48th Reloading Handbook

Thomas J. Griffin, *Editor*

Reloading Handbook, 48th Edition

IBM 9817048
UPC 011516970480

First Printing: December 2002

Published by:
Lyman Products Corporation
475 Smith Street
Middletown, CT 06457
J. Mace Thompson, President
Printed in the United States of America

WARNING

The publisher and editors have no control over how reloading is conducted by the individual or with what components and dies. Every change of equipment, procedure, and component lot will effect ballistics and/or the safety and usefulness of a load. Therefore, no warranties are implied or expressed by the data and copy contained in this book. We specifically disclaim any warranties of fitness for any and all particular purpose and specifically disclaim any and all liablilty for consequential damages of any kind.

The individual assumes all risks for the safety of reloaded ammunition. Improperly loaded ammunition, or the failure to follow all necessary precautions, may result in serious personal injury and/or death to the shooter or bystanders.

There are many precautions to which the reloader need adhere. This volume deals with many of these but cannot possibly foresee or include all possible cautions and caveats.

National Associations of Interest

Membership in an organization made up of shooters with interests similar to our own is one of the best ways to increase your enjoyment of the shooting sports. Below is a list of organizations that will be of interest to many reloaders. Please write for membership information.

American Single Shot Rifle Assoc.
709 Carolyn Drive
Delphos OH 45833

Cast Bullet Assoc.
4103 Foxcraft Drive
Traverse City, MI 49684

International Benchrest Shooters
RR1, Box 250 BB
Springville, PA 18844

International Handgun Metallic Silhouette Assoc.
PO Box 368
Burlington, IA 52601

International Practical Shooting Federation
PO Box 811
Sedro Wooley, WA 98284

National Bench Rest Shooters Assoc.
2835 Guilford Lane
Oklahoma City, OK 73120

National Reloading Manufacturers Assoc.
Suite 300, One Centerpointe Drive
Lake Oswego, OR 97035

National Rifle Assoc.
11250 Waples Mill Road
Fairfax, VA 22030

National Shooting Sports Foundation, Inc.
11 Mile Hill Road
Newtown, CT 06470

Acknowledgements

Lyman would like to thank the following companies for their assistance in the production of this handbook. Without their strong support this effort would not have been possible. Our sincere thanks and appreciation to all.

Barnes
PO Box 215
American Fork, UT 84003

Browning
One Browning Place
Morgan, UT 84050

Buffalo Arms
99 Raven Ridge
Sandpoint, ID 83864

Accurate Arms Co.
5891 Hwy. 230 W.
McEwen, TN 37101

Alliant Techsystems
PO Box 6
Radford, VA 24141

CCI/Speer
Box 856
Lewiston, ID 83501

Federal Cartridge Co.
900 Ehlen Drive
Anoka, MN 55303

Hodgdon Powder Co.
PO Box 2932
Shawnee Mission, KS 66201

Hornady Manufacturing
PO Box 1848
Grand Island, NE 68802

IMR Powder Co.
Plattsburgh, NY 12901

Kaltron-Pettibone
VihtaVuori Powders
1241 Ellis Street
Bensenville, IL 60106

Navy Arms
815 22nd Street
Union City, NJ 07087

Nosler Bullets Inc.
PO Box 671
Bend, OR 97709

Olin/Winchester
427 North Shamrock
East Alton, IL 62024

RCBS
605 Oro Dam Blvd.
Oroville, CA 95965

Redding
1089 Starr Road
Cortland, NY 13045

Remington Arms Co.
870 Remington Drive
Madison, NC 27025

Sierra Bullets
1400 Henry Street
Sedalia, MO 65301

Starline
1300 West Henry Street
Sedalia, MO 65301

Sturm, Ruger & Co.
200 Ruger Road
Prescott, AZ 86301

Swift Bullet Co.
PO Box 27
Quinter, KS 67752

Thompson/Center Arms Co.
PO Box 5002
Rochester, NH 03867

U.S. Repeating Arms Co.
275 Winchester Ave.
New Haven, CT 06511

Weatherby
3100 El Camino Real
Atascadero, CA 93422

Lyman 48th Edition Reloading Handbook

Development Staff

Publisher J. Mace Thompson

Editor: Thomas J. Griffin

Handbook Writers And Contributing Authors:
Ed Matunas
Bryce Towsley
Paul Matthews
Butch Fisher

Ed Matunas

Lyman Handbook Technical Coordinator: Jeff Henry

Lyman Staff Associates:
Jim Sheehan
Eric Fisk

Outside Contributors:
CAS Advertising
Michael Nicastre
Monique Przybysz

Photo Credits:
John Giammatteo
Bryce Towsley
Ed Matunas

Thomas J. Griffin

Jim Sheehan

Eric Fisk

Jeff Henry

Table of Contents

Editor's Forward

"This reloading handbook, our 48th Edition, is the finest Lyman has produced to date". From the very first day that we began work on this edition, it was our goal to make this statement a reality.

This was no small task or goal as the history of Lyman and Ideal handbooks spans a time frame of over 100 years. In our new handbook, we wanted to combine the best features of the past with all new information and data. In this book we have continued many of the popular features that made our past editions famous. Foremost of these is the tradition of using a variety of component brands. In our data you will find jacketed bullets made by Barnes, Hornady, Nosler, Remington, Sierra, Speer, Swift and Winchester. As with all of our previous handbooks, our cast bullet data is the most complete that can be found. Popular brands of powder used are Accurate, Alliant, Hodgdon, IMR, VihtaVuori and Winchester. A feature that we have returned to our data pages due to demand from our customers is the listing of Accuracy Loads.

Staying with our data section, a new feature appearing for the first time is the listing of ballistic coefficient and sectional densities for all bullets (jacketed and cast) used in our data. This will be a great aid for the many reloaders with ballistic computer programs. Our data section includes all the popular new calibers available at the time this handbook was printed, such as Remington's 7mm, 300, 338 and 375 Ultra Mags; Remington's 7mm and 300 Short Action Ultra Mags; and Winchester's 270, 7mm and 300 WSM calibers. The 450 Marlin also makes its first appearance in a Lyman handbook. To go along with these powerful new calibers, we have used many premium grade jacketed bullets such as Swift and

Nosler/Winchester's Combined Technology. This will produce outstanding load combinations for the reloading hunter. We have included smokeless powder data for a number of the now popular black powder cartridges such as the 40-65, 40-70, 45-90, 45-100, 45-110, and 45-120 calibers. On the handgun side, we have data for the 357 Sig and the powerful, new 480 Ruger. Overall, we have completed one of the most extensive reviews and re-shoots of the entire data section ever performed.

On the editorial side of this handbook, we have again used our old friend Ed Matunas to ably guide the reloader though all the basic reloading steps and procedures covered in chapters one through nine. This is the fourth handbook in recent years that Ed has worked with us on, in addition to several other handbooks he edited during his years with Lyman in the 1960's. While Ed is now retired from gun writing, we were happy that we could get him away from hunting and fishing long enough to again be part of the newest Lyman handbook.

Bryce Towsley makes his first appearance in a Lyman handbook and expertly writes on the subjects of reloading for accuracy, internal ballistics and external ballistics. Bryce is best known as Field Editor for the NRA's American Rifleman magazine. He also has written articles for many other magazines and has authored several books of his own.

Paul Matthews writes on reloading the 45-70 for accuracy. Paul has written eleven of his own books, including one devoted entirely to the 45-70 cartridge. In addition, he has authored many articles for such well-known publications as Gun Digest and Shooting Times magazine. Paul is also well known for his expertise in black powder cartridge reloading and is an avid and winning black powder cartridge silhouette shooter.

Butch Fisher, the inventor of Butch's Bore Shine cleaning solvent, contributes an article on the subject of barrel care and cleaning. Butch is an avid and well-known bench rest shooter who builds many of his own match winning rifles. Few can speak more authoritatively on the often overlooked and misunderstood topic of barrel care and cleaning.

The 48th Edition Reloading Handbook that you are now reading is the result of over three years work and many thousands of rounds assembled and fired by our handbook staff. We like to think that we have reached our goal of producing the best Lyman handbook ever. We certainly hope that you agree. If so, please tell your fellow shooters and reloaders about it. If not, then please tell us. As always, the Lyman staff encourages your comments and suggestions on how to improve upon this and other Lyman publications.

Section 1

1 Why Reload?

Chapter 1

The reloading of ammunition is undertaken by individual shooters for many reasons. If you shoot frequently, or want to, the first reason for getting started can be money saved. Depending upon the caliber you reload and the type of components used, you may save from 20 to 70%, with the average cost savings being 50%. For a fixed expenditure, this means the you can shoot twice as often as when using factory ammo. Yes, there is a cost for getting started, perhaps averaging about $350.00 for quality, long lasting equipment. However, it takes only about two dozen boxes of ammo (again dependant on caliber and load used) to recover this expense. At this point your equipment has cost you nothing and thereafter every round fired represents a substantial savings. Shooters using cast lead bullets can increase savings to as much as 85% of the cost of factory loaded jacketed bullet ammo.

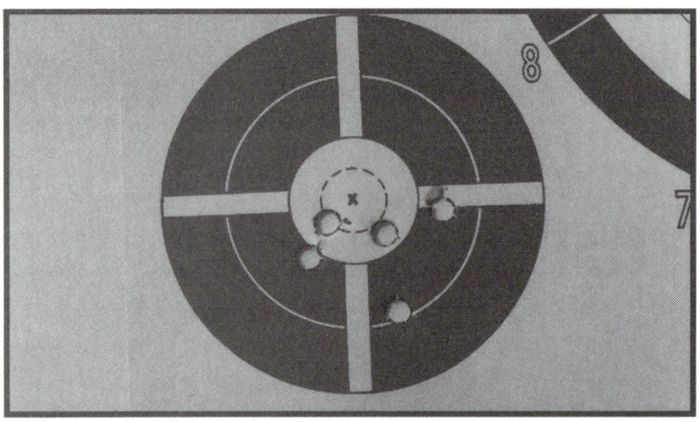

Accuracy is one of the most important reasons to reload. The target on the top was shot with factory ammunition. The target on the bottom was shot with the same firearm and bullet weight using reloads.

One of the reasons for substantial savings is that the fired case represents a large portion of the cost of a loaded round. A reloader can reuse the fired case many times. Dependant upon the caliber, load specifications, and firearm type, a case may be reloaded from 3 to as many as 15 times. Based on input from countless thousands of reloaders, on an average fired cases are reloaded about 6 times. However, when cases wear out, there is no reason to purchase new ammo. Simply buy new unfired cases. These are available from a great many component outlets and when purchased in bulk quantities are rather inexpensive.

Another reason for reloading is increased accuracy. Factory ammo seldom matches the accuracy of carefully developed and assembled reloads. The best proof of this is to look at the sport of benchrest shooting. Benchrest shooters do not consider themselves competitive unless they can put 5 bullets into a 100-yard group which, when measured from center to center of bullet hole, must be less than 0.250". To be among the best, benchrest groups must often be half this size. What kind of ammo do these folks use? It is 100% reloaded by themselves.

Long range varmint hunters are another group that demand great accuracy. Hitting a tiny prairie dog at 400 yards isn't easy. Again, the choice of ammo is nearly 100% reloads.

The author has great trust in the accuracy, reliability and performance of his reloads. Ed took this cape buffalo with a reloaded 400 grain solid fired from his 416 Rigby.

Big game hunters who pursue animals that must be shot at very long ranges, and who do some actual practice shooting, will find that many factory loads do not perform at a level to make such shooting practical. Ammo that may as easily miss as hit, or even cripple game, is not what a true sportsman wants. Reloads are often the solution to making reliable long range hits. When hunting big game, nothing will give you more confidence for a shot from this canyon rim to that canyon rim than using a reload with which you are thoroughly familiar.

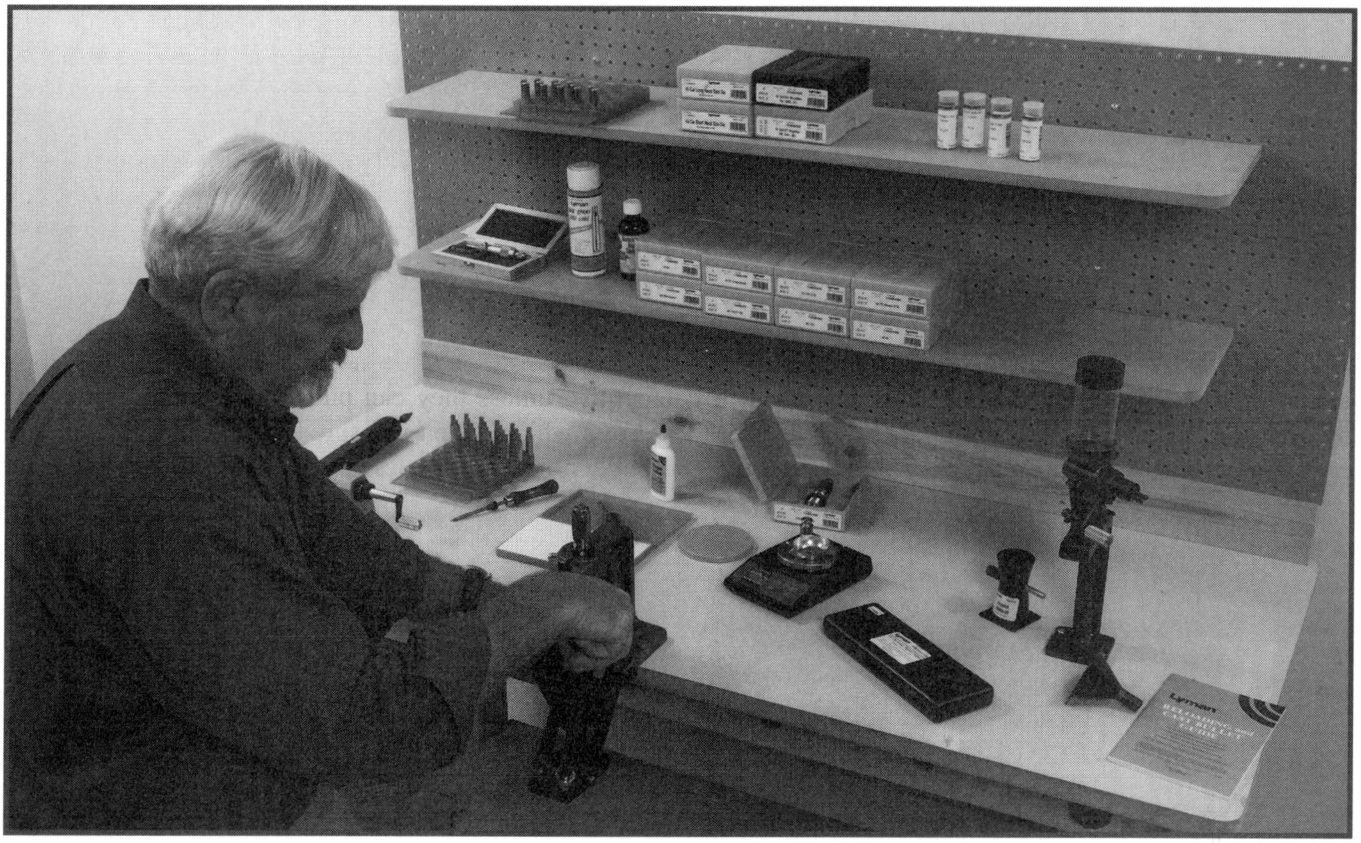

Reloading your own ammo can be very rewarding with many benefits, some of which are money saved, increased accuracy, and better marksmanship.

Despite what you may have heard to the contrary, sufficient experience with factory ammo will quickly show that it is difficult to get consistent 5-shot 100-yard groups that measure 2". Even beginning handloaders quickly learn that all groups, firearm and shooter so capable, can be under 1-1/2" and frequently down to 1". With well tuned firearms, loading to benchrest accuracy is no special feat; albeit there are a few extra handloading steps commonly employed when assembling ammo for sub minute-of-angle shooting.

The accuracy difference between factory ammo and handloaded ammo is such that most who start reloading to save money would continue to do so, regardless of cost. Still another reason to reload is the hobby aspect. Assembling ammo during leisure time can be a lot of fun. There is a sense of pride that comes from firing match winning scores, or taking game, with ammo that you assembled yourself. Increased firearm and ammunition knowledge is another benefit of reloading. Striving for better and better ammo performance will lead the reloader to countless new materials on the topic. Many serious reloaders eventually become local "gurus" on guns and ammo.

There are many other reasons to reload. Nonetheless, we caution that the undertaking of ammo assembly requires a very mature level of thinking and safety awareness. If you are willing to read all of the instruction material in this book and to follow all cautions, then you have what it takes to get started. Remember, part of being a safe and sane reloader is to review this material from time to time.

2 Getting Started Safely

Some folks getting started in reloading find it confusing. It need not be so. The short answer to the basic question of how to start is as follows. First: read all the material in sections 2 through 4 of this handbook. Second: Review the data for the cartridge you wish to reload. Third: select a bullet weight, an appropriate powder and the required primer size. Fourth: follow the suggestions in this chapter for the basic tools you will require. Finally: from all this information make a list of the items you will need and you are ready to go shopping.

Nonetheless, if you are to be a safe and sane reloader you may wish to know the long answer to the question, "How do I safely get started to assemble ammo?" Several aspects required in becoming a safe reloader cannot be purchased at the local gunshop. Most important is a good supply of common sense. Not everyone who drives is qualified to be a auto mechanic and not everyone who shoots is qualified to be a reloader. However, it is reasonably easy to develop the necessary skills and attitude to be a safe reloader.

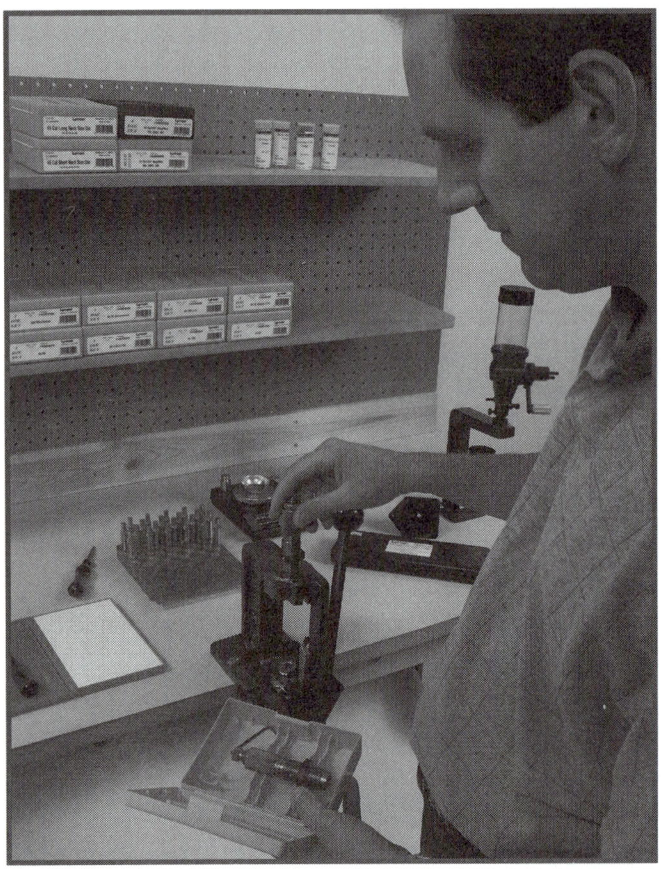

All shooters who can read, use common sense, and adhere to all the basic safety principles can become a proficient reloader.

An appropriate place to reload is another important consideration. You will require a place free from all distractions. Included among the potential distractions are radios, TVs, kids, phones, pets and a host of other items that could possibly divert your attention from the task at hand. To prevent accidents, when assembling ammo you must be able to give undivided attention to what you are doing.

The work area chosen should have adequate lighting and a suitable bench. Apartment dwellers can convert a closet to include a bench and the necessary storage space. A basement or seldom used room can also provide the basis for a suitable reloading area. Powder and primers will require safe, lockable, remote storage (more on this in the chapters dealing with components). Do not overlook potential problems such as heavy objects stored in such a manner that they could fall onto the work area and perhaps cause ignition of primers or powder.

You must follow the recipes of reloading - the data listings - exactly. This is a basic requirement. If you think published data is deliberately kept extra safe and therefore you can develop loads that use powder charges heavier than the listed maximum loads, or that you can interpret data for unlisted powders and bullet weights, then we suggest you find a different hobby.

There is never a justification for random experimentation when loading ammunition. If you cannot respect the wisdom, judgement and testing of the ballistic laboratory then you are not a candidate for reloading. It is sometimes reasonable to wish to substitute one brand of bullet for another or to want to use a different brand primer. However, such changes must always be approached with extreme caution. When doing so the load must be worked up carefully. Always remember that any change in components, procedures, or loaded round dimensions will create a measurable change in ballistics. The direction of this change is not predictable and therefore all changes require extreme caution and very careful load development.

Your first reloading efforts are best directed at duplicating the factory ammo you have been using. Start with a bullet weight that duplicates your favorite factory load. If you do everything right, you will notice an immediate improvement in accuracy.

Later, you may wish to assemble ammo for a specific purpose not well served with factory cartridges. For example, you may want to assemble an off-season varmint load for a thirty caliber big game rifle or a highly accurate target load. This is fine, but wait to do so until you have gained a bit of experience.

To insure continued savings and accuracy, the reloader should purchase the most durable equipment that can be afforded. Equipment that fails to perform as hoped will quickly need to be replaced. If you start with tools and accessories that will provide many thousands of loaded rounds of ammunition, you will never regret your purchase.

Your initial purchase need not include every conceivable accessory. You already have at least one of the basics - a reloading manual. For a good beginning you will need to equip your loading bench with a number of other specific basic items. To omit any one of them is to insure a less than 100% safe reloading operation. The essential items are as follows.

Basic and Essential Loading Equipment

1. Up-to-date Loading Manual
2. Reloading press
3. Shellholder
4. Reloading die set of appropriate caliber
5. Case lubricant
6. Case lubricant pad Note: items 5 and 6 are not required when using a carbide die with some handgun calibers.
7. Safety glasses (especially for priming)
8. Priming unit
9. Primer tray
10. Powder scale (graduated in grains avoirdupois)
11. Powder trickler
12. Powder funnel
13. Two loading blocks
14. Dial indicating caliper (graduated in 0.001" increments)
15. Case trimmer with pilot
16. Deburring tool

A brief discussion of each listed essential tool is appropriate at this point.

The first basic tools include a press, shell holder and die set.

1. Reloading Manual

Remember that all data sources become obsolete. The ballistic characteristics of specific components can and do change over time. It is possible for a component to undergo a substantial change without a name change. Thus, data changes with the passage of time. Failing to keep your data up to date can bring on serious consequences.

Data tables are only a part of the knowledge needed to assemble safe ammunition. Take the time to read all of the instructional material in each manual update. Then reread it from time to time as a refresher.

2. Reloading Press

The press selected is the foundation for the loading bench. A good beginning is to select a single station press having a "O" shaped frame, such as the Lyman Crusher II. This type tool offers precise alignment of dies and shell holder as the die station hole and shell holder ram hole are machined in line. Such tools will generally offer a lifetime of use. Lesser priced tools are available for the shooter on a tight budget.

It is often advantageous to purchase the reloading press as part of a packaged kit. Such kits include many of the essential items needed at a substantial savings when compared to buying each item individually.

3. Shellholder

The shellholder allows the cartridge case to be aligned with, pushed into, and withdrawn from the various dies. One shell holder may serve for several cartridges. For example, a 30-06 shell holder is also correct for use with calibers 22-250, 243, 25-06, 270, 308 and many others. Be sure to use the correct shell holder or the case rim may tear off during resizing. This will leave the case jammed in the die, a very difficult situation to correct, and one best handled by the die manufacturer.

4. Reloading Die Set

A die set may consist of two, three, or four separate dies. The first die is a resizing die designed to return the fired and expanded case back to factory dimensions. This die also usually removes the fired primer. Without appropriate resizing, fired cases may fail to rechamber. For beginning, as well as most advanced efforts, we suggest the purchase of a full length resizing die set. Most die sets are caliber specific.

Note: For many straight walled handgun cartridges, the use of a carbide sizing die will eliminate the need to apply sizing lubricant to the case and to remove the lubricant after sizing. This is a great time saver and prevents jammed-in-the-die cases due to improper lubrication.

WARNING: Never attempt to resize a loaded round.

Sizing dies for bottleneck cases include a spindle mounted expanding button which opens the case neck to a dimension suitable to properly accept and hold the bullet. Straight walled cases must be neck expanded in a separate die. This means that all straight case die sets have an extra die.

The bullet seating die, seats the bullet to the correct depth (adjustable). When required it can also crimp cases to bullets. A crimp is necessary for ammo used in most revolvers and in guns with tubular magazines. Ammo for use in pump or semi-automatic rifles is also normally crimped. Some shooters find crimping to be more precise if done separately from bullet seating. Thus, an extra die is sometimes used. Most bullet seat/crimp dies use a roll crimp. However, taper crimp dies are used for reloading of cases which headspace from the case mouth.

5. Case Lubricant

For all reloading efforts (except when using a carbide die for straight walled handgun cartridges), cases need to be lubricated before resizing. Failure to do so will result in a case hopelessly jammed in the resizing die.

6. Case Lubricant Pad

The proper method of applying lubricant is to use a lubricating pad to transfer sizing lubricant to your cases. As an alternative to using a pad you may purchase Lyman's Qwik Spray Case Lube.

7. Safety Glasses

Should an accident occur, this essential item can save your eyes from injury, especially when priming.

8. Priming Unit

Priming units are also called priming punches, priming rams, or priming arms. All loading presses are equipped with basic priming units that accomplish the task of seating new primers. With these, the primer is often seated as the sized case is withdrawn from the resizing die (or in the case of a straight walled case as it is withdrawn from the neck expanding die). However, it is suggested that the shooter interested in maximum primer seating uniformity will do best to replace such standard priming units with one of the Ram Prime systems.

Case lubricant and a lubricating pad are essential for all case resizing except when using a carbide die for resizing certain straight walled pistol cartridges.

Ram Prime units are mounted at the normal die station and can be adjusted to give a very uniform primer seating depth by using the press stroke stop as a means to control primer insertion depth.

9. Primer Tray

A primer tray is used to orient all primers one side up so as to insure that, as each one is picked up and placed into the primer post, it is correctly oriented. It will also help keep primers free from any contami-

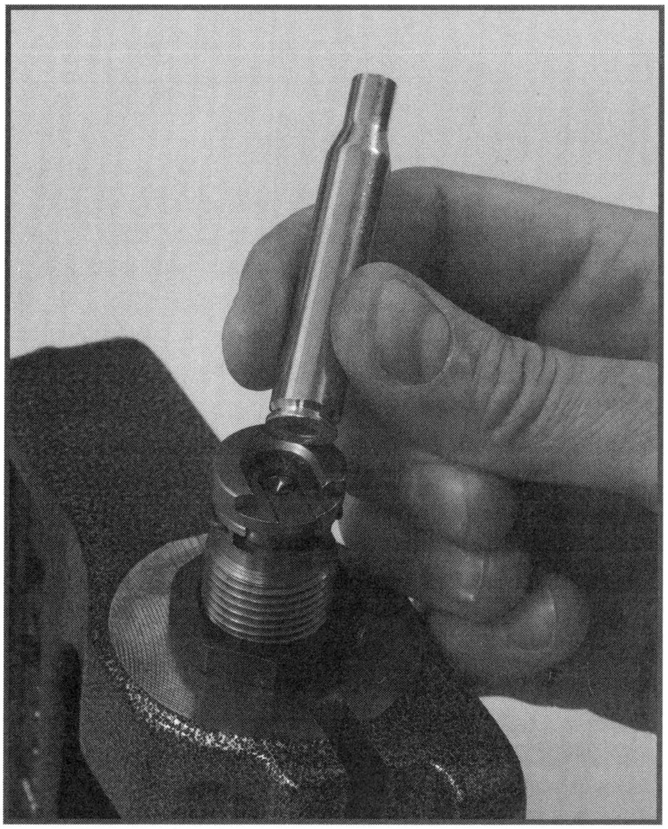

There are many styles of priming units available, but the ram prime unit shown is one of the best units for seating primers to a uniform depth. The primer tray will allow rapid orientation of primers and keep them safely in place on the bench.

nants.

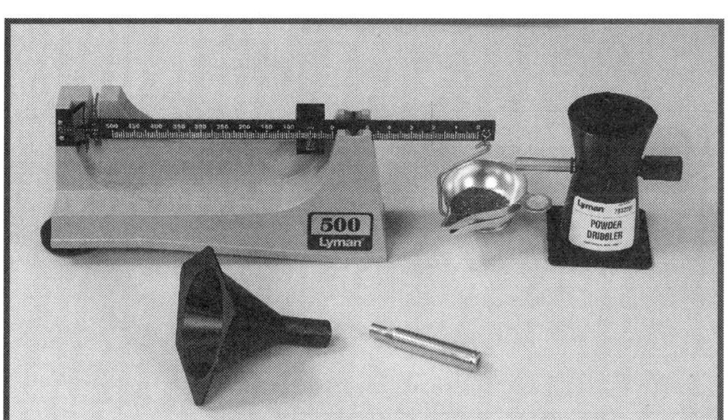

Each powder charge should be carefully weighed on an appropriate powder scale. The powder trickler will allow you to carefully add the last several granules of powder when bringing the propellant charge to exact weight. The funnel is used to pour the powder into the cartridge case.

10. Powder Scale

To insure the precise amount of powder is placed into each case, each powder charge must be carefully weighed. A powder scale is also required for the correct adjustment of any accessory powder measure.

11. Powder Trickler (aka Dribblers)

These make the job go quickly and accurately. The trickler adds powder, one granule at a time, to the scale pan in order to bring a propellant charge to exact weight. Powder tricklers are available in hand turned models for about a ten dollar bill or in the superb electric vibrator types for about forty dollars.

12. Powder Funnel

Without a powder funnel there is no practical way to get propellant from the scale pan into the case. A simple but effective funnel will work with cases as small as 22 caliber and as large as 45 caliber.

13. Loading Blocks

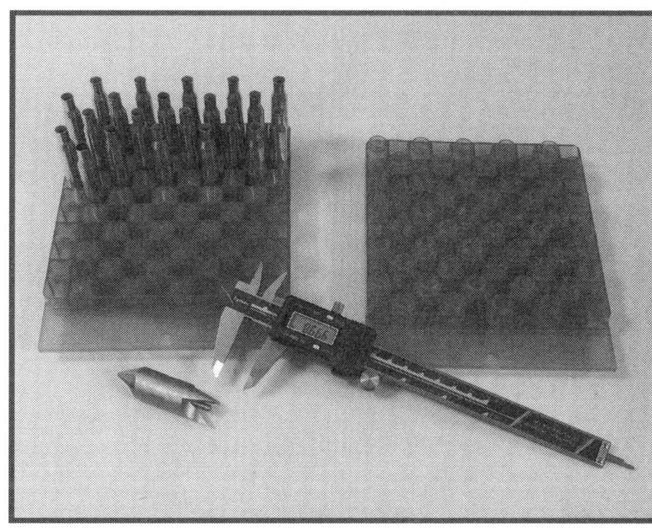

Basic and essential tools include two loading blocks, digital caliper, and a case deburring tool.

Loading blocks hold the cases on the bench in an organized manner. Always use two loading blocks. Cases are picked up, a loading step performed and then they are placed in the second loading block (positioned on the opposite side of the loading tool).

14. Dial Indicating or Digital Caliper

This tool allows the reloader to measure the depth of seated primers, to check case lengths before and after trimming, to maintain an appropriate cartridge overall length and to discover the source of many difficulties. Calipers are available as inexpensive plastic types or stainless steel models that sometimes outlast the reloader.

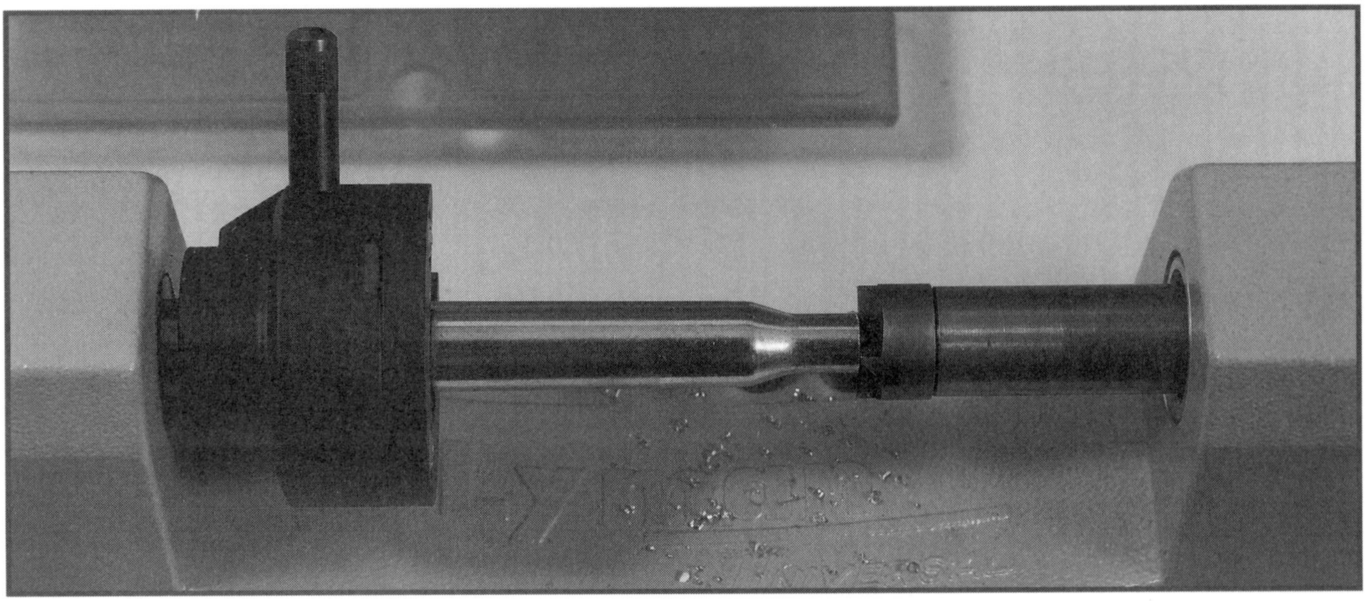

Trimming cases is essential to insuring safety, reliability and accuracy of your reloads.

15. Case Trimmer

Cases stretch when they are fired. Case bodies are often shortened when resized but their necks tend to stretch during this process. After a single firing and sizing, a case may stretch past the point of being suitable for reloading. Under some circumstances, a case may be reloaded and fired four or five times before its length exceeds the maximum allowable dimension. In any event, there will always be a point at which the case must be trimmed back to a workable dimension before it can be reloaded. Trimmers require the use of a pilot of appropriate caliber to keep cases positively aligned during the trimming process. Some case trimmers require the use of an appropriate shell holder.

16. Deburring Tool

During trimming, burrs form on the inside and outside of the case mouth. A deburring tool quickly gets the case smooth and ready for reloading. Cases should be deburred for the first reloading,especially inside the case mouth, even if they are not trimmed.

The cost of getting set up with the necessary essential tools will vary depending on where the equipment is purchased and exactly what model tools are selected. The return on the initial investment can be rapid or slow depending upon how often you shoot.

You will require a supply of primers, powder and bullets to begin reloading. The cost of these will vary dependent upon the caliber and the type of ammunition you want to assemble.

Additional Accessories

There will come a time when you may wish to add to your basic tools. Accessory items can help improve accuracy, or make the loading process faster. Some of the more useful accessories include the following:

1. Powder Measure And Optional Stand

A powder measure allows the reloader to dispense a near exact powder charge rapidly. Because charges thrown from a measure will vary, powder measures must be used in conjunction with a scale. To do so, adjust the measure to throw a charge slightly below the desired nominal. Set the measure so that no individual charge will be over the desired weight - it is nettlesome to try to remove excess propellant from the scale pan. Then, meter a charge directly onto the scale pan. Return the pan to its hanger on the scale and bring the charge into exact balance by adding the last few granules of powder with a powder trickler.

2. Primer Pocket Cleaner

After decapping, inspection of the primer pocket will reveal a hard, crusty, black deposit. If allowed to build to excessive amounts, this deposit can interfere with uniform primer seating and thus cause inaccuracy. A high primer can also be a potential hazard. This deposit is easily removed with a few twists of a primer pocket cleaner.

3. Case Cleaning Equipment

A dull looking case is no consequence. A dirty case, however, may keep a defect from being noticed or scratch reloading dies and/or chambers requiring premature die/firearm replacement.

Cases can be cleaned by hand, however, a case tumbler can make some mighty dirty brass look like new. Case tumblers are available to fit almost every need with respect to the quantity of cases to be processed and cost.

4. Scale Check Weight Sets.

Scales are delicate instruments. Should a scale be inadvertently subjected to some undue rough handling, a scale check weight set can help determine if the scales accuracy has been compromised.

5. Turret Presses

Turret presses have multiple die stations. The advantage is that a reloader can place a case into the shell holder and fully load it, by rotating each die into position. This speeds up loading and is a method sometimes favored by handgun cartridge reloaders.

Turret presses also allow the reloader who loads cases by the preferred batch method (performing the same operation on all of the cases to be loaded before moving on to the next step) to set up all the dies and leave them in position on the press. This eliminates the need to repeatedly remove and replace dies.

6. Case Neck Turning Equipment

In order to obtain better accuracy, serious accuracy buffs (such as benchrest shooters and varmint hunters) often turn the outside of case neck to make it concentric with the inside.

Neck turning is essential when making cartridges fit chambers with undersized necks often used in custom benchrest rifles. Neck turning operations are generally best left alone until after acquiring an in-depth amount of reloading experience.

7. Flashhole Uniformer

This little handheld reamer will remove the burr at the terminus of the flash hole. This is an accessory that originally saw use with serious bench rest shooters. Today, in their search for accuracy many handloaders feel it worthwhile to deburr flash holes.

8. Primer Pocket Reamer

Primer pocket reamers are required to remove the crimped material from military and some para-military style cases. Such cases have the primer locked into place by an impact-shifting of case head brass around the end of the primer pocket. It is best to use a special heavy-duty decapping rod assembly to remove crimped in-place primers. After removing fired primers from such cases, a new primer cannot be safely or correctly seated until a reamer is used to remove the crimped material.

9. Primer Pocket Uniformer

A primer pocket uniformer is used by many shooters to insure that the bottom of every primer pocket is flat. The benefits of this additional effort generally cannot be realized until groups are at or below 1/2 inch at 100-yards. It requires care to avoid overcutting the primer pocket. Removing too much material can cause gas leaks at ignition. Such leaks can quickly ruin a bolt face and are potentially dangerous. Obviously primer pocket uniforming is therefore not for everyone.

These accessories, and many others, can help make reloading faster, easier, and more accurate. When and if each item will fit in with your reloading will become self evident as you gain experience.

10. Automatic Electronic Powder Scale

While not inexpensive, this accessory can simplify the process of powder charge weighing. At a push of a button the electronic scale automatically feeds powder onto the scale pan, first at a fast rate and then at a greatly slower rate to bring the scale into balance. Once started it is a hands-off operation.

As time goes by, the reloader will find that the list of other available accessories appears near endless. Let real needs or personal goals dictate which ones should be on your loading bench.

Getting started requires only the previously discussed sixteen basic items. Indeed, one could happily load hundreds-of-thousands of rounds with only a few additional accessories - a powder measure and primer pocket cleaner being the most likely candidates. From this point on, it should be obvious when you will actually benefit from an addition to your basic tools. Lacking an obvious need, chances are you will get along nicely without further additions.

The how-to of component selection is covered in the following appropriate chapters of this manual.

Section 2

3 Reloading Rifle Cartridges

As mentioned previously, the mechanics of each reloading step are quite simple. However, there are several important considerations to be given to each step. Observing these considerations will help ensure that all ammo assembled is safe, will function reliably, and be accurate. Some of these considerations are dealt with in following chapters and in several of the cartridge comments immediately preceding specific loading data. This information is important and must be read and understood before actually beginning to assemble ammunition.

The following step by step description deals with assembling fired bottleneck (shouldered) cartridges with jacketed bullets intended for use in either rifles or handguns. Slight deviations from these steps are required when reloading ammo with lead bullets, for rifle or handgun cartridges having a straight case (no shoulder), or when loading new unfired cases. These variations are discussed at the appropriate points in the following text.

Reloading a fired cartridge generally requires 13 basic steps. The sequence or actual number of these can vary depending on the specific equipment used and reloader preference. The sequence we have chosen is based on the method most often favored by knowledgeable handloaders and our selection of tools used to illustrate the process. Again, not every step is always required, as we will indicate.

The thirteen steps to reload jacketed-bullet bottleneck cartridge ammo are as follows:

1. Selection of load and components to be used.
2. Case inspection.
3. Case cleaning.
4. Inside neck brushing
   (not required when using new unfired cases).
5. Case lubrication
6. Case resizing and fired primer removal
   (not required when using new unfired cases).
7. Lubricant removal
8. Case length measuring.
9. Case trimming and deburring
   (not always required).
10. Seating new primer.
11. Powder weighing and charging.
12. Bullet seating and if required crimping.
13. Final inspection.

CAUTION: Be sure to read all pertinent material before actually beginning to assemble ammo.

STEP ONE

SELECTING A LOAD AND COMPONENTS

At first, the selection of components may seem a difficult and confusing task. This is especially true if a general knowledge of basic ammunition details has not yet been acquired. Nonetheless, you soon will find that component selection quickly becomes an easy and fun part of reloading. The learning process can be hastened along by developing as much ammo knowledge as leisure time reading will allow. Thousands have found Lyman's Guide to: BIG GAME CARTRIDGES AND RIFLES a good book for enhancing ammo knowledge, especially with regards to bullet selection.

Selecting Cartridge Cases

The selection of fired cartridge cases, or new brass, presents no special challenge. If you have saved your fired cases you need only separate them into specific groups by brand and lot number. Lot numbers appear on the factory ammo box. This number may be on an inside flap or the box back. Keeping brass segregated by lots will maximize accuracy potential and ballistic uniformity. If you purchase new unfired cases they must, of course, be of the appropriate caliber. Purchasing bulk packaged cases (in lots of 50, 100, or more) is the least expensive way of obtaining new brass.

Test Components:

Cases	Winchester
Trim-to Length	2.484"
Primers	Winchester WLR
Primer Size	Large Rifle
Lyman Shell Holder	No. 2
Jacketed Bullets Used	Sierra HP #2110, 110 gr.

Primer size and the specific primer we used is clearly shown at the beginning of the data for each caliber.

CAUTION: Never load cartridge cases from an unknown source, i.e. cases picked up at the range or sold as once-fired brass. Use only brand new brass or cases obtained as the result of firing factory ammo in your firearm. For more on cases and their selection see Chapter Five.

Selecting Primers

The proper primer size is listed at the beginning of the data for each cartridge, for example: large rifle, large rifle magnum, small rifle. For your first loads, we suggest you use the exact primer we used for the development of our data. As an alternate, match the primer brand and correct size to the brand of the case you are reloading. Do not use magnum primers unless the data specifically calls for these as doing so can alter ballistic uniformity and the safety of the data.

Do not allow cartridge nomenclature to enter into the selection of primers. For example, the 222 Remington Magnum never requires the use of a magnum primer and the 416 Rigby always requires the use of a magnum primer. Always follow the primer size and type as listed in the data tables.

165 gr. Jacketed Ballistic Tip — **BC:** .475 **SD:** .248
3.285" OAL

Powder	Sugg Starting Grains	Velocity fps	Pressure P.S.I.	Max Load Grains	Velocity fps	Pressure P.S.I.
IMR-3031	44.0	2495	38,500	49.0	2784	56,100
H335	43.0	2474	42,600	48.0	2690	54,300
IMR-4895	45.0	2503	37,700	50.0	2815	56,200
IMR-4064	47.0	2495	38,000	**52.0**	**2851**	**58,100**
IMR-4320	47.0	2542	41,200	52.0	2843	57,500
RX15	47.0	2531	39,300	52.0	2807	53,600
H-380	48.5	2515	40,100	53.0	2804	54,000
760	52.5	2697	43,500	57.0	2959	58,600
IMR-4350	52.5	2613	45,100	57.0+	2878	56,600
RX19	54.5	2464	37,600	60.4+	2816	53,300
N160	55.5	2597	41,100	61.0+	2895	56,500
IMR-4831	54.0	2485	38,400	59.0+	2786	52,000
H-4831	56.0	2431	34,900	62.0+	2754	49,000
WXR	55.0	2401	34,200	61.0+	2720	47,800

Selection of a propellant for your first loads is easily made by referencing the bullet weight (highlighted) you are using, the accuracy powder (highlighted), and the starting load (highlighted) as shown in the data.

Selecting Propellant Powders

Propellant powders are available in about 100 different types. The burning speed of each and the ballistics obtained can vary tremendously. Powders are designed to suit specific applications such as: bullet weight, case size and shape, pressure level and other specific ballistic and firearm needs. As a result, only certain propellants are suitable for specific applications. When selecting a powder for your first reloading efforts, we suggest the use of the propellant listed in our data for the accuracy load.

CAUTION: Always start with the exact powder charge weight shown under the starting grains column. Heavier loads should not be used until the reloader has gained some experience and fully understands proper load development.

Selecting Jacketed Bullets

Bullet selection may at first seem confusing. To simplify the process, select a bullet weight to duplicate the factory ammo you favor. The appropriate diameter bullet may be determined from the caliber/bullet table in Chapter Eight. Many calibers use the same diameter bullet. For example: 300 Savage, 30-40 Krag, 308 Winchester, 30-06 Springfield, 300 Winchester Magnum, and others, all use a .308" diameter bullet (30 caliber).

There is another consideration when selecting bullets. The muzzle velocity of a cartridge may require a specific bullet jacket strength and construction. For example, the 30 MI Carbine uses a .308" bullet; but because of its very low velocity it requires a bullet having a very soft jacket in order to properly expand. The 30-30 has a modest velocity and it also needs a relatively soft bullet, albeit stronger than those for the 30 MI Carbine. Still other calibers require a bullet designed for a mid-range velocity bullet, for example the 222 Remington. Bullets that are ideal for this cartridge often are designated with nomenclature that suggest rapid bullet expansion, for example: Blitz, Expander, Super Explosive (SX) and so on. Often bullets for specific applications will be so marked on the package. Some examples are 22 Hornet, 30-30, 32 Winchester Special.

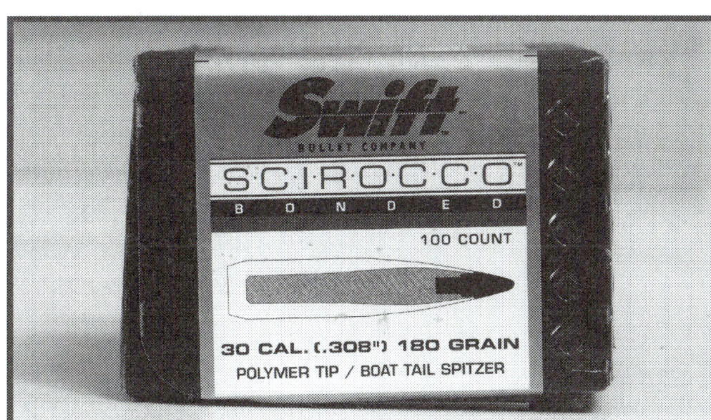

Bullets should be selected by the desired weight and correct diameter.

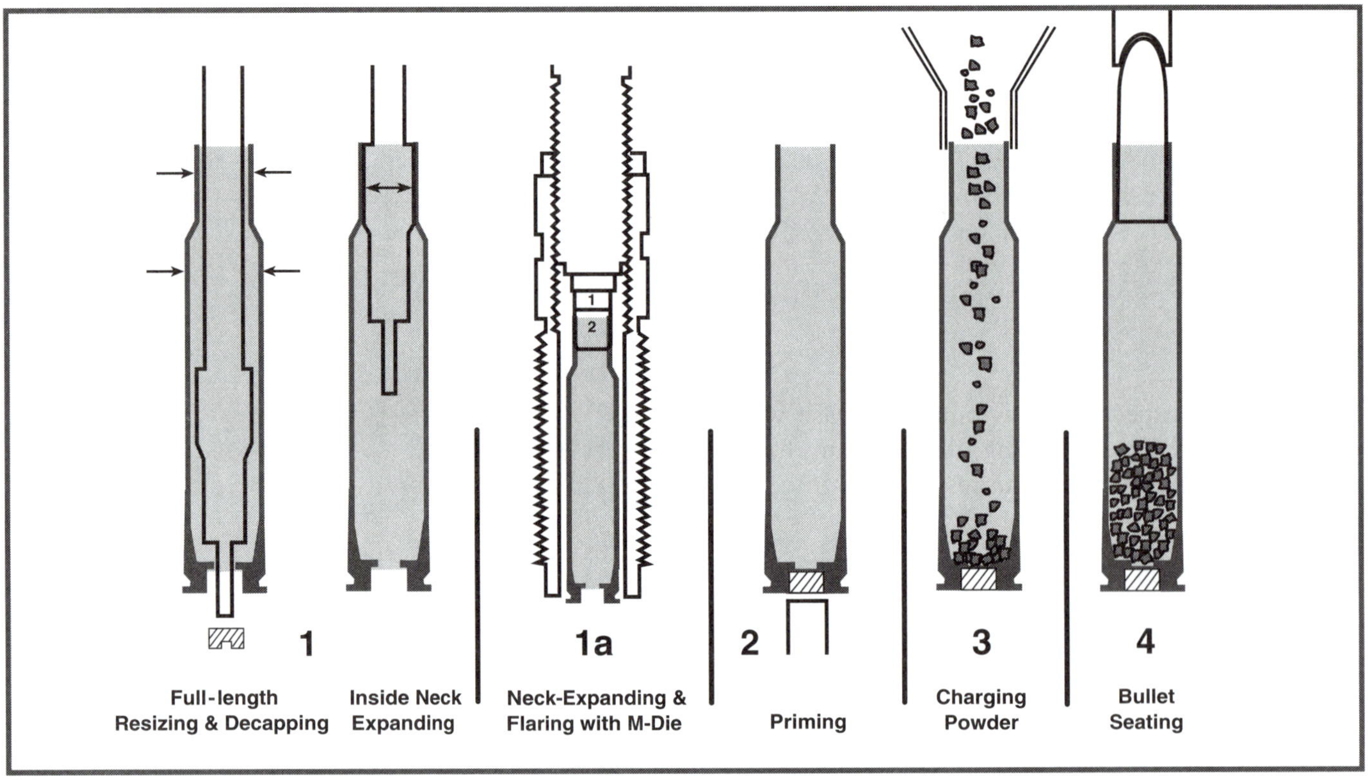

The basic steps of rifle case reloading.

I. Case is sized to return its inside neck and outside body to factory-like dimensions. The fired primer is removed automatically during this operation. II. New primer is seated III. Propellant powder is weighed and placed into case. IV. Bullet seating and crimping (when required).

All of our data specifies the bullet manufacturer's product number for each tested bullet weight. Each of the listed bullets will work fine for target shooting. However, for proper expansion on varmints or big game you will need to make certain that the bullet you select is properly designed and suitable for the velocity range of your cartridge. If you are loading for hunting, avoid the use of bullets designated as Match or Target style as these may not expand properly on game.

Often you will be able to purchase the identical bullet used in your favorite factory loads. Some of these include the Nosler Partition, Nosler Ballistic Tip, and Combined Technologies Partition Gold bullets. Many other factory bullet types are available to the handloader.

Other important bullet selection criteria are the possible need for a cannelure or the requirement for a blunt nosed bullet. These and other considerations are discussed in Chapter Eight.

As stated, Lyman's Guide To Big Game Cartridges And Rifles will help the reader with hunting bullet selection.

THE LOADING SEQUENCE

The use of two loading blocks is suggested. As each step is performed, the case should be removed from one loading block, processed, and then placed in the second loading block. This will keep the process orderly and prevent many common bench errors.

Note: We strongly suggest the reloader follow the batch method of ammo making. That is, perform a single operation on all cases to be reloaded before proceeding to the next step.

STEP TWO

CASE INSPECTION

Fired cartridge cases have a finite life. Depending upon the firearm used, caliber of the firearm, internal ballistics of the load, and other considerations, it is reasonable to expect from 2 to 15 firings from each case. Six firings are average for the typical bottleneck rifle cartridge such as the 270 Winchester or 30-06 Springfield. Belted magnum cases such as the 7mm Remington Magnum, 300 Winchester Magnum and 338 Winchester Magnum typically last for only 3 firings. Low pressure cartridges fired in strong actions, such as the 416 Rigby, generally offer the greatest number of firings. All else equal, cases fired in semi-automatic rifles will have a shorter reloading life than cases fired in bolt action rifles.

All cases reach a point when further reloading becomes unsafe. The keeping of careful reloading records and the performing of visual inspections on each case before, during, and after reloading, are essential to ensure that you use only suitably safe fired cases.

Begin your inspection by wiping each case with a cloth to remove excess fouling, dirt, and any foreign material that could scratch your resizing die or the case itself. Turn each case mouth down and tap it lightly on the bench to dislodge anything that may have entered the case after firing.

Now look for split necks or bodies, signs of incipient case separation (a bright partial or complete ring around the case at the point where the case's solid base joins the wall of the cartridge), corrosion, or burn through perforations. Also look for any signs of gas leakage around the primer pocket. Eliminate all cases with any visual defect or abnormality. (For more details refer to Chapter Five.) To prevent later inadvertent use of a rejected case, crimp its mouth shut with a pair of pliers before discarding it. Then place each case, mouth up, in a loading block.

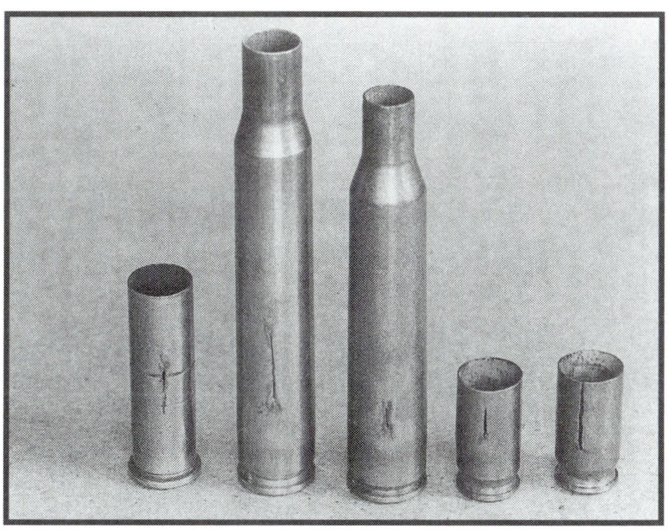

Cases must be inspected and those with any defect must be discarded.

STEP THREE

CASE CLEANING

Case cleaning is an important step to protect your reloading dies and firearm chambers. If you also want your reloads to look like new, now is the time to put all your cases into a Lyman Tumbler. Follow the instructions that come with the tumbler. After removing cases from the tumbler tap the mouth of each case on the bench to insure that no tumbling media remains in it. Then wipe each case lightly with a clean cloth.

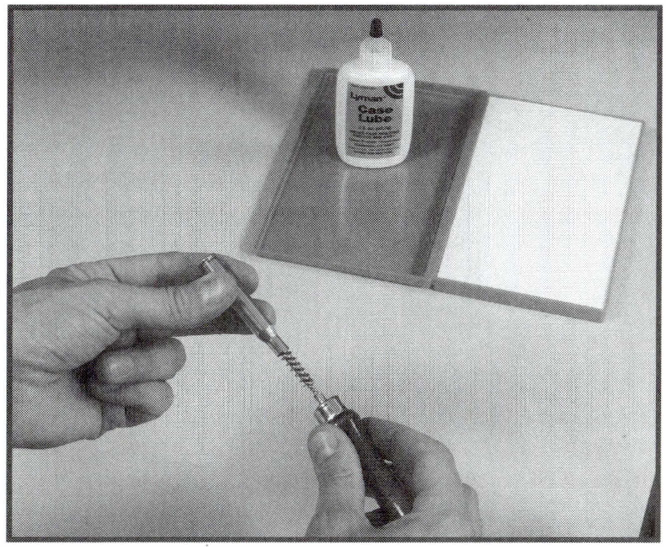

Case neck brushing will enhance accuracy.

STEP FOUR

INSIDE NECK BRUSHING

This is a step that some reloaders omit. However, it is a simple one that takes very little time and it will improve the performance of your ammunition. Brushing the inside case neck will rid it of excess firing residue, enhance the ease of pulling the expanding button through the case neck (when withdrawing the case from the sizing die) and extend the useful life of the sizing die's expanding button. It will also enhance accuracy by helping to maintain uniform bullet pull.

This operation simply involves three or four passes of a brush through the case mouth. Place the brushed case into the second loading block, mouth down, so as to allow any loosened crud to fall free of the case. To prevent the accumulation of debris in the second loading block, many reloaders tap the case mouth on the bench before placing it in the block, mouth down.

STEP FIVE

CASE LUBRICATION

When a cartridge is fired it expands. The expanded dimensions are not compatible with holding a new bullet with proper tension (bullet pull), and are not conducive to easy chambering. To avoid these, and other difficulties, all fired cases must be resized.

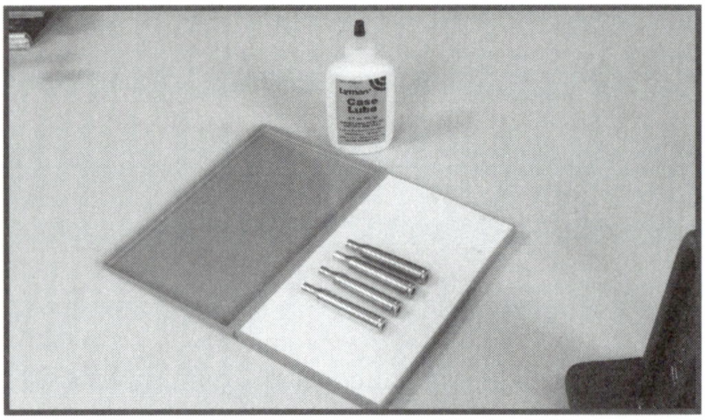

Each case must be lubricated before it is resized.

Roll a case lightly across your lubricant pad. Do not fail to lubricate each case or it will stick solidly in the resizing die creating a very difficult to correct problem. Do not use excessive lubricant as doing so will cause cases to dent during the resizing step. Use only enough lubricant to insure the case enters and leaves the resizing die without difficulty. Be neat. Do not get lubricant into the primer pocket or case mouth. Do not get lube on the case shoulder or it will dent, or possibly collapse, during resizing. Dipping the case mouth in mica, a dry powder lubricant, will help the expander button pass smoothly through the case neck. Place the case in the loading block mouth up.

STEP SIX

CASE RESIZING & FIRED PRIMER REMOVAL

Place a lubed case into the shell holder and run the case into the full length resizing die. Follow the instructions for proper die adjustment as explained in the material supplied with your die set. After the first case is resized, wipe off the lubricant with a clean cloth. Then place the case into a cartridge headspace gauge. If a die adjustment is needed, now is the time.

The resizing die generally is adjusted so that the shell holder, at the top of its travel, will contact the resizing die and create a slight cam action against the die. During this operation the fired primer will automatically be ejected from the case. Withdraw the case from the sizing die and place the case into the loading block, mouth down. Visually inspect each case as it is placed in the loading block to ensure the fired primer has been removed.

Note: The decapping rod should be adjusted just low enough to ensure the primer is pushed free of the case. If the decapping rod is too low it will impact the inside bottom of the case and be damaged. Proceed until all cases have been lubed and sized.

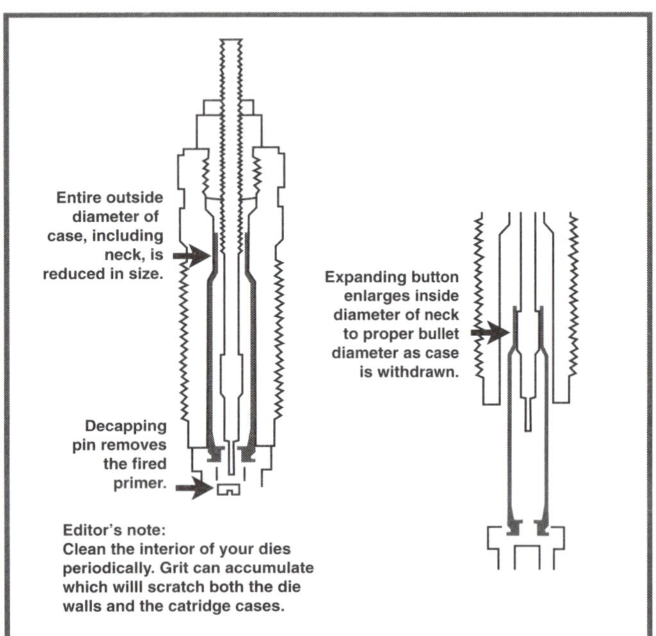

The first die usage step of reloading requires that the lubricated case be completely run into the full length resizing die. This will return the case to factory new dimensions and remove the old primer.

Note: In the past, varying sources have suggested that the primer be seated as the case is withdrawn from the resizing die. However, it is possible for case lubricant to contaminate and thus destroy primers. Therefore, we suggest that this is not the ideal time for primer seating.

STEP SEVEN

LUBRICANT REMOVAL AND SECOND INSPECTION

Carefully wipe each case with a clean cloth so as to remove all traces of sizing lubricant. Use a clean section of the cloth for each case. Next, carefully inspect the case for any flaws. Repeated case resizing and firings can cause case mouths and bodies to become brittle and split when fired or during resizing. Also watch for signs of incipient case separation (see Chapter 5). It is a good idea at this time to drop the resized case into a cartridge headspace gauge as a worthwhile inspection step. This will ensure that your sizing die is correctly adjusted to give the case the proper headspace length. Place cleaned and inspected cases in your loading block, mouth up. Should you find any defects, this is the time to discard the entire lot of cases.

IMPORTANT - Case lubricant can ruin a primer, resulting in delayed ignition or a failure to fire. To avoid potential primer contamination, this is the point to stop the reloading process and thoroughly wash and dry your hands.

SPECIAL STEP FOR STRAIGHT CASES (i.e. 444 Marlin, 45-70, 458 Win. Mag.) OR FOR WHEN USING LEAD BULLETS.

Straight cases cannot be properly neck expanded in a resizing die. Such cases must now be expanded in a special expanding die such as the Lyman M die. This extra die is supplied in all appropriate caliber die sets. If you wish to load lead bullets, the case mouth must have a two step expansion. This is also done with a M die. Follow the instructions that come with all 3 die straight case sets or with the M die when purchased as an accessory for loading lead bullets into bottleneck cases. The loading procedure for straight rifle cases and/or loading of lead bullets is the same procedure described in the next chapter for loading handgun cartridges.

STEP EIGHT

CASE LENGTH MEASURING

Case measuring is an important step both for safety and proper ammunition functioning. Cases stretch when fired and during the resizing step. If they become too long they will be difficult, if not impossible, to chamber. Excessive chamber pressure can also be caused by exceeding maximum case length. Therefore, each case must be carefully measured at this point. A dial indicating or digital caliper is the best tool for this process.

The data section drawing for each cartridge clearly indicates the maximum allowable length for the resized case. If one or more cases are found to be at maximum or greater length, trim all cases to a uniform length as described in the next step.

STEP NINE

CASE TRIMMING AND DEBURRING

As stated previously, cases must be trimmed when they exceed maximum allowable length. Additionally, case trimming is recommend whenever starting to load new or once fired brass as such cases will not be of a uniform length. Trimming cases to a uniform length will enhance accuracy and ballistic uniformity.

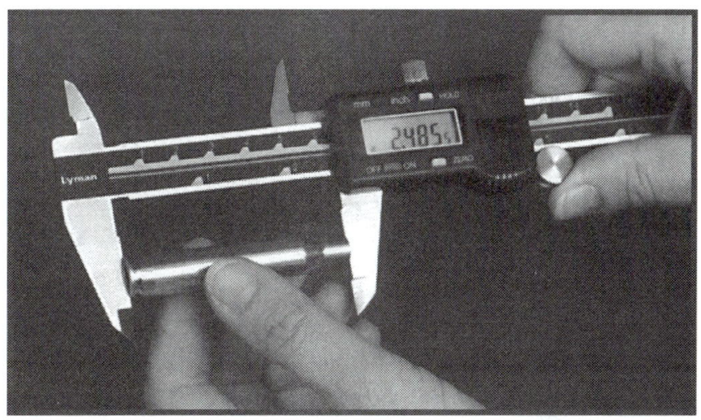

Measure all cases carefully. If one or more cases are at, or over, the maximum length shown in the cartridge drawing, trim all cases to the Trim-to-length indicated in the data.

Test Components:

Cases	Winchester
Trim-to Length	2.484"
Primers	Winchester WLR
Primer Size	Large Rifle
Lyman Shell Holder	No. 2
Jacketed Bullets Used	Sierra HP #2110, 110 gr.
	Sierra SP #2120, 125 gr.
	Hornady SP #3020, 130 gr.

Note: if you will be crimping cases to bullets (required for ammo to be used in semi-automatic and pump action rifles, or for ammo to be used in tubular magazines), the crimping process will be less than satisfactory if cases are not of a uniform length. Many reloaders, who crimp cases, trim after every

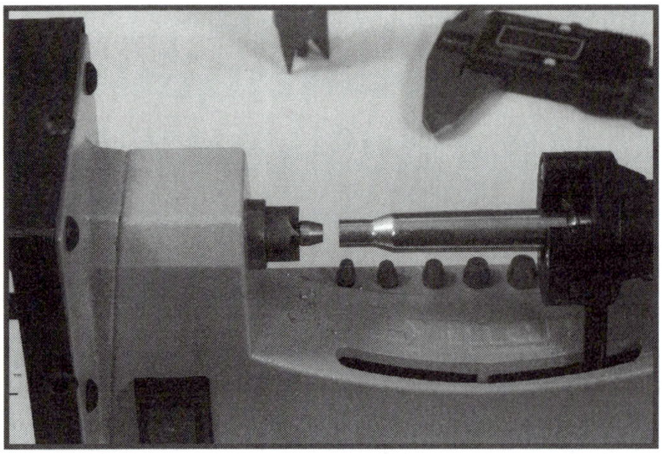

Case trimmers are miniature lathes that quickly restore a case to a safe and accurate overall length.

firing/resizing cycle.

The proper trim-to length for cases is clearly shown in the data for each cartridge. Adjust your trimmer according to the manufacture's instructions. When trimming, allow for some dwell time - that is for a number of rotations of the cutter after the case has been trimmed to length. This will help insure the maximum uniformity of finished lengths.

After trimming, remove the burrs (formed by the trimmer cutter) from both the inside and outside of the case mouth using a Lyman Deburring tool. A few twists of the tool is all that is needed. Do not deburr the case to a sharp edge. Tap the case mouth on the bench to dislodge any brass chips from inside the case. Place the case, mouth down, in your loading block.

CAUTION: The material trimmed from a case flows from the junction of the case head and wall. As brass continues to flow and is trimmed away, this section of the case becomes thinner until it reaches a point where the case is severely weakened. Therefore, never trim a case more than four times (keep careful records). When a case needs its fifth trimming it must be discarded.

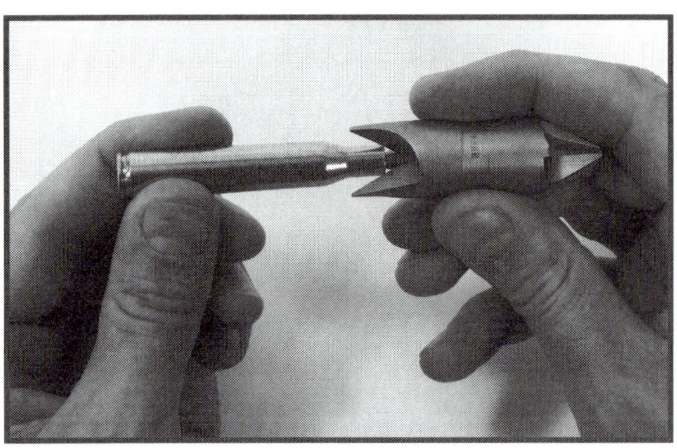

After trimming, all cases must be lightly deburred on the inside and outside of the case mouth.

Note: The initial trim of new cases is not counted when determining the number of times a case is trimmed as this trim is done not because of case stretch but rather to create a uniform length.

STEP TEN

PRIMER SEATING

While not absolutely essential, it is advisable to clean primer pockets before seating a new primer. This is a simple operation requiring only a few twists of the pocket cleaning tool. As each case's primer pocket is cleaned, place the case in the loading block, mouth up.

Bring a box of 100 primers to the bench and read the label aloud to insure that you have the correct brand and size. Then double check again. You must use the primer size and type called for in the data.

CAUTION: Safety glasses should be worn whenever handling primers.

Primer seating may be accomplished in many ways and with differing tools. One of the best methods is to use a Ram Prime unit mounted in the reloading tool's die station.

Its important to wash and thoroughly dry your hands before starting to prime. Place a quantity of primers (never more than 100 - or a lesser amount as needed) onto a primer flipper tray. Gently rotate the primer tray until all primers are anvil side up.

Most loading tools come equipped with a basic primer seating tool that primes the case at the normal shell holder position. Follow the instructions supplied with the tool. Place a primer, anvil up, into the priming post, push the post under the shell holder, and then lower the shell holder over the post to seat the primer. On many loading tools, this requires a "feel" method to seat the primer to the correct depth. Some tools will have a rudimentary stop to adjust primer seating depth.

A better method for primer seating is to use a ram-prime tool which mounts into the tool's die station. Primer seating depth then can be controlled by adjustment of the ram-prime body combined with using the press handle's solid stop at the end of the priming stroke. This method produces a very uniform primer seating depth.

Generally, primers should be seated 0.003" to 0.005" below flush of the case head - a nominal of 0.004" below flush.

CAUTION: Primers seated too high (above flush) are a needless hazard. It is possible that such primers can be ignited before the firearm action is closed, causing a serious accident. High primers are also prone to misfires. Primers seated too deeply (below flush) can become erratic in performance or misfire. For more details on primers see Chapter Six.

CAUTION: Primers are explosive and require special care in storage and handling. See Chapter Six for vital information.

Place the primed case in the loading block mouth down. When all cases have been primed, verify proper priming depth by running a finger over each case head. The novice should use a caliper to verify proper primer seating depth and then run a finger over several of these. This will teach the correct "feel" so that you can verify all remaining seating depths by "feel". After you check primer depth, return the case, mouth up, to the loading block.

After priming, return any unused primers to their original container and replace them in your storage area.

Each powder charge should be carefully weighed. The use of a powder dispenser will add speed and convenience.

STEP ELEVEN

WEIGHING POWDER AND CHARGING

Weighing powder must be done with great care and accuracy. Set up your powder scale carefully, following the instructions supplied with it. It is good practice to verify the scale's accuracy by using a weight check set.

Bring only one powder can to the loading bench. Read the label aloud. You must use the exact powder called for in the data. Then double check again. The inadvertent use of the wrong powder can cause a catastrophic accident.

Bring a box of bullets to the loading bench and read aloud the label on your bullet box to make certain the bullets are the correct weight (matched to the data you are using). Then measure the diameter and weigh a few bullets to be certain that what is in the box is the same as the label. Factory packaging errors have occurred. (A bullet will be seating immediately after a case is charged with powder.)

Pour some powder into the powder trickler and position the trickler alongside the scale pan. Also pour some powder into an open container (or preferably into a powder measure). Use a scoop of appropriate size to place a quantity of powder, somewhat less than a full charge, onto the scale pan.

If using a powder measure, adjust the measure to dispense somewhat less powder than you require. The metered charge will vary so be certain that the heaviest charges will not exceed the desired weight. (It is nettlesome to try and remove excess powder from the scale pan.) Dispense a metered charge directly onto the scale pan and then place the pan on the scale hanger.

Note: A handy powder scoop can be made by cutting off a fired case at an appropriate length and twisting a wire handle into the case rim's undercut. Straight cases such as the 30 MI Carbine, 44 Magnum and 45-70 make the best scoops.

Bring the scale into perfect balance, using the powder trickler to add one kernel of powder at a time to the scale pan. Now pour the weighed powder charge into a case using a powder funnel.

CAUTION: Make certain scale poises are not inadvertently moved during the loading process.

CAUTION: Powder is highly flammable and requires care in storage and handling. Be certain to read and follow the cautions contained in Chapter Seven.

CAUTION: Lyman lab technicians have observed a potential serious phenomena involving mechanical powder scales, plastic loading blocks, Styrofoam packaging and other objects made of plastic. These materials sometimes retain a static electric charge, enough to create an electro- static field of varying radii.

This electro-static field has proven capable of causing radical defection of uncharged and zeroed scales. Dependant upon circumstances, powder in the scale pan tends to dampen the amount of deflection by varying degrees.

We strongly urge that the loading bench be cleared before setting up the scale. Then replace equipment one piece at a time while observing the scale pointer. Any item that causes a scale deflection should be removed from the loading bench. Do this at every loading session.

Novices should avoid the use of compressed powder charges (where the powder level in the case is so high as to require compression of the powder in order to seat a bullet to the correct depth). Die adjustments for seating compressed powder charges are discussed in Chapter Seven. Chapter Seven also explains how to avoid using excessively compressed powder charges (a handloading no-no).

NOTES ON USING A POWDER MEASURE

We have instructed that after metering a powder charge, it be checked and brought into perfect balance using a scale and powder trickler. If you are loading ammunition for a non-critical application, in order to save time you may opt to pour a metered powder charge directly into a case (without the scale check and trickler balance). (**CAUTION:** This method should never be used with maximum loads or by novices at any time. When using a powder measure in this manner, ALWAYS check at least every tenth load on the scale to insure that the measure has not gone out of adjustment and that you are using a uniform metering technique.) Keep in mind that fine (small) grain or spherical powders lend themselves to more uniform metering as opposed to course (large) grain propellants. Be sure you are capable of metering uniform charges before using this method. Verify your uniformity by weighing 20 or more consecutive metered charges.

STEP TWELVE

BULLET SEATING AND CRIMPING

Next, immediately transfer the charged case to the loading tool and seat a bullet to the correct over-all length. Follow the die manufacturer's instructions to properly seat the bullet to the correct depth. The maximum overall length for a loaded round is clearly listed in the data for each cartridge. Dependant upon the bullet and equipment used, the finished individual overall cartridge length may vary by plus or minus 0.005".

Note: Generally, bullets should be seated to the overall length shown at the top of each data panel. Do check to see that ammo so assembled will feed through the magazine of your rifle and that it chambers properly. (**CAUTION:** Do this testing out of doors with the muzzle pointed at a safe backstop.) Or better yet, make a dummy round (no powder or primer) to check overall length. Circumstances, which include magazine length, chamber dimensions and bullet give, may

Immediately after adding powder to case, seat the bullet so as to have a finished cartridge of the correct overall length.

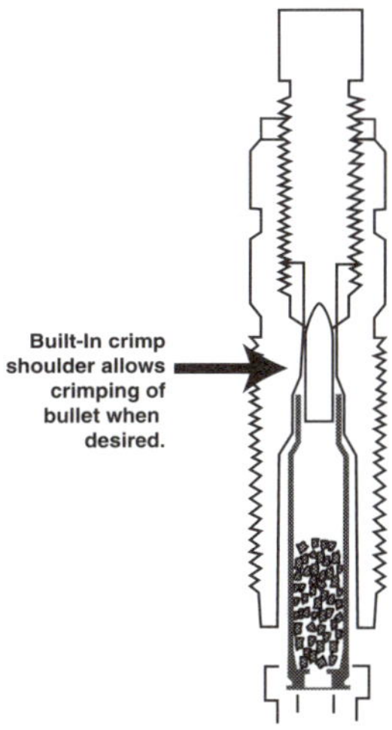

make it necessary to use a different length.

When all cases are charged and bullets seated, return all powder (from trickler, open container or powder measure) to the original container and return the container to its remote storage area.

Bullet crimping is required whenever there is a possibility of the bullet striking a firearm surface (during the feed/chamber cycle) with sufficient force to push the bullet deeper into the case. This means that ammunition for almost all semi-automatic and pump actions should have the case crimped to the bullet. Also, if there is a danger of the bullet creeping forward out of the case (loads with very heavy recoil) while in the magazine, ammunition should be crimped. All ammunition to be used in tubular magazines also must be crimped to prevent bullets from being driven deeper into the case during recoil. All ammo to be used in a revolver should be crimped.

Because crimping has a somewhat detrimental effect on accuracy, crimping should be limited to ammo intended for one or more of the just mentioned applications.

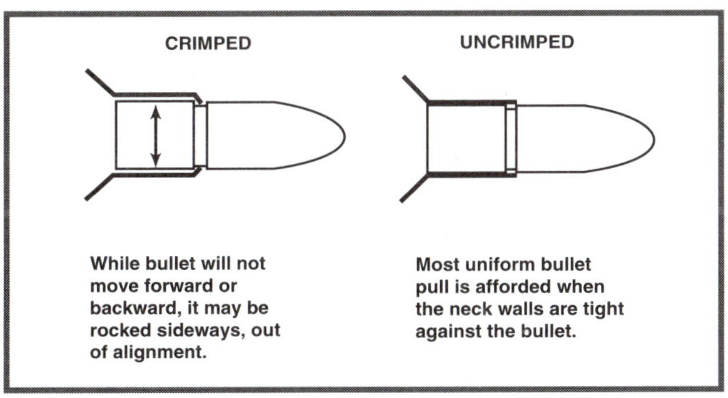

Keep in mind that your bullets must have a cannelure (a groove around the bullet) in order to crimp. Such bullets must first be seated to a depth that will align the case mouth with the center of the bullet cannelure. Adjusting your seating/crimping die requires that you first back off the bullet seating screw substantially. Then, screw the die down far enough to turn the case mouth slightly inward into the bullet cannelure when the loaded round is fully raised into the die. Be certain the bullet seating screw is backed off far enough to prevent it from touching the bullet.

Note: Crimping loads with compressed powder charges will require extra care. See Chapter Seven for details. As stated, to do away with the need to constantly re-adjust the bullet seat/crimp die, most reloaders purchase an extra die body.

Note: The crimping operation may be combined with bullet seating, but the best results are obtained when it is done as a separate operation.

STEP THIRTEEN

FINAL INSPECTION

The final inspection should be done with great care. Start by looking for imperfections. These may include, but are not limited to: case necks which split during bullet seating, and case shoulders which buckle during crimping (most often due to a poor bullet cannelure or improper die adjustment). A second check of primer depth should be made by running your finger over each case head. Dropping the loaded round into a case headspace gauge is also recommended to help insure it will chamber properly. Also, measure a sampling of the loaded rounds to check for proper over-all length. Should any round be found abnormal, discard it in a safe manner.

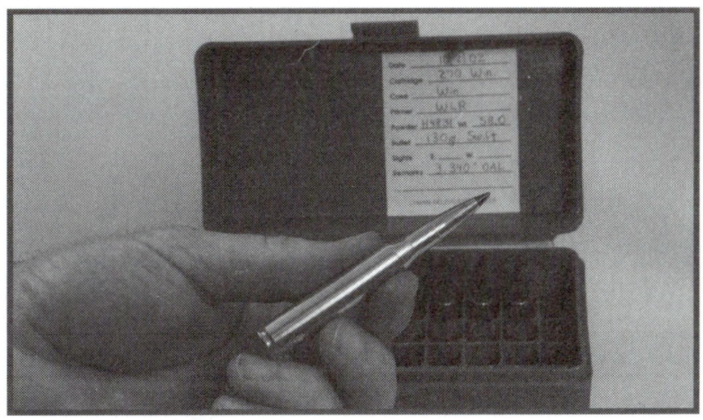

Make the final inspection of your loaded round and place it in an appropriately labeled box.

Place loaded rounds into a suitable container. Clearly mark the container with: date loaded; primer used; times trimmed; powder and charge; bullet brand, weight and type, and overall ammo length. Then, enter all this information into your reloading log. Your log should also include the lot numbers of all the components used.

As stated the foregoing outlines the basic steps to assemble bottleneck, jacketed bullet ammunition; but there are many other requirements that must be observed. BE SURE TO READ ALL THE OTHER PERTINENT CHAPTERS OF THIS BOOK BEFORE ATTEMPTING TO RELOAD AMMUNITION.

4 Reloading Handgun Cartridges

As previously stated, the mechanical steps of reloading are actually simple. Nonetheless, there are several necessary considerations to be given to each step. The observance of these considerations will help ensure that the ammo you assemble will be safe, will function reliably, and be accurate. Several of these considerations are covered in detail in sections 3 and 4 and in some of the informative lead-ins preceding the data for specific cartridges. This information is vital and must be read and understood before actually beginning to assemble ammunition.

This following material describes the actual steps involved in reloading fired handgun and rifle cartridges with straight cases. The information given applies equally to lead or jacketed bullets. Slight deviations from these steps are sometimes required. Such minor step changes are discussed as appropriate.

To reload a fired straight-cased cartridge, 13 basic operations must be performed. The sequence of these can vary slightly depending on the specific equipment employed and reloader preference. The sequence we have chosen is based on the specific tools we used to illustrate the process. This is the method most often favored by knowledgeable handloaders. Not every step is always required as we will indicate.

The thirteen steps to reload handgun ammo are as follows:

1. Selection of load and components to be used.
2. Case inspection.
3. Case cleaning (omitted when using new unfired cases).
4. Case lubrication (not required when using a carbide sizing die).
5. Case resizing and fired primer removal (not required when using new unfired cases).
6. Lubricant removal (not required when using new unfired cases or a carbide sizing die) and second inspection.
7. Case length measuring.
8. Case trimming and/or deburring (not always required).
9. Case mouth expansion.
10. Seating new primer.
11. Weighing and charging powder.
12. Bullet seating and crimping (as required).
13. Final inspection.

CAUTION: Read all pertinent material before actually beginning to assemble ammo.

STEP ONE

SELECTING A LOAD AND COMPONENTS

In the beginning, the selection of components may seem a difficult and confusing task. This is especially true for the reloader who has not yet acquired a general knowledge of basic ammunition details. Nonetheless, you will soon find that component selection quickly becomes an easy and fun part of reloading. The learning process can be hastened along by developing as much ammo knowledge as leisure time reading will allow.

Selecting Cartridge Cases

The selection of fired cartridge cases, or new brass, presents no special challenge. If you have saved your fired cases you need only separate them into specific groups by brand and lot number. Lot numbers appear on the factory ammo box. This number may be on an inside flap or the box back. Keeping brass segregated by lots will maximize accuracy potential and ballistic uniformity. If you purchase new unfired cases, of course they must be of the appropriate caliber. Bulk packaged brass (often in lots of 50, 100, or more) is the least expensive way of obtaining new cases.

Test Components:

Cases	Federal
Trim-to Length	1.149"
Primers	CCI 500
Primer Size	Small Pistol
Lyman Shell Holder	No. 1
Jacketed Bullets Used	Speer JHP #4007, 110 gr.
	Hornady JHP #35710, 125 gr.

Primer size and the specific primer we used is clearly shown at the beginning of the data for each caliber.

CAUTION: Never load cartridge cases from an unknown source, i.e. cases picked up at the range or purchased as once-fired brass. Use only brand new brass or cases obtained as the result of firing factory ammo in your firearm. For more on this topic see Chapter Five.

Selecting Primers

The correct primer size is listed at the beginning of the data for each cartridge, for example: large pistol, large pistol magnum, small pistol. In the beginning we suggest you use the exact primer we used for the development of our data. As an alternate, match the primer brand and correct size to the brand of the case you are reloading. Do not use magnum primers unless the data specifically calls for these as doing so can alter ballistic uniformity and the safety level of the data.

Do not allow cartridge nomenclature to enter into the selection of primers. For example the 32 H&R Magnum never requires the use of a magnum primer. Always, follow the primer size and type we list in our data table.

Selecting Powders

There are approximately 100 different propellant powders available to the reloader. The burning speed of each and the ballistics obtained can vary tremendously. Powders are designed to suit specific applications, such as: bullet weight, case shape and volume, pressure level and other specific ballistic and firearm needs. As a result only certain propellants are suitable for specific applications. When selecting a powder for your first reloading efforts, we suggest the use of the propellant listed in our data for the accuracy load.

158 gr. Jacketed HP 1.480" OAL — **BC:** .206 **SD:** .177

Powder	Sugg Starting Grains	Velocity fps	Pressure C.U.P.	Max Load Grains	Velocity fps	Pressure C.U.P.
Clays	2.8	648	14,500	3.2	712	16,600
Unique	4.7	645	14,400	5.0	710	16,600
				(+P)5.2	**761**	**18,200**
AA#5	5.0	674	12,100	5.8	821	16,300
Power Pistol	4.3	659	14,800	4.8	700	15,600
SR-4756	4.9	611	13,800	5.4	705	16,800
HS-6	5.5	572	12,400	6.5	726	16,600
Blue Dot	6.0	607	13,500	6.7	711	15,700
2400	8.4	645	13,500	9.4	745	16,400

Selection of the propellant chosen for your first loads is easily made by referencing the bullet weight (highlighted) you are using, the accuracy powder (highlighted), and the starting load as shown in the data.

CAUTION: Always start with the exact powder charge weight shown under the starting grains column. Heavier loads should not be used until the reloader has gained some experience and fully understands proper load development.

Selecting Bullets

Bullet selection may at first be somewhat confusing. To simplify the process, select a bullet weight to duplicate the factory ammo you favor. The appropriate diameter bullet may be determined from the diameter versus caliber bullet table in Chapter Eight. Many calibers use the same diameter bullet. For example, 38 Special, 357 Magnum and 357 Maximum all use the same jacketed bullet diameter of .357".

There is another consideration in bullet selection. This is for the need of a cannelure on any bullet to be used in a revolver. Roll crimping is essential for all such ammunition in order to prevent bullets from creeping forward under recoil. It simply is not possible to form a proper roll crimp unless the case mouth can be turned into a well formed bullet cannelure. However, ammo for any rimless semi-automatic case which headspaces on the case mouth should never be roll crimped.

For semi-automatic firearms another bullet consideration is the nose shape of the bullet. Pointed or round nose bullets are essential to proper firearm functioning in many such firearms.

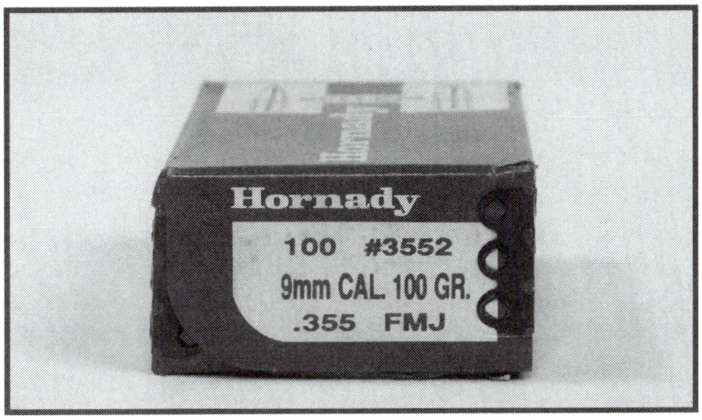

Bullets should be selected by the desired weight and correct diameter.

Our listed data specifies the bullet manufacturer's product number for each jacketed bullet we tested. We also list the bullet mould number for all cast bullet data. All of the listed bullets will work fine for target shooting. However, for proper jacketed bullet expansion on varmints or game you will need to make certain that the selected bullet is properly designed and suitable for the velocity range of your cartridge. If you are loading for a hunting purpose, avoid the use of bullets designated as Match or Target style as these may not expand properly on game.

THE LOADING SEQUENCE

The use of two loading blocks is suggested. As each step is performed, the case should be removed from one loading block, processed, and then placed in the second loading block. This will keep the process orderly and prevent many common bench errors.

Our reloading procedure follows the batch method. That is, a single operation will be performed on all cases to be reloaded before proceeding to the next operation.

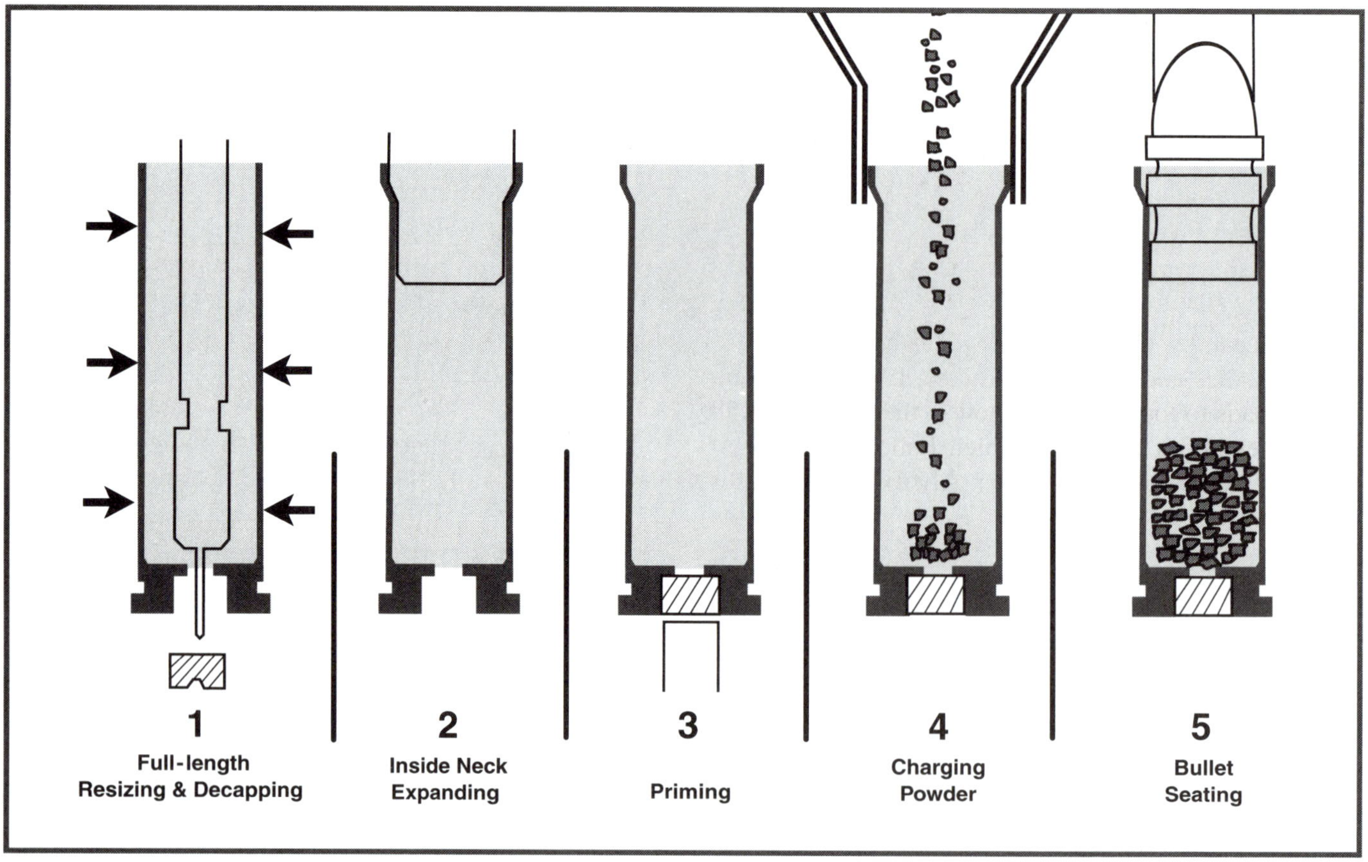

The basic steps of handgun case reloading.
I. Case is sized to return it to factory-like dimensions. The fired primer is removed during this operation. II. Case mouth is expanded III. New primer is seated IV. Propellant powder is weighed and placed into case. V. Bullet is seated and (if required) crimped.

STEP TWO

CASE INSPECTION

Fired cartridge cases have a finite life. Depending upon the firearm used, caliber of the firearm, internal ballistics of the load, and other considerations, it is reasonable to expect from 6 to 15 firings from each case. Eight firings are average for the typical handgun cartridge such as the 38 Special or 9mm Luger. Magnum cases such as the 357 Magnum, or 44 Magnum when assembled with heavy loads, typically last for only 4 or 5 firings. Low pressure cartridges fired in strong revolvers, such as the standard velocity loads for the 38 Special, generally offer the greatest number of firings.

All cases reach a point when further reloading becomes unsafe. Keep careful reloading records and perform a visual inspection of each case before, during, and after reloading to help ensure that you use only suitable fired cases.

Begin your inspection by wiping each case with a cloth to remove excess fouling, dirt, and any foreign material that could scratch your resizing die or the case itself. Turn each case mouth down and tap it lightly on the bench to dislodge anything that may have entered the case after firing.

Now look for split necks or bodies, signs of incipient case separation (a bright partial or complete ring around the case at the point where the case's solid base joins the wall of the cartridge), corrosion, or burn through perforations. Also look for any signs of gas

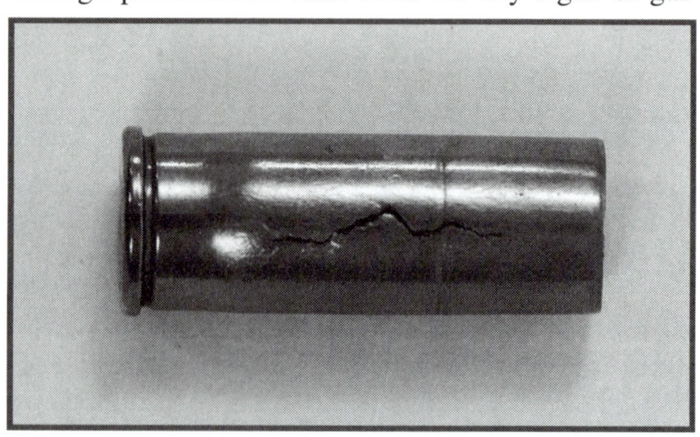

Cases must be inspected and those with any defect must be discarded.

leakage around the primer pocket. Eliminate all cases with any visual sign of defect or abnormality. (For more details refer to Chapter Five.) To prevent later inadvertent use of a rejected case, crimp its mouth shut with a pair of pliers before discarding it. Then place each case, mouth up in a loading block.

STEP THREE

CASE CLEANING

Case cleaning is an important step to protect your reloading dies and firearm chambers. If you also want your reloads to look like new, now is the time to put all your cases into a Lyman Tumbler. Follow the instructions that come with the tumbler. After removing cases from the tumbler tap the mouth of each case on the bench to insure that no tumbling media remains in it. Then wipe each case lightly with a clean cloth.

STEP FOUR

CASE LUBRICATION

When a cartridge is fired it expands somewhat. The expanded dimensions are not compatible with holding a new bullet with proper tension (bullet pull) and are not conducive to easy chambering. To avoid these and other difficulties, all fired cases must be resized.

Roll each case lightly across your lubricant pad. Do not fail to lubricate each case or it will stick solidly in the resizing die creating a very difficult to correct problem. Do not use excessive lubricant as doing so may cause cases to dent during the resizing step. Use only enough lubricant to insure the case enters and leaves the resizing die without undue difficulty. Be neat. Do not get lubricant into the primer pocket or case mouth.

We strongly recommend the use of a carbide resizing die (when available for your caliber) as this die will eliminate the need for this step as well as the later removal of the lube.

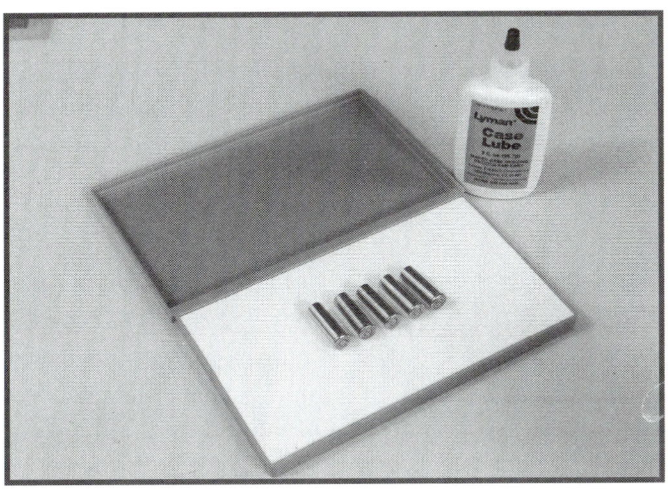

Each case must be lubricated before it is resized unless a carbide die is used.

STEP FIVE

CASE RESIZING AND FIRED PRIMER REMOVAL

Place a lubed case into the shell holder and run it into the full length resizing die. (Follow the instructions for proper die adjustment as explained in the material supplied with your die set.)

Generally, a non-carbide resizing die should be adjusted so that the shell holder, at the top of its travel, will contact the resizing die and create a slight cam action against the die. During this operation the fired primer will automatically be ejected from the case. Withdraw the case from the sizing die and place the case into the loading block, mouth down. Visually inspect each case as it is placed in the loading block to ensure the fired primer has been removed.

Note: The decapping rod should be adjusted just low enough to ensure the primer is pushed free of the case. If the decapping rod is too low it will impact the inside bottom of the case and be damaged.

Note: In the past, varying sources have suggested that the primer be seated as the case is withdrawn from the resizing die. However, it is possible for case lubricant to contaminate and thus destroy primers. Therefore, we suggest that this is not the ideal time for seating the new primer.

CAUTION: If you use a carbide resizing die the shell holder should make light contact with the bottom of the die. Heavy contact could crack the carbide sizing ring. Always follow the instructions that accompany such dies.

STEP SIX

LUBRICANT REMOVAL AND SECOND INSPECTION

Carefully wipe each case with a clean cloth so as to remove all traces of sizing lubricant. Use a clean section of the cloth for each case. Inspect the case for any flaws. Repeated case resizing and firings can cause case mouths and bodies to become brittle and split when fired or during resizing. Watch for signs of incipient case separation (see Chapter 5). It is a good idea at this time to drop the resized case into a maximum cartridge gauge as a worthwhile inspection step. Place

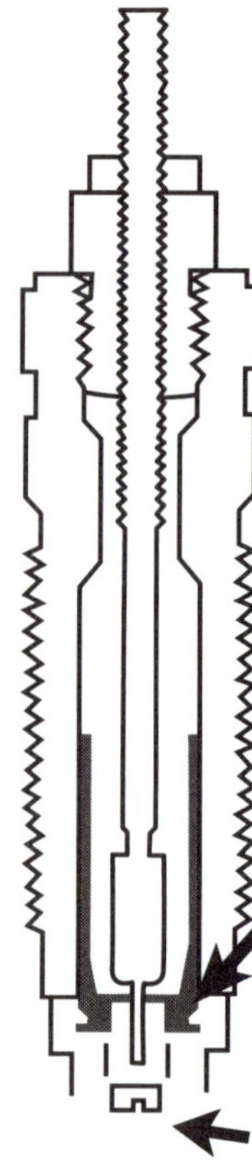

The first die usage step of reloading requires that the case be completely run into the full length resizing die. This will return the case to factory new dimensions and remove the old primer.

the cleaned and inspected case in your loading block, mouth up. Should you find any defects, this is the time to discard the entire lot of cases.

IMPORTANT - Case lubricant can ruin a primer causing delayed ignition or a failure to fire. To avoid potential primer contamination this is the point to stop the reloading process and thoroughly wash and dry your hands.

STEP SEVEN

CASE LENGTH MEASURING

Case measuring is an important step both for safety and for proper ammunition functioning. Cases stretch when fired and during resizing. If they exceed listed maximum case length, they will be difficult if not impossible to chamber. Excessive chamber pressure will also be caused by excessive case length. Therefore, each case must be carefully measured at this point. A dial indicating caliper is the best tool for this process.

The data for each cartridge clearly indicates the maximum allowable length for the resized case. If one or more cases are found to be at maximum or greater length, trim all of your cases to a uniform length.

IMPORTANT: Case length of rimless, straight cartridges is critical for proper ammo headspace. Be certain that such cases stay within the listed trim-to and maximum cartridge lengths.

STEP EIGHT

CASE TRIMMING AND DEBURRING

As stated, cases must be trimmed when they become too long. Case trimming is also recommended whenever loading new or once fired brass as they are often not of a uniform length. Trimming uneven cases to a uniform length will enhance accuracy and ballistic uniformity.

Note: if you will be crimping cases to bullets

(required for all ammo to be used in revolvers or tubular magazines), the crimping process will be less than satisfactory if cases are not of a uniform length.

The proper trim-to length for cases is clearly shown in the data for each cartridge. Adjust your trimmer according to the manufactures instructions. When trimming, allow for some dwell time - that is a number of rotations of the cutter after the case has been trimmed to length. This will help insure the maximum uniformity of finished lengths.

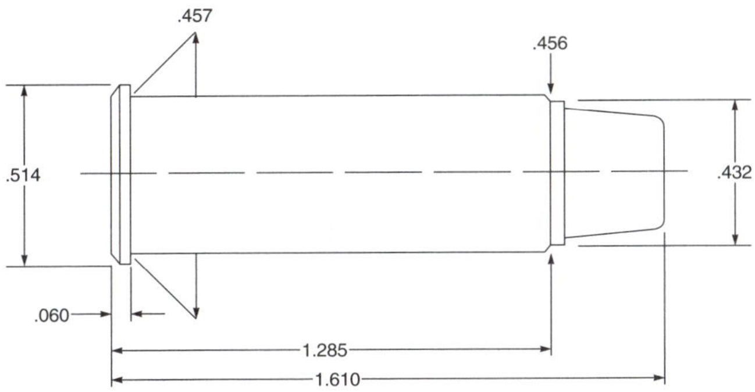

Test Components:

Cases	Remington
Trim-to Length	1.275"
Primers	CCI 300 & 350
Primer Size	Large Pistol, Std. & Magnum
Lyman Shell Holder	No. 7
Jacketed Bullets Used	Sierra JHC #8600, 180 gr.

After trimming, remove the burrs (formed by the trimmer cutter) from both the inside and outside of the case mouth using a Lyman Deburring tool. A few twists of the tool is all that is needed. Do not deburr the case to a sharp edge. Tap the case mouth on the bench to dislodge any brass chips from inside the case. Place the case, mouth down, in your loading block.

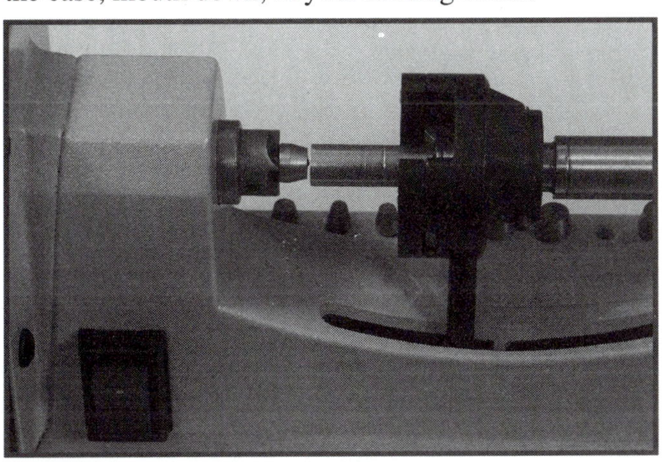

Case trimmers are miniature lathes that quickly restore a case to a safe overall length.

CAUTION: The material trimmed from a case flows from the junction of the case head and wall. As brass continues to flow and is trimmed away, this section of the case becomes thinner until it reaches a point where the case is severely weakened. Therefore, never trim a case more than four times (keep careful records). When a case needs its fifth trimming it must be discarded.

Note: Many handgun cases wear out long before any trimming is required.

Note: The initial trim of new cases is not counted when determining the number of times a case is trimmed as this trim is done not because of case stretch but rather to create a uniform length.

After trimming, all cases must be lightly deburred on the inside and outside of the case mouth.

STEP NINE

Case Mouth Expansion

Case mouths must be expanded to accept the bullet and to hold it with the proper tension (bullet pull). If you are loading lead bullets the case mouth should be both expanded and belled in order to prevent shaving lead from the bullet base during seating. Follow the instructions that come with all 3 die straight case sets or with the M die when purchased as an accessory for loading lead bullets into bottleneck cases.

Case mouth expansion in two steps is required when seating lead bullets.

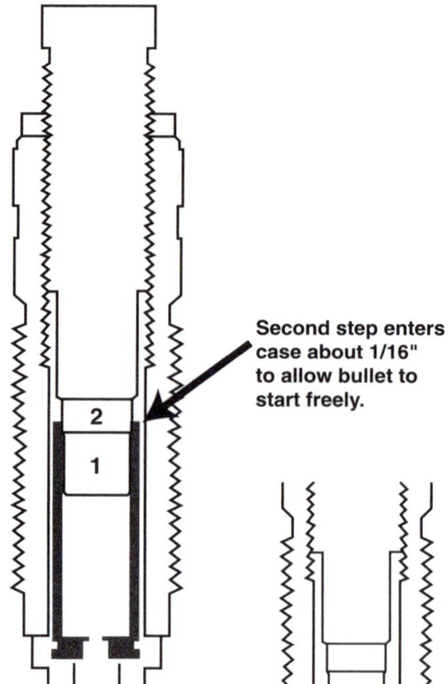

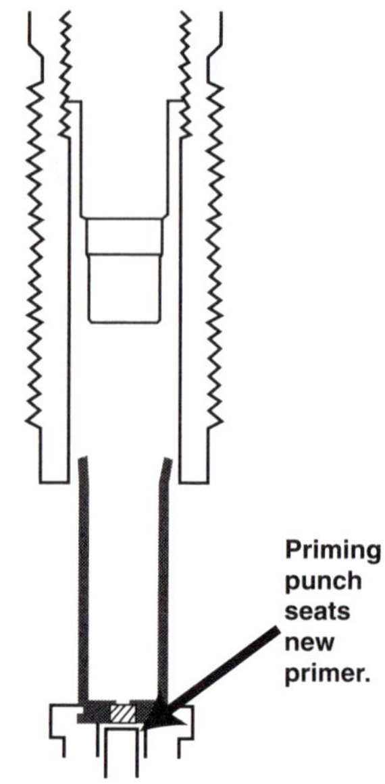

STEP TEN

PRIMER SEATING

We recommend that primer pockets be cleaned before seating a new primer. This is a simple operation requiring only a few twists of the pocket cleaning tool. (Not required when loading new cases). As each case's primer pocket is cleaned, place the case in the loading block, mouth up.

Bring a box of 100 primers to the bench and read the label aloud to insure that you have the correct brand and size. Then double check again. You must always use the primer size and type called for in the data.

Its always good practice to again wash and thoroughly dry your hands before starting to prime. Place a quantity of primers (never more than 100, or a lesser amount as needed) onto a primer flipper tray. Gently rotate the primer tray until all primers are anvil side up.

Primer seating may be accomplished in many ways and with differing tools. One of the best methods is to use a Ram Prime unit mounted in one of the reloading tool's die stations.

Most loading tools come equipped with a basic primer seating tool that primes the case at the normal shell holder position. Follow the instructions supplied with the tool. Place a primer, anvil up, into the priming post, push the post under the shell holder, and then lower the shell holder over the post to seat the primer. On many loading tools, this requires a "feel" method to seat the primer to the correct depth. Some tools will have a rudimentary stop to adjust primer seating depth.

A better method for primer seating is to use a ram-prime tool which mounts into the tool's die station. Primer seating depth can then be controlled by adjustment of the ram-prime body (using the press handle's solid stop at the end of the priming stroke). This method produces a very uniform primer seating depth.

Generally, primers should be seated 0.003" to 0.005" below flush of the case head - a nominal of 0.004" below flush.

CAUTION: Primers seated too high (above flush) are a needless hazard. It is possible that such primers can be ignited before the firearm action is closed, causing a serious accident. High primers are also prone to misfires. Primers seated too deeply (below flush) can become erratic in performance or misfire. For more details on primers see Chapter Six.

CAUTION: Primers are explosive and require special care in storage and handling. See Chapter Six for vital information.

Place the primed case in the loading block mouth down. When all cases have been primed, verify proper priming depth by running a finger over each case head. The novice should use a caliper to verify proper primer seating depth and then run a finger over several of these. This will teach the correct "feel" so that you can verify all remaining seating depths by "feel". After you check primer depth, return the case, mouth up, to the loading block.

After priming, return any unused primers to their original container and replace them in your storage area.

STEP ELEVEN

WEIGHING AND CHARGING POWDER

Weighing powder must be done with great care and accuracy. Set up your powder scale carefully, following the instructions supplied with it. It is good practice to verify the scale's accuracy by using a weight check set.

Bring only one powder can to the loading bench. Read the label aloud. You must use the exact powder called for in the data. Then double check again. The inadvertent use of the wrong powder can cause a catastrophic accident.

Bring a box of bullets to the loading bench and read aloud the label on your bullet box to make certain the bullets are the correct weight (matched to the data you are using). Then measure the diameter and weigh a few bullets to be certain that what is in the box is as described on the label. Packaging errors have occurred. (A bullet will be seated immediately after a case is charged with powder.)

Pour some powder into the powder trickler and position the trickler alongside the scale pan. Also pour some powder into an open container (or preferably into a powder measure). Using a scoop of appropriate size, place a quantity of powder, somewhat less than a full charge, onto the scale pan. (If using a powder measure, adjust the measure to dispense slightly less powder than you require. Meter a charge directly onto the scale pan and then place the scale pan on the scale hanger.)

Note: A handy powder scoop can be made by cutting off a fired case at an appropriate length and twisting a wire handle into the case rim's undercut. Straight cases such as the 357 Magnum or 44 Magnum make the best scoops.

Bring the scale into perfect balance by using the powder trickler to add one kernel of powder at a time to the scale pan. Now pour the weighed powder charge into a case using a powder funnel.

CAUTION:: Make certain scale poises are not inadvertently moved during the loading process.

CAUTION: Powder is highly flammable and requires care in storage and handling. Be certain to read and follow the cautions contained in Chapter Seven.

CAUTION: Lyman lab technicians have observed a potential serious phenomena involving mechanical powder scales, plastic loading blocks, Styrofoam packaging and other objects made of plastic. These materials sometimes retain a static electric charge, enough to create an electro-static field of varying radii.

This electro-static field has proven capable of causing radical defection of uncharged and zeroed scales. Dependant upon circumstances, powder in the scale pan tends to dampen the amount of deflection by varying degrees.

We strongly urge that the loading bench be cleared before setting up the scale. Then replace equipment one piece at a time while observing the scale pointer. Any item that causes a scale deflection should be removed from the loading bench. Do this at every loading session.

Novices should avoid the use of compressed powder charges (where the powder level in the case is so high as to require compression of the powder in order to seat a bullet to the correct depth). Die adjustments for seating compressed powder charges are discussed in Chapter Seven. Chapter Seven also explains how to avoid using excessively compressed powder charges (a handloading no-no). Happily most handgun loads avoid compressed powder charges.

Immediately after adding powder to case, seat the bullet so as to have a finished cartridge of the correct overall length.

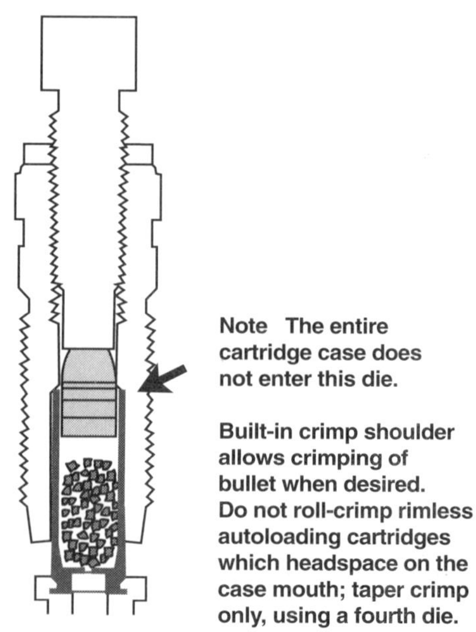

NOTES ON USING A POWDER MEASURE

We have instructed that after metering a powder charge, it be checked and brought into balance using a scale and powder trickler. If you are loading ammunition for a non-critical application, you may want to pour a metered powder charge directly into a case (without the scale check and trickler balance) in order to save time. (Note: This method should never be used with maximum loads or by novices at any time.) When using a powder measure in this manner, ALWAYS check at least every tenth load on the scale to insure that the measure has not gone out of adjustment and that you are using a uniform metering technique. Fine (small) grain powders lend themselves to more uniform metering as opposed to course (large) grain propellants. Be certain that you are capable of metering uniform charges before using this method. Verify your ability by metering and weighing at least 20 consecutive charges.

STEP TWELVE

BULLET SEATING AND CRIMPING

Next immediately transfer the charged case to the loading tool and seat a bullet to the correct over-all length. Follow the die manufacturer's instructions to properly seat the bullet to the correct depth. The maximum overall length for a loaded round is clearly listed in the data for each cartridge. Dependant upon the bullet and equipment used, the finished overall cartridge length may vary by plus or minus 0.005".

Note: Generally, bullets should be seated to the overall length shown at the top of each data panel. Do check to see that ammo so assembled will feed through the magazine of any automatic and that it chambers properly. Check revolver ammo to see that it does not come too close to the end of the cylinder. (**CAUTION:** Do this testing out of doors with the muzzle pointed at a safe backstop.) Or better yet, make a dummy round (no powder or primer) to check the overall length and function. Circumstances, which include magazine length, chamber dimensions, cannelure location, and bullet shape, may make it necessary to use a shorter than suggested overall length.

CAUTION: Excessively short overall handgun cartridge lengths can cause dangerous chamber pressure.

When all cases are charged and bullets seated, return all powder (from trickler, open container or powder measure) to the original container and return the container to its remote storage area. At this time, re-verify that the correct powder was used.

CAUTION: When loading rimless case ammo for a semi-automatic handgun (such cartridges headspace from the case mouth) never roll crimp your case. Doing so may dangerously shorten the case headspace dimension. For such ammo always use a taper crimp. Taper crimping dies are standard in some die sets or may be purchased as an accessory.

For revolvers, using a rimmed case, roll crimping is required to keep bullets from creeping out of the case during recoil. All rimmed handgun ammo to be used in tubular magazine rifles should also be roll crimped to prevent bullets from being driven deeper into the case during recoil. Only flat or blunt nosed bullets should be used in tubular magazines.

Keep in mind that when roll crimping, your bullets must have a cannelure (a groove around the bullet). Such bullets must first be seated to a depth that will align the case mouth with the center of the bullet cannelure. Adjusting your seating/crimping die requires that you first back off the bullet seating screw substantially. Then screw the die down far enough to turn the case mouth slightly inward into the bullet cannelure when the loaded round is fully raised into the die. Be certain the bullet seating screw is backed off far enough to prevent it from touching the bullet.

Note: Crimping loads which compress the powder charge will require extra care. See Chapter Seven for details. Additionally, to do away with the need to constantly re-adjust the bullet seat/crimp die, most reloaders purchase an extra die body for any crimping operation.

Note: The crimping operation may be combined with bullet seating, but the best results are obtained when it is done as a separate operation.

STEP THIRTEEN

FINAL INSPECTION

The final inspection should be done with great care. Start by looking for imperfections. These may include, but are not limited to: case necks that split during bullet seating, or cases that buckled during crimping (due most often to poor bullet cannelure or excessive die-down adjustment). A second check of primer depth should be made by running your finger over each case head. Dropping the loaded round into a maximum case gauge is recommended to insure it will chamber properly. Also measure a sampling of the loaded rounds to insure proper over-all length. Should any round be found abnormal, discard it in a safe manner.

Place loaded rounds into suitable containers. Clearly mark the containers with: date loaded; primer used; times trimmed; powder and charge; bullet brand, weight, and type, and overall length. Enter all this into your reloading log. Your log should also include the lot numbers of all components used.

The loading procedure for bottleneck handgun cases, i.e. 30 Luger and 30 Mauser, is the same as for bottleneck rifle cartridges as described in the preceding chapter.

As stated, the foregoing outlines the basic steps to assemble most handgun ammunition. However, there are many other requirements that must be observed. BE SURE TO READ ALL THE OTHER PERTINENT CHAPTERS OF THIS BOOK BEFORE ATTEMPTING TO RELOAD AMMUNITION.

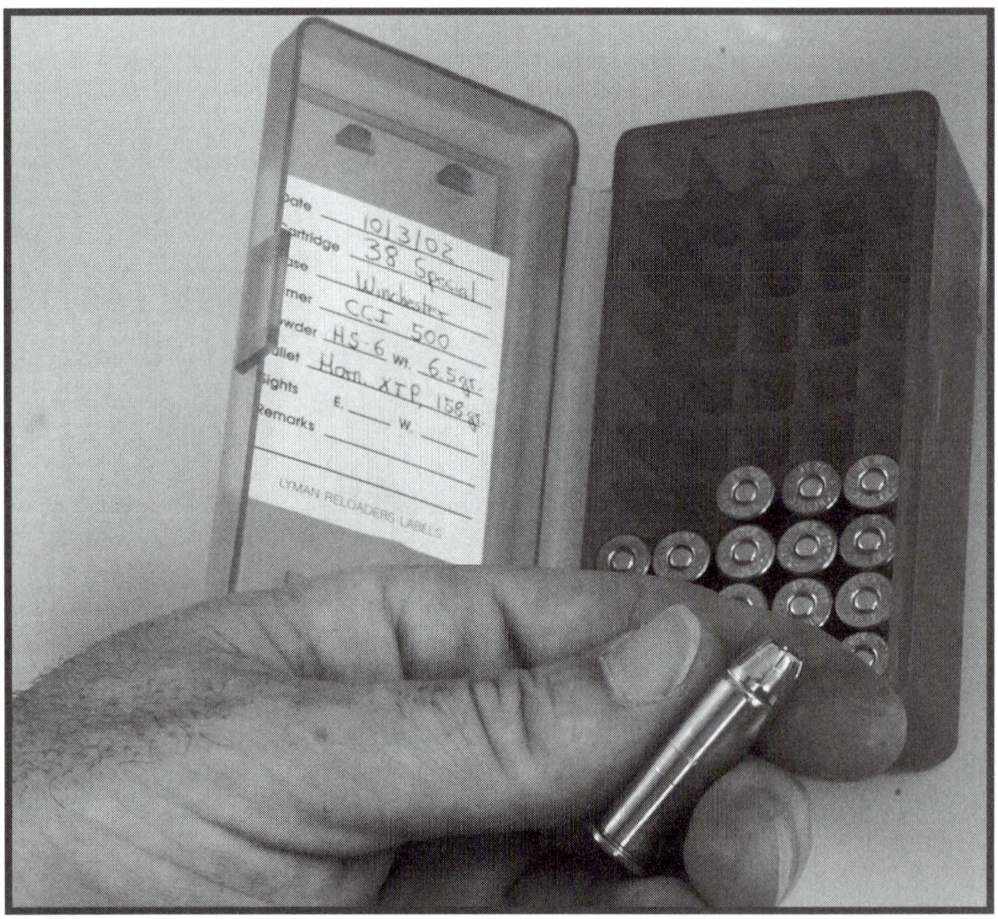

Make a final inspection of your loaded round and place it in an appropriately marked box.

Section 3

5 Cases and Their Preparation

Cartridge cases are available with two types of primer pockets. Generally, only those cases with Boxer primer pockets should be considered useful. The proximate drawing clearly shows the differences between Boxer primed and Berdan primed style cases.

Note: Berdan primed brass cases can be reloaded. However, they require difficult to find and slow to use hand tools for decapping, very hard to locate anvil-less (Berdan) primers, often special size primer seating posts, as well as special data (because of the differing ignition characteristics of the various Berdan primers). For all these reasons we feel it is impractical to reload Berdan type cases.

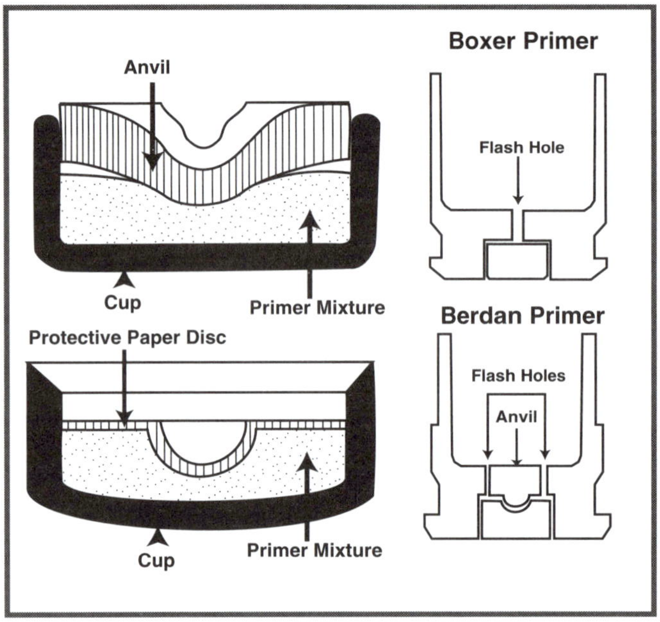

The European designed Boxer primer type case has become a standard. The American designed Berdan primer type case is almost never used in the United States — the major exception being the non-reloadable CCI Blazer ammo cases.

Case Material

Cartridge cases have been made from a wide variety of material including brass, steel, aluminum and other less encountered materials. Only cases made of brass have value for the reloader. Brass cases may sometimes be nickel plated. These are equally as reloadable as plain brass cases. The disadvantage of plated cases is that the nickel will sometimes flake off during firing/resizing cycles. This flaking can mar loading dies. The advantage of nickel plated cases is that, under adverse atmospheric conditions, they generally will not corrode as fast as brass.

CAUTION: Never reload any non-brass cases, i.e. those made of steel, aluminum, plastic or other material.

Case Configurations Rim Types

The head of a case (that portion which accepts the primer and affords a purchase for the firearm's extractor) may take several forms. These include: rimmed, semi-rimmed, rimless, rebated (reduced), and belted.

Rimmed cartridges were the first style cases developed. The large rim area allows for a substantial purchase on the case by the firearm's extractor. The rim's enlarged diameter and thickness also allows for an easy means to control headspace. Because body dimensions are less critical with this style case, chamber dimensions are often less tightly controlled as compared to chambers for rimless cases. Rimmed cases are often used in lever action and single shot rifles as well as revolvers. Typical examples of rimmed cases include the 22 Hornet, 30-30 Winchester, 45-70 Government, 38 Special and 44 Magnum. Rimmed cases may or may not have an undercut between the case body and the rim.

Case Styles

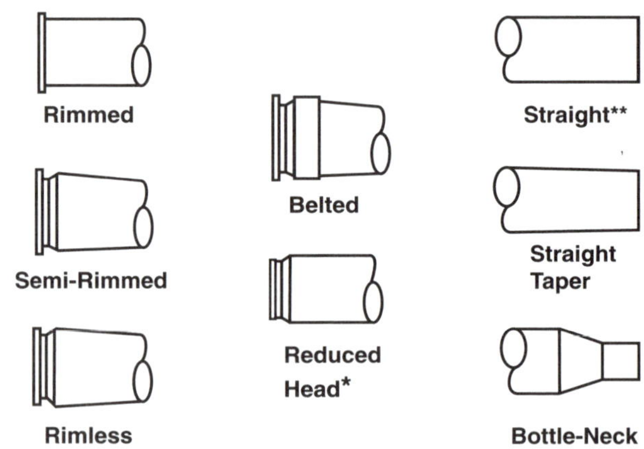

* Function is the same as rimless type.
** Straight cases do have a slight taper which is not visibly apparent.

A semi-rimmed case has a rim which is only slightly larger than the back end of the case body. This design is sometimes used to increase a smaller body rimless case's rear surface to fit a larger diameter bolt face. There are relatively few semi-rimmed cases. Those most often encountered include the 225 Winchester, 220 Swift (both designed to fit the larger 30-06 size bolt face), 25 ACP, and the 32 ACP. Semi-rimmed rifle cases are generally headspaced in the same manner as rimless cases. Semi-rimmed cases always have an undercut between the case body and rim area to ensure an adequate engaging surface for the firearm's extractor.

The rimless case design is extremely popular. This type of case has a rim of the same nominal diameter

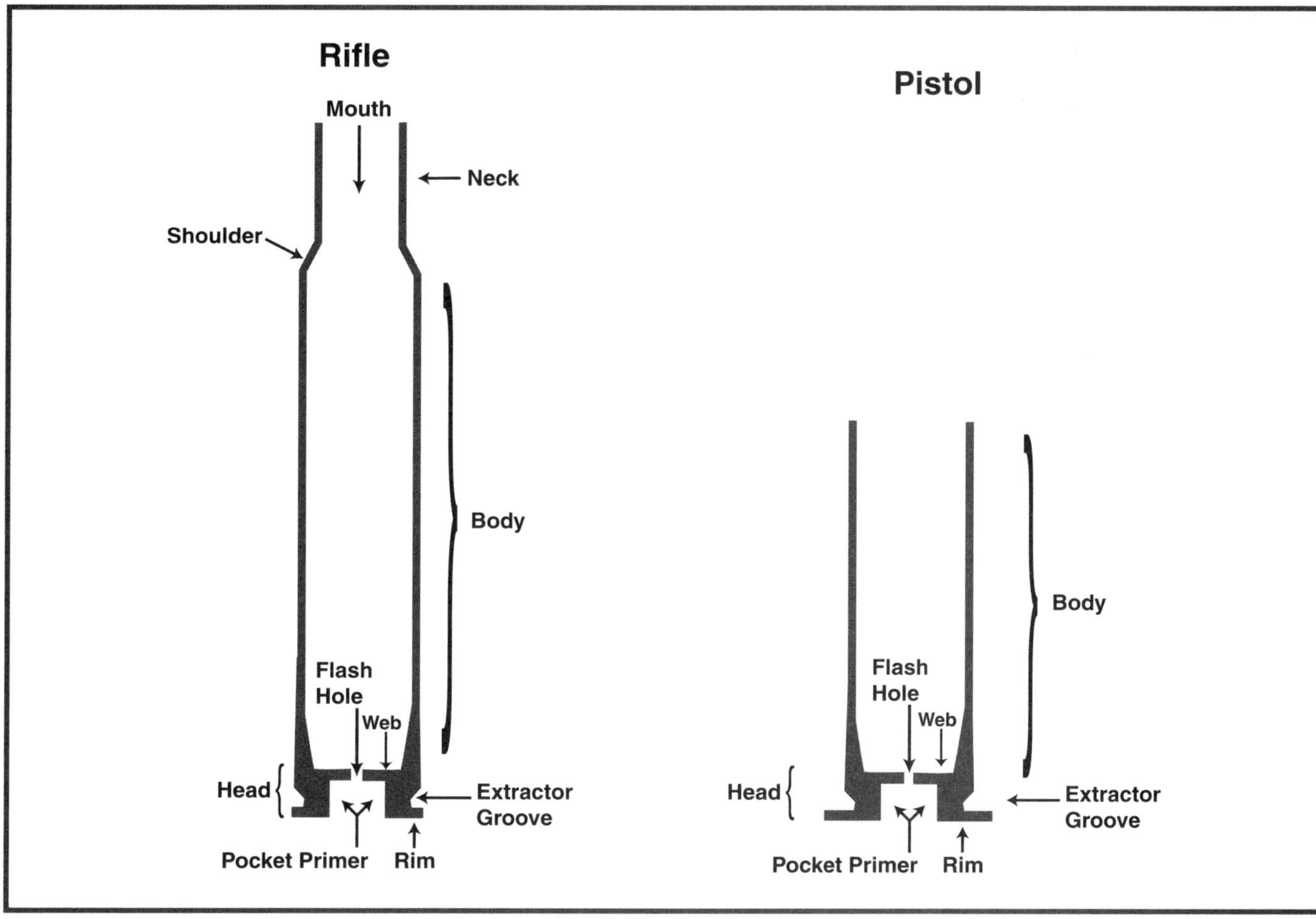

While cases vary in shape and size, the same basic nomenclature applies to all center fire cartridges.

(including any case taper) as the rear end of the case's body. There is always an under-cut between the rim and case body to afford a purchase area for the firearm's extractor. Typical of this style is the 223 Remington, 243 Winchester, 270 Winchester, 30-06 Springfield, and 45 ACP. Rimless cases are generally preferred for most firearms as they tend to feed smoothly from a magazine. The headspacing of most rimless cartridges is tightly controlled by using a diameter (datum line) around the case's shoulder and that diameter's distance from the cartridge head as the critical dimension. In cases without a shoulder, such as the 45 ACP, headspace is controlled by the dimension from the back edge of the cartridge case's head to its forward edge (case length).

A rebated (reduced) head case is basically a rimless case design that allows a large body case to be used with a bolt face designed for a smaller case. With this style case, the rim area of the case is substantially smaller than the case body. Examples of this style case include the 284 Winchester, and the 300 Winchester Short Magnum. Rebated cases always have an undercut at the rim/body juncture.

A belted case has an enlarged band at the head of the case with an undercut to allow for extractor purchase. The length of the belt, much as the rim thickness of a rimmed case, is used to control headspace. As other chamber dimensions are less critical, often they are excessively large. Therefore, as with a rimmed case, this design may suffer excessive case stretching and expansion and thereby result in shortened case reloading life.

Case Body Types

Cases come in three basic body types: straight, straight taper and bottleneck.

Straight cases originated with large caliber, low pressure cartridges designed for use with black powder. A typical cartridge of this type is the 45-70 Government. A straight case design is preferred for semiautomatic handgun cartridges. The 45 Auto is an example of this type case. Also, a straight case shape often is used for large bore cartridges to gain maximum powder space without requiring excessively large firearm actions; i.e. the 458 Winchester Magnum.

A straight taper case has a noticeable slight body taper. Examples of this type are the 38-55 Winchester and the 9mm Luger.

Bottleneck cartridges allow for the use of slower burning powders as compared to straight cases. This results in a higher level of cartridge efficiency, i.e. greater velocity with a given bullet weight. The 30-06 is a bottleneck case of the rimless type, the 30-30 is a bottleneck case of a rimmed design and the 284 Win. is a bottleneck case of a rebated rim type. Cartridges such as the 300 H&H Magnum and 375 H&H Magnum are bottleneck cases of the belted head design

.

Case Strength And Failures

Brass cartridge cases have gone through almost a century and a half of evolution. As a result, today's cases have greatly increased strength and reloadability. Earlier types had a rather thin web at the interior base. A balloon-like hump was present in the base to accept the primer pocket. Such balloon head cases should never be reloaded as they are unable to withstand the pressures developed with modern ammo.

However, the reloader must keep in mind that the brass case is still the weak link in the reloading chain. Depending upon specific caliber, cases may vary enormously in strength. Some calibers are designed to handle very modest pressure levels while others were designed to withstand working pressures of up to 65,000 psi.

Sufficient strength is a vital requirement of any case to be reloaded. For this reason the reloader must detect and eliminate all weakened or worn out cases.

Pressure Versus Case Life

Pressure development in a fired cartridge is essential. Without pressure the bullet could not be driven through the barrel. Nonetheless, pressure must be carefully controlled to insure that the safety limits of the case and firearm are not exceeded. Additionally, shot-to-shot pressure must be kept at a uniform level to insure consistent ballistic results.

Brass case reloading life is dependant on the level of pressure generated. All else equal, a cartridge that produces an average pressure of 40,000 psi. will be reloadable a great many more times than one that produces an average pressure of 60,000 psi.

Pressure Too Low

If pressure is too low, the brass case may not fully obturate (expand to fit and seal the chamber). When this happens, some gas may blow by the case mouth and leave a sooty deposit on the outside of the case body. (Note: some deposit and discoloration on the neck area is normal even when pressure levels are appropriate.) Excessively low pressure often will allow the primer to back part of the way out of the primer pocket because there is insufficient pressure to force the case head fully to the rear. This condition should be avoided as in extreme cases it could cause gas to flow into the shooters face.

Pressure Too High

High pressure causes a case to age quickly. Case stretch, and hence the frequency of case trimming, increases as pressure goes up. Primer pockets also enlarge as pressures go higher.

When pressure is too high, rapid or even sudden case failure will occur. Some failures can be catastrophic, causing firearm damage and even personal injury - or worse.

SEVERE DANGEROUS CONDITIONS

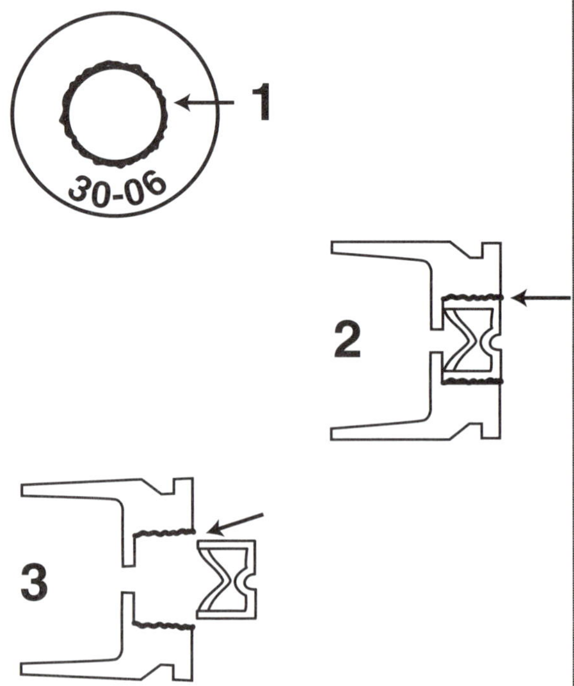

1. Dark smudge line around primer indicates gas leakage.
2. Pocket opens, leaving primer loose. A dark smudge, indicating gas leakage, is usually present.
3. Pocket opens completely, freeing the primer. A dark smudge, indicating heavy gas leakage, is usually present.

Gas leaking at the primer pocket (seen as a sooty deposit around the fired primer) or primer pockets which enlarge so much as to allow the primer to be blown partially or completely out of the case are sure signs that something is drastically wrong and creating very excessive and dangerous pressure.

Should you encounter a case in which the primer seats more easily than normal, the case must be discarded. This condition occurring on several cases, after only one or two firings, should be considered an indication that your load is developing very excessive chamber pressure. However, as cases reach the end of their useful reloading life this condition may occur naturally. When you have discarded 5% of a lot of brass because of too-easy primer seating, it is time to discard the entire lot.

CAUTION: Never reload any case in which the primer seats too easily. Such a case is a needless hazard and should be discarded. (Note: This condition can also occur with undersized primers.) Keep in mind that a few leaky primer pockets can ruin your bolt face. The escaping gases will cut multiple dimples into the bolt face. Worse, a blown primer can cause shooter injury.

The normal pressure generated in a cartridge at firing is part of the gradual deterioration of a case's useful reloading life. The actual amount of chamber pressure can cause a notable change in useful case life. Some time back, we separated a lot of brass into two portions. The first was used for a full-power hunting load with a chamber pressure of approximately 52,000 copper units of pressure (c.u.p.). These cases were fired five times before we deemed it advisable to discard the lot.

The cases in the second portion were loaded as light target loads with a chamber pressure of just over 40,000 c.u.p. These cases were fired eight times without any sign of deterioration. Indeed, they required only two trimmings during our test.

Cases From Questionable Sources

The next time you see some fired cases for sale, you may want to think twice before taking them to your reloading bench. Usually, there is no way you can be absolutely sure of: their general condition as it relates to age, number of firings, number of times trimmed, the headspace of the gun they were fired in, the pressures generated at firing, or any other abuse the cases may have suffered.

Even if the source is sure that all such cases are from the first firing of factory ammo, it will be impossible to be certain of the headspace of the firearm in which they were fired. Additionally, if but one old worn-out case was mistakenly picked up off the range with the once-fired cases, you will be reloading a potential accident. Keep in mind that modern case cleaning methods can make even old, worn out, cases look like new.

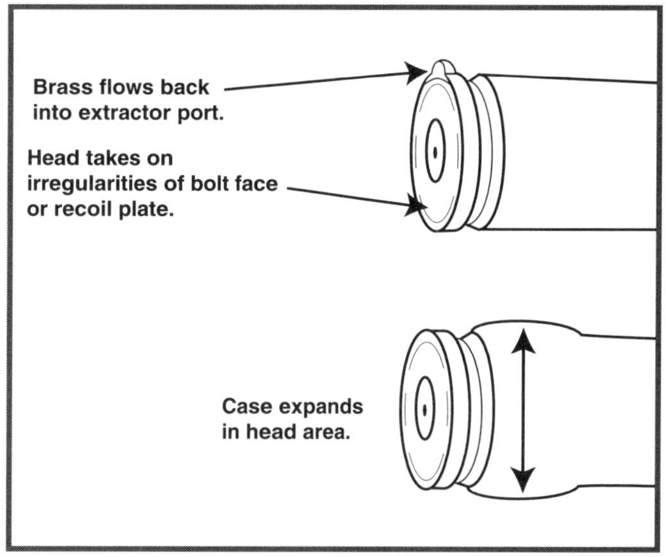

Very high and dangerous pressure can distort case shape.

CAUTION: Load only cases which you have obtained as new unfired brass or which you have on hand as the result of firing new factory ammo in the firearm for which you wish to reload. A few pennies saved on bargain so called "once-fired" brass makes little sense when one considers the cost of a firearm and/or medical treatment, or worse. Stay with new cases and keep the selection of brass worry-free and safe.

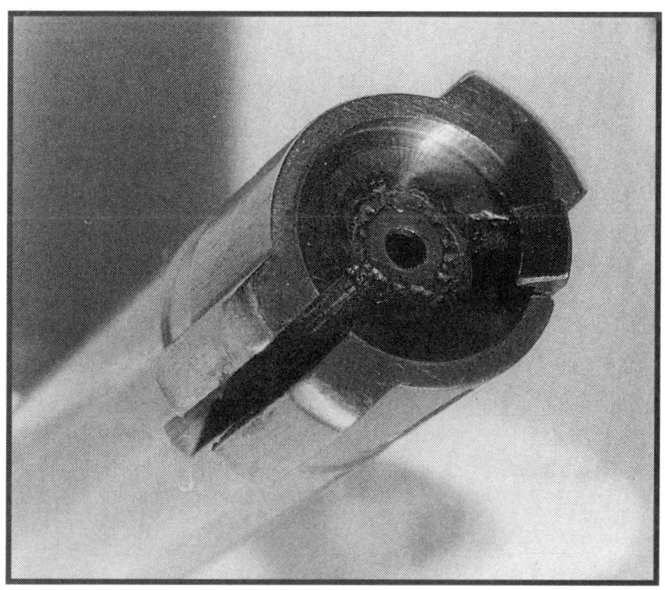

This bolt face was ruined by gas leaking around a primer pocket.

Case Deteriorating Conditions

Worn cases most be avoided. Doing so means you must keep in mind that there are many factors that wear brass out. Brass case failures can be caused by:

1. Firing in chambers having excess headspace.
2. Using loads that develop excessive pressure.
3. Too many firings.
4. Trimmed too often.
5. Becoming brittle simply due to aging.
6. Being resized in a faulty sizing die.
7. Improperly adjusted or faulty full length sizing dies.
8. Excessive crimping.
9. Other less common factors.

Case Head Separation

The separation of the case head from the body, called case separation, is a potentially dangerous condition. It is caused by the weakening of the case due to a thinning of the brass at the junction of the web and sidewall as a result of case stretching. Separation is sometimes preceded by a telltale bright ring, just in front of the case web, encircling part or all of the fired case's circumference. This condition is called incipient case separation. The next degree of failure could be a partial splitting of the case along the bright line. The final phase of failure is complete case separation.

Note: Actual separation failure of a case may not occur in steps as outlined. Incipient case separation can occur without early warning signs. In a separation situation, if the rear section of the case has sufficient flexible material to obturate and seal the chamber, no permanent damage will occur. However, if the rear portion of the case fails to seal the chamber, gases escaping rearward have the potential to damage the gun or injure the shooter.

Every reloader should learn to detect incipient case head separation the slightly lighter coloration of a ring-like area partially or fully encircling the case at the web/body junction. This discoloration often precedes the tell-tale bright ring or actual case failure.

Note: Premature case separation can also be caused by firearms having excessive headspace or faulty reloading dies.

Incipient case separation is to be avoided at all cost. Shown are cases with the bright ring that warns case separation is about to happen. These cases already have been loaded several times too often.

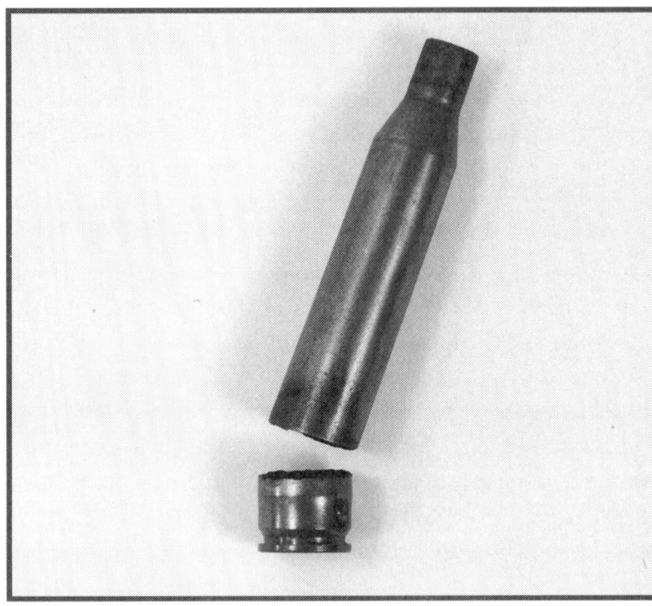

This case suffered complete separation. Such occurrences can be catastrophic to shooter and firearm.

Cartridge Case/Firearm Headspace And Headspace Gauges

The headspace of a cartridge/firearm relationship is very important. Headspace is a measurement of the distance between the headspacing surface of a cartridge (when the case is fully rearward in the chamber) and the headspacing surface of the chamber. Rimmed cartridges use the edge of the rim as the headspacing surface while rimless cartridges most often headspace from a mid-point on the case shoulder; both in relationship to a corresponding chamber surface. The headspace surfaces of a belted case system is the forward edge of the case belt and the corresponding chamber surface.

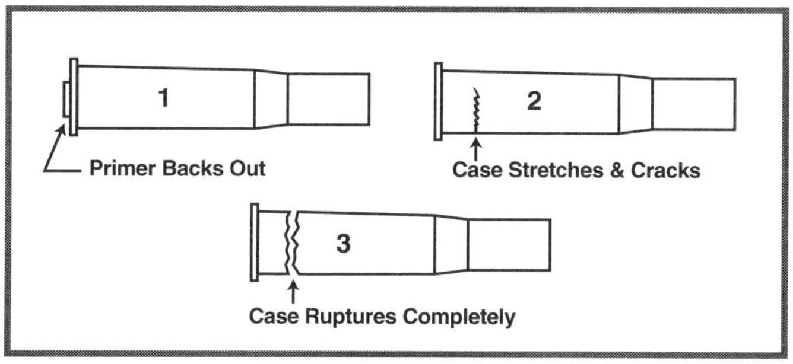

The above illustrations are indications of excessive headspace. Do not use any firearm which shows such signs of headspace problems.

If the distance from the cartridge's headspacing surface to the matching chamber surface is too great, the cartridge/firearm system is said to have excessive headspace. When this condition is present, the case is inadequately supported by the chamber and the cartridge may stretch excessively, split, or burst, upon firing.

Case headspace gauges will allow the reloader to adjust full length resizing dies so that the fired case will be shortened only a minimal amount during the reloading process. This will help prevent case separation failures. However, if a chamber is excessively deep, no amount of die adjustment will prevent premature case failure as the case will be stretched beyond reasonable limits on its first firing. Gauges made for cartridges that headspace on the shoulder, rim, or case mouth are of a fixed type. Gauges for belted cases are adjustable. Case headspace gauges are available from several manufacturers including Lyman. Ideally, every ammo assembler should have a case headspace gauge and a firearm maximum headspace gauge (a no-go gauge) for each cartridge reloaded.

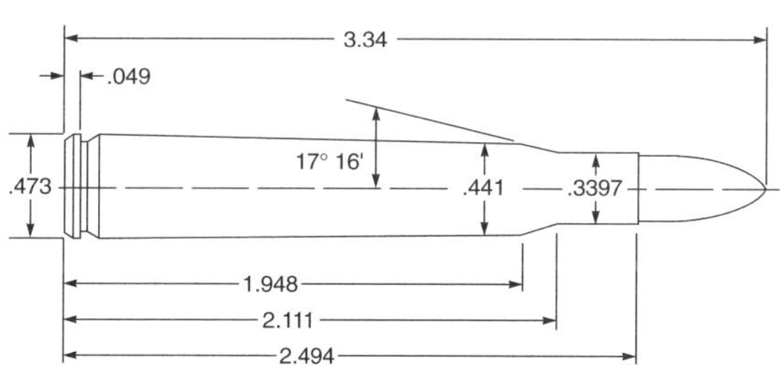

Case measurements are critical to ammo performance and safety. Checking the resized case for overall length and maximum headspace is essential during every reloading of the case.

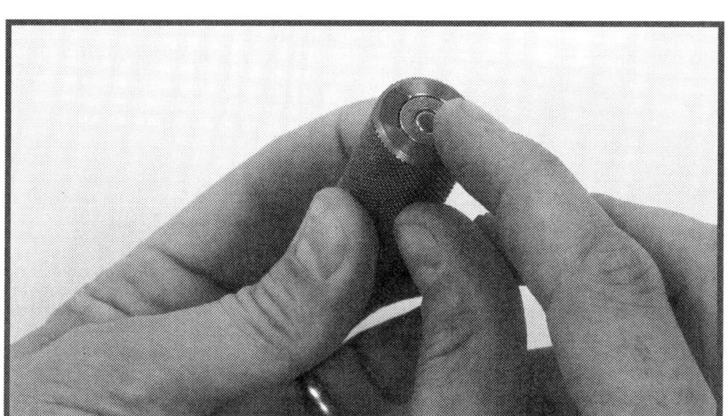

Cartridge headspace gauges can help avoid case failures.

Case Resizing And Case Life

Some firearm chambers are overly large with respect to diameters, especially older military rifles. Full-length resizing cases fired in such a chamber will work the brass excessively. This will lead to premature longitudinal splits on the case. Sometimes such a problem can be eliminated by resizing only a short portion of the case neck. (Neck sizing dies are available for most cartridges.)

Reloaders sometimes opt for neck sizing based on expected increased case life. With normal chambers, neck sizing versus full length sizing offers little case life advantage. Thus, Lyman suggests full length resizing as the norm, especially for hunting loads where smooth feeding, chambering, and extraction are important.

Case necks will eventually become work hardened and begin to split. This will begin to occur after the identical number of reloads whether the cases are neck sized or full-length sized. Some sources suggest that case life can be lengthened by annealing the neck portion of the brass. By the time a case needs annealing at the neck, other factors will have brought about the end of the case's useful safe reloading life. It is almost impossible for the reloader to control and maintain the precise temperature required for proper annealing. Therefore, Lyman cautions against attempts to anneal case necks.

Split cases must be culled from the cases to be reloaded.

Case Life is Limited

Even with the best of reloads and tight headspacing, brass cases will reach a point when they are no longer safe to reload. Thus, every reloader must have a good working knowledge of: the number of times a case has been reloaded, the factors which use up a case, and specific potential problems to be addressed during case inspections.

Experienced reloaders can sometimes obtain an incredible number of firings from each case. However, under ideal conditions, the average reloader should consider eight firings as a safety limit for rimless case life. Reloading a case more often may needlessly expose oneself to the danger of a sudden and dangerous cartridge case failure. Even the most expensive brass is comparatively cheap when used over a case life of eight firings. Don't forget to count the first firing of factory ammo as part of your case's safe reloading life.

Keep in mind that if maximum loads are routinely used, case life will be shorter - a good safety limit would then be about five firings.

Belted cases have a shorter case life. Generally, belted cases should be loaded no more than two or three times. This short useful life can sometimes be stretched a firing or two by using an adjustable cartridge headspace gauge to set up the resizing die. The short reloading life of belted cases is due to the overly long chamber dimensions that often occur in firearms so chambered and the higher pressures often used with these calibers.

Rimmed cases are often associated with firearms having large chambers. As a result, such cases most often provide a useful case life falling between rimless and belted cases.

STEP BY STEP CASE PREPARATION

To a great extent, the safety and performance of your reloads depends on how you process your cartridge cases. The following outlines many of the important aspects of processing brass cases.

Segregate Cases By Lots

All new brass, or cases saved from the firing of factory ammunition, should be segregated into homogeneous lots. Lot numbers are usually stamped, printed, or embossed on the factory box of loaded ammo. These numbers are generally found on an inside flap or the rear panel. Brass purchased in bulk may lack a lot number. In this instance, the reloader should place an assigned lot number on each box or original packaging.

Case lot segregation is essential to uniformity of ballistics and accuracy, but more significantly to allow you to know when discarding of the brass is required. Always process cases as a homogeneous lot. Never mix cases from one lot with another.

Case Inspection

The reloading operation always begins with case inspection. This is necessary to help ensure the safety and reliability of reloaded ammunition. Great care and sufficient time must be expended in the inspection process. Look carefully for any of the following:

- Cracks and splits.
- Dents.
- Perforations or burn throughs.
- Partial head separation (or any tell-tale early warning sign of eventual head separation).
- Corrosion.
- Gas leakage around primer pocket.
- Missing, backed out, or loose primer.
- Abnormal case bulging or swelling.
- Case brass extrusion into ejector or extractor cuts of bolt face.
- Any other condition that suggests less than normal firing conditions.
- Foreign material in case.

Note: Case inspections must be repeated after each loading step as cases can crack, split, suffer bent rims, or incur other damage during the loading process.

Note: Bent rims are an often overlooked cause for difficult chambering. Bent rims are usually caused by insufficient sizing lubricant resulting in undue force being applied to the rim when withdrawing the case from the sizing die.

Case Cleaning

After inspection, all cases must be cleaned. This is accomplished by simply wiping each case with an absorbent rag to remove any grit and grime that could destroy the fine polish of the resizing die. It is good practice to again inspect the case after wiping it clean.

Many reloaders prefer to restore their cases to the shiny appearance of new cases. This is accomplished by using one of Lyman's Turbo tumblers. After using a tumbler it is important to insure that all polishing media has been removed from the inside of the case.

Case tumblers are an excellent way to clean your cases and make them look factory new.

Case Resizing

The next step of the process is to restore the fired case to factory-like dimensions. Naturally, this step is not required when loading brand new cases. Case resizing is generally accomplished in one of the two types of resizing dies. Full length resizing, the method most often used and recommended, restores the entire case back to near factory dimensions. Neck sizing is a method sometimes used for benchrest shooting. When neck sizing, the entire neck, or optionally only a portion of it may be reformed, but the case body is left as fired.

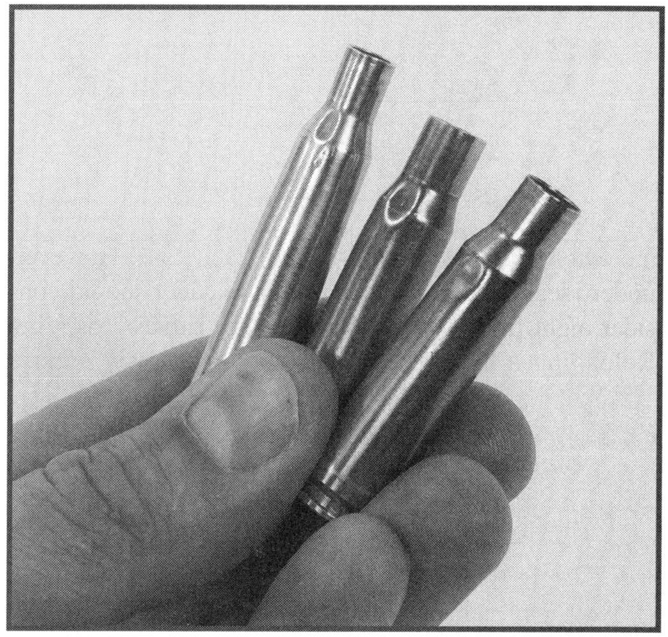

If excessively lubed, cases may dent during resizing, especially on the shoulder area.

Cases can be damaged during resizing if care is not taken. For example, the use of too much resizing lubricant can cause dents to be formed on the case, especially on the case shoulder. Or, the case mouth might be ruined if the case does not enter the sizing die correctly aligned. Case necks, and to a lesser extent case bodies, can crack and split due to the stresses of resizing. If cases are neck expanded in a separate die, splits may form during this operation.

After sizing, all traces of resizing lubricant should be removed by carefully wiping each case with a clean absorbent rag. Now is the time to perform the important second case inspection.

Case Primer Pocket Cleaning

During resizing, the fired primer will be removed. The now exposed primer pocket will show varying amounts of a carbon-like deposit. If this deposit is heavy, it will interfere with the proper seating of the new primer. A few quick twists with a primer pocket cleaning tool or brush will rid the primer pocket of any such firing residue. Omitting this important step can lead to the danger of high primers, or less than hoped-for accuracy. Take care not to use undue or excessive methods as removal of brass from the primer pocket wall must be avoided.

This is the result of resizing unlubed cases. Dies can be damaged when the stuck case is attempted to be removed improperly.

Case Trimming

Cases stretch when fired and again when resized. An overly long case will cause difficult, or actually prevent, chambering of a cartridge. Additionally an excessively long case can cause a potentially dangerous increase in chamber pressure when a cartridge is fired. Therefore, after resizing, the reloader must measure the length of every cartridge case.

Case measuring is usually accomplished by adjusting and locking a vernier or dial caliper at the maximum case length as listed in the data section for each specific cartridge. Then slip each case between the measuring jaws. Those that will not fit require trimming. Case length gauges and most cartridge headspace gauges are also used to find cartridges that need trimming.

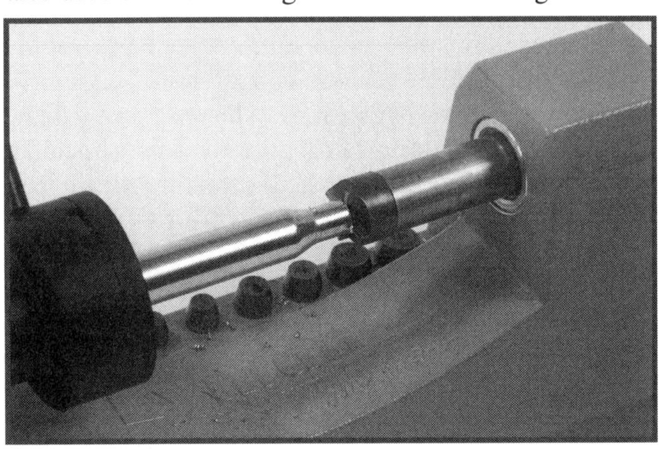

Case trimming is done with various types of case trimmers. Hand operated trimmers are the most common type used. Never trim any case more than 4 times.

It is important to keep careful records of the number of firings and trimmings of each case. Therefore, to prevent mix ups, it is best to trim the entire lot of cases when one or more show the need for trimming.

Cases are trimmed to the suggested trim-to length (as listed in the data section for each cartridge) using a small lathe-like trimming tool. After trimming, the case mouth must be deburred. A deburring tool quickly removes burrs from both the inside and outside of the case mouth. The concave end of the deburring tool cleans up the outside edge of the case while the convex end removes burrs from the inside edge. Lyman also offers a tool that will deburr both on the inside and outside surfaces in a single operation.

Lyman makes several style trimmers. The most popular of these is the Universal trimmer. This model has a patented chuck head which eliminates the need for various size collets or shell holders to retain the case. A power driven version is available for high volume users. Reloaders on a budget might prefer the economical AccuTrimmer.

CAUTION: With each trimming you may remove from 0.005" to 0.015" of material. After the fourth trimming, a total of as much as 0.060" of material will have been removed from the case. All this material has flown from the juncture of the case web and body. This leaves the case substantially weakened at the web/body juncture. Therefore, never trim a case more than four times. When a fifth trimming is needed, discard the entire lot of cases. If case stretching is excessive, fewer firings will be appropriate.

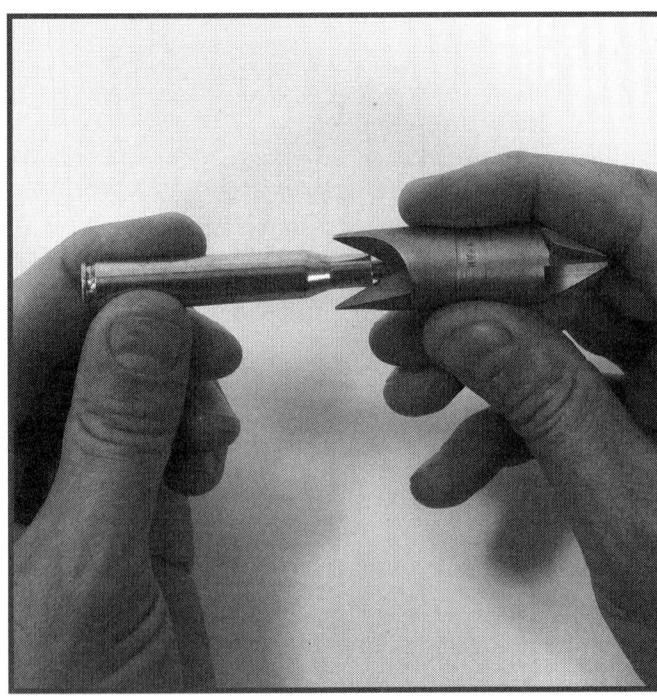

Deburring of both the inside and outside of the case neck must be done after each trimming.

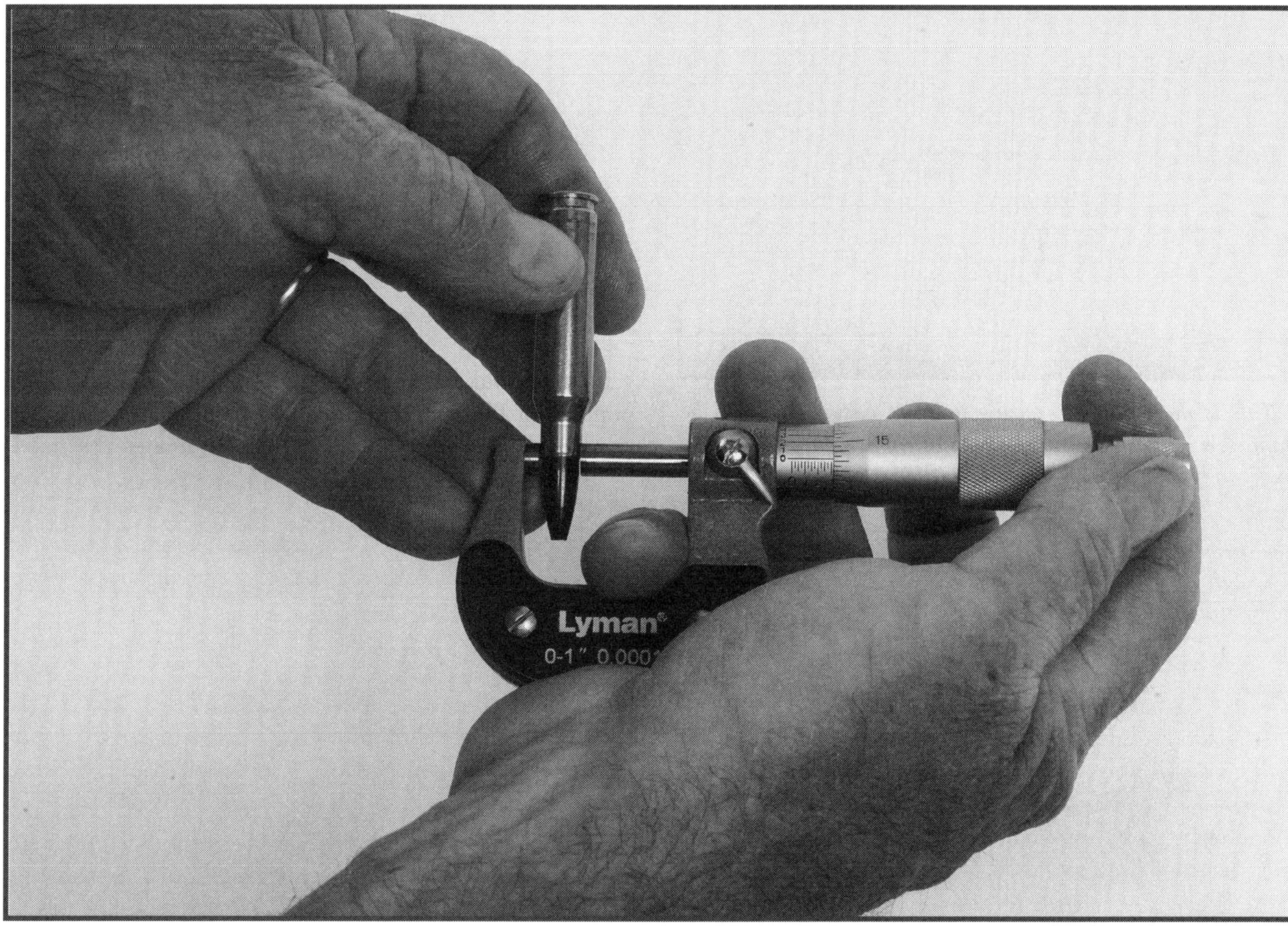

Measuring the neck diameter of a loaded cartridge is an important safety step.

Case Necks Grow Thicker

The walls of a case become thinner as it stretches from repeated firings and resizing. As a result, the neck section will thicken because of the forward brass flow. This thickening can eventually produce a loaded round with an outside neck diameter that is incompatible with the firearm chamber. When a loaded round fits snugly into the neck area of the chamber, upon firing the brass will be unable to expand sufficiently to insure an uninhibited bullet release and thus chamber pressure goes up higher than anticipated, perhaps even dangerously.

The outside case neck diameter of a loaded round should be given the same amount of consideration as case length. Loaded round maximum neck dimensions are shown on the various cartridge drawings that appear at the beginning of the data for each specific cartridge. Take the time to measure a sampling of your ammo after seating the bullet. Generally, your loaded round neck diameter will be well below the maximum. Should case neck diameter begin to get close to the maximum, it is evidence of heavy brass flow from the case body. The wise reloader will then discard the entire lot of brass.

WARNING: Case neck turning equipment is designed to be used with new cases for the improvement of accuracy or the fitting of cases to special dimension chambers that sometimes occur with custom chambered benchrest rifles. This equipment may also be used when reforming new brass of a specific caliber to another caliber (not a undertaking for any but the most experienced persons). Case neck turning equipment is not intended for use in prolonging case life by the turning of overly thick necks.

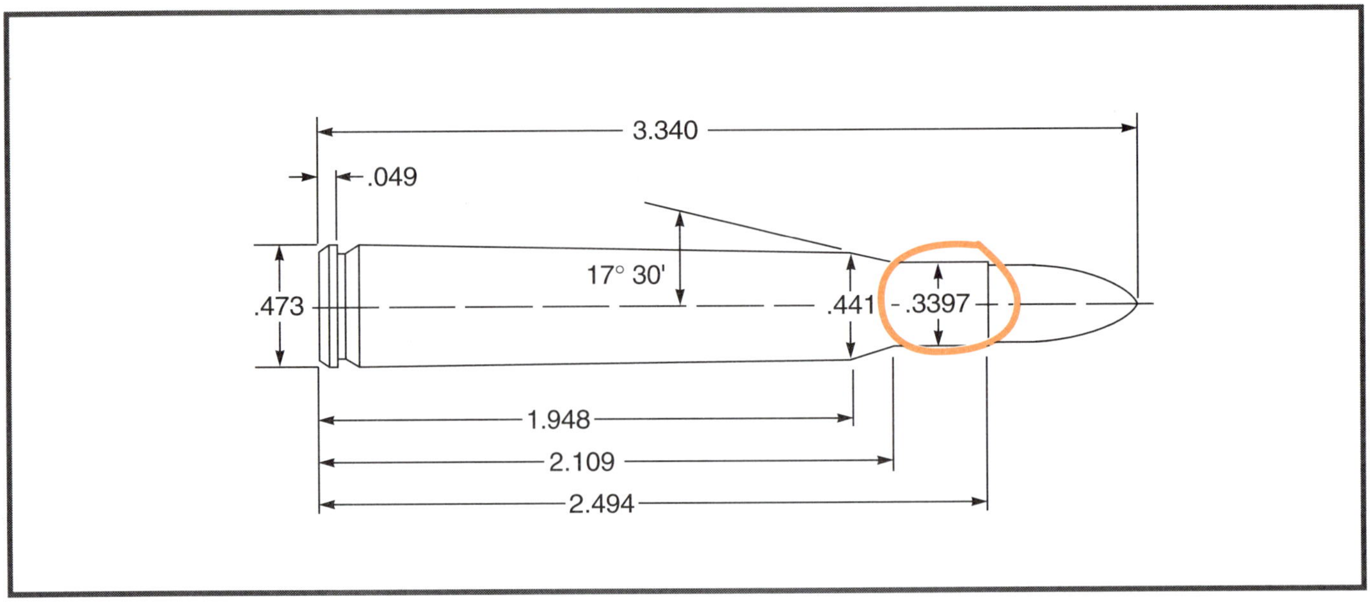

The proper reference for loaded round MAXIMUM outside-neck diameter is included in the cartridge drawing shown at the beginning of the cartridges data listing.

Case Forming

Reloaders sometimes choose to make brass of a given caliber from brass of a different caliber; for example, reforming 308 Winchester (7.62 NATO) cases into 243 Winchester cases. At first, this may appear to be a simple matter of necking down the larger case by running it into the smaller die. However, such a simplistic approach can lead to serious problems.

When reforming cases to a smaller caliber, all of the material in the larger diameter neck must be moved into the smaller diameter neck. This will invariably result in an increase in the smaller caliber case neck wall thickness as well as an increase in neck length. The extra neck length, of course, is trimmed away. But the increase in case neck wall thickness is often overlooked.

As stated, increased neck wall thickness can result in a loaded round outside neck diameter that exceeds the maximum allowable dimension. Then, the safety of the load is compromised as the case will fit the chamber so tightly that proper bullet release cannot occur at firing. When using reformed cases, careful measuring of the outside neck diameter of loaded rounds is essential. If a sample round exceeds the maximum allowable dimension, then the case necks must be turned until they are of appropriate dimension.

Case forming of a small caliber to a larger one is also sometimes undertaken by reloaders - for example necking up 30-06 cases to 35 Whelen cases. Necking up will noticeably thin the case neck. This can result in poor die performance, poor bullet retention and premature neck splitting.

Most brass can be purchased, in bulk, at a low enough price to make case alteration less than practical. If one places no value on the required time, the altering of cases may seem appealing. However, we strongly suggest that the reloader consider the hazards involved. In addition to the already mentioned hazards, there is the negative safety factor involving the use of cases with the wrong headstamp, as well as problems caused by a loss of proper case body graduated annealing (when resizing to a shorter cartridge).

Case forming sometimes can make sense, for example, when brass of a specific caliber is not generally available. Then, when undertaking the task, follow all the precautions we have mentioned and all of the instructions supplied with the reforming die set.

Military Cases Caution:

Military ammo frequently is loaded with cases having thick walls to help avoid case failures in extreme conditions such as rapid or full automatic fire. This increased case wall thickness results in a smaller case volume. Such cases will develop higher chamber pressures when loaded using data developed in standard higher volume commercial brass. This increase in chamber pressure can be notable, even dangerous. For this reason we advise strong caution be used if military cases are reloaded.

Case Savvy

Cases are the weak link in the control of chamber pressure. Every effort must be made to insure that each case used is suitable for another firing. Exacting case inspection cannot be overstressed. If there is any doubt with respect to case reusability - crimp it shut with a pair of pliers (to prevent later inadvertent usage) and throw it out. Remember that case inspection must continue throughout the loading process. After the initial inspection, cases should be looked over after each operation. Did the case crack, split, or deform during sizing? If neck expanding is done separately, did any cracks occur during this operation? Did the primer seat too easily? Or perhaps did the case mouth split during bullet seating?

Safety of a reloaded cartridge will always hinge heavily on the condition of the brass case. The reloader must always assure the case's suitability with careful and repeated inspections. The time spent so doing is part of what makes a knowledgeable and conscientious reloader.

Obviously, safe case handling demands that accurate records be kept with regards to the number of firings, the likely pressure levels (max vs. less than max), and the number of trimmings. Also, it is recommended that you keep records on the number of cases discarded due to brass failures. When 5% of the cases in a lot have been discarded it is time to summarily discard all the remaining cases in the lot. Doing so will help eliminate sudden and catastrophic case failure. Remember, when in doubt, throw it out.

Chapter 6

U.S. BOXER PRIMER CHART

	Large Rifle	Large Rifle Magnum	Small Rifle	Small Rifle Magnum	Large Pistol	Large Pistol Magnum	Small Pistol	Small Pistol Magnum
Remington	9 1/2	9 1/2M	6 1/2 +7 1/2		2 1/2		1 1/2	5 1/2
Winchester	WLR (8 1/2-120)	WLRM (8 1/2M-120)	WSR (6 1/2-116)		++WLP (7-111)		WSP (1 1/2-108)	WSPM (1 1/2M-108)
Federal	210 +210M	215 +215M	205 +205M		150 +150M	155 +155M	100 +100M	200 +200M
CCI	200 **No.34 BR2	250MAG	400 *No.41 BR4	450MAG	300	350MAG	500	550MAG

** Mil-spec small rifle primer* + *Match version* ++ *Also may be used for Magnum , Large Pistol loads.*
*** Mil-spec large rifle primer*

Primers are tiny components, requiring only brief handling and they seem to always work. Because of this, primers and the priming operation often are given no more than passing thought. However, there are aspects of primers and priming that the conscientious reloader must consider.

Mercuric Primers Are Ruinous

Until the late 1940's commercially produced ammunition often used primers containing fulminate of mercury. Upon firing, the mercury from the primer was driven deep into the brass case's grain structure causing its breakdown and a resulting embrittlement. Cases fired with mercuric primers are never suitable for reloading.

Mercuric primers were available to reloaders from retail component outlets as late as the mid-1950's. Many reloaders continued to use existing stocks of mercuric primers for ammo assembly into the early 1960's. Undoubtedly, there were mercuric primers in occasional reloading use for perhaps still another decade - until sometime in the 1970s. Since it is usually impossible to be sure that mercuric primers have not been used, brass of unknown vintage, should never be reloaded.

Primer Sizes

Boxer type primers are made in four basic sizes which include large rifle, small rifle, large pistol, and small pistol. Magnum versions of each of the basic four sizes are also available. Primers are also available as benchrest types. Substantial differences in dimensions, cup hardness, priming pellet weight, and priming pellet composition may occur with each size and type.

The selection of a primer size or type for a new cartridge is sometimes based on specific ignition or firearm characteristics. As example: very large rifle cases and their corresponding difficult to ignite heavy powder charges may demand the heaviest possible primer pellet thereby making large rifle magnum primers a must. Another example: As it was originally envisioned, the 40 S&W cartridge had a large pistol primer. When a loaded round was ejected from the chamber of the firearm to be used, the live primer would impact the ejector. This posed the problem of possible ignition of the unchambered round. Hence, a small diameter primer was substituted in order for this cartridge to safely function in the intended firearm.

Primer and firearm manufacturers have carefully worked out the nuances of primers and the reloader need not be concerned with specific differences so long as the correct size and type primer is used. The correct primer choice is listed in the data section for each specific cartridge. To show the broad range of availability, the nearby primer size chart lists the current selection from the major suppliers.

Primer Makeup

The compound used in many primers is basically made by sulfonating resorcinol with sulfuric acid. Then, nitric acid is used to nitrate the compound to form trinitroresorcinol, commonly known as styphnic acid. This acid is then treated with lead nitrate under precisely controlled conditions to form the lead styphnate salt. Several forms of this salt are used in priming compounds, the most common being "normal" structure. Another type used by one manufacturer is called "basic" structure.

The boxer type primer is used in all American metallic cartridges. The battery cup primer used in shotshells is of the same basic design, but contains an extra cup to facilitate its use in the less rigid shotshell case. The antique berdan type primer functions in the same manner as the Boxer type, but with this primer the anvil is actually part of the case.

Manufacturers add fuels (such as tetracene and barium-nitrate) and other materials including small amounts of aluminum or glass to enhance specific desired characteristics and sensitivity. Magnum primers may use heavier pellet weights and a somewhat different primer mix formulation to obtain desirable ignition characteristics for heavy difficult-to-ignite propellant charges. Obviously, a lot of thought and effort goes into the manufacturer of primers.

Lead free primers are a relatively new type. These have been developed and produced to enable the manufacturers to offer an environmentally safe primer. Such primers have special appeal for those who shoot on indoor ranges which lack ideal ventilation.

Primers Effect Ballistics

Specific primers may not always be suitable for the ignition of large powder charges. One primer may cause a misfire wherein the powder becomes scorched but fails to fully ignite and burn properly; or it may cause a hangfire (delayed ignition, sometimes with an audible pause between firing pin strike and ignition). Should either condition occur, immediately discontinue (do not attempt to fire even one more round) the use of that primer with the cartridge being used. Such ignition characteristics are extremely hazardous and can cause damage to firearms or worse.

Note: These conditions can also be caused by excessively compressed powder charges or a too-low powder loading density. Of course firearm problems can also be at fault.

The variations in primer sizes or types can create substantial variations in ballistic uniformity. For example: In one primer test, extreme velocity variation was an excellent 38 fps. With another brand primer, the extreme velocity variation was a poor 133 fps. Pressure comparisons for the same test loads showed spreads of as little as 3,500 psi. with the first primer and a very poor 9,300 psi with the second primer. Obviously, varying the primer can result in noticeable performance variation, ranging from ideal all the way to undesirable.

	Velocity fps	Pressure psi
Rem. 9 1/2	2509	52,800
Fed. 210 Match	2508	53,100
CCI 200	2526	54,000
Win. WLR	2539	54,700
CCI 34	2545	55,400
Total Spread	**37 fps**	**2,600 psi**

The above results were obtained by using the exact same 308 Win. load and only changing the primer. A change in primer will change the ballistics of a load, so we recommend that new reloaders use the exact components listed in our reloading data.

Nonetheless, is not possible to state that one specific primer does a better job than another. A given primer might perform poorly with a specific load in one caliber, satisfactorily with a second load in another caliber and outstanding with a third load in still another caliber. The point is that primer testing can be an important part of fine tuning any load.

Primer Seating

Primers are manufactured with the anvil having minimal contact with the priming pellet. In this condition primers are not sufficiently sensitized for ideal ignition. Further compression of the priming pellet by the anvil, during proper seating of the primer in the cartridge's case, completes the desired sensitizing of the primer.

Properly seated, the primer cup will be below flush of the case head and within a specific tolerance range. Such positioning will insure that the anvil is lightly pushed down into the primer cup to properly stress the priming pellet. Most primers should be seated 0.003" to 0.005" below flush. If the reloader endeavors to seat primers at a nominal depth of 0.004" below flush, normal variations found in primer heights and primer pockets will generally result in seating depths within the suggested tolerances.

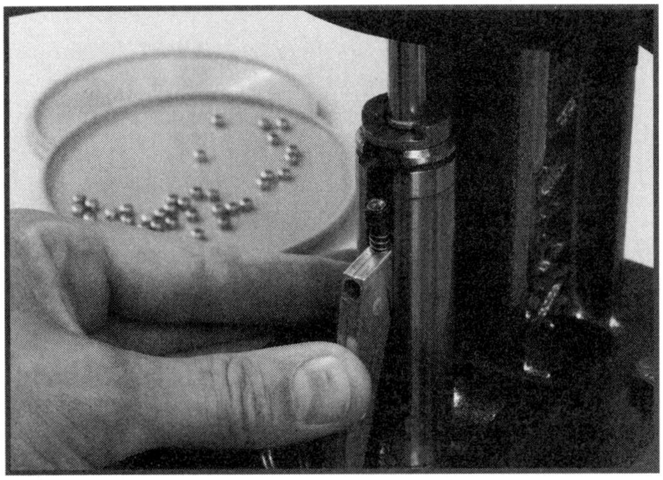

All presses come with a basic priming set-up. This turret press uses the "feel" method for priming (see text).

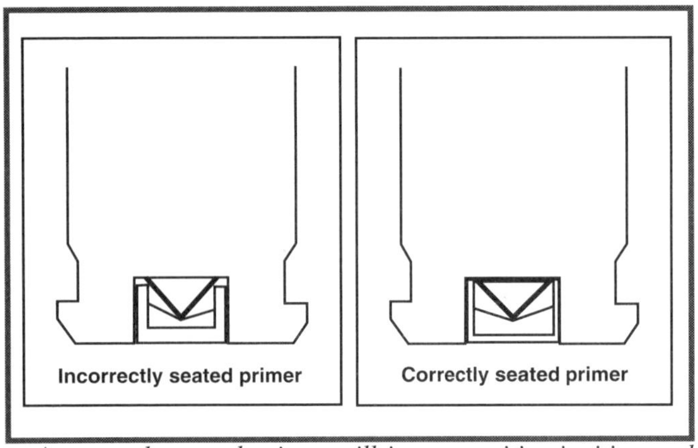

A correctly seated primer will insure positive ignition and the best ballistic uniformity. An improperly seated primer will result in unreliable ignition and poor ballistic uniformity.

CAUTION: Always wear safety glasses whenever handling primers.

Seating primers flush, or somewhat above flush, will fail to move the anvil sufficiently into the primer cup and thus the primer pellet will not be properly stressed. This sets the stage for misfires or erratic ignition.

WARNING: A primer that is above flush of the case head can be accidentally ignited by bolt or action parts during feeding. If ignition occurs before the action is closed and locked, the results can be catastrophic to firearm, shooter and bystanders.

The opposite condition, excessively deep primer seating, also can cause unwanted problems. The primer cup can collapse when seated overly deep. This condition can result in gas blow-by between cup and case. Serious bolt face erosion can then occur. Excessive stressing of the primer pellet can cause the primer to become overly sensitive. This might lead to ignition as a result of rough handling of the loaded ammo. Additionally, if a primer is seated to deeply, the priming pellet may crack as a result of excessive compression by the anvil. A cracked pellet may fail to ignite - a misfire. Excessively deep primer seating can also cause ignition during the priming operation.

Commonly, deep seating of primers causes ignition failures attributable to the firing pin not being able to strike the primer with adequate force.

Many reloaders prime by feel, that is they seat the primer until the cup is felt to bottom in the primer pocket. This method can result in unnecessary variation in primer seating depths. Due to the long chain of parts consisting of: tiny anvil, primer cup, cartridge case, priming punch, tool linkage, tool handle, and the operator's hand, it is difficult to precisely apply uniform pressure to the seating tool handle while attempting to "feel" the primer cup bottoming in the primer pocket. Seating by feel is also subject to variations in the primer cup fit in the primer pocket, and with any crud build-up in the primer pocket. The "feel" method is best left to the extensively experienced person using specialized tools designed specifically for the purpose. For most reloaders, the seating of primers by feel is simply a less than an ideal method.

Ram Priming

Generally, the best method for seating primers is to use a positive mechanical stop to arrest the motion of the tooling used to move the primer into the primer pocket. Some presses are equipped with various types of mechanical stops for primer seating.

About as fine a mechanical primer seating stop as can be found occurs in the method often called ram-priming. Successful ram-priming can be accomplished on most presses. In this method, a priming station is assembled in the normal die position at the top of a press. The ram unit is adjusted so that the normal stop of the press linkage occurs when the primer is at the correct depth.

Note: It is a wise practice to seat primers twice. That is after seating, lower the ram slightly, then rotate the case approximately 180 degrees and then complete another primer seating stroke. This helps insure that primers are squarely seated.

A dial indicating caliper makes a useful tool for checking primer depths. Start with the rod end of the

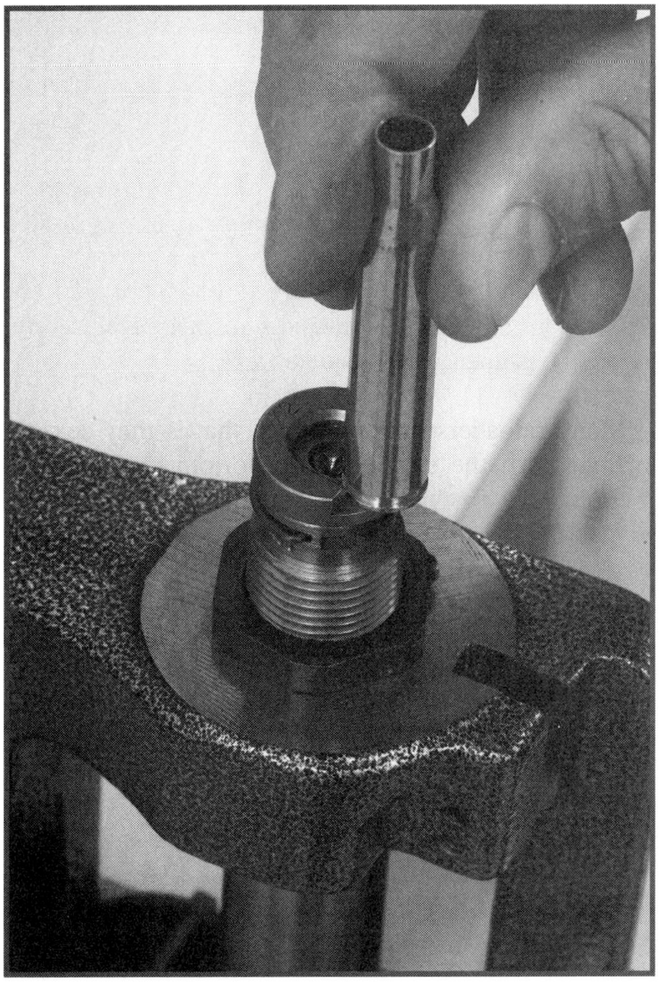

The preferred method for obtaining correct and uniform primer seating depths is to use a RAM Prime unit mounted at the die station of your press.

caliper protruding about 0.015". Then place the rod against the primer while seating the base of the vernier against the case head. The primer seating depth (below flush) can then be read from the dial. Some practice is required to obtain accurate primer depth measurements.

Running a finger over the case head, after a primer has been seated to the correct depth, will give you a feel for a correctly seated primer. With practice, you will be able to "feel" when a primer has not been seated deep enough, when it is seated within an acceptable range, or if it is too deep in the primer pocket. At the beginning of each loading session, measure a sufficient number of primers to insure they are correctly seated. "Feel" each of these to establish a norm. Then "feel" each subsequently seated primer to insure none vary from the desired depth.

Important: Before beginning to prime, make certain all sizing lubricant is removed from all cartridge cases. Then wash and dry your hands thoroughly to prevent any possibility of primer contamination. Develop a priming routine that will keep primers from becoming contaminated. It is advisable to manipulate the press handle and the cartridge cases with one hand and use the other hand solely to pick up and place primers on the priming post. This method will give 100% assurance that your primers will fire as intended.

We strongly suggest that primers be handled manually. Place no more than 100, or the lesser amount of primers actually needed, onto a primer turning tray. Rotate the tray until all primers are oriented anvil-up. Then pick up one primer at a time for transfer to the priming punch for seating into the cartridge case.

Caution: Primer Feeds Require Extreme Care

Primers, by nature, must be explosive. Despite this, handling primers one at a time is a safe procedure. However, primers handled in bulk are a potential hazard. Should you choose to use an automatic primer feed, extreme caution is required. There may be as many as 100 primers, each in contact with the next, in the automatic primer feed's magazine. Should one primer explode, all will do so spontaneously with disastrous results. Even a few primers exploding can cause serious property damage and personal injury.

Some very high volume reloaders feel an auto primer feed to be a necessary part of their equipment. If you think you number among these folks, use extreme caution when working with an automatic primer feed.

The SAAMI pamphlet "Sporting Ammunition Primers", cautions against the use of primer feeds unless adequate shielding separates the operator from the hazard of explosion. One major primer manufacturer has cautioned on their primer packaging against the use of any automatic primer feed. Obviously great care is required when using an automatic primer feed.

The use of an automatic primer feed unit demands an uncompromising light hand to prevent the crushing of a primer should a jam or malfunction of any type occur.

CAUTION: never push, bang, bump or otherwise apply any force to a primer or primer feed.

CAUTION: When using an automatic primer feed, at the first sign of a primer jam or difficult feeding - STOP! Then think about what must be done. Force nothing and do not be tempted to even slightly jiggle the tool or any part of it. Rather, carefully disassemble the primer feed. First remove the primer magazine and all primers contained therein. Clear the jam, determine the cause for the jam, and correct the problem before proceeding with use of the primer feed.

CAUTION: Extreme care must be used when filling a primer feed magazine tube with primers. Never force a primer into the pickup tube. A pressure of no more than five pounds should be required for a primer to slip into the tube. Do not use any primer pickup tube requiring more than this amount of effort. Primers must be correctly oriented in the pickup tube. It is the user's obligation to insure that this is so. Misoriented (upside-down or standing on end) primers are a serious potential hazard. Never store primers in a primer feed tube. Return all unused primers to their original packaging. Do not force any primer feed tube-follower down onto primers

Caution: Primers Can Dust

Primers can and do "dust". Minute amounts of priming mix will accumulate in the loading area, and on primer handling equipment, especially when an auto primer feed is used. An accumulation of priming dust is a fire and explosion hazard. Occasionally wipe off priming trays, priming tools and the work area with a damp cloth to reduce the likelihood of an accident. Rinse the wiping cloth thoroughly in running water to remove any primer dust it picks up. Be sure all cleaned surfaces are dry before bringing primers in contact with them.

The entire loading area and all equipment must be kept clean and free of any contamination by primer dust (as well as powder accumulations). Fired primers, separated cups and anvils, or any trace of hard or abrasive material can be the catalyst that causes primers, primer dust, or spilled powder, to ignite.

Pick up spilled primers immediately and be sure they are all recovered. Stepping on a primer may cause it to explode.

Primers are Explosive

Serious thought must be given to primers with respect to care in handling and storage. First and foremost, it must be remembered that primers are designed to explode when the priming pellet is crushed between the anvil and the primer cup. This is accomplished by a percussion blow - the firing pin striking the primer cup and crushing it against the primer anvil.

CAUTION: Primers can explode if subjected to friction, percussion, crushing, or excess heat from any source, whether open flame or not. Static electricity and many other abuses may cause primers to explode.

The reloader needs to prevent primers from exposure to such things as hammering, impacts, or percussion blows of any kind, including bullet impact. Hot tobacco ash, sparks and a great many other abuses are to be avoided. In short treat primers as an explosive. Careful handling and storage is always a must.

Never transfer primers from their original factory container to any other container. Primers stored loose or in bulk, for example in a glass jar, and having contact one with another, can, and most often will, cause a violent explosion of the entire mass if but a single primer is ignited.

When priming, never take more than a single box of 100 primers to the loading bench. Pour out only the needed amount onto the primer flipper. After each priming session, immediately return any unused primers to their original packaging. This packaging is designed to prevent accidents even if dropped.

CAUTION: never decap live primers as doing so is hazardous.

Primer Storage

Because of their explosive nature only an absolute minimum of primers should be kept in storage. The National Fire Protection Association's NFPA 495 states that not more than 10,000 primers should be stored in a private residence. This recommendation is law in most communities. Even a deeply involved reloader should have no reason to store more than 1,000 each of large rifle, large rifle magnum, small rifle, large pistol, large pistol magnum, small pistol, small pistol magnum, and shotshell primers. That's 8,000 primers. With care in replacing exhausted supplies, it is not difficult to adhere to the 10,000 primer limit. Most reloaders will be able to get along nicely with far fewer primers kept on hand.

Always store primers in a remote area away from any possible source of ignition including bullet impact. Be sure no heat, spark, electrical, percussion, or any other form of abuse can occur in the storage area.

Keep primers stored away from oxidizing agents, flammable liquids, and flammable solids (including handloading powders).

A storage cabinet is strongly recommended. A cabinet should be solidly constructed of 1" thick lumber to delay the transfer of heat to the contents in event of a fire or other mishap. Make sure the storage cabinet is remotely located with respect to trash, combustible materials, sources of heat (including the suns rays), open flames, electrical equipment, hot water heaters, mechanical equipment, furnaces, solvents, flammable gases, and so on. To avoid the potential of an unexpected or unusual accident, do not store anything else in the primer cabinet. Never remove more than 100 primers at a time from the storage area.

Naturally, never smoke around primers. No smoking signs in the storage area and at the loading bench are suggested.

A SAAMI bulletin entitled "Sporting Ammunition Primers: Properties, Handling, And Storage For Handloading" is available. Every reloader should acquire a copy from: SAAMI, 11 Mile Hill Road, Newtown, CT 06470 (www.saami.org).

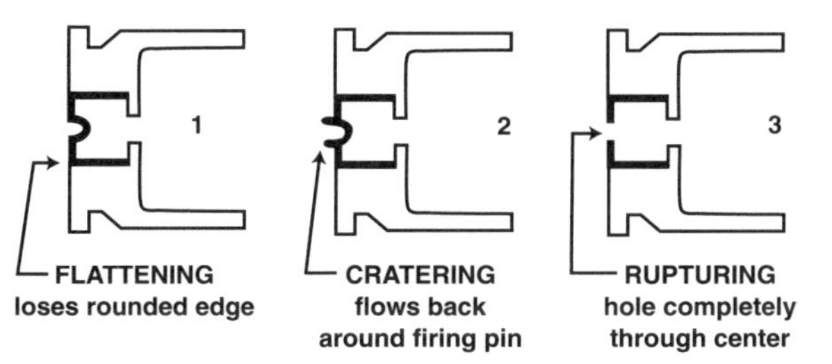

Fired Primers And Excessive Pressure

At times, the condition of a fired primer may give indications that the load being used is creating excessive and hazardous chamber pressures. A primer that shows a black smudge around it is one indication of too much chamber pressure. So is a primer that is excessively flattened when fired, a condition in which the primer loses all of its rounded edge. A primer which craters or ruptures tells of very high pressures. The safe handloader always checks the condition of fired primers immediately after firing. The nearby drawing and photo illustrate tell-tale after-firing primer conditions.

The primers in the cases on the left of each pair show a normal pressure level. The primers in the cases on the right of each pair show signs of high pressure. Primer condition is only one pressure indicator that should be monitored carefully when developing reloads.

With all of the foregoing in mind, it is obvious that giving a bit of thought to primer storage, handling, the priming operation, and primer condition after firing is essential. Proper primer consideration can result in more uniform ammo and help prevent accidents.

7 Powders

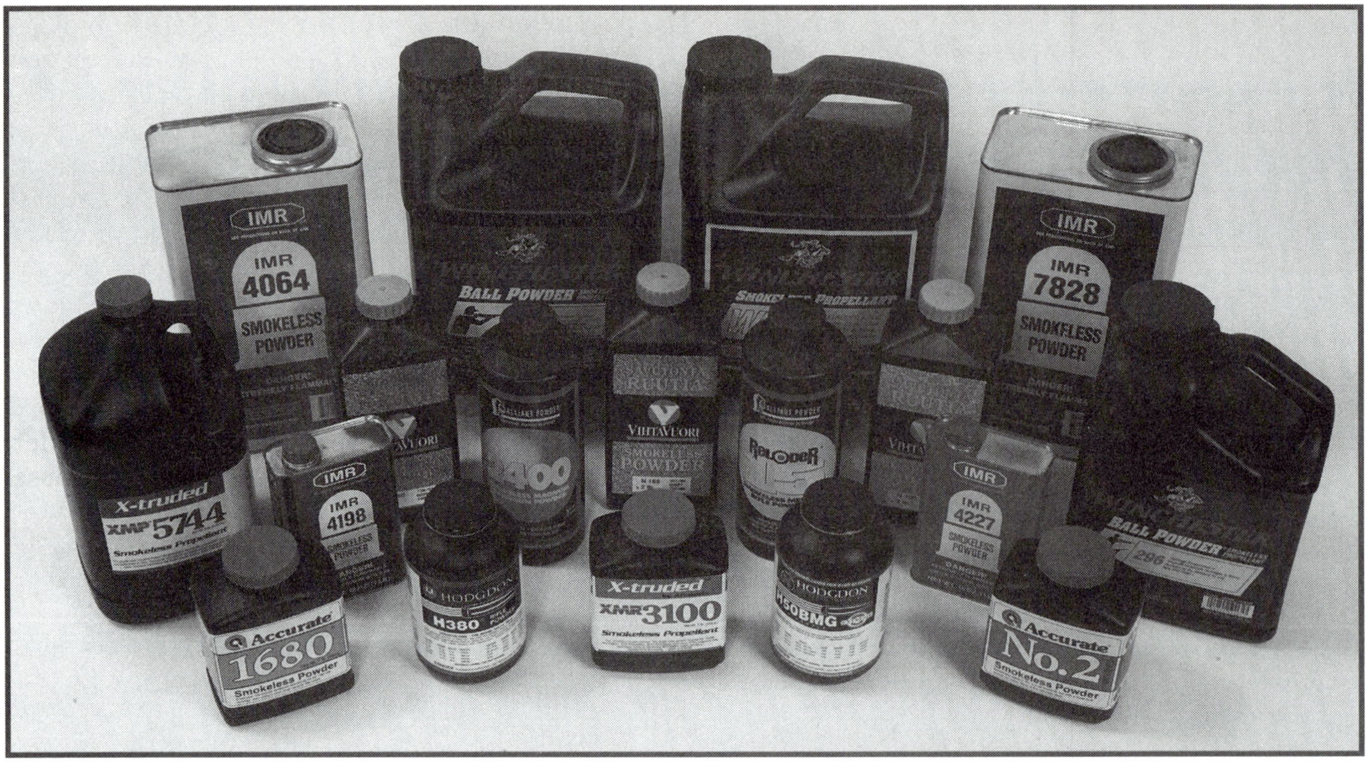

This is just a small selection of the more than 80 propellants we used to develop the data for this, our, 48th reloading handbook.

CAUTION: All smokeless powders are flammable solids. They are designed to burn rapidly and to do so without the presence of oxygen - generating all required oxygen as they burn. Therefore, treat all propellants with the respect due a highly flammable substance.

Selection Of A Propellant

There are more than 80 propellants listed in the data section of this manual. Each powder listed in our data meets our predetermined criteria for ballistic uniformity. Nonetheless, only a few of those listed will optimize ballistics and deliver tightest groups. Why then all those other powders?

One reason is reloader interest in safety. Many reloaders recognize that keeping only a few types of powder on hand helps to minimize any potential for using the wrong powder and keeps stored quantities at a minimum. These folks often select the propellant best suited to their most important application, and then use this same propellant in as many other cartridges as practical. Another reason is that not every propellant is available at every retail outlet. Individual firearms may not always deliver peak accuracy with the generally considered "best" powders. Trying several powders will often turn up one that will suit the odd performing gun. Additionally, some reloaders do not want the highest possible velocity. Others desire recoil reduction and some wish light loads for target practice. These are just some of the reasons for our extensive listings.

We list the propellant most likely to supply the best accuracy with a specific bullet in **BOLD** print. Sometimes the comments preceding the data for each cartridge will mention specific top performing propellants. Select your first powder from either the type listed in bold print or one mentioned in specific text.

168 gr. Jacketed HPBT 3.300" OAL — **BC:** .462 **SD:** .253

Powder	Sugg Starting Grains	Velocity fps	Pressure	Max Load Grains	Velocity fps	Pressure
IMR-3031	42.0	2463	34,000	48.0	2808	49,600 C
IMR-4895	**43.0**	**2444**	**35,500**	48.5	2762	47,600 C
IMR-4064	45.0	2551	34,000	50.0	2816	46,700 C
AA2520	42.5	2383	36,200	48.0	2653	48,800 C
IMR-4320	45.0	2475	36,400	51.0	2816	50,000 C
RX15	47.5	2592	38,800	51.8	2832	50,000 C
AA2700	49.5	2475	41,800	55.0	2745	54,900 P
760	50.0	2541	37,300	57.0	2839	49,700 C
IMR-4350	52.0	2590	38,000	57.0+	2873	50,400 C
RX19	55.0	2489	37,800	60.9+	2810	50,000 C
N160	55.0	2559	39,800	61.0+	2891	58,800 P
IMR-4831	54.5	2544	35,700	60.6+	2847	47,800 C
H-4831	55.0	2564	37,300	61.5+	2873	47,200 C
WXR	55.0	2489	38,100	61.0+	2799	53,300 P

Our data section lists what may be the most accurate powder for a particular bullet in bold type.

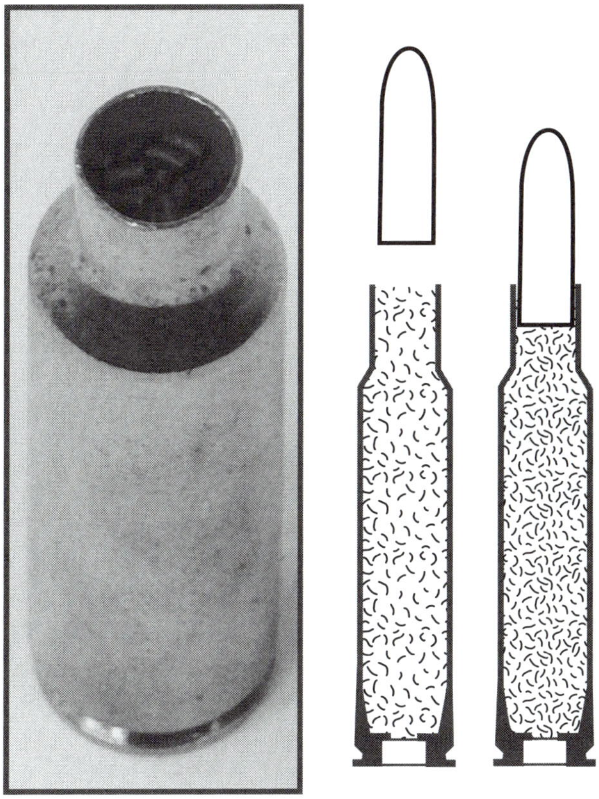

A compressed load will require that you to verify that your components, methods and seating depth do not result in an excessive compression. See text for instructions on how to proceed with this determination.

Propellant Versus Case Volume

Generally, powders that give the best ballistic uniformity and accuracy are often those that nearly fill the case to the base of the bullet. Ideally, the reloader should chose a load that fills the case to at least 90% of capacity. This insures that powder positioning in the case will not cause substantial changes in the ballistics. It is not always possible to fill up 90% of case volume. However, when it can be done this is often part of a knowledgeable reloaders approach to powder selection.

130 gr. Jacketed SP 2.660" OAL — **BC:** .394 **SD:** .230

Powder	Sugg Starting Grains	Velocity fps	Pressure P.S.I.	Max Load Grains	Velocity fps	Pressure P.S.I.
H-414	58.0	3122	58,400	61.0	3249	64,200
IMR-4350	58.0	3057	54,300	61.0	3240	63,100
N160	59.0	3095	57,000	62.5	3226	62,000
RX19	63.0	3094	54,600	66.0+	3289	64,100
WXR	62.5	3031	53,000	**66.0**	**3234**	**62,800**

+ Designates a compressed powder charge

Compressed Powder Loads

Many of the loads in this handbook are noted with a **+** sign indicating loads using compressed powder charges. This means that the propellant so fills the case that when the bullet is seated, its base packs the powder tightly into the case. If you seat bullets shorter or longer than we suggest, the amount of any compression can vary. The use of a different brand case may also alter any powder compression as some cases have larger volumes than others. The change you experience could range from no compression to excessive compression.

CAUTION: Avoid overly compressed powder charges. Excessively compressed charges will cause cartridge overall lengths to continue to increase after the loaded round is taken from the bullet seating die. This is caused by the elastic nature of the compressed charge trying to reduce compression by pushing the bullet upwards. This can lead to jammed firearms, erratic ballistics, poor accuracy and other problems.

To determine if the degree of compression is excessive, slowly seat the selected bullet in an empty case to the desired overall cartridge length. Carefully secure the seating screw to insure a uniform seating depth on succeeding rounds. Also be certain that the die body is locked in position. Make note of the overall length. Assume a potential variation of +/-0.010" as the total range to be encountered.

Then place the desired powder charge into a primed case. Let the powder charge pour into the case normally. With the previously adjusted bullet seating die, seat a bullet into the charged case. Without delay, measure and record the overall length of this round. If the charged cartridge's overall length is longer than the empty sample (not counting the tolerance of +0.010"), the loaded round has an excessively compressed powder charge.

If the loaded length is the same as the empty sample (within the tolerance of +0.010"), wait 24 hours and re-measure the loaded round. If it now exceeds the anticipated overall length, the load is excessively compressed.

Final verification requires assembling several boxes of ammo, checking each round for overall length immediately after bullet seating and again after 24 hours. If none of the loaded rounds exceeds the range of the measured empty case with bullet, the degree of propellant compression is acceptable. Keep in mind that the degree of compression can change with different lots of the same propellant.

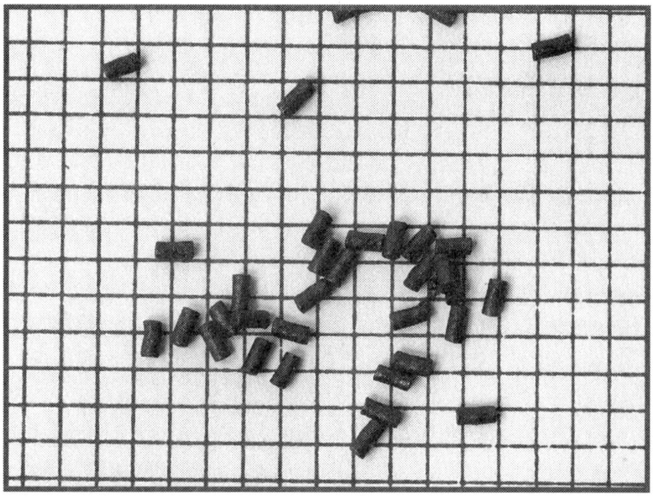

Shown above is a single base extruded powder with a tubular and perforated shape.

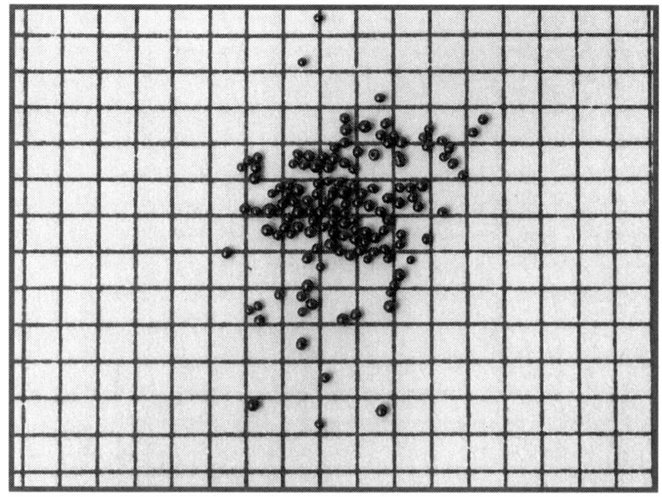

Shown above is a typical double base, spherical powder. As with most ball types it has been rolled to a somewhat flattened configuration in order to obtain the desired burning rate.

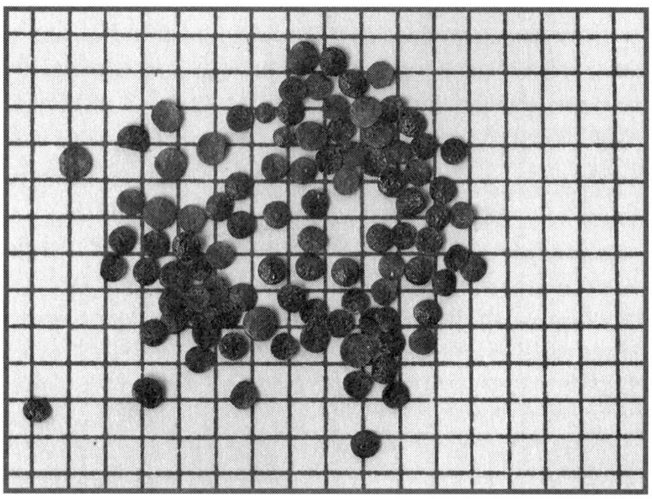

Shown above is a double base wafer (flake) shaped propellant.

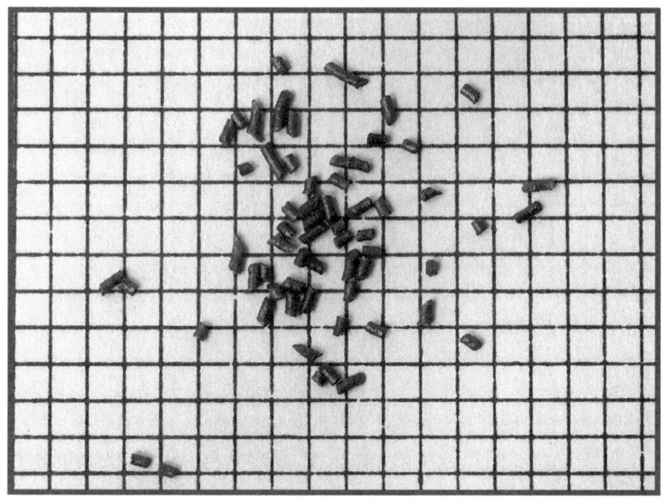

Shown above is a typical double base, extruded and perforated propellant. This one has rather short grain lengths.

Note: Do not crimp the bullet when using this method to judge compression acceptability. Also, keep in mind that a neck sized case will produce less powder compression than a full length sized case.

CAUTION: Excessive powder compression can lead to potential problems with proper propellant ignition. Sometimes this can cause dangerous conditions. Never use excessively compressed powder charges.

Powder Characteristics

Propellants are made in various configurations. Those most often encountered included: tubular granulations with or with out longitudinal perforations, wafer-like flakes, and spherical shapes.

Rifle and handgun reloading propellants are made of either nitrocellulose (called single base powder) or a combination of nitrocellulose and nitroglycerine (called double base powder). Each powder is designed to burn at a predetermined pressure level(s) in the confines of a specific cartridge case(s).

CAUTION: Powder ignition is normally caused by the heat generated by an exploding primer. However, ignition can be caused under any circumstance which causes the granulations to be heated beyond kindling temperature. Such causes can include sparks of any type or source, heat generated by any source, flame from any source, energy transfer (this includes bullet impact), and other causes. Heat transfer to the powder need not be direct. For example heat applied to the outside of the propellant container can result in ignition of the powder contained therein.

When powder burns it forms enormous quantities of

gas at an extremely high temperature. Confined, this gas will cause very high pressure. In the confines of a cartridge case, a comparatively small amount of propellant can easily cause pressure as high as 65,000 pounds per square inch. The exact amount of pressure generated is a direct result of the amount and type propellant burned and the degree of confinement.

CAUTION: If sufficient quantities of propellant burn within a confined area, such as a room, without the possibility of venting the resulting gas, pressure in that area will rise. This pressure can push out windows, doors and weak structural members. Anyone in the area could be seriously burned or suffer injury due to the force of expanding gases.

The rate of pressure generation is such that, given the means of adequate venting, burning powder possesses no undue propensity to create damage beyond that which would be caused by the high temperature of the expanding gases.

Because ignited powder can raise pressure within the storage area, there are specific requirements for powder storage. The first of these is never to transfer any propellant from its original container. The original container is designed to burst or partially open at a very low pressure. Naturally, keeping powder in the original container is also essential for proper identification of the propellant. Second, storage of propellant canisters should be in a unit that will protect the contents from outside heat sources.

CAUTION: In case of unintended ignition it is imperative that the storage container have one or more weak walls that will open out at very low pressure to vent gases before they can cause damage. Using a strong enclosure may cause a burst of considerable force causing property damage or personal injury.

Do not store propellants in the same area with solvents, flammable gases, primers, or any highly combustible material. Smoking should never be allowed in the storage area or while handling and using powder. Place conspicuous NO SMOKING signs in the storage area and at the loading bench. And remember to obey all legal requirements for storage as applicable to your community.

Many municipalities use the National Fire Protection Agency's publication NFPA 495 as law. It, in part, states that smokeless powders intended for individual use shall not exceed 20 pounds unless stored in a wooden cabinet or box with walls of at least 1" nominal thickness. So stored, the quantity shall never exceed a total of 50 pounds.

Every reloader should have, read, and adhere to the information contained in the SAAMI pamphlet entitled "Properties And Storage Of Smokeless Propellants. "This pamphlet is available from the various powder manufacturers.

Always store powder in a cool dry place away from any potential source of ignition. Do not store powder where it will be subjected to the sun's rays or be acted upon by any other source of heat. Be certain that unauthorized persons cannot gain access to the storage cabinet. Use only a brush and pan to pick up spilled powder. Dispose of spilled powder in a safe manner.

Burning Rates

The pressure generated by a specific propellant will vary with the amount of powder, the weight and type of bullet, the shape of the case, and other factors. A specific powder may burn with greater or lesser quickness (force) depending upon the specific cartridge and bullet being used. We have listed the powders in our data section in order of the apparent burning rate achieved with the specific components used. We have also included a relative burn rate chart in the reference section. The reloader, however, never should assume any specific characteristics for a given propellant based on burn rate charts. Stick to the loading data contained in this manual, neither exceeding the maximum charge nor going below the minimum charge. And never substitute one propellant for another, not even when the names are similar.

Use The Right Powder

Always read the label aloud before and after each loading session and then compare the nomenclature with your data records or this manual. Other safeguards will prove helpful. For example, some reloaders purchase their favorite powders in different size and shape cans; rifle powder is purchased in a 8 pound container, handgun powder in a 1 pound container, and so on.

CAUTION: The handloader must make absolute identification of a propellant before using it. There are many powders with similar and even identical nomenclature. Pay specific attention to manufacturer identification. For example: Accurate Arms 4350 is different from Hodgdon H-4350 and both of these are different from IMR 4350. NEVER use data developed for a specific brand powder with another brand powder with a similar identification. Each powder is different and requires specific data.

Following are brief comments on each of the more than eighty powders used in our data tables. In order to maximize space usage we have used identifying abbreviations as follows:

(S) = Single base
(D) = Double base
(B) = Spherical or ball shaped
(T) = Extruded with tubular shape
(W) = Extruded with wafer or flake shape

IMR POWDERS

IMR propellants are distributed by IMR Powder Co. of Plattsburgh, New York 12901. Any questions regarding IMR propellants should be directed to this address.

"Hi-Skor" 700-X

Developed for shotshells, it also is used in many handgun cartridge applications, especially for target loads. (D, W)

"Hi-Skor" 800-X

This propellant has applications similar to "Hi-Skor" 700-X. Its burning rate is conducive to obtaining high velocities in certain handgun cartridge applications. (D, W)

PB

PB gets its name from the initials of its basic grain structure type - porous base. The porosity of the powder is the method used to control the burning rate. It is for used for some handgun loads. (S, W)

SR 7625

We used this powder for handgun and lead bullet rifle loads. The fastest burning propellant in the SR series, the nomenclature of this and other IMR propellants using the same letters stands for Sporting Rifle - a confusing name for powders seldom so used. (S, W)

SR 4756

Designed originally as a shotshell powder, we used it for many handgun loads and some lead bullet rifle loads. (S, W)

SR 4759

This powder has limited use in small case rifle cartridges and heavy handgun cartridge loads. It is popular with cast bullet shooters and is often used with reduced loads. SR 4759 is the slowest burning of the canister SR series of powders. (S, T)

IMR 4227

The fastest burning of the IMR series, it is ideal for the 22 Hornet and other small case rifle cartridges. It is also used in many heavy handgun loads and cast bullet rifle loads. The IMR stands for Improved Military Rifle - a somewhat misleading title. (S, T)

IMR 4198

A fast burning favorite of shooters using the 17 and 222 Remingtons. It is a good selection for the 444 Marlin and has many other applications including cast bullet rifle loads. (S, T)

IMR 3031

A popular powder in many small to medium size rifle cases and a perennial favorite for the 30-30 Winchester. Once used to load 250-3000 factory ammo. It is extremely versatile and at its best in medium capacity cases. (S, T)

IMR 4895

This is a popular wide application propellant with a fine granulation. It meters well through a powder measure. Many target shooters favor it for the 308 Winchester with 168-grain bullets. (S, T)

IMR 4320

A popular, medium burning propellant, used in a wide range of rifle cartridges. Factory 308 ammo has been loaded with IMR 4320. (S, T)

IMR 4064

A versatile and accurate powder with a very wide range of applications. Favored in the 308 Winchester, the 30-06 Springfield with 150-grain or lighter bullets, and in the 375 H&H. Its applications range from the

22-250 Remington to the 416 Remington, doing quite well at both extremes. (S, T)

IMR 4350

A slow burning propellant used in cartridges from 22 to 45 caliber. It is a great choice for such popular applications as the 243 Winchester, 257 Roberts, 7x57mm Mauser, the 30-06 (with bullets of 180-grains or heavier), and 300 H&H Magnum. (S, T)

IMR 4831

This speed is ideal for many magnum style cartridges including the 338 Winchester Magnum and very large cases such as the 416 Rigby. It also is favored in the 270 Winchester and 280 Remington. Accuracy, in its best applications, is usually outstanding. It works in a wide range of cartridges having a relatively large powder capacity. (S, T)

IMR 7828

This is the slowest burning IMR powder. It is ideal for large capacity cases with heavy bullets such as the 378, 416, and 460 Weatherby Magnum cartridges. (S, T)

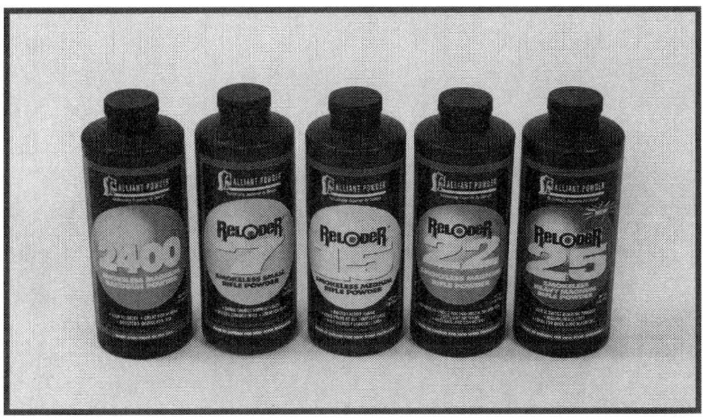

ALLIANT POWDERS

Questions regarding Alliant propellants should be addressed to Alliant Powder, P.O. Box 6, Radford, Virginia 24141 - 0096

Bullseye

This is a fast burning, popular handgun powder. It is accurate over a wide application range, often delivering near maximum, and extremely uniform, ballistics. Its reputation for performance is well deserved. Ideal from 25 ACP to the 45 ACP. Its perhaps most often used for target loadings for the 38 Special. (D, W)

Red Dot

We used Red Dot for many handgun and cast bullet rifle loads. Its name comes from the color coded granulations mixed into it. (D, W)

Green Dot

Slower burning than Red Dot this powder incorporates a small amount of green colored granulations, hence its name. Applications similar to Red Dot. (D, W)

Unique

A versatile propellant for a wide range of handgun ammo and for cast bullets in many rifle cartridges. It is the traditional choice for the 45 Colt cartridge and a good choice for handgunner's wanting slightly higher velocities in a wide range of calibers. It works from the 25 ACP to the 44 Remington Magnum. (D, W)

Power Pistol

Designed as a prime choice for semi-auto handgun cartridges, Power Pistol is popular for applications in 9mm Luger, 40 S&W and 10mm. (D, W)

Herco

This is a heavy load handgun powder, with course granulations. It is used mostly for high velocity or magnum type loadings. (D, W)

Blue Dot

A slow burning magnum handgun propellant, Blue Dot sees limited applications. This propellant contains some blue colored identification granulations. (D, W)

2400

A fine granulation propellant, 2400 is suitable for limited small cased rifle cartridges such as the 22 Hornet. It is also used for 357, 41 and 44 magnums as well as a cast rifle bullet propellant. (D, T)

Reloder 7

The fastest of the Reloder series, suitable for a wide range of cartridges from the 222 Remington to the 458 Winchester Magnum. It is perhaps the best choice for the latter with 500-grain bullets as it takes up a minimum of case space. It has a mix of red, yellow and blue identifying granulations. Abbreviated as Rx7 in our data. (D,T)

Reloder 15

A moderate burning speed, Reloder 15 fits a very wide range of applications from the 223 Remington to the 416 Remington Mag. It is ideal for the 375 H&H. Accuracy is usually superb. Abbreviated as Rx15 in our data. (D, T)

Reloder 19

This is a slow burning speed that has wide applications from the 22-250 Remington to the 340 Weatherby Magnum. A good choice for the 30-06 and many similar rounds. Abbreviated as Rx 19 in our data. (D, T)

Reloder 22

A magnum rifle propellant, Reloder 22 is loaded in many factory Weatherby Magnum cartridges. With applications from the 220 Swift to the 416 Weatherby Magnum, it is surprisingly versatile. Abbreviated as Rx 22 in our data. (D, T)

Reloder 25

This is currently the slowest burning Alliant selection. As such it is a good choice for the various Weatherby Magnum and other very large case cartridges. Abbreviated as Rx 25 in our data. (D, T)

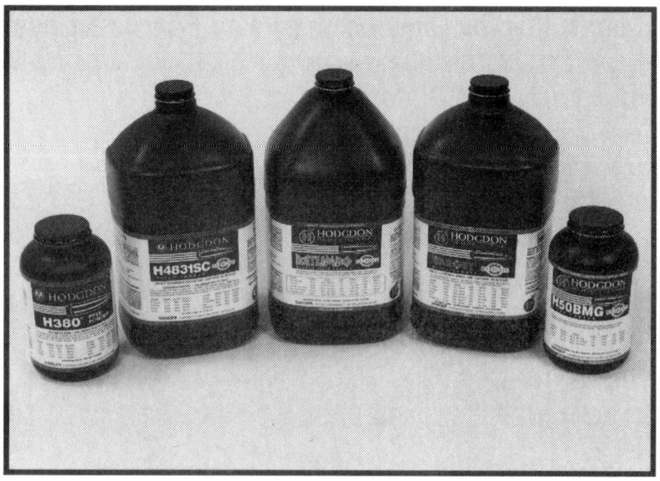

HODGDON POWDERS

Hodgdon powders are available both as newly manufactured types and as surplus types. Look for identifying statements on the powder can. For questions about Hodgdon propellants write to Hodgdon Powder Co., P.O. Box 2932, Shawnee Mission, Kansas 66201.

Clays

Designed as a porous base shotshell powder, Clays has target load applications over a wide range of handgun cartridge applications. (D, W)

Universal

A very broad spectrum of handgun cartridges can be loaded with Universal, from the 25 ACP to the 44 Mag. Also used for shotshell loads. (D, W)

Titegroup

Designed for handgun rounds, Titegroup is not sensitive to powder position in the case. Hence, it is ideal for handgun loads which occupy only a small space in the case, i.e. 45 Colt. (D, B)

HP38

This powder duplicates the performance of Winchester 231 and derives its nomenclature from Handgun Propellant ideal for the 38. It can be used in almost all handgun cartridges. (D, B)

HS6

This powder derives its nomenclature from Hodgdon Shotshell type number 6. It has handgun and cast bullet applications. (D, B)

H4227

The fastest of the Hodgdon Extreme Extruded series and works in applications similar to IMR4227, for example the 22 Hornet and magnum handgun loads. (S, T)

H110

A handgun powder that generally requires a heavy bullet and a heavy crimp to insure uniform results. Applications are similar to Winchester 296. (D, B)

H322

This powder is a universal favorite in the two most accurate cartridges ever designed - the 22PPC and the 6mm PPC. It has won more benchrest matches than any other propellant. Its short grains meter uniformly through a powder measure. (S, T)

Benchmark

A newer propellant, Benchmark is designed to produce good accuracy in a wide number of rifle cartridges. (S, T)

H335

A popular speed, H335 it enjoys an enviable reputation in cartridges like the PPC's, 222, and the 223 Remington. It has been sold as both a newly manufactured and a surplus military type. (D, B)

H4895

This speed has applications similar to IMR4895 and as such is very versatile; but it does require different data. (S, T)

Varget

An accurate choice for many cartridges, Varget has applications similar to IMR4064 and Reloder 15. (S, T)

BL-C (2)

This speed is spoken as Ball C 2 and has been available both as a newly manufactured type or as a surplus type. It is popular in diverse applications. (D, B)

H380

Named for the fact that Bruce Hodgdon found 38.0 grains to be extremely well suited to the 22-250. Because it is the only spherical type that has a true ball configuration, it meters more uniformly than any other propellant. A great favorite in the 7mm-08 and 308 cartridges, it has wide applications. (D, B)

H414

This powder is kin to Winchester 760 and as such has wide applications in diversified cartridges. It is popular in the 30-06 and similar sized rounds. (D, B)

H4831

A propellant that works superbly in many large capacity cases, H4831 is one of the best choices for the 270. It is also favored for use in most of the belted magnum calibers. Data for this propellant should not be confused with IMR 4831 data. (S, T)

H4831SC

The SC stands for "short cut" and refers to the shorter length of its granulations. The short length of its grains helps H4831SC to meter well from a powder measure. It is ballistically an exact copy of H4831. (S, T)

H1000

This propellant was named for the 1000 yard target shooters who were likely to find its ballistics attractive. It is intended for very large capacity cases with heavy bullets and is extremely slow burning. (S, T)

Retumbo

This is a very slow burning powder designed for large capacity magnum cartridges. It often produces higher velocities than other powders of this type. (S, T)

H870

This is a very slow burning speed with limited practical applications. It is used only in very large capacity cases with heavy weight bullets. (D, B)

H50BMG

Designed and named for its primary use in the 50 Browning Machine Gun round, it has limited applications because of its extremely slow burning rate. (S, T)

WINCHESTER BALL POWDERS

Winchester powders are manufactured by Olin in St. Marks, Florida. Questions regarding Winchester powders should be directed to Customer Service, Winchester, East Alton, Illinois 62024.

231

This is a fast burning powder with near universal application in handgun cartridges. It is highly favored because of its well known accuracy potential. (D, B)

WST

Winchester Super Target replaces the long discontinued 452AA. It is used by many handgun target shooters. (D, B)

WSF

Winchester Super Field has numerous applications in handgun cartridges, especially competition loads. (D, B)

296

A magnum handgun powder, 296 requires heavy bullets and heavy crimps in order to realize uniform ballistics. (D, B)

748

A relatively fast rifle powder that is used in a wide range of case sizes, 748 is popular with many benchrest and target shooters. As with all spherical powders, it meters quite uniformly through a powder measure. (D, B)

760

A slower speed, 760 is suitable for a very wide range of applications. It is accurate in many cartridges including the 270 Winchester, 280 Remington and 30-06 Springfield. (D, B)

WXR

The only Winchester extruded Rifle powder, it is slow burning and has been used in factory 7mm Mag loads. It has applications over a wide range of cartridges and is similar to both 4831 powders. (S. T)

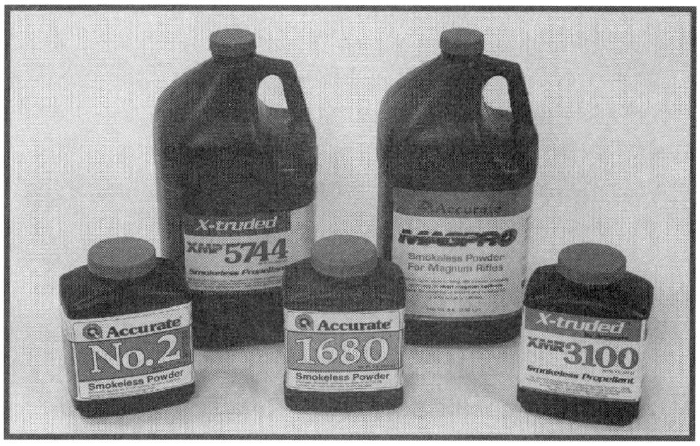

ACCURATE ARMS POWDERS

For questions concerning Accurate Powder contact Box 167, McEwen, Tennessee 37101.

No.2

This is a fast burning powder with a wide range of handgun cartridge applications including the popular 9mm Luger and 38 Special cartridges. (D, B)

No.5

Another fast burning handgun propellant, No.5 is used in a wide range of calibers including 9mm Luger, 357 Magnum, 40 S&W, 44 Remington Magnum, and 45 ACP. (D, B)

No.7

Designed for NATO 9mm carbine ammo, our use of No.7 includes 9mm Luger, 357 Magnum, 38 Super Auto., 10mm Auto., and 44 Remington calibers. (D, B)

No.9

This is a relatively slow burning handgun powder. In general, it should be used with heavy bullets and a very firm crimp. It has applications similar to Alliant 2400. (D, B)

XMP 5744

This is a rather unique powder that was designed for reduced loads. It is a bulky powder that is one of the best choices for use with cast bullet rifle loads. We have extensive data using it in this manner. (D, T)

1680

This powder was designed specifically for the 7.62x39 Russian. It is very fast burning and has limited applications. (D, B)

XMR 2015

This small grained powder is intended for small case cartridges and does well in the 22 and 6mm PPC cartridges. (S, T)

2230

Developed with the 5.56 NATO (223 Remington) in mind, 2230 is a comparatively fast burning rifle powder suitable for many medium capacity cases. (D, B)

2460

Slightly slower burning than 2230, this speed is useful from the 223 to the 30-06. It is popular for 30 caliber service rifle matches. (D, B)

2520

Designed for medium capacity rifle cartridge cases, our applications begin with the 22-250 Remington and cover a broad range of cartridges. (D, B)

2700

A relatively slow burning speed, the primary application for 2700 is with heavy bullets. (D, B)

4350

This speed was designed to compete with IMR 4350 and H4350; but data is not interchangeable. (S, T)

3100

This is another relatively slow speed propellant designed to compete with IMR 4831 and Hodgdon H4831. Thus, it is suitable as a magnum cartridge propellant. (S, T)

MAGPRO

This powder was designed especially for the new short magnum rifle cartridges. It will produce optimum velocities at nearly full case capacities. (D, B)

8700

A very slow burning propellant, 8700 is at its best in very large capacity cases. (D, B)

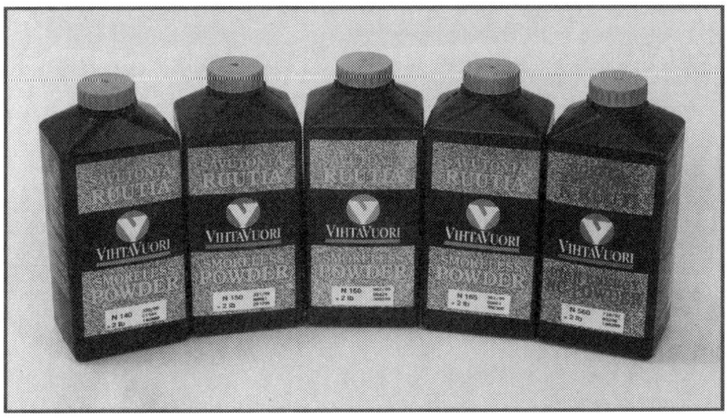

VIHTA VUORI POWDERS

Questions on Vihta Vuori powders should be directed to Kaltron - Pettibone, Vihta Vouri Powders, 1241 Ellis Street, Bensenville, Illinois 60106.

N320

We used this fast burning powder with light bullets in the 45 Colt. It also did well with cast and jacketed loads in the 45 ACP. (S, T)

N340

With a burn rate not unlike Herco, we used this speed in 9mm Luger, 40 S&W, and in the 45 Colt with heavy bullets. (S, T)

N105

This is a limited application speed which we used in the 357 Sig. (S, T)

N110

Fastest burning of the N100 series, N110 competes with both Winchester 296 and Alliant 2400. Thus, it is useful in such calibers as the 22 Hornet, 357 Magnum, and 44 Magnum. (D, T)

N120

This propellant has limited applications somewhat similar to IMR 4227. (S, T)

N130

We used this speed in the 7.62 x 39 and with some cast bullet loads. (S, T)

N133

This powder was used in cast bullet loads and applications somewhat similar to IMR4198. (S, T)

N135

Competing with IMR 4895, this speed has a wide range of applications in non-magnum cartridges. (S, T)

N140

With burning characteristics similar to Varget and Reloder 15, N140 is useful in a wide range of calibers. It is becoming popular in the 223 Rem with bullets of 69 grains and heavier. (S, T)

N150

If N140 and N160 are good in an application, N150 might prove great. (S, T)

N160

Similar in applications to IMR4350, N160 is useful in a wide range of calibers from the 243 and upward. (S, T)

N165

With a relatively slow burning rate, N165 is useful in large magnum cases and in some standard cartridges with very heavy bullets. (S, T)

N170

Extremely slow burning, this speed is limited in applications. (S, T)

N540

This speed is popular with target shooters, especially in the 223 and 260 Remingtons with heavy bullets. (D, T)

N550

Not too different from N150 in burning speed, N550 is useful in such cartridges as the 308 Win. and 30-06. (D, T)

N560

Applications for N560 include the 6,5 X 55mm Swede and the 270 Winchester. It fits some uses similar to N160. (D, T)

8 Jacketed Bullets

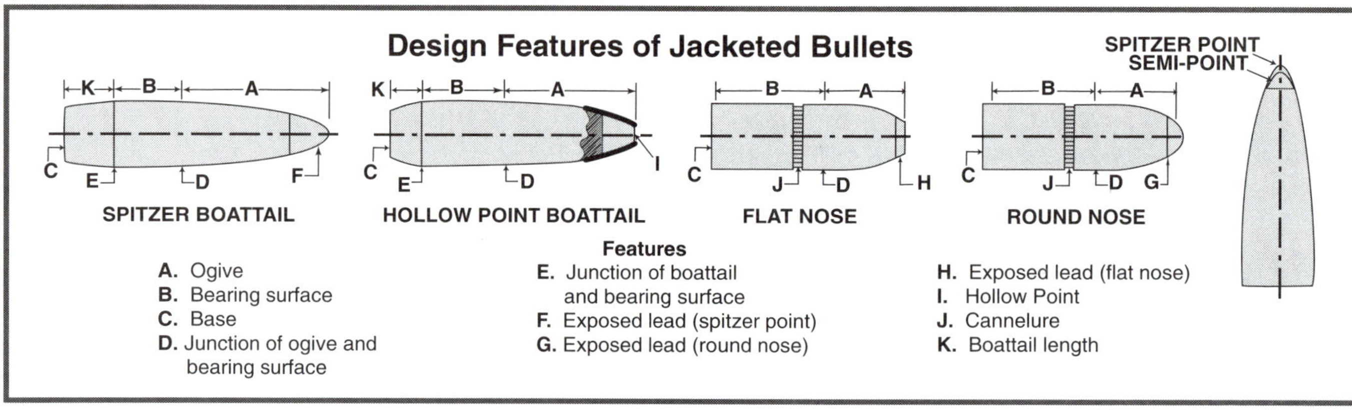

Not all jacketed bullets are created equal. Those designed specifically for target or benchrest shooting are generally capable of more accuracy than the average rifle will deliver. Nonetheless, rifle bullets of this type, should never be used for game as they may completely fail to expand or, at the other extreme, disintegrate on surface impact.

Bullets designed for target applications often carry a trade name, or generic nomenclature, suggesting this type application, i.e. "MatchKing". Some target bullets are extremely frangible being made with thin jackets using a method of assembly known as soft swaged. Such bullets also can be suitable for varmint hunting if the game is relatively small in mass. However, some target bullets are in reality non-expanding types. Target bullets may have lead exposed at the tip, or a hollow point. Neither type of construction should be interpreted to suggest the bullet is an expanding type, albeit some hunting bullets are similarly designed.

In order to insure a high degree of accuracy, target bullets are made to exacting tolerances for weight, dimensions, and concentricity. When testing bullets during production runs, it is not uncommon for some manufacturers to get groups of ten shots measuring well below half an inch. All of the effort involved in the manufacture of a good target bullet results in a slight premium in selling price.

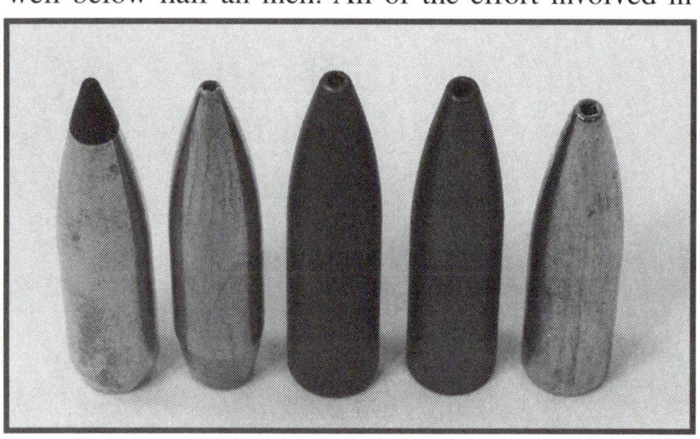

Bullets are designed for specific applications and visual inspection cannot possibly show the bullet's purpose.

Varmint Bullets

It is not easy to hit a woodchuck at 300 yards. It is even more difficult to hit a prairie dog at the same range. Obviously, varmint bullets must be extremely accurate. Additionally, rapid expansion, or even disintegration of the bullet, must occur when it strikes the limited mass of a small animal. If the bullet does not expand in this manner, it will simply pass through small quarry creating a puncture wound that will not be immediately fatal.

Frangible varmint bullets are also a safety asset. Fired at the low angles often associated with varmint hunting, strongly constructed bullets have a propensity to ricochet after striking the ground. Frangible varmint bullets will tend to break up on impact with the ground, and thus reduce potential ricochet danger.

Because of the need for a high degree of accuracy combined with a frangible construction, some target bullets are ideally suited to varmint shooting. Equally, some varmint bullets make excellent target bullets, for example the Nosler Ballistic Tip.

Big Game Bullets

A lot of consideration needs to be given to your specific needs when selecting a big game bullet. Not all big game bullets are up to the task of anchoring thick skinned or heavy boned animals. Nor, for that matter, are some up to sufficient penetration to solidly anchor even light game if any but a broad side shot is taken.

Many bullets offered for big game are quite inexpensive. Others, at first, appear very expensive. Perhaps nowhere else is the old expression, "you get what you pay for," more true than when purchasing game bullets. The run-of-the-mill, inexpensive bullets have killed a

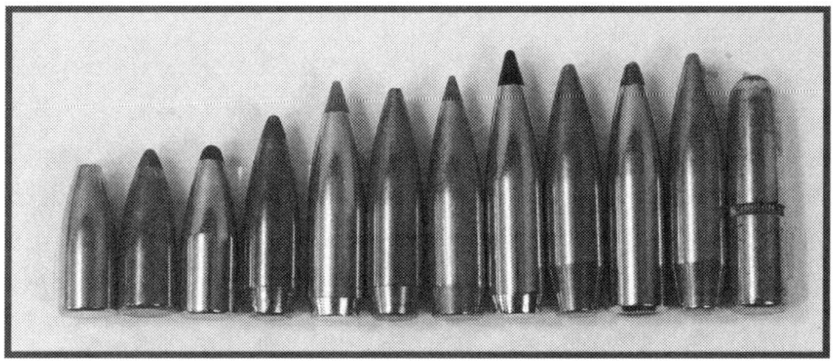

Understanding bullet design and weight applications are essential for appropriate selection. Shown from left to right are 30 caliber bullets with typical applications; a. 110 grain (varmint); b. 125 grain (varmint and light target); c. 130 grain (varmint); d. 150 grain (light big game such as deer); e. 165 grain (light big game); f. 168 grain (target); g. 178 grain (target) h. 180 grain (all around big game); i. 190 grain (target); J. 200 grain (heavy big game); k. 200 grain (target); l. 220 grain very heavy big game.

lot of game. This is especially true when the animals hunted are relatively easy to kill, i.e. antelope and deer. But, even here, many hunters have suffered lost game when the shot was at a difficult angle. Other times, hunters do not understand why game escaped after a well placed shot. Even a rare bullet failure can bring great disappointment. When the game gets tougher and heavier, the number of failures incurred with bargain priced bullets becomes notable.

Bullet expansion involves having the jacket progressively peel back allowing the lead alloy core to mushroom. During expansion of inexpensive bullets, a major portion of the jacket and core may break away from the bullet's mass. This loss of weight can be so great that the bullet's momentum will be reduced to nil and penetration will cease long before the vital organs are reached. Occasionally, an inexpensive bullet may completely fail to expand and simply punch a tiny hole through the quarry without causing a serious wound channel. Such scenarios can result in lost game.

Big game can place heavy demands upon bullet design. The requirement for outstanding accuracy is a difficult design criterion when it has to be combined with the essential need for controlled and progressive expansion. For appropriate expansion the bullet must first penetrate hide, next the bullet must begin to mushroom. The rate of expansion must be controlled so that the bullet will penetrate deeply. Ideally a bullet should expand to double its original diameter. While so doing, the bullet must reach and travel through vital organs before stopping its forward progress due to loss of momentum. To accomplish this a bullet must retain most of its mass. Passage through bone and muscle, no matter how heavy, must not create undue loss of mass and momentum.

An accurate bullet with a combination of good expansion and high weight retention is costly to manufacture. Thus, the best big game bullets carry a premium price. Happily, comparatively few bullets are required to sight-in and even fewer are actually fired when hunting. Thus, the total cost of premium bullets will go almost unnoticed. Naturally, practice shooting can be conducted with popular priced bullets; with the shooter switching to premium bullets when sighting-in immediately before hunting season.

The hunting of relatively lightweight thin-skinned game does not require as much mass retention as when hunting thick-skinned and/or very heavy game. In large and heavy game, a bullet may have to penetrate up to three times the distance before the vitals are reached. Thus, for tough game, a heavily constructed bullet is needed.

Premium hunting bullets can be placed into approximate weight retention groups. For example: Nosler Partition bullets generally retain 60% of mass. Winchester-Nosler Combined Technology Partition Gold bullets usually retain 80% of mass. Swift A-Frame bullets often retain 95% of mass. As you might expect, the price of each reflects its weight retention capability.

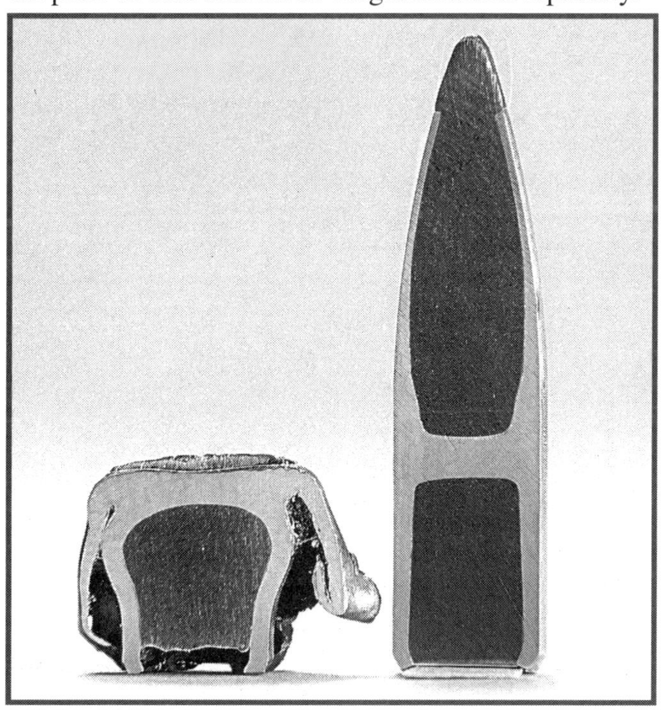

The Nosler partition bullet is shown with sectioned specimen and expanded one. The expanded bullet shows typical performance, front core gone, wings folded back close to shank, and about 65% of original weight intact.

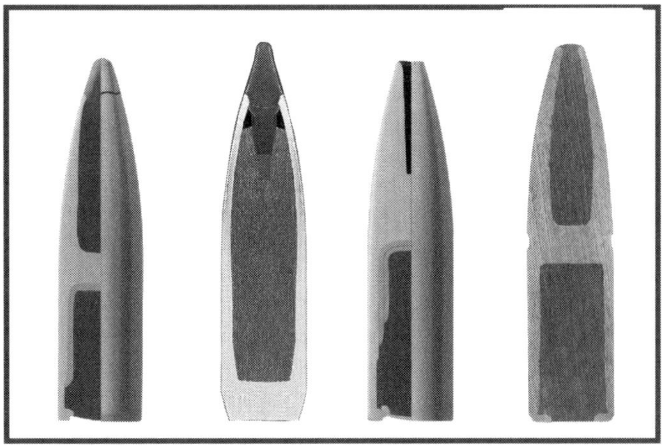

Sectioned views of some excellent premium grade bullets available to the reloader. They are left to right: Combined Technology Partition Gold, Nosler Ballistic Tip, Combined Technology Fail Safe, and Swift A-Frame.

If you are a deer hunter who takes only broadside, behind the shoulder shots you can get along with far less bullet than if you fire at a deer leaving the scene and offering only a difficult angle shot requiring perhaps three times as much penetration. As the game gets bigger, or the ranges longer, your need for the very best bullets increases. In all cases, you should always achieve a large exit hole on broadside shots. On difficult deep penetration shots, if an exit hole does not occur you should trace the wound channel and determine remaining bullet weight. If things are not as hoped, a better bullet is required. Assuming you are using an adequate cartridge, if you lose even a single animal when you absolutely know the bullet was properly placed, your choice of projectiles should be seriously questioned. Hunters with extensive experience come to realize that the bullet selected can be more important than the cartridge choice. Whatever hunting bullet you select, be sure it is up to your requirements.

Dangerous Game Solids

For the hunting of elephant, cape buffalo, or even hippo, smart hunters wisely choose a "solid" bullet to insure deep penetration through extremely thick hide, heavy fat, tough muscle and the huge bones. However, there are many "solids" which may fail to accomplish the necessary penetration.

Typically the least expensive solids are actually a "full" metal cased bullet having a lead core, a steel inner jacket and a copper outer jacket. Such bullets are closed on the nose end with the lead core exposed at the base. As many serious hunters have found, when used on truly heavy game, some bullets of this style construction sometimes bend, rivet at the nose, or even rupture. When any of these failures occurs, penetration will not be in a straight line, or of sufficient depth.

Bullets using steel jackets can result in short barrel life. We have seen several 458 Win. Mag. barrels, used exclusively with steel jacketed bullets, wear out to the point of being unable to stabilize a bullet after as few as 200 rounds were fired.

Another approach to solid bullet design is the use of a homogeneous alloy - no separate jacket and core. Bullets of this type are often made of a solid piece of bronze. These have a good reputation albeit we have experienced several failures with this type. Another type solid uses a heavy bronze jacket and a super hard tungsten core.

Still another type of solid is made with a heavy bronze front end with a hard alloyed lead rear section. The Federal Trophy Bonded Sledgehammer solid is an example of this type. These have a fine reputation for getting the job done with few failures. Such bullets may cost several dollars each, but this is a small price to add to the total cost of a safari.

Appropriate Bullet Application

Each reloader needs to consider the bullet performance required for their application and then to make a wise selection. As stated, deer and antelope hunters who take only broadside, behind the shoulder shots will place the least amount of demands on a hunting bullet. The hunter who wishes to take difficult angle shots, requiring considerable penetration, will need a better bullet. As game gets larger or more dangerous, premium bullets become essential.

Premium grade bullets are produced by several major bullet suppliers. Such bullets include the Speer Grand Slam, the Nosler Partition, the Combined Technology Partition Gold, Combined Technology Fail Safe, the Swift A-Frame and Scirocco.

Velocity Versus Bullet Selection

The reloader needs to consider impact velocity when selecting bullets and appropriate calibers. Bullets driven at greatly differing velocities do require differing construction. For a 22 caliber bullet to expand properly at 22 Hornet speeds, a very thin jacket with a good amount of exposed lead is generally required. Cartridges of approximately the velocity of a 222 Remington require the use of a slightly stronger bullet. To withstand the very high velocities of a 22-250 or 220 Swift, a 22 caliber bullet must be quite strong. The use of a 22 Hornet or 222 Remington style bullet in one of the latter could result in the bullet disintegrating in flight, turning into a puff of blue smoke due to the high rotational forces.

Larger diameter bullets require equal consideration with respect to impact velocity. For example: the very low velocity of the 30 Carbine cartridge demands very "weak" bullet construction if there is to be any bullet expansion. A 30-30 demands somewhat tougher bullets but not as tough as a 30-06 bullet. And if you are using a 300 Weatherby Magnum only the toughest bullets will penetrate properly. Lesser bullets are prone to disintegrate at the high impact velocity of super fast cartridges. As a general rule, if you are using magnum velocities, premium bullets should be considered essential.

Keep in mind that as ranges increase, terminal bullet velocity decreases. Eventually, bullet velocity will reduce to the point where proper bullet expansion is impossible. A good rule of thumb is to expect this velocity to be approximately 2,000 fps. when premium bullets are used.

Appropriate Bullet Weight

Specific bullet application will generally dictate a weight, or perhaps a range of applicable weights. For example: among 30 caliber bullets those of 110 to 130-grains are most often thin jacketed types designed for varmint hunting. In this same diameter, bullets of 150 to 165-grains are usually intended for application on light big game - i.e. antelope, deer, small black bear, goat and sheep. The heavier bullets, from 180 to 220-grains, are generally intended for medium to heavy game such as elk or moose, albeit many hunters correctly choose the 180 grain as a single weight selection for application on all big game.

In calibers from 24 to 30, the lightest bullets are usually varmint types and the heavier bullets are for progressively heavier big game animals. Bullets of 32 caliber, and larger, seldom are made in frangible varmint types. Usually, all 32 caliber and larger diameter bullets are designed for big game. The lightest of these are intended for the smaller big game animals and the progressively heavier bullets for the larger animals. Remember, using a too light bullet will result in insufficient penetration. If in doubt, use a heavier bullet weight.

Bullet Nose Configuration

Sometimes it is essential to use flat or blunt nose bullets in specific rifles. For example, cartridges to be used in a tubular magazine should always have flat or very blunt nose configuration. The use of a spitzer in a tubular magazine allows the point of one bullet to rest against the primer of the round immediately in front of it. Under the forces of recoil, a pointed bullet could ignite the primer against which it rests.

CAUTION: Never use spitzer bullets in a tubular magazine.

Some firearms are designed to feed properly only with round nose bullets. Many 458 Win. Mag. bolt action rifles are perfect examples of this need for round nose bullets. Many military handguns were also designed with only round nose bullets in mind.

However, the selection of round nose bullets sometimes occurs because of misinformation. Many handloaders (and factory ammo users too) still place credibility on the old wife's tale which states that only slow moving, heavy, round nose bullets stand any chance of plowing through heavy brush and reaching the intended quarry. Spitzer bullets, the same tale states, deflect considerably more when encountering intervening brush. The fable further acquaints us with the belief that the greater the bullet velocity, the greater the amount of deflection.

This "theory" has been debunked many times. During the mid 1960's, Lyman employees (the late Lysle Kilbourn and your editor) ran extensive deflection tests. The findings of these tests, and those of many other interested persons that followed, conclusively prove that all bullets deflect regardless of nose profile, mass, or velocity. That no one bullet form deflects more than any other has been firmly established. Those who use round nose or blunt bullets because they believe that these better penetrate through brush are working with a misleading, albeit often repeated, piece of "advice".

Others use blunt nose bullets because "they expand better than spitzers". About a hundred years ago, when smokeless powder and jacketed bullets were first coming into general use, this was indeed fact. But, modern bullet design has created spitzer bullets that expand readily, equaling or outperforming the best round nose styles.

Spitzers Are Better

At long range, a 300 Savage cartridge loaded with a spitzer bullet will outperform a 30-06 loaded with a round nose bullet. A blunt nosed 30-06 bullet of 180 grains, with a muzzle velocity of 2700 feet per second, will arrive at the 300-yard mark traveling at 1725 feet per second. A 180-grain spitzer bullet fired from a 300 Savage, with a muzzle velocity of 2350 feet per second, will arrive at 300 yards with a remaining velocity of 1750 feet per second, a tad faster than the 30-06 round nose bullet. Thus, after 300 yards, the 300 Savage spitzer overcomes the 350 feet per second advantage of the round nose 30-06 bullet.

Looking at it another way, two 30 caliber 180-grain bullets started at identical velocity will have drastically different down range ballistics dependant upon bullet profile. A very sharp nosed bullet of this type, started at 2700 feet per second, would have a 300 yard remaining velocity of 2075 feet per second versus the round nose remaining speed of 1725 feet per second - a 350 feet per second advantage for the spitzer. At 400 yards the spitzer would have approximately a 450 feet per second superiority. At either range, the difference is enough to create substantial trajectory and terminal ballistics inequality.

Spitzers are ballistically superior because their sharp profile results in less air resistance. The reloader can judge the relative performance of any bullet by referring to its ballistic coefficient. This information can be found in the data head for each bullet for which we show loading recommendations. A round nose 180-grain 30 caliber bullet might have a ballistic coefficient of .248 while a spitzer of identical weight and caliber could have a rating of .438. The larger the number the better a bullet will resist deceleration.

It makes little sense to develop a load that squeezes the last foot per second out of your favorite cartridge and then turn its down range performance into that of a lesser cartridge because a blunt nosed bullet is used. Selecting sharp bullet profiles, when appropriate, pays big ballistic dividends. The individual who never shoots past 100 yards may never notice a difference. However, as ranges increase, bullet performance variations, as a function of nose profile, become quite obvious.

Spitzers are not without a weakness. Their sharp soft lead tips deform easily during recoil forces as bullets impact with the front magazine-wall. This is most pronounced with magnum calibers. However, generally the amount of deformation normally incurred is not of major consequence.

If bullet nose damage is a serious problem due to very heavy recoil, there is an option. A "protected point" bullet will eliminate the problem. Do select a bullet with as small a meplat (the blunt tip area) as possible in order to preserve down range ballistics. Bullets such as Speer's Grand Slam or Nosler's Protected Point Partitions are examples of this type.

Certain spitzers have been designed to preclude any flattened noses. The Nosler Ballistic Tip, and the Combined Technology Ballistic Silver Tip are examples. Both use a tough insert placed into a hollow tip to form a very sharp and recoil proof bullet nose.

Bullet Base Configuration

The shape of a bullet's base can influence ballistics. Flat base bullets produce more air turbulence and hence have more drag to over come than boattail bullets. This means an increased rate of velocity loss due to a lower ballistic coefficient for the flat base types. However, this difference is generally quite small. The boattail shape is actually most effective in reducing drag when velocity has dropped below the speed of sound, about 1120 feet per second. The average reloader deals with substantially higher velocity and thus may choose flat or boattail bullet bases without concern for appreciable ballistics variation due to bullet base shape.

Bullet Seating Depth (Overall Cartridge Length)

Generally, the best accuracy occurs when the bullet must "jump" 0.005" to 0.020" from its seated-in-the-case position to the point at which its bearing surface first engages the rifling lands. The best seating depth for any specific load and firearm combination can be determined only by trial and error. Load 10 cartridges at a 0.005" jump and fire for group size. Then, repeat the test with additional groups fired with cartridges seated to 0.010", 0.015" and 0.020" bullet jump. On occasion, some firearms prove best served with a bullet jump of considerably greater than 0.020".

Handgun Bullets

Jacketed handgun bullets generally are not subjected to the extremes in velocity as rifle bullets. Most hunters will restrict handgun shots to considerably less than 100 yards, therefore a relatively narrow velocity range occurs with respect to impact velocity for a given bullet.

A minimum impact velocity of about 900 feet per second is usually required in order to obtain reliable expansion from jacketed handgun bullets. In that handgun impact velocities seldom exceed 1400 feet per second, and are often considerably slower, even the differ-

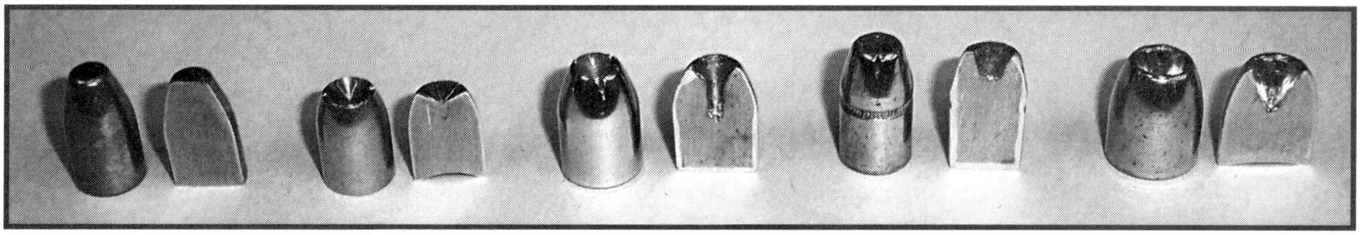

The handgun bullets pictured above show differing construction. The bullet on the left is a Speer 147gr. 9mm TMJ (Total Metal Jacket). It is fully surrounded by the copper jacket which reduces airborne lead while giving excellent feeding in semi-automatics. The others are hollow points designed for expansion. From the left: Speer 357 Sig 125gr. Gold Dot hollow point, Winchester 40 caliber 175gr. Silvertip, Hornady 38 caliber 158gr. HP/XTP, and Speer 45 caliber 200gr. Gold Dot hollow point.

ences from one caliber to another do not place extreme demands on a bullet.

Generally speaking, jacketed handgun bullets with hollow point nose configurations expand more rapidly than soft point types. There are reasons to favor both types. The soft point's slower expansion will, when taking big game, allow deeper penetration. Hollow points are often used on smaller game and varmints.

Some jacketed handgun bullets with lead exposed at the nose are not intended for expansion and are in fact "nonexpanding" target style bullets. Always follow the bullet manufacturers suggestions and guidelines.

Get The Opinion Of Experienced Persons

Unless your use is simply plinking at short ranges, bullet selection is never simply a matter of bullet weight and diameter . Requirements for accuracy, expansion, the amount of penetration and mass retention are all part of what needs to be considered. Accuracy can be determined with test shooting at the range. Controlled expansion performance can take years and several dozen big game animals to begin to provide sufficient data for an appropriate understanding. For this reason it is wise to seek out several persons who have extensive experience on a wide variety of game under differing conditions. Question them about their favored bullets. Their suggestions may provide sufficient insight to make a good first choice of a game bullet. In this way you may avoid disappointment with your first hunting loads.

Cannelures And Crimping

Roll Crimping

For some applications, it is necessary to crimp the case mouth to the bullet. Included among these applications are all ammo to be used in revolvers, all ammo used in tubular magazines, all ammo for very heavy recoiling loads i.e. ammo in calibers designed for dangerous African game.

The purpose of crimping is to prevent movement of the bullet, either forward or rearward, from the original seating depth. This is most often accomplished by roll crimping, the turning of the case mouth inward into a corresponding circular cut in the bullet circumference (the cannelure). In revolver cartridges, under the forces of recoil, bullets will tend to creep forward, even to the point of preventing cylinder rotation. In tubular magazines recoil forces tend to drive bullets deeper into the case. Ditto for firearms where the bullet forcefully impacts some surface during the feeding cycle, i.e. semi-automatic and pumps. Heavy calibers, such as the various 416s, may suffer bullet movement during recoil. For all these reasons, and others, bullet crimping may be essential to ensure proper firearm functioning and uniform ballistics.

ROLL CRIMPING

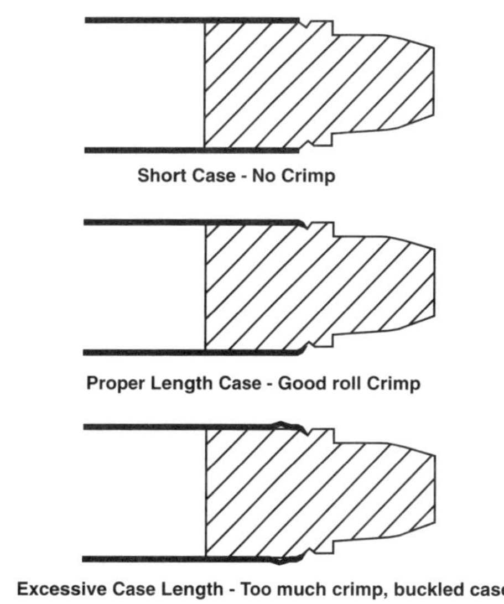

CAUTION: Never attempt to roll crimp to a bullet without a cannelure. Doing so will damage the case and bullet and will prove ineffective for the intended purpose.

Roll crimping can have a somewhat detrimental effect on accuracy. Therefore, unless called for by specific conditions, most reloading is done without crimping.

Taper Crimping

In some instances crimping may be desirable but it also may be impractical or even dangerous to roll crimp. For example, any case which headspaces from the case mouth such as the 30 Carbine, 9mm Luger, and 45 ACP, should never be roll crimped. Doing so shortens the case and alters headspacing. Yet each of these, and other cartridges, are subjected to forceful impacts with feeding surfaces of the firearm. The solution is to taper crimp. This method forces the outside of the case neck tightly against the bullet without shortening the case.

TAPER CRIMPING

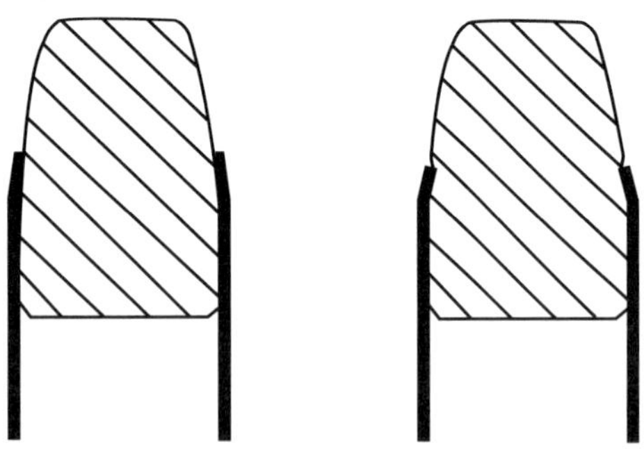

Proper taper crimp   Excessive taper crimp deforms bullet

Roll crimping is used for ammo applications in revolvers, tubular magazines, and heavy recoiling firearms. Taper crimping is used primarily with cartridges which headspace off the case mouth.

NOMINAL JACKETED RIFLE BULLET DIAMETERS

Diameter in inches	Suitable for these calibers. Please note that some calibers require bullets of specific design with respect to velocity and expansion characteristics.
.172"	17 Remington
.224"	22 Hornet (older rifles may require .223") 218 Bee; 222 Rem.; 222 Rem. Mag.; 22 PPC USA; 22 BR Rem.; 223 Rem.; 219 Zipper; 225 Win.; 224 Wea. Mag.; 22-250; 220 Swift.
.243"	6mm PPC USA; 6 x 45mm; 243 Win.; 6mm Rem.; 244 Rem.; 240 Wea. Mag.
.257"	25-20 Win.; 256 Win.; 250 Sav.; 257 Roberts; 25 -06; 257 Wea. Mag.
.264"	6.5 x50mm Jap.; 6.5mm Carcano; 6.5 x 54mm M.S.; 260 Rem.; 6.5 x 55mm Swede.; 6.5 Rem. Mag.; 264 Win. Mag.
.277"	270 Win.; 270 WSM; 270 Wea. Mag.
.284"	7-30 Waters; 7mm-08; 7mm Mauser (7 x 57mm); 284 Win.; 280 Rem.; 7mm SAUM; 7mm WSM; 7MM Rem. Mag.; 7mm Wea. Mag.; 7mm STW; 7mm Ultra.
.308"	30 M1 Carbine, 30-30 Win.; 307 Win.; 300 Sav.; 30-40 Krag; 7.5 x 55mm Swiss; 308 Win.; 30-06; 300 H&H; 300 WSM; 300 SAUM; 300 Win. Mag.; 300 Norma Mag.; 300 Wea. Mag.; 300 Ultra; 30-378 Wea. Mag.
.311"-.312"	32-20 Win.; 7.62 x 39mm; 7.62 x 54; 7.65 x 53mm Arg.; 303 Brit.; 7.7 x 58mm Jap.
.318"	8 x 57mm J
.321"	32-40 Win.; 32 Win. Spl.
.323"	8mm Mauser (8 x 57 mm JS); 8mm-06; 8mm Rem. Mag.
.338"	338-06; 338 Win. Mag.; 338 Ultra; 340 Wea. Mag.; 338-378 Wea. Mag.
.348"	348 Win.
.358"	35 Rem.; 356 Win.; 358 Win.; 350 Rem. Mag.; 35 Whelen; 358 Norma Mag.; 358 STA.
.366"	9.3 x 62mm; 9.3 x 74Rmm.
.375"	38-55 Win.; 375 Win.; 375 H&H; 375 Wea. Mag.; 375 Ultra; 378 Wea. Mag.
.400"	38-40 Win.
.416	416 Rem. Mag.; 416 Rigby; 416 Wea. Mag.
.424"-.427"	44-40 Win.
.429"	444 Marlin
.458"	45-70 Govt.; 450 Marlin; 45-90, 45-100, 45-110; 45-120; 458 Win. Mag.; 460 Wea. Mag.
.475"	470 Nitro.
.510"-.511"	50-70; 50 BMG.

NOMINAL JACKETED HANDGUN BULLET DIAMETERS

Diameter in inches	Suitable for these calibers. Please note that some calibers require bullets of specific design with respect to velocity and expansion characteristics.
.222"	22 Rem. Jet.
.224"	221 Fireball.
.251"	25 ACP.
.309"	30 Luger (7.65 Parabellum); 30 Mauser.
.311"-.312"	32 ACP.; 32 S&W; 32 S&W Long; 32 H&R Mag.
.355"	380 ACP.; 9mm Luger (9mm Parabellum); 9 x 21mm; 9 x 25mm; 38 Super; 357 Sig.
.357"	38 S&W, 38 Spl.; 357 Mag.; 357 Maximum.
.364"	9 x 18mm Makarov
.400	40 S&W.; 10mm Auto.
.410	41 Action Exp.; 41 S&W Mag.
.429"	44 S&W Spl.; 44 Rem. Mag.
.452"	45 Auto Rim; 45 ACP.; 45 Win. Mag.; 45 Colt, 454 Casull
.480"	480 Ruger
.500"	50 Action Express

ADDENDUM: MOLY COATED BULLETS

The use of bullets coated with Molybdenum disulfide or "moly" has become popular with high volume users such as varmint hunters and High Power shooters over the last ten years. Moly is a dry film lubricant that reduces friction between the bullet and the rifle's bore. Advocates of moly coating cite improved accuracy, longer barrel life, lower chamber pressure, and reduced copper fouling. The substitution of moly-coated bullets in place of an uncoated bullet over a pet load does not always yield peak accuracy. This substitution of a moly-coated bullet often requires minor tweaking of a particular load, usually increasing the powder charge by a few tenths of a grain up to a grain depending on the caliber and specific load. Use caution when working up all loads. Many shooters use Kano Kroil® to clean moly fouling. Refer to Chapter 14 for additional tips on cleaning rifle barrels fired with moly-coated bullets.

Although several bullet manufacturers offer pre-coated bullets, many shooters opt to coat their own with Lyman's Super Moly kit. The process for properly coating bullets requires that they be clean and free of any residual oils from manufacturing. Shooters using steel shot as a coating media should thoroughly clean the shot with degreaser before beginning. Lyman's Super Moly kits utilize ceramic media. This ceramic media is much lighter than steel and gives better coverage. Full instructions are enclosed with Lyman's Super Moly kits.

The Lyman technical staff fired several popular match loads with both uncoated and moly-coated bullets through our 308 Winchester electronic piezo pressure barrel. The results verify a reduction in chamber pressure as well as a slight reduction in velocity. These results are based on five-shot strings and utilized Remington cases and 9 1/2 primers. On average, moly coated bullets resulted in a three percent reduction of chamber pressure and a one to two percent reduction in velocity compared to uncoated bullets.

168-grain Sierra HPBT
41.5-grain of IMR-4895

	pressure	velocity
uncoated:	49,660 psi	2488 fps
moly-coated:	47,300 psi	2459 fps
	2,360 psi	29 fps

175-grain Sierra HPBT
42.0 grains of IMR-4064

	pressure	velocity
uncoated:	53,040 psi	2542 fps
moly-coated:	50,980 psi	2507 fps
	2,060 psi	35 fps

190-grain Sierra HPBT
41.3-grains of IMR-4895

	pressure	velocity
uncoated:	59,780 psi	2482 fps
moly-coated:	57,340 psi	2448 fps
	2,440 psi	34 fps

9 Cast Bullets

Bullet casting should always be conducted in a well ventilated area with all necessary precautions observed.

CAUTION: The making of cast lead bullets exposes the reloader to the potential risk of burns and the risk associated with accidental introduction of lead into the body. The reloader must take all the necessary steps to prevent either occurrence. It is important to read and adhere to all of the precautions listed in this chapter and those included as instructional material when you purchase lead bullet casting and reloading equipment. These and all other necessary safety steps are the responsibility of the reloader.

Cast Bullets Are Economical And Easy To Make

One advantage of cast bullets is their low cost - just pennies apiece. A box of twenty 30-06 rounds assembled with cast lead bullets will cost about $3.00; similar jacketed bullet ammunition may cost four or five times as much. Cast bullets also allow for extensive shooting without barrel wear. Barrels seem to last indefinitely when used exclusively with cast bullets.

Casting lead bullets requires the melting of a suitable alloy, pouring the melt into a mould, allowing the alloy to harden, dropping the bullet from the mould, and sizing/lubing the bullet to an appropriate diameter. The loading of cast bullets requires the edge of the case mouth to be slightly flared, allowing the cast bullet to be started into the case without shaving metal from its base. All of these steps are straightforward and not at all complex. However, as in all aspects of ammunition making, specific cautions are important. These are as follows:

CASTING PRECAUTIONS

1. Always cast in a well ventilated area - ideally outdoors. Never breath lead dust, lead fumes, or fluxing fumes.

2. If gas heat is used to melt the lead metal, it is important to take steps to prevent the inhalation of carbon monoxide.

3. Use only properly grounded electric melting pots. These are identified by a three prong plug. Do not circumvent proper grounding. Lack of grounding can cause serious electric shock.

4. Always wear heavy protective gloves, aprons, and goggles or face shield. Splatters of hot molting lead cause serious burns. Cover all parts of the body with thick protective apparel that will withstand burning through in case of an accidental spill or splatter.

5. Keep all others away from the casting area.

6. Be certain to use the melting pot only on a surface that is 100% free of any potential for the pot to be tipped or knocked over. Anchor the pot to the work surface. Never leave a pot unattended, not even for brief seconds.

7. Be cautious with fluxing materials. Avoid flammable style flux in favor of noncombustible types.

8. Place a suitable container under the pot spout to catch leaking or spilled metal to prevent unintended damage to property or people.

9. Never eat, drink, or smoke, and keep hands away from mouth, when handling lead in any manner. Thoroughly wash hands after lead handling operations.

10. Water, even minute amounts, will cause violent eruption of molten lead. Make sure all equipment is absolutely free of any moisture. Keep moisture of any source away from the casting area, including condensation on any overhead pipes.

11. Primers, loaded ammunition, and other substances, introduced into the hot melt, will cause violent eruption of the molten alloy. Keep all potential accident causing items out of the casting area. Take meticulous care to sort all lead alloy being ready for melting to eliminate any chance of introducing anything but bullet alloy into the melting pot.

Getting Started

Make certain that the apparel chosen covers all parts of the body. Don't forget the importance of protecting ankles and feet from burns. There should be no area of exposed skin to be burned in an accident.

The basics tools of bullet casting equipment include:

1. Bullet mould and handles.
2. Lead or lead alloy
3. Melting pot and dipper. (If a bottom pouring electric furnace is used, a dipper is not required.)
4. Alloy fluxing material and something to stir the flux into the molten alloy. An old large spoon works well.
5. Suitable mallet for tapping open the mould sprue-cutter plate.
6. Soft pad of nonflammable material on which to drop bullets from mould (perhaps an old ironing board cover).
7. Bullet sizing and lubricating tool.
8. Bullet lubricant.
9. Casting thermometer (optional).
10. Ingot mould (optional).

Lead Alloys

The bullet alloy used should be dependant upon the intended application. Rifle bullets loaded to near maximum velocities should be made from a hard alloy to prevent barrel leading. An ideal alloy contains 5% antimony, 5% tin and 90% lead. This alloy has a Brinell Hardness Number (BHN) of 15. Pure lead (used for all muzzleloading projectiles) has a BHN of 5, while Linotype has a BHN of 22. The hardest alloys will help prevent barrel leading but will not expand. The softest alloys expand well but may result in barrel leading especially as velocities increase.

Composition and Hardness (BHN) of Useful Bullet Alloys

	Percent			
Alloy	Lead	Tin	Antimony	BHN
Monotype	72	9	19	28
Stereotype	80	6	14	23
Linotype	84	4	12	22
Lyman No. 2*	90	5	5	15
Taracorp Magnum*	92	2	6	15
1 to 1 Lead/Lino	92	2	6	15
Electrotype	94.5	3	2.5	12
10 to 1	91	9	—	11.5
16 to 1	94	6	—	11
20 to 1	95	5	—	10
30 to 1	97	3	—	9
Wheelweights*	95.5	0.5	4	9
40 to 1	97.5	2.5	—	8.5
Pure Lead	100	—	—	5

* Responds well to heat-treatment.

CAUTION: Never attempt to salvage metal from any kind of battery. Doing so can be extremely hazardous.

Some Casting Tips

Avoid getting any lead on block interface surfaces when casting. Even minute spots of bullet metal will prevent the mould from closing properly. Wipe away such spots while the mould is hot. Do not use a synthetic rag for this purpose but rather one of a natural material such as cotton.

Keep the sprue plate screw properly tensioned. The sprue plate should turn readily but not be loose. Excessive pressure of the plate on the mould may prevent proper venting or cause wear on the blocks. A sloppy fit will result in accuracy-destroying raised sprues on the bottom of the bullet. These raised sections can prevent proper seating when gas checks are required.

Never allow a mould to rust. (Rusted moulds can sometimes be cleaned up using steel wool.) If you protect a mould with a lubricant between uses, it will be necessary to degrease the mould before each casting session. Some folks simply store cool moulds in air tight containers with a desiccant to prevent potential rusting. This eliminates the need for degreasing before each use.

The Lyman Mag 20 electric casting furnace features a thermostat for easy and precise temperature control.

Melting And Fluxing Bullet Metal

CAUTION: Be certain to wear protective clothing as previously described as well as heavy leather work gloves and goggles whenever working with molten bullet metal.

The melting pot must be capable of heating the bullet metal to 700 - 800 degrees Fahrenheit. A Lyman lead pot placed on an electric or gas stove will prove satisfactory. However the reloader will find an electric furnace made specifically for bullet casting to be faster and more convenient. Such units, properly used, are cleaner and quicker to work with. Their adjustable thermostats also make casting temperatures easier to maintain.

Whether blending raw metals into bullet alloy and forming ingots, or actually making bullets, the same melting and fluxing procedure is followed. It will take about 20 minutes for the solid metal to melt. When it is hot enough to flow freely, the melt needs to be fluxed to remove all impurities. To do so, drop a small bit of fluxing material into the pot. It is best to use Marvelux (a dry fluxing material available from Brownells). There will be no smoke or greasy fumes when using Marvelux and a good flux will be obtained. Next, stir the melt with a dipper. Stir deeply, frequently introducing a dipper full of air into the melt. The air helps with the fluxing.

The surface of a properly fluxed melt appears almost mirror bright and contains flecks of burned brown and black impurities. These must be skimmed off using a spoon like tool that has an insulated handle. During casting, as soon as any impurities are seen on top of the melt, it is time to again flux.

CAUTION: Be extremely careful when fluxing to prevent tipping over the pot or any splashing any of the molting alloy.

CAUTION: The material skimmed off is very hot, perhaps 800 or more degrees. Dispose of it in a safe fire proof container that will not be accidentally turned over.

Bullet Casting

After fluxing, the melt is ready for casting. The melt temperature must be maintained or the cast bullets will not be as expected. The nearby drawings show some common bullet problems related to temperature control. It is important to realize that proper temperature control also includes keeping the mould (and dipper if one is used) at the correct temperature. Thus, the casting pace may need to be increased or slowed, if you find the mould is too cold or too hot.

Allow the mould to preheat by resting it on the edge of the electric furnace. If you are using a cast iron pot with an independent heat source, heat the mould to casting temperature by repeatedly filling it with lead until the bullets are properly formed.

Each caster tends to develop their own routine. Beginners often find the following method works well.

Before fluxing, bullet alloy will have a dirty, lumpy appearance. After fluxing, the bullet alloy will have a bright, almost mirror-like appearance.

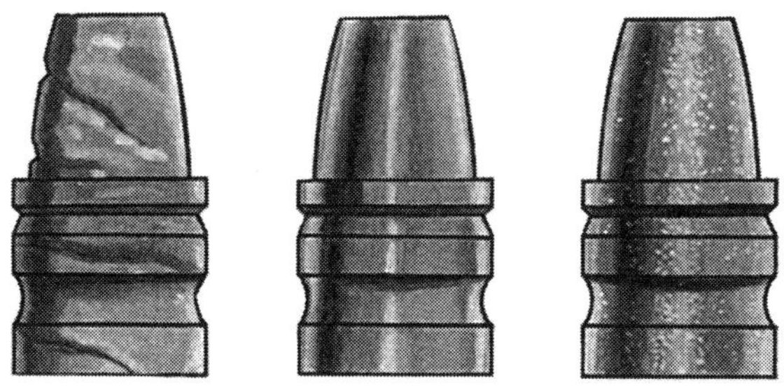

Wrinkled Bullet Temperature Too Cool | Good Bullet Temperature Correct | Frosted Bullet Temperature Too Hot

Fill the dipper with melt. Then, with the top of the bullet mould held vertically, place the dipper's poring spout into the sprue plate hole on the mould. Next, rotate mould and dipper as a unit so that the top of the mould is horizontal and the dipper is above it. Now, tip and pull the dipper slightly away to allow a small amount of alloy to fill the concave area of the sprue cutter plate. If this is not done, as the melt solidifies, a hole will form in the base of the bullet. Remove the dipper.

Wait several seconds for the bullet to harden. Small light bullets, i.e. 22 caliber bullets, will harden faster than large caliber heavy bullets, i.e. 45-70 bullets. Pace yourself so that the bullet is hard enough to prevent damage as it falls from the mould. Speed up or slow down, as required, to keep the mould at an appropriate temperature for proper bullet forming. Small diameter light bullets will require a faster pace than the larger bullets.

After the bullet cools in the mould, use a wood dowel or mallet to strike open the sprue plate, shearing the sprue from the base of the bullet. Push the sprue plate around far enough to fully clear the opening in the mould. Never use more force on the spruce plate than necessary to shear the sprue from the bullet base.

Allow the sprue to drop into a fireproof container. If bullet bases have a cavity or concave shape on the base, allow more of a sprue to form. The largest bullets will require the largest sprues.

If lead smears appear on the top of the mould, after rotating the sprue plate, allow the sprue to more completely solidify before cutting it off.

Bullets will not fill out correctly until the mould reaches the proper temperature. A dozen, or more, casts may be required before well formed bullets are obtained. Imperfect bullets can be dropped into the sprue container to be remelted at another time. Do not be tempted to heat the mould by sticking it into the melt. Such action will bring more grief than may first come to mind.

If the mould overheats do not attempt to cool it in water or use any other artificial means to lower its temperature. The only safe method is to set the mould aside and allow a bit of time for it to cool.

Wrinkled, or less than perfectly filled out bullets suggest that the temperature of either the melt or mould is too cool. Frosted bullets, or bullets which break apart easily, suggest that the either melt or mould is too hot. Properly cast bullets have sharp edges and appear clean and bright.

CAUTION: Do not drop poorly formed bullets or sprues back into the melting pot. Doing so can cause molten metal splashes which can result in serious burns or property damage.

It is a good practice to weigh all cast bullets to eliminate those with air voids. A good caster will make bullets that will vary no more than a half grain from the average weight for the lot. Those that vary more than this amount should be culled and remelted at a latter date. The actual weight of your bullets will vary with specific alloys. Bullet weights also vary from lot to lot even when using the same type of alloy. Thus, it is a good practice to segregate each lot of bullets.

Gas Checks Improve Cast Bullet Performance

The metal caps often placed on bullet bases are called gas checks. These protect the bullet base from being melted by hot firing gases. Gas checks are essential whenever using high velocities. Protecting the bullet base from having bits of lead brought to the melting point, will prevent unnecessary bore leading and avoid inaccuracy. Gas checks, attached when sizing the bullet, must be seated squarely or accuracy will be adversely affected.

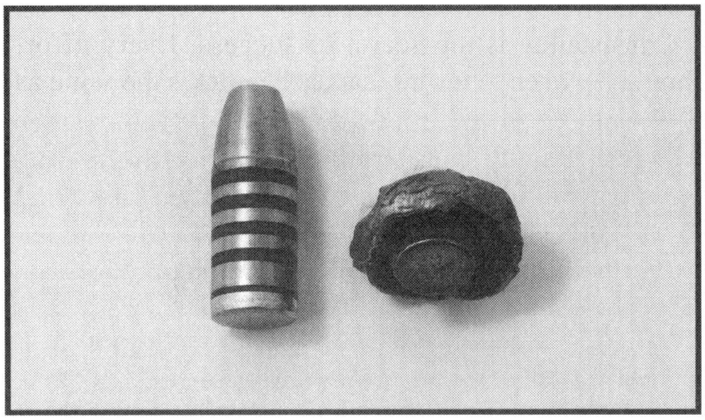

Lyman cast bullet #375449. (l) Sized, lubed, and ready to load. (r) Recovered from a Wild Boar. This bullet was cast from an alloy of 12 bhn and fired at approximately 1,750 feet per second at a distance of 80 yards. It expanded to double its original diameter with a retained weight of 264.7 grains—perfect!

When used with a high velocity bullet lubricant, rifle bullets cast of a hard alloy and protected with a gas check can withstand velocities in excess of 2,000 feet per second. Some cast bullet users find it possible to exceed even 2,500 feet per second when everything is right and the bore is perfectly smooth. However, under normal circumstances most rifle shooters do best with velocities of about 1,800 feet per second.

Gas checks can be applied only to bullets with a rebated base specifically designed for the purpose. Gas check style bullets may be used without the gas check applied, if velocity is kept modest.

CAUTION: Old style press-on gas checks should not be used when bullet bases go below the case neck. This will avoid having a loose gas check fall into the powder charge. Accuracy, with any style cast bullet, will always be best if bullet bases are completely contained within the neck of the cartridge case.

SLUGGING THE BARREL

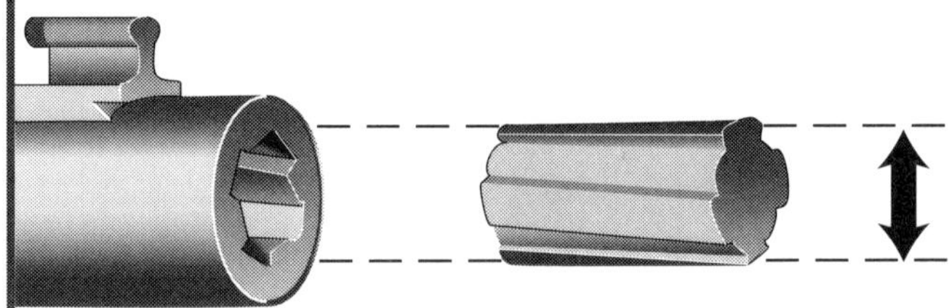

Starting at the muzzle end, a soft lead slug (slightly larger than the groove diameter) is driven through the barrel. When the slug is retrieved on the breech end, it will be perfectly engraved and clearly marked by both the lands and the grooves. Measuring (with a micrometer) the diameter of this slug at its widest point will determine the groove diameter of the rifle. Bullets should conform to this dimension.

Sizing And Lubricating Cast Bullets

Sizing cast bullets is required to ensure bullet diameters of a nominal dimension. Usually bullets are sized to, or just larger than, the actual groove diameter of the firearm to be used. To determine the actual groove diameter of any firearm, its bore should be slugged.

A slightly oversized lead slug driven through the bore and measured, will provide the barrel's smallest diameter. (Some shooters prefer to size to the diameter of the barrel at two inches in front of the chamber.) When searching for the best accuracy, first try a sizing diameter equal to the bore dimension. Then try bullet diameters of 0.001" and 0.002" larger.

Great care must be used when slugging a bore so as not to damage the bore surface. The rod used to drive the slug through the barrel must be close fitting and have a slight radius on the outside edge. If the rod buckles or scraps on the bore, irreparable damage could occur. If in doubt it is best to have a qualified and experienced gunsmith do the bore slugging.

Many shooters do not actually slug their bores, but rather start with a sizing diameter to match the diameter of jacketed bullets nominally used for their cartridge. This is a practical approach for the first series of accuracy tests. If the first tests do not provide the desired accuracy, a second series of attempts can be with bullets sized 0.001" larger. A third sequences can be with bullets 0.002" larger. It is counter productive to use any larger sizing diameter.

Note: Seldom are barrels undersized; rather when they vary from the nominal they are usually oversized. (It should be noted that some barrels of European manufacture have been found to be undersized.)

Sizing And Lubricating Tips

Lubrication of bullets demands a proper flow of lubricant. This is usually not possible when the ambient temperature is less than 70 degrees Fahrenheit, and warmer is better. This is particularly true with some of the hard, high performance lubes that are available. Note: It is possible to damage a lubricator by applying to much force in an attempt to force cold lubricant into the bullet groove(s). Heating plates are made by Lyman and others, which warm the lube/sizer press and eliminate this problem.

Note: The bullet should be pushed into the lubricator with the correct top punch. If a poor fitting top punch is used the bullet will be sized misaligned. Such a condition is contrary to the goal of accuracy. Always use the exact top punch required for the bullet to be lubricated.

If bullet lube forms on the bullet base, it may be due to excessive pressure on the lube in the reservoir. It could also be due to improper adjustment in positioning the bullet in the sizing die with respect to its relationship to the lubricant orifices. The design of the bullet can also cause this. Bevel base bullets often get lube in the bevel area. Be sure to remove any lubricant from the bullet base in order to prevent contamination of the powder charge in a loaded round.

Do not store lubricated bullets as the lubricant may pick up contaminants that would be harmful to a bore. Size and lubricate only the number of bullets that will be loaded in a single session.

If a lubed bullet is dropped on the floor, or otherwise picks up contaminants, discard it or carefully remove all lubricant and contaminants from the bullet. It may then be relubed.

Keep in mind that undersized bullets are a detriment to accuracy. If sizing at nominal bore diameter (and then increasing to 0.001" and 0.002" over nominal) does not give the desired results, slugging the bore becomes mandatory. If such slugging shows the bore to be as expected then it is reasonable to look to other areas for the cause of the inaccuracy. If the slugging shows a non-typical barrel, then the appropriate bullet sizing diameter can be addressed. Keep in mind that it is not practical to reduce a bullet's as-cast diameter by more than 0.002". Doing so negatively affects accuracy.

Consistency is the secret to success. Every effort should be taken to insure that each bullet is the same as the last. When testing alloys, lubricants, or sizing diameters do so only by changing one aspect at a time. Keep good records of the results to be certain that things do not become jumbled in memory.

Handgun Lead Bullet Tips

Handgun bullets cast of soft lead will expand if they have sufficient impact velocity (about 850 feet per second). This is equally true of flat nosed bullets as well as hollow point types. Naturally, the lead hollow points do expand somewhat faster than a lead round nose type, with other nose styles falling somewhere between. However, bullets cast with a hard alloy will not expand regardless of nose configuration.

Wadcutter style lead bullets will leave large diameter sharply cut holes in paper targets and are preferred for this reason. At normal target shooting distances (up to 50 yards) wadcutters are very accurate. However, there is a range limitation for wadcutters with respect to bullet stability. Depending upon velocity, bullet length, and base configuration, at somewhere between 60 and 100 yards most wadcutters will start to tip and eventually begin to tumble. Because of this, those wishing clean cut holes in targets at ranges past 50 yards frequently chose a semi-wadcutter bullet, for example the Lyman mould numbers 358429 and 452424 - Keith style bullets.

CAUTION: When loading hollow base lead bullets (hand cast or purchased) it is necessary to avoid very heavy loads. It is possible for the front of such a bullet to be separated from the skirt portion leaving the skirt stuck in the barrel. The separation is caused by firing pressures forcing the skirt tightly against the bore while pushing hard against the forward internal head surface. If the strength of the lead alloy is exceeded the bullet comes apart. Such a condition is very dangerous should another round be fired while the skirt is stuck in the barrel. Thus, hollow base handgun bullets should be used only for light target loads.

Suggested and Alternative Sizing Die Diameters

PISTOL Caliber	Dia. (inches)
.22 Jet, .221 Fireball	.224★, .225
25 ACP	.251★
.25 cal.	.257★, .258
.30 Luger, 30 Herritt, 30 Mauser	.310★
.32 ACP, .32/20, .32 S&W, .32 H&R Magnum	.311, .312, 313★, .314
9mm Luger*, .38, .38 Super Auto, .380 Auto	.354, .355★, .356
38 S&W, .38 Spl., 357 Mag.*, .357 Max.	.357, .358★, .359, .360
9mm Makarov	.363
.38/40, 10mm Auto, 40 S&W	.400★, .401
.41 S&W Mag., 41 Action Express	.410★
44 S&W Spl., 44 Mag.*,44 Russian	.429★, .430, .431
45 ACP, .45 Auto Rim, .45 Colt* (post-WW II), .45 Sch. .45 Win. Mag.	.450, .451★, .452
.45 Colt (pre-WW II) & .455 Webley	.454★
RIFLE **Caliber**	**Dia. (inches)**
.22 cal. (except .22 Hi-power)	.224, .225★
.243, .244, 6mm	.243, .244★
.25 cals.	.257, .258★
.264 Win. Mag., 6.5mm	.264★, .266
.270 Win.*	.277, .278★
7mm, .280 Rem., .284 Win.	.284, .285★
.30 cals.	.308, .309★, .310
7.62mm Russian*	.310
.32/20 Win.	.311, .312★
7.65mm Mauser*	.311
.303 British*, 7.7mm Jap.*	.313, 314★
.32 Win. Spec. 32 S.L., .32 Rem.	.321
8mm Mauser (S Bore)	.323, .325★
.338 Win., .33 Win.	.338★
9mm x 56, 9mm x 57	.354, .355★, .356
.35 cal.	.357, .358★, .359
.375 H&H Mag., .375 Win.	.375, .377★, .378
.38/55*	.379
.38/40	.400, 401★
.40 Cal. Rifle	.406, .410
.44/40	.427, .428★
.44 Spl., .44 Magnum	.429, .430★
.444 Marlin	.430, .431★
.43 Spanish	.439
.45/70, .458 Win.	.457, 458★, .459
.50 Cal. Rifle	.509, .512★

* Wide variations in diameter. Suggest you slug barrel.
★ Indicates suggested useful diameter, which may vary from gun to gun.

Section 4

10 The Road to Accurate Ammo by Bryce M. Towsley

Chapter 10

"Only accurate rifles are interesting."

Col. Townsend Whelen

It's at this point that most gun scribes get a little misty eyed as they blindly agree with the Colonel before launching into their thesis on how to achieve accuracy. But, with all due respect to one of the greats of gun writing, I must diverge from Colonel Whelen's famous statement. It's not that I have anything against accurate rifles; it's just that I have owned some inaccurate rifles that were pretty damn interesting.

Bryce M. Towsley. Remington Model 700 .35 Whelen. 100 yard group. Handloads with RL-15 powder and 225 grain Barnes X-Bullets.

Besides, what is accuracy? It's a term that is thrown around rather carelessly by shooters with little defined information about what exactly we are talking about. Is an accurate rifle one that will shoot a minute of angle? If so, how many shots in the group? Three, five, ten? How many groups should be fired to determine the "average" group size and establish the accuracy standard for that rifle? Should the rifle be able to do that from a cold barrel and place the first shot clustered tightly with all those that follow? Or should this rifle be "accurate" only after a fouling shot or two has warmed the barrel? If you are a long-range hunter the first shot is the most important, so accuracy here might be defined as exact predictability for the first shot. You may measure that by shooting a ten-shot group while firing one shot each day for ten days. If you are a varmint hunter you will likely define the accuracy of your rifle by how consistently it shoots with a fouled and heated barrel. If it's accurate for the first shot, or even the first five shots, but starts scattering bullets after that, it's useless to you.

Lots of rifles will group three shots in less than an inch, fewer can do five or ten and it's a special rifle indeed that will put every shot inside that magic circle. But is one MOA acceptable accuracy for a varmint rifle? Isn't a serious varmint gun that can at best keep its bullets in one inch a bit on the "inaccurate" side? What of a benchrest competition gun? Would you be competitive at the top level with a gun only capable of one MOA accuracy?

Conversely, I have seen very few lever action deer rifles that will come close to shooting to one MOA. But does that make them "inaccurate" rifles? Compared to what? What standard of accuracy is needed for hunting whitetail deer in the thick brush? What actually defines an "accurate" Winchester Model 94 in .30-30 Winchester?

Or any other rifle, for that matter?

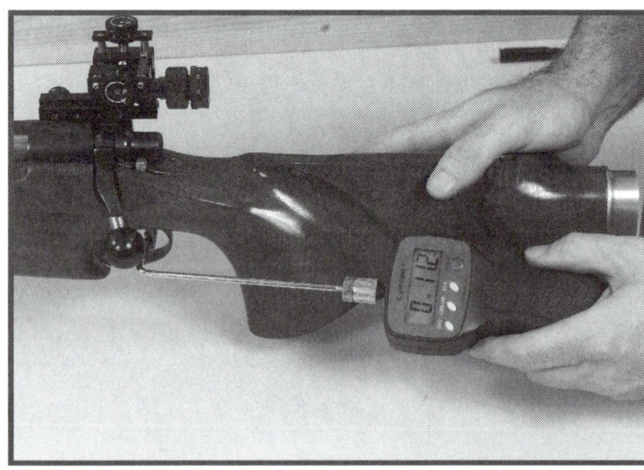

A smooth light trigger pull is important for accuracy testing. Lyman's Digital Trigger Pull Gauge allows for precise trigger pull measurement.

In the end we must accept that "accuracy" is a relative term with no single correct answer. What is considered "accurate" for one type of rifle would be unacceptable for another. I have a lever action rifle that can consistently group three shots in less than an inch and a half at 100 yards. Not just a lucky group now and then, but every time I try. That's an exceptional rifle. I also have a heavy barrel varmint rifle that shoots about the same. In this case it's a poor shooter and should be rebarreled or replaced.

In my world, accuracy is when any rifle is shooting the best it can with ammo best suited for the use that rifle is intended for. If your handloads are designed and constructed so that your rifle is running at peak performance, then you can claim you have an accurate rifle and I won't argue. How you determine that is up to you, but I'll give you my "protocol."

Big game hunting guns are fired with three shot groups and I allow the barrel to fully cool between groups. Varmint guns are fired with five shot groups and I do not let the barrel cool between groups. But I don't let it get any hotter than I would allow in the field. Usually, when it's too hot to touch it's time to let it cool off. I use a minimum of five groups, but usually ten groups, and take an average. When I can no longer achieve a smaller group average with anything I am doing with my handloads I figure I have the most accurate load for that gun.

Now, to be truthful, I don't always take this to its final conclusion. At some point you need to end the search or accept that you can wear out your rifle trying to achieve perfection. So it's best to have a "target goal" to strive for. For example, if I have a big game bolt-action rifle that is consistently shooting to less than one MOA I usually quit experimenting. Perhaps I can shave another eighth or quarter inch off the group average, but why? It will gain nothing in the hunting fields. The same with a varmint gun, if I have it shooting consistently under a half-inch I save the barrel for varmint hunting. No point in shooting it out just to try to shrink the group average a few tenths of an inch. Better those hundreds of rounds are spent on prairie dogs, which is why I have the gun. The point is I usually set an arbitrary, but fairly tough goal for the rifle and when I have achieved that I consider that my handloads are "accurate."

The key to creating accurate ammo for your rifle can be found in the three "Cs"; Consistency, Compatibility and Concentricity.

Consistency is first. Accurate ammo is no place for rebels. Individualism is frowned upon as accuracy favors conformity and to be accurate, ammo must be uniformly the same. Consistency simply means that each round of ammo is loaded exactly the same as the round before it and the round to follow. The more consistent the ammo is in every respect, the more accurate it is likely to be.

Compatibility recognizes that each rifle is an individual and that no two will be exactly the same. The best load in my rifle may not be accurate in yours. Rifles often have strong preferences about bullets, seating depth, powder selection and charge weight. Even something as simple as the primer can make a difference.

Finally concentricity. Ammo and rifle barrels are basically round. Bullets are round, cases are round, chambers are round and bores are round. So the more concentric our ammo, the more accurate it will be. The more round we keep it and the better the alignment with the round parts of the gun, and the more accurately it will shoot.

Achieving these goals is not difficult, but it does require attention to details and a thoughtful, linear approach to designing and building the ammo. As with building anything, the key is the foundation. The foundation for any cartridge is the case.

Pick a quality brand of brass from a company that pays attention to quality control. Ideally, cases should be new or once fired in the same gun you are loading for. Weigh the cases and throw out any that deviate from the average by more than a few grains. This is accomplished fast and easy with an electronic scale.

How you resize your case is important to accuracy and to reliability. The chamber in the rifle is designed to mimic the shape of the cartridge case, only slightly larger. This allows enough clearance for the unfired cartridge to enter the chamber. Brass is relatively elastic and as the pressure increases the case will expand until it encounters the chamber wall. The chamber will then support the brass, keeping it from expanding further. This directs the pressure to the base of the bullet and pushes it down the rifle bore. The expanded brass also forms a seal against the chamber walls that prevents the gas from flowing back between the cartridge case and the chamber.

If the brass were to remain in this "expanded" mode, tight against the chamber wall, it would stick and be difficult or impossible to extract from the rifle. However, because of its elasticity, once the bullet exits the muzzle and the pressure drops, the memory in the brass will cause it to contract. This again creates enough clearance to allow the empty case to exit the rifle. However, it does not contract as much as it had expanded. And while the case should fit easily back into the chamber, the neck will not hold a bullet.

A rifle chamber must be designed to allow the neck area of the cartridge to expand enough to easily release the bullet. If it cannot because the chamber's neck is not sufficiently larger than the outside diameter of the cartridge case neck, pressure will be too high as the bullet fights its way free from the vise-like grip of the too-tight neck. In order for this to work correctly, the amount the brass must expand is enough that the rebounding case neck's inside diameter will still be too large to correctly hold a bullet in a fired case. It's a rather fine line, as the neck must expand enough to release the bullet, but too much space will degrade accuracy.

Other than the neck being too large to hold a new bullet, the case is now at the maximum dimension for reuse in that rifle and a perfect fit for that chamber. This is because it has been fire formed to that rifle's chamber, which can be a good thing. When a rifle is made, the chamber will be cut to fit within a given set of tolerances. If the chamber reamer is new, it might be on the large side of this tolerance scale. If the reamer is old and worn, it might hit the small end of the chart. Millions of rifles are chambered for a given cartridge with thousands of different reamers. Throw in the variations in worker skill or quality control, and no two chambers in a given caliber selection will be exactly alike. To make factory ammo work in this wide diversity of rifle chambers, the cartridge must also fall under specific tolerance guidelines. If it's too big, it will not chamber in some rifles. So, to play the odds, ammo makers are going to favor the smaller end of the tolerance scale. As long as it chambers in all rifles and doesn't have any problems with headspace, that is the smart approach in making factory ammo. But, as Colonel Whelen himself said in a 1919 *Outdoor Life* article, "Tolerance is the antithesis of accuracy."

Because of that pesky concentricity thing, accuracy with factory ammo is not always all it can be. The closer a bullet is aligned with the rifle's bore when a cartridge is chambered, the more accurate that ammo is going to be. When ammo is smaller than the chamber (because of the tolerances), it lies on the bottom of the chamber, which tilts the bullet slightly out of alignment with the bore. Brass that has been fired in the same chamber is then fire formed to that chamber. The case fills all the space except what is needed to allow the cartridge to enter or exit. The shoulder of the case fits against the shoulder of the chamber and the end result is that the bullet is more closely aligned with the bore. In theory, the ammo should be inherently more accurate.

Full length resizing will bring the outside dimensions of the case back to "factory" specifications. For big game hunting where reliability is probably more important than accuracy, full length resizing makes sense. Hunting ammo should have a little "tolerance" built in to allow for changes in conditions. The last thing you want is ammo that chambers hard, or not at all when you are hunting in snow or the gun is dirty. While full-length sizing may shorten the life of the brass a bit and cut the accuracy slightly, it also increases the reliability of the ammo.

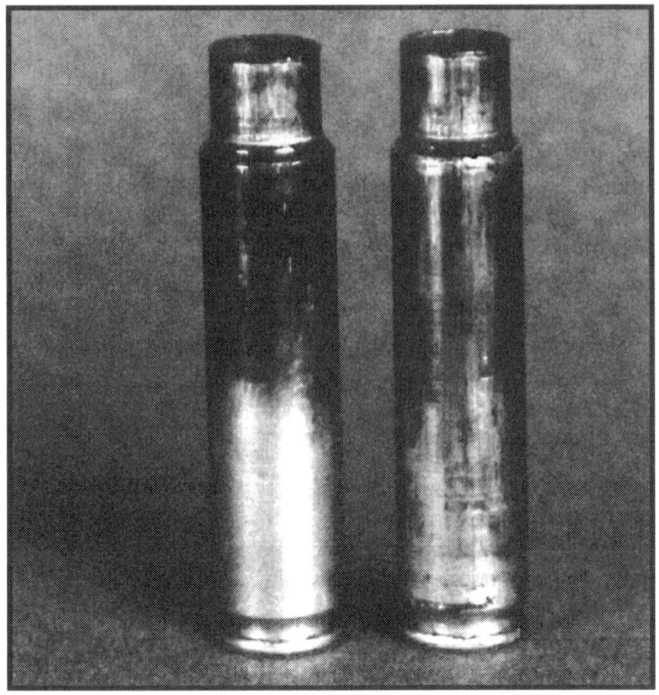

Two 7mm TCU cases that have been smoked in a candle flame and sized. Case on left is neck sized only, case on right has been full length sized. Also note the hard contact with the shoulder on the right case.

For varmint or target ammo that will be used in a bolt-action rifle, consider neck sizing only with a "neck-size only" die. This will bring the neck back down to the proper size to hold the bullet, but leave the rest of the case formed to fit the rifle's chamber, creating more accurate ammo.

Neck sizing should only be used if the case is going to be fired in the same chamber it was used in previously. Chamber tolerances are enough that a cartridge that has been fire formed in one gun may not fit in a different rifle. To work best, the case should also be indexed in exactly the same position in the chamber. Some shooters make a "witness mark" on the brass so that it can be oriented exactly the same way every time. This mark usually will be placed at the top each time the case is chambered. If the chamber is not exactly centered with the bore, or if it's out of round (as it is in

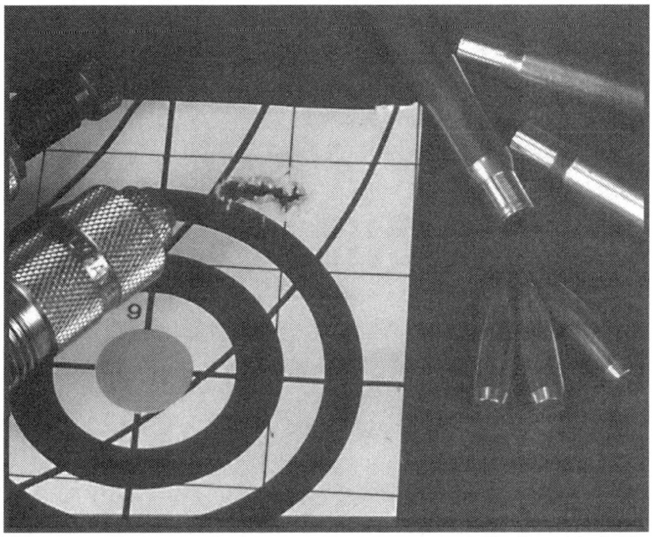

These cartridges had the necks turned and sized. the goal is better accuracy.

one rifle I own) this insures that the brass shape mates correctly with the chamber shape.

The brass will "spring back" a little less each time it is used and after a few firings it may become difficult to close the rifle's action. At this time the case should be run through a full length sizing die that is adjusted to bring it back to the minimum dimension needed for your rifle.

This is actually a viable alternative to neck sizing. If once-fired brass was used in the same gun you will be shooting the handloads in, you might also consider adjusting the sizing die until it just contacts the shoulder. This will resize the brass as fully as possible without pushing the shoulder back. It will enhance accuracy, reduce any possibility of misfires from excessive headspace and extend the life of the brass. It also saves the price of an extra neck-sizing die and creates ammo that is still more accurate than full length resizing, but more reliable than neck sizing only. The trick is to adjust the full length resizing die so that it resizes the neck, partially resizes the body and pushes the shoulder back only when it becomes necessary. This is done by adjusting the die so that is just contacts the shoulder on a once fired case, but does not move it. The easiest way I have found is by smoking the lubricated case in a candle flame and watching how the die rubs the soot off the case as it's sized. Start with the die backed out in the press and turn it in a little at a time, sizing a marked case each time, until it is contacting the case correctly. Then lock it in place.

This works well for any bottlenecked cartridge that is designed to headspace off the shoulder. It will also provide more accurate ammo with belted magnum cases, as it will cause the case to headspace off the shoulder instead of the belt and will align the cartridge with the bore more precisely. A cartridge headspace gauge also is easy to use and accomplishes the same task.

Neck turning insures that the brass in the neck is the same thickness all the way around, which helps to keep bullet tension and release consistent, which can aid in accuracy. It's all part of "concentricity." This can be a bit time consuming, but is often worth the trouble.

Inside neck reaming will reduce the thickness, but this process will not insure the thickness of the brass remains the same around the perimeter of the neck. Outside neck turning is usually a better solution. Because the inside of the neck in a sized case is round and neck turning removes metal from the outside of the case, the brass thickness will be "trued-up" all around the neck.

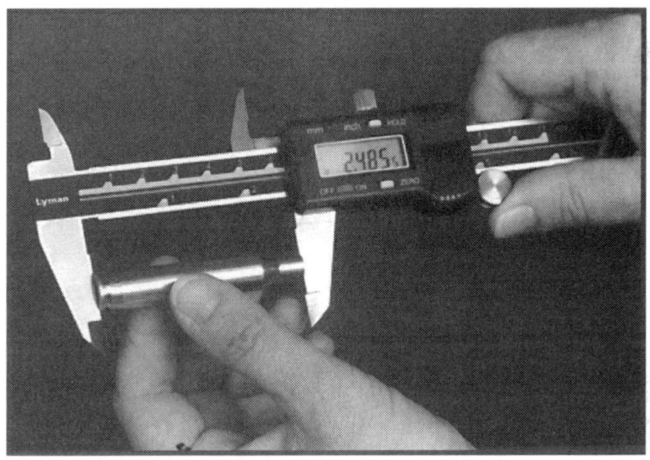

Lyman Stainless Electronic Digital Caliper measuring a case.

Clean the cases in a vibrating or tumbling case cleaner. This removes all lubricant as well as polishes the brass. Use a dial caliper to measure the length of each of the cases to find the shortest of the lot. If it is less than the maximum case length listed in the loading manual, adjust your case trimmer so that the cutter will barely "kiss" the neck. Make sure that the cutter has removed metal all the way around the neck so that the entire circumference is shiny. If the case is longer than the maximum case length, trim it to the correct length. Once you have the trimmer set up, trim all the cases to this same length. Deburr the outside of the case neck and chamfer the inside slightly. For most handloading, a hand-held tool will work fine for this. If you are going to be loading lots of ammo, the electric Lyman Power Deburring Kit takes some of the work out of it.

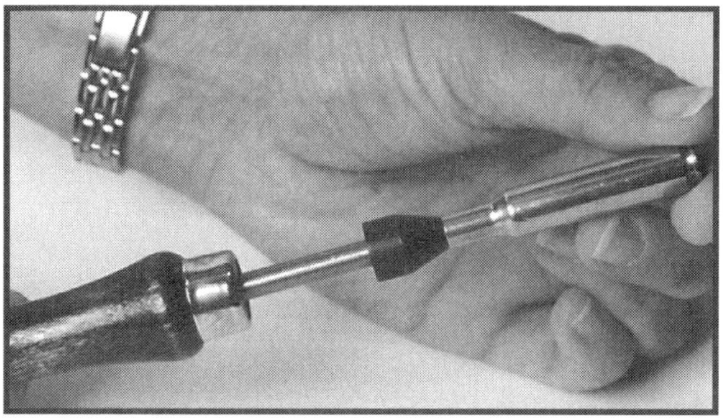

Lyman Flash Hole Uniformer being used on case.

The burr inside the case made by punching the flash hole through the head should be removed to insure an even flow of the flame from the primer to the powder. The Lyman Flash Hole Uniformer is designed for this job. Then square the bottom of the primer pocket with a Primer Pocket Uniformer. These last two steps are "one-time" only and will not need to be repeated for any subsequent loadings.

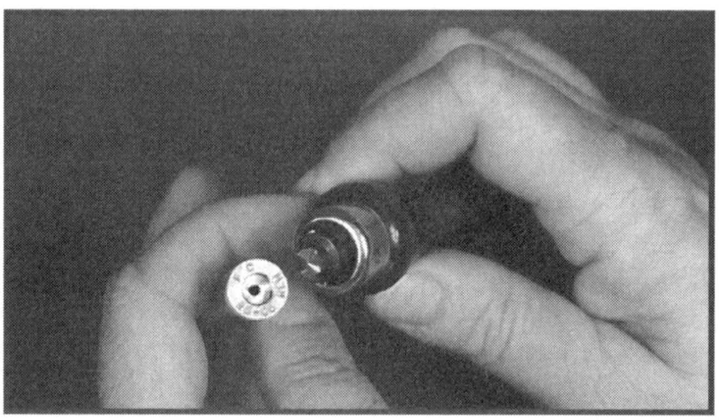

Lyman Primer Pocket Uniformer was used to square this primer pocket.

Each primer must be seated fully in the primer pocket without crushing or distorting the primer. This is done best with a hand tool that allows the user to "feel" the primer as it contacts the bottom of the primer pocket. This keeps the primer from being crushed or damaged from too much pressure and still insures that the legs of the primer anvil are all fully supported. Seating the primer with all three legs of the anvil on the squared bottom of the primer pocket insures that none of the firing pin energy will be wasted by pushing a partially seated primer further into the pocket. Also, a high primer, one that's not fully seated, can cause feeding or chambering problems.

One area of confusion for any handloader and particularly for one new to handloading is the question of which powder to use? Most manuals don't help much. Instead of saying, "use this powder" they list several for each bullet and often dozens for each cartridge. This manual is different than most because it lists the most accurate powder tested for each bullet and cartridge and that's always a good place to start. Often a gun will show a preference for one powder over the others, but it's not always the same from gun to gun. That's that "compatibility" thing mentioned earlier. It's likely that to find the best load for a particular gun and bullet you will need to test more than one powder.

Bullets, primers, cases, powder and dies for .220 Swift and .243 Winchester.

When trying to decide on a powder in a cartridge that you are unfamiliar with, another good rule of thumb is to choose the powder that is giving the highest velocity listed in the loading manual for the bullet you are using. Usually that, or the most accurate powder tested in the Lyman labs will be your best choice. With all that said it's likely that more than one powder is going to produce good results in your ammo, so the search shouldn't be as complicated or as expensive as it sounds.

So how do you know if you made the right choice? Accuracy is one place to start the search. If you have tried several powder charge weights, experimented with a couple of different bullets and various bullet seating depths and still can't make your handloads shoot accurately, it's time to try another propellant. More on bullet seating depths later.

If you don't have a chronograph, beg, borrow or steal one. No handloader can accurately evaluate their work without shooting through a chronograph. They are not expensive and without one you are mostly spinning your wheels. One of the pieces of information it will give you is the standard deviation. This is a measure of the uniformity of velocity with your loads. The lower the number, the more accurate the ammo should be. If you simply cannot get the standard deviation number down to a low number, take that as another indication that you probably are not using the best powder.

One other consideration is the style of powder. Loading for high volume shooting such as prairie dog shooting might call for a compromise in performance in return for ease of loading. Most extruded powders do not flow well through a powder measure and for best results every charge should be checked and adjusted on a scale. This takes time. Not a problem when you are loading 20 cartridges, but if you are loading 500 it can become a big bottleneck in the process. A ball powder will flow through the powder measure consistently and you can simply charge your cases from the adjusted powder measure and spot-check every ten cases. If the ball powder happens to shoot groups that are a 1/4-inch larger than an extruded powder, the ease and speed in loading is a fair trade off.

With most rifle powders other than ball style, it's best to weigh every powder charge. Adjust the powder measure to throw a charge that is about a half a grain under your desired weight. Put this on the scale and use a powder trickler to bring the charge up to the correct weight. This insures that each powder charge is consistent from cartridge to cartridge. Use a powder funnel to charge the case and then stand the charged case up in a loading block. Once all the cases are charged with powder, hold the block under a light to visually check to see that every case has powder and that all the powder charges are the same volume in the case.

On thing to remember when testing powders is to never have more than one can of powder on your loading bench at any time. When you have finished with that powder, pour the unused portion back into the can, seal it and put it in storage away from the loading bench before bringing another powder to the bench. This prevents mix-ups in which powder is being used and prevents inadvertently pouring the unused powder into the wrong container and contaminating the powder in that can.

While the options can be narrowed down by doing your homework ahead of time, the answer to the "what's the best powder" question is impossible to say for sure before loading and shooting. That answer can only be found by burning some of the powder under controlled conditions and assessing the results.

The keystone of any handload is the bullet. It's the only physical connection with the target and considerable thought should be given to its selection. A quality game bullet that is also very accurate used to be as rare as a sober Elvis sighting, but bullet technology has

Lyman Crusher II Reloading Press. With .22-250 cartridge. Barnes XLC bullet.

advanced enough so that most hunting bullets will deliver excellent accuracy. However, the designs that make them good hunting bullets can also make them a bit temperamental. A "hard" bullet is not as forgiving of a rifle's idiosyncrasies as a thin jacketed, soft-core bullet and occasionally an individual rifle may not like a particular brand of hunting bullet. So, try another one.

If accuracy is the primary goal then of course you should concentrate your efforts on the various target bullets. But here again your rifle might not like a particular brand, so be prepared to try another.

One of the biggest reasons many shooters handload is the option to "tune" a specific load to a rifle. The handloader can improve on factory ammo because of the ability to tweak the handloads for the best accuracy from a given rifle. One variable that often makes a huge difference in accuracy is bullet-seating depth.

Seating bullets to the maximum overall cartridge length listed in the handloading manuals is always a safe, conservative approach, but it will rarely produce the best accuracy results. It is better to match the overall cartridge length to the gun for which you are loading the ammo. By experimenting with bullet seating depth, a rifle's accuracy can almost always be improved. Sometimes rather dramatically.

One graphic illustration of this can be found in my .280 Ackley Improved. When I was working up the initial handloads I simply seated the bullets to the suggested overall cartridge length for the parent .280 Remington. It would be an understatement to say the accuracy was disappointing. In investigating the problem, I found that the chamber had an unusually long throat. As a result the bullets were making an almost epic journey before they encountered any rifling. With the next batch of handloads I seated the bullets out in the case neck until they were .025" from contacting the lands. The average of several three-shot groups with this new ammo shrunk to a bit less than one inch. The ammo looks a little goofy, but the increase in accuracy was enough that I'll endure the jeers and laughter from my shooting pals.

The best starting point for any rifle cartridge is to seat the bullet until it has a gap of .010-inch to .030-inch before it engages the rifling. After shooting the ammo to test for function and accuracy, this can be fine tuned even more by experimenting with small changes in the seating depth and then shooting groups to judge the effect the change has on the rifle's accuracy.

The method of finding the seating depth I find easiest is to first resize a case, but do not seat a primer or charge it with powder. Seat a bullet in the case just far enough to hold it securely and gently try to chamber it in the rifle. The bolt likely will not close. Don't force it or the bullet will stick in the rifling and pull out of the case. One trick if that happens and you don't have a cleaning rod handy, is to drop a smaller diameter bullet down the barrel. This will usually dislodge a lightly stuck bullet.

Now, turn the seating adjustment stem on the seating die down one-quarter turn and run the case through the die again to push the bullet a little deeper into the case neck. Try the case in the gun again. Repeat this process until the bolt will just close without forcing it. The overall length of this cartridge is the "to the lands" measurement. Mark this dummy cartridge with a felt tipped pen and put it with your dies for future reference use. To adjust for the bullet to be off the lands, you would adjust the seating die to seat the bullet the selected amount deeper in the case neck than the "to the lands" dummy cartridge. For example, if the dummy case is 3.450-inch and you want a .020-inch gap between the bullet and the lands, you would adjust your seating die so the resulting overall cartridge length is 3.430-inch.

Once you have determined the bullet seating depth that is best for that bullet and load in your rifle, make up another "dummy" round with no powder or primer and the bullet seated to that depth. Mark it with the information and put it with your dies for that cartridge. If you load several bullets, the bullet-seating die will need to be adjusted each time it is used and that's where this "dummy" will earn its keep. Simply screw the seating stem out in the seating die until it will not contact the bullet, run the case fully into the die, then screw the seating stem down until it makes firm contact with the bullet and lock it in place. Now each succeeding bullet will be seated to the correct depth.

All bullets are not the same and you must determine the correct seating depth for each make, style and weight of bullet individually. The ogive shape and design will determine where the bullet will contact the rifling and this will vary from bullet to bullet. Some bullets will shoot best with a longer or shorter gap and again the only way to tell for sure is by shooting for groups. Remember, too, that some "hard" bullets like the Barnes X-Bullet require a larger gap to the rifling.

Because rifle throats will vary a great deal, the only way to find the bullet seating depth in relation to the lands is to measure it in the rifle and with the bullet you are going to load. The resulting ammo is not acceptable for any other rifle. If you have more than one firearm chambered for the same cartridge make certain that the ammo is labeled properly and is only fired in the gun it was intended to be used in.

A limiting factor for bullet seating depth can be the length of the rifle's magazine. The bullet must be seated deep enough so that the ammo will feed through the magazine without binding. With some rifles, seating the bullet close to the rifling lands will create a cartridge that is too long to work in the magazine. In that case, if you wish to use the rifle as a repeater you must seat the bullet deep enough to insure reliable functioning in the magazine and hope for the best in accuracy. Or switch to a bullet with a blunter ogive profile.

Cartridges intended for use in guns with tubular

magazines such as many lever actions must have their bullets seated to the cannelure and the case mouth crimped into the groove. This keeps the bullet from sliding further into the case from the pressure of the magazine spring combined with the forces of recoil. This limits the options for seating depth and may result in a longer bullet jump than is optimum for accuracy. That's a point that is probably moot in this style of rifle as they are rarely used in any way that would require the last little bit of accuracy potential be extracted.

Make a careful visual inspection of each cartridge as you wipe them clean with a soft cloth. It's a good idea to check the concentricity. Place the case in a concentricity gauge and spin it while checking for bullet run out or a neck that is out of center with the case. Everything should spin on the same axis center with no wobble.

Place the loaded ammo in a cartridge box and label it with all the load data and the date. Never assume that you will remember the data used and do not rely on notes in your loading manual or log. Label each box with all the important information.

All this simply gets you through the door to the party. Now the fun starts and we get to make some noise. Shoot this ammo carefully from a bench rest to evaluate accuracy. Try moving the powder charge up a little at a time to see if groups tighten or open up. If you are not at maximum, keep going because sometimes they will open up with a change and then close tighter as you move the charge weight higher. Watch the standard deviations on your chronograph to see if they are getting smaller or larger. Smaller is good as smaller standard deviations usually indicate more accurate ammo. But larger, particularly a rapid jump with just a small change in powder charge, might indicate that pressure is at maximum. Sometimes large standard deviations can be reduced by changing primers. Particularly if you go from a standard to a magnum primer. However, reduce your powder charge 5% if you do because the magnum primers will cause a jump in pressure.

Once you have the best powder charge, experiment with bullet seating depth to see if accuracy is improved. Move the bullet in .005-inch increments and shoot to see if accuracy gets better or worse.

If you simply can't make a load shoot, you should probably switch powders or bullets. But always change only one thing at a time. Check the results before moving on to change the next one. This applies to components as well as variations in charge weights or bullet seating depth.

In following these steps you are on a straight road to accurate ammo. It may take a few trips to the range, but range trips are what it's all about. In truth, aren't we really all in this for the shooting?

11 The Mystery of Pressure by Bryce M. Towsley

Like any technical pursuit, handloading has its own language. Included in that are a lot of catch phrases that are sure to be strewn through any conversation between one or more devotees. "Pressure" is one term that comes up a lot with people who create their own ammunition. But, it's often misunderstood. We speak of pressure as a bad thing, but without it our bullets wouldn't fly. Pressure is what forces them down the barrel and out into the wide world of free flight. Without pressure they would just continue to sit, fat, happy and content, inside the case neck.

In simple terms, the higher the pressure the harder it will push on the base of the bullet and the faster it will make that bullet go because there is a direct correlation between pressure levels and velocity. Of course, a lot of factors come into play and a 40-grain .22 bullet with the same pressure as a 300-grain .375 bullet is not going to have the same muzzle velocity. But, with all else being equal, on any given bullet the higher the pressure the higher the velocity.

Of course, there are limits. In a modern rifle the cartridge case is probably the limiting factor. They can only withstand so much pressure without failing, so our ammo must be kept below that limit by a reasonable safety margin. In some older designs the gun is the limiting factor. In either case, that "failure" point minus the safety margin is the "maximum" pressure limit for a given cartridge.

Maximum loads are based on the upper pressure limit listed for that specific cartridge by the Sporting Arms and Ammunition Manufacturers' Institute, Inc. (SAAMI.) This regulatory body determines voluntary industry performance standards for pressure and velocity in ammunition for commercial manufacturers.

By looking at a number of factors they determine what the maximum pressure level for a given rifle cartridge should be. Those factors include case design and the type of rifle the cartridge is used in. For example, newly designed cartridges to be used in bolt-action rifles will likely have a higher allowable maximum pressure level than new cartridges designed to be used in a lever action rifle. Primarily because the bolt-action rifle design is stronger than a lever action rifle and can withstand higher pressure.

The age of the cartridge is another big factor. For example, the new rifle cartridges being introduced to the market will only be used in new rifles made with modern materials and manufacturing techniques. The rifles also should be constructed to SAAMI specifications. Because of those factors most of the recent sporting rifle cartridges designed for bolt action and similar rifles are listed with a maximum average pressure of (MAP) 65,000 psi, which is currently the highest SAAMI allows in any cartridge. Examples of these cartridges would include all the new short action magnums, as well as the Remington Ultra Mag family.

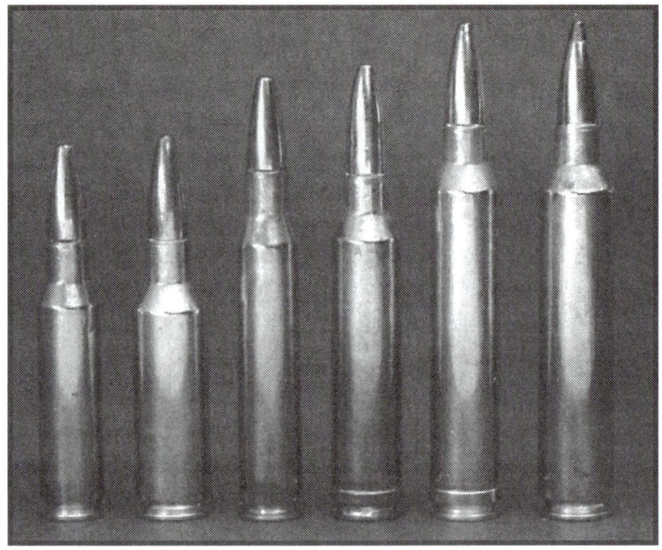

L to R: 7mm-08 Rem, 7mm Remington Short Action Ultra Mag, 280 Rem, 7mm Rem Mag, 7mm STW & 7mm Ultra Mag.

However, popular old cartridges like the .45-70 Government will be used in a wide range of rifles from new and very strong guns to old trapdoor Springfield rifles that were never very strong even before age and metal fatigue are considered. As a result, the SAAMI MAP for factory loaded .45-70 ammo is only 28,000 psi.

The rest will fall someplace in between. Cartridges with a lot of old guns of unknown quality around might be low, for example the .300 Savage is listed at 47,000 psi. While at the other end of the spectrum is the .300 Weatherby, which has been around a while, but has always operated at the ragged leading edge of pressure and is commonly chambered in strong firearms. The MAP for the Weatherby is 65,000 psi.

There are basically two laboratory pressure-measuring systems in use to gauge chamber pressure for modern rifle cartridges. The copper crusher system, for which the resulting pressure data is expressed as "copper units of pressure" or CUP. The other is the piezoelectric transducer system. The pressure data from this system is listed in pounds per square inch or PSI. (Unless the information is not available, the pressure references in this article are listed in psi.)

This is a Universal Receiver shown with a copper crusher resting on the piston.

Of the two, the copper crusher system is older and is slowly being phased out in favor of the more modern piezoelectric transducer system. However, both are recognized by SAAMI and both continue to see a lot of use.

The two systems result in differing numbers for the same loads and, for mathematical reasons that are beyond the normal range of understanding for anybody not a double carrier of the nerd gene, *there is no direct correlation between the two that would allow a simple mathematical formula to convert from one to the other. So don't even try.*

SAAMI lists the information from both systems for most, but not all cartridges. In general terms the older cartridges are listed for CUP and the newer cartridges listed in terms of PSI, with a lot of overlap that have both listed.

The copper crusher system which measures peak pressure in copper units of pressure uses a special barrel with a cylinder bored perpendicular to the cartridge axis. (Without the gunny "geek speak", that means there is a hole drilled into the sidc of the chamber.) A piston is inserted into the hole with a copper gas check between it and the cartridge. A copper cylinder is held firmly against the piston. When the gun fires, the pressure in the case will push against the piston which in turn crushes the copper cylinder. The cylinder is then removed and measured to see how much it was crushed. The results are expressed in copper units of pressure.

These are copper crusher cylinders. The one on the left has been compressed from a pressure test. The one on the right is new and shows the original size.

Each lot of crusher cylinders are of a known metallurgy and are calibrated to a tarage table which is matched with each lot to show the pressure for the amount of crush measured. The operator will remove the crushed cylinder and measure it, then find the corresponding measurement on the chart which will have the pressure listed in CUP on the same line.

A similar measuring device is used for shotshells, but because of the lower pressure levels encountered, it uses lead rather than copper cylinders. The resulting pressure measurement is listed as lead units of pressure (LUP.)

The piezoelectric method uses a transducer sensor mounted in the chamber wall. This contains a piezoelectric crystal that generates a small electrical charge when compressed. The strength of the electrical charge is directly related to the pressure applied to the transducer. This information is automatically fed into a computer which will read the pressure out in pounds per square inch.

Both systems are calibrated using reference ammunition. Reference ammunition is loaded to exacting standards and carefully evaluated for assessed pressure and velocity levels. The reference ammo is fired in the test barrel and the resulting pressure is checked against the reference lot's assessed level. The difference is then used to "zero" the test barrel being used against the known qualities of the reference ammo.

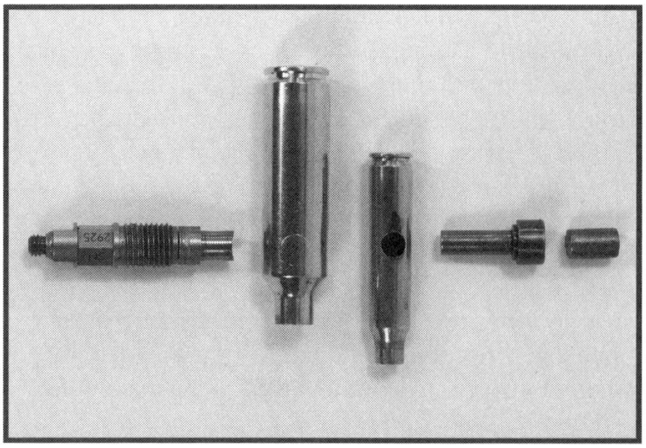

Cases are shown after pressure testing with the two systems. The transducer on the left leaves a distinct circle imprint on the case, however the case is intact. The copper crusher system on the right depends on the case rupturing and releasing the pressure against the piston and copper crusher cylinder.

Test barrels are made to very exacting standard set out by SAAMI, but of course no two are exactly the same. The reference ammo allows them to be calibrated to a common dominator.

These test barrels are used to generate the handloading data for this and most other loading manuals. Which, of course, would be fine if we all were shooting our ammo only in SAAMI spec test barrels and under controlled laboratory conditions. But, of course things are different in the "real world." Our guns are usually production quality, made on an assembly line to tolerance specifications that are by necessity much different than test barrels. The chambers and barrels may vary quite a bit in dimension as well as in the design and finish. A barrel that is on the outside of the tolerance level is not going to generate the same pressure levels with the same ammo as a "tight" barrel that is on the small end of the tolerance scale. Also, a rough bore with lots of tool marks is not going to act the same as a smooth, hand lapped barrel. Also, we rarely shoot under the climate controlled conditions of a ballistic laboratory and we may be using the ammo in conditions that range from blistering heat to brutal cold and both have an effect on the internal ballistics.

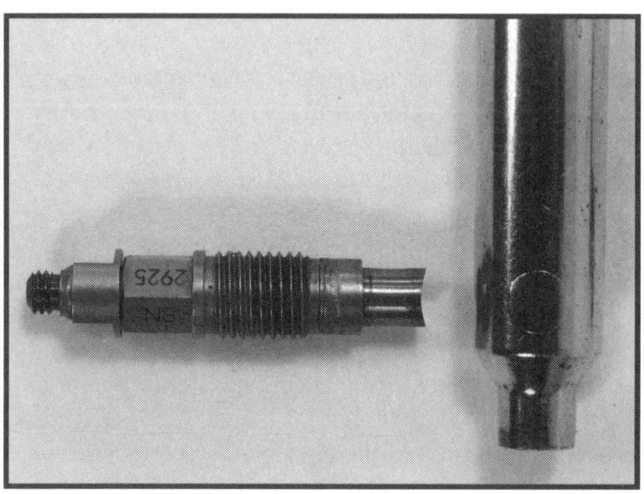

An imprint is left on cases after pressure testing with a conformal transducer. This occurs as the case forms itself against the face of the transducer, which matches the shape of the chamber wall.

What that means is that our ammo requirements might be a little different than what the test barrels indicate. It's our responsibility as handloaders to make sure our ammo is safe in our guns. There are some easy ways to do this.

Today's approach is vastly improved from the "wild and woolly" days of shooting when a different philosophy prevailed. I remember about twenty-five years ago when a shooting buddy called a famous old gun writer to ask about his handgun loads. They were well into and beyond the "hot" region of any reasonable loading information, but this was the early years of IHMSA competition when everybody was trying to find the "magic" bullet to knock over the 50-pound rams. Several of us had told my buddy his loads were too extreme, but he liked them because they knocked the targets down. He didn't believe us, so he called this world famous expert. Once he got him on the phone my pal explained in detail what the load was and how it was acting in the gun and what pressure signs he was getting from the fired cases. He had data on how many loads he would get on average out of each case, the average velocity and standard deviation of those loads and a lot of other technical information. The old guy listened patiently until he was done explaining it all in great detail, then asked gruffly, "anything blow up or break yet?"

"No."

"Probably all right then."

A week later the forcing cone of his expensive revolver cracked, as did the forcing cone on his wife's identical gun a week after that.

A better approach than using gun blow-ups and failures to gauge your handloads would be to follow the guidelines in this loading manual. As is prudent, always start with the suggested starting load and work up until you reach the "maximum" safe pressure level in your gun, or the maximum load listed in the manual, whichever comes first.

But without a pressure barrel and fancy, expensive pressure measuring equipment, work up to what? How do you know when your loads are at the "maximum" safe powder charge? Well, obviously you should never exceed the maximum load listed, but the maximum for some rifles may be well below that. I once had a Ruger Number One rifle chambered for 7 x 57 Mauser that had so tight a barrel that all the loads I tried maxed out well before reaching the top powder charge listed any reloading manual. That was likely the worst example I have encountered, but a lot of rifles I have worked with over the years have started showing signs of excessive pressure before the maximum powder charges listed in the loading manual were reached.

It's also important to know what to look for because you will not always be using the loading manual's recipe exactly as it's listed. You may be using a different brand of case, primer or bullet than what is listed. When you deviate from the recipe listed in the book, things can change. Changing brands of brass can affect chamber pressure because some cases have different internal capacity, which at maximum load pressures can be enough to get you into trouble.

Bullets are not all created equal and they can have a big effect on chamber pressure. Don't assume that just because the book lists a 180-grain bullet that any 180-grain bullet is fine. The length of bearing surface of the bullet will have an effect on the pressure because it causes friction and that friction will vary with how much of the bullet is contacting the barrel. Also the hardness of the bullet can affect chamber pressure. A "hard" bullet like the solid copper Barnes X-Bullet can cause pressures to rise because it is more resistant to shaping itself to the contours of the rifle's bore. Also, because copper is less dense than lead, a bullet of a given weight will be longer when it uses more copper or gilding metal. Or, if the core is lead alloy rather than pure lead, it will require a bit more length to achieve the same weight because alloy is less dense than lead. The longer bullets have a greater bearing surface as a rule, which means more friction. This is true of X-Bullets, Winchester Fail Safe, Trophy Bonded and many other premium hunting bullets. These hard bullets can also be sensitive to bore diameter and rifling style and cause higher pressures than expected if the bore is on the tight side of tolerances. But, don't assume any rule of thumb to be true, especially this one. Sometimes the "hard" bullet shows lower pressure than expected and a "soft" bullet will cause higher pressure than you thought it would. A soft bullet with a thin jacket and soft lead core may cause higher pressures because the bullet is so malleable that the sides will push hard against the bore of the rifle when high pressure is exerted on its base. Even the bullet's jacket material, copper or gilding metal can change pressure. So, the rule is simply this, assume nothing. If you deviate from the known load or change any component, back off the powder charge weight by 15% and work up again to the maximum safe level.

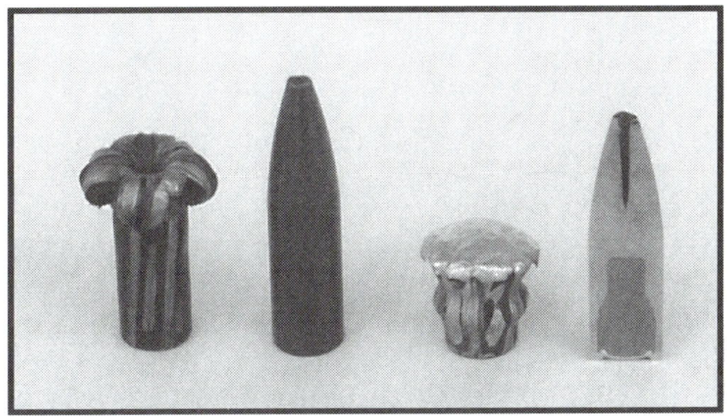

Winchester Fail Safe .30 bullets. L to R: 180 grain .30-06 recovered from a bear. 180 grain unfired. 165 grain fired at 3,000 fps into blocks of ballistic gelatin with a large elk leg bone molded in. Sectioned 165 grain.

Anytime you change something in your load, the pressure can change. Something as small as a primer change can send the pressure over the safe limit. This is particularly true if you switch from a standard primer to a magnum primer. Even shortening the gap from the bullet to the rifling by altering the seating depth can change pressure levels. Always back off until you see what the results of the change are.

There are several signs of excessive pressure to look for and these tried and true methods have worked for generations of handloaders. Several of the older loading manuals were actually compiled using these methods. The amazing thing is that when the industry started using carefully controlled pressure barrels and scientific measuring devices, it was discovered that most of the old data was correct. That was due mostly to the skill of the technicians who developed the load data.

Perhaps the best method is to measure the case head expansion. The solid brass in the head of the case will expand a small amount with each firing. This will vary with the thickness of the brass, the composition of the metal and the pressure encountered. However, it is generally accepted that most brass will expand about .0005-inch with a pressure of about 50,000 CUP, or 60,000 psi which is the max listed for the .30-06 Springfield and .280 Remington.

It's best to use once fired brass, as the first firing usually will give a larger reading. After two firings the brass may start to work harden and skew the results. But by using once fired brass you have established a constant. The down side is you must have a blade micrometer that is capable of reading to one ten thousandths of an inch. You must measure the solid web between the case body and the extractor groove and the anvils of a standard micrometer will not fit, so you need a blade mic with thinner anvils. A caliper is not accurate enough, even if it reads to one ten thousandths. So a blade micrometer is the only answer. Always measure at exactly the same location, using a point on the headstamp as a reference. When you reach .0003 to .0005-inch expansion over the once fired diameter, you are probably approaching the maximum load. Again, this is for calibers operating at approximately 60,000 psi.

Most reloaders know about "reading the primer," but that's not always an accurate measurement of pressure. A lot of things can have an affect on the primer condition in a fired case, including the condition of the gun and the firing pin. Also, how a primer reacts to pressure will vary by brand and style. A pressure level that causes changes in one brand of primer that indicate excessively high pressure may appear fine with another brand of primer. You simply cannot look at the primer of a fired case and know what is happening in terms of pressure.

That's not to say that "reading" a primer is not useful, it is, but you must do it correctly. If you are constantly monitoring the primer condition from the starting loads on up, using the same primers, as you increase powder charge weight you will be able to see changes in the primer condition that might be indications of pressure levels that are approaching maximum. From the early loads you will know what the primer looks like with mild loads and as you increase the charge weight watch carefully for any changes in that appearance. If the primer is losing its radius along the rounded top edge and starting to fill the gap to the top edge of the primer pocket you are at or near maximum pressure.

Bear in mind that primers can also flatten for other reasons. For example, if the headspace is a bit long in that rifle it can cause the primer to flatten. What happens is the firing pin will push the cartridge forward until it fires. The primer can then back out of the case and into the space between the case and the bolt face. As the pressure builds, the head of the case will move back against the bolt and the extruded primer may flatten out to fill the primer pocket. Pressure might well still be within safe limits, but the uninformed will look at the flat primer and assume that the load is too hot. That's one reason why it's important to watch the primers with starting loads that you know are low pressure (relatively speaking) and see how the primer reacts as the powder charge is increased. If it flattens with the mild load, you probably have other problems with the gun. But if the primer appearance follows a predictable pattern as you increase the powder charges until it starts to show signs of high pressure, then you can use it as one indicator of when your loads are approaching maximum.

Primer cratering can occur from a weak firing pin spring or a firing pin that fits poorly in the bolt. But, it can also indicate that pressures are getting too high. If your mild starting loads also showed cratering, then chances are it's a gun problem. But if the cratering suddenly started appearing when the charge weights were increased, especially if other signs of excessive pressure are also appearing, then pay attention, you are getting into dangerous territory.

Watch for sudden changes in the primer condition and for erratic primer appearance that is different from the rest of that lot of ammo. For example, if the first five shots are showing a slight loss of the radius, but suddenly the fifth shot has a primer that is flat and extruded to fill the entire primer pocket, you probably had a spike in pressure with that load. It also indicates that the powder charge is too hot in general, you should back off a little.

Keep a close eye out for black smudges between the primer and the case that indicates gas leaking. This is a sign to back off on the powder charge. Of course any loose primer is an indication that the load is way too hot and a clear danger sign. Note how the primers seat as you reload the case. If they seat easily with little pressure the primer pocket is probably stretching and the load is too hot. The general rule of thumb with high intensity rifle cartridges is if on the third load the primer pocket is still tight, the load is likely safe. Cases with loose primer pockets should be discarded.

Watch the case head condition. If a shiny spot on the head appears, the case brass is likely extruding into the ejector cut on the bolt and is "shaved" off as the bolt rotates to open. That load is far too hot.

Watch for signs of sticky extraction. A sticky bolt is often the first sign of excessive pressure. If the bolt opens hard or the case balks at exiting the chamber, the load is too hot. If you have to beat the bolt open with a mallet, it's way too hot!

Often lever action rifles and other rifles with rear locking actions and bottleneck cartridges may start to show stretching in the case ahead of the web. This will show up as a shiny ring around the outside of the fired case just ahead of the solid head. It can also be checked by bending the end of a straightened paper clip to 90 degrees. Insert this to the bottom of the case and scrape the sharp edge of the end along the inside wall of the case. If the paper clip hesitates or catches just up from the bottom of the case it indicates a thin spot caused by case stretch. That case will probably separate on the next firing, which is not only dangerous, but will render the gun useless until the front section of the case is extracted from the chamber.

Every serious handloader should have a chronograph. They are relatively inexpensive and without one you are just throwing bullets downrange with no idea what they are really doing. Until I got my first one I had used the same .30-06 load for years, smug in my ignorance that it was all the book said it would be. I thought my new chronograph must have been broken because the load showed a muzzle velocity that was several hundred ft/s lower than I expected. I had worked the load up carefully and knew it was at the safe limit for this gun. Several other guns were hitting the velocity expected and when I checked that .30-06 with some factory loads they were close to the velocity expected, so the problem was with my handloads, not the chronograph. The problem was corrected by changing powder and the bullets were soon zipping out the barrel at the speed I had anticipated. As an added benefit, the new load was more accurate than the slower load. Without the chronograph I would have continued to use that underachieving load without a clue to how poor it really was. But, that's only one of many reasons that you should own a chronograph.

A chronograph can also help to identify when you are approaching the maximum load. Of course, if you are hitting your target velocity then chances are you are at the maximum powder charge. But, another good indicator is the standard deviation of your handloads. If you have been getting consistent readings (particularly if they have been low) and a small increase in the powder charge causes the standard deviation to spike up, you have exceed the maximum charge.

Remember that powders react differently as pressure changes and they can become erratic at higher pressure. A one-grain change in the powder charge does not always result in the same percentage of change in pressure. Often a small increase can cause pressure to spike up dramatically.

When you see any of these pressure signs start to appear, back off about 5% on the powder charge. If the pressure signs recede, your load is likely the maximum for your rifle, even if it's still below the maximum powder charge listed in the manual. If the velocity is lower than what is indicated for that cartridge and bullet weight you can consider going to the next slower burning powder and working up again. But, be reasonable in your demands. Just because the book shows 3,000 ft/s and your load is only getting 2,950 ft/s; like the song said, "don't worry - be happy." There are a lot of variables in this game and sometimes they just don't favor you; besides 50 ft/s or even 100 ft/s is not a significant difference. Also, don't forget to allow for differences in barrel length. Most test barrels are 24-inch, while your rifle may be 22-inches and velocity will be lower.

Remember, pressure is a shooter's friend. It's what causes the bullet to propel down the barrel and out into this big, wide, world. Pressure is a necessary thing and within certain limits the higher the pressure the higher the velocity will be, but recognize that there are limits. Learn to accept pressure, but always fear it a little bit too. Like fire, it's only your friend if you understand it, respect it, and never try to cheat on its rules.

12 The Flight of a Bullet by Bryce M. Towsley

It all changed the day that one of our hairy-legged ancestors picked up a rock and instead of bashing his four-legged lunch fixins' in the head up close and dangerously personal, threw the rock to accomplish the same result from a safer distance. Since then, man has been trying to perfect the flight of deadly projectiles.

The path of today's modern hunting rifle bullet is invisible and a mystery to most hunters. Yet it is an incredible feat when an accurate rifle is able to fire bullets with consistency enough that they can travel through the mass of assorted gasses we call air and strike in a predictable location, often a very long distance from the shooter.

We have progressed a great deal from guns of the 1300s that fired almost anything that was handy, including rocks or gravel. Those primitive arms were charged with black powder that was often of an unknown quality and power. One reference says that the advice was to fill the bore "three-fifths full with powder before inserting a wad and bullet." The reasoning was that even with weak powder it would be enough but, with luck, not so much as to blow up the gun if it turned out to be "strong" powder.

Those crude hand-held guns then were fired by inserting a glowing wire into a touchhole at the breech end or by using a hot coal to light a small pile of powder on top of the touchhole. (Talk about long "lock time!") Of course, to do this the shooter must watch the ignition device, so sights were rather frivolous.

These guns were deemed the "Devil's Invention" by some clergy and were often outlawed for battle. (The crossbow met with a similar designation when Pope Innocent II forbade their use in 1139. He said they were "deathly and hateful to God and unfit to be used among Christians." However, he noted that he didn't have a problem with skewering non-Christians with a long bow.)

At any rate, from these humble beginnings have evolved today's rifles. We now have rifles capable of firing projectiles to 4,000 ft/s with accuracy that will keep all the bullets in less than one inch at two hundred yards or more. The technology that has made this possible has come from our understanding of how bullets react during their flight. The witchery of it all remains, though, that during this flight they are out of our direct control and at the mercy of physics and fate.

That alone is enough to still keep it interesting. It is an amazing thing to observe the modern hunter and his priorities are at times puzzling. Any seminar from a well-known whitetail guru will pack the house. Let any "expert", real or pretender, expound on the latest way to hunt a scrape line, place a tree stand or blow a grunt call and the adoring fans will fall at his feet. But, if the biggest expert in the field gave an outdoor show seminar on today's rifle bullets and ballistics, they likely wouldn't need much in the way of crowd control.

That's a little troubling because the bullet is the one physical connection with the game the hunter is pursuing. The bullet is the sole determiner of the outcome of any hunt, as you touch a live animal with nothing else. Yet, the rifle bullet and its flight remains the wallflower of the dance, sitting in a corner while everybody is caught up with the glamour and paying it no attention.

The bottom line is - everything you do on any hunt, all the expense, the time, the sweat and the tears, all comes down to one thing; the bullet. If it does its job you are happy and it's forgotten. If it does not, you have nobody but yourself to blame.

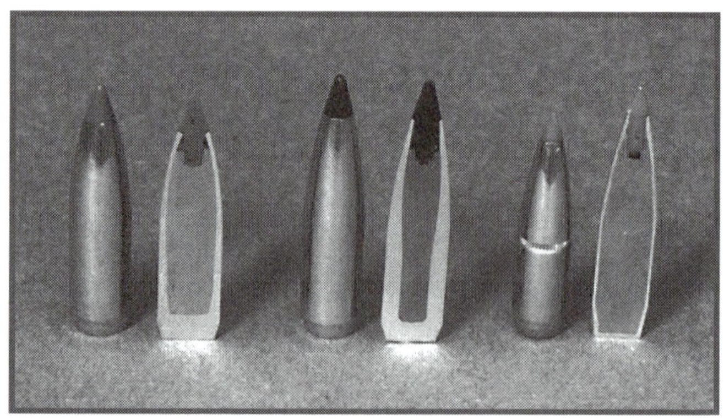

Plastic tipped bullets for big game. L to R: Nosler Ballistic Tip, Swift Scirocco, Hornady SST

The bullet is truly the unsung hero of the modern big game rifle hunter. The demands placed on it are incredible. An engineering student not aware of today's bullets and given an assignment to design one, would likely change his major to modern dance rather than accept. We expect our bullets to be accurate enough in flight to allow us to hit very small targets at very long ranges. Then we demand that they work correctly over a range of impact velocity and target variations that would be deemed unreasonable by any objective outside observer. Yet they do it so well these days that most hunters virtually ignore the specifics.

When smokeless powder made velocities of 2,000 ft/s and higher possible around the turn of the century, bullet makers cringed. But they rose to the call and jacketed bullets arrived on the scene. Then a few years later, as technology allowed higher and higher velocities and brought the bullet's speed up past 3,000 ft/s with regularity, they changed bullet designs to keep up. These higher velocities coupled with a parallel trend in the development of optical sights and hunters started demanding long-range accuracy, flatter trajectories and predictable terminal performance over a wider range of impact speeds.

In recent years we have seen a revolution of hunting bullets designed to accommodate the American hunter's insatiable appetite for more velocity. Even now, as big game rifle velocities approach the next plateau of 4,000 ft/s, today's super bullets can handle impact velocity ranges that designers dared not even dream about a generation back. These bullets remain stable and accurate at these high velocities and they have designs and shapes that cut through the air slicker than a politician at a fundraiser.

The Swift Scirocco bullets.

Has it reached a pinnacle? Is this the limit? Some thought it was with each of the milestones of the past, and they were wrong. To think today's advances have taken bullet design as far as it can possibly go would be fool's thinking. There is always one more hill to climb, one more frontier to conquer and one more problem to solve. As long as we are free to own and shoot rifles and as long as those rifles continue to use bullets as projectiles, we will continue to witness both subtle and amazing technological advances in their design.

And we as hunters will continue to all but ignore the bullets we shoot, demanding perfection and expecting and receiving nothing less.

The three common types of ballistics often referred to by shooters can be a little confusing. But, it's really very simple. *Internal Ballistics* is what happens inside the gun. It refers to the powder burning, pressure, the bullet's path down the barrel, etc. *External Ballistics* is what happens during the bullet's flight through the air. It deals with trajectory, velocity, energy, etc. Finally, *Terminal Ballistics* is what happens after the bullet has hit the target. It's the study of such things as expansion, weight retention, penetration, etc. Internal Ballistics is covered in another article, so we will deal here with the last two; what occurs after the bullet leaves the rifle's barrel.

But first, let's look at the barrel and what it does. It's been long known that by imparting a spin on a projectile it will become more stable in flight and the rifling in a barrel is designed to make the bullet spin at a predetermined rate. Just exactly how the concept of rifling in a gun's bore was developed is a bit unclear, but it's thought that archers who attached feathers to their arrows noticed that the curved wing feathers caused the arrow to spin in flight and made them more accurate.

Rifling in guns can be traced to the 15th century. Some say it was Gaspard Kollner, a Vienna gun maker, who first put spiral groves in musket barrels. Others claim that his rifling was straight and that it was Augustus Kotter of Nuremberg who first used spiral rifling in 1520. Another text says that the earliest known rifled gun was a matchlock owned by Emperor Maximillian in 1500. Regardless of the exact date when rifling was first used to spin a bullet in flight, it remained a seldom-used concept for centuries. Rifled barrels were simply too difficult and expensive to make and only the mega-wealthy could afford such a luxury.

Nobody back then knew exactly how rifling worked to make a gun more accurate, but the popular theory was that spinning the bullet kept the Devil from riding it and steering it away from its intended target. Today we know that spinning a bullet stabilizes it with a gyroscopic effect. The gyroscopic effect causes a spinning object to resist a change in direction and that's the real reason the Devil can't steer the spinning bullet away from the target.

This is the same effect that keeps a child's top standing upright as long as it's spinning. A bullet spinning on its axis will resist outside forces that might cause a non-rotating projectile to veer off course. This also minimizes the influences of different densities within the bullet and the slight irregularities on the surface that change air friction. Without the spinning induced by rifling, these may eventually cause the bullet to veer off course. But with it spinning, they are in constant motion and so their influence on the bullet's course is minimized. When combined with the much more powerful gyroscopic effect, it will result in a more accurate projectile.

Rifling is simply a series of parallel "ridges" that spiral within the bore of the rifle barrel. These ridges are called "lands" and the valleys in between them are called the "grooves." The distance between the opposing lands is the bore diameter. The distance between opposing grooves is the groove diameter, which can vary depending on the depth of the rifling grooves in the barrel.

The amount of spin is dependent on the rate of twist the rifling has. This twist rate is expressed in the number of inches of forward travel needed to complete one rotation. For example a bullet traveling down a barrel with a 1:10 twist rate will make one complete rotation in ten inches of forward movement. The general rule of thumb is that a longer bullet in relation to its diameter will require a faster rate of twist than a shorter bullet. Also, a more pointed bullet will need a faster twist than a blunt bullet (because it's longer.) Finally, velocity can affect the rate of twist needed to stabilize a bullet. A given bullet will stabilize with a slower twist rate if the velocity is higher.

While the barrel propels the bullet and adds spin to it, once it leaves the muzzle it's on its own. The laws of gravity have not yet been "interpreted" by the federal courts, so they still dictate that a bullet will begin to drop toward the earth from the moment it leaves the rifle barrel.

The reason that some rifles shoot "flatter" than others is simply because that bullet is traveling faster and is able to cover more distance in a given amount of time. The bullet is still dropping to the earth at a rate of acceleration of 32 feet per second, but it simply covers more ground than a slower bullet before dropping a given distance.

Because the bullet begins dropping as soon as it is free from the constraints of the rifle, if it were simply fired with the bore of the rifle pointed at the target, it would hit below the target every time. It is necessary to elevate the rifle's bore slightly above the line of sight so that the bullet's path rises on a curve that has it crossing the line of sight twice.

The line of sight is a straight line from the sights and to the target. Vision is not subject to gravity so your eye and the bullet take different paths to the target. While your eye's path is as straight as a laser, the bullet's path will be an arch to compensate for the effects of gravity.

The first time the bullet crosses your line of sight will be close to the rifle. The exact distance will depend on several factors such as how high the sights or scope are above the bore, the distance to your zeroed point of impact and the relative flatness of the trajectory for your bullet. However, it is usually somewhere around 25 to 50 yards. The next time the bullet crosses your line of sight will be at the distance you have selected to zero your rifle for. At any point in the bullet's path other than these two, the bullet will always be above or below your line of sight.

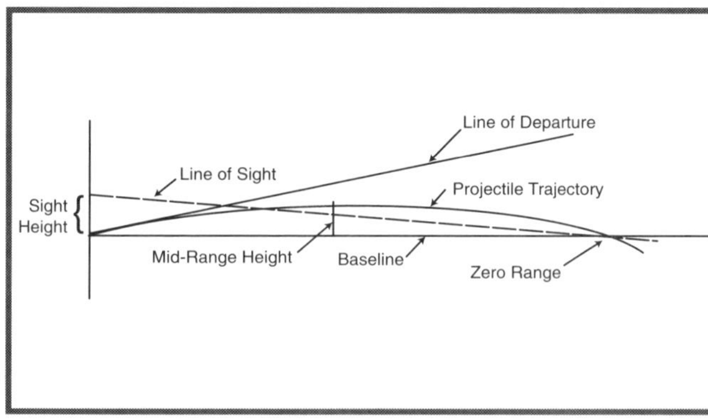

Because a slower bullet takes longer to get to a given distant point, it must arch higher above the line of sight than a bullet that is traveling much faster. Let's take an extreme example and compare the .45-70 and the .300 Remington Ultra Mag, both sighted for 200 yard zero.

The mid-range trajectory is the apex of the bullet's arch or the highest point that the bullet will be above the line of sight during its flight. With the .300 Remington Ultra Mag and a 200 yard zero, the mid-range trajectory of the bullet occurs at about 120 yards and is only 1.17 inches. The mid-range trajectory for a .45-70 with a 400-grain bullet and a 200 yard zero is 13.37 inches and occurs at 110 yards.

Total bullet drop at 200 yards for the .300 Remington

Ultra Mag is 6.99 inches. That is, if the bullet were fired perfectly level it would have dropped 6.99 inches by the time it reached the 200-yard mark. Total 200-yard drop for the .45-70 is 51.23 inches.

The difference is that the .300 Remington Ultra Mag has a muzzle velocity of 3,300 ft/s and uses a streamlined bullet. It takes it only .19467 seconds in flight to reach 200 yards and the bullet is still traveling at 2,881 ft/s at that distance. The .45-70 uses a bullet that is much less streamlined so it loses velocity much quicker and it starts out at a much slower 1,300 ft/s. The .45-70 bullet crosses the 200-yard mark with a time in flight of .53846 seconds and with 991 ft/s velocity remaining.

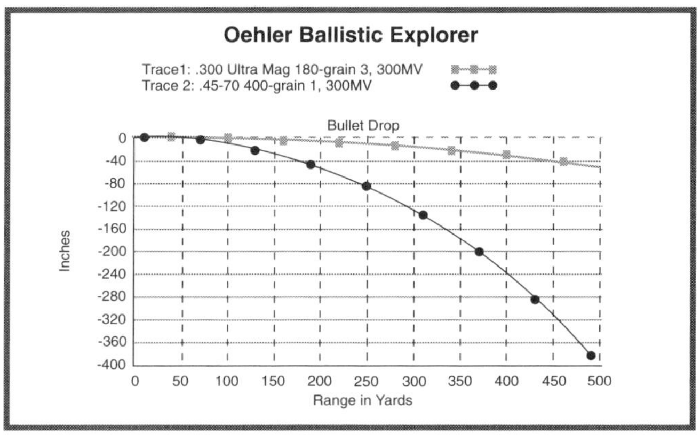

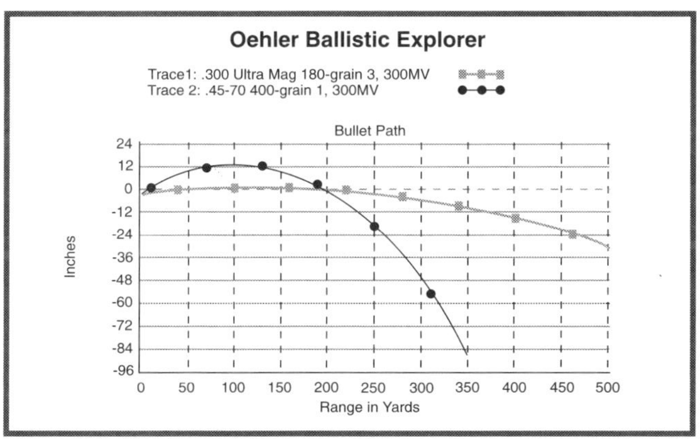

Graphs Courtsey of Oehler Research Inc.

Of course, the .300 Ultra Mag is much faster to start out than the .45-70, but notice too that it retains a higher percentage of its velocity down range, 87% vs. 76%. That is because of bullet design. The ability of a bullet to retain its velocity is expressed as the ballistic coefficient. This is a measure of how aerodynamic a bullet is and how well it cuts through the air. The technical explanation is "the ratio of the sectional density of a bullet to its coefficient of form." The higher the number, the less the bullet is slowed by air resistance over a given distance at a given velocity. That means a higher BC bullet retains its velocity better than a low BC bullet. This results in flatter trajectories and increased down range energy retention. Sectional density is a measure of a bullet's mass relative to its cross-section. In simple terms, a relationship of its weight to its diameter. The SD of a bullet is found by taking the bullet's weight in pounds and dividing it by the square of its diameter in inches. The higher the SD of a bullet in a given diameter, the heavier and longer that bullet will be. This is important to several factors, including the BC of the bullet. However, from a big game hunter's standpoint the most relevant measure of SD is that with all else being equal, the higher the SD of a bullet, the better its ability to penetrate.

There are several factors that affect the BC of a bullet, but shape is the one that is most easily identified. A pointed bullet will have a higher BC than a flat-nose or round-nose bullet of the same diameter and weight, simply because it offers less wind resistance. In this case the .300 Ultra Mag's bullet has a BC of .474 while the .45-70 bullet's BC is .214.

Simply put, any bullet must arch in its path to the target relative to the line of sight. The faster the bullet gets to the target, the flatter that arch or trajectory will be. Also, the further from the gun that the sights are adjusted to zero the point of impact the higher the arch will be. If the gun is zeroed for 300 yards the apex of the arch will be higher above the line of sight than it will be if the gun is zeroed for 100 yards with the same load.

Also, with a slower velocity and lower ballistic coefficient, the amount of drop at any point beyond the zeroed range will be greater. At three hundred yards the .300 Ultra Mag will impact 5.27 inches below the line of sight while the .45-70 will impact 48.07 inches below the line of sight.

Even when we compare apples to apples and use bullets of the same weight and fired from the same gun the long range benefits of a bullet with a higher BC are still apparent. Comparing identical loads made with both round nose and pointed bullets shows the impact that BC has down range. Let's compare 180-grain .30-06 factory loads - one pointed, the other round nose. The following charts show the differences between the two designs.

Pointed Bullet	Muzzle	100 yards	200 yards	300 yards	400 yards	500 yards
Velocity ft/s	2,700	2,469	2,250	2,042	1,846	1,663
Energy ft.-lbs.	2,913	2,436	2,023	1,666	1,362	1,105
Bullet Path in inches	-1.5	+2.1	zero	-9.0	-26.3	-54.0

Round Nose Bullet	Muzzle	100 yards	200 yards	300 yards	400 yards	500 yards
Velocity ft/s	2,700	2,348	2,023	1,727	1,466	1,215
Energy ft.-lbs.	2,913	2,203	1,635	1,192	859	625
Bullet Path in inches	-1.5	+2.4	zero	-11.0	-33.8	-72.8

Another confusing aspect of shooting is what happens to the bullet's path when the shot is fired at a sharp uphill or downhill angle.

I think I have heard every theory on earth, but the most common is that when you are shooting downhill the bullet will strike higher and when you are shooting uphill the bullet will strike lower. Technically, both are wrong.

Gravity doesn't care if the bullet is flying up or down; it works its magic on it in exactly the same way. That is, if we disregard air friction, it pulls the bullet toward the center of the earth at a rate of acceleration of 32 ft/s. This is what causes a trajectory curve in every bullet fired. If we could fire a bullet in space where there is no gravity then it would not drop a bit, but here on earth gravity is a fact of life. Gravity pulls on the bullet as it travels in relation to the center of the gravitational force, or the center of the earth. The longer gravity has to work on the bullet, the faster it will drop. A bullet that is fifty yards from the muzzle is dropping much more slowly than a bullet that is 300 yards from the muzzle. This is because the 300-yard bullet has had more time for gravity to act on it. That's one reason why trajectories curve much more sharply down as the distance progresses.

In theory, a bullet that is fired from a perfectly level rifle barrel will strike the earth at the same time as a bullet that is dropped from the end of the barrel at precisely the same time the fired bullet exits the muzzle. The difference is that one bullet will hit the ground near the rifle while the other uses the time to fly through the air and will strike the earth at some distant point.

No bullet can actually go up in defiance of the law of gravity and they all begin dropping toward the earth as soon as they exit the muzzle. The distance a bullet travels in relation to the earth's center, or gravitational pull, from the muzzle to the target is what determines how far a bullet will drop in a given time span. With a steep up or down angle, the distance to the target will be longer than the distance the bullet travels in relation to the earth's center. Rather than dropping the amount we expect for the distance traveled to the target, the bullet actually will drop for the distance it travels in relation to the earth's center. The steeper the angle, the shorter that distance becomes.

For example, let's assume that you are shooting at a sheep that is 400 actual yards away, but on a very steep angle. The gun is a 7mm Remington Magnum with a 140-grain bullet and sighted for a 200 yard zero. You will expect the bullet to impact eighteen inches below the line of sight for a 400-yard shot. However, the path of the bullet's flight in relation to the earth's center or the center of gravitational pull is in reality 300 yards, so the bullet is only 6 inches below the line of sight when it gets to the sheep. If you hold for 400 yards, it will result in shooting over the ram you just paid $15,000 to hunt.

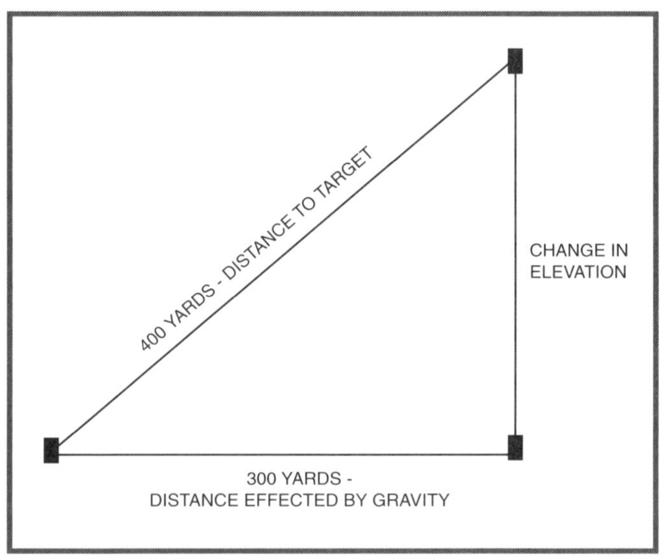

Of course, as the distance shortens this becomes less of a factor because the bullet is moving faster and has had less time in flight for gravity to act. The result is that, due to the shorter range, the bullet has dropped less and the trajectory curve is much flatter. Obviously, the flatter the rifle shoots the less the difference will be. With a flat shooting modern rifle and a sight-in that keeps the bullet within three inches of the line of sight out to a given distance; if the target is less than that distance from you, the bullet will always strike within three inches of your point of aim, regardless of the angle. Which should not cause you to miss a big game animal.

The problem comes in when you cannot close the distance and must take a long shot, compounded by an angle that's steep. Suppose the angle is very steep and the distance to the sheep is 300 yards, while the distance in relation to gravitational pull is 200 yards. With that same 7mm rifle you will be expecting the bullet to drop 6 inches from the line of sight for a 300 yard shot, when in reality it will hit right where the crosshairs are centered because the rifle is zeroed at 200 yards, which is the actual bullet travel in relation to gravitational pull. With luck, if you held for a six-inch bullet drop you might spine the sheep instead of shooting over his back.

Angle degrees	Bullet path in inches below line of sight at 400 yards	Variance in inches in the bullet path from level
Level	-23.14	NA
10 Degrees	-22.47	.67
20 Degrees	-20.46	2.68
30 Degrees	-17.16	5.98
40 Degrees	-12.69	10.45
50 Degrees	-7.20	15.94
60 Degrees	-0.81	22.33

Difference in bullet path with varying angles.
.30-06 Springfield 165 grain Speer Spitzer
MV 2,800 ft/s - Zero range 200 yards.
Target distance - 400 yards.

The effect of wind on a bullet's flight has been a subject of heated discourse for as long as shooters have been trying to hit targets at long distances. We have developed charts and computer programs in attempts to make predictions, but at best they are only a guideline. All the mathematical theorists from Einstein to my cousin Philip can't predict exactly what a bullet will do when it is fired through a real-world moving air mass.

However, the computer programs and ballistic tables are useful in learning about the wind's effects and it is hardly a waste of time to study them. For example, the accompanying table will give you some idea of the difference of wind direction on a bullet's path. It will also help to illustrate the degree of wind deflection on differing bullets and velocities. But the flaw in these tables and programs is that the predictions are built on laboratory conditions and assume a constant force and direction of wind.

Things are different in the atmosphere we shoot through. Winds gust and wane with currents and flows affected by terrain, vegetation and a million other factors. While it may be blowing one speed where you are, your bullet can travel through wind that is moving several different speeds before it reaches a distant target. The same with wind direction; over a given distance it can curl around, double back, eddy, switch and change its attitude more times than a six year old with a sugar buzz. If there is one constant in a wind's influence on a bullet it is the inconsistency.

The effect as predicted in the charts and computer programs uses the velocity and ballistic coefficient of the bullet to determine the time of flight, which is in effect the amount of time the wind has to act on the bullet. The more time in flight, the more the wind will blow it off course. Other factors include the bullet's size and shape or the amount of area available for the wind to blow against. Also the bullet's weight, as it is easier to push around a light object than it is a heavy one. Who would you rather wrestle, Hulk Hogan or Bill Gates? Same principle.

Cartridge	Drop & drift with 20 mph wind from 06:00 o'clock	Drop & drift with 20 mph wind from 03:00 o'clock	Drop & drift with 20 mph wind from 12:00 o'clock
.44 Rem. Mag. 240 gr. 1,200 ft/s			
100 yds	13.55/00	13.73/8.54	13.93/00
200 yds	59.52/00	60.96/31.07	62.51/00
300 yds	145/00	149.7/65.17	154.8/00
400 yds	276.9/00	287.8/110.2	299.9/00
500 yds	462.7/00	483.8/166.4	507.6/00
.220 Swift 55 gr. 3,900 ft/s			
100 yds	1.24/00	1.24/1.68	1.24/00
200 yds	5.38/00	5.39/7.05	5.41/00
300 yds	13.21/00	13.26/16.69	13.32/00
400 yds	25.77/00	25.93/31.39	26.09/00
500 yds	44.50/00	44.88/52.21	45.27/00
.30-06 Springfield 180 gr. 2,520 ft/s			
100 yds	2.87/00	2.87/1.59	2.88/00
200 yds	12.09/00	12.11/6.58	12.14/00
300 yds	28.68/00	28.77/15.40	28.86/00
400 yds	53.87/00	54.12/28.48	54.37/00
500 yds	89.19/00	89.73/46.34	90.30/00
12 ga. sabot slug 437.5 gr. 1,350 ft/s			
100 yds	12.13/00	12.37/15.56	12.62/00
200 yds	57.51/00	59.55/54.88	61.78/00
300 yds	148.1/00	155.2/113.6	163/00
400 yds	297.2/00	314.7/192.3	334.6/00
500 yds	521.1/00	558/293.2	601.3/00

Wind direction is another factor. A ninety-degree side wind will blow against the full profile of the bullet, but any angle that is different will have a differing percentage of the bullet's surface to act on. A full head or tail wind will have either the point or the base of the bullet to blow against, which affects retained velocity and consequently bullet drop. For example, a .44 Magnum 240-grain bullet with a 20 mph tail wind will have almost 100 ft/s more retained velocity at 500 yards than the same bullet with a 20 mph head wind. This will change the bullet path even though there is no theoretical wind deflection and the difference in drop is almost 45 inches. Of course this is an extreme example and nobody really considers the .44 Magnum a 500-yard cartridge, but every bullet from any cartridge will be affected in some way.

Now suppose the wind is gusting from a 45 degree angle, or 20 degrees or perhaps ten degrees. Maybe it will be from all three before the bullet makes it from your gun to the target. How much will that affect the lateral wind deflection? How about the bullet drop?

I don't have a clue.

Neither does anybody else, not with all the variables.

Shooting in the wind is far more an art than a science and the only real way to learn about reading the wind is to get out and shoot in it. A couple of days in a prairie dog town will teach you more than all the theoretical computer programs or printed ballistic tables in the world. If there is a pronounced lack of prairie dogs where you are, simply find a windy place with some good distance and shoot at clay pigeons, plastic soda bottles full of water or even paper targets and do it with a variety of firearms. Have a pal watch your hits through a spotting scope and try to evaluate each shot. After a few thousand rounds or so you may finally start to get a handle on how to shoot in the wind.

The greatest debate on bullets these days might well be in the terminal ballistics, what happens after the bullet hits the game. With varmints or targets it's pretty well a non-issue, but with big game things change. There is a popular theory that a bullet should stay in a big game animal's body, "dumping all its energy." It is being promoted by several gun writers and unfortunately many hunters are accepting it as gospel.

Personally, I find it irresponsible.

A .22 LR will always stay in a deer, dumping all its energy and often resulting in the termination of the deer. Does that make it a deer caliber? Of course not.

When applied with a modern rifle caliber this theory can produce some spectacular results, but those hunters who persist in using it as a guide for choosing their bullets and cartridges are courting disaster. There are far too many variables such as impact velocity, bullet path, game size, material penetrated, physiological variables in the target animal and many more for this to ever work.

Consider that a .220 Swift with a 40-grain bullet at 4,100 ft/s generates 1,493 ft-lbs of energy, far more than any generally accepted 1,000 ft-lbs minimum for whitetail deer. If you place that bullet precisely in the lungs the deer will likely drop as if hit by lightning. Move the impact a few inches forward to the shoulder and he may run off. If he is lucky, the coyotes will find him soon and end his suffering. Both shots dumped all their energy into the deer's body, so what happened?

You must remember that death is not a result of energy transfer, but of tissue destruction. Dumping energy is useless if it does not result in damage to vital tissue. The bullet that hit the shoulder likely fragmented and failed to penetrate through anything important.

L to R: Barnes 300 grain .375 solid, Swift 300 grain A-Frame, .375 Remington Ultra Mag, Swift 300 grain A-Frame recovered from buffalo after hitting spine. Two Swift 300 grain A-Frame bullets recovered from buffalo.

Whitetail deer are our most popular big game animal and as a result are the basis for most hunters' experiences. The fact that whitetails are relatively easy to dispatch and that most hunters today use a rifle chambered for a cartridge that is far more powerful than is necessary, allows even the poorest bullets to drop the deer most of the time. But with bad bullet and/or cartridge choices, sooner or later one of the variables will enter into the equation and the result will be a wounding loss. Try it with bigger game, particularly on something inherently tough like elk or bears and sooner or later it will bite you in the backside. (The bear might too!)

L to R: Win. Fail Safe, Win. Partition Gold, Win. Ballistic Silvertip, Barnes X-Bullet, Nosler Partition, Speer Grand Slam, Swift A-Frame, Hornady Interlock, Speer Hot-Cor, Nosler Ballistic Tip, Swift Scirocco & Hornady SST

The Flight of a Bullet

The other end of the equation is a bullet that passes completely through the animal, but due to its design does not expand and causes very little tissue damage with results that are just as poor.

An ethical hunter will choose a caliber of adequate power for the game hunted and match it with a quality bullet. For most North American hunting this bullet should be designed to expand very soon after initial impact and to hold together enough to continue through the animal completely, under all but the most extreme conditions.

Wound channels are relative to energy transfer and the frontal area of the bullet. Consider how it would look if the animal were a homogeneous material, like ballistic gelatin. As a bullet first contacts, the wound channel is bullet diameter. Then as it penetrates, the wound diameter will grow relative to the energy transferred and the contact damage caused by the expanding bullet diameter and shape. If the bullet is going to stay in the game and velocity slows as energy levels drop, the wound channel diameter will again begin to taper until it reaches current bullet diameter at the point where the bullet stops. This results in a misshapen football shaped wound channel, with a large maximum diameter, fine if it's in the right place, but if it ends short of vital tissue, problems arise. Or the more likely scenario is that the bullet tears apart and scatters small pieces which lose energy and stop penetrating. This results in a large, but very short wound channel.

With proper bullet design and initial energy the bullet exits with energy remaining. The resulting wound channel has a minimal decreasing profile and will remain near the maximum diameter all the way to the exit.

These bullets expanded even at reduced velocities. Left is a Barnes X-Bullet and right is a Winchester Fail Safe.

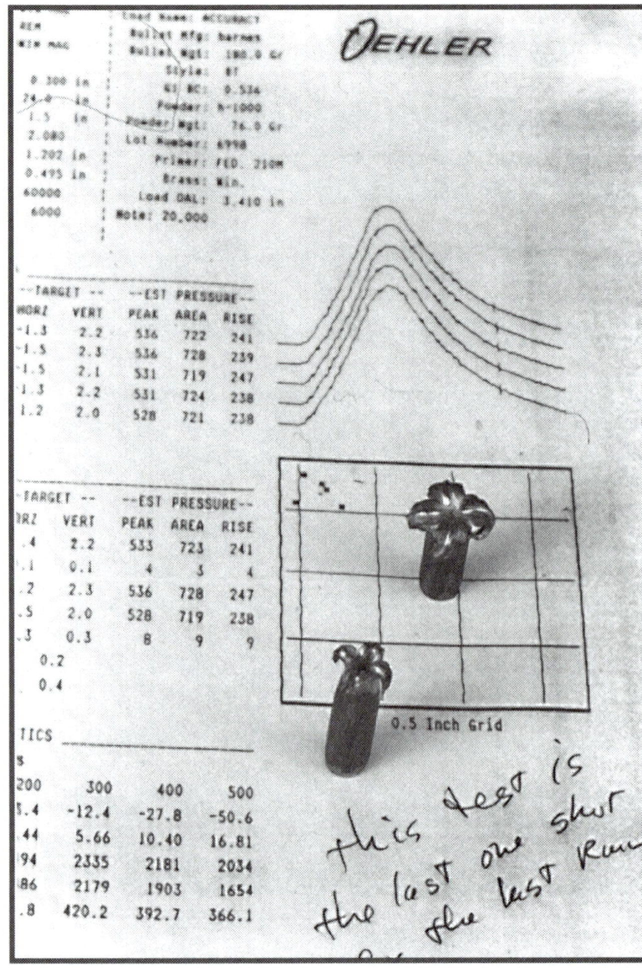

Two Barnes X-Bullets show the results of test for expansion at the extremes of impact velocity.

So what if the bullet exits with remaining energy?

The important thing is that it created a sufficient wound channel completely through the target. Even though that wound channel diameter may be less at its peak than that created by a bullet that dumps all its energy and fails to exit, the wound channel is much longer and has a better chance of passing through something vital. It probably also has more total area because of its length.

The results may not be as spectacular as a double lung hit with a bullet that tears itself apart, dumps its energy in a short wound channel and fails to exit, but it is much more predictable, reliable and consistent. These are far more desirable traits for an ethical hunter than dramatic impact results with a high failure potential.

As velocity increases, the terminal demands on a bullet increase exponentially. Even though they may take the same bullets, not all cartridges are going to extract the same performance from them. You simply can't expect a bullet to act the same way from a .308 Winchester as it does from the .30-378 Weatherby. It may be traveling nearly 1,000 ft/s faster from the big gun and that changes things radically. With the current trend to faster and faster cartridges it becomes more and more important to use a high quality game bullet. A bullet that performs well on deer from a .30-06 might tear apart like a varmint bullet when pushed from a .300 Ultra Mag. With their extreme velocity, the new generation "super magnum" cartridges all but demand one of the highly engineered "super bullets" to work well on big game.

Today's bullets are so well designed that they truly are a modern day ballistic miracle. But you have to do your part to make it happen. Pick the right bullet, work up a load that is accurate and learn its trajectory. Otherwise the Devil might steer it into trouble.

13 Reloading the 45-70 for Accuracy by Paul A. Matthews

Chapter 13

The 45-70 is more popular than ever and is a favorite of reloaders.

Few shooters today use the word accuracy and the term .45-70 in the same sentence. Their concept of rifle accuracy usually stirs visions of one of the smaller calibers from a bolt action rifle wearing a scope sight. To them the .45-70 is comparable to a bulldozer competing at the Indy 500.

That is too bad, because the .45-70 is one of our classic cartridges dating back to 1873 when it was first put into the hands of our troops on the western frontier. Since that date over a century and a quarter ago, the cartridge itself has never been out of production. And since the 1960s when Navy Arms started offering their rolling block rifle chambered for this venerable cartridge, and Ruger came out with the same chambering in their No. 3 single shot, the production of rifles chambered for the .45-70 has skyrocketed. Indeed, for many years since the 1960s, the sale of reloading dies for the .45-70 has been very near the top of the popularity list.

I shot my first .45-70 cartridge over fifty years ago and have been using that cartridge ever since. For many years, right up through 1993 when I switched from hunting to competitive shooting, the .45-70 was the rifle I reached for when I took to the deer woods of Pennsylvania or went for the wild boar of Tennessee or North Carolina. And when I went into competitive shooting from 200 yards on out to 1000 yards, it was a .45-70 that I wrapped my fist around. To me, it is one of the greatest cartridges ever designed, one of the most reliable and versatile cartridges in existence today. If there is any fault with the .45-70, it lies in the shooter who has failed to understand the cartridge and its limitations. Because when you understand the .45-70 and use it within its limitations - and all cartridges have limitations of one type or another - you will find it to be superbly accurate with either jacketed or lead alloy bullets, smokeless or black powder, and with whisper loads developed for tin can plinking or full-throttle black powder loads for use on the silhouettes out to 547 yards and on paper out to 1000 yards. It is just plain and simple a great cartridge, but you have to understand it in order to get the most out of it.

Before getting into the details of reloading, I want to point out that different reloading handbooks often list entirely different charges of the same powder as a maximum recommended load for the same bullet. This does not mean that one handbook is right and the other handbook is wrong, but that there may be a variation in the cartridge case, bullet, primer or test barrel used, all of which may have an effect on the chamber pressure developed. So I beg you, implore you and urge you to study the reloading handbooks and carefully follow the instructions as to the powder charge and bullet used as well as the rifle for which that load is intended. Do not use loading data intended for the Ruger No. 1 or No. 3 in a Winchester 1886 lever action, and if you are using a rifle originally built for black powder, my advice is to use only black powder in that rifle. If you are using smokeless powder and find you have an airspace between the powder charge and the base of the bullet, DO NOT use a filler to fill up that airspace. More than one fine rifle has had a barrel bulged or been totally destroyed by the improper use of a filler over smokeless powder. Finally, if you must use a maximum recommended load, work up to that load a half a grain at a time. More often than not, you will find your best accuracy with a load that is less than the recommended maximum.

WARNING: The loads appearing in this article were found to be safe in the rifles in which they were used. However, since neither the author nor Lyman Products Corporation has any control over the rifles in which you may use these loads, nor do we have any control over your loading procedures, neither the author nor Lyman Products Corporation will accept any responsibility for damage to equipment or injury or death of any person as a result of using any load shown in this article. If you choose to use one or all of the loads described in this article, you assume responsibility for any and all liability.

During the years that I hunted deer and boar with the .45-70, my favorite rifles were the Ruger single shots and the new Marlin 1895 lever action. The last two deer I killed in 1993 were with the Marlin 1895. That year I took a 9-point buck on the first day and a chunky button buck on the last day. Both deer were taken with the Hornady 350 grain flat nose jacketed bullet backed by 55 grains of IMR-3031. One shot apiece was all that was required. It should be noted that this load was worked up from the 3rd edition of the Hornady handbook which lists 55.8 grains of IMR-3031 as a recommended load for the new Marlin 1895 lever action rifle.

Although I much prefer the 350 grain Hornady bullet because of its accuracy and performance, many shooters using the .45-70 opt for the 300 grain Hornady hollow point. There is nothing wrong with this. The 300 grain bullet is a superb killer on deer when backed with a recommended charge of IMR-4198 or IMR-3031, but because of its short body length, I was never able to get quite the accuracy with the 300 grain bullet that I got with the longer 350 grain bullet.

One point that I believe should be made relative to accuracy from the .45-70 is that bullets, whether cast or jacketed, having a long body to grasp the rifling always gave better accuracy than the shorter bullets. That is to say that where the 350 grain Hornady always gave a bit better accuracy than the 300 grainer, a 400 or 405 grain bullet with a recommended charge of IMR-3031 fired from my little Ruger No. 3 would give better accuracy than the 350 grain bullet. In fact, from my Ruger No. 3, five rounds of a 400 or 405 grain jacketed bullet would fit in one large ragged hole at 100 yards.

But what kind of accuracy do we need from a .45-70

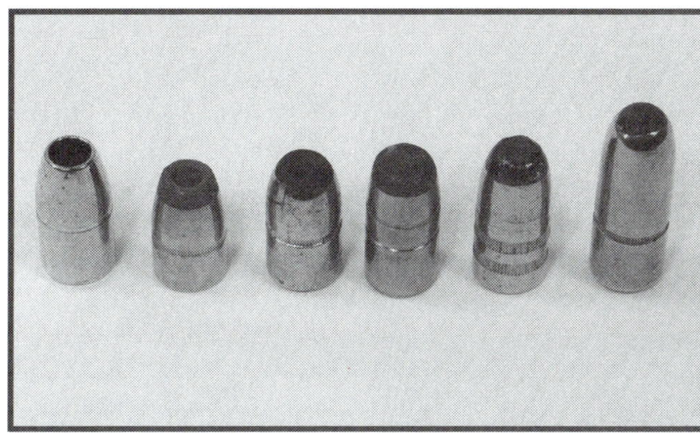

A large variety of jacketed bullets are available for reloading the 45-70. Shown above are: 250 grain Barnes X, 300 grain Sierra flat nose hollow point, 350 grain Hornady flat nose soft point, 400 grain Speer flat nose soft point and 500 grain Hornady round nose soft point.

for hunting purposes? Anyone who has ever used a .45-70 for serious hunting or competitive shooting knows that the vulnerability of that cartridge lies in it high trajectory and low velocity. Knowing this to be true, any sensible hunter using the .45-70 will restrict his shots to no further than 125 yards, and he will avoid all running shots. Hunting deer within those limitations - and most woodlot deer are shot at less than 50 yards - one would have to say that 2-inch grouping capability at 100 yards is acceptable, and is surely well within the capability of the jacketed bullets mentioned thus far.

For those who wish to reload with jacketed bullets in the .45-70, I strongly recommend that you first purchase 100 new cartridge cases, all of the same make and lot number. By doing this, you will be starting with cartridge cases that are as near alike in powder capacity and wall thickness as is mechanically possible to get them. This is the first step in assuring uniformity from cartridge to cartridge. Keep in mind that of the three major domestic cartridge cases available - Winchester, Remington and Federal - the Winchester cases have the greatest powder capacity, followed by Remington. Federal cartridge cases have the thickest wall section and thus the least powder capacity.

Once you get your new cartridge cases, run them through a full-length resizing die to assure roundness, and then trim them to a length of 2.095 inches to square up the mouth of each case and to assure the same amount of neck tension on each bullet. Following this, you should remove any internal flash hole burr with a Lyman Flash Hole Uniformer, and deburr the inside and outside of the mouth of the case with a Lyman deburring tool . All of this is to get your cartridge cases as uniformly alike as possible.

As far as primers are concerned, it makes little difference within hunting ranges which brand of primer you use just so they are all of the same brand and lot number and are either a large rifle or a large rifle magnum primer as recommended in your loading data.

When it comes to the choice of powder for smokeless loads in the .45-70, always select a powder that bulks up well within the case. For heavy, full-throttle smokeless loads in a suitable rifle using either a 405 or 350 grain jacketed bullet, IMR-3031 has been a great favorite for well over fifty years. For the lighter 300 grain bullet, IMR-4198 is often the preferred powder.

If you want to load at less than full-throttle velocities, perhaps in the 1400 to 1600 fps category, SR-4759 is an excellent powder made especially for use in original black powder cartridges. It bulks up well within the case and ignites easily.

Obviously there are other smokeless powders that will bulk up in the case and give good velocities at safe pressures. These are listed in the loading tables for the .45-70 along with the expected velocity and corresponding chamber pressure.

Finally we come to crimping the bullet. Crimping serves two functions in the smokeless powder .45-70 cartridge. First, it assists in the ignition and burning of the powder. It adds more resistance to the bullet to help build up the pressures necessary to make the smokeless powder ignite and burn uniformly from cartridge to cartridge. And secondly, in a tubular magazine rifle a heavy crimp in the cannelure on the bullet is necessary to keep the bullet from being hammered further back into the cartridge case under heavy recoil. With single shot rifles, this is not a problem and only a light crimp or no crimp at all is necessary. If the bullet is crimped for use in a single shot rifle, it should be for the purpose of improving ignition and burning of the powder.

Once your new cases have been fired, it is advisable not to full-length resize them any more than necessary to permit easy chambering in the rifle in which they were originally fired. This eliminates a lot of wear and tear on the cartridge case and greatly extends its life. Not only that, but the less resizing done to the cartridge case, the better that case will fit the chamber and the more concentric the bullet will be with the bore before the cartridge is fired. This helps promote accuracy.

The best method for treating your cartridge cases once they have been fire-formed to fit the chamber, is to not full-length resize them at all, but to resize only the neck of the case with a special neck sizing die. This gives the cartridge a perfect fit with the chamber for several firings. Of course, cartridge cases that have not been resized, often will not chamber in other rifles of the same caliber, and after several firings may require full-length resizing in order to chamber easily in the

Lyman offers Neck Size dies and Taper Crimp dies for the 45-70

parent rifle.

While most hunters using the .45-70 rely on a jacketed bullet to get the job done, there is a large percentage of hunters who realize the .45-70 was originally developed with a lead alloy bullet and is still an excellent cartridge with such bullets. The first deer I ever shot with the .45-70 was with the Lyman No. 457122 (originally 456122) Gould hollow point bullet weighing about 330 grains when cast of a 16 to 1 lead-tin alloy. That doe was 137 paces away from me and the bullet went through the rib cage shattering the heart so that she left a highly visible blood trail for the short distance she ran.

I used a Winchester Model 1886 lever action rifle with a P.O. Ackley barrel for that shot, and had the cartridge loaded with a stout charge of World War II vintage 4895 powder. If I were to try to duplicate that load today, I would start with 51.5 grains of IMR-3031 as recommended in an IMR Handloader's Guide for a 405 grain jacketed bullet.

The Lyman No. 457122 hollow point bullet when cast of a 16 to 1 or the softer 20 to 1 lead-tin alloy makes an excellent bullet for deer. Its reputation for reliability reaches back over 100 years to the late 1800s when that bullet was especially designed by J. H. Barlow of the old Ideal Manufacturing Co. for Mr. A. C. Gould, editor of *Shooting and Fishing*.

Two other Lyman bullets which should give excellent performance on deer are Nos. 457643 and 457193. Both

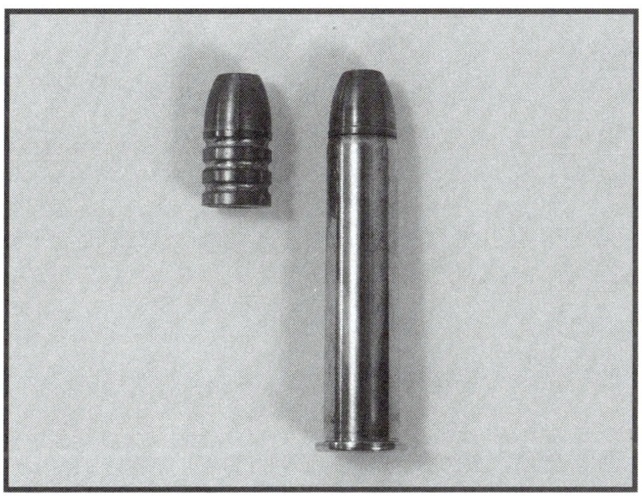

The Lyman #457122 hollow point has been a popular hunting bullet for many years.

of these are flat-nose bullets weighing about 400 and 405 grains respectively. Of special significance relative to accuracy is the fact that at least 60% of the overall length of each bullet is made up of driving bands and lubrication grooves. As stated earlier, bullets having a long body to grasp the rifling and to better align the bullet within the bore usually prove the most accurate. And when it comes to hunting with cast bullets, a flat nose always seems to deliver more bullet energy to the game than does a round nose or semi-spitzer.

To get the best performance from cast bullets in the .45-70, cast the bullets of a 20 to 1 or 16 to 1 lead-tin alloy and size them to 0.459 or 0.458 inch diameter. These alloys are soft enough to promote expansion without breaking up, and the 0.459 or 0.458 inch diameter is large enough to seal the bore and help prevent leading due to hot powder gases rushing around the bullet. Lubricate your bullets with either the Lyman Black Powder Gold or Super Moly bullet lubricant, both of which will work well with smokeless loads.

When it comes to smokeless powder for cast bullets in the .45-70, SR-4759 is an old time favorite especially developed for that job. IMR-3031 is another excellent powder that bulks up well within the case and seems to burn a bit cooler than SR-4759.

Since lead alloy cast bullets are not as strong structurally as a jacketed bullet, it usually pays to start at the lower end of the velocity scale and work up, checking the load for accuracy and barrel leading as you go. And if you are unable to obtain the velocity you want without an undue amount of lead in the barrel, switch to one of the slower burning powders listed in the loading tables.

Loading .45-70 ammunition with cast bullets is little different than loading with jacketed bullets except that one must be careful not to shave the side of the bullet or otherwise deform it during the bullet seating process. The best way to accomplish this is to open the mouth of the cartridge case with a Lyman M-die. This is a two-diameter neck expanding die that allows a jacketed or cast bullet to be started perfectly centered in the case without shaving the circumference of the bullet.

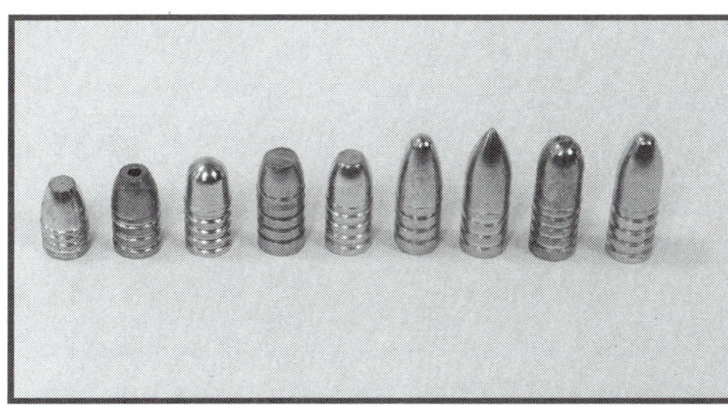

Lyman offers a large variety of bullet moulds for the 45-70. Shown from left to right are: #457191, #457122, #457124, #457643, #457193, #457671, #457658, #457125 and #457132.

If you are going to hunt with cast bullets in the .45-70, do not hesitate to make them of a soft lead-tin alloy and hold your velocity down to 1300 to 1500 fps. If you do your part in placing the bullet properly, the bullet will take care of the rest. After all, the .45-90 with its 300 grain bullet at 1480 fps was considered one of the great hunting cartridges of the past. Today, the .45-70 with smokeless loads can easily duplicate and far surpass the black powder vintage .45-90.

One of the big problems in obtaining good accuracy with cast bullets is in the depth of the rifling. Marlin rifles having shallow Micro-Groove barrels are particularly difficult in the accuracy department with soft alloy cast bullets unless velocities are held to 1500 fps or below. Hardening the bullets with antimony and/or cold water quenching may permit somewhat higher velocity, but at the expense of bullet expansion on game. And don't ever think for a moment that just because you are using a .45-caliber bullet, you can consider it a pre-expanded .30-caliber and forget all about expansion. It doesn't work that way.

While we have thus far talked primarily about accuracy in the hunting field with the .45-70, for the past several years there has been a fast growing community of shooters using this venerable cartridge in vintage single shot rifles with cast bullets, black powder and iron sights. Not only are they duplicating in the accuracy department what some of our grandfathers and great-grandfathers did, they are improving upon it with accuracy levels that would probably astound the Old Timers. These shooters strive to grasp the final degree of accuracy from their rifle.

How do they do it? First, they adhere to the cardinal rule of accuracy that uniformity begets uniformity. That is, everything they do to make every shot is just exactly as it was done on the last shot and the shot before that and the shot before that. Secondly, and of equal importance, they understand the black powder cartridge and what it takes to get precision accuracy from it.

To begin with, the .45-70 was developed as a black powder cartridge, and as such was always loaded to 100% loading density. That is, with black powder it was always loaded with NO airspace between the top of the powder charge and the base of the bullet. A full 70 grain charge of black powder was usually compressed a considerable amount to accommodate the bullet, and in those government cartridges containing only 55 grains of powder in order to reduce recoil, the space between the powder charge and the base of the bullet was filled with wads. But whatever the load, when black powder was used there was never any airspace between the top of the powder charge and the base of the bullet.

The competitive black powder cartridge rifle shooter using the .45-70 usually starts out by purchasing three or four hundred cartridge cases of the same make and lot number. He goes through the same resizing, trimming and deburring processes described for the smokeless powder shooter, and then he fire-forms his cartridge cases so that they are an exact fit in the chamber of his rifle. Beyond that he does as little resizing as possible, often using only a neck die that has been honed out so as to leave the inside diameter of the mouth of the case only one or two thousandths of an inch smaller than the diameter of the cast bullet to be seated. Some shooters have the neck die honed so that a bullet can just be finger seated in the cartridge case. Whichever way it is done, any crimping is usually accomplished with a taper crimp die just sufficiently enough to remove any flare or irregularity from the mouth of the case.

When it comes to primers, the competitive black powder shooter does a lot of testing to determine which primer gives the best accuracy with the powder charge being used. Two very popular primers are the Winchester WLRM large rifle magnum primer and the Federal No. 215 large rifle magnum primer. Some shooters prefer a standard large rifle primer as opposed to the magnum primer, while others make good use of a large pistol primer. Whichever primer is used, it is selected only after considerable testing.

Black powder selection for the .45-70 usually involves GOEX FFg or Cartridge, Elephant FFg or Swiss FFg or 1-1/2 Fg. But again, the shooter has to do considerable testing to determine which powder make and granulation will provide the best accuracy in his rifle. If I were just starting out, I would select the GOEX FFg and work from there.

If you are going to load your black powder cartridges with precision accuracy as the goal, then your powder charges should be weighed to one tenth of a grain. While this may be contrary to the standard practice of loading a muzzleloader with a volumetric measure, it is vital to precision accuracy in the cartridge rifle. Whereas the amount of powder in a muzzleloader can vary by plus or minus half a grain, as long as the shooter exerts the same pressure on the ramrod each time when seating the ball, the powder will be compressed the same and velocity will be the same even though the height of the powder column will vary.

But in a black powder cartridge, the powder column has to be compressed to an exact depth in order to maintain a precise overall length of the cartridge. Any variation in the amount of powder charge makes a variation in the amount of compression as well as velocity, and this variation will show up on your target.

Dribble the powder charge into the cartridge case through a 24-inch drop tube to settle the powder within the cartridge case. Top it off with a .45-caliber wad cut from a milk carton, vegetable fiber gasket material or low density polyethylene. Thickness of the wad can range from about 0.023 inch for the milk carton to 0.030 or 0.060 inch for the vegetable fiber and the low density polyethylene. Whichever wad you choose, always use a wad of the same type and thickness for every cartridge. (Wads and wad material are available from Buffalo Arms Co.)

While there are many different opinions on how much the powder should be compressed, I personally like to keep it between 0.060 and 0.100 inch. This provides a sufficient amount of compression to make the powder burn cleaner and more efficiently, and yet not so much compression as to start fracturing the individual kernels of powder and thus change their burning rate.

Most long-time black powder cartridge rifle shooters have tried a dozen or more different style cast bullets in an effort to obtain the finest degree of accuracy. But for the beginner just getting into the game and wanting to keep expenses down, I can think of no better bullet than the Lyman No. 457132 Postell design. Cast the bullet from a 25 to 1 or 20 to 1 lead-tin alloy, size it to 0.458 or 0.459 inch diameter and lubricate it with Black Powder Gold bullet lube. Finally, seat the bullet in the cartridge case and run the cartridge into a taper crimp die just far enough to remove any flare or irregularity on the mouth of the case.

Now is all this fuss and bother worth it? Does it really pay off with superb accuracy from a black powder cartridge using lead alloy bullets?

Back in November 2001, Dan Theodore of Mountain View, California, using a .40-65 single shot Browning rifle, fired twenty consecutive shots at 100 yards to make four 5-shot groups, each of which measured well under 1-inch from center-to-center of the widest apart holes. Not only did he accomplish this fine bit of shooting with black powder and lead alloy bullets, he

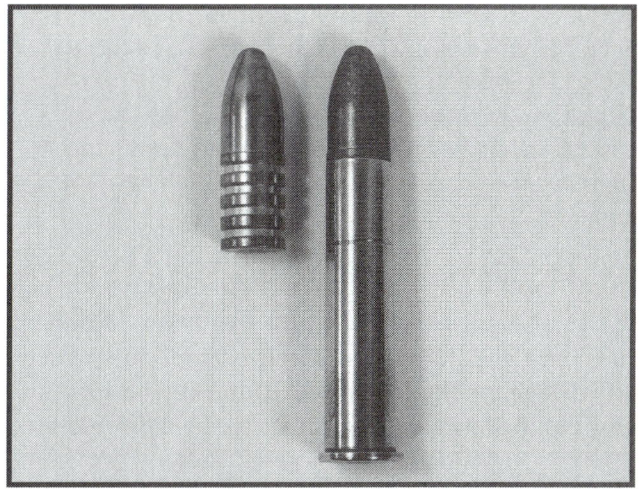

The Lyman #457132 Postell bullet is very popular with black powder cartridge silhouette shooter due to its excellent long-range accuracy.

did it three different times - a total of 60 rounds - just to prove it was no fluke!

Superb accuracy from a black powder cartridge rifle, whether it be a .40-65 or a .45-70, is never accidental. It is the result of precision casting of premium bullets and paying careful attention to details in the reloading process. It is the result of the study of black powder ballistics and knowing what has to be done to accomplish certain goals.

The fine accuracy accomplished by Dan Theodore at 100 yards does not necessarily translate into fine accuracy out at 385 meters (421 yards) for the turkey silhouettes nor at 500 meters (547 yards) where the rams stand. When our bullet gets out to these targets, it has lost a significant amount of linear velocity and with it some rotational velocity. And when rotational velocity drops off, bullet stability starts to deteriorate and the bullet becomes more vulnerable to wind.

When loading for accuracy on the turkeys and rams and beyond, shooters normally use heavy bullets for the caliber involved. In the case of the .45-70, this means bullets ranging from 500 to 550 grains at sufficient velocity from an 18-inch twist barrel to assure bullet stability all the way out to the target. Obviously, this means recoil and plenty of it, and lots and lots of practice so that recoil does not interfere with concentration on sight picture and trigger squeeze.

Probably the two most critical elements of accuracy with the .45-70, or any other rifle, is total concentration of sight picture and trigger squeeze. Without this, the inherent accuracy of your rifle and ammunition is of no consequence. They can both be the best on the firing line and you will still loose out to some shooter carrying a beat up relic of dubious ancestry, but who knows how to freeze onto a sight picture and keep applying even pressure to the trigger at one and the same time.

Now what kind of accuracy does one need to hit a steel ram 547 yards away? First, shooting at silhouettes is not like shooting at a paper target with a perfectly round black aiming point that nicely fits inside your aperture front sight. In silhouette shooting the targets are of an irregular shape - chickens, pigs, turkeys and rams - and you shoot at a different target with every shot for score. The only thing these targets have in common is that their major aiming area covers a tad more than two minutes of angle at the distance at which the target is placed.

For the rams, this means that the ram target measures a shade over 12 inches from back line to belly line behind the foreleg. With this in mind, it is my opinion that to be competitive the rifle, ammunition and shooter must all be capable of minute and a half angle of accuracy or better all the way out to 547 yards. At that distance, a minute and a half angle of accuracy translates into a group measuring just under 8-5/8 inches, leaving less than two inches top and bottom for a margin of error. When you consider the influence of wind, mirage, light change and human error, it seems remarkable that any shooter ever hits a ram using a cartridge with a mid-range trajectory of around twelve feet from a rifle equipped with iron sights!

Yet shooters using the .45-70 with carefully crafted loads commonly hit eight out of ten rams, and often get all ten of them. On July 25 and 26, 1998 during a two-day 120 round match, using a .45-70 built by Ron Snover on a Remington rolling block action, I tipped 28 out of 30 rams. And during that same match on the 26th, I hit 59 pigs without a miss to set a new national record. For my load during that match, I used a 547 grain original design Postell bullet backed by 67 grains of FFg GOEX powder fired by a Winchester WLRM primer. A 0.060-inch thick vegetable fiber wad from John Walters was used over the powder and the powder was compressed about 0.090 inch.

The .45-70 has a lot of potential in the accuracy department for the shooter who has the savvy and persistence to work it all out. Not only is it accurate, but with modern loads in suitable rifles it can be adequate for any big game on the North American continent as well as for some of the critters in other places. When it comes to long range accuracy, the .45-70 is right at home, but because of its high trajectory and low velocity demands a lot more from the shooter than does a high power smokeless rifle.

This is the challenge that makes the .45-70 so great. Over fifty years ago I wrote Elmer Keith a letter criticizing him because he had said some unkind things about the .30-06. Elmer wrote back and said that more people agreed with me than with him, but if I ever had the opportunity to try a big bore rifle, try it and make up my own mind.

I did. And the .45-70 has been with me ever since.

The author just after making a bullseye hit at 1000 yards on July 3, 1993 near Williamsport, Pennsylvania. Rifle is a Navy Arms rolling block .45-70 that was rechambered with a match reamer by John Korzinek of Canton, Pennsylvania. The rifle is equipped with a long-range windgauge tang sight and a spirit level front sight made by Ron Snover of IXL Enterprises. The load that day consisted of 52 grains of RS Pyrodex, 475 grain Lyman No. 457132 original design Postell bullet, Winchester large rifle magnum primer and a grease cookie. The throat of the barrel was wire brushed after every fifth shot.

Photo by Sam Saxton

14 Barrel Care and Cleaning for Accuracy by Butch Fisher

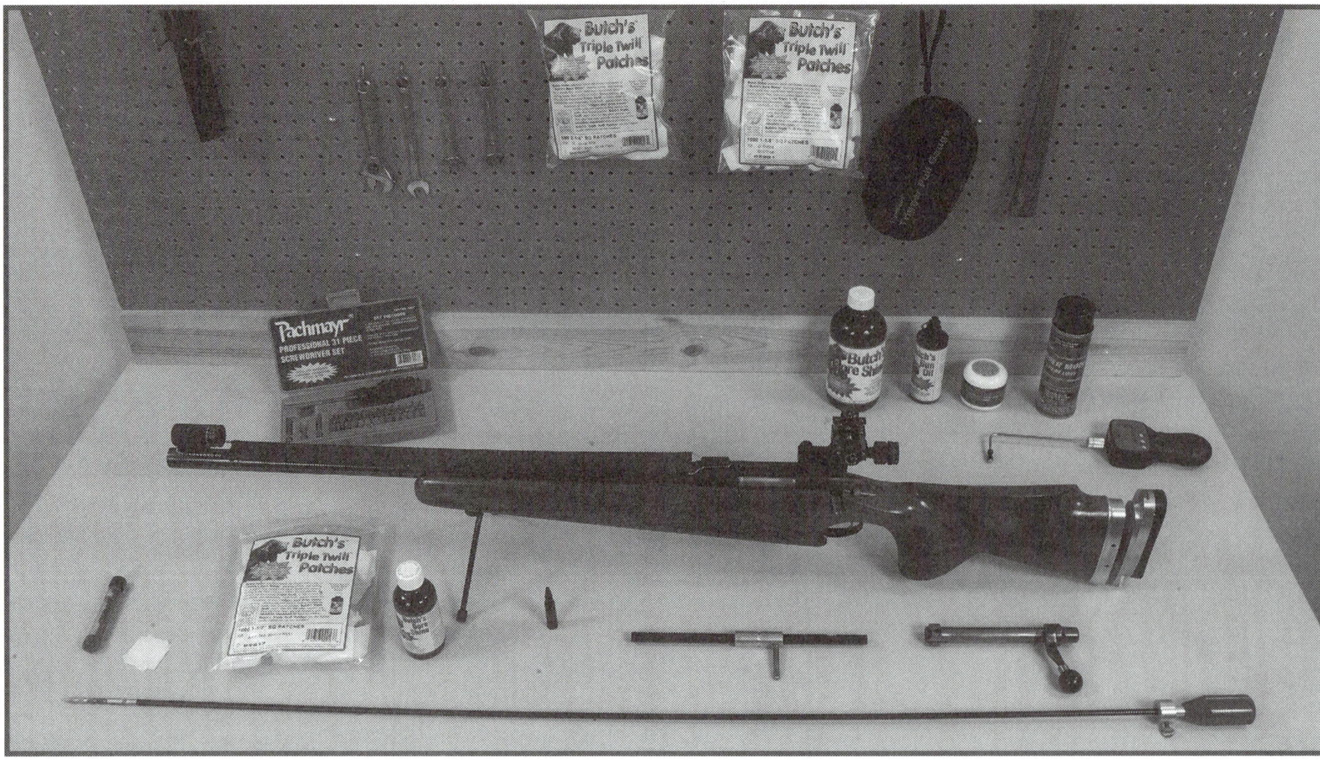

A good quality cleaning rod and a bore guide are important tools necessary to properly clean a rifle.

I have been involved in some form of competitive shooting for over 40 years. I currently shoot in precision bench rest competitions all around the US and Canada. Obviously, as a competitive shooter, I have always been concerned with firearm accuracy. There are certainly many different and varied factors which can contribute to the accuracy of any firearm. However, despite the type of firearm being utilized, and the many variable factors involved, there is always one accuracy problem which is common to all firearms. That problem is termed very simply as "bore fouling." Any build-up of residues in the bore of a firearm will very quickly begin to cause accuracy deterioration. The rate of that deterioration can and will vary from gun to gun, but it universally affects all types of firearms. On many occasions over the years I have had individuals come to me with complaints about firearms they claim used to group extremely well. However, these firearms were now performing poorly. A few of these guns had actual problems, but the majority again fired decent groups after I performed nothing more than a proper, thorough cleaning on the firearm.

Growing up as a youngster in Pennsylvania I learned at an early age the negative effects that fouling can produce. My father had a nice collection of original Pennsylvania Rifles and he also built muzzle loading rifles as a hobby. My first experience with firearms consisted of shooting targets at 50 yards with a muzzle loading rifle he had made. I vividly remember being very disconcerted at the difficulty I encountered trying to ram a patched ball down the muzzle of that weapon after only a handful of shots. Any sort of group I might be shooting deteriorated very quickly after a just a few rounds. Subsequently, most of my shooting time was spent with patches, a cleaning jag and a bucket of soapy water trying to keep that rifle clean. Certainly the fouling created by an old can of DuPont FF black powder cannot be compared to the fouling problems we see in the modern firearms of today. But bore fouling in today's weapons is still a major contributor to accuracy deterioration.

The types and rate of fouling one encounters in modern weapons can vary greatly due to a variety of factors. Those factors can include the ammunition a shooter might utilize, the type of projectile, the bore diameter, the type, depth and twist of the rifling, and many other variables. There are also several very distinct and diverse fouling problems. **Fouling is simply the build-up of any residues in a weapon's bore.** Most shooters are familiar with powder fouling. However, fouling residues can also include several other substances. Shooters who utilize cast lead bullets will experience a build-up of lead. Copper wash from copper jacketed bullets can be a very difficult and stubborn substance to remove. It also builds up quickly. Some small caliber factory ammunition can cause a build-up of wax from the wax coatings on that ammunition. Scattergun users will often experience fouling from plastic wads which will eventually cause a deterioration in the patterns those guns shoot. In recent years, the introduction of bullet lubricating substances such as Moly and Danzac have spawned their own unique fouling problems. Additives such as teflon in some lubricating oils can also create problems. Understanding and controlling all these types of fouling residues is absolutely paramount to the top performance of any firearm.

How Fouling Affects Accuracy

To understand how fouling affects accuracy one must first have an understanding of the various types of fouling and how they build in your barrel. Powder fouling is common to all firearms. Powder fouling is primarily burnt carbon residue. When you pull the trigger on any firearm, the heat produced by the combustion of the burning powder is tremendous. It is not unlike taking a blow torch and blasting it up your barrel. Hot gases from the burning powder create immense heat and pressure. As they begin to erupt and roll out of the case behind the projectile, they first blast what is called the "lead" of the barrel. The "lead" is the small area between the end of the chamber and the "throat" of the barrel. The "throat" is where the projectile first contacts the lands of the barrel. As the pressure builds, the projectile is pushed forward through the "throat" and then spins on up through the barrel. Since the peak pressure level is concentrated in the "lead," the "throat," and a few inches beyond, the majority of the powder fouling is also concentrated in these areas. Often in the "lead" area of a firearm you will find what is called "fire cracking." These are actual microscopic cracks in the steel caused by the tremendous heat created by the cartridge. This "lead" area of the barrel usually becomes caked with residue. The heat and pressure also puts a lot of wear and tear on the "throat" area as well. Some carbon fouling will usually extend the entire length of the bore but it is much more concentrated in this lower end. As more rounds are fired, the carbon begins to build. At some point the build-up of burnt carbon residue begins to actually shrink the diameter of the bore, causing bullet pressures to rise. As more pressure is applied the projectile's ability to properly engage the lands and smoothly move down the bore becomes impaired. You now have a powder fouling problem.

Shooters who use copper jacketed bullets will also experience what is known as copper fouling. As the heat and pressure build, those gases create tremendous torque behind the projectile. This pushes the jacketed projectile forward and spinning up the barrel at high velocities. As the rifling cuts into the projectile, the high amount of friction taking place shears off microscopic fragments of copper. These microscopic fragments of copper create a "copper wash" which is blown on up the barrel. The immense amount of friction taking place causes a sort of copper plating effect in the upper portions of the barrel. As more rounds are fired this plating process continues to build. At some point the lands of the rifling can no longer penetrate the jacket and this begins to affect the movement of the projectile. You now have a case of copper fouling. This copper plating effect can be very difficult to remove if it is allowed to build.

Shooters who use cast lead bullets will also experience a build-up of lead for the same reasons as the users of copper jacketed bullets. Lead is even more malleable than copper. If you take a piece of lead and scratch it across your concrete garage floor, it will leave a distinct mark or residue due to the friction involved in the action of the movement. In the barrel of a firearm the friction involved causes cast bullets to readily shed lead. As the lead builds it will also begin to impair the smooth movement of the projectile and you will start to have a lead fouling problem. Lead can also be difficult to remove, but not nearly as difficult as copper.

Wax coatings in some factory ammunition and certainly plastic wads in shotgun shells will also cause similar fouling problems in the upper ends of barrels. However, they are usually relatively easy to remove. Bullets coated with substances such as Molybdenum Disulfide and Danzac have become popular in recent years. If allowed to go unchecked, they can also be a problem

The Rate of Build-Up

The rate at which any form of fouling builds up in a firearm can and does vary from gun to gun. This is primarily because of the wide diversity in barrels. Most factory produced barrels fashioned by the larger firearms companies today are usually made from less expensive chrome moly varieties of steel. The mass production tooling process utilized by these manufacturers often leaves a substantial amount of tooling marks and imperfections in these barrels. These tooling scratches and imperfections are places which will catch and hold fouling. Subsequently, these barrels usually tend to foul rather quickly. Custom barrels produced by smaller companies like Lilja, Hart, Shilen, Krieger, and others around the country are usually made from better varieties of stainless steels. The tooling imperfections in these barrels are then hand lapped away by the manufacturer. The rate of fouling build-up in these custom barrels will often be retarded, to some extent, but even quality custom barrels will still foul reasonably fast. The rifling in a barrel always plays a role. The number of lands, their depth, the rate of twist, and the type of rifling will all have an effect on the rate at which the barrel will foul. For example, in my experience cut rifling tends to foul a bit more quickly than button or mandrel forged rifling. The more lands a barrel has and the deeper the grooves, the quicker it will foul. Faster twist barrels often foul a bit quicker than slower twist barrels. Bullet caliber and case size also have an effect. Smaller caliber firearms with big cases will tend to foul faster than firearms with more average size cases. This is simply because you are using a large amount of powder to push a small diameter projectile. The more powder packed into a case, the faster the fouling. The high velocities these cartridges produce will also increase the amount of copper wash and therefore cause faster copper build-up.

Cleaning Frequency

In competitive bench rest shooting the better shooters are usually fastidious cleaners and often clean after every five to ten rounds. Cleaning every few rounds might seem a bit excessive to the average shooter, but most competitive shooters will tell you that bore fouling starts to have a noticeable effect on accuracy after only a few rounds. How many rounds that might be depends on the firearm and the variables I discussed above. I have owned firearms that could handle 25 to 30 rounds before I began to notice any real accuracy deterioration. I have also had some that would foul after only five or six shots. The frequency of fouling will certainly vary from gun to gun but you can be assured that fouling will definitely begin to affect accuracy in all firearms at some point. Keeping a firearm free from all forms of fouling is probably the easiest chore a shooter can perform to assure that the weapon is performing at its best. Any shooter concerned with accuracy needs to clean frequently and correctly.

It is always difficult to ascertain the exact extent of fouling that is present in any firearm without the use of a bore scope. A bore scope is a tool which can be inserted down the barrel of a firearm. It magnifies the inside of the bore so as to enable the user to examine the entire length of that barrel. These wonderful instruments allow a close examination of the lands and grooves, revealing any residue build-up, barrel wear, and the barrel's overall health and condition. Obviously, since these tools are rather expensive, few shooters have the luxury of access to such an instrument. However, there are some things you can do without a bore scope to estimate the condition of your firearm. First, taking a long look through the bore can reveal a lot. Often one can see fouling with the naked eye alone. Another trick which works well when endeavoring to ascertain the extent of copper fouling is to take the barrel out into the sunlight. Hold the barrel out away from your body a foot or two and a couple feet below eye level. Put the sun over your shoulder and view the crown of the barrel from that angle. You can usually see down the first inch of the barrel. Move the barrel around until you get the sunlight to reflect off that first inch of the inside of your barrel. Bright sunlight will reflect off any copper residue below the barrel crown. If you have a fair amount of copper just below the crown, you can be certain you have it further down the bore. Another method of judging fouling is simply by feel. When working your cleaning rod through the barrel you can often feel "rough spots" where the rod does not move through smoothly. These are more than likely places that have a heavy build-up of residues. One of the easiest ways to check for severe copper fouling is an examination of your cleaning patches. If you are using a solvent like my "Butch's Bore Shine" or another copper cleaner with ammonia, a lot of blue coloration on the patch will indicate a substantial amount of copper residue.

Cleaning Accessories

So far we have some idea of how fouling works and the problems it creates. So the question now becomes, "How do we correctly remove it?" Before cleaning your firearm, I always recommend the purchase of a few cleaning tools and accoutrements. The first purchase should always be a quality cleaning rod. The rod should be plastic coated, since metal rods can scratch a barrel. Some poorer quality plastic coated rods can also lose pieces of their coating and can then pick up particles of fouling residue while being worked through the bore. If this happens they can actually act like a file, doing damage to your barrel. So purchase a good quality rod! Your second purchase is a rod guide. A rod guide is an instrument which slides into the receiver of a bolt action rifle and is inserted into the chamber, replacing the bolt which is removed. Some guides are actually equipped with a bolt handle which can be locked into place. The front of the guide is equipped with a rubber gasket to seal the chamber. The back end of the guide extends out the back of the receiver. Some guides have a convenient solvent port cut into the top in order to allow cleaning fluids to be applied through the opening. A cleaning rod can be inserted through the rear of the guide and then pushed through into the bore. Guides come in various sizes designed to fit any chamber, and are reasonably priced. The primary purpose for using a rod guide is to allow the rod to slip smoothly into the bore. Without a guide, rods can scratch the throat area as they are inserted into the barrel. They also help eliminate some of the

A good quality patch is saturated with solvent and is ready to be inserted into the bore guide.

mess involved in the cleaning process by keeping solvents and residues from dripping back into your receiver.

You will also need the correct size jag, a chamber mop, and some brushes to go along with your rod. Patches are another must. Here you will probably need to experiment a bit in order to find patches of the correct thickness and consistency in order to work well in your particular barrel. What you are looking for is a patch that fits snuggly yet can be easily pushed through the bore. If your patch is a bit too tight, try placing your jag in the corner of the patch rather than the center. This often will help a thicker patch work well in a smaller barrel. I have what I consider an excellent patch out on the market that is specially treated for absorbency and corrugated for scrubbing. It will actually hold more solvent. However, any good cotton flannel patch will work.

A gun cradle is another tool I like to use. Cleaning can be vigorous and it's great to have your gun sitting in a sturdy cradle rather than trying to man-handle it on the kitchen table. Your wife might also appreciate the purchase! If you are handy, a cradle can be easily constructed or you can purchase one for a reasonable price.

It is also nice to have an old sock or some type of cover that you can slip over the rear of your stock. Some of the composite gun stocks have beautiful paint jobs; a cover will help protect them. Frankly, I've always been somewhat amazed by individuals who will pay thousands of dollars for a custom-built rifle, yet not spend a few extra dollars to purchase the equipment necessary to keep it adequately maintained. Once you have your cleaning accessories in place, you are now ready to clean your firearm.

A Normal Cleaning Regimen

For purposes of this discussion, I am going to assume that your firearm is ready for what I call a normal maintenance cleaning. By normal maintenance I am referring to a firearm that has been fired no more than 15 to 25 times. If more rounds have been fired you will probably have more severe fouling and it must be treated differently. The first thing you need is a quality solvent. I am definitely going to recommend my product, Butch's Bore Shine solvent. I formulated BBS to be what I call a one-step patch solvent. That means that it is designed to work, in most cases, with only a few patches in order to eliminate the need for brushing. I call it a one-step product because unless your barrel is severely fouled, BBS will remove all forms of fouling, including copper, quickly and without too much effort. Most products on the market today will only work on one form of fouling and you need several different products to properly clean a barrel. As I stated above, I generally recommend a maintenance cleaning every 15 to 25 rounds. This figure will likely vary from gun to gun but can be used as a good rule of thumb or starting point. I have found that accuracy generally starts to be affected by bore fouling somewhere around that figure in most firearms. Just remember, the more you shoot the more the fouling build-up. My regimen only requires a few patches and can be performed very quickly so there is no excuse not to be a diligent cleaner.

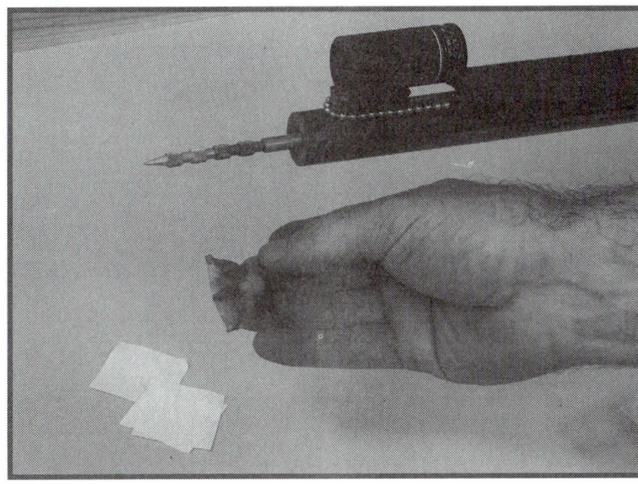

The first patch run through a dirty barrel will come out completely soiled from powder fouling.

To begin, place a patch on the point of your rod jag and push it into the rear of the rod guide until it reaches a point directly below the solvent port. Liberally dispense enough solvent into the port to thoroughly saturate your patch. Push the patch on through the bore slowly until it exits the muzzle. This first patch will be black and very dirty. Most of the residue on the patch will be from powder fouling. When pushing the cleaning rod through a bore be careful that you do not put undo pressure on the rod in an upward or downward direction, as this can cause your rod to bow and rub against the barrel. This can cause scratches. Follow the first patch with a second liberally saturated patch. However, be sure to wipe your rod with a towel or piece of cloth between each patch. Wiping the rod removes any grit that might have been picked up so that it won't scratch the barrel if the rod accidentally comes into contact with it. Inspect the second patch, it shouldn't be nearly as dirty as the first. Continue with another saturated patch. Short stroke this patch back and forth as you push it through the bore, utilizing a little vigor. Carefully examine the third patch for signs of blue coloration which indicates copper fouling. If you find a fair amount of blue color, short stroke a couple more patches. When a patch finally exits reasonably clean you can follow with a couple dry patches to remove any solvent that might remain in the bore. A very small amount of blue color can show up on a clean patch. This is caused by the brass jag (which contains copper) rather than copper fouling in the

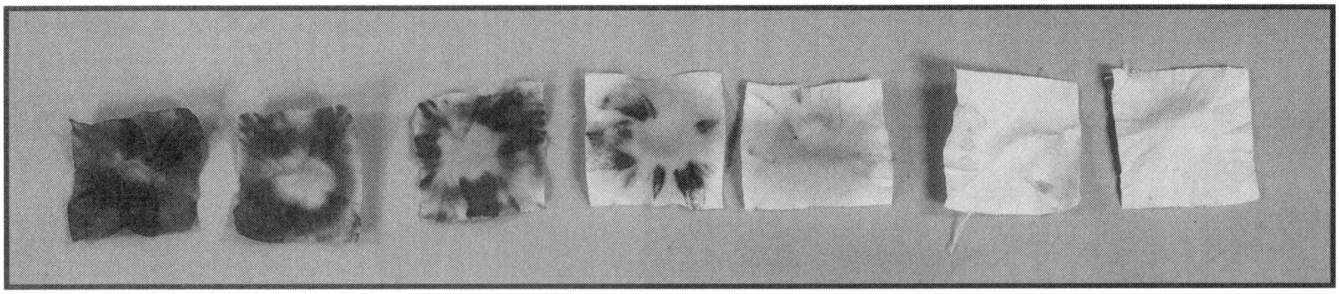

A line-up of patches shows the cleaning process from start to finish. Watch for any blue discoloration on the patches, which would indicate copper fouling in the barrel.

barrel. Always conclude a cleaning with lubrication. I will discuss lubrication later. This simple cleaning regimen, if done frequently before fouling has a chance to become severe, should keep most weapons performing as intended.

Severe Fouling

Sometimes shooters will fire 100 rounds or more without cleaning. And some individuals utilizing jacketed bullets will clean only with a powder remover. In these instances, you can be sure that bore fouling has been allowed to become much more severe. If you clean with only a powder solvent, copper or lead fouling can really become a problem since most powder solvents, unlike BBS, won't work well on copper or lead. Therefore, your cleaning techniques will have to be more complex and diligent. You will now have to resort to brushing. Never use a brush with steel bristles. Steel brushes can scratch and damage a bore. I much prefer a bronze brush. There are nylon brushes available but I just don't think they do a decent enough job. Before brushing try another couple of patches liberally soaked with my BBS. Run them through the bore, then sit back and let it soak. You can soak your barrel with BBS as long as you like. BBS will not harm barrel steel. Letting it soak will allow the solvent to work on the fouling. Now screw the brush into your rod and push the rod into the guide. Liberally soak the brush with solvent and push it on through your barrel. Pull the brush back through on the reverse stroke. Continue this process one stroke for every round you think was fired. One round per stroke is just something I use as a gauge, or rule of thumb, it's not any hard and fast rule. Severe fouling may require additional brushing or might clean up with just a few strokes. It depends on how severely fouled the barrel is. Lead fouling and heavy carbon will usually clean out with a good brushing but copper can at times be a much more difficult problem to alleviate. After you have completed brushing, switch back to the jag and run a solvent soaked patch through the bore. You might now see a lot of blue color on the patch. This is because there is copper in a bronze brush as well as copper in your barrel. Much of the blue color on your patch is from the brush, not the barrel. BBS solvent removes all forms of fouling, including copper, and therefore will remove some of the copper from the brush as you are cleaning. Therefore, always rinse your brush well after using BBS because any solvent left on the brush will continue to eat away at the bristles and turn the brush black. I use lacquer thinner for rinsing but even water will help if you have nothing else available. After rinsing the brush, squeeze it with a rag or an old towel to remove any excess. This helps give your brush a longer life. However, brushes are very reasonably priced and the cost is well worth a clean gun. Now try another patch or two soaked with solvent. Closely examine these patches for blue color. There should be little or no blue color present if your brushing did the job. Once you get a color free patch, dry the bore and follow with lubrication.

If after a thorough brushing you are still finding blue color on your patches, or perhaps feeling tight spots in the barrel which you think might still be fouling, you will probably have to resort to the use of a bore paste. Bore pastes are products that contain abrasives. They can be found in most sporting goods stores. Wrap a patch around your jag, or you can utilize an old used brush. Heavily coat it with the bore paste. Vigorously work your rod back and forth through the bore five to ten times. Remember, bore pastes are abrasives, and if overused can do damage to the barrel. Use them sparingly! With a bore paste you are actually sanding the fouling out of a barrel. However, the use of a bit of bore paste every several hundred rounds shouldn't harm a barrel. Again, follow with solvent soaked patches to clean out the paste and check for blue. Once a patch comes through clean, dry the bore and lubricate it.

Cleaning a Moly/Danzac Fouled Barrel

Since I created BBS, many of the questions I field from shooters concern cleaning techniques for use with firearms that utilize coated bullets. Bullets coated with substances such as Molybdenum Disulfide and Danzac have become popular in recent years. Actually, variations of these coatings have been around since the 1950's. Users and manufacturers now tout longer barrel life, higher velocities, and reduced copper fouling among the benefits of coated bullets. I find that Moly build-up, if allowed to go unchecked, can create a real mess in a barrel. I have observed severe Moly fouling in

Butch's gun care products include Bore Shine cleaning solvent, Black Powder Bore Shine cleaning solvent, Gun Oil and "Triple Twill" cleaning patches.

barrels using a bore scope. The Moly build-up is usually concentrated several inches above the throat, although it can also extend further up the barrel. It appears very irregular and knobby looking, not unlike snake or lizard skin.

In examining cleaning procedures for firearms that utilize Moly/Danzac, I think there are some variables we need to look at. There are probably some definite benefits with Moly/Danzac such as reduced copper fouling. Logic dictates to me that shooters should not have to clean as often when using these substances. However, when comparing the cleaning frequency of firearms using standard bullets to those using coated bullets, at what point do we set that variable? This can be a real problem and one that will not be uniform to all firearms. It requires experimentation with the individual firearm involved. In my experience, firearms using Moly/Danzac that are cleaned infrequently (after 50 or more rounds), usually tend to have fouling build-up problems in varying degrees. The more rounds fired between cleaning, the bigger the problem. Therefore, one should experiment, but certainly adjust the cleaning frequency downward if experiencing any form of Moly/Danzac build-up. Since I am a fastidious cleaner, I recommend my cleaning regimen every 15 to 25 rounds for regular bullets. I would extend that out to perhaps 40 rounds for shooters using Moly/Danzac and start from there. Some knowledgeable individuals who shoot with Moly and clean with BBS feel their cleaning results have been effective with that frequency. Since BBS is primarily a patch solvent it does not require brushing if used frequently. Brushing can remove the Moly coating on the barrel surface and defeat the purpose of the product. If you do get a case of severe Moly/Danzac build-up, you will have to resort to the brushes or possibly a bore paste. Just follow the same procedures I discussed in the severe fouling segment. Remove all fouling and start again from scratch.

Lubrication

Once you have a barrel that is clean of all forms of fouling you should always coat the barrel with a light gun oil. Shooting over a dry barrel will produce immediate copper streaking and can also cause excess wear to the bore. Some individuals are leery about using oil because they are overly concerned about first shot accuracy. They think an oiled barrel will cause a flyer on the first shot after cleaning. This can be true to some extent; however, if your barrel is squeaky clean and you follow your oiled patch with a dry patch in order to remove excess oil, your first point of impact should be quite close to your target aim. The benefits of a lightly lubricated barrel, in my opinion, far outweigh the risk of a flyer on that first round. Oil will fill those minute pores and tool marks in your barrel, creating a smoother surface to shoot over. Oil will also retard fouling to some extent, allowing more rounds of accurate shooting. Think of your barrel as the bore of a gasoline engine and the bullet as the piston. You would never start that engine without oil for lubrication. The same is true of a firearm.

I experimented for over two years with various types of oils in developing my Butch's Gun Oil. My oil is a special blend of several natural oils that hold up very well under the high pressures and the tremendous heat created by a cartridge. It also contains a rust inhibitor for the maximum in corrosion protection. But any light gun oil will certainly work better than no oil to protect your barrel and to extend its life and accuracy. The only lubricants I would avoid are oils containing synthetic additives such as teflon. I have found that some additives can create a fouling problem of their own. Once you have lubricated the barrel, remove your rod guide. You can now take a chamber mop and give the chamber a quick turn or two so that you remove any dirt, drips or residue from that area. Whenever you do an extensive cleaning, you should always use a little bolt grease on the lugs and on the bolt cam. There are several quality high pressure greases available at your gun store. You are now ready to shoot again and your firearm should perform with the accuracy it was designed to provide.

Good Luck & Good Shooting !
Butch Fisher

The author, Butch Fisher, is a well known bench rest competitor and is the originator of Butch's Gun Care products.

Section 5

INTRODUCTION TO DATA
Important: Please Read

The data in this section cannot be used without a full understanding of what it represents or without complete reloading knowledge. Any attempts to use this data without the necessary cautions, procedures and background knowledge could be extremely
dangerous causing personal and property injury, even death.

The Data listed in this section has been tested by Lyman technicians and found safe when loaded with their test components, tooling, procedures and finished cartridge dimensions and when fired under laboratory controlled conditions in the test firearms. The publisher and editors have no control over how reloading is conducted by the individual. Every change in equipment, procedure, component lot, finished cartridge dimensions and firearm used will effect ballistics and/or the safety and usefulness of a load. Therefore, no warranties are implied or expressed for the data and copy contained in this book. We specifically disclaim any warranties of fitness for any and all particular purposes and specifically disclaim any and all liability for consequential damages of any kind.

The individual assumes all risks for the safety of reloaded ammunition. Improperly loaded ammunition, or the failure to follow all necessary precautions, may result in serious personal injury and/or death to the shooter or bystanders.

There are many precautions to which the reloader must adhere. This introduction deals with many of these but cannot possibly foresee, or include, all
possible cautions or procedures.

Maximum listed loads must always be worked up to carefully. **ALWAYS START WITH THE SUGGESTED STARTING LOAD** and develop the load slowly, carefully, firing at least ten shots at each incremental propellant charge weight.

Test Components

Keep in mind that the test components are not of Lyman manufacture. Therefore, it is possible that production changes which would effect ballistics or load safety could occur without Lyman's knowledge. If there is any doubt, created for any reason, including unexpected results from even a single shot, immediately stop. Do not proceed with reloading or fire another shot until you have contacted the component manufacturer and cleared up the difficulty.

Test Specifications

These are self explanatory and can sometimes help the shooter form some judgment as to performance in their firearm.

Cartridge Drawing

This information is based on the maximum allowable loaded cartridge dimensions according to accepted industry standards. With the exception of bullet diameters, it is unlikely that the reloader will ever encounter a cartridge with these dimensions unless a problem is present. The listed dimensions, when compared with the reloader's ammo dimensions can be a help in making certain maximum dimensions are not exceeded. For example, if your loaded cartridges have a neck diameter equal or greater than the listed dimension for your cartridge the ammo should not be fired.

Comments

The information given here is to help the reloader with component selection and/or to impart specific information about the cartridge. The comments should always be read for specific loading suggestions which may make the reloader's work go easier.

Bullets

Not every possible bullet has been tested. We have used bullets which we feel are of a popular weight and type. The listed data for each bullet is all of the data we can supply at this time. Keep in mind that seating depths can cause noticeable changes in ballistics. This is especially true for handgun ammunition. It is strongly suggested that the specific cartridge overall length be used as shown.

Powders

We have used powders that were generally popular at the time the specific testing was conducted. Not every cartridge was shot for this edition. Thus, some of the newer propellants may not be listed with some of the less popular loadings or bullets. Where a single or a few specific powders have proven the best possible choices with respect to ballistic uniformity, we have identified these loads as "accuracy loads". While not actually fired at targets, all accuracy loads have high potential for producing outstanding accuracy since uniform internal ballistics is critical to accuracy on the target. You cannot have one without the other.

Suggested Starting Grains

Loads shown in this column are the place to begin load development. Use the exact charge shown. Never go below the starting charge as to do so can sometimes cause dangerous conditions. Lighter charges can sometimes result in dangerously high pressure. Work up from the starting charge to the maximum charge slowly and in small increments. Be sure you have sufficient knowledge to interpret the results of your progressively heavier charges. Of course, you may choose not to use a maximum load and stay at the starting charge or some point between it and the maximum load.

As a beginning handloader you may want to confine all loads to the starting charges. When you are familiar with all the aspects of ammo assembly then you can begin to work up loads.

Maximum Load Grains

Maximum charges are never to be exceeded. THEY ARE NOT DELIBERATELY HELD TO CONSERVATIVE LEVELS. They are the exact charge that produced the maximum pressure level allowed or otherwise proved to be the maximum charge we could list. Never use any maximum charge without carefully working up to it from the Suggested Starting Grains.

It is quite possible that listed maximum loads may not be safe in your firearm, with your components, with your loading procedures or with your finished cartridge dimensions. Use caution when developing a load. If even a single cartridge shows evidence of excessive pressure stop using the load and propellant. Switch to another propellant and begin the process again.

Should trouble continue, stop all loading and contact the component manufacturer.

Maximum load listings are dependent on a number of criteria beyond average pressures of the loads. Ballistics must be uniform or individual high and low pressures could be dangerous. The amount of powder that will fit into the case without excess compression (this will vary with case lots, powder lots, and bullet seating depth), performance at high or low temperatures, muzzle flash and other important factors are considered when we develop maximum loads. Never assume listed data can be exceeded.

Compressed Loads

All compressed loads are indicated with a + sign. Depending upon the volume of the specific case lot being used the load may or may not fit into the case. Never attempt to force feed the powder into the case. If the bullet cannot be started into the case reduce the powder charge to allow for 1/10" of free space in the case neck. Cases which bulge after bullet seating should not be used. Never compress powder so that it causes case bulging. For more information on determining the suitability of a compressed load refer to the chapter on powder.

Universal Receiver set-up for C.U.P. pressure testing

17 Remington

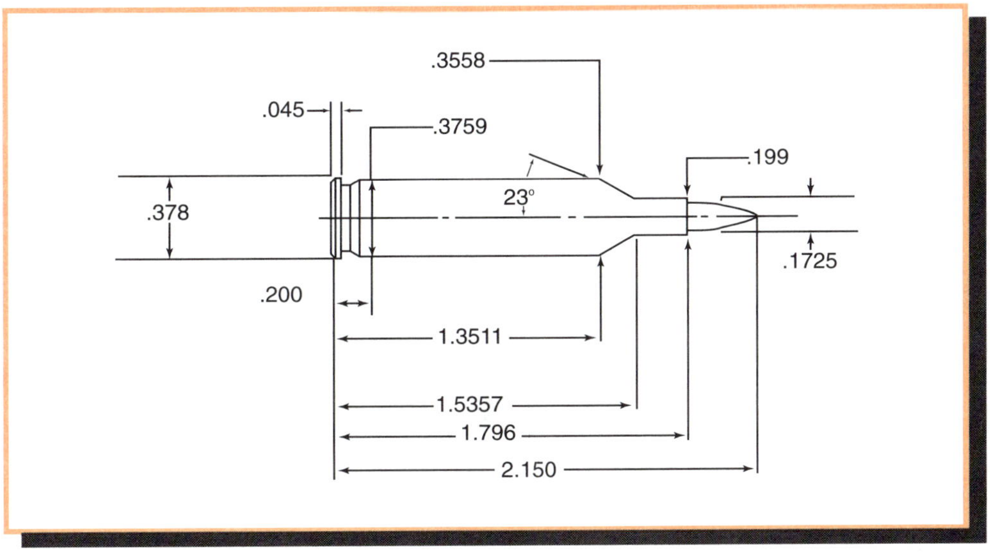

Comments:

Introduced in 1971 by Remington this case, based on the 222 Remington has never been very popular. Most of the lack of popularity stems from practical reasons.

The 17 caliber cartridges tend to foul barrels quickly. With most rifles, if the bore is not thoroughly cleaned every 15 shots or so, accuracy quickly disappears. It is not every gun shop that stocks 17 caliber bullets or the skinny cleaning rods required.

Factory ammo achieves a velocity of about 4,000 fps. The reloader can duplicate this speed using IMR 4320.

Test Components:

Cases . . . . .Remington
Trim-to Length . . . . .1.786"
Primers . . . . .Remington 7½
Primer Size . . . . .Small Rifle
Lyman Shell Holder . . . . .No. 26
Jacketed Bullets Used . . . . .Hornady HP #1710, 25 gr.

Test Specifications:
(Velocity Only)

Firearm Used . . . . .Remington Model 700
Barrel Length . . . . .24"
Twist . . . . .1-10"
Groove Dia. . . . . .172"

25 gr. Jacketed HP
2.150" OAL
BC: .187
SD: .121

Powder	Sugg Starting Grains	Velocity fps	Pressure C.U.P.	Max Load Grains	Velocity fps	Pressure C.U.P.
IMR-4198	17.0	3478	—	**19.3**	**3830**	—
RX7	17.0	3473	—	19.0	3759	—
IMR-3031	20.0	3452	—	22.0	3857	—
BL-C (2)	20.3	3450	—	21.9	3731	—
748	20.0	3418	—	22.8	3761	—
IMR-4895	21.5	3616	—	23.3	3934	—
IMR-4064	21.0	3455	—	23.0	3819	—
IMR-4320	22.5	3692	—	24.2	4023	—

Note: Loads shown in shaded panels are maximum. Loads shown in bold designate potentially most accurate load.

22 Hornet

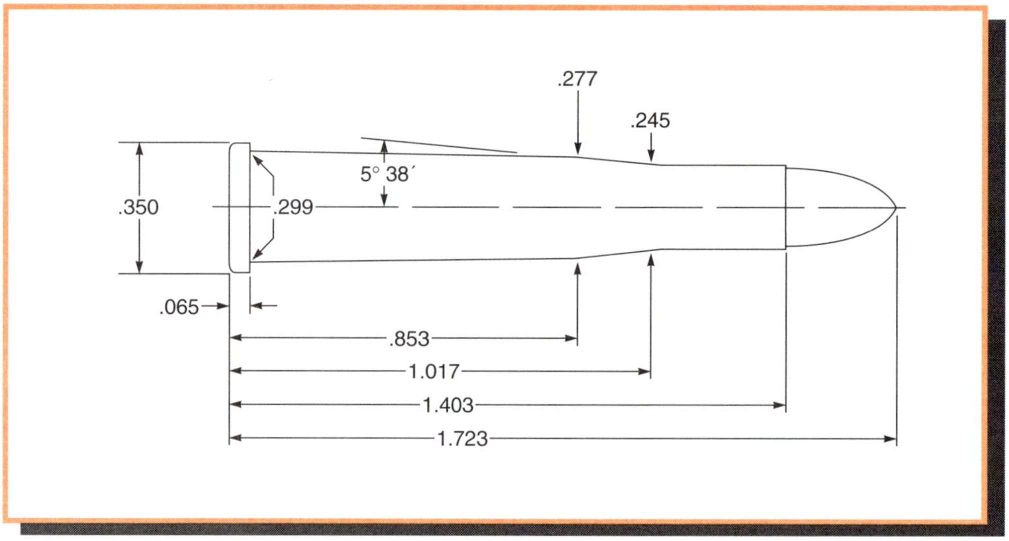

Comments:

The 22 Hornet evolved out of experiments with the old 22 WCF cartridge at Springfield Armory during the 1920s. Any shooters loading for an older rifle should be aware that two different groove diameters exist for the Hornet. All currently produced 22 Hornets utilize the .224″ diameter groove common to all .22 caliber center-fire rifles. Most rifles manufactured up to the World War Two period have a .223″ diameter groove. If in doubt, slug the bore and select the appropriate diameter bullet. The Hornet is a long-time favorite among handloaders and is a fine cartridge out to 150 yards. Its accuracy, mild recoil, and wide-selection of quality rifles and pistols add to its appeal.

The Hornet is not similar to any other current cartridge. As such, several precautions are necessary. Case walls of the Hornet are quite thin and easily crush if the die is not properly adjusted. The expander button installed in the full-length sizing die must be adjusted as low as possible without contacting the inside bottom of the case. Improper adjustment of the expander button can shear off the upper portion of the neck leaving it stuck within the die body. Some shooters prevent any such problems by simply removing the expander button completely and using an M-die to neck expand as in a three-die set. Suitable powders are limited due to the shape and capacity of the cartridge case. Powders regarded as slow pistol powders or very fast rifle powder fall within this rather narrow burn rate. 40 and 45-grain bullets usually work best with the Hornet's 1-16″ twist. Most manufacturers produce bullets in both diameters constructed specifically for the Hornet's performance level.

Test Components:

Cases . . . . .Winchester
Trim-to Length . . . . .1.393"
Primers . . . . .WSR
Primer Size . . . . .Small Rifle
Lyman Shell Holder . . . . .No. 4
Jacketed Bullets Used . . . . . .Sierra JSP #1200, 40 gr.
Hornady JSP #2230, 45 gr.
Hornady V-Max #22261, 50 gr.
Speer JHP #1035, 52 gr.
Hornady SPSX #2260, 55 gr.
Cast Bullets Used . . . .(sized to .224" dia)
*gas check bullet *#225438, 44 gr.
*#225415, 55 gr.

Test Specifications: (Velocity & Pressure)

Firearm Used . .Ruger K77/22VHZ, Universal Receiver
Barrel Length . . . . .24"
Twist . . . . .1-16"
Groove Dia. . . . . .Ruger; .224"
Universal Receiver; .223"

40 gr. Jacketed JSP 1.722" OAL — **BC:** .122 **SD:** .114

Powder	Sugg Starting Grains	Velocity fps	Pressure C.U.P.	Max Load Grains	Velocity fps	Pressure C.U.P.
2400	7.2	2007	—	8.0	2235	—
N110	8.2	2384	—	9.2	2627	—
H110	9.9	2592	—	11.0	2805	—
IMR 4227	10.5	2534	—	11.7+	2671	—
AA1680	12.6	2547	—	**14.0+**	**2768**	—

45 gr. Jacketed JSP 1.722" OAL — **BC:** .202 **SD:** .128

Powder	Sugg Starting Grains	Velocity fps	Pressure C.U.P.	Max Load Grains	Velocity fps	Pressure C.U.P.
2400	7.9	2235	—	**8.8**	**2506**	—
N110	7.9	2322	—	8.8	2472	—
H110	9.1	2383	—	10.2	2664	—
IMR 4227	10.3	2406	—	11.5+	2652	—
AA1680	11.2	2323	—	12.5+	2537	—

Note: Loads shown in shaded panels are maximum.
Loads shown in bold designate potentially most accurate load.
+Designates a compressed powder charge.

22 Hornet

50 gr. Jacketed V-Max
1.723" OAL

BC: .242
SD: .142

Powder	Sugg Starting Grains	Velocity fps	Pressure C.U.P.	Max Load Grains	Velocity fps	Pressure C.U.P.
2400	6.3	1775	—	7.0	2014	—
N110	7.3	2174	—	8.2	2303	—
H110	8.5	2292	—	9.5	2549	—
IMR-4227	**9.9**	**2350**	—	11.0	2576	—
AA 1680	9.9	2181	—	11.0	2341	—

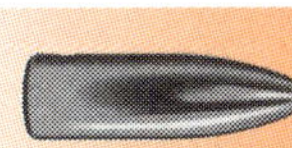

52 gr. Jacketed JHP
1.723" OAL

BC: .225
SD: .147

Powder	Sugg Starting Grains	Velocity fps	Pressure C.U.P.	Max Load Grains	Velocity fps	Pressure C.U.P.
2400	6.1	1653	—	6.8	1955	—
N110	7.2	2103	—	8.0	2217	—
H110	**8.1**	**2193**	—	9.0	2425	—
IMR-4227	9.0	2233	—	10.0	2404	—
AA 1680	9.7	2094	—	10.8	2300	—

55 gr. Jacketed SPSX
1.723" OAL

BC: .235
SD: .157

Powder	Sugg Starting Grains	Velocity fps	Pressure C.U.P.	Max Load Grains	Velocity fps	Pressure C.U.P.
2400	5.8	1591	—	6.5	1752	—
N110	6.9	1989	—	7.7	2092	—
H110	7.6	1945	—	**8.5**	**2190**	—
IMR-4227	8.1	1941	—	9.0	2171	—
AA 1680	9.1	1963	—	10.2	2098	—

#225438
44 gr. (#2 Alloy) 1.675" OAL

BC: .094
SD: .125

Powder	Sugg Starting Grains	Velocity fps	Pressure C.U.P.	Max Load Grains	Velocity fps	Pressure C.U.P.
Unique	4.2	1945	—	4.7	2038	—
2400	5.4	1585	—	6.0	1742	—
N110	6.3	1707	—	7.0	2069	—
H110	7.2	1913	—	8.0	2173	—
IMR-4227	7.2	1711	—	8.0	1909	—
AA 1680	9.4	2076	—	10.5	2325	—

#225415
55 gr. (#2 Alloy) 1.694" OAL

BC: .116
SD: .157

Powder	Sugg Starting Grains	Velocity fps	Pressure C.U.P.	Max Load Grains	Velocity fps	Pressure C.U.P.
Red Dot	2.0	1060	—	3.7	1645	—
700X	2.0	1050	—	3.7	1665	—
Green Dot	2.5	1230	—	4.2	1765	—
PB	2.5	1185	—	4.2	1705	—
Unique	3.0	1320	—	4.6	1805	—
SR-7625	3.0	1350	—	4.4	1745	—
H110	4.2	1124	—	6.5	1722	20,700
SR-4759	5.6	1266	13,500	7.5	1821	25,700
Rx7	6.5	1159	—	9.5	1785	15,300

Note: Loads shown in shaded panels are maximum.
Loads shown in bold designate potentially most accurate load.

218 Bee

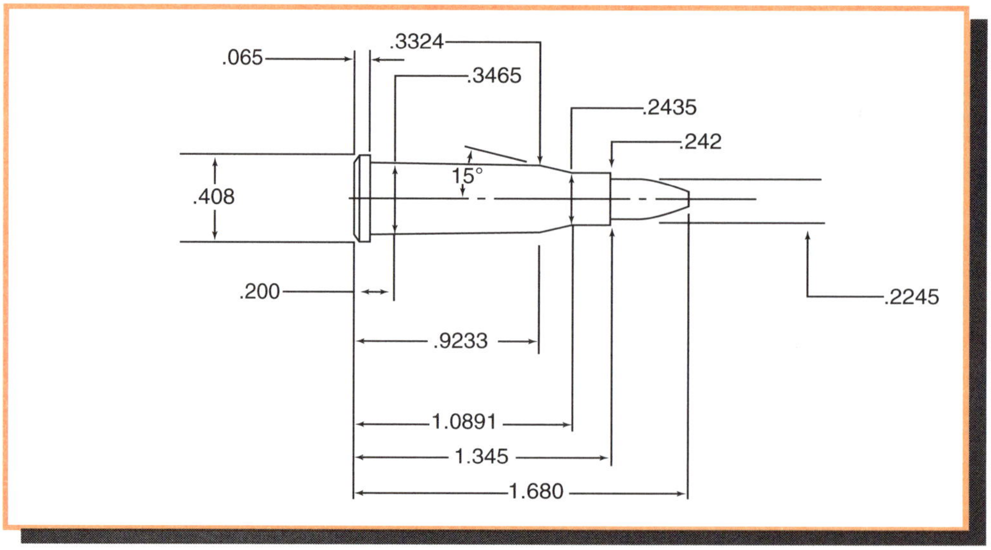

Comments:

The reloader cannot generally duplicate advertised factory ballistics with canister powders currently available. When loading ammo for use in tubular magazines DO NOT USE ANY SHARP NOSED bullets. Cartridges for tubular magazines must be loaded only with bullets having blunt noses.

Hornet style bullets must be used if expansion is to be certain. IMR 4227 seems the best powder to begin with, along with bullets of 45-grains.

Accuracy in lever action rifles is seldom impressive but when used in a bolt or single shot action the 218 Bee often gives accuracy approaching that of the 22 Hornet.

Test Components:

Cases . . . . .Remington
Trim-to Length . . . . .1.335"
Primers . . . . .Remington 6½
Primer Size . . . . .Small Rifle
Lyman Shell Holder . . . . .No. 10
Jacketed Bullets Used . . . . .Speer Sp. SP #1017, 40gr.
Sierra Varminter #1210, 45gr.
Sierra SMP #1320, 50gr.
Speer HP #1035, 52 gr.
Sierra SMP #1350, 55gr.
Cast Bullets Used . . . .(sized to .223" dia)
*gas check bullets *#225415, 55gr.

Test Specifications: (Velocity Only)

Firearm Used . . . . .Winchester Low Wall
Marlin Over/Under
Barrel Length . . . . .Winchester; 20"
Marlin; 26"
Twist . . . . .1-16"
Groove Dia. . . . . .223"

40 gr. Jacketed Sp. SP 1.680" OAL — **BC:** .144 **SD:** .114

Powder	Sugg Starting Grains	Velocity fps	Pressure C.U.P.	Max Load Grains	Velocity fps	Pressure C.U.P.
2400	11.0	2645	—	12.5	2906	—
IMR-4227	11.0	2457	—	**13.5**	**2932**	—
IMR-4198	14.0	2564	—	15.0+	2762	—

45 gr. Jacketed Varminter 1.680" OAL — **BC:** .137 **SD:** .128

Powder	Sugg Starting Grains	Velocity fps	Pressure C.U.P.	Max Load Grains	Velocity fps	Pressure C.U.P.
2400	10.5	2463	—	12.0	2777	—
IMR-4227	11.5	2481	—	**13.0**	**2754**	—
IMR-4198	13.0	2309	—	14.0	2564	—

Note: Loads shown in shaded panels are maximum.
Loads shown in bold designate potentially most accurate load.
+Designates a compressed powder charge.

218 Bee

50 gr. Jacketed SMP
1.680" OAL

BC: .190
SD: .142

Powder	Sugg Starting Grains	Velocity fps	Pressure C.U.P.	Max Load Grains	Velocity fps	Pressure C.U.P.
2400	10.0	2331	—	11.5	2617	—
IMR-4227	11.0	2331	—	**12.5**	**2617**	—
IMR-4198	12.0	2105	—	13.0	2325	—

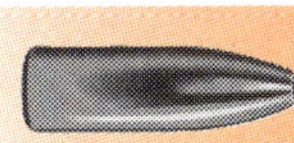

52 gr. Jacketed HP
1.680" OAL

BC: .225
SD: .147

Powder	Sugg Starting Grains	Velocity fps	Pressure C.U.P.	Max Load Grains	Velocity fps	Pressure C.U.P.
2400	9.5	2178	—	10.5	2487	—
IMR-4227	10.5	2159	—	**11.5**	**2506**	—
IMR-4198	12.0	2083	—	13.5	2267	—

55 gr. Jacketed SMP
1.680" OAL

BC: .212
SD: .157

Powder	Sugg Starting Grains	Velocity fps	Pressure C.U.P.	Max Load Grains	Velocity fps	Pressure C.U.P.
2400	9.5	2183	—	10.5	2369	—
IMR-4227	10.0	2087	—	**11.5**	**2380**	—
IMR-4198	12.0	2145	—	13.2	2309	—

#225415
55 gr. (#2 Alloy) 1.595" OAL

BC: .116
SD: .157

Powder	Sugg Starting Grains	Velocity fps	Pressure C.U.P.	Max Load Grains	Velocity fps	Pressure C.U.P.
*Red Dot	2.8	1295	—	4.7	1805	—
***700x**	2.8	1310	—	**4.7**	**1805**	—
*Green Dot	3.3	1410	—	5.2	1915	—
*PB	3.3	1365	—	5.2	1845	—
*Unique	3.7	1450	—	5.7	1960	—
*SR-7625	3.8	1480	—	6.0	2005	—

Note: Loads shown in shaded panels are maximum.
Loads shown in bold designate potentially most accurate load.
*Tested in a Winchester Low Wall

222 Remington

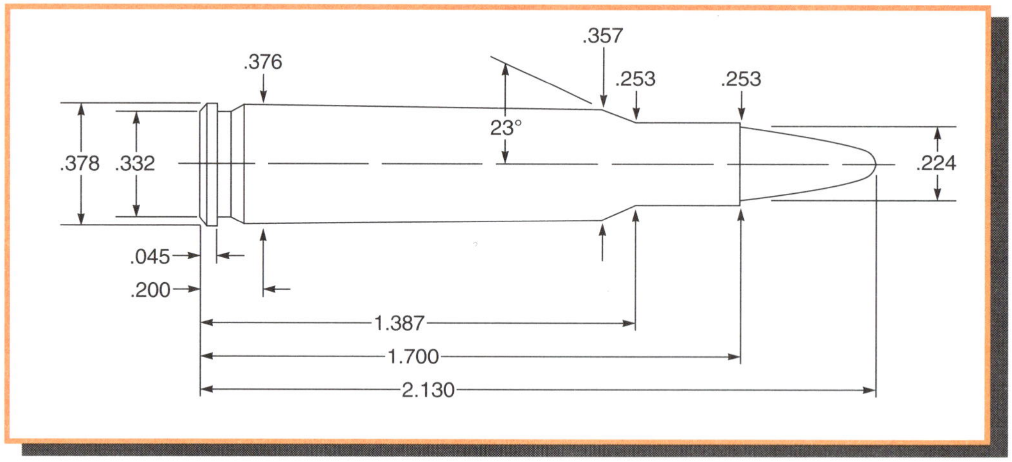

Comments:

Remington introduced their 222 in 1950 as an entirely new design not based on any existing case. The .378" diameter rimless cartridge constituted something entirely new to the American shooter. Its inherent accuracy helped establish the 222 Remington as a top-notch varmint cartridge which soon dethroned the older, rimmed 219 Zipper and Wasp cartridges on the Benchrest circuit. The "triple-deuce" has itself been overtaken in Benchrest by the PPC family of cartridges and has lost ground to the 223 Remington elsewhere. Despite this, the cartridge still has its die-hard followers and offers a lot to those who load for it. 50 or 55-grain bullets with a proper charge of H-322 make a potent combination out to 250 yards. Cast bullets should be cast hard, at least 15 BHN, and kept below 2300 feet per second. The 222 Remington is notable as the basis of several latter cartridges including the 222 Remington Magnum, 223 Remington, 221 Fireball, and 17 Remington along with numerous wildcats.

Test Components:

Cases . . . . .Remington
Trim-to Length . . . . .1.690"
Primers . . . . .Remington 7½
Primer Size . . . . .Small Rifle
Lyman Shell Holder . . . . .No. 26
Jacketed Bullets Used . . . .Speer Sp. SP #1017, 40 gr.
Sierra SPT #1310, 45 gr.
Sierra Blitz #1340, 50 gr.
Sierra HP Match King #1400, 53 gr.
Sierra SPT #1360, 55 gr.
Hornady HP #2275, 60 gr.
Sierra SMP #1370, 63 gr.
Cast Bullets Used . . . .(sized to .224" dia)
*gas check bullets *#225415, 55 gr.
*#225646, 55 gr.

Test Specifications:
(Velocity & Pressure)

Firearm Used . . . . .Universal Receiver
Barrel Length . . . . .26"
Twist . . . . .1-14"
Groove Dia. . . . .224"

40 gr. Jacketed Sp. SP
2.040" OAL
BC: .144
SD: .114

Powder	Sugg Starting Grains	Velocity fps	Pressure C.U.P.	Max Load Grains	Velocity fps	Pressure C.U.P.
IMR-4198	18.0	2688	24,000	22.0+	3460	44,600
XMR-2015	21.6	2963	34,800	24.0+	3277	45,300
RX7	19.8	2919	29,900	22.0	3295	44,100
IMR-3031	20.0	2283	19,200	23.5+	3076	33,500
H322	22.1	2883	30,000	24.5+	3264	44,300
AA2230	**24.5**	**3250**	42,000	27.2+	3586	45,100
H335	23.0	2884	29,400	25.5	3234	43,000
IMR-4895	21.0	2590	26,400	25.0+	3125	36,400
IMR-4064	21.0	2481	25,200	23.6+	2848	30,500
IMR-4320	22.0	2583	27,600	26.0+	3067	36,900

45 gr. Jacketed SPT
2.125" OAL
BC: .210
SD: .128

Powder	Sugg Starting Grains	Velocity fps	Pressure C.U.P.	Max Load Grains	Velocity fps	Pressure C.U.P.
IMR-4198	18.0	2672	24,600	21.5+	3333	45,400
XMR2015	21.3	2810	31,800	23.7+	3152	44,700
RX7	18.9	2896	32,700	21.0	3106	42,200
IMR-3031	20.0	2538	27,600	23.5+	3134	39,400
H322	21.6	2811	29,400	24.0+	3280	44,400
AA2230	23.4	3037	41,500	26.0+	3342	45,500
H335	**23.6**	**2925**	**34,200**	26.2+	3234	45,300
IMR-4895	21.0	2583	27,000	25.0+	3115	39,400
IMR-4064	20.0	2331	24,600	23.6+	2857	32,000
IMR-4320	22.0	2570	27,000	26.0+	3039	37,700

Note: Loads shown in shaded panels are maximum.
Loads shown in bold designate potentially most accurate load.
+Designates a compressed powder charge.

50 gr. Jacketed Blitz

2.125" OAL — **BC:** .222 **SD:** .142

Powder	Sugg Starting Grains	Velocity fps	Pressure C.U.P.	Max Load Grains	Velocity fps	Pressure C.U.P.
IMR-4198	18.0	2717	28,200	20.5	3205	45,900
XMR-2015	20.7	2875	32,600	23.9	3267	46,200
RX7	19.4	2703	28,600	21.5	3060	43,900
IMR-3031	20.0	2538	26,400	23.5+	3115	42,000
H322	21.6	2852	32,700	**24.0+**	**3090**	**45,600**
H335	23.0	2797	33,000	25.5	3143	45,300
IMR-4895	22.0	2762	31,500	25.0+	3105	42,400
AA2460	21.0	2749	32,400	23.5	3050	43,000
748	22.0	2691	26,800	25.0+	3032	38,900
IMR-4064	20.0	2314	23,400	23.6+	2849	34,000
Varget	22.5	2648	30,500	25.0+	2947	37,500
IMR-4320	23.0	2754	34,000	26.0+	3048	42,400

53 gr. Jacketed HP Match King

2.130" OAL — **BC:** .224 **SD:** .151

Powder	Sugg Starting Grains	Velocity fps	Pressure C.U.P.	Max Load Grains	Velocity fps	Pressure C.U.P.
IMR-4198	17.1	2705	31,700	19.5	3025	45,500
XMR-2015	19.2	2703	32,200	22.1	3084	45,500
IMR-3031	20.0	2544	27,600	23.5+	3039	41,100
H-322	19.8	2738	34,300	**22.5+**	**3083**	**45,500**
AA2230	20.2	2601	29,200	23.5	3060	46,200
H-335	22.1	2744	34,200	24.5	3051	45,600
IMR-4895	22.0	2747	34,000	25.0+	3058	45,900
AA2460	19.0	2410	23,900	23.6	3004	45,900
748	21.5	2651	29,000	24.5	3024	44,600
IMR-4064	20.0	2341	25,800	23.6+	2840	36,900
Varget	22.1	2590	29,100	24.5+	2930	40,900
IMR-4320	20.3	2409	27,800	23.0+	2807	39,500

55 gr. Jacketed Spt

2.130" OAL — **BC:** .237 **SD:** .157

Powder	Sugg Starting Grains	Velocity fps	Pressure C.U.P.	Max Load Grains	Velocity fps	Pressure C.U.P.
IMR-4198	17.0	2538	27,000	19.5	2980	45,100
XMR-2015	19.0	2605	26,200	22.1	3098	46,700
IMR-3031	20.0	2531	28,800	23.5+	3048	45,400
H-322	19.3	2609	27,800	22.0	2990	45,500
AA-2230	19.6	2550	27,700	23.0	2975	45,900
H-335	21.6	2639	32,200	**24.0**	**2974**	**44,600**
IMR-4895	20.7	2454	28,000	23.5+	2837	38,800
AA2460	20.2	2555	30,200	23.2	2943	46,100
748	21.8	2539	31,000	24.2+	2885	44,200
IMR-4064	20.0	2347	25,800	23.6+	2849	37,700
Varget	22.1	2651	31,500	24.5+	2863	35,700
IMR-4320	20.2	2394	26,300	23.0+	2744	39,000

60 gr. Jacketed HP

2.130" OAL — **BC:** .271 **SD:** .171

Powder	Sugg Starting Grains	Velocity fps	Pressure C.U.P.	Max Load Grains	Velocity fps	Pressure C.U.P.
IMR-4198	16.3	2473	30,000	18.5	2787	44,300
XMR-2015	18.5	2527	30,300	21.5	2907	46,000
H-322	19.8	2539	33,200	22.0	2818	45,200
AA-2230	19.6	2462	31,000	23.0	2923	45,400
H-335	20.5	2528	31,400	**23.0**	**2889**	**45,400**
IMR-4895	20.5	2469	32,000	23.0+	2748	38,700
AA-2460	19.0	2366	30,000	22.1	2749	43,600
748	21.0	2423	32,300	23.0	2745	45,000
IMR-4064	19.5	2267	28,200	23.6+	2785	41,600
Varget	22.1	2572	32,000	24.5+	2830	42,400
IMR-4320	21.5	2450	34,000	23.0+	2680	37,200
RX15	21.2	2487	32,200	23.5+	2779	43,500

63 gr. Jacketed SMP

2.130" OAL — **BC:** .235 **SD:** .179

Powder	Sugg Starting Grains	Velocity fps	Pressure C.U.P.	Max Load Grains	Velocity fps	Pressure C.U.P.
IMR-4198	17.5	2629	35,700	19.3	2798	43,020
XMR-2015	17.8	2394	28,800	21.0	2836	45,200
IMR-3031	19.0	2325	29,400	21.0+	2648	35,500
H-322	20.3	2598	32,200	22.0+	2926	45,500
AA-2230	19.0	2384	28,800	22.3	2777	45,400
IMR-4895	20.0	2364	30,000	23.0+	2747	40,000
AA-2460	**20.8**	**2542**	**36,700**	23.1	2824	45,800
IMR-4064	19.0	2141	25,200	23.6+	2785	40,700
Varget	22.7	2682	33,800	25.2+	2950	42,600
RX15	21.8	2591	34,300	24.2+	2898	44,700

#225415

55 gr. (#2 Alloy) 2.107" OAL — **BC:** .116 **SD:** .157

Powder	Sugg Starting Grains	Velocity fps	Pressure C.U.P.	Max Load Grains	Velocity fps	Pressure C.U.P.
Red Dot	5.1	1785	25,200	7.4	2217	39,000
700x	4.7	1709	27,000	6.4	2053	39,900
Green Dot	5.3	1779	26,400	7.7	2232	40,700
PB	5.2	1718	25,800	7.0	2024	39,400
Unique	6.0	1972	25,800	8.0	2358	36,000
SR-7625	5.7	1785	24,000	7.0	2016	35,500
Herco	6.5	1956	27,000	8.8	2336	40,700
SR-4756	6.0	1769	21,600	8.7	2267	40,300
2400	11.0	2153	15,500	15.8	2835	31,300
SR-4759	**12.2**	**2185**	**15,300**	16.5	2895	36,700
IMR-4227	13.0	2314	19,700	15.9	2751	24,200
748	17.0	2019	12,100	25.0	2881	25,400

Note: Loads shown in shaded panels are maximum.
Loads shown in bold designate potentially most accurate load.
\+ Designates a compressed powder charge.

222 Remington

#225646
55 gr. (#2 Alloy) 2.130" OAL

BC: .155
SD: .157

Powder	Sugg Starting Grains	Velocity fps	Pressure C.U.P.	Max Load Grains	Velocity fps	Pressure C.U.P.
Red Dot	5.5	1658	22,600	7.5	2033	38,300
700x	5.6	1728	23,700	7.6	2109	42,200
Green Dot	5.9	1728	21,800	7.7	2077	42,100
PB	5.7	1639	21,800	7.5	1958	41,100
Unique	6.2	1754	21,000	8.0	2115	36,400
SR-7625	5.6	1607	23,900	8.0	1988	41,000
Herco	6.2	1741	20,000	8.5	2172	39,700
SR-4756	6.0	1630	19,000	8.3	2045	41,800
2400	10.5	2170	21,800	12.1	2436	35,800
SR-4759	12.0	2261	26,200	14.0	2569	42,800
748	20.0	2261	17,100	26.0	3058	42,700

Note: Loads shown in shaded panels are maximum.
Loads shown in bold designate potentially most accurate load.

22 PPC

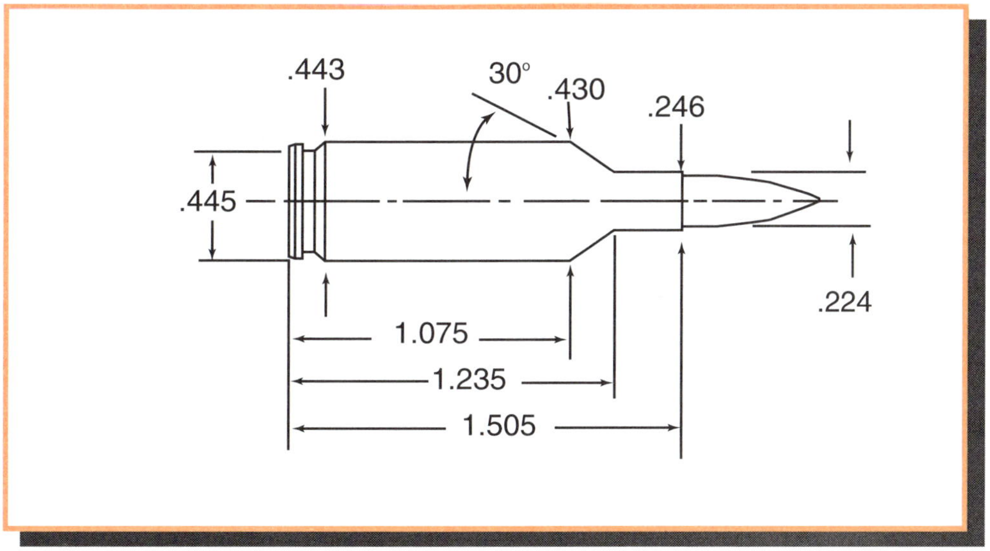

Comments:

The 22 PPC is the necked down version of the 6 mm PPC. It is identical to the 6mm PPC except for the cartridge neck and bullet diameters. This is without doubt the second most accurate cartridge available, being bested only by its 6mm near twin.

The PPC nomenclature is taken from the initials of its designers: Pindell, Palmisano Cartridge. It has won many benchrest shoots, but the 6mm version is the real champ in competition. This round makes a good selection for varmint shooting. Hodgdon's H322 is the single best propellant selection for jacketed bullets, though all listed types will shoot very well indeed. Match grade bullets are required to get all that this cartridge can deliver.

If you use cases other than Sako PPC, USA, approach all loads with extreme caution.

Test Components:

Cases . . . . .SAKO
Trim-to Length . . . . .1.495"
Primers . . . . .Remington 7½
Primer Size . . . . .Small Rifle
Lyman Shell Holder . . . . .No. 3
Jacketed Bullets Used . . . . . . .Speer HP #1030, 50 gr.
Sierra HPBT #1410, 52 gr.
Sierra SPT #1360, 55 gr.
Hornady HP #2275, 60 gr.

Test Specifications:
(Velocity & Pressure)

Firearm Used . . . . .Universal Receiver
Barrel Length . . . . .22"
Twist . . . . .1-14"
Groove Dia. . . . . .2245"

50 gr. Jacketed HP
2.025" OAL
BC: .223
SD: .142

Powder	Sugg Starting Grains	Velocity fps	Pressure C.U.P.	Max Load Grains	Velocity fps	Pressure C.U.P.
N-130	22.7	3030	35,900	25.3	3385	47,800
IMR-4198	21.0	2959	31,400	23.0	3295	44,200
XMR-2015	22.0	2991	29,200	25.5	3391	41,300
H-322	23.0	3030	30,200	26.0	3359	41,200
AA2230	25.0	3039	29,700	28.0	3388	41,500
IMR-4895	25.0	2979	30,300	28.0+	3353	42,400
BL-C(2)	27.0	3051	32,600	30.0	3350	43,200
748	27.0	3028	29,400	30.0	3396	42,500
Varget	27.0	3122	36,600	**30.0+**	**3456**	**46,400**

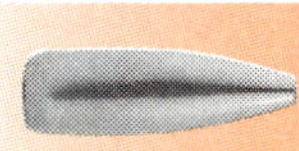

52 gr. Jacketed HPBT
2.025" OAL
BC: .225
SD: .148

Powder	Sugg Starting Grains	Velocity fps	Pressure C.U.P.	Max Load Grains	Velocity fps	Pressure C.U.P.
N130	22.0	2989	37,500	24.5	3260	47,300
IMR-4198	20.0	2995	33,000	23.0	3338	48,200
XMR-2015	22.0	2957	31,200	25.5	3382	43,700
H-322	23.0	3038	34,400	26.0	3369	45,900
AA-2230	24.5	3007	32,600	27.5	3340	44,100
IMR-4895	25.0	2993	32,200	28.0+	3378	47,200
BL-C(2)	27.0	3043	35,200	30.0	3335	46,800
748	27.0	3053	33,200	30.0	3405	45,900
Varget	26.1	3029	35,400	**29.0+**	**3358**	**46,900**

Note: Loads shown in shaded panels are maximum.
Loads shown in bold designate potentially most accurate load.
+Designates a compressed powder charge.

22 PPC

55 gr. Jacketed SPT
2.025" OAL

BC: .237
SD: .157

Powder	Sugg Starting Grains	Velocity fps	Pressure C.U.P.	Max Load Grains	Velocity fps	Pressure C.U.P.
N130	**22.0**	**2934**	**37,500**	24.5	3229	47,500
IMR-4198	20.0	2979	35,400	23.0	3307	49,500
XMR-2015	23.0	2914	31,600	26.0	3313	43,700
H-322	23.0	3040	38,100	25.5	3281	45,400
AA-2230	24.0	2854	30,200	27.5	3311	45,700
IMR-4895	25.5	2923	32,400	28.5+	3338	46,300
BL-C(2)	27.5	3030	36,800	30.0	3320	48,600
748	26.5	2913	30,400	29.5	3262	41,700
Varget	26.1	2997	36,600	29.0	3311	46,200

60 gr. Jacketed HP
2.115" OAL

BC: .271
SD: .171

Powder	Sugg Starting Grains	Velocity fps	Pressure C.U.P.	Max Load Grains	Velocity fps	Pressure C.U.P.
N130	**21.1**	**2784**	**36,700**	23.5	3052	47,800
IMR-4198	19.0	2668	34,200	22.0	2998	46,400
XMR-2015	21.5	2767	34,000	25.3	3194	48,000
RX7	19.0	2617	35,100	22.2	2920	45,800
H-322	22.5	2814	37,000	26.0	3176	49,200
AA-2230	23.5	2760	34,200	27.2	3162	46,200
IMR-4895	23.5	2681	29,200	27.3	3187	48,500
BL-C (2)	24.7	2739	33,400	29.5	3205	48,100
748	25.5	2827	32,800	29.4	3184	45,600
Varget	25.0	2854	35,600	27.8	3167	47,200

Note: Loads shown in shaded panels are maximum.
Loads shown in bold designate potentially most accurate load.
+Designates a compressed powder charge.

222 Remington Magnum

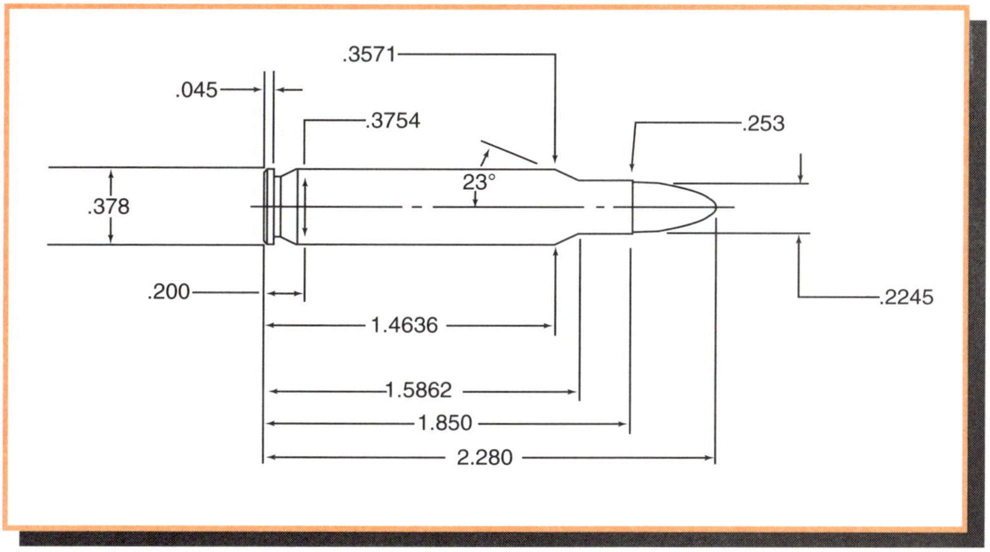

Comments:

This cartridge is all but obsolete, having been the forerunner to the 223 Remington. Owners of rifles chambered for both the Remington 223 and 222 Magnum cartridges must use extreme care not to use the wrong ammo in a rifle.

Powders such as IMR 4198, IMR 4895, Reloder 7, and H335 are all good choices with jacketed bullets from 52 to 60 grains.

Cast bullets driven at 2,000 to 2,300 fps. perform best, especially when lubed with Lyman Alox and gas checked.

Test Components:

Cases . . . . .Remington
Trim-to Length . . . . .1.840"
Primers . . . . .Remington 7½
Primer Size . . . . .Small Rifle
Lyman Shell Holder . . . . .No. 26
Jacketed Bullets Used . . . .Speer Sp. SP #1017, 40 gr.
Sierra SPT # 1310, 45 gr.
Sierra SPT # 1330, 50 gr.
Speer HP #1035, 52 gr.
Hornady SP # 2265, 55 gr.
Sierra SMP #1370, 63 gr.
Cast Bullets Used . . . .(sized to .224" dia)
*gas check bullets *#225415, 55 gr.

Test Specifications:
(Velocity Only)

Firearm Used . . . . . . .Remington 700, Jacketed Bullets
Remington 722, Cast Bullets
Barrel Length . . . . .24"
Twist . . . . .1-14"
Groove Dia. . . . .224"

40 gr. Jacketed Sp. SP 2.200" OAL — **BC:** .144 **SD:** .114

Powder	Sugg Starting Grains	Velocity fps	Pressure C.U.P.	Max Load Grains	Velocity fps	Pressure C.U.P.
IMR-4198	22.0	3322	—	24.5	3787	—
RX7	22.0	3333	—	24.0	3598	—
IMR-3031	23.0	3012	—	26.5+	3636	—
H-335	26.0	3205	—	29.0	3597	—
BL-C (2)	26.0	3125	—	29.0	3497	—
IMR-4895	25.0	3086	—	27.0+	3427	—
IMR-4064	24.0	2754	—	27.0+	3322	—
IMR-4320	26.0	3144	—	28.0+	3448	—

45 gr. Jacketed SPT 2.220" OAL — **BC:** .210 **SD:** .128

Powder	Sugg Starting Grains	Velocity fps	Pressure C.U.P.	Max Load Grains	Velocity fps	Pressure C.U.P.
IMR-4198	22.0	3278	—	24.0	3663	—
RX7	22.0	3236	—	23.5	3427	—
IMR-3031	23.0	3039	—	26.5+	3546	—
H-335	25.0	2906	—	28.5	3484	—
BL-C (2)	25.0	2958	—	28.5	3333	—
IMR-4895	**24.0**	**2840**	—	27.0+	3333	—
IMR-4064	24.0	2739	—	27.0+	3247	—
IMR-4320	25.0	2873	—	28.0+	3367	—

Note: Loads shown in shaded panels are maximum.
Loads shown in bold designate potentially most accurate load.
+Designates a compressed powder charge.

222 Remington Magnum

50 gr. Jacketed SPT
2.250" OAL

BC: .222
SD: .142

Powder	Sugg Starting Grains	Velocity fps	Pressure C.U.P.	Max Load Grains	Velocity fps	Pressure C.U.P.
RX7	21.0	3030	—	23.0	3300	—
IMR-3031	23.0	2967	—	26.5+	3497	—
H-335	25.0	2976	—	28.0	3378	—
BL-C (2)	25.0	2906	—	28.0	3300	—
IMR-4895	24.0	2857	—	**27.0+**	**3311**	—
IMR-4064	24.0	2770	—	27.0+	3225	—
IMR-4320	25.0	2873	—	27.2+	3236	—

52 gr. Jacketed HP
2.250" OAL

BC: .225
SD: .147

Powder	Sugg Starting Grains	Velocity fps	Pressure C.U.P.	Max Load Grains	Velocity fps	Pressure C.U.P.
RX7	20.0	2890	—	22.8	3268	—
IMR-3031	23.0	2941	—	25.5+	3367	—
H-335	24.0	2652	—	27.5	3236	—
BL-C (2)	24.0	2680	—	27.5	3125	—
IMR-4895	24.0	2808	—	**27.0+**	**3279**	—
IMR-4064	24.0	2564	—	27.0+	3194	—
IMR-4320	24.0	2754	—	26.2+	3225	—

55 gr. Jacketed SP
2.310" OAL

BC: .235
SD: .157

Powder	Sugg Starting Grains	Velocity fps	Pressure C.U.P.	Max Load Grains	Velocity fps	Pressure C.U.P.
RX7	20.0	2890	—	22.5	3194	—
IMR-3031	23.0	2949	—	25.0+	3290	—
H-335	24.0	2793	—	27.2	3194	—
BL-C (2)	24.0	2724	—	27.0	3125	—
IMR-4895	24.0	2840	—	**26.5**	**3247**	—
IMR-4064	24.0	2762	—	27.0+	3236	—
IMR-4320	23.0	2617	—	25.5	3076	—

63 gr. Jacketed SMP
2.280" OAL

BC: .231
SD: .179

Powder	Sugg Starting Grains	Velocity fps	Pressure C.U.P.	Max Load Grains	Velocity fps	Pressure C.U.P.
RX7	20.0	2816	—	22.0	3039	—
IMR-3031	21.0	2659	—	23.0	2915	—
H-335	23.0	2702	—	26.5	3086	—
BL-C (2)	23.0	2659	—	26.5	3021	—
IMR-4895	23.0	2672	—	25.5	3303	—
IMR-4064	22.0	2506	—	25.5+	3012	—
IMR-4320	23.0	2631	—	24.5	2865	—

#225415
55 gr. (#2 Alloy) 2.150" OAL

BC: .116
SD: .157

Powder	Sugg Starting Grains	Velocity fps	Pressure C.U.P.	Max Load Grains	Velocity fps	Pressure C.U.P.
Red Dot	5.6	1800	—	9.1	2335	—
700X	5.6	1800	—	8.0	2175	—
Green Dot	6.2	1870	—	9.2	2295	—
PB	6.3	1800	—	9.0	2190	—
Unique	6.8	1925	—	9.7	2325	—
SR-7625	6.7	1855	—	9.1	2205	—

Note: Loads shown in shaded panels are maximum.
Loads shown in bold designate potentially most accurate load.
+Designates a compressed powder charge.

223 Remington

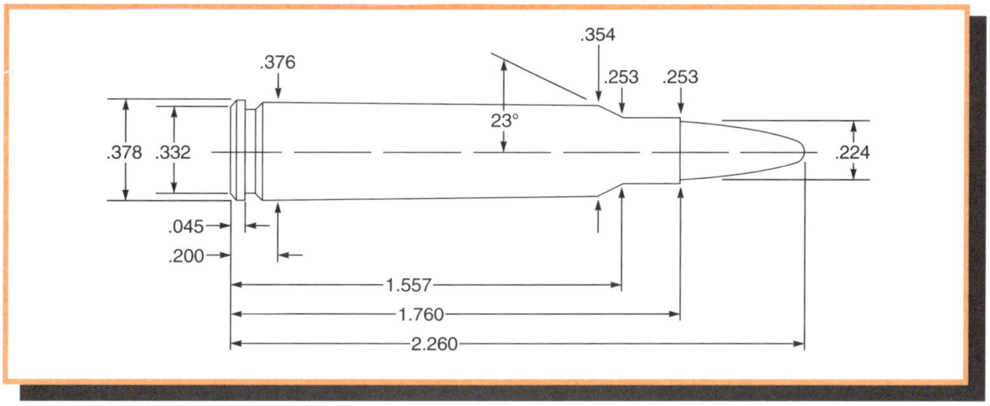

Comments:

The 223 Remington/5.56 NATO has become one of the most popular and versatile commercial cartridges in the United States over the last forty years. It works well with a wide number of components and is easy and economical to reload. Remington introduced this cartridge to the commercial market in 1964 soon after its adoption by the U.S. Army as the 5.56 NATO. Much like the 7.62/308 and the earlier 30-06, ready availability of surplus GI brass nurtured its popularity. The 223 Remington is similar to the older 222 Remington Magnum but is .090" shorter in length and has the shoulder set back approximately .030" Although the 223 Remington cartridges will chamber in 222 Magnum rifles, ***shooters should not fire the 223 Remington cartridge in any rifle chambered in 222 Magnum.*** The 223 Remington has been chambered for a wide variety of firearms over the years including the T/C Contender pistol. In addition to its military service, the cartridge has been used for varmints, small game and target shooting with great success along with use by law enforcement agencies. The 223 Remington should not be considered a deer cartridge. Many states mandate the 6mm/.243" bore size as the minimum caliber for whitetails.

Part of the versatility enjoyed by this cartridge revolves around the wide range of usable bullet weights. Best results with a particular bullet however requires use of a barrel with the appropriate rate of twist. This cartridge was originally introduced by different manufacturers in both 1-12" and 1-14" twist barrels. These twist rates work well for bullets weighing 40 to 55-grains. IMR-3031, H-322, H-335, 748, and XMR-2015 are all great powders in this category. This cartridge has also come a long way in recent years due to development of heavy weight bullets featuring high ballistic coefficients for long-range match competition. The introduction of 80-grain bullets by several different manufacturers allowed the AR-15/M16 series of rifles to compete head to head (shooters of the "black gun" have demonstrated) with the older 30-caliber M14/M1A rifles on the 600 yard stage of the National Match course.

Use of the 77 or 80-grain Sierra MatchKings require a 1-7" or 1-8" twist barrel for proper stabilization. The 80-grain MatchKing has a long, slender ogive and will need to be seated to an overall length longer than the 2.260" specified as maximum. The 80-grain bullet should not be seated to the 2.260" magazine length. Excessive pressures will result from the reduction in available case volume with an 80-grain bullet seated so deeply. Shooters are best advised to use a bullet length comparator (such as those made by Stoney Point) to determine distance of the bullet ogive's engagement to the lands. Bullets should be seated off the lands at least .005"–sometimes more–for best accuracy. The shooter should begin with the bullet approximately .025" off the lands and work closer to the lands of the rifling in .005" increments. This will produce a cartridge OAL around 2.550" in Colt manufactured rifles, shorter in most custom barrels. HBar rifles as manufactured by Colt have a throat approximately .100" longer than other commercially manufactured rifles. Cartridges loaded with 80-grain bullets must be loaded and fired singly. This is no handicap as the 600 yard stage is a slow fire event. These heavier bullets works best with slower burning propellants than what are usually associated with the 223 Remington. H-4895 often gives excellent accuracy with the 80-grainer while N140 and Varget also work well. The 77-grain MatchKing's ogive is more blunt and can be loaded to 2.260" magazine length as can their 69-grain MatchKing. Some shooters do report good results with 75-grain bullets in 1-9" twist barrels, while many shooters have successfully used the 52-grain Sierra MatchKing in 1-7" barrels on the reduced 200-yard course. Sierra's 69-grain MatchKing is suitable for 1-7" to 1-10" barrels.

Cast bullets velocities should be kept between 1,900 and 2,300 feet per second for best accuracy. Bullets should be cast of a hard alloy measuring at least 15 bhn such as Lyman's # 2 alloy. Bullet #225646 gave good results around 2,100 feet per second.

Test Components:

Cases	Remington
Trim-to Length	1.750"
Primers	Remington 7½
Primer Size	Small Rifle
Lyman Shell Holder	No. 26
Jacketed Bullets Used	Hornady V-Max, #22241, 40 gr.
	Sierra SPT #1310, 45 gr.
	Sierra Blitz #1340, 50 gr.
	Sierra HPBT #1410, 52 gr.
	Sierra SPT #1360, 55 gr.
	Hornady V-Max #22281, 60 gr.
	Sierra SMP #1370, 63 gr.
	Sierra HPBT #1380, 69 gr.
	Hornady A-Max #22792, 75 gr.
	Sierra HPBT #9377, 77 gr.
	Sierra HPBT #9390, 80 gr.
Cast Bullets Used	(sized to .224" dia.)
*gas check bullets	*#225415, 55 gr.
	*#225646, 55 gr.

Test Specifications: (Velocity & Pressure)

Firearm Used	Universal Receiver
	Colt AR15
Barrel Length	Universal Receiver: 24"
	Colt AR15: 20"
Twist	Universal Receiver: 1-12"
	Colt AR15: 1-7"
Groove Dia.	.224

223 Remington

40 gr. Jacketed V-Max
2.215" OAL
BC: .200
SD: .114

Powder	Sugg Starting Grains	Velocity fps	Pressure C.U.P.	Max Load Grains	Velocity fps	Pressure C.U.P.
IMR-4198	20.6	3170	34,300	22.9	3545	46,900
XMR-2015	24.4	3253	32,300	27.0	3760	49,700
RX7	21.6	3163	32,300	24.0	3583	47,400
IMR-3031	23.4	2957	31,600	26.0+	3395	44,500
Benchmark	25.2	3276	38,800	28.0+	3627	50,400
H322	24.3	3095	32,000	27.0+	3619	49,000
AA-2230	**24.3**	**3156**	**36,800**	27.0	3649	50,200
IMR-4895	25.0	3001	34,500	27.8+	3444	47,700
H335	25.5	3135	33,200	28.3	3640	50,000
BL-C (2)	27.3	3207	37,800	30.3	3685	50,400
Varget	25.6	3042	31,000	28.0+	3383	38,900

45 gr. Jacketed SPT
2.240" OAL
BC: .210
SD: .128

Powder	Sugg Starting Grains	Velocity fps	Pressure C.U.P.	Max Load Grains	Velocity fps	Pressure C.U.P.
IMR-4198	20.6	3171	36,600	22.9	3542	51,800
XMR-2015	23.8	3083	35,100	25.5	3558	50,900
RX7	21.7	3114	35,000	24.2	3537	51,700
IMR-3031	22.9	2798	31,200	25.5+	3346	46,700
Benchmark	24.7	3173	40,200	27.5	3480	48,700
H322	**24.4**	**3274**	**40,300**	27.2+	3571	50,900
AA-2230	24.3	3166	36,600	27.6	3659	51,700
IMR-4895	25.2	3009	32,600	28.1+	3420	47,700
H335	26.1	3200	37,300	29.4	3678	51,700
BL-C (2)	26.0	3218	37,800	29.0	3641	51,300
748	27.0	3211	36,300	30.0	3592	47,600
Varget	25.2	3023	28,600	28.0+	3369	40,700

50 gr. Jacketed Blitz
2.235" OAL
BC: .222
SD: .142

Powder	Sugg Starting Grains	Velocity fps	Pressure C.U.P.	Max Load Grains	Velocity fps	Pressure C.U.P.
IMR-4198	20.0	2739	—	22.0	3115	—
XMR-2015	**23.2**	**3069**	**32,200**	25.0	3480	49,200
RX7	21.7	3009	34,200	24.1	3330	47,600
IMR-3031	22.0	2688	—	25.5+	3257	—
Benchmark	23.8	3043	36.400	26.5	3332	48,400
AA2230	23.8	3080	36,800	26.0	3453	50,300
IMR-4895	23.0	2570	—	26.5+	3115	—
H335	24.8	3025	34,300	27.7	3459	51,600
BL-C (2)	24.0	2666	—	27.5	3076	—
AA2460	23.9	2964	38,800	26.6	3303	49,900
748	26.1	2977	33,300	29.0	3401	46,500
IMR-4064	23.0	2538	—	26.0+	2967	—
Varget	25.4	3010	32,800	28.2+	3376	45,400
IMR-4320	24.0	2638	—	27.5	3134	—

52 gr. Jacketed HPBT
2.250" OAL
BC: .225
SD: .148

Powder	Sugg Starting Grains	Velocity fps	Pressure C.U.P.	Max Load Grains	Velocity fps	Pressure C.U.P.
IMR-4198	19.0	2666	—	21.6	3039	—
XMR-2015	22.8	2965	32,300	25.3	3396	48,600
IMR-3031	22.0	2645	—	25.0+	3125	—
Benchmark	23.1	2932	37,100	**25.7**	**3251**	**49,400**
AA2230	22.9	2868	34,300	25.5	3299	48,800
IMR-4895	23.0	2538	—	26.5+	3086	—
H335	24.5	2950	34,200	27.2	3361	51,500
BL-C (2)	24.0	2564	—	27.0	2915	—
AA2460	23.9	2907	39,500	26.5	3228	50,200
748	25.2	2974	36,300	28.0	3318	49,200
IMR-4064	23.0	2512	—	26.0+	2941	—
Varget	25.6	3019	35,500	28.0+	3377	48,700
IMR-4320	24.0	2659	—	27.5+	3125	—

55 gr. Jacketed SPT
2.260" OAL
BC: .237
SD: .157

Powder	Sugg Starting Grains	Velocity fps	Pressure C.U.P.	Max Load Grains	Velocity fps	Pressure C.U.P.
IMR-4198	19.0	2645	—	21.7	3067	—
XMR-2015	22.5	2970	33,800	25.0	3353	50,800
IMR-3031	21.0	2506	—	24.5+	3076	—
Benchmark	22.7	2793	33,100	25.3	3137	48,000
AA2230	22.5	2891	37,000	25.0	3272	50,600
IMR-4895	23.0	2564	—	26.0+	3030	—
H335	24.3	3142	35,200	27.0	3270	49,100
BL-C (2)	23.0	2525	—	26.5	2949	—
AA2460	23.7	2853	36,700	26.3	3182	49,900
748	25.0	2849	33,500	27.8	3228	49,600
IMR-4064	23.0	2531	—	26.0+	2949	—
Varget	**25.0**	**2977**	**34,700**	27.8+	3346	51,400
IMR-4320	24.0	2672	—	27.5+	3144	—

60 gr. Jacketed V-Max
2.260" OAL
BC: .265
SD: .171

Powder	Sugg Starting Grains	Velocity fps	Pressure C.U.P.	Max Load Grains	Velocity fps	Pressure C.U.P.
XMR-2015	22.0	2882	35,500	**24.5**	**3217**	**52,000**
IMR-3031	21.4	2648	35,000	23.8	3047	48,300
Benchmark	22.0	2678	34,900	24.5	3018	48,200
AA2230	22.0	2749	37,300	24.5	3134	51,800
IMR-4895	22.9	2900	37,300	25.5	3023	49,500
H335	23.2	2728	34,700	25.8	3121	51,200
BL-C (2)	23.4	2673	33,200	26.0	3035	47,800
AA-2460	22.5	2678	38,700	25.0	2997	50,600
748	23.4	2608	32,200	26.0	3030	49,400
IMR-4064	23.8	2725	35,500	26.0+	3074	49,900
Varget	24.4	2854	36,300	27.2+	3170	49,300
IMR-4320	23.3	2663	36,300	25.9	2995	49,900
RX15	23.8	2783	36,800	26.5+	3098	47,300

Note: Loads shown in shaded panels are maximum.
Loads shown in bold designate potentially most accurate load.
+Designates a compressed powder charge.

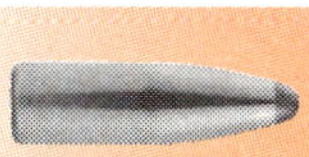

63 gr. Jacketed SMP

2.260" OAL

BC: .231
SD: .179

Powder	Sugg Starting Grains	Velocity fps	Pressure C.U.P.	Max Load Grains	Velocity fps	Pressure C.U.P.
IMR-4198	18.0	2487	—	20.0	2544	—
XMR-2015	22.0	2773	35,100	24.5	3088	50,500
IMR-3031	20.0	2403	—	22.5	2777	—
Benchmark	21.8	2648	34,400	24.3	2943	46,400
AA2230	**23.5**	**2926**	**37,300**	**26.1**	**3250**	**50,400**
IMR-4895	22.0	2444	—	25.0	2865	—
H335	23.7	2800	32,900	26.4	3139	48,700
BL-C (2)	23.0	2544	—	26.0	2840	—
AA2460	23.4	2739	34,700	26.0	3103	50,400
748	24.7	2768	32,600	27.5	3142	49,300
IMR-4064	22.0	2409	—	25.0+	2808	—
Varget	25.0	2891	36,000	27.0	3231	49,700
IMR-4320	22.0	2421	—	25.5	2881	—
RX15	24.7	2892	36,500	27.5	3210	51,200

69 gr. Jacketed HPBT

2.260" OAL

BC: .301
SD: .196

Powder	Sugg Starting Grains	Velocity fps	Pressure C.U.P.	Max Load Grains	Velocity fps	Pressure C.U.P.
XMR-2015	21.6	2677	36,600	24.0	2958	49,000
AA2230	22.5	2717	34,200	25.0	3063	48,400
IMR-4895	22.5	2614	38,400	25.0	2896	50,800
H335	22.9	2593	32,700	25.5	2994	51,400
BL-C (2)	23.4	2525	32,100	26.5	3018	50,700
AA-2460	22.5	2613	36,400	25.0	2928	49,300
748	23.6	2584	34,500	26.3	2994	51,100
IMR-4064	22.9	2603	37,100	25.5	2930	48,800
Varget	23.4	2654	33,700	**26.0+**	**2922**	**40,500**
IMR-4320	23.9	2682	37,200	26.5+	2946	47,100
N140	23.4	2650	34,300	26.0+	2926	45,800
RX15	23.9	2721	36,700	26.5+	2975	45,000

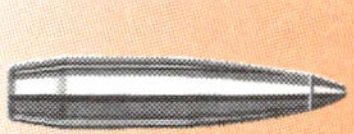

75 gr. Jacketed A-Max

2.390" OAL

BC: .435
SD: .214

Powder	Sugg Starting Grains	Velocity fps	Pressure C.U.P.	Max Load Grains	Velocity fps	Pressure C.U.P.
XMR-2015	20.3	2548	36,500	22.6	2797	49,100
AA-2230	20.7	2462	35,300	23.0	2758	48,300
IMR-4895	21.6	2390	33,800	24.0	2720	49,100
H335	22.5	2577	37,000	25.1	2850	52,000
BL-C (2)	23.0	2545	36,300	25.6	2845	48,800
AA-2460	21.6	2495	39,300	24.0	2719	49,200
748	23.0	2490	35,700	25.5	2810	50,600
IMR-4064	22.2	2458	37,500	24.7	2767	49,000
Varget	**23.0**	**2606**	**37,200**	25.6	2843	48,600
IMR-4320	22.1	2408	36,800	24.6	2724	50,000
N140	22.7	2557	38,700	25.2	2773	49,000
RX15	23.4	2652	39,300	26.1	2923	52,400

77 gr. Jacketed HPBT

2.260" OAL

BC: .362
SD: .219

Powder	Sugg Starting Grains	Velocity fps	Pressure C.U.P.	Max Load Grains	Velocity fps	Pressure C.U.P.
XMR-2015	20.2	2443	35,000	22.5+	2724	48,600
AA2230	21.7	2589	38,800	24.2	2880	49,600
IMR-4895	22.0	2469	35,400	24.5+	2771	46,100
H335	22.5	2551	36,500	25.0	2851	51,200
BL-C (2)	23.0	2473	31,600	24.7	2858	50,400
AA-2460	21.9	2470	36,100	24.3	2788	51,000
748	22.6	2417	31,900	24.1	2800	50,300
IMR-4064	22.5	2475	34,600	25.0+	2772	46,200
Varget	22.5	2591	39,000	25.0+	2788	44,500
IMR-4320	22.0	2497	34,600	24.5+	2742	46,900
N140	22.2	2463	35,800	24.7+	2758	48,500
RX15	**22.0**	**2544**	**36,900**	24.5+	2803	46,400

80 gr. Jacketed HPBT

2.550" OAL

BC: .420
SD: .228

Powder	Sugg Starting Grains	Velocity fps	Pressure C.U.P.	Max Load Grains	Velocity fps	Pressure C.U.P.
XMR-2015	19.8	2495	38,200	22.2	2769	51,100
AA-2230	21.5	2518	32,700	24.0	2849	47,500
IMR-4895	22.0	2467	33,200	24.5+	2789	47,500
H4895	21.8	2536	36,600	**23.2+**	**2757**	**43,700**
H335	22.5	2464	32,400	25.0	2834	48,600
BL-C (2)	23.4	2522	33,800	25.0	2860	48,600
AA-2460	21.7	2467	33,900	24.1	2744	48,500
748	21.8	2443	31,900	24.0	2799	50,100
IMR-4064	22.0	2451	33,700	24.0+	2833	49,500
Varget	23.0	2595	38,300	25.0+	2823	46,500
IMR-4320	22.5	2460	34,000	25.0+	2777	48,400
N140	22.0	2453	34,000	24.5+	2734	46,200

*69 gr. Jacketed HPBT

2.260" OAL

BC: .301
SD: .196

Powder	Sugg Starting Grains	Velocity fps	Pressure C.U.P.	Max Load Grains	Velocity fps	Pressure C.U.P.
XMR-2015	21.6	2596	—	24.0	2936	—
AA-2230	22.5	2629	—	25.0	2979	—
IMR-4895	22.5	2475	—	25.0	2786	—
H335	22.9	2498	—	25.5	2839	—
BL-C (2)	23.4	2442	—	26.5	2861	—
AA-2460	22.5	2511	—	25.0	2883	—
748	23.6	2468	—	26.3	2834	—
IMR-4064	22.4	2344	—	25.5	2743	—
Varget	23.4	2589	—	**26.0+**	**2832**	—
IMR-4320	23.9	2595	—	26.5+	2922	—
N140	23.4	2581	—	26.0+	2897	—
RX15	23.9	2632	—	26.5+	2933	—

Note: Loads shown in shaded panels are maximum.
Loads shown in bold designate potentially most accurate load.
+Designates a compressed powder charge.
*Fired for velocity in 20" barrel, 1-7" twist.

223 Remington

***77 gr. Jacketed HPBT** 2.260" OAL — **BC:** .362 **SD:** .219

Powder	Sugg Starting Grains	Velocity fps	Pressure C.U.P.	Max Load Grains	Velocity fps	Pressure C.U.P.
XMR-2015	20.2	2395	—	22.5	2680	—
AA-2230	21.7	2512	—	24.2	2827	—
IMR-4895	22.0	2441	—	24.5+	2721	—
H335	22.5	2458	—	25.0	2821	—
BL-C (2)	23.0	2342	—	24.7	2745	—
AA-2460	21.9	2433	—	24.3	2716	—
748	22.6	2348	—	24.1	2697	—
IMR-4064	22.5	2393	—	25.0+	2736	—
Varget	22.5	2460	—	25.0+	2705	—
IMR-4320	22.0	2334	—	24.5+	2663	—
N140	22.2	2422	—	24.7+	2696	—
RX15	**22.0**	**2472**	—	24.5+	2741	—

***80 gr. Jacketed HPBT** 2.550" OAL — **BC:** .420 **SD:** .228

Powder	Sugg Starting Grains	Velocity fps	Pressure C.U.P.	Max Load Grains	Velocity fps	Pressure C.U.P.
XMR-2015	19.8	2348	—	22.2	2742	—
AA2230	21.5	2548	—	24.0	2795	—
IMR-4895	22.0	2483	—	24.5+	2669	—
H4895	21.8	2430	—	**23.2+**	**2723**	—
H335	22.5	2471	—	25.0	2752	—
BL-C (2)	23.4	2417	—	25.0	2791	—
AA-2460	21.7	2383	—	24.1	2681	—
748	21.8	2332	—	24.0	2697	—
IMR-4064	22.0	2413	—	24.0+	2654	—
Varget	23.0	2499	—	25.0+	2761	—
N140	22.0	2348	—	24.5+	2644	—

#225415 55 gr. (#2 Alloy) 2.060" OAL — **BC:** .116 **SD:** .157

Powder	Sugg Starting Grains	Velocity fps	Pressure C.U.P.	Max Load Grains	Velocity fps	Pressure C.U.P.
Red Dot	5.6	1795	23,400	9.0	2280	40,700
700X	5.6	1790	23,400	7.7	2080	39,400
Green Dot	6.2	1855	23,400	9.2	2270	39,000
PB	6.3	1775	24,000	8.7	2120	39,400
Unique	6.8	1900	20,400	9.5	2300	35,000
SR-7625	6.6	1820	24,000	8.8	2120	39,900
SR-4759	9.3	1762	18,000	11.3	2195	24,200
XMP-5744	**11.0**	**1765**	**18,000**	14.5	2234	26,600
IMR-4227	11.0	1762	18,000	14.0	2193	23,200

#225646 55 gr. (#2 Alloy) 2.260" OAL — **BC:** .155 **SD:** .157

Powder	Sugg Starting Grains	Velocity fps	Pressure C.U.P.	Max Load Grains	Velocity fps	Pressure C.U.P.
Unique	6.2	1753	26,700	9.0	2209	41,400
SR-7625	6.7	1763	26,800	9.5	2205	43,900
AA9	7.8	1749	17,500	11.0	2198	28,900
2400	8.0	1753	18,000	11.5	2206	27,100
SR-4759	8.5	1751	17,500	11.5	2206	27,300
XMP-5744	10.3	1757	17,000	14.0	2186	27,000
IMR-4227	10.0	1766	17,000	13.0	2180	26,200
AA-1680	10.9	1756	16,000	14.3	2181	23,500
IMR-4198	10.7	1756	17,000	14.5	2223	25,200
RX7	**11.5**	**1758**	**17,000**	15.6	2226	25,300

Note: Loads shown in shaded panels are maximum.
Loads shown in bold designate potentially most accurate load.
*Fired for velocity in 20" barrel, 1–7" twist.

225 Winchester

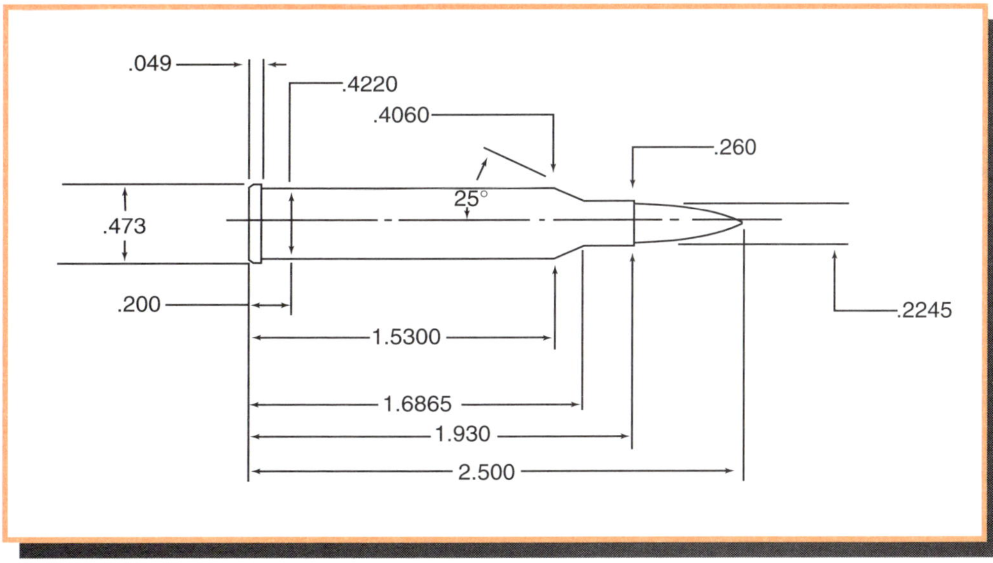

Comments:

Winchester announced their 225 Winchester in 1964 as a replacement for the 220 Swift. Remington's simultaneous introduction of the long popular 22-250 wildcat as a factory offering grabbed the shooting public's attention and the 225 never really took off. Ironically, the Swift has made a comeback while many Model 70s chambered in 225 have been re-barreled to other calibers. Although given up for dead as a commercial cartridge, the 225 Winchester has served as the parent case for several wildcats in recent years. The 225 is a semi-rimmed case but it headspaces on the shoulder. Those loading for this cartridge should be served well by bullets weighing between 50 and 55-grains with IMR-4064 or IMR-4320.

Test Components:

Cases . . . . .Winchester
Trim-to Length . . . . .1.920"
Primers . . . . .WLR
Primer Size . . . . .Large Rifle
Lyman Shell Holder . . . . .No. 5
Jacketed Bullets Used . . .Speer Spire SP #1017, 40 gr.
Sierra SPT #1310, 45 gr.
Sierra SPT #1340, 50 gr.
Speer HP #1035, 52 gr.
Hornady SP #2265, 55 gr.
Sierra SMP #1370, 63 gr.
Cast Bullets Used . . . .(sized to .224" dia)
*gas check bullets *#225415, 55 gr.

Test Specifications:
(Velocity Only)

Firearm Used . . . . .Winchester Model 70
Barrel Length . . . . .22"
Twist . . . . .1-14"
Groove Dia. . . . . .224"

40 gr. Jacketed SP
2.370" OAL
BC: .144
SD: .114

Powder	Sugg Starting Grains	Velocity fps	Pressure C.U.P.	Max Load Grains	Velocity fps	Pressure C.U.P.
IMR-4198	23.0	3086	—	26.0	3484	—
IMR-3031	26.0	3086	—	**29.0**	**3497**	—
IMR-4895	27.0	2849	—	30.0	3268	—
IMR-4064	31.0	3333	—	34.0	3846	—
IMR-4320	29.0	3067	—	31.0	3345	—
H-380	34.0	3356	—	38.0+	3786	—

45 gr. Jacketed SPT
2.390" OAL
BC: .210
SD: .128

Powder	Sugg Starting Grains	Velocity fps	Pressure C.U.P.	Max Load Grains	Velocity fps	Pressure C.U.P.
IMR-4198	22.0	2865	—	25.0	3215	—
IMR-3031	25.0	2849	—	28.0	3257	—
IMR-4895	26.0	2710	—	29.0	3144	—
IMR-4064	30.0	3164	—	33.5	3745	—
IMR-4320	28.0	2857	—	**31.0**	**3300**	—
H-380	33.0	3184	—	37.5+	3571	—

Note: Loads shown in shaded panels are maximum.
Loads shown in bold designate potentially most accurate load.
+Designates a compressed powder charge.

225 Winchester

50 gr. Jacketed SPT
2.400" OAL

BC: .222
SD: .142

Powder	Sugg Starting Grains	Velocity fps	Pressure C.U.P.	Max Load Grains	Velocity fps	Pressure C.U.P.
IMR-4198	21.0	2710	—	23.0	2941	—
IMR-3031	25.0	2865	—	28.0	3225	—
IMR-4895	26.0	2732	—	29.0	3125	—
IMR-4064	30.0	3174	—	33.0	3636	—
IMR-4320	27.0	2808	—	**30.0**	**3194**	—
H-380	33.0	3174	—	37.0+	3497	—

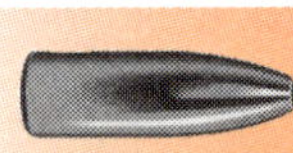

52 gr. Jacketed HP
2.500" OAL

BC: .225
SD: .147

Powder	Sugg Starting Grains	Velocity fps	Pressure C.U.P.	Max Load Grains	Velocity fps	Pressure C.U.P.
IMR-4198	20.0	2645	—	22.5	2906	—
IMR-3031	24.0	2793	—	27.0	3030	—
IMR-4895	25.0	2666	—	28.0	3012	—
IMR-4064	29.0	3067	—	32.5	3610	—
IMR-4320	26.0	2710	—	**29.0**	**3048**	—
H-380	32.0	3076	—	36.5+	3436	—

55 gr. Jacketed SP
2.420" OAL

BC: .235
SD: .157

Powder	Sugg Starting Grains	Velocity fps	Pressure C.U.P.	Max Load Grains	Velocity fps	Pressure C.U.P.
IMR-4198	20.0	2638	—	22.0	2857	—
IMR-3031	23.0	2652	—	26.0	2976	—
IMR-4895	25.0	2659	—	28.0	2976	—
IMR-4064	29.0	3048	—	**32.0**	**3534**	—
IMR-4320	25.0	2610	—	28.0	2906	—
H-380	32.0	3039	—	36.0+	3401	—

63 gr. Jacketed SMP
2.440" OAL

BC: .231
SD: .179

Powder	Sugg Starting Grains	Velocity fps	Pressure C.U.P.	Max Load Grains	Velocity fps	Pressure C.U.P.
IMR-4198	19.0	2487	—	21.0	2695	—
IMR-3031	22.0	2544	—	25.0	2849	—
IMR-4895	24.0	2557	—	27.0	2832	—
IMR-4064	28.0	2923	—	31.5	3378	—
IMR-4320	24.0	2518	—	**27.0**	**2816**	—
H-380	31.0	2915	—	35.0	3215	—

#225415
55 gr. (#2 Alloy) 2.235" OAL

BC: .116
SD: .157

Powder	Sugg Starting Grains	Velocity fps	Pressure C.U.P.	Max Load Grains	Velocity fps	Pressure C.U.P.
Red Dot	6.0	1770	—	9.0	2235	—
700X	**6.0**	**1775**	—	9.0	2185	—
Green Dot	6.5	1845	—	9.5	2280	—
PB	6.5	1760	—	9.5	2170	—
Unique	7.0	1880	—	10.0	2315	—
SR-7625	7.0	1825	—	10.0	2195	—

Note: Loads shown in shaded panels are maximum.
Loads shown in bold designate potentially most accurate load.
+Designates a compressed powder charge.

224 Weatherby Magnum

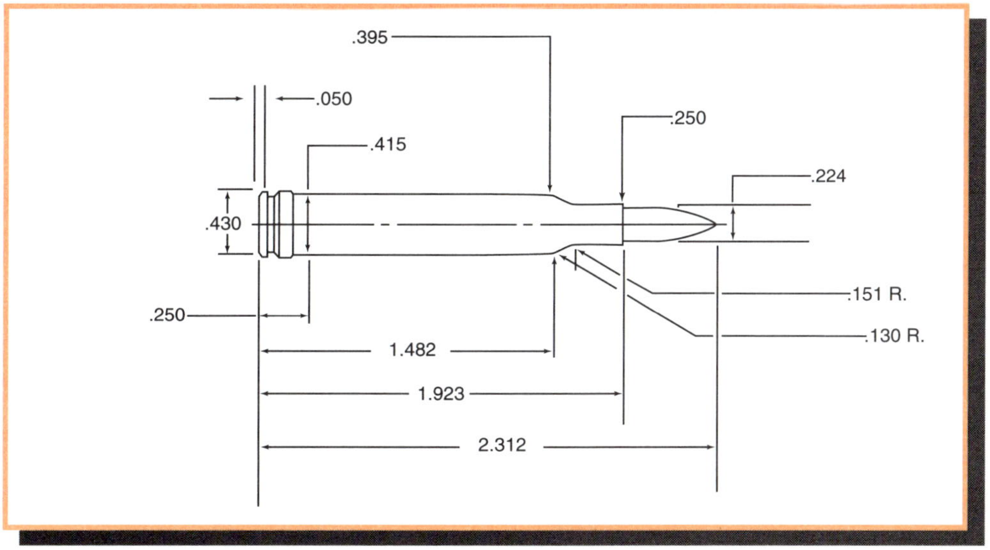

Comments:

This is a very accurate cartridge which has never gained a great deal of popularity. Perhaps this is due to the fact that only Weatherby rifles have generally been chambered for the round and only Weatherby supplies ammo or brass. Do not use magnum style primers with this case as dangerous pressures will result. Jacketed bullets of 50 to 60 grains are the best choices. Use only bullets designed for high velocities to prevent premature bullet break up. Our recommended powder choice is IMR-4198.

Test Components:

Cases . . . . .Weatherby
Trim-to Length . . . . .1.915"
Primers . . . . .Remington 9½
Primer Size . . . . .Large Rifle
Lyman Shell Holder . . . . .No. 3
Jacketed Bullets Used . . .Speer Spire SP #1017, 40 gr.
Sierra SPT #1310, 45 gr.
Sierra SPT #1340, 50 gr.
Speer HP #1035, 52 gr.
Hornady V-Max #22271, 55 gr.
Sierra SMP #1370, 63 gr.

Test Specifications: (Velocity Only)

Firearm Used . . . . .Weatherby Varmint Master
Barrel Length . . . . .26"
Twist . . . . .1-14"
Groove Dia. . . . . .224"

40 gr. Jacketed SP 2.310" OAL — **BC:** .144 **SD:** .114

Powder	Sugg Starting Grains	Velocity fps	Pressure C.U.P.	Max Load Grains	Velocity fps	Pressure C.U.P.
IMR-4198	**25.0**	**3521**	—	28.5	4098	—
IMR-3031	29.0	3597	—	32.5+	4149	—
IMR-4895	30.0	3497	—	33.5+	4000	—
IMR-4064	30.0	3448	—	33.0+	3875	—
IMR-4320	32.0	3663	—	35.0+	4115	—

45 gr. Jacketed SPT 2.310" OAL — **BC:** .210 **SD:** .128

Powder	Sugg Starting Grains	Velocity fps	Pressure C.U.P.	Max Load Grains	Velocity fps	Pressure C.U.P.
IMR-4198	**25.0**	**3448**	—	28.0	3891	—
IMR-3031	29.0	3584	—	32.0+	3984	—
IMR-4895	30.0	3546	—	33.0+	3875	—
IMR-4064	30.0	3460	—	33.0+	3816	—
IMR-4320	31.0	3509	—	34.5+	3891	—

Note: Loads shown in shaded panels are maximum.
Loads shown in bold designate potentially most accurate load.
+Designates a compressed powder charge.

224 Weatherby Magnum

50 gr. Jacketed SPT
2.310" OAL

BC: .222
SD: .142

Powder	Sugg Starting Grains	Velocity fps	Pressure C.U.P.	Max Load Grains	Velocity fps	Pressure C.U.P.
IMR-4198	24.0	3247	—	**27.5**	**3708**	—
IMR-3031	28.0	3356	—	31.5+	3831	—
IMR-4895	29.0	3322	—	32.5+	3745	—
IMR-4064	30.0	3484	—	33.0+	3802	—
IMR-4320	31.0	3460	—	34.0+	3802	—

52 gr. Jacketed HP
2.310" OAL

BC: .225
SD: .147

Powder	Sugg Starting Grains	Velocity fps	Pressure C.U.P.	Max Load Grains	Velocity fps	Pressure C.U.P.
IMR-4198	**24.0**	**3215**	—	27.0	3623	—
IMR-3031	28.0	3356	—	31.0+	3773	—
IMR-4895	29.0	3247	—	32.0+	3650	—
IMR-4064	30.0	3413	—	33.0+	3759	—
IMR-4320	30.0	3322	—	33.5+	3717	—

55 gr. Jacketed V-Max
2.350" OAL

BC: .255
SD: .157

Powder	Sugg Starting Grains	Velocity fps	Pressure C.U.P.	Max Load Grains	Velocity fps	Pressure C.U.P.
IMR-4198	23.0	3095	—	26.5	3509	—
IMR-3031	27.0	3225	—	30.5+	3677	—
IMR-4895	28.0	3164	—	31.5	3597	—
IMR-4064	29.0	3279	—	**32.8+**	**3717**	—
IMR-4320	30.0	3311	—	33.0+	3650	—

63 gr. Jacketed SMP
2.310" OAL

BC: .231
SD: .179

Powder	Sugg Starting Grains	Velocity fps	Pressure C.U.P.	Max Load Grains	Velocity fps	Pressure C.U.P.
IMR-4198	**23.0**	**2967**	—	26.0	3356	—
IMR-3031	26.0	3039	—	29.5	3472	—
IMR-4895	27.0	3003	—	30.5	3413	—
IMR-4064	29.0	3247	—	32.0+	3559	—
IMR-4320	29.0	3144	—	32.0+	3484	—

Note: Loads shown in shaded panels are maximum.
Loads shown in bold designate potentially most accurate load.
+Designates a compressed powder charge.

22-250 Remington

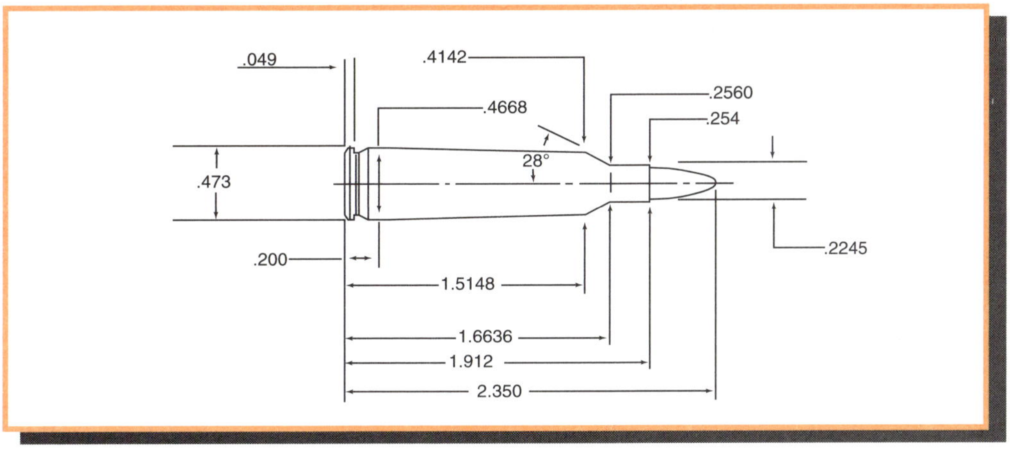

Comments:

This is one of the best choices for a long range 22 caliber varmint rifle. Accuracy is generally excellent and its flat trajectory makes hitting small targets easier - even up to 300 yards.

Jacketed bullets of 50 grains or heavier are best. First recommendation on powder would be Hodgdon H380. In fact, a charge of 38.0 grains with a 55 grain bullet has become legendary. It began when Bruce Hodgdon first tried it and found it so accurate that he named the powder after the 38.0 grain charge - H380.

The data listed are intended only for standard chambers as manufactured to current industry specifications. Early chambers marked 22-250 or Varminter may differ from current standards by a substantial amount, to the point that this data may prove unsafe in such firearms.

Cast bullet #225646 at about 2,100 fps. is the best choice for lead bullet use in this cartridge, as it is in several other 22 caliber cartridges.

Test Components:

Cases . . . . Winchester
Trim-to Length . . . . 1.902"
Primers . . . . WLR
Primer Size . . . . Large Rifle
Lyman Shell Holder . . . . No. 2
Jacketed Bullets Used . Hornady V-Max #22241, 40 gr.
Sierra SPT #1310, 45 gr.
Sierra Blitz #1340, 50 gr.
Hornady A-Max #22492, 52 gr.
Hornady V-Max #22271, 55 gr.
Hornady HP #2275, 60 gr.
Sierra SMP 1370, 63 gr.
Cast Bullets Used . . . . (sized to .224" dia)
*gas check bullets *#225415, 55 gr.
*#225646, 55 gr.

Test Specifications:
(Velocity & Pressure)

Firearm Used . . . . Universal Receiver
Barrel Length . . . . 24"
Twist . . . . 1-14"
Groove Dia. . . . . .224"

40 gr. Jacketed V-Max
2.350" OAL
BC: .200
SD: .114

Powder	Sugg Starting Grains	Velocity fps	Pressure C.U.P.	Max Load Grains	Velocity fps	Pressure C.U.P.
RX7	29.0	3709	39,400	33.0	4130	52,800
IMR-3031	32.5	3367	32,800	36.0	3982	50,500
AA-2230	33.3	3510	34,500	37.0	4106	50,900
IMR-4895	34.2	3402	35,500	37.0	3919	51,300
AA-2460	34.7	3496	35,700	38.0	3981	49,600
IMR-4064	34.7	3365	34,300	**38.5+**	**3920**	**49,900**
Varget	36.0	3690	38,600	40.0	4130	52,000
IMR-4320	34.0	3566	41,600	38.5	4003	53,200
N140	35.6	3552	37,600	39.0	3953	49,900
RX15	34.5	3493	36,000	38.0+	4017	50,800
H380	38.2	3394	32,300	42.0+	3770	51,900
IMR-4350	36.0	2992	28,800	40.0+	3458	39,000

45 gr. Jacketed SPT
2.310" OAL
BC: .210
SD: .128

Powder	Sugg Starting Grains	Velocity fps	Pressure C.U.P.	Max Load Grains	Velocity fps	Pressure C.U.P.
IMR-3031	31.0	3545	40,300	35.5	3965	50,800
AA-2230	33.3	3512	35,800	36.5	3973	49,100
IMR-4895	32.5	3353	36,400	36.5	3901	52,000
AA-2460	34.0	3486	37,400	37.2	3899	51,500
IMR-4064	33.0	3373	34,500	37.5+	3945	52,400
Varget	36.0	3538	36,400	39.5+	3982	51,300
IMR-4320	33.0	3390	38,600	37.5	3836	50,000
N140	35.0	3424	35,500	38.5	3887	51,800
RX15	33.6	3465	35,300	**37.0**	**3947**	**51,500**
H380	37.0	3741	41,100	41.0	4055	51,600
IMR-4350	36.0	3239	33,500	40.0+	3613	43,300
**SR-4759	15.0	2476	24,700	19.5	3052	33,700
**XMP-5744	18.0	2436	23,200	23.5	3007	30,900

Note: Loads shown in shaded panels are maximum.
Loads shown in bold designate potentially most accurate load.

+Designates a compressed powder charge.
**Designates a reduced load.

22-250 Remington

50 gr. Jacketed Blitz 2.310" OAL — **BC:** .222 **SD:** .142

Powder	Sugg Starting Grains	Velocity fps	Pressure C.U.P.	Max Load Grains	Velocity fps	Pressure C.U.P.
RX7	28.0	3478	41,100	31.5	3779	52,400
XMR-2015	29.7	3229	37,000	34.0	3716	50,600
IMR-3031	30.0	3181	33,000	35.3	3861	52,400
AA2230	31.8	3290	40,500	35.4	3626	47,000
IMR-4895	**33.5**	**3230**	**34,800**	37.0+	3704	50,900
AA-2460	31.4	3154	37,300	36.2	3693	52,400
748	34.3	3391	36,800	39.0	3811	50,700
IMR-4064	32.0	3210	32,000	37.0+	3854	52,400
AA-2520	32.5	3319	39,100	37.0	3724	51,200
IMR-4320	33.0	3408	40,700	37.0	3742	51,200
H-380	36.0	3569	39,400	40.0	3912	52,000
N150	33.6	3433	37,400	37.5+	3814	49,800
H414	37.8	3200	36,500	41.5+	3693	50,100
IMR-4350	36.0	3182	34,000	40.0+	3591	45,000

52 gr. Jacketed A-Max 2.350" OAL — **BC:** .247 **SD:** .148

Powder	Sugg Starting Grains	Velocity fps	Pressure C.U.P.	Max Load Grains	Velocity fps	Pressure C.U.P.
RX7	27.0	3333	39,900	31.2	3561	52,200
XMR-2015	28.9	3151	36,900	33.8	3631	52,000
IMR-3031	31.7	3206	34,000	35.2	3692	50,800
AA-2230	31.0	3330	35,500	34.0	3789	50,100
IMR-4895	32.6	3157	34,000	36.2	3604	50,000
AA-2460	32.4	3285	37,200	35.0	3648	50,900
IMR-4064	32.3	3123	33,500	36.0	3657	51,200
Varget	33.5	3367	37,400	**38.0**	**3764**	**50,500**
IMR-4320	32.0	3153	36,400	36.7	3645	51,200
RX15	32.3	3239	35,200	36.0	3666	48,900
H380	36.9	3218	34,200	41.0+	3584	43,800
N150	33.0	3414	40,900	37.0	3794	52,400
760	37.8	3267	37,300	41.0+	3701	51,300
IMR-4350	36.0	3024	32,600	40.0+	3396	45,000

55 gr. Jacketed V-Max 2.345" OAL — **BC:** .255 **SD:** .157

Powder	Sugg Starting Grains	Velocity fps	Pressure C.U.P.	Max Load Grains	Velocity fps	Pressure C.U.P.
XMR-2015	29.0	3131	38,300	33.9	3602	52,500
AA2230	31.1	3096	37,400	35.0	3536	51,000
IMR-4895	32.0	3152	34,500	35.5	3576	50,200
AA2460	30.0	3298	39,800	34.8	3588	49,500
IMR-4064	32.0	3221	36,200	35.5	3634	51,500
Varget	34.0	3325	39,100	37.0	3696	51,600
AA-2520	32.0	3276	41,900	36.0	3613	51,800
IMR-4320	32.0	3163	36,400	36.5	3630	52,000
RX15	31.7	3306	38,500	35.5	3694	51,400
H380	36.9	3213	36,900	41.0+	3553	47,000
N150	32.7	3343	42,400	**36.5**	**3663**	**51,900**
IMR-4350	36.0	3021	33,500	40.0+	3368	43,900
**SR-4759	16.0	2472	28,900	20.0	2933	42,700
**XMP-5744	18.5	2406	25,000	24.5	3034	36,800

60 gr. Jacketed HP 2.350" OAL — **BC:** .271 **SD:** .171

Powder	Sugg Starting Grains	Velocity fps	Pressure C.U.P.	Max Load Grains	Velocity fps	Pressure C.U.P.
XMR-2015	27.2	2936	37,100	32.4	3460	53,000
AA2230	30.1	2973	37,900	34.8	3460	52,400
IMR-4895	33.0	3141	36,300	36.0+	3581	51,500
AA2460	28.9	2859	34,600	35.0	3467	52,100
IMR-4064	31.0	3128	36,400	35.5	3589	52,000
Varget	32.5	3165	37,700	36.0	3493	48,800
AA2520	29.8	2882	35,700	34.7	3392	52,300
IMR-4320	31.0	3064	37,300	36.0+	3503	51,600
RX15	31.8	3242	39,600	35.0	3610	51,100
H-380	34.5	3362	44,200	38.0	3622	52,400
N150	32.0	3186	40,400	**35.5**	**3507**	**50,800**
H-414	35.4	3136	38,400	38.5+	3540	51,700
IMR-4350	36.0	3136	38,600	40.0+	3497	50,400
RX19	36.9	2922	34,000	41.0+	3268	44,600

63 gr. Jacketed SMP 2.350" OAL — **BC:** .231 **SD:** .179

Powder	Sugg Starting Grains	Velocity fps	Pressure C.U.P.	Max Load Grains	Velocity fps	Pressure C.U.P.
XMR-2015	26.8	2814	37,600	32.0	3311	52,000
IMR-4895	30.0	3056	36,900	34.5	3479	52,400
AA2460	27.7	2729	35,300	33.8	3309	51,800
IMR-4064	31.0	3090	36,400	35.0	3490	50,800
Varget	32.6	3139	40,500	36.0	3447	51,000
IMR-4320	31.0	3011	36,900	35.8	3480	52,800
RX15	30.5	3162	40,100	34.5	3508	51,400
H380	**34.0**	**3268**	**43,700**	37.5	3554	52,800
N150	31.4	3096	42,800	35.0	3380	51,400
H414	35.0	3128	38,500	38.0+	3510	52,100
IMR-4350	36.0	3149	40,700	40.0+	3464	51,200
RX19	37.4	2999	36,300	41.0+	3350	47,300

#225415 55 gr. (#2 Alloy) 2.325" OAL — **BC:** .116 **SD:** .157

Powder	Sugg Starting Grains	Velocity fps	Pressure C.U.P.	Max Load Grains	Velocity fps	Pressure C.U.P.
Red Dot	7.2	1981	22,800	9.5	2323	35,000
700X	7.0	1850	18,000	9.0	2268	33,500
Green Dot	7.5	1992	21,000	10.0	2356	34,500
PB	7.5	1920	18,000	9.5	2188	31,000
SR-7625	8.0	1981	20,400	10.0	2237	36,900
SR-4756	8.0	1915	18,000	11.0	2308	32,000
SR-4759	14.5	2176	13,000	19.0	2820	26,900
XMP-5744	**15.3**	**2032**	13,000	17.0	2219	20,500
IMR-4198	16.9	2173	12,100	22.1	2867	23,900
748	22.3	2175	13,600	28.5	2814	21,500

Note: Loads shown in shaded panels are maximum.
Loads shown in bold designate potentially most accurate load.

+Designates a compressed powder charge.
**Designates a reduced load.

22-250 Remington

#225646
55 gr. (#2 Alloy) 2.350" OAL

BC: .155
SD: .157

Powder	Sugg Starting Grains	Velocity fps	Pressure C.U.P.	Max Load Grains	Velocity fps	Pressure C.U.P.
Red Dot	7.7	1797	22,100	10.4	2165	33,900
700X	8.0	1887	24,500	11.0	2286	40,200
Green Dot	8.2	1862	23,200	11.5	2292	39,700
PB	9.0	1892	27,300	11.0	2128	38,900
SR-7625	9.2	1905	26,700	11.7	2219	42,700
SR-4756	10.2	1986	26,700	12.6	2271	38,100
SR-4759	16.5	2218	22,000	18.5	2543	35,200
IMR-4227	16.5	2274	23,500	21.0	2754	36,200
XMP-5744	16.2	2065	18,000	**18.0**	**2243**	**22,400**
IMR-4198	18.5	2172	18,400	24.6	2913	37,100
748	28.5	2380	19,000	34.0	3100	38,000

Note: Loads shown in shaded panels are maximum.
Loads shown in bold designate potentially most accurate load.

220 Swift

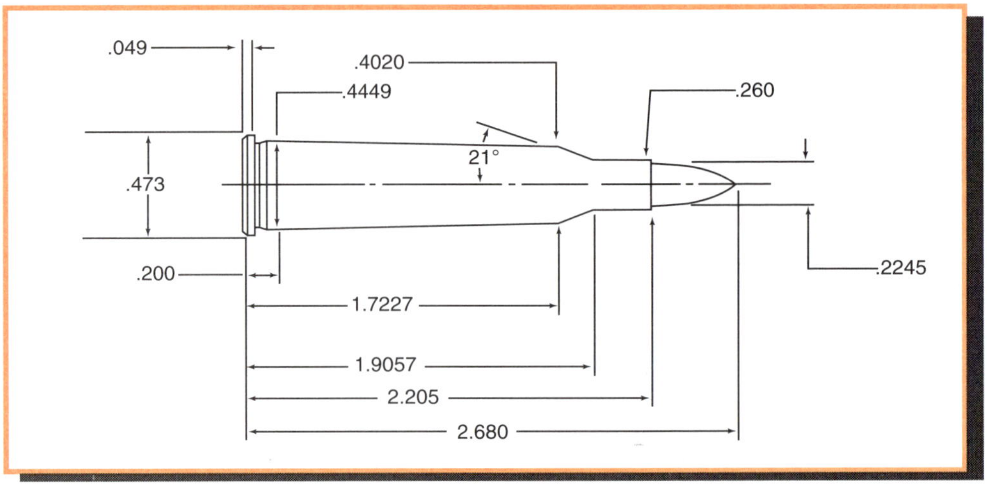

Comments:

Winchester introduced the 220 Swift in 1935 as a high performance varmint cartridge. Nearly 70 years later, it is still the fastest factory offering among the 22 caliber center-fire cartridges. The Swift is based on the old 6mm Lee Navy necked down to accept .224" bullets. Controversy has followed the Swift throughout its history. Most of this controversy has focused around barrel life. Technology at the time of its introduction handicapped the full potential of the Swift. However, latter-day developments of better quality steel and slower burning powders combined with sensible handloading have improved the life expectancy of a Swift barrel. Shooters should not overheat the barrel with sustained fire and avoid fouling by frequent cleaning—preferably with a coated rod and a good bore-guide. Some shooters opt to keep loads below maximum velocities to prolong barrel life. Handloaders seeking 4,000 feet per second with 50-grain bullets are not doing their Swift any favors. Winchester discontinued the Swift in favor of a new cartridge, the 225 Winchester. While the 225 only lasted for several years before fading away, the 220 Swift remains fairly popular with varmint hunters. Remington and Ruger both currently offer rifles chambered for the Swift. Several different VihtaVuori powders produced uniform results during our testing but IMR-4064 has been a favorite among Swift shooters for many years.

Test Components:

Cases . . . . Winchester
Trim-to Length . . . . 2.195"
Primers . . . . Remington 9½, Winchester WLR, CCI 200
Primer Size . . . . Large Rifle
Lyman Shell Holder . . . . No. 5
Jacketed Bullets Used . Speer Spire Point #1017, 40 gr.
Sierra SPT #1310, 45 gr.
Sierra SP Blitz #1340, 50 gr.
Sierra SPT #1360, 55 gr.
Hornady V-Max #22281, 60 gr.
Sierra SMP #1370, 63 gr.
Speer SP #1053, 70 gr.
Cast Bullets Used . . . . (sized to .224" dia)
*gas check bullet *#225415, 55gr.

Test Specifications: (Velocity & Pressure)

Firearm Used . . . . Winchester Model 70
Universal Receiver
Barrel Length . . . . Winchester, 26"
Universal Receiver, 24"
Twist . . . . 1-14"
Groove Dia. . . . . .224"

****40 gr. Jacketed SP** 2.650" OAL — **BC:** .144 **SD:** .114

Powder	Sugg Starting Grains	Velocity fps	Pressure C.U.P.	Max Load Grains	Velocity fps	Pressure C.U.P.
IMR-3031	**35.0**	**3906**	—	39.0	4385	—
IMR-4064	37.0	3891	—	41.0+	4347	—
IMR-4895	35.0	3663	—	39.0	4166	—
IMR-4320	37.0	3861	—	41.0	4291	—
H-380	40.0	3831	—	44.0+	4166	—
IMR-4350	39.0	3546	—	43.0+	3968	—

45 gr. Jacketed SPT 2.660" OAL — **BC:** .210 **SD:** .128

Powder	Sugg Starting Grains	Velocity fps	Pressure C.U.P.	Max Load Grains	Velocity fps	Pressure C.U.P.
**IMR-3031	34.0	3717	—	38.5	4219	—
**IMR-4895	34.0	3559	—	38.5	4065	—
**IMR-4064	36.0	3626	—	40.5+	4184	—
**IMR-4320	36.0	3663	—	40.5	4166	—
***N140	36.0	3638	42,000	40.0	4042	51,900
***RX15	36.5	3708	43,300	40.5	4121	52,200
**H-380	39.0	3663	—	43.0	3984	—
*****N150**	**36.5**	**3689**	**43,100**	40.5	4044	53,200
***AA2700	40.5	3681	44,500	45.0	3974	53,900
**IMR-4350	39.0	3521	—	43.0+	3861	—

Note: Loads shown in shaded panels are maximum.
Loads shown in bold designate potentially most accurate load.
+ Designates a compressed powder charge.
** Tested in a Winchester Model 70 with Remington 9½ primers.
*** Tested in a Universal Receiver with Remington 9½ primers.

220 Swift

50 gr. Jacketed Blitz
2.670" OAL

BC: .222 **SD:** .142

Powder	Sugg Starting Grains	Velocity fps	Pressure C.U.P.	Max Load Grains	Velocity fps	Pressure C.U.P.
**IMR-3031	34.0	3650	—	38.0	4065	—
**IMR-4895	34.0	3559	—	38.0	3952	—
**IMR-4064	36.0	3650	—	40.0	4081	—
**IMR-4320	36.0	3597	—	40.0	4000	—
***N140	35.0	3537	44,400	39.0	3878	53,500
**H-380	38.0	3571	—	42.5	3906	—
*****N150**	**34.5**	**3519**	**42,900**	37.5	3839	53,300
***AA2700	39.5	3519	44,200	44.0	3865	53,500
***760	38.7	3107	33,900	44.0	3934	53,000
**IMR-4350	39.0	3509	—	43.0+	3921	—
***RX19	40.5	3205	39,000	45.0+	3639	48,700

55 gr. Jacketed SPT
2.675" OAL

BC: .237 **SD:** .157

Powder	Sugg Starting Grains	Velocity fps	Pressure C.U.P.	Max Load Grains	Velocity fps	Pressure C.U.P.
**IMR-3031	33.0	3509	—	37.0	3921	—
**IMR-4895	33.0	3427	—	37.0	3787	—
**IMR-4064	35.0	3472	—	39.0	3906	—
**IMR-4320	35.0	3460	—	39.0	3861	—
**H-380	38.0	3497	—	42.0	3773	—
***N150	31.5	3353	44,100	35.0	3597	51,800
***AA2700	39.2	3457	44,900	43.5	3729	53,400
***760	38.0	3382	42,800	42.0	3683	52,700
*****N160**	39.5	3388	42,500	**44.0**	**3764**	**51,100**
**IMR-4350	38.0	3356	—	42.5+	3787	—
***RX19	40.0	3334	43,800	44.5+	3643	53,300

***60 gr. Jacketed V-Max
2.680" OAL

BC: .265 **SD:** .171

Powder	Sugg Starting Grains	Velocity fps	Pressure C.U.P.	Max Load Grains	Velocity fps	Pressure C.U.P.
IMR-4064	33.3	3296	45,800	37.0	3572	53,600
Varget	32.5	3288	45,700	36.0	3507	53,800
IMR-4320	33.0	3227	43,800	37.5	3550	53,000
H-380	36.0	3173	45,500	40.0	3389	53,800
AA2700	37.8	3314	46,500	42.0	3466	53,000
760	37.5	3279	43,200	41.5	3585	52,700
N160	37.8	3291	44,100	**42.0**	**3568**	**52,300**
IMR-4350	37.8	3236	43,500	42.0+	3559	53,100
RX19	40.0	3255	45,800	44.5+	3540	54,000
XMR-3100	39.5	3163	43,300	44.0+	3497	53,800

63 gr. Jacketed SMP
2.680" OAL

BC: .231 **SD:** .179

Powder	Sugg Starting Grains	Velocity fps	Pressure C.U.P.	Max Load Grains	Velocity fps	Pressure C.U.P.
**IMR-4895	30.0	3067	—	33.5	3378	—
**IMR-4064	31.0	3144	—	35.0	3472	—
**IMR-4320	31.0	3105	—	35.0	3436	—
**H-380	37.0	3356	—	41.0	3623	—
***AA2700	36.2	3264	48,300	40.2	3417	53,000
***760	36.0	3170	41,500	41.0	3488	52,300
****IMR-4350**	**37.0**	**3322**	—	40.0	3584	—
***RX19	39.5	3241	46,200	44.0+	3530	54,000

***70 gr. Jacketed SP
2.600" OAL

BC: .214 **SD:** .199

Powder	Sugg Starting Grains	Velocity fps	Pressure C.U.P.	Max Load Grains	Velocity fps	Pressure C.U.P.
IMR-4895	29.5	2987	45,000	33.0	3284	53,000
IMR-4064	29.5	3025	47,700	33.5	3282	53,000
IMR-4320	29.9	2937	43,800	34.0	3251	52,500
AA2700	33.3	2880	42,200	37.0	3198	53,000
760	34.3	3002	44,300	39.0	3324	52,900
IMR-4350	**34.3**	**3112**	**44,000**	39.0	3399	53,000
XMR-3100	37.8	3046	46,300	42.0+	3260	54,000
RX22	38.7	3143	49,200	43.0+	3327	53,500

#225415
55 gr. (#2 Alloy) 2.546" OAL

BC: .116 **SD:** .156

Powder	Sugg Starting Grains	Velocity fps	Pressure C.U.P.	Max Load Grains	Velocity fps	Pressure C.U.P.
+++Red Dot	6.5	1730	—	10.0	2270	—
+++700X	6.5	1710	—	10.0	2275	—
+++Green Dot	7.0	1800	—	10.5	2305	—
+++PB	7.0	1680	—	10.5	2210	—
+++Unique	7.5	1805	—	11.0	2320	—
+++SR-7625	7.5	1765	—	11.0	2280	—
++2400	10.5	1772	16,500	17.4	2723	31,500
++SR-4759	**12.2**	**1748**	**15,500**	18.5	2683	30,300

Note: Loads shown in shaded panels are maximum.
Loads shown in bold designate potentially most accurate load.
\+ Designates a compressed powder charge.
** Tested in a Winchester Model 70 with Remington 9½ primers.
*** Tested in a Universal Receiver with Remington 9½ primers.
+++Tested in a Winchester Model 70 with Win. WLR primers.
++Tested in a Universal Receiver with CCI 200 primers.

6mm PPC

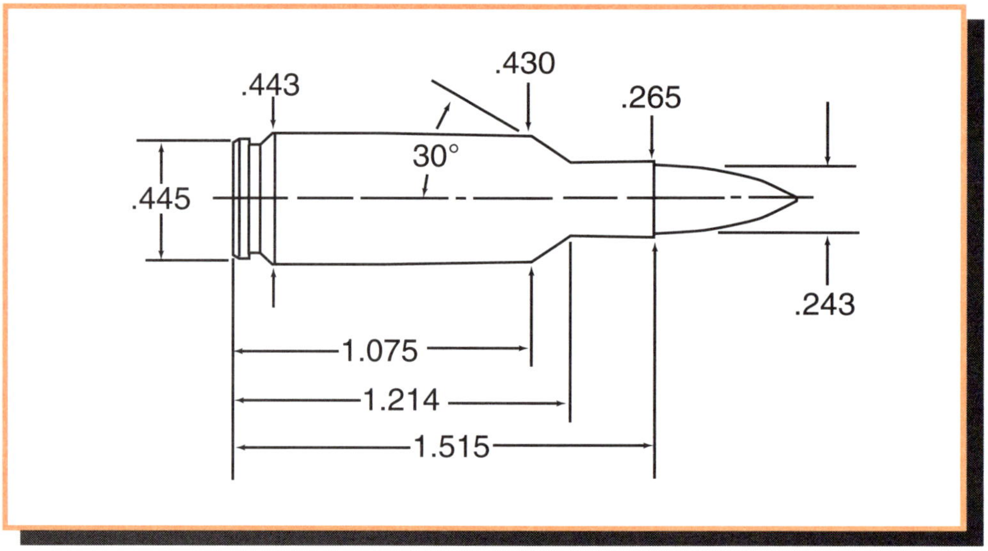

Comments:

This cartridge has proven itself the most accurate round ever developed. Its originators, Ferris Pindell and Dr. Louis Palmisano were the first of many bench rest shooters to so prove. Today the "Six" holds more records than any other cartridge.

A group of 10 shots at 100 yards going into a 1/4" center to center group would not get the attention of a serious bench rest shooter with 6mm PPC experience. The cartridge will do better. Naturally such accuracy demands the very best in bullets, loading technique, and the appropriate charge of H322 or perhaps H335. Still, we have fired a production Sako single shot rifle with factory Sako ammunition and obtained frequent 1/4" to 1/2" groups. Nothing else has ever come close to this kind of out-of-the-box performance.

The twist rate of most 6mm PPC's works best with bullets of 68 to 75 grains. Heavier bullets will not always stabilize and poor results may occur.

Test Components:

Cases . . . . .SAKO
Trim-to Length . . . . .1.505"
Primers . . . . .Remington 7½
Primer Size . . . . .Small Rifle
Lyman Shell Holder . . . . .No. 3
Jacketed Bullets Used . . . . . . .Sierra HP #1500, 60 gr.
Sierra HPBT #1505, 70 gr.
Hornady HP #2420, 75 gr.
Sierra HPBT #1530, 85 gr.

Test Specifications:
(Velocity & Pressure)

Firearm Used . . . . .Universal Receiver
Barrel Length . . . . .22"
Twist . . . . .1-14"
Groove Dia. . . . . .2435"

60 gr. Jacketed HP
2.050" OAL
BC: .182
SD: .145

Powder	Sugg Starting Grains	Velocity fps	Pressure C.U.P.	Max Load Grains	Velocity fps	Pressure C.U.P.
N130	21.1	2634	29,500	26.0	3168	47,900
IMR-4198	21.0	2734	32,200	23.5	3076	46,000
XMR-2015	25.0	2961	34,400	27.5	3272	45,300
H-322	24.8	2846	33,000	27.5	3200	47,200
AA2230	26.2	2876	35,400	30.0+	3212	47,900
H-335	26.5	2874	34,500	29.5	3208	48,500
BL-C(2)	28.0	2803	34,400	31.2+	3092	46,400
AA2460	26.5	2846	34,600	29.0	3133	46,600
748	27.6	2819	32,800	31.7	3204	48,900
Varget	**25.4**	**2672**	**29,900**	30.0+	3144	42,900

70 gr. Jacketed HPBT
2.075" OAL
BC: .259
SD: .169

Powder	Sugg Starting Grains	Velocity fps	Pressure C.U.P.	Max Load Grains	Velocity fps	Pressure C.U.P.
N130	22.9	2753	35,900	25.5	3016	48,600
IMR-4198	21.0	2611	32,400	23.5+	2962	48,700
XMR-2015	24.5	2801	35,000	27.0	3138	48,900
H-322	24.5	2742	34,200	26.5	2999	45,500
AA-2230	26.0	2732	34,800	29.0	3050	48,200
H-335	**26.1**	**2674**	**33,800**	**29.0**	**2992**	**48,200**
BL-C(2)	27.5	2675	35,000	30.5	2962	47,400
AA2460	26.0	2705	35,200	28.5	2980	46,300
748	27.5	2697	34,000	31.0+	3032	47,700
Varget	25.3	2561	30,400	29.0+	2986	43,500

Note: Loads shown in shaded panels are maximum.
Loads shown in bold designate potentially most accurate load.
\+ Designates a compressed powder charge.

6mm PPC

75 gr. Jacketed HP
2.120" OAL

BC: .294
SD: .181

Powder	Sugg Starting Grains	Velocity fps	Pressure C.U.P.	Max Load Grains	Velocity fps	Pressure C.U.P.
N130	22.7	2639	34,500	25.2	2900	45,800
IMR-4198	20.5	2555	34,000	23.0+	2835	46,800
XMR-2015	23.2	2629	33,800	26.0	2990	48,300
H-322	**23.8**	**2645**	**35,400**	26.0	2930	47,200
AA2230	25.0	2620	33,800	28.0	2932	47,100
H-335	26.3	2657	34,900	29.3	2978	48,500
BL-C(2)	26.2	2533	32,400	29.5	2880	47,400
AA2460	25.3	2570	33,200	28.5	2930	48,100
748	26.8	2649	33,600	30.2+	2976	48,100
Varget	26.5	2587	31,900	29.5+	2974	44,500

85 gr. Jacketed HPBT
2.035" OAL

BC: .282
SD: .206

Powder	Sugg Starting Grains	Velocity fps	Pressure C.U.P.	Max Load Grains	Velocity fps	Pressure C.U.P.
N130	21.8	2484	35,300	**24.2**	**2716**	**47,300**
IMR-4198	19.5	2366	35,000	21.5+	2622	47,300
XMR-2015	22.0	2483	34,400	24.5	2776	46,400
H-322	22.5	2465	35,500	24.5	2702	45,800
AA230	24.3	2473	34,000	27.0	2739	46,700
H-335	25.5	2494	34,000	28.4	2809	49,000
BL-C(2)	25.2	2429	34,400	28.5	2729	47,300
AA2460	24.0	2440	34,600	27.2	2757	48,900
748	25.8	2500	34,400	29.0+	2790	48,000
Varget	25.6	2560	36,100	28.5+	2809	45,200

Note: Loads shown in shaded panels are maximum.
Loads shown in bold designate potentially most accurate load.
\+ Designates a compressed powder charge.

243 Winchester

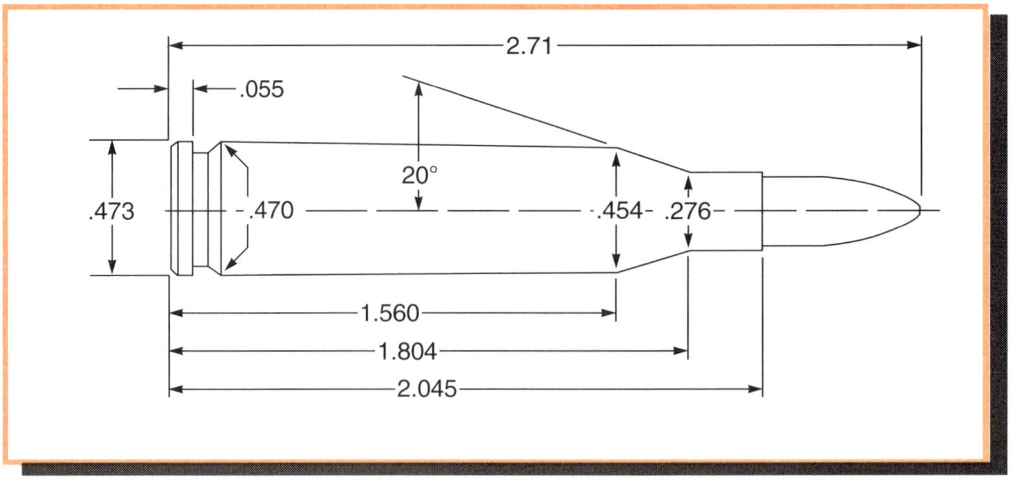

Comments:

The 243 Winchester is one of the first cartridges to evolve from the 308 Winchester /7.62 NATO. The late Warren Page is the gunwriter most closely associated with the 243 as a result of his experiments with necking down the 308 during the early 1950s. Winchester's 243 immediately proved suitable not only for smaller varmints but also for deer sized game due to its 1 in 10" twist rate. The 1-12" twist of Remington's concurrently introduced 244 Remington limited it to lighter weight bullets generally unsuitable for deer. Proper shot placement is crucial as the 243/6mm bore size is the minimum allowed for pursuing whitetails in many states. The fairly wide range of .243" diameter bullets available make this a versatile cartridge for the handloader. It is a good all-around cartridge for the more recoil-sensitive shooter. The 243 has also found appeal in recent years with competitors on the High Power Rifle circuit. Its "overbored" nature require use of slower powders normally associated with larger magnum rifles. IMR-4350 and Reloder 22 often produce best results, especially with bullets over 90 grains. Shooters using cast bullets should keep velocities around 2,000 feet per second or below.

Test Components:

Cases . . . . .Remington
Trim-to Length . . . . .2.035"
Primers . . . . .Remington 9½
Primer Size . . . . .Large Rifle
Lyman Shell Holder . . . . .No. 2
Jacketed Bullets Used .Hornady V-Max #22411, 58 gr.
Sierra HP #1500, 60 gr.
Hornady V-Max #22415, 65 gr.
Sierra HPBT #1505, 70 gr.
Speer HP #1205, 75 gr.
Speer HP #1211, 80 gr.
Sierra HPBT #1530, 85 gr.
Sierra FMJBT #1535, 90 gr.
Hornady, BTSP #2453, 100 gr.
Hornady A-Max #24562, 105 gr.
Cast Bullets Used . . . .(sized to .243" dia)
*gas check bullet *245496, 84 gr

Test Specifications:
(Velocity & Pressure)

Firearm Used . . . . .Universal Receiver
Barrel Length . . . . .24"
Twist . . . . .1-10"
Groove Dia. . . . .243"

58 gr. Jacketed V-Max
2.625" OAL
BC: .250
SD: .140

Powder	Sugg Starting Grains	Velocity fps	Pressure P.S.I.	Max Load Grains	Velocity fps	Pressure P.S.I.
RX7	33.3	3404	45,900	37.0	3701	58,600
IMR-3031	38.7	3482	42,000	43.0	3884	56,900
IMR-4895	40.5	3490	45,000	45.0	3877	59,700
Varget	40.9	3526	44,400	45.5	3891	58,800
AA2520	38.7	3423	43,300	43.0	3747	55,500
IMR-4320	40.9	3531	46,800	45.5	3876	59,800
H-380	44.6	3437	42,800	49.6	3850	58,300
H-414	45.0	3488	44,530	50.0	3824	56,900
IMR-4350	**45.0**	**3331**	**41,200**	50.0+	3776	57,000
H-4831	47.7	3267	38,100	53.0+	3662	52,600

60 gr. Jacketed HP
2.600" OAL
BC: .182
SD: .145

Powder	Sugg Starting Grains	Velocity fps	Pressure C.U.P.	Max Load Grains	Velocity fps	Pressure C.U.P.
IMR-3031	36.0	3365	36,400	41.0	3773	49,600
BL-C(2)	37.0	3323	37,300	41.5	3731	50,000
IMR-4064	38.0	3409	38,100	42.5	3774	49,200
IMR-4895	38.0	3488	43,300	42.0	3772	52,000
IMR-4320	39.0	3385	41,100	43.5	3707	50,000
H-380	40.0	3448	42,000	45.0	3778	51,200
IMR-4350	42.0	3291	37,300	47.0+	3726	51,600
H-4831	43.0	3057	34,000	49.0+	3596	48,400

Note: Loads shown in shaded panels are maximum.
Loads shown in bold designate potentially most accurate load.
\+ Designates a compressed powder charge.

243 Winchester

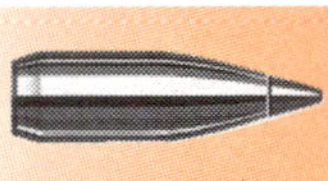

65 gr. Jacketed V-Max
2.625" OAL

BC: .280
SD: .157

Powder	Sugg Starting Grains	Velocity fps	Pressure P.S.I.	Max Load Grains	Velocity fps	Pressure P.S.I.
RX7	30.7	3175	45,800	34.2	3445	58,000
IMR-3031	36.4	3360	45,900	40.5	3636	57,000
IMR-4895	38.7	3328	45,800	43.0	3645	58,000
Varget	38.2	3342	45,100	42.0	3644	57,100
AA2520	36.2	3329	48,000	40.5	3603	58,800
IMR-4064	39.6	3369	45,900	44.0	3696	58,400
IMR-4320	39.6	3363	47,600	44.0	3676	59,400
H-380	43.2	3346	46,500	48.0	3670	59,700
760	43.7	3306	42,800	48.5	3685	58,300
H414	44.0	3294	43,300	49.0	3664	57,600
IMR-4350	**43.2**	**3209**	**42,700**	48.0	3602	57,400
H4831 SC	47.2	3260	44,100	52.5+	3607	58,500

70 gr. Jacketed HPBT
2.625" OAL

BC: .259
SD: .169

Powder	Sugg Starting Grains	Velocity fps	Pressure	Max Load Grains	Velocity fps	Pressure
XMR-2015	31.0	3081	36,800	37.0	3448	51,900 C
IMR-3031	35.0	3281	41,600	39.5	3482	52,000 C
H-335	32.6	3011	42,300	37.0	3291	51,900 C
BL-C(2)	35.0	3142	41,100	39.0	3419	50,800 C
IMR-4895	36.0	3257	41,100	40.0	3539	51,200 C
IMR-4064	37.0	3316	42,900	41.0	3610	52,000 C
Varget	37.5	3301	47,900	**41.7**	**3553**	**58,700 P**
AA2520	34.0	3031	38,600	39.0	3345	50,500 C
IMR-4320	37.0	3177	38,600	42.0	3526	49,600 C
H-380	37.0	3136	39,900	41.0	3440	50,800 C
760	41.6	3146	39,300	47.3	3492	52,000 C
H-414	43.1	3263	43,100	49.0+	3592	52,000 C
IMR-4350	40.0	3193	39,900	45.5	3578	52,400 C
H4831SC	43.0	3142	38,100	49.0+	3552	52,000 C

75 gr. Jacketed HP
2.600" OAL

BC: .234
SD: .181

Powder	Sugg Starting Grains	Velocity fps	Pressure	Max Load Grains	Velocity fps	Pressure
XMR-2015	30.0	2937	40,600	34.5	3224	51,900 C
IMR-3031	34.0	3097	38,600	38.0	3407	49,000 C
H-335	32.6	2930	46,800	37.0	3193	52,000 C
BL-C(2)	34.0	2974	41,100	38.0	3264	50,400 C
IMR-4895	34.0	3041	37,700	39.0	3407	50,800 C
IMR-4064	36.0	3127	39,000	40.5	3460	50,400 C
Varget	37.5	3181	47,900	**41.0**	**3446**	**59,700 P**
IMR-4320	37.0	3155	42,000	41.0	3437	50,000 C
H-380	36.0	3020	42,000	40.0	3294	52,000 C
760	40.5	3027	38,000	46.0	3375	51,200 C
H-414	41.4	3077	41,300	47.0	3435	51,900 C
IMR-4350	40.0	3160	44,200	45.0	3425	51,200 C
H4831	43.0	3099	41,100	48.0+	3486	51,600 C
RX22	42.1	3048	44,000	47.0+	3474	59,400 P

80 gr. Jacketed SP
2.655" OAL

BC: .365
SD: .193

Powder	Sugg Starting Grains	Velocity fps	Pressure	Max Load Grains	Velocity fps	Pressure
IMR-4895	32.0	2873	35,000	38.5	3308	50,400 C
IMR-4064	34.0	2958	36,400	39.5	3378	49,600 C
AA-2520	31.9	2861	46,900	**35.5**	**3118**	**58,000P**
IMR-4320	36.0	3057	40,700	40.5	3329	49,200 C
RX15	33.0	2951	43,100	37.0	3151	51,000 C
H-380	35.0	2956	41,600	39.0	3210	51,600 C
760	38.2	3019	48,600	42.5	3285	59,100 P
H-414	40.0	3058	46,300	45.5	3314	52,000 C
IMR-4350	39.0	3035	39,400	41.5	3194	52,000 C
RX19	43.0	3087	42,600	48.0	3373	51,400 C
H4831	42.0	3059	41,600	47.0+	3406	51,600 C
RX22	42.1	3064	46,000	47.0+	3413	59,400 P

85 gr. Jacketed HPBT
2.615" OAL

BC: .282
SD: .206

Powder	Sugg Starting Grains	Velocity fps	Pressure	Max Load Grains	Velocity fps	Pressure
IMR-4895	33.0	2883	38,600	38.0	3211	49,600 C
IMR-4064	34.0	2824	33,500	39.0	3236	49,200 C
Varget	35.2	2991	48,600	39.0	3222	58,900 P
IMR-4320	35.0	2920	38,600	40.0	3255	50,000 C
AA2700	39.0	2937	46,700	42.5	3224	59,900 P
RX15	32.0	2787	36,800	38.0	3147	50,000 C
H-380	33.0	2686	36,400	38.5	3094	50,400 C
760	39.2	2907	41,000	44.5	3176	50,800 C
H414	40.0	2924	41,600	45.5	3234	51,200 C
IMR-4350	38.0	2844	36,900	44.0	3283	51,200 C
RX19	41.0	2847	42,100	46.0	3210	57,300 P
XMR3100	42.7	2813	41,200	47.5	3198	57,700 P
H-4831	**41.0**	**2832**	**36,400**	46.5+	3266	49,600 C
RX22	41.2	2837	43,900	45.5	3250	59,000 P

90 gr. Jacketed FMJBT
2.650" OAL

BC: .387
SD: .218

Powder	Sugg Starting Grains	Velocity fps	Pressure P.S.I.	Max Load Grains	Velocity fps	Pressure P.S.I.
IMR-4064	34.5	2887	48,600	38.5	3107	57,400
Varget	34.0	2901	49,300	38.0	3121	59,380
IMR-4320	35.5	2952	53,200	39.0	3143	59,900
AA2700	39.5	2830	44,900	44.0	3153	59,800
RX15	34.2	2891	49,400	38.1	3095	58,340
H-380	35.0	2741	45,100	38.0	2996	56,900
H-414	40.0	2953	49,900	44.0	3168	57,500
N160	40.5	2953	52,500	45.0	3147	58,800
IMR-4350	40.5	2934	48,400	45.0	3215	59,700
RX19	42.7	2922	46,100	47.3	3243	60,000
XMR-3100	41.0	2788	42,100	46.0	3179	59,700
H4831	43.0	2916	47,100	**48.0**	**3174**	**58,600**
RX22	40.3	2807	44,800	45.0	3146	56,700

Note: Loads shown in shaded panels are maximum.
Loads shown in bold designate potentially most accurate load.
\+ Designates a compressed powder charge.
In pressure column, C=Copper Units of Pressure (CUP); P=Pounds per Square Inch (PSI)

243 Winchester

100 gr. Jacketed BTSP
2.630" OAL
BC: .405
SD: .242

Powder	Sugg Starting Grains	Velocity fps	Pressure P.S.I.	Max Load Grains	Velocity fps	Pressure P.S.I.
Varget	31.0	2636	52,200	**34.5**	**2845**	**58,800**
H-414	35.1	2613	48,800	39.0	2841	56,500
IMR-4350	35.5	2579	48,500	39.5	2858	57,000
XMR-4350	36.4	2613	48,200	40.5	2876	56,000
RX19	37.3	2600	48,100	41.5	2898	57,400
XMR-3100	37.8	2471	46,200	42.0	2787	55,800
RX22	39.5	2612	44,200	44.0	2931	57,000

105 gr. Jacketed A-Max
2.710" OAL
BC: .500
SD: .254

Powder	Sugg Starting Grains	Velocity fps	Pressure P.S.I.	Max Load Grains	Velocity fps	Pressure P.S.I.
Varget	29.7	2545	52,300	33.0	2708	57,200
H-414	34.2	2492	47,800	38.0	2753	56,400
N160	36.0	2566	49,500	40.0	2796	56,800
IMR-4350	35.2	2468	37,300	**39.2**	**2749**	**55,500**
XMR-3100	37.3	2430	46,700	41.5	2738	56,500
RX22	37.8	2501	47,200	42.0	2793	56,400

#245496
84 gr. (#2 Alloy) 2.480" OAL
BC: .202
SD: .203

Powder	Sugg Starting Grains	Velocity fps	Pressure C.U.P.	Max Load Grains	Velocity fps	Pressure C.U.P.
Red Dot	10.0	1845	33,000	12.0	2032	40,300
700X	8.5	1684	30,500	10.5	1904	39,000
PB	9.0	1653	29,400	11.0	1836	36,400
SR-7625	9.5	1703	30,500	11.5	1909	39,400
SR-4759	14.9	1668	12,800	20.0	2156	21,400
XMP-5744	**13.0**	**1605**	**12,800**	17.5	2006	19,600
IMR-4227	14.5	1623	11,900	19.7	2156	23,200
IMR-4198	16.0	1638	10,800	22.7	2175	19,200
RX7	16.0	1676	12,400	21.0	2092	20,200
748	26.1	2114	18,200	39.0	2906	43,800
H-335	25.8	2191	21,200	37.8	2926	46,000
H-4895	25.0	2126	17,300	37.2	2970	47,400
IMR-4320	28.2	2178	20,800	40.2	2996	45,700

Note: Loads shown in shaded panels are maximum.
Loads shown in bold designate potentially most accurate load.

6mm Remington (244 Remington)

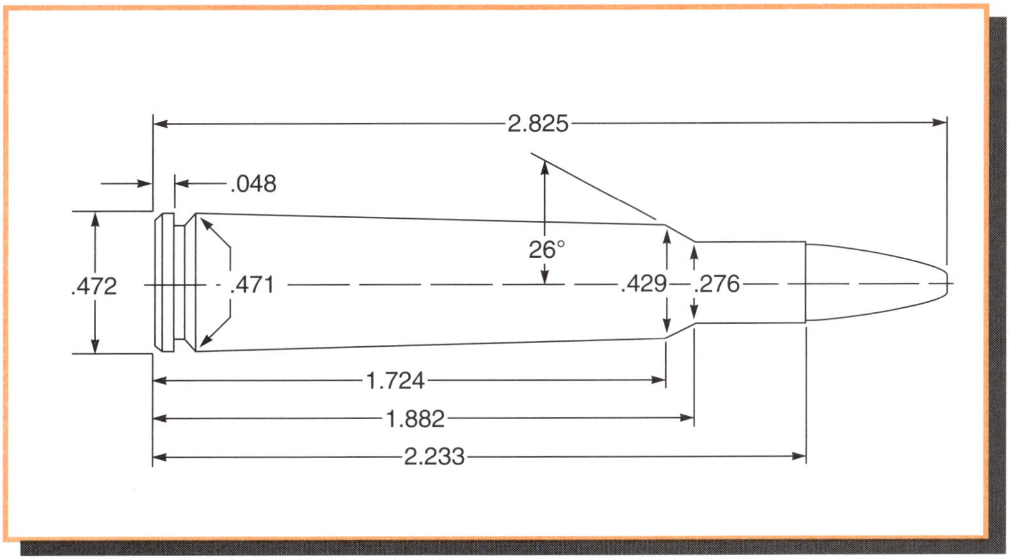

Comments:

The 6mm Remington has a slightly larger case capacity than the 243 suggesting a bit more velocity. However since most shooters have found it slightly less accurate than the Winchester round it has never been extremely popular.

Early rifles marked 244 Remington had a 1 in 12" twist and shot best with bullets of 90 grains or less. Rifles with the 6mm Remington markings usually have a 1 in 9" twist and shoot well with bullets up to 100 grains.

IMR 4350 is the ideal first propellant selection for all jacketed bullet weights.

Cast bullet #245496 at 1,700 fps., or less, is the best choice for lead bullet shooting.

Test Components:

Cases . . . . .Remington
Trim-to Length . . . . .2.225"
Primers . . . . .Remington 9½
Primer Size . . . . .Large Rifle
Lyman Shell Holder . . . . .No. 2
Jacketed Bullets Used . . . . . . .Sierra HP #1500, 60 gr.
Sierra HPBT #1505, 70 gr.
Speer HP #1205, 75 gr.
Speer SP #1211, 80 gr.
Sierra HPBT #1530, 85 gr.
Hornady V-Max #22440, 87 gr.
Hornady BTSP #2453, 100 gr.
Cast Bullets Used. . . . .(sized to .243" dia)
*gas check bullets *#245496, 84 gr.

Test Specifications: (Velocity & Pressure)

Firearm Used . . . . .Universal Receiver
Barrel Length . . . . .24"
Twist . . . . .1-9"
Groove Dia. . . . . .243"

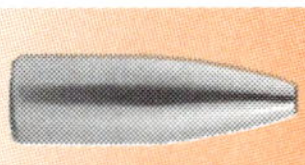

60 gr. Jacketed HP
2.750" OAL
BC: .182
SD: .145

Powder	Sugg Starting Grains	Velocity fps	Pressure C.U.P.	Max Load Grains	Velocity fps	Pressure C.U.P.
XMR-2015	35.4	3374	42,700	39.5	3682	51,360
IMR-3031	37.0	3290	39,400	41.5	3677	50,400
IMR-4895	38.0	3279	42,400	42.5	3623	51,200
IMR-4064	38.0	3164	36,900	44.0	3759	52,800
Varget	39.0	3456	40,200	43.5	3700	49,600
AA-2520	38.0	3454	42,100	42.5	3700	47,300
IMR-4320	39.0	3194	39,900	44.5	3663	52,400
N140	38.5	3328	41,000	43.0	3673	49,000
H-380	40.0	3279	37,700	46.5	3759	52,000
IMR-4350	42.0	3021	36,400	**49.0+**	**3623**	**52,000**
H-4831	46.0	3154	39,000	50.5+	3497	48,800

70 gr. Jacketed HPBT
2.825" OAL
BC: .259
SD: .169

Powder	Sugg Starting Grains	Velocity fps	Pressure C.U.P.	Max Load Grains	Velocity fps	Pressure C.U.P.
IMR-3031	36.0	3105	36,400	41.0	3509	49,600
IMR-4895	37.0	3067	38,600	42.0	3448	49,600
H-335	35.0	3308	45,900	40.0	3598	52,000
IMR-4064	37.5	3067	35,000	43.5	3584	51,600
Varget	36.7	3261	39,300	41.0	3555	49,300
AA2520	33.0	2875	37,700	41.0	3423	49,300
IMR-4320	38.5	3105	40,300	44.0	3509	52,000
N140	37.0	3248	42,300	42.3	3530	51,200
H-380	39.0	3095	36,400	45.0	3588	52,000
760	40.0	3222	42,800	46.0	3611	50,700
H-414	40.0	3299	40,300	46.0	3643	51,400
IMR-4350	42.0	2976	38,600	**48.5+**	**3484**	**51,200**
H-4831	45.0	2967	36,400	50.5+	3413	48,800

Note: Loads shown in shaded panels are maximum.
Loads shown in bold designate potentially most accurate load.
\+ Designates a compressed powder charge.

6mm Remington (244 Remington)

75 gr. Jacketed HP
2.825" OAL

BC: .234
SD: .181

Powder	Sugg Starting Grains	Velocity fps	Pressure C.U.P.	Max Load Grains	Velocity fps	Pressure C.U.P.
IMR-3031	36.0	3105	40,300	40.5	3485	50,700
IMR-4895	37.0	3105	40,300	39.0	3385	51,400
H-335	36.3	3168	44,600	40.5	3424	50,800
IMR-4064	37.0	3048	37,300	42.5	3448	51,600
Varget	36.3	3164	40,600	40.5	3457	49,400
AA-2520	32.0	2790	40,000	40.5	3333	52,000
IMR-4320	37.5	2994	37,700	41.0	3400	52,000
N140	36.7	3092	40,600	**41.0**	**3398**	**50,300**
H-380	36.0	2865	34,500	43.5	3423	48,100
IMR-4350	41.5	2976	37,300	48.0+	3448	52,000
XMR-3100	42.0	2761	37,700	48.5+	3249	50,700
H-4831	44.0	2941	37,700	48.0+	3288	45,000

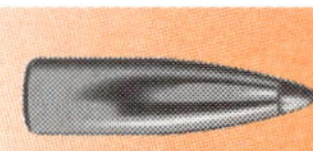

80 gr. Jacketed SP
2.825" OAL

BC: .365
SD: .194

Powder	Sugg Starting Grains	Velocity fps	Pressure C.U.P.	Max Load Grains	Velocity fps	Pressure C.U.P.
IMR-3031	35.0	2994	37,300	39.5	3412	50,600
IMR-4895	35.8	3107	44,000	40.0	3325	49,000
IMR-4064	36.5	2915	36,000	42.0	3390	50,400
Varget	**35.4**	**3188**	**44,500**	39.6	3364	49,400
IMR-4320	36.7	3122	43,700	41.0	3376	49,900
N140	35.8	3044	42,900	40.0	3301	50,400
H-380	35.0	2785	34,500	43.0	3290	50,800
760	41.3	3152	40,700	47.0	3467	51,800
H-414	40.4	3209	42,600	46.0	3447	48,900
IMR-4350	41.0	2873	36,000	47.0+	3367	52,000
H-4831	44.0	2890	36,400	50.5+	3322	48,400
RX22	40.0	2922	40,600	49.0+	3294	49,600

85 gr. Jacketed HPBT
2.825" OAL

BC: .282
SD: .206

Powder	Sugg Starting Grains	Velocity fps	Pressure C.U.P.	Max Load Grains	Velocity fps	Pressure C.U.P.
IMR-3031	34.0	2906	38,100	39.0	3354	52,000
IMR-4895	35.0	2915	39,900	40.0	3236	51,600
IMR-4064	36.0	2865	37,300	41.0	3279	51,200
Varget	35.8	3075	43,200	40.0	3295	50,300
IMR-4320	36.0	2849	36,900	42.0	3268	51,200
N140	**36.0**	**3020**	**43,600**	39.5	3270	51,400
H-380	35.0	2770	37,300	41.0	3362	52,000
760	39.0	3055	41,100	45.0	3417	51,700
H414	38.0	3029	41,800	45.0	3436	52,000
IMR-4350	40.0	2808	35,500	46.5+	3300	52,000
RX19	42.0	2836	39,100	46.5	3211	50,800
XMR3100	42.5	2715	39,500	47.5	3130	49,900
H-4831	44.0	2958	38,600	47.5+	3394	51,600
RX22	44.0	2879	38,000	49.4	3301	51,900

87 gr. Jacketed V-Max
2.825" OAL

BC: .400
SD: .210

Powder	Sugg Starting Grains	Velocity fps	Pressure C.U.P.	Max Load Grains	Velocity fps	Pressure C.U.P.
IMR-3031	35.4	3031	42,600	39.0	3276	50,100
IMR-4064	36.0	2948	37,900	40.0	3266	48,800
Varget	35.8	3034	41,800	40.0	3278	50,000
IMR-4320	35.8	2954	39,500	40.0	3253	49,800
760	40.8	2993	37,200	45.5	3335	50,200
H-414	40.5	2979	38,300	45.2	3291	48,900
N160	42.0	2987	40,000	47.0	3320	50,000
IMR-4350	40.0	2882	36,400	44.7	3256	49,100
XMR-3100	42.6	2882	37,900	**47.5+**	**3251**	**51,200**
H4831	44.0	2979	39,000	49.0+	3300	49,300
RX22	45.0	3091	40,400	50.0+	3435	50,700

100 gr. Jacketed BTSP
2.820" OAL

BC: .405
SD: .242

Powder	Sugg Starting Grains	Velocity fps	Pressure C.U.P.	Max Load Grains	Velocity fps	Pressure C.U.P.
IMR-4064	33.8	2811	38,500	38.5	3089	50,100
AA2520	31.0	2457	39,500	37.2	2869	51,000
H-380	34.0	2570	39,000	38.5	2865	51,600
760	37.8	2800	39,600	43.0	3093	48,000
H-414	36.9	2769	36,300	43.5	3122	50,500
N160	39.0	2852	45,500	43.5	3049	50,300
IMR-4350	37.4	2811	39,500	42.5	3143	51,100
RX19	38.0	2559	38,100	46.0	3058	51,700
XMR-3100	40.0	2576	39,400	45.0	2955	51,000
H-4831	42.0	2883	44,200	47.0+	3131	50,900
RX22	41.5	2723	38,300	46.6+	3084	50,300
IMR-7828	**43.0**	**2841**	**40,900**	48.0+	3149	50,300
H-1000	45.7	2869	40,400	51.0+	3141	49,300

#245496
84 gr. (#2 Alloy) 2.690" OAL

BC: .202
SD: .200

Powder	Sugg Starting Grains	Velocity fps	Pressure C.U.P.	Max Load Grains	Velocity fps	Pressure C.U.P.
Red Dot	9.0	1726	30,500	11.5	1973	40,700
700X	8.5	1635	29,400	10.7	1879	40,700
Green Dot	9.0	1703	27,600	11.5	1947	39,900
PB	9.0	1642	27,600	11.0	1829	38,600
SR-7625	9.0	1628	29,400	11.0	1818	38,600
SR-4756	10.0	1725	29,400	12.5	1974	40,700
SR-4759	13.0	1661	16,100	19.5	2202	34,000
IMR-4227	**13.7**	**1711**	**17,300**	19.7	2181	28,200
IMR-4198	16.8	1828	17,000	21.0	2144	22,500
RX7	14.5	1651	17,100	20.0	2117	25,800
748	20.8	1772	14,700	26.0	2173	23,400
H-4895	19.6	1760	16,300	24.5	2132	23,300

Note: Loads shown in shaded panels are maximum.
Loads shown in bold designate potentially most accurate load.
\+ Designates a compressed powder charge.

240 Weatherby Magnum

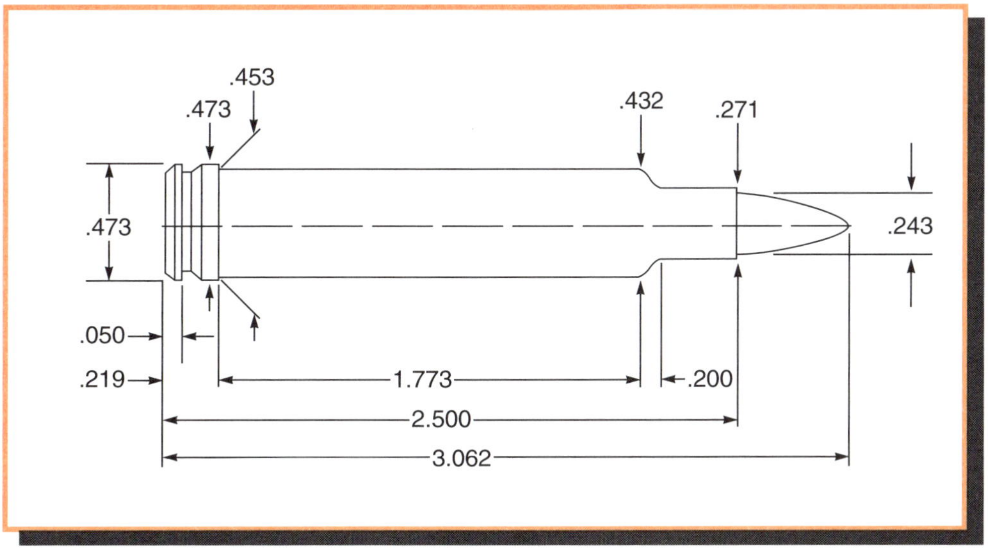

Comments:

The 240 Weatherby is very fast, very accurate, and not very popular. Weatherby introduced their 240 Magnum in 1968 as the fastest of the 6mm/243 bore size available in factory form. This cartridge can be thought of as a 30-06 necked down to 6mm, with a belt added to the case. The fact that it is an original case design, which cannot be formed from any other cartridge, has not helped its appeal. The 240 requires slower burning powders for top performance.

IMR-4350 produced best results with bullets up to 90-grains, H-4831 performed best with the 100-grainers. Despite the good availability of .243" diameter varmint bullets, the 240 shouldn't be considered for such as extended shooting would be rough on barrel life. This data is intended for commercially produced and chambered rifles. It is not for use in custom guns which may lack the free bore found in standard Weatherby chambers.

Test Components:

Cases . . . . .Weatherby
Trim-to Length . . . . .2.490"
Primers . . . . .Federal 215
Primer Size . . . . .Large Rifle Magnum
Lyman Shell Holder . . . . .No. 2
Jacketed Bullets Used . .Hornady V-Max #22420, 75 gr.
Speer SP #1211, 80 gr.
Nosler Partition #16314, 85 gr.
Speer SP #1217, 90 gr.
Hornady BTSP #2453, 100 gr.

Test Specifications:
(Velocity Only)

Firearm Used . . . . .Weatherby Mk V
Barrel Length . . . . .24"
Twist . . . . .1-10"
Groove Dia. . . . .243"

75 gr. Jacketed V-Max 3.062" OAL — **BC:** .330 **SD:** .181

Powder	Sugg Starting Grains	Velocity fps	Pressure	Max Load Grains	Velocity fps	Pressure
H-414	46.8	3091	—	52.0	3508	—
N160	52.0	3444	—	54.5	3677	—
IMR-4350	50.0	3413	—	**52.5**	**3629**	—
RX19	48.1	3004	—	53.5	3389	—
H4831SC	53.0	3358	—	56.0	3522	—

80 gr. Jacketed SP 3.062" OAL — **BC:** .365 **SD:** .193

Powder	Sugg Starting Grains	Velocity fps	Pressure	Max Load Grains	Velocity fps	Pressure
N160	50.0	3288	—	53.5	3536	—
IMR-4350	**48.0**	**3257**	—	51.0	3462	—
RX19	46.8	2848	—	52.0	3264	—
XMR-3100	50.4	3004	—	56.0	3422	—
H4831SC	52.0	3246	—	55.0	3431	—

Note: Loads shown in shaded panels are maximum.
Loads shown in bold designate potentially most accurate load.

240 Weatherby Magnum

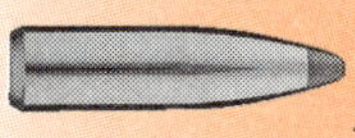

85 gr. Jacketed Partition 3.062" OAL — **BC:** .315 **SD:** .206

Powder	Sugg Starting Grains	Velocity fps	Pressure	Max Load Grains	Velocity fps	Pressure
N160	49.0	3195	—	51.5	3360	—
IMR-4350	48.0	3213	—	51.0	3409	—
RX19	46.8	2924	—	**52.0**	**3296**	—
XMR-3100	48.6	2868	—	54.0	3280	—
H4831SC	52.0	3219	—	54.5	3371	—

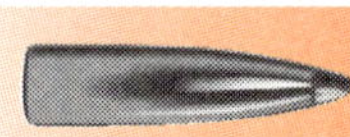

90 gr. Jacketed SP 3.062" OAL — **BC:** .385 **SD:** .217

Powder	Sugg Starting Grains	Velocity fps	Pressure	Max Load Grains	Velocity fps	Pressure
N160	47.0	3024	—	50.0	3257	—
IMR-4350	**47.0**	**3076**	—	49.5	3265	—
RX19	46.0	2783	—	51.0	3142	—
XMR3100	48.5	2864	—	54.0	3260	—
H4831SC	52.0	3193	—	54.5	3355	—

100 gr. Jacketed BTSP 3.062" OAL — **BC:** .405 **SD:** .242

Powder	Sugg Starting Grains	Velocity fps	Pressure	Max Load Grains	Velocity fps	Pressure
N160	47.0	3012	—	49.0	3125	—
IMR-4350	44.0	2835	—	47.0	3047	—
RX19	45.0	2725	—	50.0	3072	—
XMR-3100	47.0	2733	—	52.0	3103	—
H4831SC	**50.0**	**3082**	—	53.0	3241	—
RX22	48.0	2930	—	51.0	3135	—

Note: Loads shown in shaded panels are maximum.
Loads shown in bold designate potentially most accurate load.

25-20 Winchester (25WCF)

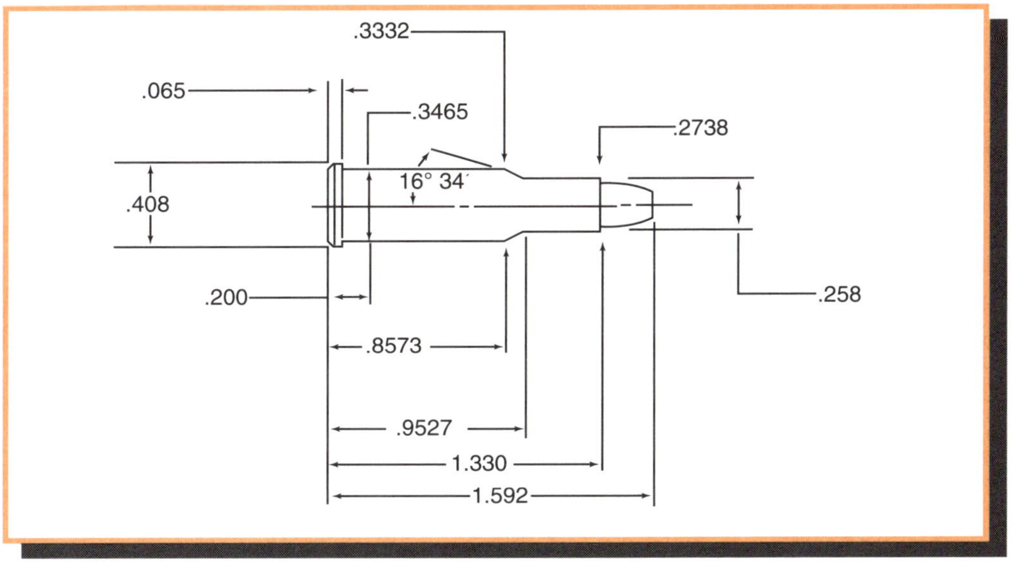

Comments:

This data is not for use in old firearms designed for black powder. It is intended only for use in firearms designed for smokeless powder loads and in good condition.

When loading for tubular magazines you must use flat, blunt or round nose bullets. It is suggested that a light crimp be employed in ammo assembly for cartridges that will be used in a tubular magazine.

IMR 4198 is a good first powder choice. Data is brief due to the limited popularity of the cartridge.

Test Components:

Cases . . . . .Remington, Winchester
Trim-to Length . . . . .1.320"
Primers . . . . .Remington 6½, Win. WSR
Primer Size . . . . .Small Rifle
Lyman Shell Holder . . . . .No. 10
Jacketed Bullets Used . . .Remington SP #22735, 86 gr.

Cast bullets used . . . .(sized to .257)
*gas check bullet *257420, 65 gr.

Test Specifications: (Velocity Only)

Firearm Used . . . . .Winchester Model 92
Barrel Length . . . . .20"+24"
Twist . . . . .1-14"
Groove Dia. . . . . .256"

86 gr. Jacketed SP
1.592" OAL
BC: .191
SD: .186

Powder	Sugg Starting Grains	Velocity fps	Pressure	Max Load Grains	Velocity fps	Pressure
Unique	4.5	1219	—	5.5	1508	—
2400	7.0	1150	—	9.0	1610	—
IMR-4227	8.0	1213	—	10.7	1713	—
IMR-4198	**10.5**	**1283**	—	13.0+	1751	—
H-110	7.0	1060	—	8.0	1294	—

***257420**
65 gr. (#2 Alloy) 1.560" OAL
BC: .129
SD: .140

Powder	Sugg Starting Grains	Velocity fps	Pressure C.U.P.	Max Load Grains	Velocity fps	Pressure C.U.P.
Red Dot	3.5	1320	—	4.5	1560	—
700X	3.5	1345	—	4.5	1580	—
Green Dot	4.0	1400	—	5.0	1665	—
PB	4.0	1360	—	5.0	1605	—
Unique	4.5	1455	—	6.0	1805	—
SR-7625	4.5	1480	—	5.5	1700	—

Note: Loads shown in shaded panels are maximum.
Loads shown in bold designate potentially most accurate load.
+ Designates a compressed powder charge.
* Fired in a 24" barrel with Winchester cases and primers.

250 Savage (250/3000 Savage)

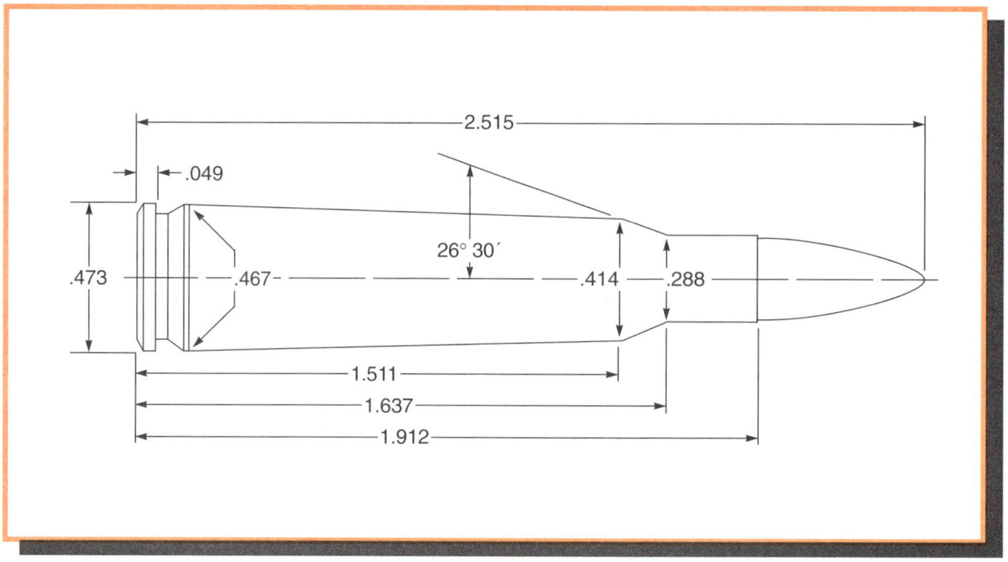

Comments:

This cartridge is extremely accurate, but the proper length (weight) bullets must be selected. Early firearms had a 1 in 14" twist because the only factory bullet weight was 87 grains (at 3000 fps.) for the cartridges original name (250/3000). In these barrels 100 grain bullets seldom perform well. When the 100 grain bullet was introduced a twist rate of 1 -10" was adopted. This twist often works with bullets as light as 75 grains and as heavy as 117 grains.

Due to the cases small volume the 100 grain bullet is the best for light big game. Be sure to select a big game style bullet as some 100 grain bullets are of the varmint type. Due to the limited ballistics of the cartridge it would be wise to select a premium grade bullet to optimize performance on light big game.

IMR 3031 is the best choice for accuracy in most rifles with all bullet weights.

When loading for lever action rifles use extra caution in developing a maximum load. Many lever actions will not allow the listed maximum load to be used safely.

Test Components:

Cases . . . . .Remington
Trim-to Length . . . . .1.902"
Primers . . . . .Remington 9½
Primer Size . . . . .Large Rifle
Lyman Shell Holder . . . . .No. 2
Jacketed Bullets Used . . . . . . .Sierra HP #1600, 75 gr.
Speer SP #1241, 87 gr.
Sierra SP #1620, 100 gr.
Sierra SP #1640, 117 gr.

Test Specifications: (Velocity & Pressure)

Firearm Used . . . . .Universal Receiver Custom Mauser 98
Barrel Length . . . . .24"
Twist . . . . .1-14"
Groove Dia. . . . . .257"

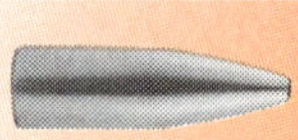

75 gr. Jacketed HP 2.450" OAL — **BC:** .189 **SD:** .162

Powder	Sugg Starting Grains	Velocity fps	Pressure C.U.P.	Max Load Grains	Velocity fps	Pressure C.U.P.
IMR-3031	32.0	3003	—	36.0	3472	—
BL-C (2)	35.0	2985	—	39.0	3300	—
H-335	33.0	2793	—	37.0	3058	—
IMR-4895	**32.0**	**2612**	**35,700**	36.0	3023	44,400
IMR-4064	34.0	2906	—	38.0+	3322	—
Varget	34.0	2734	33,300	38.0+	3175	44,300
IMR-4320	32.0	2636	36,500	36.0	2983	44,600
N140	32.0	2669	34,500	36.0	3034	43,900
RX15	33.0	2701	33,700	37.0	3051	43,900
H-380	38.0	2967	—	42.0+	3164	—

87 gr. Jacketed SP 2.450" OAL — **BC:** .300 **SD:** .188

Powder	Sugg Starting Grains	Velocity fps	Pressure C.U.P.	Max Load Grains	Velocity fps	Pressure C.U.P.
IMR-3031	31.0	2840	—	35.0	3247	—
BL-C(2)	33.5	2625	34,900	37.3	2972	43,900
H-335	33.0	2645	—	36.0	2923	—
IMR-4895	31.0	2467	34,500	34.5	2882	44,200
IMR-4064	33.0	2816	—	37.0+	3205	—
Varget	32.4	2576	35,100	36.0	2961	43,600
IMR-4320	31.0	2567	36,600	34.5	2835	44,300
N140	32.0	2670	36,600	36.0	2965	44,200
RX15	**32.0**	**2681**	**38,100**	35.7	2956	44,300
H-380	37.0	2824	—	41.0+	3012	—

Note: Loads shown in shaded panels are maximum.
Loads shown in bold designate potentially most accurate load.
\+ Designates a compressed powder charge.

250 Savage (250/3000 Savage)

100 gr. Jacketed SP
2.515" OAL

BC: .330
SD: .216

Powder	Sugg Starting Grains	Velocity fps	Pressure C.U.P.	Max Load Grains	Velocity fps	Pressure C.U.P.
IMR-3031	30.0	2695	—	33.0	3012	—
BL-C(2)	31.0	2463	—	35.0	2801	—
H-335	32.0	2544	—	34.0	2702	—
IMR-4895	31.0	2563	36,200	34.5	2823	44,100
IMR-4064	31.0	2577	—	35.5	3021	—
Varget	32.0	2587	37,900	35.5	2864	45,000
IMR-4320	31.0	2522	37,500	34.5	2768	43,600
RX15	**31.0**	**2530**	**38,100**	34.6	2831	45,000
H-380	35.0	2645	—	39.0	2865	—

117 gr. Jacketed SP
2.515" OAL

BC: .388
SD: .253

Powder	Sugg Starting Grains	Velocity fps	Pressure C.U.P.	Max Load Grains	Velocity fps	Pressure C.U.P.
IMR-3031	28.0	2444	—	31.0	2672	—
BL-C(2)	29.0	2283	—	32.0	2481	—
H-335	30.0	2369	—	32.0	2506	—
IMR-4895	**28.0**	**2304**	**35,200**	31.0	2577	45,000
IMR-4064	30.0	2463	—	33.5	2710	—
Varget	29.0	2317	38,400	32.2	2534	44,700
IMR-4320	28.0	2246	35,500	31.8	2506	44,900
H-380	32.0	2427	—	36.0	2702	—
RX19	34.0	2254	36,100	38.0	2537	44,800
H4831SC	34.6	2228	36,800	38.5+	2489	44,500

Note: Loads shown in shaded panels are maximum.
Loads shown in bold designate potentially most accurate load.
\+ Designates a compressed powder charge.

257 Roberts

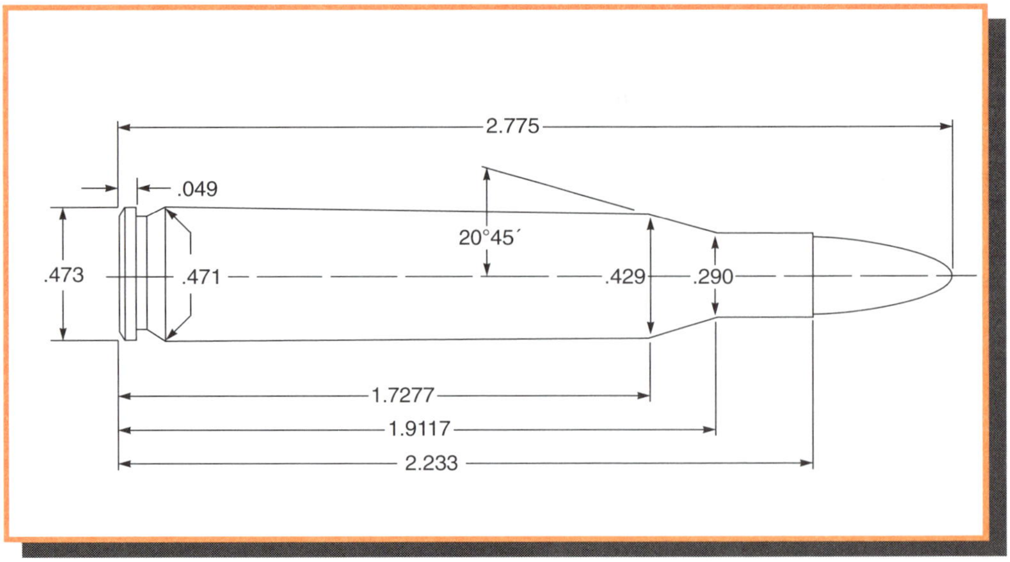

Comments:

Noted gun-writer Ned Roberts had experimented with the 7x57mm Mauser cartridge necked down to 25-caliber during the late 1920s and early thirties. Remington introduced the 257 Roberts in 1934 as a standardized version of this wildcat. This was one of the most common deer cartridges before being overshadowed by later developments such as the 243 Winchester. While its popularity has waned over the past few decades, it has had some renewed interest in recent years.

The Roberts is an excellent all-around cartridge, which offers low recoil even in a lightweight rifle. Handloaders can tailor a load for everything from varmints on up to deer sized game. Unfortunately, many older rifles have a long throat while using a short magazine. This can make peak accuracy difficult to attain. Most new rifles have corrected this condition. Shooters may also encounter brass that has two different rim/extractor dimensions. If so, it may be necessary to use two different shell holders. Medium to slower burning powders provide the best accuracy with Varget being perhaps the best choice for all bullet weights.

Any +P loads listed here should not be used in older guns. They should be used only in modern, recently made rifles approved by the manufacturer for high-pressure +P loads.

Test Components:

Cases . . . . .Remington
Trim-to Length . . . . .2.223"
Primers . . . . .Remington 9½
Primer Size . . . . .Large Rifle
Lyman Shell Holder . . . . .No. 2
Jacketed Bullets Used . . . . . . .Sierra HP #1600, 75gr.
Speer SP #1241, 87gr.
Sierra SP #1620, 100gr.
Sierra SP #1640, 117gr.

Test Specifications:
(Velocity & Pressure)

Firearm Used . . . . .Remington M700 and Universal Receiver
Barrel Length . . . . .24"
Twist . . . . .1-10"
Groove Dia. . . . . .257"

75 gr. Jacketed HP
2.735" OAL
BC: .189
SD: .162

Powder	Sugg Starting Grains	Velocity fps	Pressure C.U.P.	Max Load Grains	Velocity fps	Pressure C.U.P.
IMR-3031	36.0	2976	—	40.0	3311	—
BL-C(2)	36.0	2801	—	42.0	3311	—
IMR-4895	38.0	2890	—	42.0	3268	—
IMR-4064	40.0	3012	—	44.0	3509	—
Varget	39.0	2920	—	41.0	3084	—
IMR-4320	38.0	2808	—	42.0	3144	—
AA2700	44.0	2982	—	46.0	3143	—
RX15	**39.0**	**2967**	—	41.8	3190	—
H-380	41.0	2906	—	45.0	3236	—
760	40.2	2849	33,500	45.7	3153	41,200
H-414	42.5	3016	37,800	47.0	3330	44,900
XMR-4350	45.0	2728	—	47.0	2963	—

87 gr. Jacketed SP
2.750" OAL
BC: .300
SD: .188

Powder	Sugg Starting Grains	Velocity fps	Pressure C.U.P.	Max Load Grains	Velocity fps	Pressure C.U.P.
IMR-3031	35.0	2840	—	39.0	3134	—
BL-C(2)	37.0	2824	—	41.0	3030	—
IMR-4895	36.0	2724	—	40.0	3030	—
IMR-4064	39.0	2932	—	43.0	3311	—
Varget	**38.0**	**2893**	—	40.2	3029	—
IMR-4320	36.0	2659	—	40.0	2932	—
AA2700	39.5	2708	—	42.5	2853	—
RX15	39.0	2953	—	41.0	3101	—
H-380	40.0	2849	—	44.0	3086	—
760	41.1	2879	38,700	46.7	3092	44,200
H-414	40.5	2896	38,500	46.0	3140	44,200
XMR-4350	43.0	2642	—	45.5	2814	—

Note: Loads shown in shaded panels are maximum.
Loads shown in bold designate potentially most accurate load.

257 Roberts

100 gr. Jacketed SP
2.775" OAL

BC: .330
SD: .216

Powder	Sugg Starting Grains	Velocity fps	Pressure C.U.P.	Max Load Grains	Velocity fps	Pressure C.U.P.
IMR-3031	34.0	2717	—	38.0	2985	—
BL-C(2)	36.0	2747	—	40.0	2898	—
IMR-4895	34.0	2557	—	36.5	2711	—
IMR-4064	36.0	2631	—	40.0	2958	—
Varget	**36.0**	**2712**	—	38.0	2841	—
IMR-4320	34.0	2463	—	38.5	2793	—
H-380	38.0	2564	—	42.0	2857	—
760	38.5	2591	35,700	44.7	2928	43,200
H-414	38.3	2683	38,000	43.5	2973	45,200
XMR-4350	41.0	2480	—	43.5	2654	—
				(+P)46.0	2885	—
RX19	43.0	2644	—	45.0	2777	—
				(+P)47.5	2967	—
XMR-3100	43.7	2552	—	46.0	2736	—

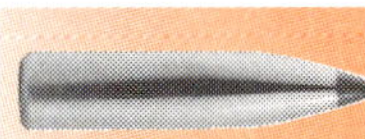

117 gr. Jacketed SP
2.775" OAL

BC: .388
SD: .253

Powder	Sugg Starting Grains	Velocity fps	Pressure C.U.P.	Max Load Grains	Velocity fps	Pressure C.U.P.
IMR-3031	30.0	2341	—	33.0	2551	—
BL-C(2)	32.0	2380	—	35.0	2564	—
IMR-4895	31.0	2288	—	34.0	2493	—
IMR-4064	32.0	2347	—	36.0	2610	—
Varget	33.0	2449	—	**35.0**	**2565**	—
IMR-4320	33.0	2369	—	36.0	2595	—
H-380	34.0	2325	—	38.0	2544	—
760	36.3	2319	34,300	41.2	2688	44,800
H-414	35.0	2394	36,500	39.7	2706	45,100
XMR-4350	40.0	2444	—	42.5	2587	—
				(+P)44.0	2729	—
XMR-3100	41.8	2445	—	44.0	2640	—
RX22	42.0	2468	—	44.0	2580	—
				(+P)45.5	2655	—

Note: Loads shown in shaded panels are maximum.
Loads shown in bold designate potentially most accurate load.

25-06 Remington

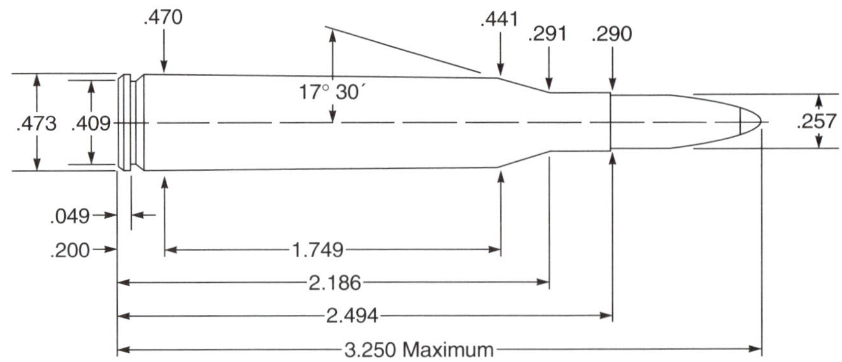

Comments:

Gunmaker Adolph O. Niedner first necked down the 30-06 to accept .257" diameter bullets during the 1920s. However, the slow burning powders necessary to maximize the performance of this round were simply not available at the time. The lack of such powders handicapped many wildcat cartridges based on 30-06 cartridges necked down to smaller calibers. The post-World War Two availability of propellants such as H-4831 and IMR-4350 changed all this. Such "over-bore" cartridges now became viable. The 25-06 was a rather popular wildcat for many years before Remington legitimized it as a factory offering in 1969. This data is intended for use in standard commercial rifles originally chambered for this round. Shooters may occasionally encounter custom chambered rifles built before the 25-06 became standardized. If in doubt, a chamber cast by a qualified gunsmith is in order to verify chamber dimensions.

Test Components:

Cases . . . . .Remington
Trim-to Length . . . . .2.484"
Primers . . . . .Winchester WLR
Primer Size . . . . .Large Rifle
Lyman Shell Holder . . . . .No. 2
Jacketed Bullets Used . . . . . . .Sierra HP #1600, 75 gr.
Hornady SP, #2530, 87 gr.
Sierra SP #1620, 100 gr.
Sierra SP #1640, 117 gr.
Hornady HP #2560, 120 gr.

Test Specifications:
(Velocity & Pressure)

Firearm Used . . . . .Universal Receiver
Barrel Length . . . . .26"
Twist . . . . .1-10"
Groove Dia. . . . . .257"

75 gr. Jacketed HP
3.065" OAL
BC: .189
SD: .162

Powder	Sugg Starting Grains	Velocity fps	Pressure C.U.P.	Max Load Grains	Velocity fps	Pressure C.U.P.
IMR-3031	43.0	3304	42,400	48.0	3632	49,600
IMR-4895	43.5	3170	42,000	49.0	3595	50,800
IMR-4064	45.0	3304	42,000	**51.0**	**3740**	**52,800**
Varget	45.5	3288	40,300	50.5	3667	46,200
IMR-4320	45.0	3199	40,700	51.5	3671	52,400
AA2700	49.5	3375	41,200	55.0	3689	50,100
H-380	47.0	3373	42,900	52.0	3663	51,200
760	50.6	3160	39,300	57.5	3608	51,800
H-414	51.3	3477	45,000	57.0	3671	48,600
IMR-4350	50.0	3128	41,600	56.5	3575	52,000
H4831	53.0	3105	41,100	59.0+	3559	52,400

87 gr. Jacketed SP
3.096" OAL
BC: .322
SD: .188

Powder	Sugg Starting Grains	Velocity fps	Pressure C.U.P.	Max Load Grains	Velocity fps	Pressure C.U.P.
IMR-3031	42.0	3113	41,600	47.0	3452	50,800
IMR-4064	44.0	3153	42,900	49.0	3485	50,800
Varget	43.0	3304	42,800	48.0	3598	50,400
IMR-4320	44.0	3080	41,100	49.5	3420	51,200
RX15	43.0	2943	35,000	48.0	3554	48,800
H-380	45.0	3111	41,600	50.0	3436	52,400
H-414	46.4	3210	43,400	51.5	3447	48,700
IMR-4350	**49.0**	**3078**	**41,100**	55.0	3492	52,800
RX19	50.5	2963	36,000	58.0	3509	50,400
XMR-3100	52.0	3011	37,200	58.0	3454	50,600
H-4831	52.0	3015	40,300	58.0+	3407	50,400

Note: Loads shown in shaded panels are maximum.
Loads shown in bold designate potentially most accurate load.
\+ Designates a compressed powder charge.

25-06 Remington

100 gr. Jacketed SP
3.155" OAL

BC: .330
SD: .216

Powder	Sugg Starting Grains	Velocity fps	Pressure C.U.P.	Max Load Grains	Velocity fps	Pressure C.U.P.
IMR-4064	40.0	2856	40,300	44.5	3158	51,200
IMR-4320	41.0	2822	41,600	46.5	3139	50,800
H-380	43.0	2951	45,900	47.5	3141	51,600
IMR-4350	46.5	2622	37,300	**52.0**	**3202**	**50,400**
RX19	51.0	2956	40,600	57.0	3307	51,100
XMR-3100	51.0	2878	40,000	57.0	3243	50,700
RX22	52.7	3024	42,800	58.5+	3309	50,100
H1000	55.8	2933	38,300	62.0+	3260	48,500

117 gr. Jacketed SP
3.155" OAL

BC: .388
SD: .253

Powder	Sugg Starting Grains	Velocity fps	Pressure C.U.P.	Max Load Grains	Velocity fps	Pressure C.U.P.
IMR-4064	37.1	2566	40,300	42.4	2866	49,600
H-380	41.0	2681	43,300	44.5	2843	51,200
IMR-4350	46.0	2735	41,600	51.0	3013	50,400
RX19	48.2	2807	43,200	53.5	3058	50,100
XMR-3100	49.0	2675	40,500	54.3	3013	49,600
H-4831	49.0	2772	42,600	53.5	2996	50,000
RX22	48.2	2863	46,800	53.5	3014	50,400
IMR-7828	49.5	2773	43,800	55.0	3019	50,600
H1000	52.2	2758	40,000	**58.0**	**3038**	**49,600**

120 gr. Jacketed HP
3.250" OAL

BC: .394
SD: .260

Powder	Sugg Starting Grains	Velocity fps	Pressure C.U.P.	Max Load Grains	Velocity fps	Pressure C.U.P.
IMR-4064	40.0	2715	43,500	44.5	2916	50,200
IMR-4350	44.4	2631	39,200	50.5	2963	50,600
RX19	**48.0**	**2793**	**41,900**	53.5	3051	50,500
XMR-3100	49.0	2715	42,000	54.0	3009	52,000
H-4831	48.0	2781	43,500	53.5	2962	50,700
RX22	46.8	2826	47,000	52.0	2950	49,000
IMR-7828	49.5	2723	41,400	55.0	3009	50,500
H1000	51.3	2787	42,600	57.0	2996	49,900

Note: Loads shown in shaded panels are maximum.
Loads shown in bold designate potentially most accurate load.
\+ Designates a compressed powder charge.

257 Weatherby Magnum

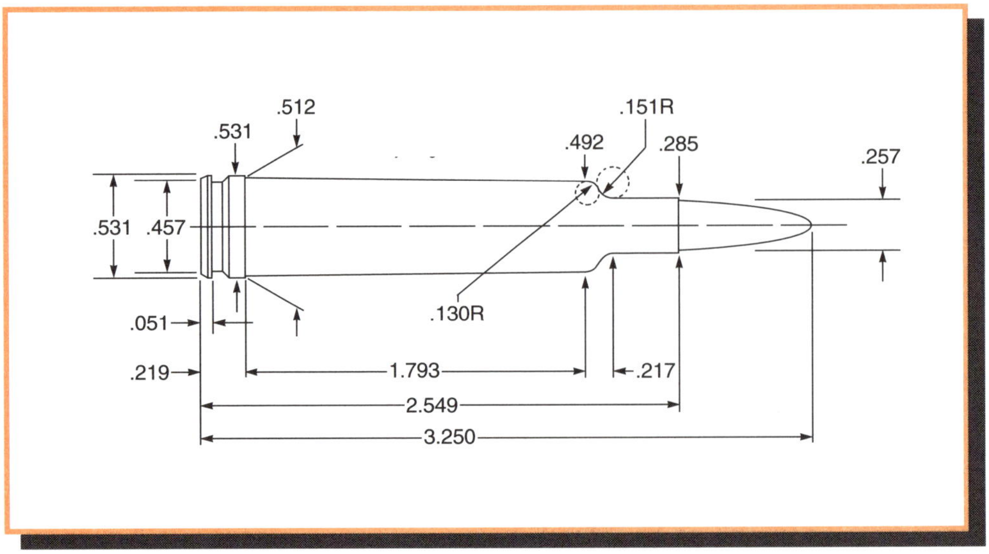

Comments:

The 257 Weatherby Magnum is about as much power as one can get in the 25 caliber rifle. It is one of the original Weatherby cartridges developed around 1944 based on the 300 H&H case. Careful handloading can wring around 200 feet per second over the 25-06. It has never been one of Weatherby's more popular offerings despite its impressive performance. This performance level predictably requires use of premium bullets such as the 115-grain Ballistic Silvertip for larger game. The 75 and 87-grain bullets are best left to varmints and smaller game. The "overbored" nature of the 257 mandates the use of slow burning powders. H-4831 has long been the powder of choice and Reloder 25 provided best results with the 115-grain bullet. Shooters should not overheat barrels with extended firing and prevent fouling from accumulating through proper cleaning. This data is intended for commercially produced and chambered rifles. It is not for use in custom guns that may lack the free bore found in standard Weatherby chambers.

Test Components:

Cases . . . . .Weatherby
Trim-to Length . . . . .2.540"
Primers . . . . .Federal 215
Primer Size . . . . .Large Rifle Magnum
Lyman Shell Holder . . . . .No. 13
Jacketed Bullets Used . . . . . . .Sierra HP #1600, 75 gr.
Hornady SP #2530, 87 gr.
Sierra SBT #1620, 100 gr.
Nosler Ballistic Silvertip #51050, 115 gr.

Test Specifications:
(Velocity Only)

Firearm Used . . . . .Weatherby Mk V
Barrel Length . . . . .26"
Twist . . . . .1-12"
Groove Dia. . . . . .257"

75 gr. Jacketed HP
3.025" OAL
BC: .189
SD: .162

Powder	Sugg Starting Grains	Velocity fps	Pressure	Max Load Grains	Velocity fps	Pressure
IMR-4064	55.0	3367	—	61.0	3745	—
IMR-4320	57.0	3460	—	63.0	3875	—
H-380	62.0	3472	—	68.0	3952	—
IMR-4350	64.0	3390	—	**71.0**	**4048**	—
XMR-3100	69.0	3582	—	73.0	3883	—
H-4831	68.0	3367	—	75.0	3891	—
RX22	72.0	3661	—	75.5	3975	—

87 gr. Jacketed SP
3.185" OAL
BC: .322
SD: .188

Powder	Sugg Starting Grains	Velocity fps	Pressure	Max Load Grains	Velocity fps	Pressure
IMR-4064	52.0	3125	—	58.0	3521	—
IMR-4320	54.0	3290	—	60.0	3597	—
H-380	60.0	3401	—	66.0	3690	—
IMR-4350	62.0	3279	—	69.0	3759	—
XMR-3100	65.0	3258	—	69.0	3541	—
H-4831	66.0	3184	—	**73.0**	**3731**	—
RX22	69.0	3452	—	73.0	3689	—
IMR-7828	71.0	3550	—	74.5+	3762	—

Note: Loads shown in shaded panels are maximum.
Loads shown in bold designate potentially most accurate load.
\+ Designates a compressed powder charge.

257 Weatherby Magnum

100 gr. Jacketed SBT
3.250" OAL
BC: .355
SD: .216

Powder	Sugg Starting Grains	Velocity fps	Pressure	Max Load Grains	Velocity fps	Pressure
IMR-4064	50.0	2898	—	56.0	3279	—
IMR-4320	52.0	3021	—	58.0	3390	—
H-380	58.0	3086	—	63.0	3367	—
IMR-4350	59.0	3048	—	**66.0**	**3472**	—
XMR-3100	65.5	3227	—	69.0	3467	—
H-4831	63.0	2923	—	70.0	3436	—
RX22	65.0	3237	—	69.0	3430	—
IMR-7828	70.0	3445	—	73.0	3659	—
RX25	71.0	3322	—	75.0	3630	—

115 gr. Jacketed BT
3.250" OAL
BC: .453
SD: .249

Powder	Sugg Starting Grains	Velocity fps	Pressure	Max Load Grains	Velocity fps	Pressure
IMR-4831	60.0	3039	—	64.0	3247	—
XMR-3100	63.2	3099	—	66.5	3294	—
H4831	63.0	3036	—	66.0	3209	—
RX22	61.5	3049	—	64.5	3158	—
IMR-7828	66.5	3212	—	70.0	3392	—
RX25	67.0	3069	—	**71.0**	**3327**	—
H870	72.0	2775	—	76.0	3058	—

Note: Loads shown in shaded panels are maximum.
Loads shown in bold designate potentially most accurate load.

260 Remington

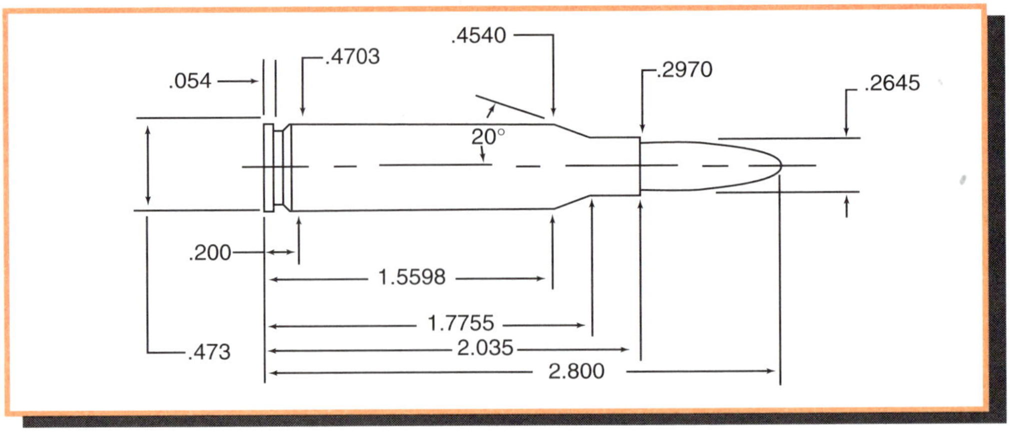

Comments:

Many High-Power shooters have used a number of 6.5mm wildcats over recent years to great effect. Various 6.5 wildcats based on the 308 Winchester case had been around for years before Remington standardized the 260 as a factory round in 1997. While ballistically similar to the 6.5x55 Swedish Mauser cartridge, the 260 enjoys the advantage of being loaded to higher pressures due to the absence of older, weaker actions in this chambering. An ideal package of low recoil, high accuracy potential, plus a wide selection of match grade and hunting bullets with good ballistic coefficients offer a very useful combination. The ability to launch the 120-grain Nosler Ballistic Tip at 2,900 feet per second makes a handy whitetail cartridge at the closer "non-magnum" distances found along the east coast. High Power shooters should find the 142-grain Sierra Match King with IMR-4350 a winner. Shooters loading for custom rifles marked 6.5-08, 6.5-308, or any similarly designated rifle — especially semi-autos — should exercise caution as minor dimensional variations can be encountered in such custom chambers. If in doubt, a chamber cast by a competent gunsmith is recommended. Laboratory tests showed good results with most of the VihtaVuori powders as well as Varget and IMR-4350.

Test Components:

Cases . . . . .Remington
Trim-to Length . . . . .2.025"
Primers . . . . .Remington 9½
Primer Size . . . . .Large Rifle
Lyman Shell Holder . . . . .No. 2
Jacketed Bullets Used . . . . . . .Sierra HP #1700, 85 gr.
Hornady V-Max #22601, 95 gr.
Sierra HP #1710, 100 gr.
Nosler Ballistic Tip #26120, 120 gr.
Hornady A-Max #26332, 140 gr.
Sierra HPBT #1742, 142 gr.
Hornady RN #2640, 160 gr.
Cast Bullets Used . . . .(sized to .264" dia)
*gas check bullets *#266469, 140 gr.
*#266673, 150 gr.

Test Specifications: (Velocity & Pressure)

Firearm Used . . . . .Universal Receiver
Barrel Length . . . . .24"
Twist . . . . .1-9"
Groove Dia. . . . . .2645"

85 gr. Jacketed HP 2.670" OAL — **BC:** .225 **SD:** .174

Powder	Sugg Starting Grains	Velocity fps	Pressure P.S.I.	Max Load Grains	Velocity fps	Pressure P.S.I.
IMR-4064	39.0	3022	40,000	44.0	3432	58,400
Varget	39.5	3097	43,600	44.0	3443	59,700
RX15	41.0	3084	42,800	45.5	3447	59,000
N150	40.5	3106	46,100	**45.0**	**3376**	**58,300**
AA2700	44.0	2965	38,300	49.0	3384	57,900
760	43.5	2988	37,200	48.5	3406	55,400
H-414	45.0	2966	38,100	50.0	3387	57,800
IMR-4350	44.1	2813	35,600	49.0+	3190	49,000
RX19	46.0	2777	34,600	51.0+	3159	48,200
IMR-4831	44.0	2797	31,800	49.0+	3079	44,600
XMR-3100	44.0	2603	30,200	49.0+	2944	40,900
WXR	45.0	2716	32,000	50.0+	3055	43,500

95 gr. Jacketed V-Max 2.785" OAL — **BC:** .365 **SD:** .195

Powder	Sugg Starting Grains	Velocity fps	Pressure P.S.I.	Max Load Grains	Velocity fps	Pressure P.S.I.
IMR-4064	38.5	2899	40,100	42.5	3309	59,500
Varget	**38.0**	**2987**	**44,800**	42.5	3287	58,900
RX15	40.0	2964	43,400	44.0	3307	59,700
N150	39.8	2928	42,900	44.2	3217	57,800
AA2700	43.5	2958	46,600	48.2	3228	58,900
760	44.0	2944	40,600	48.0	3312	58,400
H-414	44.5	2882	39,400	49.5	3275	57,500
IMR-4350	44.0	2832	40,000	48.0+	3197	55,200
RX19	46.0	2804	39,400	51.0+	3172	54,100
IMR-4831	44.0	2736	36,000	49.0+	3114	50,600
XMR-3100	43.0	2542	31,300	48.0+	2914	44,600
WXR	44.0	2595	31,800	49.0+	2958	44,500

Note: Loads shown in shaded panels are maximum.
Loads shown in bold designate potentially most accurate load.
\+ Designates a compressed powder charge.

260 Remington

100 gr. Jacketed HP
2.710" OAL

BC: .259
SD: .205

Powder	Sugg Starting Grains	Velocity fps	Pressure P.S.I.	Max Load Grains	Velocity fps	Pressure P.S.I.
IMR-4064	37.5	2846	41,500	41.5	3175	57,500
Varget	37.3	2895	46,000	41.7	3161	58,200
RX15	39.0	2905	45,400	43.0	3193	59,000
N150	**39.3**	**2877**	**46,100**	**43.7**	**3126**	**59,300**
AA2700	42.0	2831	43,300	46.5	3116	58,600
760	42.7	2886	42,300	47.5	3221	58,500
H-414	44.0	2864	42,600	49.0	3213	59,700
IMR-4350	43.7	2766	40,000	47.5+	3156	58,100
RX19	45.0	2745	39,300	50.0+	3106	54,700
IMR-4831	44.0	2695	37,000	49.0+	3095	53,300
XMR-3100	44.0	2582	35,400	48.0+	2950	50,600
WXR	44.0	2610	34,300	48.5+	2958	47,500

120 gr. Jacketed Ballistic Tip
2.785" OAL

BC: .458
SD: .246

Powder	Sugg Starting Grains	Velocity fps	Pressure P.S.I.	Max Load Grains	Velocity fps	Pressure P.S.I.
Varget	36.5	2619	47,500	40.5	2844	58,500
RX15	36.5	2648	48,900	40.5	2875	59,400
N150	**36.5**	**2593**	**48,800**	**40.5**	**2799**	**58,900**
AA2700	38.7	2472	41,400	43.0	2799	58,900
760	40.5	2603	42,300	45.0	2902	58,000
H-414	40.7	2560	41,400	45.2	2892	58,700
IMR-4350	41.0	2589	43,900	45.7	2900	59,100
RX19	43.0	2594	42,500	47.7+	2894	56,800
IMR-4831	42.0	2570	41,500	46.5+	2889	56,600
XMR-3100	42.3	2458	39,600	47.0+	2754	52,900
WXR	42.0	2498	39,000	47.0+	2758	49,200
IMR-7828	40.5	2288	33,400	45.0+	2552	43,200

140 gr. Jacketed A-Max
2.785" OAL

BC: .550
SD: .287

Powder	Sugg Starting Grains	Velocity fps	Pressure P.S.I.	Max Load Grains	Velocity fps	Pressure P.S.I.
RX15	34.7	2399	45,800	38.5	2635	58,000
AA2700	37.8	2383	45,400	42.0	2625	58,400
760	38.7	2412	42,500	44.0	2714	58,700
H-414	39.0	2413	42,800	43.0	2705	58,100
RX19	38.3	2295	38,500	42.5+	2513	46,500
N160	**37.5**	**2315**	**42,100**	**41.5+**	**2520**	**50,000**
IMR-4350	37.5	2298	39,400	41.5+	2557	51,200
IMR-4831	37.5	2263	37,300	41.5+	2500	46,900
XMR-3100	37.0	2131	34,500	41.0+	2344	42,400
WXR	38.7	2276	38,500	43.0+	2485	46,000
IMR-7828	36.0	2006	30,000	40.0+	2220	36,200
H1000	38.3	2049	31,100	42.5+	2229	36,000

142 gr. Jacketed HPBT
2.800" OAL

BC: .580
SD: .291

Powder	Sugg Starting Grains	Velocity fps	Pressure P.S.I.	Max Load Grains	Velocity fps	Pressure P.S.I.
Varget	34.5	2457	47,300	38.5	2669	57,400
N540	35.5	3475	46,000	39.5	2712	58,000
N550	37.3	2469	42,900	41.5	2729	56,000
AA2700	37.7	2477	45,200	42.0	2718	55,700
H-414	39.0	2454	42,800	43.5	2755	59,300
IMR-4350	39.5	2546	49,000	44.0	2760	57,800
N160	40.5	2517	45,500	45.0	2763	58,000
N560	**42.3**	**2473**	**43,100**	**47.0**	**2749**	**56,600**
RX19	41.5	2557	47,200	46.0	2787	57,000
IMR-4831	40.5	2516	45,100	45.0	2767	57,200
H4831SC	42.3	2489	45,600	47.0	2723	58,200
RX22	43.2	2523	44,500	48.0+	2819	58,800

160 gr. Jacketed RN
2.840" OAL

BC: .283
SD: .328

Powder	Sugg Starting Grains	Velocity fps	Pressure P.S.I.	Max Load Grains	Velocity fps	Pressure P.S.I.
N160	35.5	2246	52,600	39.5	2409	58,900
RX19	39.5	2334	47,300	44.0	2561	58,500
IMR-4831	38.3	2301	47,400	42.5	2513	58,700
XMR-3100	39.3	2242	47,100	43.7	2466	58,700
RX22	**41.0**	**2341**	**47,600**	**45.5**	**2566**	**58,300**
WXR	39.5	2347	50,800	44.0	2533	59,500
IMR-7828	**40.5**	**2287**	**47,100**	**45.0+**	**2514**	**58,800**
H1000	43.0	2310	47,600	48.0+	2524	59,000

#266469
140gr. (#2 Alloy) 2.600" OAL

BC: .323
SD: .286

Powder	Sugg Starting Grains	Velocity fps	Pressure P.S.I.	Max Load Grains	Velocity fps	Pressure P.S.I.
SR-4759	15.0	1633	22,300	24.0	2227	55,500
XMP-5744	17.0	1666	20,600	25.0	2194	39,000
IMR-4198	**17.0**	**1641**	**18,700**	28.0	2326	45,900
RX7	18.0	1651	19,900	27.0	2198	37,900
IMR-3031	21.5	1661	17,500	30.0	2201	31,900
IMR-4895	22.0	1609	16,300	32.0	2261	34,400

Note: Loads shown in shaded panels are maximum.
Loads shown in bold designate potentially most accurate load.
\+ Designates a compressed powder charge.

260 Remington

#266673
150gr. (#2 Alloy) 2.800" OAL

BC: .305
SD: .307

Powder	Sugg Starting Grains	Velocity fps	Pressure P.S.I.	Max Load Grains	Velocity fps	Pressure P.S.I.
SR-4759	**18.0**	**1760**	**30,300**	25.0	2167	56,300
XMR-5744	18.0	1645	21,100	28.0	2243	46,400
IMR-4198	18.0	1624	19,700	28.0	2213	45,300
RX7	20.0	1698	21,400	30.0	2247	45,700
IMR-3031	23.0	1659	17,900	33.0	2266	40,200
IMR-4895	24.0	1658	18,200	33.0	2199	35,400

Note: Loads shown in shaded panels are maximum.
Loads shown in bold designate potentially most accurate load.

6.5 x 55mm Swedish Mauser

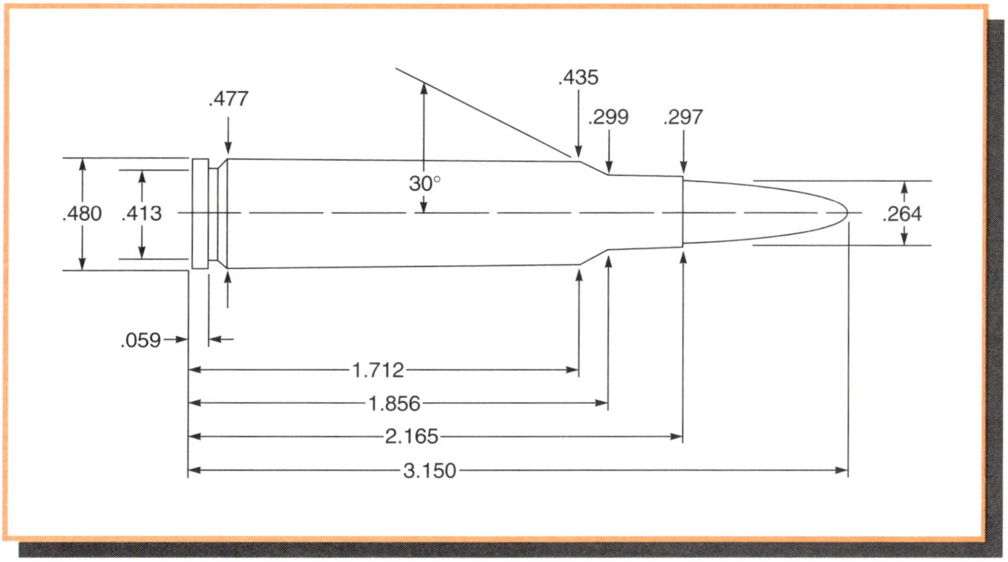

Comments:

This was for many years the least known of the Mauser military cartridges to American shooters. Recent importation of large numbers of inexpensive — and well made — Swedish Mausers have popularized this fine old cartridge as never before. The accuracy of many of these rifles rivals that of currently produced factory models. Indeed, most American rifle manufacturers have added this chambering in the past few years. Scandinavian hunters have used this cartridge for many years on game as large as moose as well as a match cartridge. Its combination of excellent accuracy and low recoil have found favor with American shooters as well. The 6.5x55 suffered for many years due to the lack–or great expense–of factory ammo. Shooters loading for the Swede should not under any circumstance attempt to form brass from 30-06 cases. Case diameter of the 30-06 runs .008" to .010" smaller than the Swedish cartridge. Such reformed cases will bulge dramatically when fired, or worse.

All major ammunition manufacturers currently offer loadings for this caliber as well as bulk brass so such case forming operations are unnecessary. Although well made, shooters loading for any of the M94, M96, or M38 Swedish military rifles should work up loads cautiously, observe for pressure signs, and back off accordingly should any arise. The same warning applies to any Norwegian issue Krag-Jorgensen rifle one may be loading for. SAAMI established a Maximum Average Pressure (MAP) of 46,000 CUP in deference to design of these older military actions. Powders in the medium to medium-slow range usually give the best results. IMR-4895 and IMR-4064 usually produce excellent results with bullets up to 140-grains. Many surplus rifles show variation in the groove diameter and the Swedish Mausers are no exception. Shooters should slug their bore and size accordingly.

Test Components:

Cases	Federal
Trim-to Length	2.155"
Primers	Federal 210
Primer Size	Large Rifle
Lyman Shell Holder	No. 27
Jacketed Bullets Used	Sierra HP #1710, 100 gr.
	Nosler Ballistic Tip #26120, 120 gr.
	Hornady SP, #2620, 129 gr.
	Sierra HPBT, #1740, 140 gr.
	Hornady RN, #2640, 160 gr.
Cast Bullets Used	(sized to .264" dia)
*gas check bullets	*#266469, 140 gr.
	*#266673, 150 gr.

Test Specifications:
(Velocity Only)

Firearm Used	Swedish Mauser M38
Barrel Length	24"
Twist	1-7½"
Groove Dia.	.264"

6.5 x 55mm Swedish Mauser

100 gr. Jacketed HP
2.850" OAL

BC: .259
SD: .205

Powder	Sugg Starting Grains	Velocity fps	Pressure	Max Load Grains	Velocity fps	Pressure
IMR-4895	39.0	2671	—	43.5	3027	—
IMR-4064	39.5	2654	—	44.0	2990	—
Varget	35.0	2351	—	39.0	2693	—
N150	37.0	2419	—	41.0	2745	—
N550	39.5	2399	—	44.0	2728	—
AA2700	41.5	2441	—	46.0	2790	—
H-414	45.0	2724	—	50.0	3034	—
XMR-4350	43.0	2407	—	48.0	2784	—
RX19	**46.0**	**2612**	—	51.0+	2939	—

120 gr. Jacketed Ballistic Tip
3.000" OAL

BC: .458
SD: .246

Powder	Sugg Starting Grains	Velocity fps	Pressure	Max Load Grains	Velocity fps	Pressure
IMR-4895	37.0	2474	—	41.0	2762	—
IMR-4064	36.0	2406	—	40.0	2643	—
Varget	35.0	2383	—	39.0	2606	—
N150	36.5	2431	—	**40.5**	**2652**	—
N550	39.5	2497	—	44.0	2747	—
AA2700	40.0	2413	—	44.0	2652	—
H-414	40.0	2379	—	44.5	2664	—
XMR-4350	41.5	2374	—	46.0	2657	—
RX19	43.0	2459	—	48.0	2740	—

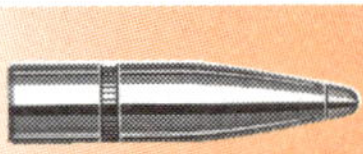

129 gr. Jacketed SP
3.035" OAL

BC: .445
SD: .264

Powder	Sugg Starting Grains	Velocity fps	Pressure	Max Load Grains	Velocity fps	Pressure
IMR-4895	35.0	2301	—	39.0	2567	—
IMR-4064	**36.5**	**2355**	—	40.5	2602	—
H-414	39.5	2343	—	44.0	2570	—
N160	41.5	2353	—	46.0	2600	—
N560	43.5	2352	—	48.5	2603	—
XMR-4350	41.0	2235	—	45.5	2547	—
RX19	41.5	2355	—	46.5	2615	—

140 gr. Jacketed HPBT
3.050" OAL

BC: .526
SD: .287

Powder	Sugg Starting Grains	Velocity fps	Pressure	Max Load Grains	Velocity fps	Pressure
IMR-4895	32.5	2177	—	**36.5**	**2370**	—
IMR-4064	33.5	2188	—	37.5	2410	—
Varget	33.5	2251	—	37.0	2440	—
H-414	37.5	2222	—	41.5	2455	—
N160	39.5	2273	—	44.0	2512	—
N560	42.0	2300	—	47.0	2574	—
IMR-4350	37.5	2180	—	42.0	2455	—
XMR-4350	39.5	2163	—	44.0	2445	—
RX19	40.0	2234	—	44.5	2521	—
XMR-3100	41.0	2162	—	45.5	2475	—
H4831SC	41.5	2258	—	46.0	2508	—
RX22	42.3	2275	—	47.0	2576	—

160 gr. Jacketed RN
3.035" OAL

BC: .283
SD: .328

Powder	Sugg Starting Grains	Velocity fps	Pressure	Max Load Grains	Velocity fps	Pressure
H-414	37.0	2108	—	41.0	2314	—
N160	38.7	2125	—	43.0	2347	—
N560	41.5	2162	—	46.0	2370	—
XMR-4350	39.5	2065	—	44.0	2330	—
RX19	38.7	2061	—	43.0	2317	—
IMR-4831	38.7	2100	—	43.0	2339	—
XMR-3100	39.5	2003	—	44.0	2269	—
H-4831SC	41.5	2166	—	46.0	2388	—
RX22	41.5	2129	—	**46.0**	**2361**	—

#266469
140gr. (#2 Alloy) 2.925" OAL

BC: .323
SD: .286

Powder	Sugg Starting Grains	Velocity fps	Pressure	Max Load Grains	Velocity fps	Pressure
AA#9	12.5	1479	—	16.5	1765	—
2400	12.5	1435	—	17.0	1741	—
SR-4759	13.0	1375	—	25.0	2199	—
IMR-4227	**17.0**	**1630**	—	25.0	2098	—
XMP-5744	18.5	1674	—	25.5	2111	—
IMR-4198	15.0	1443	—	26.0	2091	—
RX7	15.0	1425	—	27.0	2182	—

#266673
150gr. (#2 Alloy) 3.025" OAL

BC: .305
SD: .307

Powder	Sugg Starting Grains	Velocity fps	Pressure	Max Load Grains	Velocity fps	Pressure
AA#9	13.0	1472	—	17.0	1724	—
2400	13.0	1444	—	18.0	1726	—
SR-4759	**16.5**	**1608**	—	24.0	2044	—
IMR-4227	17.5	1619	—	25.0	2044	—
XMP-5744	18.0	1605	—	27.0	2136	—
IMR-4198	18.0	1627	—	26.0	2079	—
RX7	18.0	1619	—	27.0	2108	—

Note: Loads shown in shaded panels are maximum.
Loads shown in bold designate potentially most accurate load.
\+ Designates a compressed powder load.

264 Winchester Magnum

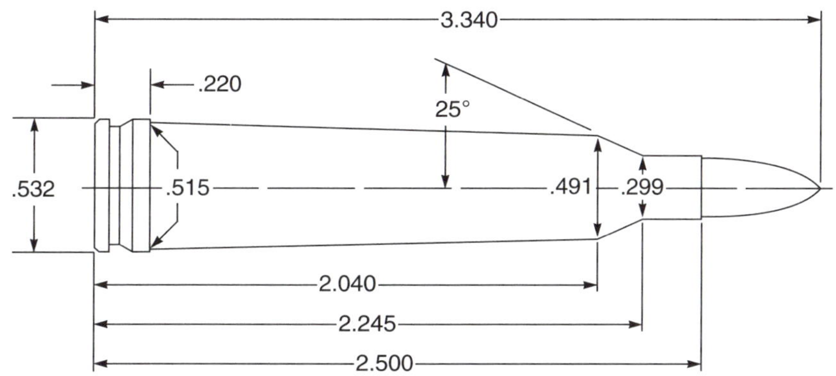

Comments:

The 264 Magnum was the smallest of several belted magnums offered by Winchester during the late 1950s. The 264 Winchester Magnum has been regarded by some as ahead of its time. Most American shooters had known the 6.5mm bore size only through ballistically lackluster, foreign military surplus. Its high velocity and flat trajectory necessitated the 26-inch barrel of Winchester's Model 70 Westerner to achieve peak performance. Factory ammunition utilized a unique two-step projectile similar in design to a bore-riding cast bullet.

As with any overbore rifle such as the 264 Winchester, barrel life is an issue. The availability of slow-burning powders for reloading at the time of the 264s introduction was limited. Fortunately, the current selection of powder suitable for the 264 Magnum is quite good and allows the handloader to bring out its full potential. Bullets from after market suppliers are of conventional design and will need to be seated deep in order to clear the rifling.

Remington's introduction of their 7mm Magnum in 1962 immediately outshone the 264 and it never quite recovered. Shooters loading for the 7mm enjoyed a better selection of bullets weighing up to 175-grains. Winchester currently catalogs the 264 as a Custom Shop item.

Test Components:

Cases . . . . .Winchester
Trim-to Length . . . . .2.490"
Primers . . . . .Remington 9½ Magnum
Primer Size . . . . .Large Rifle
Lyman Shell Holder . . . . .No. 13
Jacketed Bullets Used . . . . .Sierra SBT #1730, 140 gr.

Test Specifications:
(Velocity & Pressure)

Firearm Used . . . . .Universal Receiver
Barrel Length . . . . .24"
Twist . . . . .1-9"
Groove Dia. . . . .264"

140 gr. Jacketed SBT
3.340" OAL

BC: .495
SD: .287

Powder	Sugg Starting Grains	Velocity fps	Pressure C.U.P.	Max Load Grains	Velocity fps	Pressure C.U.P.
IMR-4831	57.5	2869	48,500	60.8	2975	51,700
H-4831SC	58.0	2824	47,500	61.0	2943	54,000
RX22	56.0	2770	41,600	**59.0**	**2903**	**46,300**
IMR-7828	54.5	2662	38,600	**60.5**	**2909**	**46,200**
H1000	60.0	2753	44,400	63.0	2849	46,400
RX25	60.5	2854	45,200	64.0	2959	47,200
H50BMG	66.5	2734	43,300	70.0	2874	47,600
AA8700	69.5	2732	39,400	73.5	2883	45,800

Note: Loads shown in shaded panels are maximum.
Loads shown in bold designate potentially most accurate load.

270 Winchester

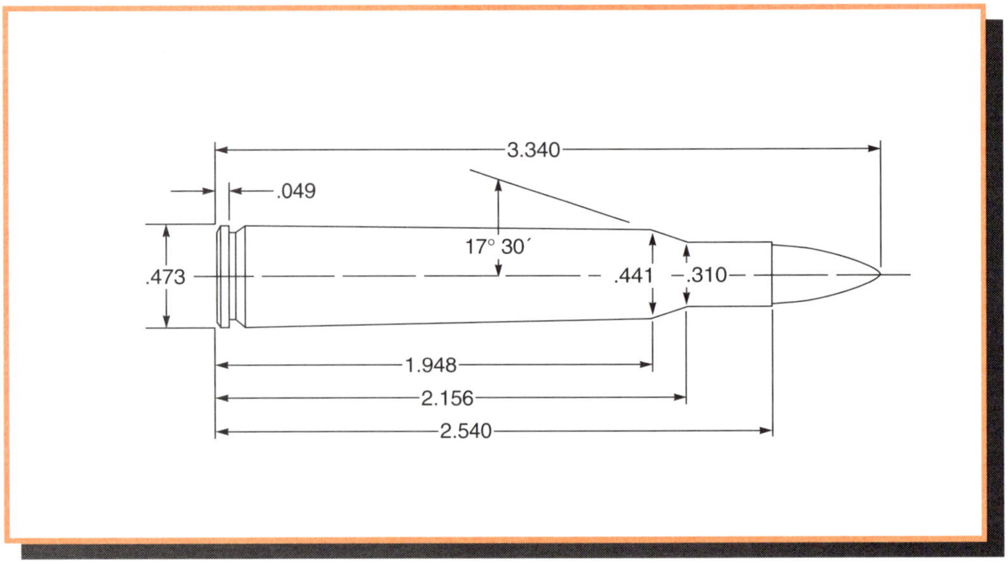

Comments:

The 270 Winchester has become one of the classic American rifle cartridges over the last 75 years. It has taken virtually every thin-skinned game imaginable and provides endless debate as one half of the "30-06 Springfield versus 270 Winchester" argument. Indeed, 270 Winchester dies have consistently been among the top three in Lyman's rifle die sales for many years. Winchester introduced the 270 in 1925 in their Model 54 rifle. The cartridge is essentially a necked down 30-06 with minor dimensional changes. The .277" diameter bullet was entirely new to the shooting public. Some sources indicate Winchester may have based this bullet size on an obscure prototype for a Chinese military cartridge dating to the turn of the century.

The 270 Winchester has become synonymous with the late Jack O'Connor of *Outdoor Life* magazine. His untiring advocacy of the 270 Winchester loaded with 130-grain bullets established it as a potent, flat-shooting cartridge suitable most North American hunting situations. Properly loaded, the 270 will do much of what the 30-06 will do on all but the larger species in North America.

Test Components:

Cases . . . . Winchester
Trim-to Length . . . . 2.530"
Primers . . . . Winchester WLR
Primer Size . . . . Large Rifle
Lyman Shell Holder . . . . No. 2
Jacketed Bullets Used . . . . Sierra HP #1800, 90 gr.
Hornady SP #2710, 100 gr.
Sierra SPT #1810, 110 gr.
Sierra SBT #1820, 130 gr.
Hornady BTSP #2735, 140 gr.
Hornady SP #2740, 150 gr.
Nosler SP, #16324, 160 gr.
Cast Bullets Used . . . . (sized to .278" dia)
*gas check bullet *280642, 150 gr.

Test Specifications: (Velocity & Pressure)

Firearm Used . . . . Universal Receiver
Barrel Length . . . . 26"
Twist . . . . 1-10"
Groove Dia. . . . . .277"

90 gr. Jacketed HP
3.090" OAL
BC: .195
SD: .168

Powder	Sugg Starting Grains	Velocity fps	Pressure C.U.P.	Max Load Grains	Velocity fps	Pressure C.U.P.
IMR-4064	48.5	3032	37,700	53.8	3532	51,500
Varget	48.5	3112	38,000	54.0	3536	50,300
RX15	48.5	3142	38,700	54.0	3527	49,100
AA2700	50.5	3020	39,500	**56.0**	**3401**	**50,500**
IMR-4350	52.0	2883	36,800	58.0+	3376	49,900
IMR-4831	54.0	2907	37,400	60.0+	3416	50,200

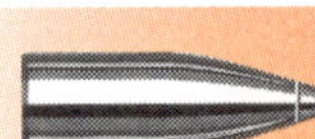

100 gr. Jacketed SP
3.175" OAL
BC: .307
SD: .186

Powder	Sugg Starting Grains	Velocity fps	Pressure C.U.P.	Max Load Grains	Velocity fps	Pressure C.U.P.
IMR-4895	46.5	3134	44,600	50.0	3356	53,600
IMR-4064	46.0	3067	40,300	51.5	3413	53,600
Varget	47.5	3071	41,200	53.0	3416	50,900
IMR-4320	46.0	2976	39,900	51.5	3333	53,200
RX15	48.0	2991	38,700	53.5	3364	50,300
H-380	48.0	3086	39,900	54.0	3436	54,000
AA2700	50.5	3039	45,200	**56.0**	**3353**	**52,000**
IMR-4350	50.0	2949	41,100	57.0+	3345	54,400
IMR-4831	53.5	2763	37,000	59.5+	3254	48,000
H4831	55.0	2702	40,300	62.0+	3356	51,600
**SR-4759	23.0	2279	33,900	27.3	2607	39,900
**XMP-5744	26.5	2326	31,200	31.3	2614	36,900

Note: Loads shown in shaded panels are maximum.
Loads shown in bold designate potentially most accurate load.
\+ Designates a compressed powder charge.
** Designates a reduced load.

270 Winchester

110 gr. Jacketed SPT
3.285" OAL

BC: .318
SD: .205

Powder	Sugg Starting Grains	Velocity fps	Pressure C.U.P.	Max Load Grains	Velocity fps	Pressure C.U.P.
IMR-4064	45.0	2920	40,000	**50.5**	**3271**	**50,600**
Varget	46.0	2922	40,000	51.0	3213	50,800
RX15	45.0	2861	38,000	52.7	3296	51,400
AA2700	48.0	2881	43,200	53.5	3128	50,600
H-414	49.2	2833	35,500	56.0	3266	50,400
IMR-4350	49.2	2873	38,700	56.0	3274	49,600
IMR-4831	54.0	2846	37,500	61.0+	3376	51,900
XMR-3100	52.0	2655	36,100	61.5+	3230	51,800
H4831	54.0	2824	38,600	61.0+	3174	49,200

130 gr. Jacketed SBT
3.250" OAL

BC: .436
SD: .242

Powder	Sugg Starting Grains	Velocity fps	Pressure C.U.P.	Max Load Grains	Velocity fps	Pressure C.U.P.
IMR-4064	44.5	2695	39,500	49.5	3014	50,600
Varget	42.0	2723	41,500	47.0	2970	52,000
H-380	44.0	2680	41,100	50.0	2994	54,000
H-414	48.5	2726	40,700	54.0	3033	50,500
IMR-4350	49.0	2654	38,200	54.5	3032	51,100
N160	50.0	2652	37,700	55.5	2999	49,800
RX19	48.5	2486	35,000	57.0	3066	51,900
IMR-4831	53.0	2823	40,800	57.0+	3127	52,100
XMR-3100	50.0	2536	36,300	59.0+	2939	51,900
H4831	52.0	2726	40,200	**58.0+**	**2990**	**50,900**
RX22	51.0	2564	35,600	60.0+	3173	51,800
IMR-7828	53.5	2600	39,200	59.5+	3042	51,700
**SR-4759	24.0	2210	39,000	28.5	2502	50,200
**XMP-5744	27.0	2223	35,100	31.0	2469	39,900

140 gr. Jacketed BTSP
3.335" OAL

BC: .486
SD: .261

Powder	Sugg Starting Grains	Velocity fps	Pressure C.U.P.	Max Load Grains	Velocity fps	Pressure C.U.P.
IMR-4064	41.3	2612	39,000	47.0	2929	52,000
H-414	46.2	2684	44,200	52.5	2914	51,700
N160	50.0	2588	38,900	**55.5**	**2929**	**50,900**
IMR-4350	49.0	2604	38,800	54.0	2982	51,000
RX19	51.3	2651	40,200	57.0	3010	51,400
IMR-4831	52.0	2703	39,900	56.3	3010	51,500
XMR-3100	48.8	2453	36,000	57.5	2979	51,900
H-4831	51.0	2603	35,000	58.0+	3004	51,000
RX22	52.5	2587	37,100	58.5+	2980	50,400
IMR-7828	53.0	2544	37,300	59.0+	2958	50,400

150 gr. Jacketed SP
3.285" OAL

BC: .462
SD: .279

Powder	Sugg Starting Grains	Velocity fps	Pressure C.U.P.	Max Load Grains	Velocity fps	Pressure C.U.P.
IMR-4064	42.5	2510	41,800	47.0	2750	50,900
H-380	40.4	2434	41,300	46.0	2655	51,500
H-414	46.0	2547	44,200	51.0	2777	50,400
N160	48.0	2494	41,600	53.5	2766	50,900
IMR-4350	48.0	2524	40,400	52.0	2833	51,100
RX19	46.9	2386	36,300	55.2	2857	51,600
IMR-4831	50.3	2614	42,000	54.3	2860	52,000
XMR-3100	48.0	2354	36,300	56.6	2852	51,900
H-4831	50.5	2440	39,400	**56.0**	**2788**	**51,200**
RX22	48.8	2466	36,300	57.5+	2852	51,000
IMR-7828	51.5	2465	38,300	57.0+	2810	48,800
RX25	56.0	2569	39,400	62.0+	2907	49,600

160 gr. Jacketed SP
3.340" OAL

BC: .434
SD: .298

Powder	Sugg Starting Grains	Velocity fps	Pressure C.U.P.	Max Load Grains	Velocity fps	Pressure C.U.P.
IMR-4350	47.0	2470	40,500	51.0	2755	50,600
IMR-4831	48.0	2403	37,000	53.5	2741	49,000
XMR-3100	49.5	2344	38,100	55.0	2685	50,000
RX22	51.5	2447	38,300	**57.0+**	**2798**	**49,900**
IMR-7828	51.5	2412	39,900	57.0+	2781	49,800
H1000	54.0	2407	35,700	60.0+	2776	49,300
RX25	55.5	2478	37,800	61.5+	2853	50,800

#280642
150 gr. (#2 Alloy) 3.073" OAL

BC: .260
SD: .279

Powder	Sugg Starting Grains	Velocity fps	Pressure C.U.P.	Max Load Grains	Velocity fps	Pressure C.U.P.
Unique	12.0	1519	26,300	19.0	2025	48,200
SR-7625	13.0	1533	29,800	18.0	1876	50,100
SR-4756	13.0	1500	26,100	19.0	1911	47,800
SR-4759	16.0	1696	23,500	23.0	2153	40,400
AA-1680	20.5	1753	22,500	28.5	2160	35,300
XMP-5744	18.0	1685	22,300	24.5	2059	33,900
IMR-4198	18.0	1675	21,600	26.0	2147	34,000
XMR-2015	30.0	2058	27,300	34.0	2254	33,500
RX7	19.0	1711	20,700	**27.0**	**2155**	**34,700**

Note: Loads shown in shaded panels are maximum.
Loads shown in bold designate potentially most accurate load.
+ Designates a compressed powder charge.
**Designates a reduced load.

270 WSM (Winchester Short Magnum)

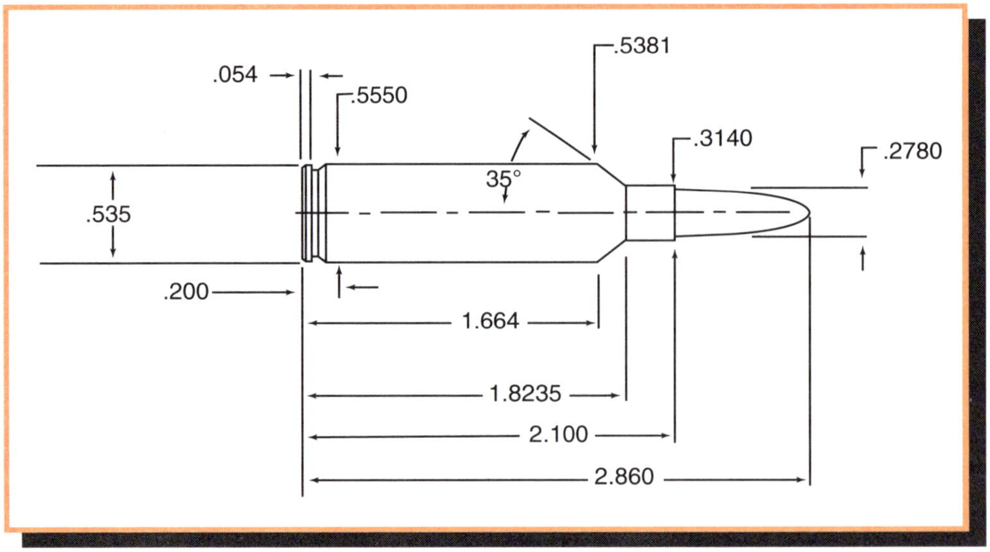

Comments:

The late Jack O'Connor spent much of his long career spreading the gospel of the 270 Winchester in the pages of *Outdoor Life* magazine. Along with Weatherby's 270 Magnum developed in 1943, they constituted the only two factory produced chamberings in the .277" bullet size for many years. Aficionados of the .277" diameter bullet now have their own Short Magnum cartridge. U.S. Repeating Arms/Winchester introduced the 270 WSM in 2002 as the first new factory offering in the 270 caliber in nearly 60 years and the first designed for a short action. The 270 WSM is a necked down version of the 300 WSM sharing its 35-degree shoulder and 2.100" length. Ballistic performance of the WSM tops the time-proven 270 Winchester by around 200 feet per second and is about on par to that of the 270 Weatherby Magnum. The overall length of many premium bullets may require that they be seated deeper into the case than they would be in a longer action. Unfortunately, the configuration of Lyman's current 270 caliber cast bullet prevented its use in the WSM. Powders that gave best results in laboratory tests included Alliant's Reloder 19 and Reloder 22, IMR-7828, and VihtaVuori N165.

Test Components:

Cases . . . . .Winchester
Trim-to Length . . . . .2.090"
Primers . . . . .Winchester WLRM
Primer Size . . . . .Large Rifle Magnum
Lyman Shell Holder . . . . .No. 34
Jacketed Bullets Used . . . . . . .Sierra HP #1800, 90 gr.
Sierra SPT #1810, 110 gr.
Speer Grand Slam #1465, 130 gr.
Swift Scirocco, 130 gr.
Hornady BTSP #2735, 140 gr.
Barnes X #27735, 150 gr.
Hornady RN #2745, 150 gr.
Nosler SP #16324, 160 gr.

Test Specifications:
(Velocity & Pressure)

Firearm Used . . . . .Universal Receiver
Barrel Length . . . . .24"
Twist . . . . .1-10"
Groove Dia. . . . . .277"

90 gr. Jacketed HP
2.700" OAL
BC: .195
SD: .168

Powder	Sugg Starting Grains	Velocity fps	Pressure P.S.I.	Max Load Grains	Velocity fps	Pressure P.S.I.
IMR-4350	66.5	3544	50,900	**70.0**	**3786**	**61,400**
RX19	71.0	3545	51,900	75.0+	3777	60,900

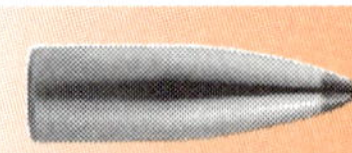

110 gr. Jacketed SPT
2.750" OAL
BC: .318
SD: .205

Powder	Sugg Starting Grains	Velocity fps	Pressure P.S.I.	Max Load Grains	Velocity fps	Pressure P.S.I.
760	61.5	3316	56,800	65.0	3431	62,100
IMR-4350	62.5	3274	52,000	66.0	3467	61,200
N160	64.0	3378	59,000	67.5	3481	62,100
RX19	67.5	3312	52,700	**71.0+**	**3515**	**62,100**

Note: Loads shown in shaded panels are maximum.
Loads shown in bold designate potentially most accurate load.
\+ Designates a compressed powder charge.

270 WSM (Winchester Short Magnum)

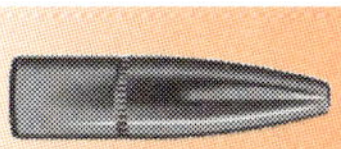

130 gr. Jacketed GSSP
2.765" OAL — **BC:** .345 **SD:** .241

Powder	Sugg Starting Grains	Velocity fps	Pressure P.S.I.	Max Load Grains	Velocity fps	Pressure P.S.I.
760	56.5	2977	53,500	59.5	3102	58,700
H414	57.0	2979	53,300	60.0	3155	61,500
IMR-4350	59.5	3074	54,700	63.0	3222	61,600
IMR-4831	63.0	3105	54,000	66.5	3280	62,400
H-4831	64.0	3054	55,900	**67.5**	**3183**	**62,200**
RX22	64.5	3196	58,000	68.0	3302	61,600
WXR	64.5	3033	50,800	68.0	3234	60,000
MagPro	69.5	3133	53,000	73.5	3290	60,500

130 gr. Jacketed Scirocco
2.860" OAL — **BC:** .450 **SD:** .242

Powder	Sugg Starting Grains	Velocity fps	Pressure P.S.I.	Max Load Grains	Velocity fps	Pressure P.S.I.
H414	56.0	2990	53,400	59.0	3138	60,400
IMR-4350	59.0	3046	53,600	62.0	3199	61,200
N160	60.0	3053	54,600	63.5	3203	62,000
H-4831	63.0	3011	54,400	66.5	3158	61,900
RX19	63.5	3074	52,900	**67.0**	**3257**	**62,000**
WXR	65.5	3100	53,700	69.0+	3284	62,400
MagPro	68.5	3092	51,400	72.0	3252	60,100

140 gr. Jacketed BTSP
2.785" OAL — **BC:** .486 **SD:** .261

Powder	Sugg Starting Grains	Velocity fps	Pressure P.S.I.	Max Load Grains	Velocity fps	Pressure P.S.I.
H414	54.5	2880	53,000	57.5	3042	62,000
IMR-4350	58.5	2992	54,400	61.5	3141	62,400
IMR-4831	58.0	2954	52,900	61.0	3104	60,800
XMR-3100	61.5	2953	52,800	65.0	3137	62,700
RX22	60.0	3021	55,100	63.0	3160	62,100
WXR	61.0	3009	55,200	64.5	3138	61,500
IMR-7828	60.5	2966	53,000	64.0	3114	60,600
N165	61.5	2998	56,300	65.0	3111	61,800
RX25	64.0	2960	51,600	67.5+	3138	60,400
MagPro	65.5	2925	48,300	69.0	3154	60,400
Retumbo	68.0	2984	52,300	**71.5**	**3133**	**60,700**

150 gr. Barnes X
2.775" OAL — **BC:** .491 **SD:** .279

Powder	Sugg Starting Grains	Velocity fps	Pressure P.S.I.	Max Load Grains	Velocity fps	Pressure P.S.I.
RX22	55.5	2758	53,500	58.5	2893	59,900
WXR	60.5	2851	55,800	64.0	2994	62,400
IMR-7828	60.5	2842	56,200	**64.0**	**2982**	**62,500**
RX25	63.5	2856	55,600	67.0	3004	62,200
MagPro	64.5	2837	54,000	68.0	2996	62,000

150 gr. Jacketed RN
2.775" OAL — **BC:** .269 **SD:** .279

Powder	Sugg Starting Grains	Velocity fps	Pressure P.S.I.	Max Load Grains	Velocity fps	Pressure P.S.I.
H414	55.0	2846	56,700	58.0	2963	62,400
IMR-4831	58.0	2887	54,900	61.0	3020	62,300
RX22	59.5	2944	56,500	63.0	3058	62,600
WXR	60.5	2887	53,700	63.7	3034	61,600
N165	61.5	2894	56,100	**65.0**	**3010**	**62,300**
IMR-7828	61.5	2900	53,600	65.0	3048	62,000
RX25	65.0	2961	55,700	68.0	3088	62,300
MagPro	64.5	2896	53,200	68.0	3049	62,000
H1000	67.0	2868	53,800	70.5+	3004	61,100

160 gr. Jacketed SP
2.830" OAL — **BC:** .434 **SD:** .298

Powder	Sugg Starting Grains	Velocity fps	Pressure P.S.I.	Max Load Grains	Velocity fps	Pressure P.S.I.
RX22	55.0	2809	57,600	**58.0**	**2905**	**61,900**
IMR-7828	58.0	2798	54,800	61.0	2929	61,700
RX25	61.0	2821	54,000	64.5	2953	60,700
N170	62.5	2806	56,200	66.0	2917	61,900
H1000	62.5	2769	55,100	66.0	2913	62,900
MagPro	63.0	2832	55,500	66.5	2952	61,200
Retumbo	64.5	2852	56,100	68.0	2978	62,700

Note: Loads shown in shaded panels are maximum.
Loads shown in bold designate potentially most accurate load.
\+ Designates a compressed powder charge.

270 Weatherby Magnum

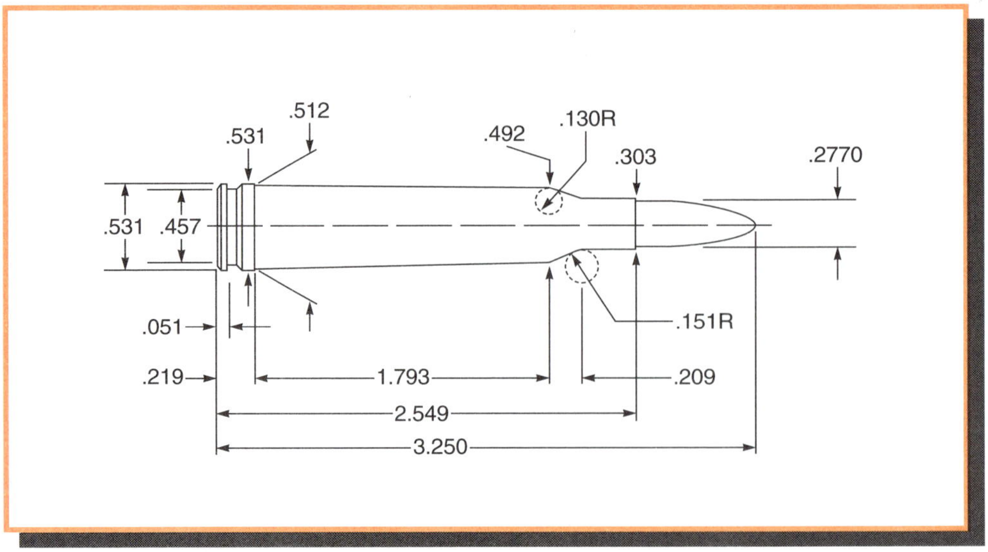

Comments:

Roy Weatherby developed his 270 Magnum as the first in his long and distinguished line of high velocity rifle cartridges in 1943. Based on a blown out 300 Holland and Holland case, its instantly recognizable double radius shoulder became the trademark of all future Weatherby designs. This was for many years the last word in high performance in the .277 bore size giving on average, some 200 feet per second over the 270 Winchester. Winchester's recent introduction of their 270 WSM however has given the Weatherby some competition. This cartridge works best with slower powders. IMR-7828 and H-1000 gave good results in our tests although H-4831 is a longtime favorite. Weatherby recommends exclusive use of the Federal 215 Magnum primers for use with all jacketed bullets. Shooters loading cast bullets should use standard primers. This data is intended for commercially produced and chambered rifles. This data is not for use in custom guns that may lack the free bore found in standard Weatherby chambers.

Test Components:

Cases . . . . .Federal
Trim-to Length . . . . .2.540"
Primers . . . . . . .Federal 215 Magnum(Jacketed bullets)
Federal 210 (Cast bullets)
Primer Size . . . . .Large Rifle
Lyman Shell Holder . . . . .No. 13
Jacketed Bullets Used . . .Speer Grand Slam #1465, 130 gr.
Hornady BTSP #2735, 140 gr.
Barnes X #27735, 150 gr.
Nosler Partition #16324, 160 gr.
Cast Bullets Used . . . .(sized to .277" dia)
*gas check bullet *#280642, 150 gr.

Test Specifications: (Velocity Only)

Firearm Used . . . . .Weatherby Mk V
Barrel Length . . . . .26"
Twist . . . . .1-10"
Groove Dia. . . . . .277"

130 gr. Jacketed GSSP 3.210" OAL — **BC:** .345 **SD:** .242

Powder	Sugg Starting Grains	Velocity fps	Pressure	Max Load Grains	Velocity fps	Pressure
IMR-4831	65.5	3078	—	69.0	3288	—
XMR-3100	66.5	3059	—	70.0	3302	—
H-4831	68.5	3136	—	72.0	3276	—
RX22	66.5	3145	—	70.0	3297	—
IMR-7828	67.5	3070	—	71.0	3321	—
H1000	72.0	3047	—	**76.0**	**3300**	**—**

140 gr. Jacketed BTSP 3.240" OAL — **BC:** .486 **SD:** .261

Powder	Sugg Starting Grains	Velocity fps	Pressure	Max Load Grains	Velocity fps	Pressure
IMR-4831	62.5	2897	—	66.0	3117	—
XMR-3100	63.5	2945	—	67.0	3141	—
H-4831	65.5	2951	—	69.0	3134	—
RX22	64.5	2960	—	68.0	3148	—
IMR-7828	65.5	2959	—	69.0	3145	—
H1000	69.0	2896	—	**73.0**	**3119**	**—**

Note: Loads shown in shaded panels are maximum.
Loads shown in bold designate potentially most accurate load.

270 Weatherby Magnum

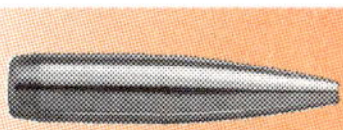

150 gr. Barnes X
3.250" OAL

BC: .491
SD: .279

Powder	Sugg Starting Grains	Velocity fps	Pressure	Max Load Grains	Velocity fps	Pressure
IMR-4831	62.5	2870	—	66.0	3055	—
XMR-3100	64.5	2928	—	**68.0**	**3108**	**—**
H-4831	64.5	2867	—	68.0	3064	—
RX22	63.5	2944	—	67.0	3097	—
IMR-7828	63.5	2830	—	67.0	3046	—
H1000	70.0	2915	—	74.0+	3113	—

160 gr. Jacketed SP
3.250" OAL

BC: .434
SD: .298

Powder	Sugg Starting Grains	Velocity fps	Pressure	Max Load Grains	Velocity fps	Pressure
IMR-4831	61.5	2767	—	65.0	2969	—
XMR-3100	61.5	2752	—	65.0	2943	—
H-4831	63.5	2809	—	67.0	2966	—
RX22	62.5	2867	—	66.0	3002	—
IMR-7828	63.5	2846	—	**67.0**	**2995**	**—**
H1000	67.5	2767	—	71.0	2968	—

#280642**
150 gr. (#2 Alloy) 3.200" OAL

BC: .260
SD: .279

Powder	Sugg Starting Grains	Velocity fps	Pressure	Max Load Grains	Velocity fps	Pressure
SR-4759	20.0	1725	—	28.0	2136	—
IMR-4227	22.0	1723	—	30.0	2112	—
XMP-5744	22.5	1707	—	30.0	2105	—
IMR-4198	**22.0**	**1696**	—	30.5	2103	—
RX7	23.0	1695	—	31.0	2109	—

Note: Loads shown in shaded panels are maximum.
Loads shown in bold designate potentially most accurate load.
** Designates use of Federal 210 primers.
+Designates a compressed powder charge.

7-30 Waters

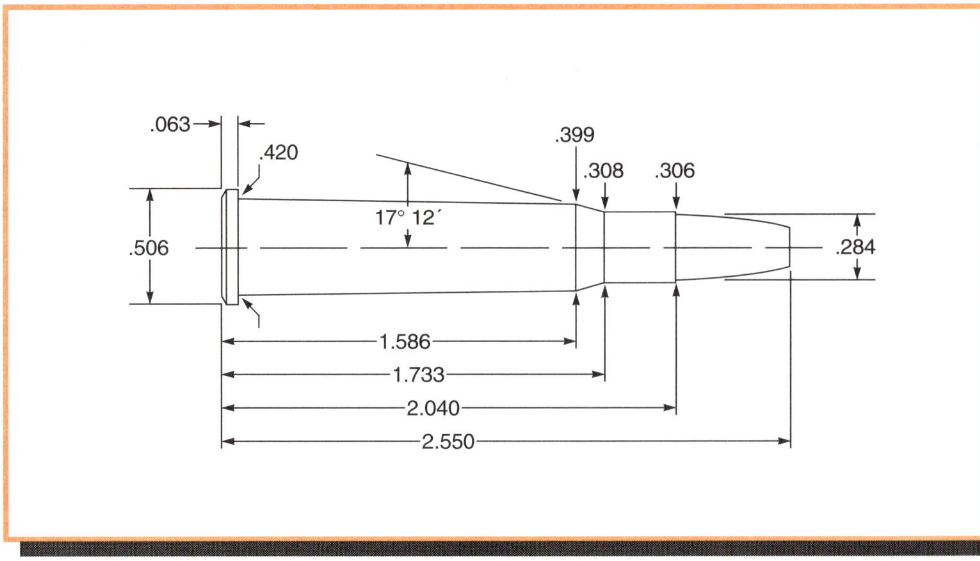

Comments:

Attempts to improve the ballistics of the 30-30 over the years resulted in various "Improved" or wildcat chamberings. Gunwriter Ken Waters developed this cartridge during the 1970s to give shooters of lever-action, tubular magazine rifles a flatter shooting, higher velocity cartridge than what was generally available. The 7-30 is basically a necked down and blown out 30-30 Winchester featuring a shorter neck and approximately five percent more powder capacity. U.S. Repeating Arms commenced production of Model 94 rifles chambered for the Waters in 1983. As the only tubular magazine rifle built for .284" diameter bullets, bullet selection is limited. Lyman offers data to duplicate both the 120-grain factory load as well as the 139-grain load originally envisioned by Mr. Waters. Shooters can expect an 80 to 100 foot per second increase over the listed data if they're using a 24-inch barrel. In addition to our listed accuracy loads, IMR-3031 and 748 will often give good results with jacketed bullets. *Caution: Shooters should not use any pointed bullets in rifles built with tubular magazines*

Test Components:

Cases . . . . .Federal
Trim-to Length . . . . .2.030"
Primers . . . . .CCI 200
Primer Size . . . . .Large Rifle
Lyman Shell Holder . . . . .No. 6
Jacketed Bullets Used . . . .Nosler FP #28121, 120 gr.
Hornady FP #2822, 139 gr.
Cast Bullets Used . . . .(sized to .285" dia)
*gas check bullet *#287346, 135 gr.

Test Specifications: (Velocity & Pressure)

Firearm Used . . . . .Universal Receiver
Barrel Length . . . . .20"
Twist . . . . .1-9.9"
Groove Dia. . . . .284"

120 gr. Jacketed FP 2.530" OAL — **BC:** .195 **SD:** .213

Powder	Sugg Starting Grains	Velocity fps	Pressure C.U.P.	Max Load Grains	Velocity fps	Pressure C.U.P.
IMR-3031	29.0	2056	24,200	32.5	2398	35,500
IMR-4895	31.0	2155	28,400	**34.5**	**2443**	**38,000**
H-335	31.3	2144	24,200	34.8	2453	36,700
BL-C (2)	33.8	2214	26,500	37.5	2520	35,400
AA2460	30.0	2142	28,000	33.6	2441	39,200
748	34.0	2183	25,200	38.0	2486	33,000
AA-2520	31.5	2310	30,000	35.0	2554	39,000
RX15	32.5	2247	29,400	36.0	2497	36,900

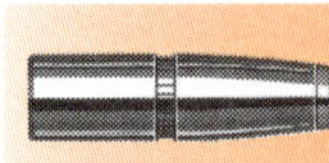

139 gr. Jacketed FP 2.545" OAL — **BC:** .196 **SD:** .246

Powder	Sugg Starting Grains	Velocity fps	Pressure C.U.P.	Max Load Grains	Velocity fps	Pressure C.U.P.
IMR-3031	28.2	1986	24,300	31.9	2334	37,700
IMR-4895	29.3	2005	25,000	34.0+	2370	39,700
H-335	29.5	2093	28,900	33.7	2352	38,600
BL-C (2)	31.4	2019	26,000	35.7	2327	38,000
AA-2460	29.0	1965	25,300	32.8	2277	37,400
748	32.0	2132	27,500	36.0	2415	39,600
AA-2520	30.5	2129	27,600	34.0	2412	39,400
RX15	30.5	2074	28,500	**34.0**	**2321**	**36,900**

Note: Loads shown in shaded panels are maximum.
Loads shown in bold designate potentially most accurate load.
\+ Designates a compressed powder charge.

#287346
135 gr. (#2 Alloy) 2.475" OAL

BC: .235
SD: .239

Powder	Sugg Starting Grains	Velocity fps	Pressure C.U.P.	Max Load Grains	Velocity fps	Pressure C.U.P.
SR-4759	16.5	1746	18,900	20.0	2017	34,600
XMP-5744	20.3	1884	27,000	**22.5**	**2053**	**32,300**
RX7	23.0	1859	19,200	26.9	2145	31,000
IMR-3031	25.5	1816	16,800	31.0+	2277	34,900
748	29.5	1975	20,000	33.5+	2264	32,800
AA-2460	26.0	1882	19,500	32.0	2273	34,600
H-335	26.2	1904	18,500	32.0	2300	34,500

Note: Loads shown in shaded panels are maximum.
Loads shown in bold designate potentially most accurate load.
\+ Designates a compressed powder charge.

7mm-08 Remington

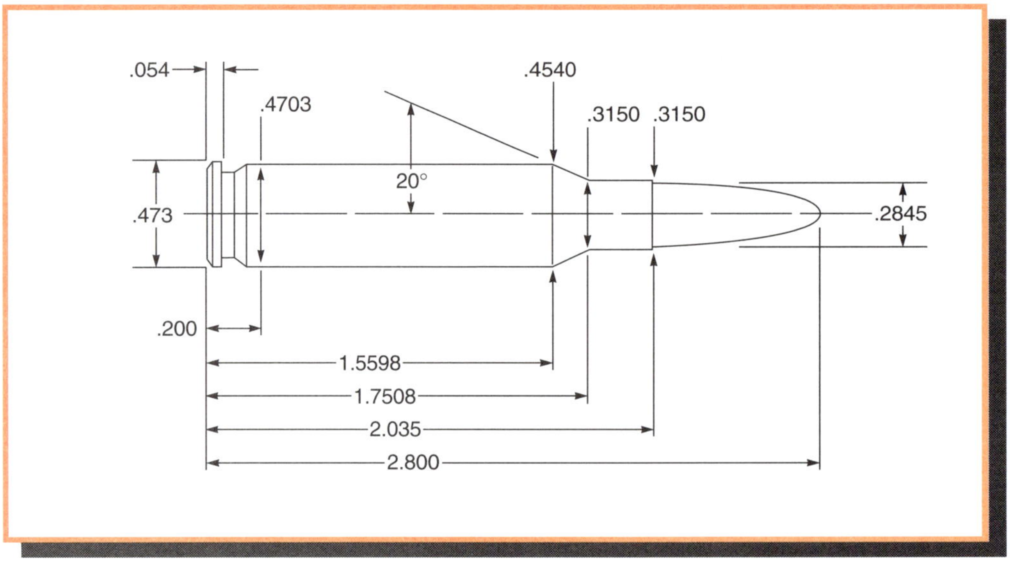

Comments:

This cartridge was designed to fill a need for silhouette shooting. It is a necked down version of the 308 and has the advantage of less recoil as compared to its parent round. The 7mm-08 ranks right near the top among deer hunting cartridges. It has a flat trajectory and great accuracy. It is somewhat handicapped with the heavier bullets due to the short magazines used in rifles chambered for this round.

The factory ammo we have looked at was loaded with a Ball Powder that was just a tad faster burning than Winchester 760. Thus, 760 is as close as possible to duplicating factory specs. Nonetheless best accuracy often occurs with all weight bullets when using Hodgdon H380. Second choice would be IMR 4350 but this will mean the inconvenience of compressed powder loads.

Test Components:

Cases	Remington
Trim-to Length	2.025"
Primers	Remington 9½
Primer Size	Large Rifle
Lyman Shell Holder	No. 2
Jacketed Bullets Used	Sierra SPT #1900, 120 gr.
	Hornady SP #2820, 139 gr.
	Hornady SP #2830, 154 gr.
	Hornady BTSP #2845, 162 gr.
	Sierra HPBT #1930, 168 gr.
	Speer Grand Slam #1643, 175 gr.
Cast Bullets Used	(sized to .284" dia)
*gas check bullet	*#287346, 135 gr.
	*#287641, 160 gr.

Test Specifications: (Velocity & Pressure)

Firearm Used	Universal Receiver
Barrel Length	24"
Twist	1-9½"
Groove Dia.	.284"

120 gr. Jacketed SPT
2.790" OAL
BC: .328
SD: .213

Powder	Sugg Starting Grains	Velocity fps	Pressure C.U.P.	Max Load Grains	Velocity fps	Pressure C.U.P.
RX7	33.1	2655	38,300	36.0	2811	50,000
IMR-3031	37.4	2706	34,800	42.6	3012	47,000
748	41.0	2704	35,700	46.0	3037	49,900
IMR-4064	39.0	2657	37,000	44.5+	3023	48,100
IMR-4895	37.8	2705	38,700	43.0	2999	48,100
H-380	43.5	2765	36,500	49.5+	3100	50,000
H-414	44.0	2680	36,500	50.0+	3007	48,400

139 gr. Jacketed SP
2.750" OAL
BC: .392
SD: .246

Powder	Sugg Starting Grains	Velocity fps	Pressure C.U.P.	Max Load Grains	Velocity fps	Pressure C.U.P.
IMR-3031	37.5	2564	39,000	40.0	2765	47,200
AA2460	37.0	2433	36,700	41.4	2766	50,900
748	41.5	2591	38,200	44.5	2929	51,200
IMR-4895	**38.7**	**2534**	**36,100**	43.0	2812	50,800
IMR-4064	38.0	2456	34,000	43.2	2854	50,200
IMR-4320	37.5	2381	33,500	42.6	2727	49,300
RX15	37.3	2467	35,400	43.0	2824	50,200
H-380	41.7	2513	36,200	47.0	2815	50,500
760	43.0	2500	34,900	48.0	2818	48,500
H-414	42.3	2516	33,500	47.6	2840	46,100
IMR-4350	45.0	2600	39,400	48.0+	2858	50,500
RX19	46.0	2512	37,500	52.0+	2857	49,300
**SR-4759	20.0	1879	33,700	26.0	2304	50,300
**XMP-5744	23.0	1904	30,700	28.5	2306	39,700

Note: Loads shown in shaded panels are maximum.
Loads shown in bold designate potentially most accurate load.
\+ Designates a compressed powder charge.
**Designates a reduced load.

154 gr. Jacketed SP
2.800" OAL

BC: .433
SD: .273

Powder	Sugg Starting Grains	Velocity fps	Pressure C.U.P.	Max Load Grains	Velocity fps	Pressure C.U.P.
IMR-3031	34.3	2335	34,000	39.0	2632	47,200
AA-2460	35.3	2262	38,500	40.5	2598	51,100
IMR-4895	37.0	2371	37,200	40.0	2644	49,500
IMR-4064	**36.7**	**2388**	**36,300**	41.8+	2702	51,600
IMR-4320	36.5	2315	37,000	41.6+	2641	50,100
RX15	38.7	2478	40,800	42.0	2736	51,000
H-380	38.5	2388	38,300	43.8	2668	51,000
760	40.5	2315	33,300	46.0	2702	50,800
H-414	40.5	2357	34,800	46.0	2731	51,700
IMR-4350	39.5	2324	35,000	45.0+	2622	46,900
RX19	44.0	2394	37,800	49.5+	2676	47,400
**SR-4759	22.0	1910	40,900	25.5	2121	50,900
**XMP-5744	24.0	1911	33,000	27.5	2142	38,600

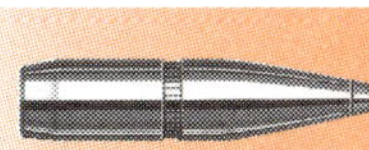

162 gr. Jacketed BTSP
2.800" OAL

BC: .514
SD: .287

Powder	Sugg Starting Grains	Velocity fps	Pressure C.U.P.	Max Load Grains	Velocity fps	Pressure C.U.P.
IMR-3031	34.3	2339	35,700	39.0	2631	51,300
AA2460	33.3	2181	37,000	39.0	2533	50,700
IMR-4895	36.5	2350	37,500	**39.5**	**2604**	**50,000**
IMR-4064	35.5	2317	35,000	40.5	2626	49,700
IMR-4320	36.0	2301	36,200	40.9	2589	51,200
RX15	38.5	2421	39,100	41.5	2690	51,000
H-380	38.7	2325	35,800	44.0+	2627	50,900
760	41.8	2395	37,200	45.5+	2663	49,400
H-414	40.0	2312	34,500	45.4	2677	51,300
IMR-4350	39.5	2343	36,000	45.0+	2622	47,300
RX19	42.0	2264	34,800	49.0+	2702	50,200

168 gr. Jacketed HPBT
2.800" OAL

BC: .494
SD: .298

Powder	Sugg Starting Grains	Velocity fps	Pressure C.U.P.	Max Load Grains	Velocity fps	Pressure C.U.P.
IMR-3031	33.5	2290	36,500	38.2	2575	51,500
AA2460	35.0	2347	41,200	38.0	2607	51,900
IMR-4895	34.2	2306	39,200	38.9	2555	51,300
IMR-4064	35.6	2339	39,600	40.5+	2605	51,800
IMR-4320	34.7	2251	37,300	39.5	2510	49,100
RX15	37.3	2361	38,100	**40.5**	**2628**	**51,700**
H-380	38.2	2256	37,000	43.5+	2555	51,600
760	39.5	2280	36,200	43.5	2632	51,800
H-414	39.0	2287	36,800	44.5	2628	51,900
IMR-4350	39.5	2357	38,300	45.0+	2605	50,700
RX19	43.0	2467	42,500	48.0+	2718	52,000

175 gr. Jacketed GSSP
2.735" OAL

BC: .465
SD: .310

Powder	Sugg Starting Grains	Velocity fps	Pressure C.U.P.	Max Load Grains	Velocity fps	Pressure C.U.P.
IMR-3031	32.3	2112	37,200	36.8	2408	50,400
AA2460	31.0	1997	39,800	37.3	2323	51,300
IMR-4895	34.3	2207	40,300	39.0	2473	50,600
IMR-4064	35.5	2260	41,900	39.2	2487	51,400
IMR-4320	34.3	2171	40,200	39.0	2425	51,400
RX15	**36.2**	**2275**	**40,900**	40.0	2476	50,900
H-380	35.8	2103	38,000	40.7	2361	50,200
760	38.5	2208	37,500	43.8	2485	48,500
H-414	39.0	2301	38,000	43.8	2589	52,000
IMR-4350	38.7	2251	39,000	44.0+	2529	51,300
RX19	40.3	2191	38,700	47.0+	2539	51,400

#287346
135 gr. (#2 Alloy) 2.600" OAL

BC: .235
SD: .239

Powder	Sugg Starting Grains	Velocity fps	Pressure C.U.P.	Max Load Grains	Velocity fps	Pressure C.U.P.
SR-4759	15.3	1529	13,300	23.5	2189	36,900
IMR-4227	15.8	1584	14,600	24.0	2183	34,400
IMR-4198	16.8	1546	11,500	25.5	2161	32,300
XMP-5744	**16.5**	**1558**	**12,600**	25.0	2128	32,200
RX7	17.4	1559	12,200	26.5	2164	31,600
IMR-3031	19.8	1495	10,900	30.0	2171	27,300
748	21.1	1509	9,600	32.0	2179	28,000

#287641
160 gr. (#2 Alloy) 2.705" OAL

BC: .382
SD: .283

Powder	Sugg Starting Grains	Velocity fps	Pressure C.U.P.	Max Load Grains	Velocity fps	Pressure C.U.P.
Unique	12.0	1497	24,000	15.5	1769	42,100
SR-4759	20.7	1761	21,500	26.0	2154	45,100
IMR-4227	21.0	1748	20,800	27.0	2105	38,500
IMR-4198	23.0	1805	19,500	29.5	2181	36,100
XMP-5744	**18.0**	**1601**	**25,000**	26.0	2108	36,400
RX7	22.0	1716	18,200	28.0	2080	33,300
IMR-3031	28.7	1906	18,800	33.0	2238	32,000
XMR-2015	30.2	2060	26,900	34.0	2322	41,000
H-335	31.5	1999	23,800	36.0	2266	34,400
IMR-4895	29.5	1871	19,200	35.3	2296	36,000
748	34.5	1949	20,000	39.0	2299	33,500

Note: Loads shown in shaded panels are maximum.
Loads shown in bold designate potentially most accurate load.
\+ Designates a compressed powder charge.
** Designates a reduced load.

7mm Mauser (7 x 57mm)

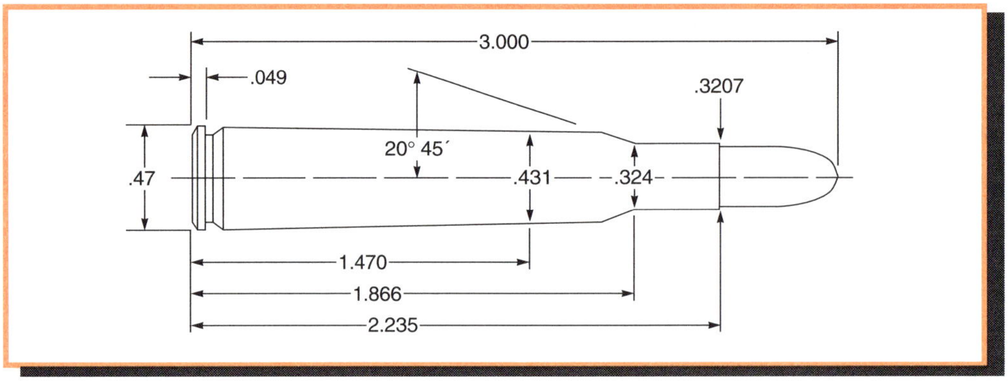

Comments:

The 7x57mm Mauser enjoys a longevity that few rifle cartridges can match. The 7mm Mauser made its name on both the battlefields and hunting grounds around the world. Even though it is one of the earliest smokeless cartridges, it's outlasted most all of its contemporaries and is still popular after 110 years. This cartridge first earned its reputation in the hands of Spanish forces in Cuba during the Spanish-American War in 1898. Its range, flat trajectory, and rapid rate of fire from the clip-fed Mauser Model 93 rifle was an unpleasant surprise for American troops armed with the archaic Trapdoor Springfield chambered in 45-70. This harrowing experience helped lead to adaptation of the M1903 Springfield and the 30-06 cartridge several years later.

Military sales of rifles chambered in the 7x57 were wide spread, particularly in Central and South America. As a result, it has been one of the most popular—and useful—of the military cartridges on the surplus market. The original military load consisted of a 173-grain bullet with a muzzle velocity of approximately 2,300 feet per second. Several other military loads featuring lighter bullets at higher velocities also appeared. The 7x57 loaded with 175-grain bullets became legendary as a hunting cartridge and eventually served as the parent case for the 257 Roberts. Rifles chambered for this cartridge often have a relatively fast twist, usually around 1 in 8 1/2." In light of this, shooters may not always get the best results with lighter weight bullets. IMR-4350 has long been a favorite for loading the 7x57. SAAMI established the Maximum Average Pressure (MAP) of 46,000 CUP due to the large numbers of rifles in circulation built upon the older M93 and M95 style actions. Most American manufacturers have produced rifles chambered in 7x57 in recent years while loaded ammo and bulk brass is in ample supply. As with most any foreign military surplus, a variation in groove diameters can be expected. Shooters should slug their bore and size cast bullets accordingly. Cast bullet # 287641 has produced excellent results in our testing.

Test Components:

Cases . . . . .Winchester
Trim-to Length . . . . .2.225"
Primers . . . . .Winchester WLR
Primer Size . . . . .Large Rifle
Lyman Shell Holder . . . . .No. 2
Jacketed Bullets Used . . . . .Sierra SPT #1900, 120 gr.
Hornady SP #2820, 139 gr.
Sierra, HPBT #1915, 150 gr.
Hornady BTSP #2845, 162 gr.
Sierra HPBT #1930, 168 gr.
Speer Grand Slam #1643, 175 gr.
Cast Bullets Used . . . .(sized to .284" dia)
*gas check bullet *#287641, 160 gr.

Test Specifications:
(Velocity & Pressure)

Firearm Used . . . . .Mauser 95
Universal Receiver
Barrel Length . . . . .Mauser 29"
Universal Receiver 24"
Twist . . . . .Mauser 1-10"
Universal Receiver 1-8¾"
Groove Dia. .Mauser .2865"
Universal Receiver .284"

120 gr. Jacketed SPT
2.945" OAL

BC: .328
SD: .213

Powder	Sugg Starting Grains	Velocity fps	Pressure C.U.P.	Max Load Grains	Velocity fps	Pressure C.U.P.
IMR-3031	39.0	2717	—	43.0	2958	—
H-335	37.3	2693	39,100	41.6	2901	45,700
BL-C(2)	41.0	2688	—	45.0	2958	—
IMR-4895	41.0	2824	—	45.0	3067	—
IMR-4064	41.0	2770	—	45.0	3048	—
Varget	39.5	2661	32,630	44.0	2958	44,700
IMR-4320	42.0	2816	—	47.0	3067	—
H-380	43.0	2652	—	48.0	2949	—
760	46.7	2669	34,700	52.0	3001	44,500
IMR-4350	45.0	2652	—	50.0+	2932	—
H-4831	50.0	2666	—	53.5+	2865	—

Note: Loads shown in shaded panels are maximum.
Loads shown in bold designate potentially most accurate load.
\+ Designates a compressed powder charge.

7mm Mauser (7 x 57mm)

139 gr. Jacketed SP
2.970" OAL

BC: .392
SD: .246

Powder	Sugg Starting Grains	Velocity fps	Pressure C.U.P.	Max Load Grains	Velocity fps	Pressure C.U.P.
H-335	36.0	2457	37,200	40.5	2687	45,300
BL-C(2)	39.0	2427	—	43.0	2695	—
IMR-4895	34.5	2420	32,800	38.0	2724	44,800
IMR-4064	39.0	2557	—	43.0	2824	—
Varget	36.0	2432	33,000	**40.5**	**2741**	**43,700**
AA2520	31.4	2050	32,700	42.0	2605	44,500
IMR-4320	41.0	2666	—	45.0	2898	—
H-380	41.0	2525	—	46.0	2785	—
N150	39.0	2444	32,600	44.0	2703	42,800
760	44.0	2430	32,700	49.0	2721	39,700
IMR-4350	43.0	2475	—	48.0+	2785	—
RX19	41.0	2040	31,000	50.8+	2681	45,400
H-4831	49.0	2583	—	53.5+	2816	—
RX22	44.8	2180	32,700	54.8+	2796	45,700

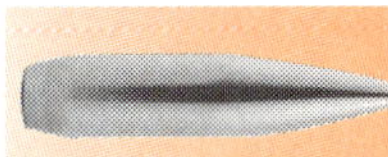

150 gr. Jacketed HPBT
3.025" OAL

BC: .429
SD: .266

Powder	Sugg Starting Grains	Velocity fps	Pressure C.U.P.	Max Load Grains	Velocity fps	Pressure C.U.P.
IMR-4895	32.0	2331	34,000	36.0	2619	45,500
IMR-4064	36.0	2381	—	40.5	2640	—
Varget	34.5	2370	35,000	38.5	2617	45,100
AA-2520	29.0	1894	31,200	38.4	2418	44,300
IMR-4320	37.5	2334	33,300	41.5	2555	40,900
N150	38.7	2342	35,200	**43.0**	**2576**	**45,100**
760	41.3	2390	34,300	47.0	2664	43,900
H-414	41.5	2316	29,700	46.0	2590	42,200
IMR-4350	42.0	2427	—	47.0	2739	—
RX19	43.5	2202	31,600	48.7	2529	44,900
H-4831	48.0	2538	—	53.0+	2816	—
RX22	45.0	2119	28,600	50.0+	2478	43,200

162 gr. Jacketed BTSP
3.037" OAL

BC: .514
SD: .287

Powder	Sugg Starting Grains	Velocity fps	Pressure C.U.P.	Max Load Grains	Velocity fps	Pressure C.U.P.
IMR-4064	36.3	2358	35,000	41.3	2597	44,200
Varget	33.5	2281	34,500	37.5	2514	44,400
AA-2520	33.5	2185	30,900	37.5	2415	42,200
IMR-4320	35.5	2181	32,800	40.5	2498	42,800
H-380	36.0	2236	33,800	40.5	2446	43,000
760	40.0	2272	30,500	45.5	2560	44,800
H-414	39.5	2296	32,900	44.0	2551	44,500
N160	40.8	2258	30,900	**48.7**	**2563**	**43,000**
IMR-4350	39.4	2188	29,200	44.8	2564	44,700
RX19	42.7	2234	31,400	47.5	2543	44,800
H-4831	44.0	1905	24,600	48.0+	2560	44,900
RX22	45.5	2158	28,300	49.5+	2492	43,600

168 gr. Jacketed HPBT
3.050" OAL

BC: .494
SD: .298

Powder	Sugg Starting Grains	Velocity fps	Pressure C.U.P.	Max Load Grains	Velocity fps	Pressure C.U.P.
IMR-4064	35.9	2327	24,600	40.8	2531	45,000
Varget	33.5	2253	35,500	37.0	2498	44,500
IMR-4320	35.0	2178	32,300	39.8	2475	45,400
H-414	38.0	2179	29,200	42.5	2484	43,500
N160	43.0	2201	31,200	**48.0**	**2509**	**44,200**
IMR-4350	39.0	2199	31,300	44.4	2537	44,900
RX19	38.0	1940	32,700	47.0+	2460	45,100
XMR-3100	38.2	1859	32,000	47.8+	2427	45,300
H-4831	42.2	2165	29,800	47.0+	2433	39,400
RX22	40.2	2017	32,500	49.2+	2537	45,400

175 gr. Jacketed GSSP
2.935" OAL

BC: .465
SD: .310

Powder	Sugg Starting Grains	Velocity fps	Pressure C.U.P.	Max Load Grains	Velocity fps	Pressure C.U.P.
IMR-4064	35.0	2178	—	39.0	2439	—
H-380	39.0	2320	—	43.0	2500	—
760	39.0	2169	33,000	44.5	2478	45,700
H-414	39.0	2165	32,500	44.4	2465	44,300
IMR-4350	41.0	2079	—	45.0	2531	—
XMR-3100	38.0	1854	31,200	47.6+	2365	45,000
H-4831	41.5	2134	31,400	46.0	2403	41,600
RX22	38.6	1948	33,000	48.6+	2473	45,800
IMR-7828	44.0	2107	31,300	**48.0+**	**2398**	**43,700**

#287641
160 gr. (#2 Alloy) 2.900" OAL

BC: .382
SD: .283

Powder	Sugg Starting Grains	Velocity fps	Pressure C.U.P.	Max Load Grains	Velocity fps	Pressure C.U.P.
Unique	9.7	1258	20,500	14.2	1625	33,400
SR-7625	10.0	1153	20,200	11.7	1351	28,100
SR-4759	19.7	1581	20,500	24.0	1919	35,200
IMR-4227	22.0	1723	22,700	27.0	2066	38,600
XMP-5744	**20.0**	**1672**	**21,600**	29.0	2268	45,500
AA-1680	23.2	1686	23,100	30.0	2094	38,200
RX7	22.4	1647	21,900	26.5	1869	26,000
IMR-3031	27.0	1687	20,500	34.8	2195	32,600
AA-2230	29.4	1832	23,600	35.0	2155	33,100
748	33.5	1869	22,800	37.0	2104	32,700

Note: Loads shown in shaded panels are maximum.
Loads shown in bold designate potentially most accurate load.
\+ Designates a compressed powder charge.

284 Winchester

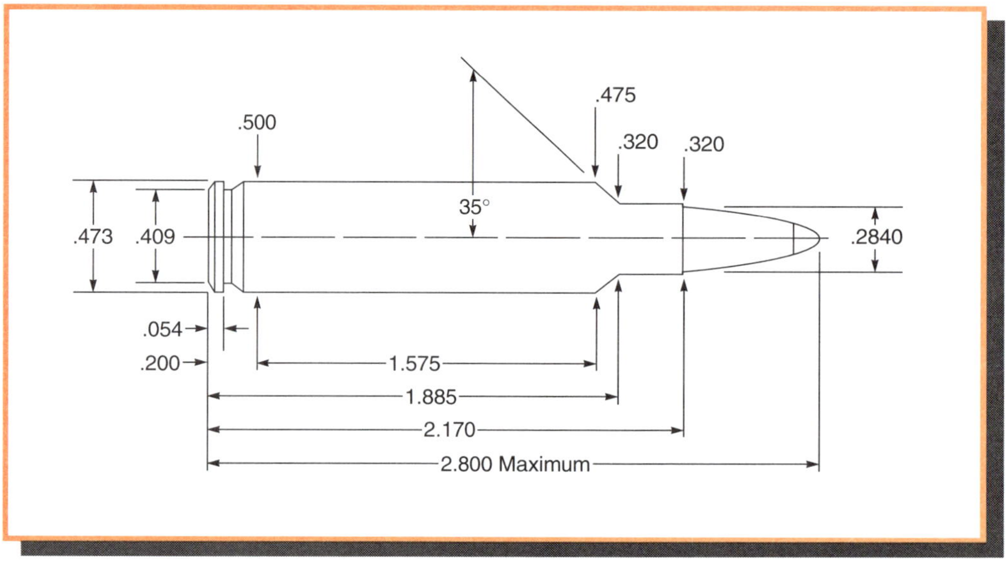

Comments:

The 284 Winchester was a departure in cartridge design for American shooters. Its rebated rim, larger diameter body, and 35-degree shoulder were designed to obtain greater ballistics than previously possible with a short action cartridge while using a standard bolt face. Winchester introduced the cartridge in 1963 in their Models 88 and 100 rifles while a few other manufacturers offered it at different times. Winchester currently lists a 150-grain factory loading although the 284 is pretty much extinct as a commercial cartridge. However, the idea of rebated rims lives on through the current trend of the Remington Ultra Mag and Winchester Short Magnum families. The 284 is presently most notable as the basis of the 475 Wildey pistol cartridge, the 6.5x284 Norma, and a few other wildcats popular among the varmint and long-range shooting fraternity.

Test Components:

Cases . . . . .Winchester
Trim-to Length . . . . .2.160"
Primers . . . . .Winchester WLR
Primer Size . . . . .Large Rifle
Lyman Shell Holder . . . . .No. 2
Jacketed Bullets Used . . . . .Sierra SPT #1900, 120 gr.
Hornady SP #2820, 139 gr.
Sierra SBT #1913, 150 gr.
Cast Bullets Used . . . .(sized to .284" dia)
*gas check bullet *#287346, 135 gr.

Test Specifications: (Velocity & Pressure)

Firearm Used . . . . .Winchester Model 88, Universal Receiver
Barrel Length . . . . .22"
Twist . . . . .1-10"
Groove Dia. . . . .284"

120 gr. Jacketed SPT 2.800" OAL — **BC:** .328 **SD:** .213

Powder	Sugg Starting Grains	Velocity fps	Pressure C.U.P.	Max Load Grains	Velocity fps	Pressure C.U.P.
IMR-3031	41.0	2645	—	45.0	2932	—
IMR-4895	42.0	2544	—	47.0	2890	—
IMR-4064	44.0	2660	—	49.0	3003	—
IMR-4320	44.0	2610	—	49.0	2967	—
H-380	47.0	2590	—	53.0	2932	—
IMR-4350	**51.0**	**2710**	—	57.0+	3076	—
RX19	52.5	2738	42,700	58.3	3113	53,700
XMR-3100	53.0	2660	46,000	59.0+	3026	54,000
H-4831	53.0	2557	—	59.0+	2949	—

139 gr. Jacketed SP 2.800" OAL — **BC:** .392 **SD:** .246

Powder	Sugg Starting Grains	Velocity fps	Pressure C.U.P.	Max Load Grains	Velocity fps	Pressure C.U.P.
IMR-3031	40.0	2525	—	44.5	2793	—
IMR-4895	41.0	2427	—	46.0	2785	—
IMR-4064	42.0	2506	—	47.5	2832	—
IMR-4320	42.0	2444	—	47.5	2777	—
H-380	47.0	2613	46,500	52.0	2820	53,200
IMR-4350	**50.0**	**2617**	—	55.0+	2906	—
RX19	49.5	2533	42,000	55.0	2819	52,900
XMR-3100	50.5	2441	43,200	56.0+	2766	53,100
H-4831	52.0	2444	—	58.0+	2793	—

Note: Loads shown in shaded panels are maximum.
Loads shown in bold designate potentially most accurate load.
\+ Designates a compressed powder charge.

284 Winchester

150 gr. Jacketed SBT
2.800" OAL

BC: .436
SD: .266

Powder	Sugg Starting Grains	Velocity fps	Pressure C.U.P.	Max Load Grains	Velocity fps	Pressure C.U.P.
IMR-3031	37.0	2298	—	41.0	2518	—
IMR-4895	40.0	2488	42,100	45.5	2776	53,500
IMR-4064	41.0	2487	—	46.0	2762	—
IMR-4320	41.0	2525	44,900	45.5	2737	53,300
H-380	43.0	2506	46,100	47.5	2651	53,300
IMR-4350	49.0	2610	—	**54.0+**	**2816**	—
RX19	48.5	2548	42,100	54.0	2850	53,000
XMR-3100	48.5	2373	40,100	54.0	2711	52,600
H-4831	51.0	2439	—	57.0+	2724	—

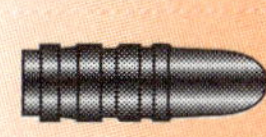

#287346
135 gr. (#2 Alloy) 2.800" OAL

BC: .235
SD: .239

Powder	Sugg Starting Grains	Velocity fps	Pressure C.U.P.	Max Load Grains	Velocity fps	Pressure C.U.P.
SR-4759	15.0	1528	23,000	22.0	2059	35,600
XMP-5744	**16.0**	**1499**	**22,000**	26.0	2114	34,600
IMR-4198	16.0	1512	22,500	25.0	2063	30,500
RX7	17.0	1535	23,000	26.0	2073	32,200

Note: Loads shown in shaded panels are maximum.
Loads shown in bold designate potentially most accurate load.
\+ Designates a compressed powder charge.

280 Remington (7mm Express Remington)

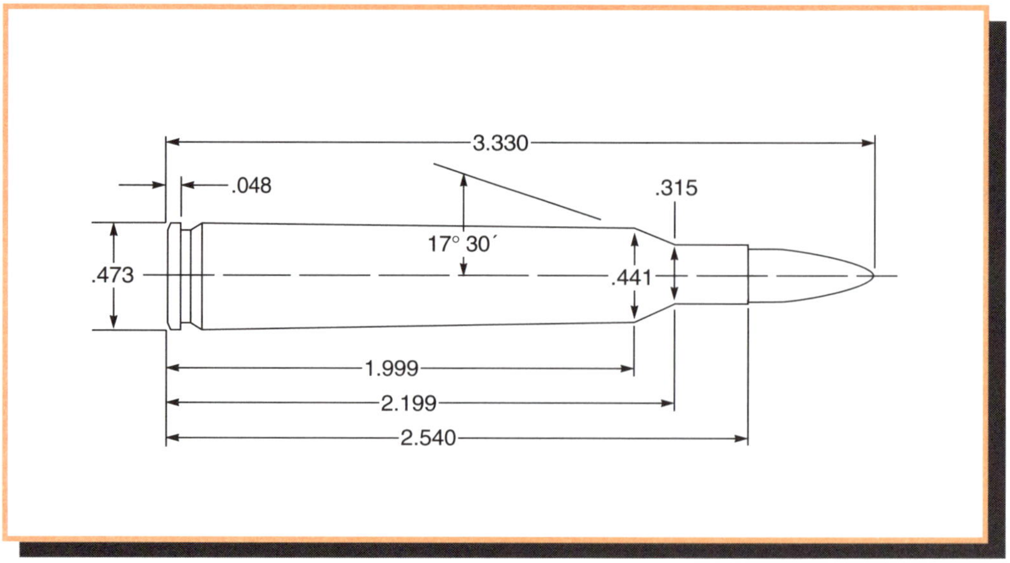

Comments:

This cartridge started life as the 280 Remington. When it proved less than popular it was subjected to a name change - 7mm Express Remington. With the name change came some modifications of the factory loads. Eventually, Remington went back to the original designation.

Ballistically the 280 fills a very narrow gap between the 30-06 Springfield and the 270 Winchester. It seems best with bullets of about 160 to 162 grains, though it does well with a wide range of weights.

Hodgdon H4831 is particularly well suited to balancing uniform ballistics and accuracy in this cartridge. IMR 4350 is also a fine performer with this round.

Test Components:

Cases . . . . .Remington
Trim-to Length . . . . .2.530"
Primers . . . . .Remington 9½
Primer Size . . . . .Large Rifle
Lyman Shell Holder . . . . .No. 2
Jacketed Bullets Used . . . . .Sierra SPT #1900, 120 gr.
Hornady SP #2820, 139 gr.
Nosler SP #16326, 150 gr.
Hornady BTSP #2845, 162 gr.
Sierra HPBT #1930, 168 gr.
Speer Grand Slam #1643, 175 gr.
Cast Bullets Used . . . .(sized to .284" dia)
*gas check bullet *#287346, 135 gr.
*#287641, 160 gr.

Test Specifications: (Velocity & Pressure)

Firearm Used . . . . .Remington Model 700, Universal Receiver
Barrel Length . . . . .Model 700 22", Universal Receiver 24"
Twist . . . . .1-9½"
Groove Dia. . . . . .284"

120 gr. Jacketed SPT 3.237" OAL — **BC:** .328 **SD:** .213

Powder	Sugg Starting Grains	Velocity fps	Pressure C.U.P.	Max Load Grains	Velocity fps	Pressure C.U.P.
IMR-4064	44.0	2680	—	49.0	2976	—
Varget	43.5	2754	39,300	48.5	3002	47,400
H-380	49.0	2754	—	54.0	3039	—
H-414	46.0	2693	35,700	52.0	3003	44,200
IMR-4350	51.0	2747	—	**57.0+**	**3144**	—
H-4831	53.0	2597	—	59.0+	3003	—

139gr. Jacketed SP 3.270" OAL — **BC:** .392 **SD:** .246

Powder	Sugg Starting Grains	Velocity fps	Pressure C.U.P.	Max Load Grains	Velocity fps	Pressure C.U.P.
IMR-4064	43.0	2531	—	48.5	2857	—
Varget	43.0	2604	38,300	46.0	2783	46,100
RX15	43.0	2526	40,000	48.0	2780	47,900
H-414	**47.0**	**2501**	**34,800**	52.5	2797	47,400
IMR-4350	48.5	2616	39,100	54.0	2912	49,000
RX19	52.5	2631	37,800	58.2	2989	49,500
H-4831	52.0	2518	—	58.0+	2873	—

Note: Loads shown in shaded panels are maximum.
Loads shown in bold designate potentially most accurate load.
+ Designates a compressed powder charge.

280 Remington (7mm Express Remington)

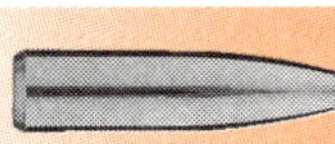

150gr. Jacketed SP
3.280" OAL

BC: .456
SD: .266

Powder	Sugg Starting Grains	Velocity fps	Pressure C.U.P.	Max Load Grains	Velocity fps	Pressure C.U.P.
IMR-4064	43.0	2551	—	48.0	2840	—
Varget	40.0	2385	35,200	44.5	2646	46,600
RX15	40.0	2396	37,800	44.5	2612	45,600
760	46.5	2546	39,600	53.0	2829	48,800
H-414	46.5	2555	38,900	52.5	2871	49,000
IMR-4350	46.0	2480	38,500	**52.5**	**2787**	**48,800**
XMR-3100	50.5	2473	38,700	56.3+	2761	49,700
H-4831	51.0	2550	—	57.0+	2816	—

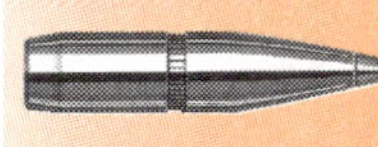

162gr. Jacketed BTSP
3.320" OAL

BC: .514
SD: .287

Powder	Sugg Starting Grains	Velocity fps	Pressure C.U.P.	Max Load Grains	Velocity fps	Pressure C.U.P.
IMR-4064	41.0	2426	38,000	47.0	2707	48,800
Varget	39.5	2347	38,900	43.0	2526	47,200
760	45.5	2441	38,300	51.7	2745	48,300
H-414	44.0	2417	37,500	50.3	2729	49,900
IMR-4350	45.0	2379	34,700	**51.5**	**2718**	**48,400**
RX19	48.0	2425	37,300	53.8	2751	48,800
XMR-3100	49.5	2446	38,700	54.4	2722	49,100
H-4831	51.0	2493	37,700	56.5+	2784	46,900
RX22	49.0	2465	37,300	54.3	2747	47,900

168gr. Jacketed HPBT
3.290" OAL

BC: .494
SD: .298

Powder	Sugg Starting Grains	Velocity fps	Pressure C.U.P.	Max Load Grains	Velocity fps	Pressure C.U.P.
IMR-4064	40.0	2345	37,800	46.0	2632	49,300
Varget	39.0	2280	37,800	42.5	2474	45,900
760	44.0	2365	39,200	50.0	2653	49,400
H-414	43.5	2377	37,700	49.5	2667	49,100
IMR-4350	46.0	2451	38,700	**52.2**	**2735**	**50,000**
RX19	45.5	2375	38,000	52.6	2675	49,800
XMR-3100	46.0	2298	37,300	52.0	2575	49,400
H-4831	47.5	2338	36,000	54.0	2658	49,500
RX22	47.0	2402	38,300	53.0	2692	49,800

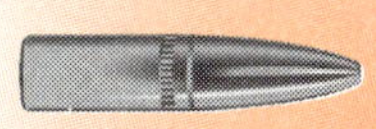

175gr. Jacketed GSSP
3.300" OAL

BC: .465
SD: .310

Powder	Sugg Starting Grains	Velocity fps	Pressure C.U.P.	Max Load Grains	Velocity fps	Pressure C.U.P.
IMR-4064	40.0	2252	—	43.0	2463	—
Varget	36.5	2176	38,200	40.5	2381	46,700
760	41.5	2219	38,200	47.2	2493	49,100
H-414	40.5	2227	39,000	46.3	2502	49,800
IMR-4350	41.5	2247	37,500	**47.5**	**2520**	**47,700**
H-4831	48.0	2212	—	53.0	2518	—

#287346
135 gr. (#2 Alloy) 3.176" OAL

BC: .235
SD: .239

Powder	Sugg Starting Grains	Velocity fps	Pressure C.U.P.	Max Load Grains	Velocity fps	Pressure C.U.P.
SR-4759	27.5	2213	40,500	30.5	2374	44,000
XMP-5744	**29.5**	**2221**	**34,200**	33.0	2364	37,400
AA-1680	28.5	2160	33,900	31.5	2359	42,700
RX7	31.5	2273	31,700	35.0	2388	34,500
IMR-3031	34.0	2331	28,300	38.0	2383	32,100
748	37.5	2188	27,700	41.5	2373	30,400
IMR-4895	34.5	2209	29,700	38.5	2364	33,300

#287641
160 gr. (#2 Alloy) 3.205" OAL

BC: .382
SD: .283

Powder	Sugg Starting Grains	Velocity fps	Pressure C.U.P.	Max Load Grains	Velocity fps	Pressure C.U.P.
Unique	15.5	1596	26,800	19.0	1815	40,400
SR-7625	13.0	1349	21,500	17.0	1637	41,200
SR-4756	15.0	1483	23,500	19.0	1773	41,700
SR-4759	**24.5**	**1872**	**26,000**	29.6	2145	39,800
AA-1680	28.5	1864	21,300	34.4	2155	34,300
IMR-3031	33.5	1984	20,000	38.6	2301	32,600
XMR-2015	31.0	1991	22,500	36.0	2206	31,600
RX7	25.5	1792	18,800	30.5	2040	27,900
AA-2230	34.0	2017	24,800	39.5	2298	35,800
748	38.0	1930	20,000	44.0	2362	35,000
IMR-4895	32.0	1908	21,000	35.3	2020	22,700

Note: Loads shown in shaded panels are maximum.
Loads shown in bold designate potentially most accurate load.
\+ Designates a compressed powder charge.

7mm Remington Magnum

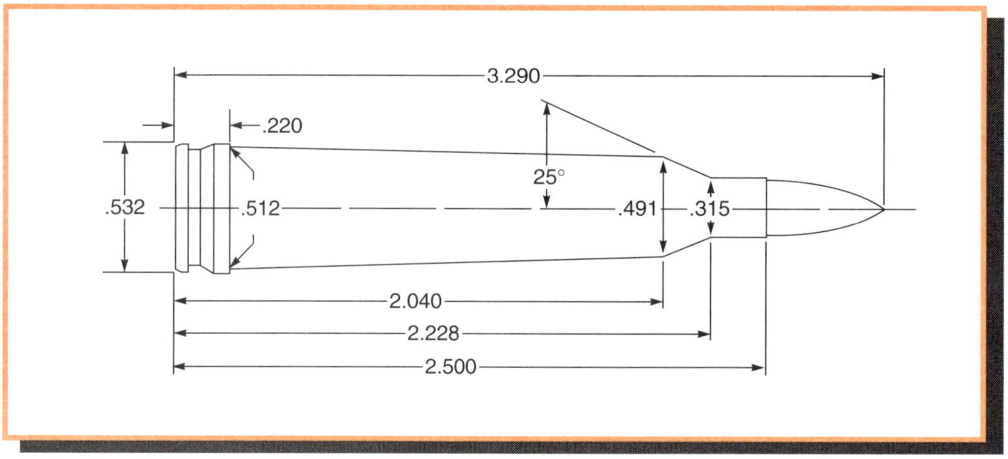

Comments:

This is easily the third most popular big game cartridge in this country. It started with the need for a cartridge with more punch than the 30-06 but not more recoil.

The 7mm Remington Magnum has been used to take every big game animal that walks on this continent. It really peaks out however with game as large as elk, although some would disagree. Bullets of 160 to 175 grains are best when the game gets tough.

Magnum primers and slow burning propellants such as Hodgdon H4831 and IMR 7828 are naturals for this application. Hercules Reloder 22 would also be a good choice.

Cast bullets need to be kept under 1800 fps for acceptable accuracy.

Test Components:

Cases . . . . .Remington
Trim-to Length . . . . .2.490"
Primers . . . . .Remington 9½M
Primer Size . . . . .Large Rifle Magnum
Lyman Shell Holder . . . . .No. 13
Jacketed Bullets Used . . .Hornady V-Max #22810, 120 gr.
Speer SP #1623, 130 gr.
Nosler SP #16325, 140 gr.
Comb. Tech. Bal. Silvertip #51110, 150 gr.
Hornady SP #2830, 154 gr.
Combined Tech. Fail Safe #53160, 160 gr.
Sierra HPBT #1930, 168 gr.
Speer Grand Slam #1643, 175 gr.
Cast Bullets Used . . . .(sized to .284" dia)
*gas check bullet *#287641, 160 gr.

Test Specifications:
(Velocity & Pressure)

Firearm Used . . . . .Universal Receiver
Barrel Length . . . . .24" (PSI), 26" (CUP)
Twist . . . . .1-9½"
Groove Dia. . . . . .284"

120gr. Jacketed V-Max
3.290" OAL
BC: .365
SD: .213

Powder	Sugg Starting Grains	Velocity fps	Pressure P.S.I.	Max Load Grains	Velocity fps	Pressure P.S.I.
IMR-4064	55.5	3037	45,900	61.5	3302	58,900
H-414	57.0	2938	44,600	63.0	3267	59,700
N160	61.0	3013	47,400	**67.5**	**3290**	**60,300**
IMR-4350	61.0	2990	44,300	68.0	3340	59,500
WXR	66.0	2990	42,800	73.0	3348	59,900
XMR-3100	63.5	2929	43,900	70.5	3264	59,300
H-4831	65.5	2957	43,600	73.0	3283	59,000
RX22	66.0	2990	43,700	73.0	3366	59,800
IMR-7828	65.0	2927	43,000	72.0+	3290	58,800
H-1000	70.0	3011	44,500	77.0+	3318	59,000

130gr. Jacketed SP
3.235" OAL
BC: .394
SD: .230

Powder	Sugg Starting Grains	Velocity fps	Pressure	Max Load Grains	Velocity fps	Pressure
IMR-4064	53.0	2914	42,400	59.5	3185	52,000 C
H-414	56.0	2859	47,000	62.0	3120	59,200 P
N160	57.0	2909	52,300	63.5	3125	60,700 P
IMR-4350	58.5	2946	49,400	65.0	3186	59,200 P
WXR	61.0	2916	47,400	68.0	3186	58,900 P
XMR-3100	**62.0**	**2902**	**48,600**	69.0	3133	58,100 P
H-4831	64.0	2887	38,600	71.0	3310	52,400 C
RX22	65.0	2992	48,000	72.0	3270	59,600 P
IMR-7828	64.5	2941	47,800	70.5+	3225	60,900 P
H-1000	68.5	2996	49,800	75.0+	3228	60,600 P

Note: Loads shown in shaded panels are maximum.
Loads shown in bold designate potentially most accurate load.
\+ Designates a compressed powder charge.
In pressure column, C=Copper Units of Pressure (CUP); P=Pounds per Square Inch (PSI)

7mm Remington Magnum

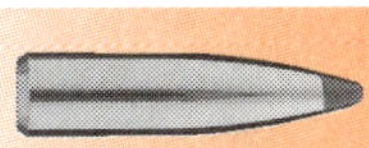

140 gr. Jacketed SP 3.290" OAL — **BC:** .434 **SD:** .248

Powder	Sugg Starting Grains	Velocity fps	Pressure	Max Load Grains	Velocity fps	Pressure
IMR-4064	46.3	2674	37,700	54.5	3038	51,700 C
H-414	49.5	2702	39,100	58.0	3017	51,500 C
N160	55.0	2788	49,800	61.0	2993	58,400 P
IMR-4350	58.0	2842	42,400	66.0	3133	52,000 C
WXR	60.0	2846	47,700	67.0	3110	59,600 P
XMR-3100	61.0	2819	48,800	68.0	3083	60,900 P
H-4831	62.0	2791	39,000	67.5	3146	50,600 C
RX22	64.5	3037	42,900	70.5	3275	59,000 P
IMR-7828	62.0	2834	47,300	68.5	3105	59,100 P
H-1000	66.0	2885	48,200	**72.7+**	**3139**	**60,600 P**

150 gr. Jacketed Silvertip 3.290" OAL — **BC:** .493 **SD:** .266

Powder	Sugg Starting Grains	Velocity fps	Pressure P.S.I.	Max Load Grains	Velocity fps	Pressure P.S.I.
H-414	54.0	2639	44,400	60.0	2940	60,000
N160	54.0	2642	44,200	60.6	2918	59,100
IMR-4350	57.5	2700	45,000	64.0	3013	59,800
IMR-4831	59.5	2729	44,200	66.0	3053	59,800
WXR	60.0	2735	45,800	67.0	3041	59,900
XMR-3100	61.0	2591	41,700	67.5	2955	58,900
H-4831	60.0	2701	45,200	67.0	2975	58,700
RX22	60.0	2742	43,900	67.0	3106	60,400
IMR-7828	61.0	2701	44,300	68.0	3035	59,800
H-1000	65.0	2713	42,600	71.5	3030	58,700
RX25	64.5	2682	40,600	**72.0**	**3086**	**58,900**

154 gr. Jacketed SP 3.290" OAL — **BC:** .433 **SD:** .273

Powder	Sugg Starting Grains	Velocity fps	Pressure	Max Load Grains	Velocity fps	Pressure
H-414	53.5	2570	47,000	59.2	2856	60,700 P
N160	52.0	2627	52,700	58.0	2797	59,600 P
IMR-4350	55.0	2723	40,300	63.0	3012	51,400 C
IMR-4831	57.5	2738	50,400	64.0	2952	59,800 P
WXR	60.0	2764	52,300	66.5	2966	60,700 P
XMR-3100	59.5	2669	50,400	**66.0**	**2891**	**59,900 P**
H-4831	59.0	2710	37,700	66.0	3051	51,200 C
RX22	60.0	2755	48,200	67.0	3014	59,800 P
IMR-7828	61.0	2743	50,400	68.0	2972	60,500 P
H-1000	64.0	2818	56,100	71.0	2953	59,600 P
RX25	64.5	2842	53,200	71.5	3021	59,600 P

160 gr. Jacketed Fail Safe 3.290" OAL — **BC:** .382 **SD:** .283

Powder	Sugg Starting Grains	Velocity fps	Pressure P.S.I.	Max Load Grains	Velocity fps	Pressure P.S.I.
N160	52.0	2690	50,300	57.0	2801	59,300
IMR-4350	55.0	2598	48,000	61.1	2839	59,500
IMR-4831	56.5	2626	48,100	63.0	2867	59,300
WXR	56.5	2594	47,400	63.0	2853	59,300
XMR-3100	59.5	2543	45,200	65.0	2832	59,500
H-4831	59.5	2622	48,700	66.0	2848	59,800
RX22	60.0	2647	46,100	**66.0**	**2933**	**59,400**
IMR-7828	60.5	2595	45,800	67.0	2893	60,500
H1000	63.5	2572	45,000	70.5	2867	60,600
RX25	62.5	2603	43,200	69.2	2919	57,600

168 gr. Jacketed HPBT 3.290" OAL — **BC:** .488 **SD:** .298

Powder	Sugg Starting Grains	Velocity fps	Pressure	Max Load Grains	Velocity fps	Pressure
N160	50.0	2600	49,400	56.5	2817	59,500 P
IMR-4350	53.0	2571	37,700	61.0	2925	52,000 C
IMR-4831	58.0	2638	45,900	64.0	2915	60,500 P
WXR	58.0	2659	47,800	64.0	2919	60,600 P
H-4831	59.0	2685	38,600	**65.0**	**2971**	**52,000 C**
RX22	57.0	2594	38,500	66.0	2978	50,400 C
IMR-7828	61.5	2622	37,500	69.0	2946	52,100 C
H-1000	65.5	2645	38,500	73.0	2971	51,100 C
RX25	61.5	2594	41,900	68.5	2943	59,200 P
AA8700	70.0	2629	37,800	78.5	2948	49,900 C

175 gr. Jacketed GSSP 3.260" OAL — **BC:** .465 **SD:** .310

Powder	Sugg Starting Grains	Velocity fps	Pressure	Max Load Grains	Velocity fps	Pressure
N160	50.5	2449	53,300	56.0	2629	60,300 P
IMR-4350	52.0	2498	50,300	58.0	2706	60,200 P
IMR-4831	53.5	2515	49,600	59.5	2732	60,200 P
WXR	55.5	2616	53,400	62.5	2787	60,300 P
H-4831	53.7	2521	40,200	60.0	2748	46,400 C
RX22	56.0	2566	49,400	**62.5**	**2797**	**60,300 P**
IMR-7828	61.3	2620	40,200	68.5	2910	51,100 C
H-1000	61.0	2559	40,200	71.0	2897	51,400 C
RX25	62.0	2687	53,300	68.5	2871	59,900 P
AA8700	71.0	2481	42,700	77.0+	2799	57,600 P

Note: Loads shown in shaded panels are maximum.
Loads shown in bold designate potentially most accurate load.
\+ Designates a compressed powder charge.
In pressure column, C=Copper Units of Pressure (CUP); P=Pounds per Square Inch (PSI)

7mm Remington Magnum

#287641
160 gr. (#2 Alloy) 3.170" OAL

BC: .382
SD: .283

Powder	Sugg Starting Grains	Velocity fps	Pressure	Max Load Grains	Velocity fps	Pressure
Unique	17.0	1680	24,300	21.5	1917	36,200 C
SR-4756	18.5	1694	26,000	22.0	1880	36,700 C
SR-4759	29.0	2023	25,500	38.0	2440	48,000 C
IMR-4227	31.0	2067	25,800	36.0	2282	35,000 C
IMR-4198	32.0	2082	24,000	41.0	2462	39,300 C
XMP-5744	**22.5**	**1611**	**15,800**	40.0	2432	43,500 P
RX7	35.0	2151	26,000	46.0	2613	45,000 C

Note: Loads shown in shaded panels are maximum.
Loads shown in bold designate potentially most accurate load.
C=Copper Units of Pressure (CUP); P=Pounds per Square Inch (PSI)

7mm Weatherby Magnum

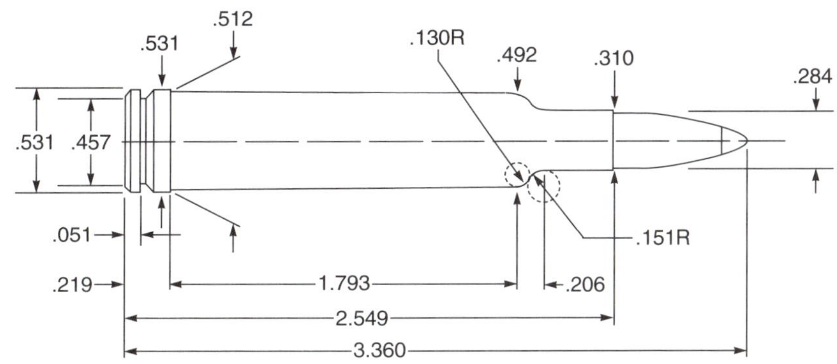

Comments:

This is one of Roy Weatherby's original cartridges developed in 1943. The 7mm shares the double-radius shoulder and 300 H&H parent case of its siblings, the 270 and 300 Weatherby Magnums. The 7mm has never been one of the more poplar calibers in the Weatherby line. It has been largely overshadowed by the 7mm Remington Magnum despite a slight ballistic advantage. What the future holds for the 7mm Weatherby remains to be seen due to the recent introduction of several high performance 7mm Magnum cartridges by the major manufacturers. Nevertheless, it is a capable cartridge suitable for thin-skinned game most anywhere in the world.

Like other Weatherbys, slow burning powders and premium bullets are in order. Shooters should also refrain from overheating the barrel with sustained firing. Hodgdon's H-4831 has been extensively used by reloaders over the years. Reloder 22 and H-1000 also gave good ballistic uniformity in our tests. Weatherby recommends the use of Federal 215 Magnum primers for all jacketed loads. Those loading cast bullet #287641 should use standard primers as indicated. This data is intended for commercially produced and chambered rifles. It is not for use in custom guns that may lack the free bore found in standard Weatherby chambers.

Test Components:

Cases . . . . .Federal
Trim-to Length . . . . .2.540"
Primers . . . . .Federal 215 Mag and 210
Primer Size . . . . .Large Rifle, Magnum & Standard
Lyman Shell Holder . . . . .No. 13
Jacketed Bullets Used . . .Nosler Partition #16325, 140 gr.
Nosler Ballistic Silvertip, #51110, 150 gr.
Combined Tech. Fail Safe #53160, 160 gr.
Speer Grand Slam #1643, 175 gr.
Cast Bullets Used . . . .(sized to .284" dia)
*gas check bullet *#287641, 160 gr.

Test Specifications:
(Velocity Only)

Firearm Used . . . . .Weatherby Mk V
Barrel Length . . . . .26"
Twist . . . . .1-10"
Groove Dia. . . . .284"

140 gr. Jacketed SP 3.300" OAL — **BC:** .434 **SD:** .248

Powder	Sugg Starting Grains	Velocity fps	Pressure	Max Load Grains	Velocity fps	Pressure
IMR-4831	67.0	3043	—	71.0	3259	—
XMR-3100	67.0	3001	—	70.5	3197	—
H-4831	69.0	3028	—	73.0	3219	—
RX22	68.0	3044	—	72.0	3317	—
IMR-7828	68.0	2979	—	72.0	3248	—
H1000	73.0	3021	—	**77.0**	**3245**	—

150 gr. Jacketed Bal. Silvertip 3.250" OAL — **BC:** .493 **SD:** .266

Powder	Sugg Starting Grains	Velocity fps	Pressure	Max Load Grains	Velocity fps	Pressure
IMR-4831	65.5	2939	—	69.0	3148	—
XMR-3100	66.5	2976	—	70.0	3176	—
H-4831	67.0	3011	—	71.0	3139	—
RX22	66.5	2978	—	70.0	3203	—
IMR-7828	66.5	2905	—	70.0	3138	—
H1000	73.0	3058	—	**77.0**	**3197**	—

Note: Loads shown in shaded panels are maximum.
Loads shown in bold designate potentially most accurate load.

7mm Weatherby Magnum

160 gr. Jacketed Fail Safe
3.250" OAL

BC: .382
SD: .283

Powder	Sugg Starting Grains	Velocity fps	Pressure	Max Load Grains	Velocity fps	Pressure
IMR-4831	64.0	2802	—	68.0	2994	—
XMR-3100	65.5	2812	—	69.0	3007	—
H-4831	66.5	2834	—	70.0	3017	—
RX22	65.5	2811	—	**69.0**	**3008**	—
IMR-7828	66.0	2787	—	69.5	3014	—
H1000	70.0	2826	—	74.5	3004	—

175 gr. Jacketed GSSP
3.350" OAL

BC: .465
SD: .310

Powder	Sugg Starting Grains	Velocity fps	Pressure	Max Load Grains	Velocity fps	Pressure
XMR-3100	64.0	2730	—	67.5	2911	—
H-4831	65.5	2766	—	69.0	2915	—
RX22	64.0	2738	—	67.5	2865	—
IMR-7828	65.5	2751	—	69.0	2931	—
H1000	**69.0**	**2720**	—	73.0	2902	—

***#287641**
160 gr. (#2 Alloy) 3.250" OAL

BC: .382
SD: .283

Powder	Sugg Starting Grains	Velocity fps	Pressure	Max Load Grains	Velocity fps	Pressure
SR-4759	19.0	1620	—	26.0	2008	—
IMR-4227	20.5	1609	—	28.5	2013	—
XMP-5744	21.0	1604	—	28.5	1998	—
IMR-4198	20.0	1591	—	**28.5**	**1996**	—
RX7	21.0	1622	—	29.5	2002	—

Note: Loads shown in shaded panels are maximum.
Loads shown in bold designate potentially most accurate load.
* Designates use of 210 primers.

7mm WSM (Winchester Short Magnum)

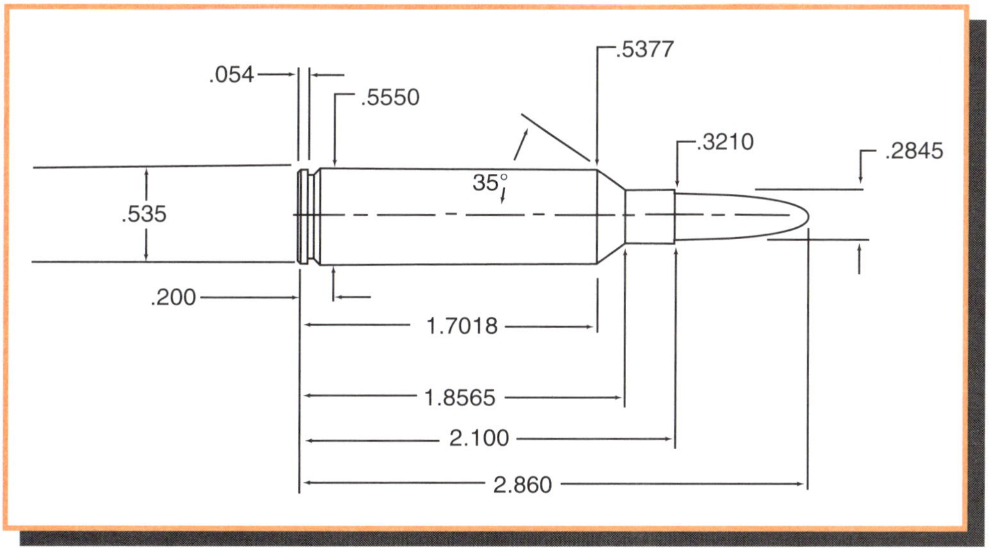

Comments:

The 7mm WSM is Winchester's first development in the .284" bore size since their 284 nearly forty years ago. Like its 300 and 270 WSM brethren, the 7mm WSM is designed to take advantage of the short stubby position of powder for more efficient ignition. The WSM is a non-belted design and will function in short actions unlike most other high performance 7mm chamberings. The 7mm WSM features a longer body and shorter neck to prevent the cartridge from being mistakenly chambered in 270 WSM chambered rifles. On paper, ballistics of the 7mm WSM are on par with the time honored 7mm Remington Magnum. Data is provided for shooters to duplicate the 160-grain Fail-Safe® factory load. Barnes Bullets does not recommend data for their coated XLC bullets be used for uncoated bullets. The 7mm WSM displayed unusually good ballistic uniformity with nearly every powder tested. IMR-4350, IMR-4831, Reloder 19, and WXR performed especially well. The short neck found on the 7mm WSM prevented effective use of our current cast bullet designs.

Test Components:

Cases . . . . Winchester
Trim-to Length . . . . 2.090"
Primers . . . . Winchester
Primer Size . . . . Large Rifle Magnum
Lyman Shell Holder . . . . No. 34
Jacketed Bullets Used . . . . . Sierra SPT #1900, 120 gr.
Nosler Ballistic Tip #28140, 140 gr.
Nosler Bal. SilverTip #51110, 150 gr.
Comb. Tech. Fail Safe #53160, 160 gr.
Barnes XLC #28458, 160 gr.

Test Specifications: (Velocity & Pressure)

Firearm Used . . . . Universal Receiver
Barrel Length . . . . 24"
Twist . . . . 1-9½"
Groove Dia. . . . . .284"

120 gr. Jacketed SPT 2.760" OAL — **BC:** .328 **SD:** .213

Powder	Sugg Starting Grains	Velocity fps	Pressure	Max Load Grains	Velocity fps	Pressure
IMR-4350	61.5	2982	45,200	68.5	3407	61,500
760	62.0	3170	52,100	69.0	3433	61,900
XMR-4350	62.5	3025	45,500	69.5	3431	62,100
IMR-4831	64.0	2997	44,900	71.0	3437	62,200
RX19	66.5	3060	47,200	**74.0+**	**3443**	**61,900**

140 gr. Jacketed Ballistic Tip 2.810" OAL — **BC:** .485 **SD:** .248

Powder	Sugg Starting Grains	Velocity fps	Pressure	Max Load Grains	Velocity fps	Pressure
H-414	58.5	2941	51,600	65.0	3198	62,200
760	58.5	2896	49,000	65.0	3196	61,500
IMR-4350	58.5	2833	46,400	65.0	3197	61,500
XMR-4350	60.0	2835	45,200	66.5	3226	62,100
IMR-4831	61.0	2878	47,100	**68.0+**	**3243**	**62,600**
N560	62.5	2832	45,800	69.5+	3188	59,700
XMR-3100	62.5	2825	47,000	69.5+	3181	61,300
RX19	64.0	2912	47,700	71.0	3276	62,800
WXR	64.0	2868	46,700	71.0+	3231	61,300
MAGPRO	70.0	3013	49,700	74.5+	3256	60,400

Note: Loads shown in shaded panels are maximum.
Loads shown in bold designate potentially most accurate load.
\+ Designates a compressed powder charge.

7mm WSM (Winchester Short Magnum)

150 gr. Jacketed Silvertip
2.850" OAL

BC: .493
SD: .266

Powder	Sugg Starting Grains	Velocity fps	Pressure P.S.I.	Max Load Grains	Velocity fps	Pressure P.S.I.
H-414	56.5	2804	51,900	63.0	3041	60,200
760	57.0	2807	51,100	63.5	3058	60,700
IMR-4350	58.0	2755	47,400	**64.5**	**3100**	**61,900**
XMR-4350	59.5	2752	46,100	66.0	3119	62,600
IMR-4831	61.0	2775	47,000	67.5	3133	62,000
XMR-3100	61.5	2722	46,400	68.5+	3073	60,900
RX19	63.0	2805	47,700	70.0	3155	62,300
WXR	63.0	2761	46,900	70.0+	3119	61,600
MAGPRO	70.0	2943	51,800	74.0+	3135	59,900

160 gr. Jacketed Fail Safe
2.770" OAL

BC: .382
SD: .283

Powder	Sugg Starting Grains	Velocity fps	Pressure P.S.I.	Max Load Grains	Velocity fps	Pressure P.S.I.
H-414	53.0	2601	48,000	59.0	2894	61,800
760	56.5	2770	57,200	59.5	2899	63,200
IMR-4350	57.0	2670	48,800	63.5	2977	62,700
XMR-4350	58.0	2672	47,900	64.5	2978	61,900
IMR-4831	59.5	2695	48,200	**66.0**	**2996**	**61,800**
XMR-3100	60.5	2621	46,700	67.5+	2952	60,700
RX19	61.5	2715	48,600	68.5	3034	62,500
WXR	62.5	2688	48,100	69.5+	3013	60,900
MAGPRO	69.0	2783	49,400	75.0+	3017	62,400

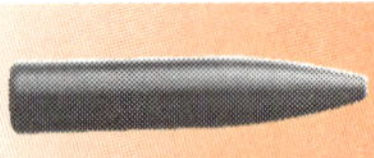

160 gr. Barnes XLC
2.775" OAL

BC: .508
SD: .283

Powder	Sugg Starting Grains	Velocity fps	Pressure P.S.I.	Max Load Grains	Velocity fps	Pressure P.S.I.
IMR-4350	57.0	2583	46,400	63.5+	2944	61,200
RX19	62.0	2685	48,200	69.0	3002	61,400
WXR	61.5	2662	47,400	**68.5+**	**2956**	**59,000**
IMR-7828	62.0	2705	49,000	69.0+	2997	60,400
MAGPRO	72.0	2920	54,300	75.0+	3014	60,500

Note: Loads shown in shaded panels are maximum.
Loads shown in bold designate potentially most accurate load.
\+ Designates a compressed powder charge.

7mm Rem. Short Action Ultra Mag

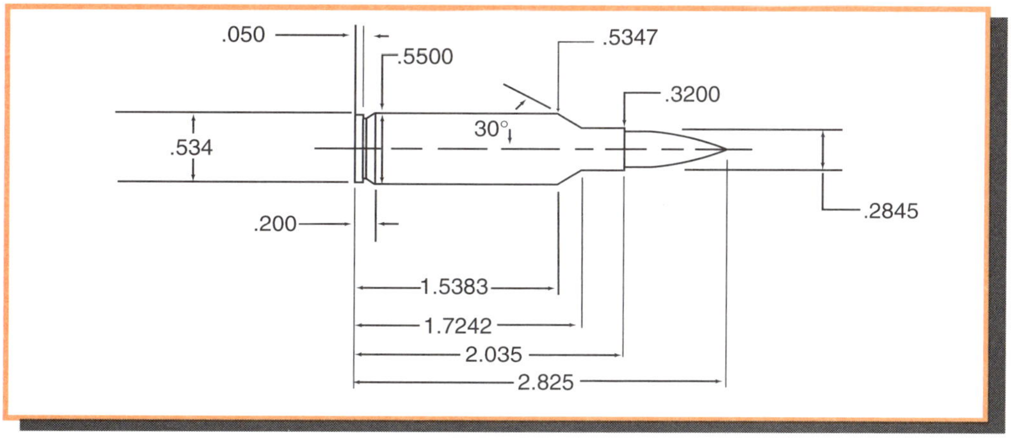

Comments:

Newly introduced for 2002, this is the second short action version of the non-belted Ultra Mag family of cartridges from Remington. The 7mm SAUM shares the 30-degree shoulder of its parent, the 300 SAUM, but has a nearly 1/16"longer neck than its 7mm WSM counterpart. Factory data from Remington shows performance of this cartridge neck in neck with their old standby, the 7mm Remington Magnum. Lyman offers data to duplicate the factory loaded 160-grain Nosler Partition as well as several other premium hunting bullets. Also included is data for the 115-grain Speer HP for those with the urge to take their SAUM varmint hunting. The 120-grain Sierra and 130-grain Speer bullets are best for plinking or coyote-sized game. All powders listed with cast bullet # 287641 gave good ballistic uniformity, particularly SR-4759 and IMR-4198. Overall length with the cast bullet allows the lube grooves and gas check to both remain within the neck of the case. Those shooters loading this cast bullet may want to run the bore-riding nose portion through a .277" diameter H & I die, as this bullet will engage the rifling. Best performance with jacketed bullets came from moderate to slow burning powders such as WXR and MagPro.

Test Components:

Cases . . . . .Remington
Trim-to Length . . . . .2.025"
Primers . . . . .Remington 9½ M & 9½ (cast bullets)
Primer Size . . . . .Large Rifle, Magnum & Standard
Lyman Shell Holder . . . . .No. 34
Jacketed Bullets Used . . . . . .Speer HP #1617, 115 gr.
Sierra SPT #1900, 120 gr.
Speer SP #1623, 130 gr.
Nosler Partition #16325, 140 gr.
Barnes X #28425, 140 gr.
Swift Scirocco, 150 gr.
Nosler Partition #16327, 160 gr.
Barnes XLC #28458, 160 gr.
Sierra HPBT #1930, 168 gr.
Cast Bullets Used . . . .(sized to .284" dia)
*gas check bullet *#287641, 160 gr.

Test Specifications: (Velocity & Pressure)

Firearm Used . . . . .Universal Receiver
Barrel Length . . . . .24"
Twist . . . . .1-9¼"
Groove Dia. . . . . .284"

115 gr. Jacketed HP
2.590" OAL
BC: .257
SD: .204

Powder	Sugg Starting Grains	Velocity fps	Pressure P.S.I.	Max Load Grains	Velocity fps	Pressure P.S.I.
H-414	60.0	3192	51,900	63.0	3381	61,700
IMR-4350	61.0	3203	53,600	64.0	3415	63,600
RX19	64.5	3208	53,900	68.0+	3401	62,700
WXR	65.5	3190	53,800	69.0+	3401	64,000
MAGPRO	71.0	3315	55,200	**75.0+**	**3449**	**59,800**

120 gr. Jacketed SPT
2.675" OAL
BC: .328
SD: .213

Powder	Sugg Starting Grains	Velocity fps	Pressure P.S.I.	Max Load Grains	Velocity fps	Pressure P.S.I.
H-414	59.5	3224	55,800	62.7	3381	63,100
IMR-4350	60.0	3153	52,700	63.5	3383	63,800
N160	62.0	3218	54,500	65.5	3393	63,400
RX19	64.5	3201	55,000	68.0+	3382	63,600
WXR	65.0	3200	55,700	**68.5+**	**3381**	**63,900**

Note: Loads shown in shaded panels are maximum.
Loads shown in bold designate potentially most accurate load.
\+ Designates a compressed powder charge.

7mm Rem. Short Action Ultra Mag

130 gr. Jacketed SP
2.660" OAL
BC: .394 **SD:** .230

Powder	Sugg Starting Grains	Velocity fps	Pressure P.S.I.	Max Load Grains	Velocity fps	Pressure P.S.I.
H-414	58.0	3122	58,400	61.0	3249	64,200
IMR-4350	58.0	3057	54,300	61.0	3240	63,100
N160	59.0	3095	57,000	62.5	3226	62,000
RX19	63.0	3094	54,600	66.0+	3289	64,100
WXR	62.5	3031	53,000	**66.0**	**3234**	**62,800**

140 gr. Jacketed SP
2.800" OAL
BC: .434 **SD:** .248

Powder	Sugg Starting Grains	Velocity fps	Pressure P.S.I.	Max Load Grains	Velocity fps	Pressure P.S.I.
H-414	55.0	2993	57,500	58.0	3120	63,400
IMR-4350	57.0	3003	56,000	60.0	3137	62,700
XMR-4350	58.0	3050	57,100	61.5	3184	63,500
IMR-4831	59.0	2979	53,600	62.5	3162	61,900
N560	60.5	3037	57,800	64.0	3170	63,900
RX19	61.0	3000	54,400	64.5	3165	62,700
H4831SC	61.0	2957	56,400	**64.0**	**3089**	**63,000**
WXR	62.0	3007	54,600	65.5	3198	64,400
MAGPRO	67.0	3082	54,500	70.5+	3231	61,500

140 gr. Barnes X
2.755" OAL
BC: .436 **SD:** .248

Powder	Sugg Starting Grains	Velocity fps	Pressure P.S.I.	Max Load Grains	Velocity fps	Pressure P.S.I.
IMR-4350	56.5	2885	56,200	59.5	3059	64,200
XMR-4350	57.0	2885	54,700	60.0	3054	62,700
IMR-4831	59.0	2922	56,300	62.0	3098	64,100
RX19	61.0	2947	56,000	64.0+	3106	63,300
H-4831SC	61.0	2880	56,900	64.0	3019	64,000
WXR	61.5	2910	54,900	**65.0+**	**3100**	**64,000**
MAGPRO	66.5	3034	56,000	70.0	3197	64,200

150 gr. Jacketed Scirocco
2.825" OAL
BC: .515 **SD:** .266

Powder	Sugg Starting Grains	Velocity fps	Pressure P.S.I.	Max Load Grains	Velocity fps	Pressure P.S.I.
H-414	54.0	2908	57,400	57.0	3043	63,800
IMR-4350	54.0	2885	56,500	57.0	3024	63,800
IMR-4831	56.0	2874	55,200	59.0	3025	63,100
RX19	58.0	2885	55,100	61.0	3047	63,200
H-4831SC	59.0	2875	57,700	62.0	3003	64,400
WXR	59.0	2892	55,900	62.5	3053	63,800
MAGPRO	65.5	2928	54,400	**69.0+**	**3083**	**62,300**

160 gr. Jacketed SP
2.825" OAL
BC: .475 **SD:** .283

Powder	Sugg Starting Grains	Velocity fps	Pressure P.S.I.	Max Load Grains	Velocity fps	Pressure P.S.I.
IMR-4350	52.5	2782	57,000	55.5	2908	63,100
IMR-4831	56.0	2858	60,700	59.0	2952	64,100
H-4831SC	56.5	2776	58,200	59.5	2884	64,000
RX22	56.0	2902	61,000	59.0	2972	63,300
N165	58.5	2821	56,500	61.5	2940	62,400
WXR	59.0	2838	57,400	**62.0**	**2970**	**63,900**
IMR-7828	58.5	2826	56,100	61.5	2957	62,500
MAGPRO	62.5	2863	57,200	65.7	2966	61,000
H-1000	63.0	2781	56,300	66.5+	2924	64,400

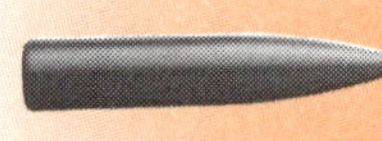

160 gr. Barnes XLC
2.665" OAL
BC: .508 **SD:** .283

Powder	Sugg Starting Grains	Velocity fps	Pressure P.S.I.	Max Load Grains	Velocity fps	Pressure P.S.I.
IMR-4350	55.5	2709	54,300	58.5+	2878	62,500
IMR-4831	58.0	2787	55,800	61.0	2925	63,700
RX19	58.0	2787	55,700	61.5+	2902	63,000
H-4831SC	59.0	2696	56,000	62.0+	2846	63,400
WXR	61.0	2788	57,700	64.5+	2927	64,300
IMR-7828	61.0	2742	54,300	64.0+	2881	61,400
N165	62.5	2792	53,300	**66.0+**	**2931**	**60,600**
H-1000	64.5	2679	52,200	68.0+	2828	59,900
MAGPRO	67.0	2828	54,600	70.5+	2989	62,900

168 gr. Jacketed HPBT
2.825" OAL
BC: .488 **SD:** .298

Powder	Sugg Starting Grains	Velocity fps	Pressure P.S.I.	Max Load Grains	Velocity fps	Pressure P.S.I.
IMR-4350	54.0	2775	56,400	57.0	2895	62,600
IMR-4831	56.0	2789	56,200	59.0	2929	63,900
RX19	56.5	2789	55,500	59.5	2936	63,800
H-4831SC	58.0	2731	56,000	61.0	2855	63,500
WXR	58.0	2807	57,100	**61.5**	**2941**	**63,900**
RX22	58.5	2800	55,500	61.5	2949	63,900
H-1000	63.5	2742	54,500	67.0+	2873	61,800
MAGPRO	63.0	2821	54,700	66.5	2949	61,500

*#287641
160 gr. (#2 Alloy) 2.740" OAL
BC: .382 **SD:** .283

Powder	Sugg Starting Grains	Velocity fps	Pressure P.S.I.	Max Load Grains	Velocity fps	Pressure P.S.I.
SR-4759	17.5	1606	23,100	**26.0**	**2088**	**41,600**
XMP-5744	19.5	1596	21,500	29.0	2118	34,700
IMR-4198	20.0	1626	21,200	29.0	2092	33,400
RX7	20.5	1596	20,500	30.0	2095	33,300
N133	24.0	1615	19,900	32.5	2096	31,800
H-4895	25.0	1609	20,400	34.5	2089	29,500

Note: Loads shown in shaded panels are maximum.
Loads shown in bold designate potentially most accurate load.

+Designates a compressed powder charge.
*Designates the use of Rem. 9 1/2 primers.

7mm STW (Shooting Times Westerner)

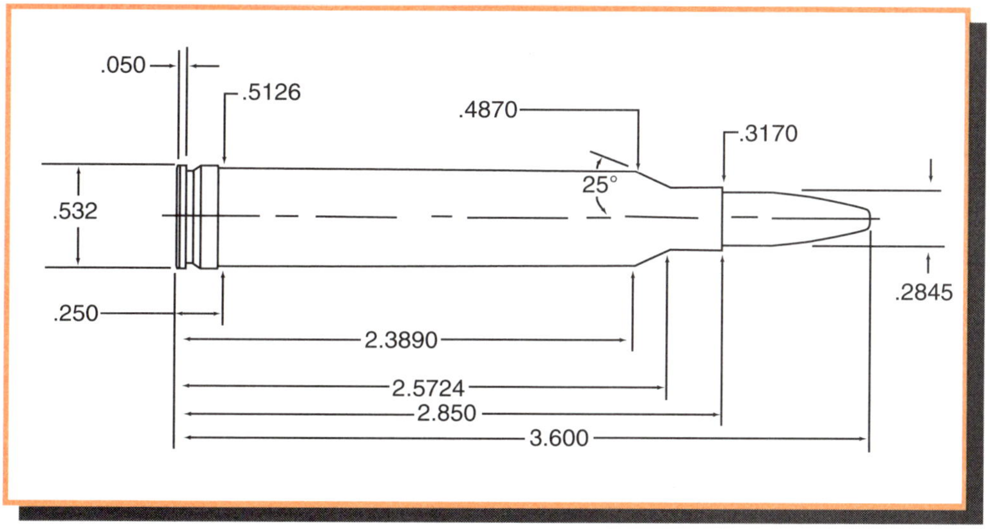

Comments:

While many cartridges languish in wildcat status for years before being legitimized as factory offerings, the STW's performance gained early—and widespread—acceptance. Gun writer Layne Simpson developed the STW in 1989 based on a necked down 8mm Remington Magnum. Flat trajectory, hard-hitting power, excellent accuracy, and a wide selection of .284" diameter bullets led to a popularity level never enjoyed by the parent cartridge. Remington standardized the STW in 1997 and other manufacturers including Ruger and Winchester soon followed. This high intensity cartridge offers approximately 200 feet per second over the vaunted 7mm Remington Magnum and 50 to 100 feet over the 7mm Weatherby. One-shot kills out to over four hundred yards and beyond are not unheard of with this cartridge. The STW works best with slower burning powders including, IMR-7828, H-1000, Reloder 22, Reloder 25, and Federal 215 Magnum primers. Shooters loading this cartridge should stick with premium grade hunting bullets designed to withstand the STW's velocity levels. While the pressure test barrel held up reasonably well in our lab, shooters should refrain from shooting long strings to prolong barrel life.

Test Components:

Cases . . . . Federal
Trim-to Length . . . . 2.840"
Primers . . . . Federal 215
Primer Size . . . . Large Rifle Magnum
Lyman Shell Holder . . . . No. 13
Jacketed Bullets Used . . . . . Sierra SPT #1900, 120 gr.
Nosler SP #16325, 140 gr.
Combined Tech. Bal. Silvertip #51110, 150 gr.
Combined Tech. Fail Safe #53160, 160 gr.
Speer Grand Slam # 1643, 175 gr.
Cast Bullets Used . . . . (sized to .285" dia)
*gas check bullet *#287641, 160 gr.

Test Specifications:
(Velocity & Pressure)

Firearm Used . . . . Universal Receiver
Barrel Length . . . . 24"
Twist . . . . 1-9½"
Groove Dia. . . . . .284"

120 gr. Jacketed SP 3.585" OAL — **BC:** .328 **SD:** .213

Powder	Sugg Starting Grains	Velocity fps	Pressure P.S.I.	Max Load Grains	Velocity fps	Pressure P.S.I.
IMR-4350	68.0	3208	50,300	75.8	3522	63,700
H-4831	72.0	3180	49,800	80.0	3478	63,100
N160	73.0	3260	56,400	77.0	3423	63,400
WXR	77.5	3253	53,400	82.0	3455	62,000
RX22	75.5	3146	45,700	84.5	3536	63,100
IMR-7828	75.0	3123	46,700	83.0	3507	63,400
H-1000	**79.0**	**3114**	**45,100**	86.2	3498	62,600
AA8700	85.0	2832	37,600	94.5+	3317	55,800
H-870	88.5	3023	45,000	98.4+	3503	61,800

140 gr. Jacketed SP 3.585" OAL — **BC:** .434 **SD:** .248

Powder	Sugg Starting Grains	Velocity fps	Pressure P.S.I.	Max Load Grains	Velocity fps	Pressure P.S.I.
H-4831	70.5	3054	54,600	78.0	3257	63,100
WXR	74.0	3123	56,100	78.0	3279	62,800
MAGPRO	75.0	3116	58,100	79.0	3209	61,000
RX22	74.0	3023	47,500	**82.0**	**3382**	**63,700**
IMR-7828	72.0	2981	48,100	79.3	3313	63,200
H-1000	75.5	3074	51,700	83.0	3299	63,200
AA 8700	82.5	2938	45,400	91.5	3330	63,200
H-870	86.5	2959	44,100	96.0	3375	63,100

Note: Loads shown in shaded panels are maximum.
Loads shown in bold designate potentially most accurate load.
\+ Designates a compressed powder charge.

7mm STW (Shooting Times Westerner)

150 gr. Jacketed Silvertip
3.585" OAL
BC: .493
SD: .266

Powder	Sugg Starting Grains	Velocity fps	Pressure P.S.I.	Max Load Grains	Velocity fps	Pressure P.S.I.
H-4831	70.0	2853	47,700	**76.0**	**3155**	**62,800**
WXR	73.5	2999	54,500	77.5	3191	63,100
N165	73.5	2909	54,100	77.5	3102	62,600
MAGPRO	74.0	2986	54,700	78.0	3156	62,900
RX22	71.0	2884	45,800	79.0	3246	62,800
IMR-7828	70.0	2801	44,000	77.8	3192	62,600
H-1000	74.0	2840	44,700	80.0	3212	63,600
RX25	74.0	2844	43,500	82.0	3304	63,700
Retumbo	79.0	3047	56,600	83.5+	3209	63,200
H50BMG	85.5	2810	47,900	90.0+	2989	55,400
AA8700	81.0	2825	43,900	90.0	3209	62,700
H-870	86.0	2932	45,800	94.0	3290	62,400

160 gr. Jacketed Fail Safe
3.585" OAL
BC: .382
SD: .283

Powder	Sugg Starting Grains	Velocity fps	Pressure P.S.I.	Max Load Grains	Velocity fps	Pressure P.S.I.
H-4831	69.0	2775	49,800	76.0	3054	63,200
MAGPRO	73.0	2902	57,400	77.0	3025	62,500
Retumbo	76.0	2919	54,200	80.0	3084	62,600
RX22	70.0	2815	47,500	**77.5**	**3145**	**63,000**
IMR-7828	69.0	2730	45,900	77.0	3076	63,400
H-1000	73.0	2781	48,000	81.5	3088	63,100
RX25	71.0	2778	46,000	79.0	3147	63,600
AA8700	81.5	2778	46,200	90.5	3125	62,200
H-870	86.0	2819	46,500	92.5	3170	62,800

175 gr. Jacketed GSSP
3.585" OAL
BC: .465
SD: .310

Powder	Sugg Starting Grains	Velocity fps	Pressure P.S.I.	Max Load Grains	Velocity fps	Pressure P.S.I.
H-4831	72.5	2751	54,900	72.5	2943	63,400
N170	70.0	2782	57,700	74.0	2902	63,200
MAGPRO	72.0	2840	59,500	76.0	2945	63,100
RX22	**68.0**	**2753**	**49,900**	75.5	3020	63,000
IMR-7828	67.5	2746	52,100	75.0	2985	63,700
H-1000	69.0	2772	54,400	77.0	2954	62,800
RX25	70.0	2763	50,900	78.0	3051	63,500
Retumbo	74.0	2840	54,800	78.0	2999	62,900
H50BMG	84.5	2858	58,300	89.0+	2987	64,100
AA8700	81.0	2715	46,100	90.0	3064	63,200
H-870	82.0	2722	45,200	90.0	3083	63,400

#287641
160 gr. (#2 Alloy) 3.380" OAL
BC: .382
SD: .283

Powder	Sugg Starting Grains	Velocity fps	Pressure P.S.I.	Max Load Grains	Velocity fps	Pressure P.S.I.
SR4759	27.0	1881	25,000	35.0	2222	41,000
XMP-5744	**30.0**	**1959**	**22,300**	42.0	2444	40,300
IMR-4831	50.0	2020	20,500	60.0	2397	31,100

Note: Loads shown in shaded panels are maximum.
Loads shown in bold designate potentially most accurate load.
\+ Designates a compressed powder charge.

7mm Remington Ultra Mag

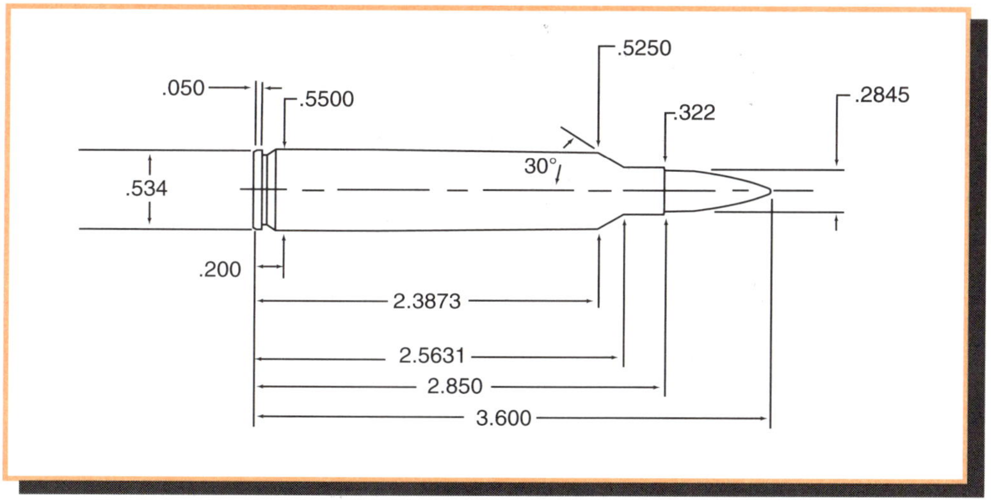

Comments:

Remington introduced the 7mm Ultra Mag in 2001 as the smallest caliber in the non-belted Ultra Mag family of cartridges. Advertised by the factory as 20 percent flatter shooting than the 7mm Remington Magnum and approximately 150 feet per second faster than the 7mm STW, the 7mm RUM is about as big as one can get in current 7mm factory rifles. Recent developments in slow burning powders and premium bullets such as the Swift Scirocco bring out the full potential of the 7mm RUM. Perhaps the most highly over-bored of current Magnum rifle cartridges, the 7mm RUM has an insatiable appetite for slow burning powder. Unfortunately, one of them, H-870, has recently been discontinued by Hodgdon. However, Hodgdon's H50BMG produced good results with almost all bullet weights. Initial feedback also reports excellent results with Alliant's Reloder 25. This cartridge has what one would expect in a long-range hunting cartridge; more speed, more downrange energy, more recoil, and more barrel wear. Shooters should take care not to overheat the barrel with sustained firing and not allow excessive fouling to accumulate. Three-shot strings are about maximum before excessive barrel heat necessitates cooling off. The use of a bore guide and a coated cleaning rod should give years of service to the long range hunter who fires only a few shots per season.

Test Components:

Cases . . . . Remington
Trim-to Length . . . . 2.840"
Primers . . . . Federal 215 Magnum and 210
Primer Size . . . . Large Rifle
Lyman Shell Holder . . . . No. 13
Jacketed Bullets Used . . . . . . Speer SP #1623, 130 gr.
Nosler Partition #16325, 140 gr.
Swift Scirocco, 150 gr.
Hornady SP #2845, 162 gr.
Cast Bullets Used . . . . (sized to .284" dia)
*gas check bullet *#287641, 160 gr.

Test Specifications: (Velocity & Pressure)

Firearm Used . . . . Universal Receiver
Barrel Length . . . . 26"
Twist . . . . 1-9½"
Groove Dia. . . . . .284"

130 gr. Jacketed SP
3.600" OAL
BC: .394 **SD:** .230

Powder	Sugg Starting Grains	Velocity fps	Pressure P.S.I.	Max Load Grains	Velocity fps	Pressure P.S.I.
IMR-4831	82.0	3355	57,200	86.5	3495	62,600
XMR-3100	86.0	3311	54,200	90.5	3500	62,100
H-4831SC	85.5	3303	57,100	90.0	3433	62,200
RX22	87.0	3365	55,600	**92.0**	**3560**	**63,400**
IMR-7828	87.0	3353	55,700	92.0	3534	63,200
H-1000	93.0	3312	56,000	98.0	3472	62,700
RX25	90.0	3325	53,500	95.5	3556	62,300
H50BMG	102.5	3270	55,100	**107.0+**	**3429**	**62,500**
AA8700	104.5	3369	55,000	110.0+	3560	62,900
H870	102.0	3164	47,700	107.5	3521	58,700

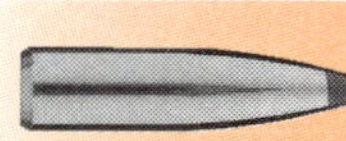

140 gr. Jacketed Partition
3.600" OAL
BC: .434 **SD:** .248

Powder	Sugg Starting Grains	Velocity fps	Pressure P.S.I.	Max Load Grains	Velocity fps	Pressure P.S.I.
IMR-4831	80.0	3273	57,500	84.0	3398	62,100
XMR-3100	81.0	3267	55,600	85.5	3428	63,000
H-4831SC	83.0	3249	58,200	87.5	3360	63,000
RX22	85.0	3328	56,700	89.5	3487	63,000
IMR-7828	84.5	3347	57,700	89.0	3509	65,000
H-1000	90.0	3256	57,300	95.0	3408	63,000
RX25	88.5	3316	56,200	**93.5**	**3493**	**63,400**
H50BMG	99.0	3228	55,300	**104.5**	**3398**	**63,300**
AA8700	101.0	3253	53,300	106.0	3457	61,400
H870	100.0	3218	52,200	105.0	3492	64,100

Note: Loads shown in shaded panels are maximum.
Loads shown in bold designate potentially most accurate load.
\+ Designates a compressed powder charge.

7mm Remington Ultra Mag

150 gr. Jacketed Scirocco
3.600" OAL

BC: .515
SD: .266

Powder	Sugg Starting Grains	Velocity fps	Pressure P.S.I.	Max Load Grains	Velocity fps	Pressure P.S.I.
IMR-4831	77.5	3144	56,000	82.0	3300	63,100
XMR-3100	79.0	3132	53,800	84.0	3294	61,500
H-4831SC	80.0	3126	56,300	85.0	3286	63,300
RX22	83.0	3196	55,300	87.5	3380	63,800
IMR-7828	82.5	3150	50,800	87.0	3349	63,000
H1000	88.0	3063	52,300	92.5	3291	62,600
RX25	86.5	3139	52,900	91.0	3344	61,500
H50BMG	99.0	3156	56,500	**104.0+**	**3308**	**63,100**
AA8700	100.0	3180	53,100	105.0	3396	62,500
H870	100.0	3159	52,500	105.5	3362	61,500

162 gr. Jacketed SP
3.650" OAL

BC: .514
SD: .287

Powder	Sugg Starting Grains	Velocity fps	Pressure P.S.I.	Max Load Grains	Velocity fps	Pressure P.S.I.
IMR-7828	80.0	2940	54,200	85.0	3146	61,500
H1000	84.5	2763	49,800	89.0	3008	59,100
RX25	85.0	2980	55,000	89.5	3189	62,400
H50BMG	95.0	2901	54,900	100.0	3166	65,000
AA8700	95.5	3014	56,400	**100.5**	**3251**	**65,000**
H870	97.0	3024	56,800	102.0	3223	64,300

***#287641**
160 gr. (#2 Alloy) 3.620" OAL

BC: .382
SD: .283

Powder	Sugg Starting Grains	Velocity fps	Pressure P.S.I.	Max Load Grains	Velocity fps	Pressure P.S.I.
SR-4759	24.5	1659	20,200	33.5	2111	35,200
XMP-5744	26.0	1661	19,000	**36.0**	**2101**	**28,800**
IMR-4198	25.0	1670	19,400	38.0	2142	30,700
RX7	26.0	1652	18,700	38.0	2110	29,000
Varget	35.0	1756	19,500	47.0	2120	26,300

Note: Loads shown in shaded panels are maximum.
Loads shown in bold designate potentially most accurate load.
* Designates use of 210 primers.
\+ Designates a compressed powder charge.

30 M1 Carbine

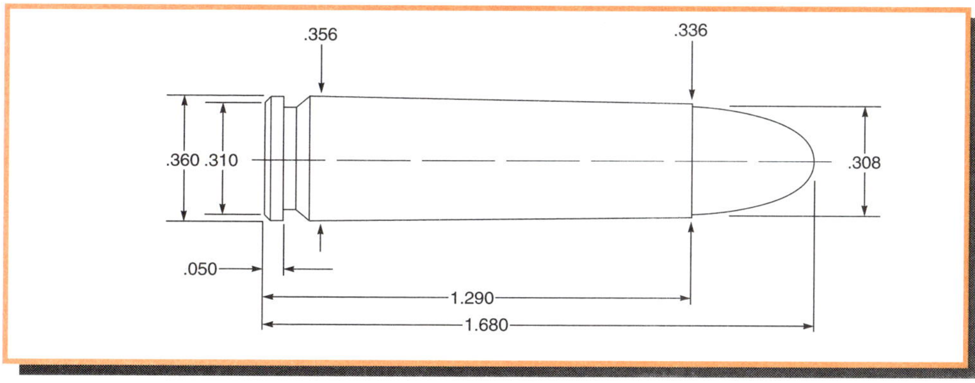

Comments:

The United States Army developed the M1 Carbine just prior to the outbreak of World War Two. The Carbine was originally intended as a small, lightweight rifle for engineers, mortar crews, and other personnel for whom the M1 Garand would be too unwieldy, but would have been "under-gunned" with the .45 caliber pistol. The M1 Carbine saw widespread use in World War II, and Korea, and was widely distributed throughout the world via military aid programs in subsequent years. Its first major civilian use occurred during the 1960s through surplus sales. The M1 Carbine's low recoil and ease of handling make it a fun cartridge to shoot and well suited to introducing new shooters to center-fire rifles. Large numbers of M1 Carbines have returned from overseas in recent years, many in rather worn condition.

The 30 Carbine closely resembles a lengthened semiautomatic pistol cartridge. Several manufacturers have in fact chambered pistols for this round over the years. Winchester based the 30 Carbine round on their short-lived 32 SL cartridge. The 30 Carbine has its limits and they should be adhered to. Original specifications call for a 110-grain round nose bullet at just over 1,900 feet per second...not especially impressive for 30-caliber ballistics even 60 years ago. This cartridge has a straight-walled, rimless case which headspaces on the case mouth. Case length should be uniform and should not be trimmed below the "trim-to" length. The internal ballistics of the cartridge and its required overall length limits bullet selection. Most major manufacturers produce bullets designed expressly for the little Carbine. Much like the 22 Hornet, the shape and small capacity of the case requires powders considered slow pistol or fast rifle propellants. VihtaVuori N110 produced good ballistic uniformity in our tests.

Test Components:

Cases . . . . .Remington
Trim-to Length . . . . .1.286"
Primers . . . . .Remington 6½
Primer Size . . . . .Small Rifle
Lyman Shell Holder . . . . .No. 19
Jacketed Bullets Used . . . . .Sierra RN #2100, 110 gr.
Cast Bullets Used . . . .(sized to .308" dia)
*gas check bullet *#311359, 115 gr.
#311410, 130 gr.

Test Specifications: (Velocity & Pressure)

Firearm Used . . . . .Universal Receiver
Barrel Length . . . . .20"
Twist . . . . .1-20"
Groove Dia. . . . . .308"

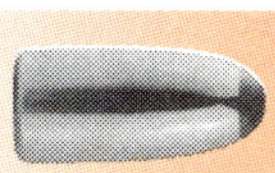

110 gr. Jacketed RN 1.680" OAL — **BC:** .170 **SD:** .166

Powder	Sugg Starting Grains	Velocity fps	Pressure C.U.P.	Max Load Grains	Velocity fps	Pressure C.U.P.
SR-4756	6.5	1355	27,000	7.3	1509	38,600
AA#9	11.7	1805	29,900	13.0	1935	38,800
2400	12.0	1765	28,800	13.5+	1953	36,500
N110	**12.5**	**1858**	**29,200**	13.8	2011	38,000
H110	14.0	1935	31,600	15.5	2079	38,800
IMR-4227	13.5	1642	28,200	15.0+	1845	39,000
AA-1680	15.5	1642	22,300	17.3+	1897	34,400

#311359 115 gr. (#2 Alloy) 1.635" OAL — **BC:** .181 **SD:** .173

Powder	Sugg Starting Grains	Velocity fps	Pressure C.U.P.	Max Load Grains	Velocity fps	Pressure C.U.P.
Unique	5.5	1388	27,600	6.5	1555	38,600
SR-7625	4.6	1194	22,800	5.6	1368	39,000
SR-4756	5.5	1298	23,400	6.5	1463	38,100
2400	11.0	1633	31,500	12.5+	1828	40,000
N110	**10.0**	**1635**	**31,300**	11.3	1754	39,700
IMR-4227	11.5	1497	27,000	13.3+	1733	39,000
AA1680	14.0	1576	29,000	15.5	1768	38,300

#311410 130 gr. (#2 Alloy) 1.680" OAL — **BC:** .239 **SD:** .195

Powder	Sugg Starting Grains	Velocity fps	Pressure C.U.P.	Max Load Grains	Velocity fps	Pressure C.U.P.
Unique	4.9	1260	23,400	6.1	1468	35,000
SR-4756	5.0	1180	22,000	6.0	1371	37,300
2400	10.5	1564	28,800	12.0+	1733	38,600
N110	**10.0**	**1659**	**29,700**	11.2	1768	38,500
IMR-4227	10.6	1366	24,600	12.7+	1647	38,600
AA1680	13.5	1563	27,900	15.0+	1735	38,100

Note: Loads shown in shaded panels are maximum.
Loads shown in bold designate potentially most accurate load.
\+ Designates a compressed powder charge.

30-30 Winchester (30 WCF) (7.62 x 51R mm)

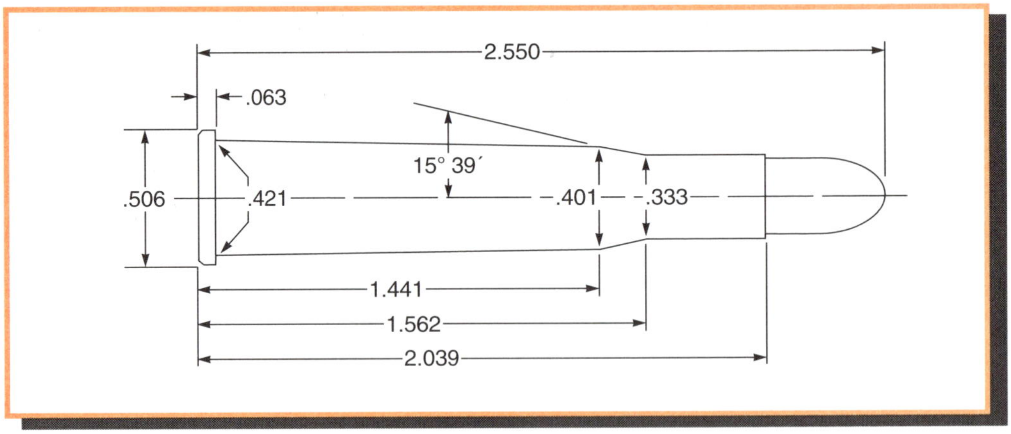

Comments:

Winchester introduced the 30 Winchester Center Fire (WCF) cartridge in their new Model 94 lever-action in 1895. The rest is history. Probably no other cartridge in North America has put as much venison on the table as the venerable old "thirty-thirty". Although ballistically unimpressive by today's standards, the lightweight, fast-handling rifles often chambered for the 30-30 proved well suited to hunting the brushy wood lots of the east coast and New England. Even if it had faded into obsolescence, the 30-30 would have earned its place in history as the first sporting cartridge to be loaded exclusively with smokeless powder.

The vast majority of rifles chambered for the 30-30 have tubular magazines. Such rifles require the use blunt or flat nosed bullets with a cannelure, and should be crimped in place. Loaded cartridges must be kept within the maximum overall length or they will not cycle through the rifle's action. All major bullet manufacturers offer projectiles designed especially for the 30-30. Shooters should trim all cases to uniform length prior to crimping. Many opt to purchase an additional seat body and seat and crimp in two separate operations.

Despite the indelible association with lever action rifles, there have been a few bolt action rifles chambered in 30-30 over the years. Shooters loading for bolt-action rifles or any of the Thompson/Center Contenders should refrain from exceeding maximum loads due to strength limitations of the cartridge case itself. The 30-30 works well with a number of powders including IMR-3031, IMR-4064, and 748. Cast bullet # 311041 was originally designed for the 30-30 and has a good reputation for accuracy. Reloder 7, XMP-5744, and IMR 4198 should work well with any of the listed cast bullets. Best results with cast bullets usually occur between 1,500 and 2,100 feet per second. Shooters loading for Marlin rifles with Micro-Groove® rifling should keep cast bullet velocities below 1,600 feet per second for best accuracy.

Test Components:

Cases . . . . .Winchester
Trim-to Length . . . . .2.028"
Primers . . . . .Winchester WLR and CCI 200
Primer Size . . . . .Large Rifle
Lyman Shell Holder . . . . .No. 6
Jacketed Bullets Used . . .Hornady RN #3015, 110 gr.
Sierra FP #2020, 125 gr.
Hornady RN #3035, 150 gr.
Hornady FN #3060, 170 gr.
Cast Bullets Used . . . .(sized to .308" dia)
*gas check bullet *#311291, 170 gr.
*#311041, 173 gr.

Test Specifications: (Velocity & Pressure)

Firearm Used . . . . .Universal Receiver
Barrel Length . . . . .24"
Twist . . . . .1-12"
Groove Dia. . . . . .308"

110 gr. Jacketed RN
2.470" OAL
BC: .150
SD: .166

Powder	Sugg Starting Grains	Velocity fps	Pressure C.U.P.	Max Load Grains	Velocity fps	Pressure C.U.P.
IMR-4198	27.0	2358	28,200	29.5	2617	38,100
IMR-3031	26.5	2085	28,400	33.5	2492	37,900
IMR-4895	31.0	2258	28,200	34.0	2545	37,700
H-335	34.0	2278	28,000	37.8	2587	37,400
BL-C(2)	35.0	2246	26,200	38.7	2542	35,600
748	36.0	2280	25,600	41.0+	2631	31,400
IMR-4064	31.0	2168	27,000	35.0+	2568	37,300
IMR-4320	28.0	2044	27,700	33.0	2418	37,100

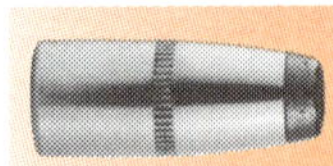

125 gr. Jacketed FN
2.420" OAL
BC: .153
SD: .188

Powder	Sugg Starting Grains	Velocity fps	Pressure C.U.P.	Max Load Grains	Velocity fps	Pressure C.U.P.
IMR-4198	25.0	2171	28,000	28.0	2406	35,700
RX7	27.0	2214	27,400	30.0	2428	34,600
IMR-3031	29.5	2012	24,100	33.0+	2385	36,700
AA2230	29.0	2075	25,400	32.0	2381	35,700
IMR-4895	30.5	2074	26,500	34.0	2331	32,800
H335	34.0	2398	32,800	37.5	2626	37,600
BL-C(2)	34.5	2312	29,700	38.0	2543	34,000
748	35.0	2338	29,500	39.0+	2531	30,900
IMR-4064	31.5	2070	27,300	35.0+	2356	34,400
IMR-4320	31.5	2090	27,400	35.0	2349	34,400
760	34.5	2056	25,200	38.5+	2265	28,700
RX15	32.0	2200	28,300	**36.0+**	**2441**	**35,000**

Note: Loads shown in shaded panels are maximum.
Loads shown in bold designate potentially most accurate load.
\+ Designates a compressed powder charge.

30-30 Winchester (30 WCF) (7.62 x 51R mm)

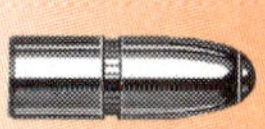

150 gr. Jacketed RN
2.540" OAL

BC: .186
SD: .266

Powder	Sugg Starting Grains	Velocity fps	Pressure C.U.P.	Max Load Grains	Velocity fps	Pressure C.U.P.
IMR-4198	20.5	1884	30,000	23.5	2092	37,600
RX7	25.0	2154	30,500	28.0	2350	38,100
IMR-3031	26.5	1876	25,200	28.5	2145	38,000
AA2230	30.0	2108	27,800	31.7	2244	36,400
IMR-4895	26.5	1944	30,500	30.0	2145	37,400
H-335	30.5	2151	29,600	34.0	2382	38,000
BL-C(2)	33.0	2050	29,500	**36.4**	**2292**	**36,000**
748	32.0	2153	31,600	36.5	2335	36,400
IMR-4064	29.0	2055	32,000	31.5	2199	37,800
IMR-4320	29.0	1956	31,000	31.5	2171	37,800
760	34.0	2027	29,200	38.0	2244	36,300
IMR-4350	32.0	1980	31,000	36.0+	2218	38,100

170 gr. Jacketed FP
2.540" OAL

BC: .189
SD: .256

Powder	Sugg Starting Grains	Velocity fps	Pressure C.U.P.	Max Load Grains	Velocity fps	Pressure C.U.P.
IMR-3031	25.0	1786	27,600	28.5	2054	36,900
IMR-4895	24.5	1807	31,200	28.0	1973	36,800
H-335	29.5	1913	26,700	33.0	2135	35,900
BL-C(2)	32.0	1996	29,000	35.8	2256	38,000
748	32.0	1961	28,800	35.6	2167	35,300
IMR-4064	27.0	1845	28,200	**30.5**	**2150**	**38,100**
AA-2520	29.0	1974	27,500	32.4	2170	36,000
IMR-4320	24.5	1708	30,200	28.0	1948	37,100
RX15	29.0	1831	27,200	32.0+	2110	37,400
IMR-4350	30.0	1816	28,200	34.5+	2104	38,100

#311291
170 gr. (#2 Alloy) 2.521" OAL

BC: .202
SD: .256

Powder	Sugg Starting Grains	Velocity fps	Pressure C.U.P.	Max Load Grains	Velocity fps	Pressure C.U.P.
Unique	7.0	1211	19,800	10.6	1568	36,000
Herco	9.0	1328	28,200	11.5	1538	38,100
SR-4756	7.0	1123	19,800	10.5	1431	35,000
2400	15.0	1624	25,500	18.5	1853	37,500
*SR-4759	14.8	1530	19,900	18.5	1862	36,100
IMR-4227	**16.5**	**1575**	**26,500**	22.0	1959	35,800
XMP-5744	18.0	1607	28,200	24.0	2016	38,600
IMR-4198	18.0	1606	25,200	24.0	2013	34,800
N130	19.0	1581	23,000	24.5	2007	32,800
*RX7	19.0	1635	20,200	28.0	2152	34,400
*IMR-3031	22.5	1599	18,300	28.5	2095	32,500
*H-335	20.3	1638	18,500	32.5	2230	32,800
*748	23.8	1604	16,900	37.3+	2355	35,100

#311041
173 gr. (#2 Alloy) 2.510" OAL

BC: .220
SD: .260

Powder	Sugg Starting Grains	Velocity fps	Pressure C.U.P.	Max Load Grains	Velocity fps	Pressure C.U.P.
Unique	7.0	1240	21,600	10.6	1555	35,500
Herco	8.5	1281	25,800	11.0	1505	33,000
SR-4756	7.0	1102	19,800	10.5	1423	36,400
2400	14.5	1586	27,000	19.0	1909	36,600
*SR-4759	15.5	1613	23,700	17.7	1803	32,600
IMR-4227	17.0	1615	26,100	22.0	1983	34,800
XMP-5744	18.0	1605	26,100	24.0	2019	36,000
IMR-4198	**18.0**	**1602**	**23,200**	23.5	1999	31,400
N130	19.5	1608	22,800	24.3	1994	29,900
*RX7	20.0	1632	19,800	28.6	2165	34,500
*IMR-3031	21.5	1555	20,100	27.0	2016	31,300
*H-335	22.0	1596	23,200	30.0	2254	33,800
*748	25.0	1613	22,500	35.0+	2270	32,900

Note: Loads shown in shaded panels are maximum.
Loads shown in bold designate potentially most accurate load.
+ Designates a compressed powder charge.
* Designates use of CCI 200 primers.

300 Savage

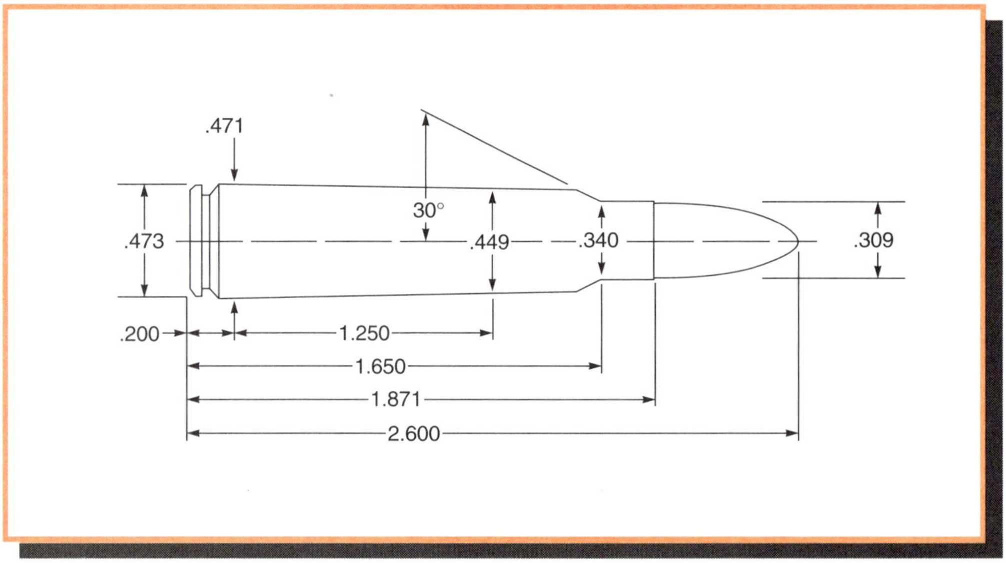

Comments:

Savage first offered this caliber in their lever action Model 99 beginning in 1920. It has been chambered in several different firearms over the years. This cartridge remained quite popular until it gradually lost favor to the 308 Winchester. The 300 Savage enjoyed a ballistic advantage over the 30-30 Winchester due to its ability to fire spitzer bullets without the problems normally associated with tubular magazines. The cartridge looks very similar to the 308 Winchester when compared side by side until the shoulder. The Savage has a sharper 30-degree shoulder and shorter (by .082") neck. The 300 Savage is best loaded with flat-based bullets because of this short neck. IMR-4064 and Varget gave best results in our testing while IMR-4320 has also been a longtime favorite. Cast bullets #311359 and #311291 will need to be seated with the gas check well below the neck for proper overall length.

Test Components:

Cases . . . . .Remington
Trim-to Length . . . . .1.865"
Primers . . . . .Remington 9½ and Federal 210
Primer Size . . . . .Large Rifle
Lyman Shell Holder . . . . .No. 2
Jacketed Bullets Used . . . . . .Sierra HP #2110, 110 gr.
Hornady SP #3020, 130 gr.
Hornady SP #3031, 150 gr.
Sierra RN #2170, 180 gr.
Cast Bullets Used . . . .(sized to .309" dia)
*gas check bullet *#311359, 115 gr.
*#311672, 160 gr.
*#311291, 170 gr.

Test Specifications: (Velocity & Pressure)

Firearm Used . . . . .Savage 99 and Universal Receiver
Barrel Length . . . . .Savage 99 22", Universal Receiver 26"
Twist . . . . .1-12"
Groove Dia. . . . . Savage 99 .3075", Universal Receiver .309"

110 gr. Jacketed HP 2.425" OAL — **BC:** .188 **SD:** .166

Powder	Sugg Starting Grains	Velocity fps	Pressure C.U.P.	Max Load Grains	Velocity fps	Pressure C.U.P.
IMR-3031	38.0	2672	—	42.0	2967	—
IMR-4895	41.0	2659	—	45.0+	2967	—
AA2460	39.0	2591	31,500	43.0	2897	44,400
N135	40.5	2765	34,200	45.0+	3081	42,800
IMR-4064	41.0	2624	—	45.0+	2890	—
Varget	42.5	2740	34,600	46.0+	3080	44,800
RX15	42.5	2677	34,400	47.0+	2954	43,700

130 gr. Jacketed SP 2.565" OAL — **BC:** .295 **SD:** .196

Powder	Sugg Starting Grains	Velocity fps	Pressure C.U.P.	Max Load Grains	Velocity fps	Pressure C.U.P.
IMR-3031	37.0	2525	—	41.0	2808	—
IMR-4895	39.0	2412	34,600	43.0	2707	44,800
AA2460	36.5	2348	33,300	40.5	2614	41,000
748	41.5	2514	33,300	46.0	2808	44,100
IMR-4064	40.0	2493	—	44.0+	2754	—
Varget	40.0	2637	33,900	**44.5+**	**2943**	**45,600**
N140	39.5	2476	32,400	44.0	2778	45,000
RX15	41.0	2494	33,500	45.5+	2854	45,400

Note: Loads shown in shaded panels are maximum.
Loads shown in bold designate potentially most accurate load.
\+ Designates a compressed powder charge.

300 Savage

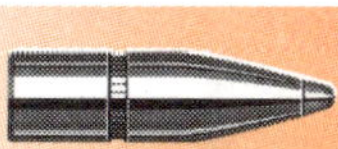

150 gr. Jacketed SP
2.600" OAL

BC: .338
SD: .226

Powder	Sugg Starting Grains	Velocity fps	Pressure C.U.P.	Max Load Grains	Velocity fps	Pressure C.U.P.
IMR-3031	35.0	2271	31,700	39.0	2615	45,600
IMR-4895	37.0	2325	—	41.5+	2652	—
AA2460	35.5	2289	34,200	39.5	2521	42,200
748	39.0	2337	28,300	43.0	2676	45,600
IMR-4064	38.0	2326	34,500	42.2	2609	43,300
Varget	38.0	2438	33,000	**42.0+**	**2741**	**44,300**
N140	38.0	2326	34,000	42.0+	2612	44,900
RX15	38.0	2371	33,900	42.5+	2657	44,400

180 gr. Jacketed RN
2.560" OAL

BC: .280
SD: .271

Powder	Sugg Starting Grains	Velocity fps	Pressure C.U.P.	Max Load Grains	Velocity fps	Pressure C.U.P.
IMR-4895	35.0	2145	32,900	39.0	2384	41,700
AA2460	35.0	2130	32,600	38.7	2393	43,400
748	37.5	2181	31,900	41.7	2480	45,500
IMR-4064	36.0	2159	—	40.0+	2375	—
Varget	37.0	2273	34,300	**41.0+**	**2500**	**44,100**
IMR-4320	37.0	2192	—	41.0+	2439	—
N140	35.5	2175	34,100	39.7	2388	42,400
RX15	37.5	2258	35,700	41.5+	2498	44,500

*#311359
115 gr. (#2 Alloy) 2.280" OAL

BC: .181
SD: .173

Powder	Sugg Starting Grains	Velocity fps	Pressure C.U.P.	Max Load Grains	Velocity fps	Pressure C.U.P.
Unique	9.0	1530	12,500	13.0	1960	36,400
SR-7625	9.5	1550	28,200	12.5	1815	42,900
SR-4756	10.0	1555	23,400	14.0	1920	41,600
SR-4759	**17.0**	**1731**	**14,700**	24.5	2448	40,400
XMP-5744	21.0	1897	24,400	29.0	2445	38,500
IMR-4198	18.0	1513	12,500	29.0	2373	32,700
RX7	26.0	1884	16,200	33.0	2489	32,400

*#311672
160 gr. (#2 Alloy) 2.550" OAL

BC: .245
SD: .239

Powder	Sugg Starting Grains	Velocity fps	Pressure C.U.P.	Max Load Grains	Velocity fps	Pressure C.U.P.
SR-4759	**17.0**	**1655**	**18,100**	24.0	2191	44,300
XMP-5744	20.0	1637	19,100	31.0	2361	45,700
IMR-4198	21.0	1629	15,500	32.0	2441	43,700
RX7	22.0	1617	15,200	35.0	2455	42,700
IMR-3031	27.0	1615	14,700	37.0	2452	39,700

*#311291
170 gr. (#2 Alloy) 2.358" OAL

BC: .202
SD: .256

Powder	Sugg Starting Grains	Velocity fps	Pressure C.U.P.	Max Load Grains	Velocity fps	Pressure C.U.P.
SR-4759	20.0	1888	28,800	22.5	2047	40,200
XMP-5744	**20.0**	**1694**	**26,300**	27.5	2142	44,800
IMR-4198	20.0	1563	16,100	30.5	2279	43,700
RX7	22.0	1812	19,300	32.5	2339	39,500
IMR-3031	26.5	1823	18,700	35.0	2388	38,400
748	29.0	1817	17,000	41.5	2456	36,600

Note: Loads shown in shaded panels are maximum.
Loads shown in bold designate potentially most accurate load.
* Designates use of Federal 210 primers.
\+ Designates a compressed powder charge.

30-40 Krag

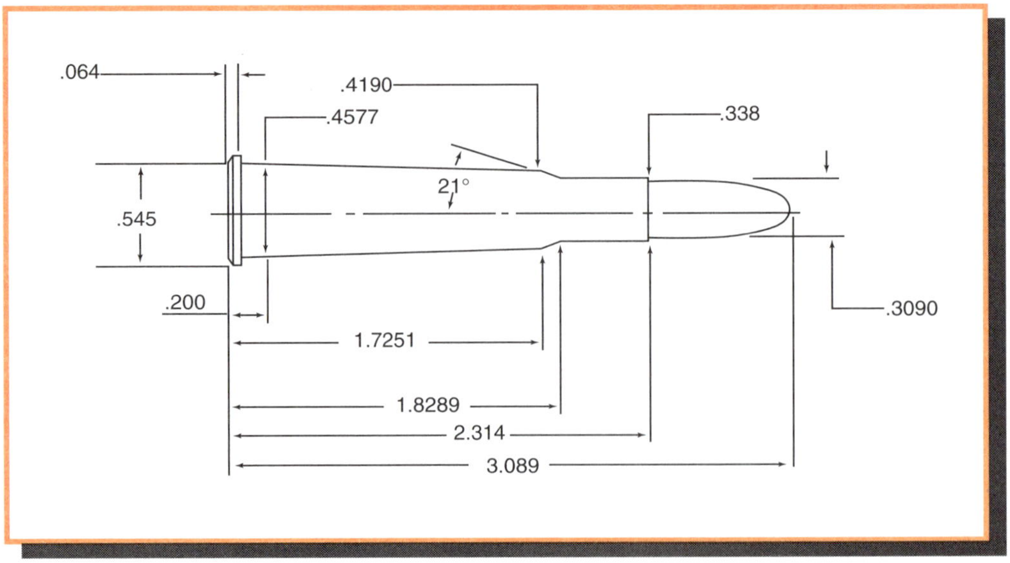

Comments:

The 30-40 Krag enjoys the distinction of being the first smokeless rifle cartridge adopted by the United States Army. Technical advances in small arms cartridges came fast and furious in the last fifteen years of the nineteenth century. The Model 92 Krag-Jorgensen rifle chambered in 30-40 became standardized—if not widely issued—in 1892 in response to developments in high-velocity, smokeless cartridges by several European powers. The Krag was originally a Norwegian design chambered in 6.5x55mm and featured an unusual side-loading magazine.

The new American cartridge was loaded with a 220-grain round nose bullet with a muzzle velocity of 2,200 feet per second. Many shooters regard these rifles as the smoothest operating bolt-action rifles ever produced. However, its single locking lug does not qualify it as the strongest action around. SAAMI has established a Maximum Average Pressure (MAP) of 40,000 CUP for this cartridge. Shooters loading for a Krag should have it inspected by a competent gunsmith for cracks in the receiver and bolt. The 30-40 earned an excellent reputation on game as large as elk with the 220-grain bullet despite its modest (on paper anyway) ballistics. The chamber throats of Krag-Jorgensen rifles are configured for this rather long, heavy bullet. Those loading lightweight bullets may have indifferent results. Shooters can also encounter the 30-40 chambering in the Winchester Model 1885, Model 95 and Ruger Number 3.

Like any elderly military rifle, variations in bore sizes of Krag rifles do exist. Fortunately, the Krag-Jorgensen is a great candidate for a cast bullet gun. Shooters will need to slug their bores and size bullets accordingly. Cast bullet #311284 was originally designed for the 30-40 and closely replicates both the weight and profile of the 220-grain jacketed loading. Shooters wishing to load cast bullet #311332 may use the data for bullet #311644 and seat the bullet to an overall length of 3.025". Reloder 7, SR-4759, and XMP-5744 provided the best results in our cast bullet testing

Test Components:

Cases . . . . .Remington
Trim-to Length . . . . .2.304"
Primers . . . . .Remington 9½
Primer Size . . . . .Large Rifle
Lyman Shell Holder . . . . .No. 7
Jacketed Bullets Used . . . . .Sierra SBT #2125, 150 gr.
Speer SP #2053, 180 gr.
Hornady RN #3090, 220 gr.
Cast Bullets Used . . . .(sized to .309" dia)
*gas check bullet
*#311359, 115 gr.
*#311672, 160 gr.
*#311291, 170 gr.
*#311041, 173 gr.
*#311644, 190 gr.
*#311284, 210 gr.

Note: Loads shown in shaded panels are maximum. Loads shown in bold designate potentially most accurate load.
\+ Designates a compressed powder charge.

Test Specifications:
(Velocity & Pressure)

Firearm Used . . . . .Universal Receiver
Barrel Length . . . . .24"
Twist . . . . .1-10"
Groove Dia. . . . . .309"

150 gr. Jacketed SBT
3.000" OAL
BC: .380
SD: .226

Powder	Sugg Starting Grains	Velocity fps	Pressure C.U.P.	Max Load Grains	Velocity fps	Pressure C.U.P.
IMR-3031	38.5	2339	29,000	43.0	2654	39,100
IMR-4895	39.0	2265	28,000	**43.5**	**2609**	**37,300**
IMR-4064	39.5	2252	29,300	44.0	2600	39,100
Varget	41.0	2488	32,900	45.0	2740	39,600
N140	41.0	2393	29,700	45.0	2672	39,000
RX15	40.0	2376	30,200	44.5	2624	38,500
H380	43.0	2306	28,400	48.0	2584	37,900
IMR-4350	45.0	2316	33,900	50.5	2598	39,900
XMR-4350	46.0	2182	30,000	51.0+	2548	39,900

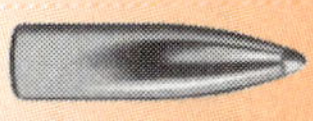

180 gr. Jacketed SP
3.080" OAL

BC: .483
SD: .271

Powder	Sugg Starting Grains	Velocity fps	Pressure C.U.P.	Max Load Grains	Velocity fps	Pressure C.U.P.
IMR-3031	36.0	2151	31,300	40.0	2408	38,700
IMR-4895	38.0	2172	30,800	42.0	2432	39,900
IMR-4064	38.0	2148	29,700	42.0	2389	39,300
Varget	**36.0**	**2077**	**28,900**	40.0	2341	38,900
RX15	37.0	2167	30,100	41.5	2386	38,400
N150	38.0	2181	33,200	42.0	2389	39,400
H380	42.0	2164	31,000	46.0	2402	39,700
IMR-4350	42.0	2031	29,700	46.0	2293	38,800
XMR-4350	41.5	1921	28,400	46.0	2247	38,600

#311359
115 gr. (#2 Alloy) 2.725" OAL

BC: .181
SD: .172

Powder	Sugg Starting Grains	Velocity fps	Pressure C.U.P.	Max Load Grains	Velocity fps	Pressure C.U.P.
AA#9	**14.5**	**1666**	—	19.0	2049	—
2400	14.0	1622	—	18.5	1989	—
SR-4759	15.0	1613	—	20.0	2049	—
IMR-4227	17.0	1630	—	22.0	2028	—
XMP-5744	19.0	1708	—	25.0	2102	—
IMR-4198	19.0	1691	—	25.0	2112	—

#311291
170 gr. (#2 Alloy) 2.825" OAL

BC: .202
SD: .256

Powder	Sugg Starting Grains	Velocity fps	Pressure C.U.P.	Max Load Grains	Velocity fps	Pressure C.U.P.
AA#9	16.0	1619	—	22.0	1991	—
2400	16.0	1610	—	23.0	2038	—
SR-4759	17.0	1620	—	23.5	2022	—
IMR-4227	17.5	1549	—	**24.0**	**1961**	—
XMP-5744	20.0	1638	—	26.5	2020	—
IMR-4198	20.0	1648	—	27.0	2043	—
RX7	20.0	1599	—	27.5	2044	—
N135	24.0	1578	—	31.5	2010	—
Varget	27.0	1693	—	34.0	2059	—

#311644
190 gr. (#2 Alloy) 2.925" OAL

BC: .272
SD: .286

Powder	Sugg Starting Grains	Velocity fps	Pressure C.U.P.	Max Load Grains	Velocity fps	Pressure C.U.P.
SR-4759	16.0	1489	—	24.0	1997	—
IMR-4227	17.0	1481	—	25.0	1967	—
AA-1680	19.0	1475	—	26.5	1927	—
XMP-5744	19.0	1547	—	27.0	2015	—
IMR-4198	19.0	1516	—	27.5	2023	—
RX7	**19.5**	**1522**	—	28.5	2062	—
N135	24.0	1504	—	32.5	2043	—
Varget	25.0	1514	—	35.0	2065	—

220 gr. Jacketed RN
3.080" OAL

BC: .300
SD: .331

Powder	Sugg Starting Grains	Velocity fps	Pressure C.U.P.	Max Load Grains	Velocity fps	Pressure C.U.P.
IMR-4895	35.0	1896	30,700	38.5	2132	39,500
IMR-4064	35.5	1910	31,100	39.5	2120	39,000
N150	34.0	1877	33,500	37.0	2055	39,000
IMR-4350	40.0	1874	29,500	45.0	2177	39,200
XMR-3100	41.0	1728	29,400	45.5	2042	38,700
RX19	42.0	1937	30,000	**46.5**	**2204**	**38,500**
H-4831SC	44.0	2109	36,600	48.5	2169	39,700

#311672
160 gr. (#2 Alloy) 2.900" OAL

BC: .245
SD: .239

Powder	Sugg Starting Grains	Velocity fps	Pressure C.U.P.	Max Load Grains	Velocity fps	Pressure C.U.P.
AA#9	15.0	1571	—	21.0	1975	—
2400	15.0	1561	—	20.5	1936	—
SR-4759	15.0	1493	—	23.5	2052	—
IMR-4227	17.0	1494	—	25.0	2020	—
XMP-5744	18.0	1513	—	27.0	2039	—
IMR-4198	18.0	1519	—	27.0	2048	—
RX7	20.0	1608	—	**26.0**	**1992**	—
N135	24.0	1528	—	32.0	2035	—
Varget	25.0	1540	—	35.0	2067	—

#311041
173 gr. (#2 Alloy) 2.835" OAL

BC: .220
SD: .258

Powder	Sugg Starting Grains	Velocity fps	Pressure C.U.P.	Max Load Grains	Velocity fps	Pressure C.U.P.
AA#9	15.5	1579	—	21.5	1993	—
2400	15.5	1563	—	21.5	1942	—
SR-4759	16.0	1562	—	**23.5**	**2051**	—
IMR-4227	17.5	1541	—	23.0	1914	—
XMP-5744	19.0	1590	—	26.5	2035	—
IMR-4198	19.0	1566	—	27.0	2054	—
RX7	20.5	1613	—	27.0	2034	—
N135	24.0	1547	—	32.0	2024	—
Varget	27.5	1644	—	34.0	2026	—

#311284
210 gr. (#2 Alloy) 3.025" OAL

BC: .332
SD: .314

Powder	Sugg Starting Grains	Velocity fps	Pressure C.U.P.	Max Load Grains	Velocity fps	Pressure C.U.P.
SR-4759	15.5	1426	—	24.5	1945	—
IMR-4227	16.5	1405	—	22.0	1731	—
AA-1680	19.0	1425	—	25.0	1810	—
XMP-5744	18.0	1429	—	**28.0**	**1995**	—
IMR-4198	18.0	1441	—	28.0	1978	—
RX7	18.5	1437	—	29.0	2015	—
N135	23.0	1442	—	33.0	2000	—
Varget	25.0	1515	—	35.0	2046	—

Note: Loads shown in shaded panels are maximum.
Loads shown in bold designate potentially most accurate load.

308 Winchester (7.62 x 51mm) (7.62 NATO)

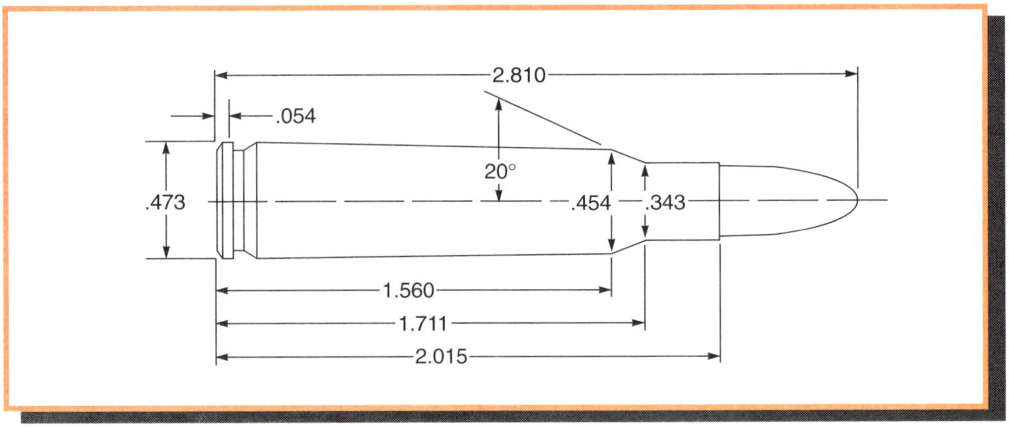

Comments:

The 308 Winchester/7.62 NATO is one of the most versatile and popular center-fire rifle cartridges ever made. It is the end result of the U.S. military's experiment with shortened 30-06 cartridges during the late 1940s. It will do much of what the 30-06 will do with bullets up to 180-grains. Winchester introduced the 308 as the civilian counterpart to the 7.62 NATO in 1952, several years before its adoption by the U.S. military. It works well with a wide range of powders although those in the medium range such as IMR-4895, IMR-4064, Varget, N140, and Reloder 15 usually give best results with most common bullets weights. The 308 has been chambered in every conceivable type of rifle as well as the T/C Encore pistol. It has spawned a family of standard cartridges and wildcats too numerous to list. The cartridge enjoys a high level of inherent accuracy and its 2.810" overall length functions in short actions.

The 308 has proven effective for both hunting and competitive shooting. 110, 125, and 130-grain bullets should be relegated to varmints and smaller game. The 150, 165, and 180-grain bullets are suitable for whitetails on up to elk with proper shot placement. The 308 loaded with the 168-grain Sierra MatchKing has been a standard match load in High Power Rifle competition for many years. Sierra's newer 175-grain MatchKing is suitable for the same situations but offers a higher ballistic coefficient. Sierra's 190-grain MatchKing will shoot accurately out to 1,000 yards. Many shooters loading for M1A/M14 rifles find best results with IMR-4895 or IMR-4064. Shooters reloading military cases will need to remove the primer crimp in order to re-prime cartridge cases. Shooters should also stay one to two grains below the listed maximum charges due to the smaller volume of GI brass.

The 308 Winchester also works well with cast bullets. Best accuracy will usually occur in the velocity ranges of 1,600 to 2,000 feet per second. Cast bullet #311672, #311644, and #311299 have proven popular with metallic silhouette shooters. Reloder 7 and SR-4759 will often give best results with cast bullets.

Test Components:

Cases . . . . .Remington
Trim-to Length . . . . .2.005"
Primers . . . . .Remington 9½
Primer Size . . . . .Large Rifle
Lyman Shell Holder . . . . .No. 2
Jacketed Bullets Used . . . . . .Sierra HP #2110, 110 gr.
Sierra SP #2120, 125 gr.
Hornady SP #3020, 130 gr.
Hornady SP #3031, 150 gr.
Nosler J4 Match #53155, 155 gr.
Nosler Partition #16330, 165 gr.
Sierra HPBT #2200, 168 gr.
Sierra HPBT #2275, 175 gr.
Hornady A-Max #30712, 178 gr.
Nosler Ballistic Tip #30180, 180 gr.
Sierra HPBT #2210, 190 gr.
Sierra HPBT #2230, 200 gr.
Cast Bullets Used . . . .(sized to .308" dia)
*gas check bullet
*#311359, 115 gr.
*#311672, 160 gr.
*#311291, 170 gr.
*#311041, 173 gr.
*#311332, 180 gr.
*#311644, 190 gr.
*#311299, 200 gr.

Test Specifications: (Velocity & Pressure)

Firearm Used . . . . .Universal Receiver
Barrel Length . . . . .24"
Twist . . . . .1-10"
Groove Dia. . . . . .308"

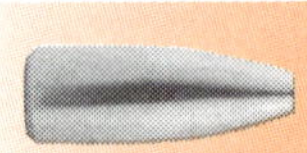

110 gr. Jacketed HP
2.603" OAL
BC: .177
SD: .166

Powder	Sugg Starting Grains	Velocity fps	Pressure C.U.P.	Max Load Grains	Velocity fps	Pressure C.U.P.
RX7	37.5	2898	36,900	42.0	3194	49,200
IMR-3031	39.6	2604	30,300	45.0+	3017	42,200
748	44.0	2656	27,200	50.0+	3023	36,200
H-335	42.0	2823	31,400	49.0	3282	49,400
BL-C(2)	44.9	2765	31,000	51.0+	3115	42,000
IMR-4064	45.0	2849	36,900	50.0+	3184	49,200
IMR-4895	44.0	2890	39,000	50.0+	3225	51,600
IMR-4320	44.0	2702	34,500	51.5+	3154	51,600
H-380	49.0	2994	41,600	55.5+	3268	50,800

Note: Loads shown in shaded panels are maximum. Loads shown in bold designate potentially most accurate load.
+ Designates a compressed powder charge.

308 Winchester (7.62 x 51mm) (7.62 NATO)

125 gr. Jacketed SP
2.622" OAL

BC: .279 **SD:** .188

Powder	Sugg Starting Grains	Velocity fps	Pressure C.U.P.	Max Load Grains	Velocity fps	Pressure C.U.P.
RX7	35.0	2624	31,000	40.2	3021	51,200
IMR-3031	38.7	2656	41,200	44.0+	2996	52,100
748	45.4	2687	29,500	**51.2+**	**3034**	**41,600**
H-335	41.0	2685	31,600	47.0	3108	49,400
BL-C(2)	45.0	2793	36,400	50.0	3086	50,000
IMR-4064	42.5	2702	36,300	48.0+	3017	50,500
IMR-4895	43.0	2747	38,600	47.0	3003	49,600
IMR-4320	45.0	2724	38,600	49.5+	2994	51,200
H-380	48.0	2890	42,400	54.0+	3115	49,600

130 gr. Jacketed SP
2.690" OAL

BC: .295 **SD:** .196

Powder	Sugg Starting Grains	Velocity fps	Pressure	Max Load Grains	Velocity fps	Pressure
XMR-2015	41.0	2830	38,800	45.0	3123	51,200 C
IMR-3031	40.0	2755	39,600	45.0+	3085	57,100 P
Benchmark	41.5	2743	41,400	46.0	3077	60,500 P
AA 2230	41.5	2665	35,400	47.5	3056	51,600 C
IMR-4895	43.0	2754	40,700	47.0	2958	50,000 C
H-335	41.0	2664	33,400	47.0	3085	48,800 C
BL-C(2)	45.0	2770	36,400	50.0	3058	50,800 C
748	44.0	2645	30,200	50.6+	2992	44,200 C
N135	42.0	2813	43,100	**46.7+**	**3096**	**59,800 P**
IMR-4064	44.0	2770	39,000	47.0+	2983	49,000 C
Varget	46.0	2892	44,600	50.0+	3174	59,900 P
IMR-4320	45.0	2702	37,300	49.5	2941	50,000C
H-380	48.0	2873	43,300	54.0+	3125	50,400C

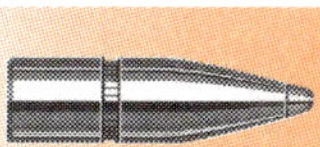

150 gr. Jacketed SP
2.735" OAL

BC: .338 **SD:** .226

Powder	Sugg Starting Grains	Velocity fps	Pressure	Max Load Grains	Velocity fps	Pressure
XMR-2015	38.0	2588	39,300	42.0	2846	49,800 C
IMR-3031	40.0	2544	36,000	43.5+	2942	49,100 C
AA2230	40.5	2538	39,300	44.5	2803	51,100 C
IMR-4895	40.0	2403	32,000	45.0	2777	50,000 C
H-335	38.0	2457	32,300	45.0	2923	52,000 C
BL-C(2)	45.0	2717	40,700	49.0	2915	52,400 C
AA2460	40.0	2533	40,500	44.5	2786	51,800 C
748	44.0	2690	32,700	50.0+	2996	43,200 C
IMR-4064	43.0	2645	41,100	48.0+	2890	52,000 C
Varget	42.5	2632	42,200	47.0	2891	58,000 P
IMR-4320	42.5	2544	41,200	47.0	2874	59,900 P
N140	41.5	2487	39,400	46.2	2795	56,700 P
RX15	41.0	2521	36,300	46.2	2847	50,000 C
H-380	**48.0**	**2607**	**38,700**	52.0+	2787	45,200 P

155 gr. Jacketed J4
2.800" OAL

BC: .450 **SD:** .233

Powder	Sugg Starting Grains	Velocity fps	Pressure P.S.I.	Max Load Grains	Velocity fps	Pressure P.S.I.
Benchmark	39.5	2577	42,800	**44.0**	**2861**	**61,100**
IMR-4895	42.0	2636	43,600	46.0+	2914	59,000
748	43.0	2546	34,700	48.0	2910	57,700
N135	39.5	2574	41,800	44.0	2845	59,700
IMR-4064	41.0	2566	39,600	46.0+	2885	56,700
Varget	43.0	2601	40,700	48.0+	2905	58,300
AA2520	41.0	2551	35,600	45.7	2924	56,700
N140	39.5	2566	40,800	44.0	2843	59,200
N540	43.0	2590	39,600	48.0+	2923	58,200

165 gr. Jacketed Partition
2.780" OAL

BC: .410 **SD:** .248

Powder	Sugg Starting Grains	Velocity fps	Pressure	Max Load Grains	Velocity fps	Pressure
IMR-3031	37.0	2331	35,000	42.0	2652	49,600 C
H-335	36.0	2334	32,300	42.0	2674	47,000 C
IMR-4895	38.0	2426	39,300	43.2	2704	51,800 C
BL-C(2)	41.0	2398	35,000	46.0	2732	52,000 C
AA-2460	39.0	2377	38,700	44.0	2645	50,600 C
748	42.0	2449	33,800	47.5	2756	49,800 C
IMR-4064	39.0	2406	35,300	44.5+	2698	50,300 C
Varget	41.0	2502	43,900	46.0+	2756	59,800 P
AA2520	40.0	2466	39,800	46.0	2724	51,200 C
IMR-4320	42.0	2463	42,400	45.0	2632	51,200 C
RX15	41.0	2424	34,800	44.5	2684	47,900 C
H-380	45.5	2441	38,800	50.5+	2641	45,900 P
N150	**42.0**	**2512**	**45,400**	47.0	2745	59,400 P
760	45.0	2328	31,800	51.5+	2646	43,500 C
IMR-4350	45.5	2360	40,400	50.5+	2637	53,100 P

168 gr. Jacketed HPBT
2.775" OAL

BC: .462 **SD:** .253

Powder	Sugg Starting Grains	Velocity fps	Pressure	Max Load Grains	Velocity fps	Pressure
IMR-3031	37.0	2336	35,000	42.0	2645	49,600 C
Benchmark	37.0	2381	41,200	41.0	2639	56,300 P
IMR-4895	38.0	2331	35,500	42.5	2624	51,200 C
BL-C(2)	41.0	2444	37,700	45.0	2695	48,800 C
AA2460	38.0	2336	36,900	42.0	2608	51,000 C
748	40.0	2371	32,800	45.7+	2714	50,200 C
IMR-4064	40.0	2415	35,500	45.0+	2717	52,000 C
Varget	41.0	2491	42,200	45.7+	2766	61,100 P
AA2520	40.5	2511	46,800	45.0	2732	60,800 P
IMR-4320	42.0	2475	40,700	45.0	2659	50,800 C
N140	39.0	2339	39,000	43.5	2638	57,200 P
RX15	40.0	2411	41,000	44.8	2724	61,200 P
N150	41.0	2438	41,900	45.5	2690	57,300 P
N550	**44.5**	**2524**	**43,300**	48.5+	2819	59,300 P
IMR-4350	45.0	2339	38,600	50.0+	2666	55,100 P

Note: Loads shown in shaded panels are maximum.
Loads shown in bold designate potentially most accurate load.
\+ Designates a compressed powder charge.
In pressure column, C=Copper Units of Pressure (CUP); P=Pounds per Square Inch (PSI)

308 Winchester (7.62 x 51mm) (7.62 NATO)

175 gr. Jacketed HPBT
2.800" OAL — **BC:** .505 **SD:** .264

Powder	Sugg Starting Grains	Velocity fps	Pressure P.S.I.	Max Load Grains	Velocity fps	Pressure P.S.I.
IMR-4895	39.0	2429	43,100	43.5	2692	59,400
748	40.5	2360	35,700	45.0	2701	56,600
IMR-4064	39.5	2370	38,600	44.0	2688	58,000
Varget	40.0	2459	43,900	45.2+	2708	59,300
AA2520	38.5	2375	37,900	43.0	2673	55,100
IMR-4320	39.0	2370	41,600	43.5	2659	59,800
N140	39.0	2370	40,800	43.5	2651	58,700
RX15	40.0	2428	41,500	44.3	2697	58,600
N150	41.0	2430	41,900	45.5	2701	59,900
N550	**43.5**	**2405**	**38,700**	48.5+	2753	58,900

178 gr. Jacketed A-Max
2.780" OAL — **BC:** .495 **SD:** .268

Powder	Sugg Starting Grains	Velocity fps	Pressure P.S.I.	Max Load Grains	Velocity fps	Pressure P.S.I.
IMR-3031	37.0	2292	39,700	41.0	2591	58,500
IMR-4895	37.5	2267	39,500	42.0	2600	60,400
H-335	38.5	2253	39,700	42.8	2604	57,600
BL-C(2)	40.0	2325	39,200	44.8	2687	59,300
AA2460	37.0	2262	39,000	42.0	2596	59,400
748	40.5	2346	39,900	45.0	2662	60,500
IMR-4064	39.5	2373	43,600	44.3	2669	61,100
Varget	39.5	2401	43,700	44.0	2661	60,500
AA2520	38.0	2357	45,000	42.3	2593	59,500
IMR-4320	38.5	2273	40,100	42.6	2585	58,700
N140	39.5	2340	44,500	43.8	2614	61,700
N150	40.5	2390	44,600	45.2	2637	60,800
760	46.0	2432	44,100	50.7	2691	59,000
H-414	**44.0**	**2357**	**43,700**	49.3	2632	58,900
IMR-4350	45.0	2326	41,600	50.0	2631	58,400

180 gr. Jacketed Ballistic Tip
2.800" OAL — **BC:** .507 **SD:** .271

Powder	Sugg Starting Grains	Velocity fps	Pressure P.S.I.	Max Load Grains	Velocity fps	Pressure P.S.I.
IMR-4895	38.0	2320	45,400	**42.5**	**2550**	**58,500**
IMR-4064	38.5	2298	44,700	43.0	2550	57,500
Varget	39.0	2273	44,800	43.5	2521	56,500
AA2520	38.0	2192	37,400	42.2	2549	58,100
IMR-4320	38.0	2234	42,900	42.0	2505	58,400
N140	38.0	2204	40,900	42.0	2503	58,800
RX15	39.0	2278	41,900	43.5	2548	57,800
N540	39.5	2276	41,300	44.0	2578	58,500
N150	39.5	2289	42,600	**43.7**	**2524**	**57,900**
760	43.0	2345	44,600	48.0+	2549	54,700

190 gr. Jacketed HPBT
2.775" OAL — **BC:** .533 **SD:** .286

Powder	Sugg Starting Grains	Velocity fps	Pressure	Max Load Grains	Velocity fps	Pressure
IMR-4895	37.0	2217	35,500	41.5	2500	51,200 C
AA2460	37.5	2300	45,600	41.8	2546	61,700 P
748	39.0	2281	35,500	44.1	2573	51,100 C
IMR-4064	35.5	2192	37,100	41.0+	2470	48,100 C
Varget	38.5	2322	43,900	**42.8+**	**2575**	**60,200 P**
AA2520	37.5	2291	45,100	41.7	2522	59,300 P
IMR-4320	39.0	2247	39,000	43.0	2481	50,800 C
N140	37.5	2231	40,900	42.0	2500	58,300 P
RX15	38.5	2332	38,500	42.5	2549	50,500 C
H-380	43.0	2409	42,400	48.0+	2624	51,200 C
N150	39.5	2338	45,500	44.0+	2571	60,700 P
N550	42.5	2396	44,300	47.0+	2667	60,900 P
760	42.0	2225	36,300	47.2+	2522	50,600 C
H-414	40.5	2166	32,400	46.0+	2426	44,300 C
IMR-4350	38.5	2101	35,000	44.0+	2366	45,800 C

200 gr. Jacketed HPBT
2.780" OAL — **BC:** .565 **SD:** .301

Powder	Sugg Starting Grains	Velocity fps	Pressure	Max Load Grains	Velocity fps	Pressure
IMR-4895	37.0	2150	36,900	40.5	2392	51,600 C
IMR-4064	38.5	2290	46,400	43.0	2521	60,400 P
Varget	38.0	2291	46,600	42.5	2503	60,200 P
AA2520	34.5	2105	36,300	40.5	2396	51,200 C
IMR-4320	39.0	2178	37,300	42.5	2409	52,400 C
N140	**37.5**	**2218**	**44,800**	42.0	2465	62,000 P
RX15	37.5	2257	38,500	42.2	2507	51,800 C
H-380	39.0	2178	33,200	45.0+	2447	43,800 C
N150	38.5	2259	45,700	43.0	2476	60,200 P
AA2700	42.0	2211	40,700	47.5+	2458	51,300 C
760	41.0	2124	35,000	46.2+	2423	47,500 C
H-414	39.5	2075	32,500	45.0+	2347	43,200 C
IMR-4350	43.0	2145	36,000	48.0+	2457	51,200 C
H-4831	45.0	2061	32,500	49.5+	2309	42,000 C

#311359
115 gr. (#2 Alloy) 2.430" OAL — **BC:** .181 **SD:** .173

Powder	Sugg Starting Grains	Velocity fps	Pressure C.U.P.	Max Load Grains	Velocity fps	Pressure C.U.P.
Red Dot	10.0	1623	24,000	14.0	2000	39,400
700X	9.0	1540	24,600	12.5	1879	40,300
Green Dot	11.0	1692	25,200	15.0	2049	40,300
PB	10.5	1601	24,600	13.5	1879	38,600
Unique	**12.0**	**1831**	**24,600**	17.0	2272	40,700
SR-7625	12.0	1704	26,400	15.0	1953	39,400
Herco	14.0	1886	25,800	17.0	2159	38,600

Note: Loads shown in shaded panels are maximum.
Loads shown in bold designate potentially most accurate load.
\+ Designates a compressed powder charge.
In pressure column, C=Copper Units of Pressure (CUP); P=Pounds per Square Inch (PSI)

308 Winchester (7.62 x 51mm) (7.62 NATO)

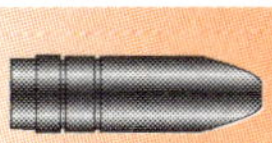

#311672 BC: .272 SD: .241
160 gr. (#2 Alloy) 2.635" OAL

Powder	Sugg Starting Grains	Velocity fps	Pressure P.S.I.	Max Load Grains	Velocity fps	Pressure P.S.I.
Unique	9.8	1386	20,700	13.0	1646	31,800
SR-7625	**10.7**	**1391**	**20,000**	14.0	1655	31,800
SR-4756	11.3	1404	17,400	14.5	1656	27,900
2400	15.5	1501	14,600	19.5	1805	23,200
SR-4759	18.0	1638	15,000	27.5	2317	49,900
IMR-4227	20.0	1596	13,000	30.0	2328	40,600
XMP-5744	**20.0**	**1672**	**17,000**	31.0	2337	40,700
IMR-4198	21.0	1674	14,200	31.7	2378	37,800
XMR-2015	25.0	1602	12,300	35.0	2318	30,900
RX7	25.0	1703	13,600	34.7	2356	34,700
IMR-3031	28.0	1634	13,400	38.0	2347	30,900

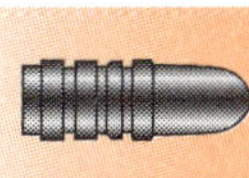

#311291 BC: .202 SD: .256
170 gr. (#2 Alloy) 2.510" OAL

Powder	Sugg Starting Grains	Velocity fps	Pressure	Max Load Grains	Velocity fps	Pressure
Unique	10.0	1368	23,400	15.0	1816	37,700 C
SR-7625	10.0	1290	28,800	13.0	1560	39,900 C
SR-4756	12.0	1416	30,000	14.5	1607	37,700 C
2400	16.5	1606	21,600	23.0	2010	37,700 P
SR-4759	19.5	1847	22,300	26.0	2255	47,800 C
IMR-4227	21.5	1867	22,400	27.8	2270	49,500 C
XMP-5744	**19.0**	**1615**	**18,100**	31.0	2356	51,600 P
IMR-4198	24.0	1897	22,700	30.5	2341	48,200 C
RX7	**23.0**	**1870**	**21,200**	41.5	2602	49,700 C
IMR-3031	28.5	1868	20,200	39.5+	2653	49,800 C

#311041 BC: .220 SD: .260
173 gr. (#2 Alloy) 2.617" OAL

Powder	Sugg Starting Grains	Velocity fps	Pressure	Max Load Grains	Velocity fps	Pressure
Unique	10.0	1377	21,600	15.0	1802	37,700 C
SR-7625	10.0	1272	24,600	13.0	1499	38,600 C
SR-4756	12.0	1392	27,000	14.5	1587	37,700 C
2400	16.0	1595	20,300	22.5	2008	37,700 P
SR-4759	21.5	1897	26,100	27.7	2345	50,300 C
IMR-4227	20.5	1712	21,100	28.0	2183	42,600 P
XMP-5744	**20.0**	**1681**	**19,200**	32.0	2361	47,400 P
IMR-4198	25.0	1898	22,700	34.0	2460	38,700 C
RX7	27.5	1924	24,100	39.0	2627	48,900 C
IMR-3031	29.0	1858	23,500	41.8	2709	51,200 C
H-335	30.0	1847	19,800	44.2	2787	50,200 C

#311332 BC: .320 SD: .271
180 gr. (#2 Alloy) 2.705" OAL

Powder	Sugg Starting Grains	Velocity fps	Pressure	Max Load Grains	Velocity fps	Pressure
Unique	12.2	1425	23,000	15.5	1692	38,000 C
SR-7625	11.5	1270	21,500	14.0	1492	36,400 C
SR-4756	13.0	1378	22,700	15.5	1587	36,400 C
2400	17.0	1599	20,000	24.0	2034	40,100 P
SR-4759	23.5	1827	25,000	26.3	2035	38,600 C
IMR-4227	25.0	1867	24,300	29.0	2126	36,900 C
XMP-5744	**21.0**	**1667**	**19,300**	32.5	2308	49,200 P
IMR-4198	26.5	1856	22,000	32.0	2229	40,200 C
RX7	28.0	1827	20,800	32.0	2172	35,800 C
XMR-2015	31.0	1891	19,500	36.0	2324	37,900 C
IMR-3031	33.0	2001	20,500	37.5	2361	37,800 C
H-335	32.0	1902	21,800	39.0	2316	36,800 C
IMR-4895	33.5	1909	20,000	39.0	2372	39,100 C

#311644 BC: .272 SD: .286
190 gr. (#2 Alloy) 2.675" OAL

Powder	Sugg Starting Grains	Velocity fps	Pressure	Max Load Grains	Velocity fps	Pressure
Unique	10.5	1289	21,000	14.5	1614	40,700 C
SR-7625	10.0	1174	21,300	14.0	1501	42,200 C
SR-4756	11.5	1292	22,000	14.5	1531	37,900 C
2400	17.0	1596	21,500	24.0	2022	42,800 P
SR-4759	20.0	1682	21,600	26.0	2029	47,100 C
IMR-4227	**21.0**	**1720**	**23,100**	28.0	2117	42,300 P
XMP-5744	19.5	1578	18,200	32.5	2349	55,800 P
IMR-4198	23.5	1760	21,900	31.0	2192	45,400 C
RX7	25.0	1738	18,500	33.0	2222	41,700 C
XMR-2015	28.0	1833	21,200	38.0	2340	47,700 C
IMR-3031	32.0	1943	22,700	40.0+	2435	48,100 C
H-335	31.0	1875	22,300	40.0	2409	46,800 C
IMR-4895	33.0	1923	22,800	42.0+	2447	49,200 C

#311299 BC: .377 SD: .301
200 gr. (#2 Alloy) 2.875" OAL

Powder	Sugg Starting Grains	Velocity fps	Pressure P.S.I.	Max Load Grains	Velocity fps	Pressure P.S.I.
Unique	11.5	1400	28,300	16.0	1699	45,400
AA#9	17.0	1616	23,600	23.0	1983	43,400
2400	18.0	1638	23,600	24.0	1995	43,700
SR-4759	17.0	1590	22,300	24.0	1993	45,700
IMR-4227	19.5	1617	19,400	26.0	1994	38,100
XMP-5744	20.0	1608	19,700	**27.0**	**2016**	**35,300**
N130	22.0	1598	17,200	28.5	2005	30,900
IMR-4198	20.5	1615	18,300	27.3	1999	30,900
RX7	22.0	1622	19,300	29.0	2014	33,800
Varget	28.0	1608	15,700	34.0	1994	25,800

Note: Loads shown in shaded panels are maximum.
Loads shown in bold designate potentially most accurate load.
\+ Designates a compressed powder charge.
In pressure column, C=Copper Units of Pressure (CUP); P=Pounds per Square Inch (PSI)

30-06 Springfield (7.62 x 63mm)

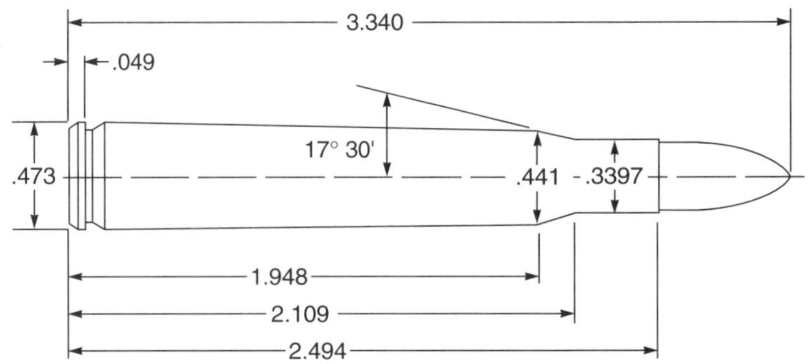

Comments:

The 30-06 has reigned for 100 years as probably the most popular American center-fire rifle cartridge. Its ability to shoot bullets from 110 to 220-grains and digest a wide number of smokeless powders gives it a flexibility that few rifle cartridges can match. As a military cartridge, it served the United States in both World Wars and Korea. In civil guise, it has hunted all of North America and most — if not all — of the rest of the world, and dominated High-Power rifle competition for years. The 30-06 has served as the basis for many of our current rifle cartridges as well as the parent of innumerable wildcats. The 30-06 is still going strong despite the best efforts by some gunwriters over the years to declare it obsolete. Reloading dies for the 30-06 have consistently been one of the three best selling Lyman die sets for decades.

The 30-06 evolved from the short-lived 30-03 cartridge originally chambered in M1903 Springfield Rifle. In spite of the thoroughly modern design of the new '03 rifle, the ballistics of the 30-03 showed little improvement over the 30-40 Krag, which it replaced. The German military's refinement of their 7.92x57mm cartridge with a 154-grain spitzer bullet in 1905 caught the attention of the world's major powers. The just-adopted 30-03 with its 220-grain round nose bullet at 2,300 feet per second became obsolete overnight. The U.S. military refined the 30-03 into the 30-06 in short order. The newly altered cartridge featured a case shortened by .070" and fired a 152-grain bullet at 2,740 feet per second. This cartridge served in the U.S. military in numerous loadings into the early 1960s. The original 30-03 round survived a few more years as a chambering in Winchester's Model 95 before finally fading into history.

While the 30-06 is an extremely flexible cartridge, shooters loading for the M1 Garand rifle must follow certain precautions. The Garand's gas system is designed around military issue ammunition with bullets usually weighing either 152 or 173-grains. The best powders for loading in the M1 Garand are IMR-4895, or IMR-4064. Garand shooters should stick with spitzer bullets weighing between 150 to 168-grains. Round nosed bullets may not feed properly through the M1's mechanism. Moreover, the use of heavier bullets can eventually damage the rifle's operating system. Shooters reloading military cases should keep maximum loads one to two grains below those listed here. Cases should be full-length sized for proper functioning as in any semiautomatic rifle.

The wide-range of bullet weights usable in the 30-06 mandate some discretion in powder selection. Mid-range powders such as IMR-4895 usually work well with all bullet weights. Lighter bullets will work with faster powders such as, IMR-3031 or H-335. Heavier 200 and 220-grain bullets should respond well with IMR-4350, Reloder 19 and other slower burning powders.

The 30-06 also works extremely well with cast bullets. Results can vary from one gun to the next and shooters will often need to experiment with sizing diameters and seating depths but the results are well worth it. Shooters loading for the M1903A3 rifles with two-groove barrels have had particularly good accuracy from Lyman cast bullets #311291 and #311644. Best powder choices for cast bullets include Reloder 7, XMP-5744, and SR-4759.

Test Components:

Cases . . . . .Winchester
Trim-to Length . . . . .2.484"
Primers . . . . .Winchester WLR
Primer Size . . . . .Large Rifle
Lyman Shell Holder . . . . .No. 2
Jacketed Bullets Used . . . . . .Sierra HP #2110, 110 gr.
Sierra SP #2120, 125 gr.
Hornady SP #3020, 130 gr.
Hornady SP # 3031, 150 gr.
Barnes XLC #30854, 150 gr.
Nosler Ballistic Tip #30165, 165 gr.
Sierra HPBT #2200, 168 gr.
Barnes X #30835, 180 gr.
Sierra SBT #2160, 180 gr.
Sierra HPBT #2210, 190 gr.
Sierra HPBT #2230, 200 gr.
Hornady RN #3090, 220 gr.

Cast Bullets Used . . . .(sized to .309" dia)
*gas check bullet
*#311359, 115 gr.
*#311672, 160 gr.
*#311291, 170 gr.
*#311041, 173 gr.
*#311644, 190 gr.
*#311299, 200 gr.
*#311284, 210 gr.

Test Specifications:
(Velocity & Pressure)

Firearm Used . . . . .Universal Receiver
Barrel Length . . . . .24"
Twist . . . . .1-10"
Groove Dia. . . . . .308"

Note: Loads shown in shaded panels are maximum. Loads shown in bold designate potentially most accurate load.
+ Designates a compressed powder charge.
In pressure column, C=Copper Units of Pressure (CUP); P=Pounds per Square Inch (PSI)

30-06 Springfield (7.62 x 63mm)

110 gr. Jacketed HP

3.000" OAL

BC: .177
SD: .166

Powder	Sugg Starting Grains	Velocity fps	Pressure	Max Load Grains	Velocity fps	Pressure
RX7	37.0	2736	35,500	42.0	3073	50,000 C
IMR-3031	47.0	2930	35,000	52.0	3257	46,700 C
H-335	45.0	2923	40,400	51.0	3203	48,900 C
IMR-4895	49.0	3012	39,400	53.5	3266	50,000 C
BL-C(2)	51.0	3144	41,100	55.0	3378	50,400 C
AA2460	53.0	3122	41,900	58.5	3453	58,600 P
748	51.0	3060	39,000	60.0	3387	49,000 C
IMR-4064	50.0	3030	39,000	**55.5**	**3356**	**48,800 C**
IMR-4320	50.0	2906	38,600	55.5	3184	49,600 C
760	55.5	2922	33,800	63.0+	3206	42,900 C
IMR-4350	56.0	2898	37,300	61.0+	3194	45,400 C
H-4831	55.0	2557	28,800	61.5+	2985	39,400 C

125 gr. Jacketed SP

3.080" OAL

BC: .279
SD: .188

Powder	Sugg Starting Grains	Velocity fps	Pressure	Max Load Grains	Velocity fps	Pressure
RX7	38.5	2710	38,500	43.0	2932	47,600 C
IMR-3031	45.0	2732	33,000	50.0	3095	45,000 C
H-335	45.5	2815	37,500	51.5	3144	48,500 C
IMR-4895	48.0	2815	36,500	53.5	3096	47,300 C
BL-C(2)	49.0	2890	35,000	54.0	3215	48,000 C
AA-2460	50.5	2737	35,600	56.0	3222	57,000 P
748	51.5	2890	35,300	58.0	3241	48,300 C
IMR-4064	48.0	2762	31,500	55.0	3215	47,200 C
IMR-4320	49.0	2770	36,000	**55.0**	**3105**	**46,700 C**
H-380	51.0	2967	39,000	56.5	3194	49,600 C
760	56.0	2801	34,200	63.0+	3215	47,600 C
IMR-4350	55.0	2754	34,000	61.0+	3105	46,000 C
IMR-4831	57.0	2700	33,600	63.0+	3017	43,200 C
H-4831	55.0	2544	30,000	61.5+	2906	40,000 C

130 gr. Jacketed SP

3.090" OAL

BC: .295
SD: .196

Powder	Sugg Starting Grains	Velocity fps	Pressure	Max Load Grains	Velocity fps	Pressure
IMR-3031	45.0	2793	36,400	50.0	3048	45,000 C
H-335	46.0	2813	40,800	52.0	3081	50,000 C
IMR-4895	48.0	2890	42,000	51.0	3036	49,900 C
BL-C(2)	49.0	2932	38,100	53.5	3174	48,400 C
AA-2460	49.5	2856	42,000	55.0	3196	58,800 P
748	53.0	2967	46,800	**58.0**	**3214**	**56,500 P**
IMR-4064	48.0	2808	36,000	54.0	3184	49,600 C
IMR-4320	49.0	2793	39,000	54.0	3076	48,400 C
H-380	51.0	2967	42,000	54.5	3125	50,000 C
760	55.5	2822	35,000	63.0+	3111	44,200 C
IMR-4350	51.0	2623	34,200	58.0+	2949	44,200 C
IMR-4831	57.0	2686	32,900	62.0+	2984	44,400 C
H-4831	55.0	2570	32,500	61.5+	2923	41,600 C

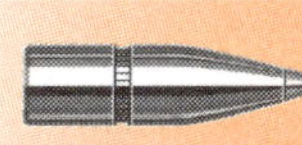

150 gr. Jacketed SP

3.200" OAL

BC: .338
SD: .226

Powder	Sugg Starting Grains	Velocity fps	Pressure	Max Load Grains	Velocity fps	Pressure
IMR-3031	42.0	2450	29,400	50.0	2932	48,000 C
AA2230	46.0	2597	39,800	51.0	2870	49,300 C
IMR-4895	46.0	2680	36,900	51.5	2958	49,200 C
BL-C(2)	46.0	2770	44,600	51.0	2932	48,000 C
AA-2460	46.5	2597	38,500	51.7	2869	48,500 C
IMR-4064	**48.0**	**2695**	**36,000**	53.0	3012	49,600 C
IMR-4320	47.0	2624	37,300	53.0	2923	48,400 C
RX15	50.5	2777	40,000	54.0	2987	49,300 C
H-380	48.0	2724	40,300	53.0	2976	50,000 C
760	48.0	2535	34,200	56.0	2878	49,600 C
N160	54.5	2588	35,900	60.5+	2919	50,500 P
IMR-4350	49.0	2545	37,300	57.0+	2887	47,800 C
RX19	58.5	2690	40,800	64.3+	2925	49,600 C
IMR-4831	55.0	2571	33,800	61.0+	2903	47,100 C
H-4831	55.0	2551	33,000	61.5+	2890	43,300 C

150 gr. Jacketed XLC

3.230" OAL

BC: .428
SD: .226

Powder	Sugg Starting Grains	Velocity fps	Pressure P.S.I.	Max Load Grains	Velocity fps	Pressure P.S.I.
IMR-3031	44.0	2538	37,600	49.2	2874	55,400
H-335	46.0	2607	43,300	51.2	2918	59,500
IMR-4895	47.0	2624	41,400	52.5	2952	58,400
748	48.0	2680	43,800	53.0	2963	59,000
BL-C(2)	48.5	2646	42,700	54.0	2970	59,400
IMR-4064	47.0	2515	36,500	52.0	2900	55,900
IMR-4320	47.0	2598	41,800	52.5	2916	58,400
RX15	48.0	2569	38,700	**52.5**	**2940**	**57,900**
N150	49.5	2636	41,600	55.0	2945	59,300
H-380	52.5	2664	43,000	58.5	2931	56,400
N550	53.0	2696	43,100	59.0	3002	57,300
760	53.5	2797	47,600	59.5	3005	57,900

165 gr. Jacketed Ballistic Tip

3.285" OAL

BC: .475
SD: .248

Powder	Sugg Starting Grains	Velocity fps	Pressure P.S.I.	Max Load Grains	Velocity fps	Pressure P.S.I.
IMR-3031	44.0	2495	38,500	49.0	2784	56,100
H335	43.0	2474	42,600	48.0	2690	54,300
IMR-4895	45.0	2503	37,700	50.0	2815	56,200
IMR-4064	47.0	2495	38,000	**52.0**	**2851**	**58,100**
IMR-4320	47.0	2542	41,200	52.0	2843	57,500
RX15	47.0	2531	39,300	52.0	2807	53,600
H-380	48.5	2515	40,100	53.0	2804	54,000
760	52.5	2697	43,500	57.0	2959	58,600
IMR-4350	52.5	2613	45,100	57.0+	2878	56,600
RX19	54.5	2464	37,600	60.4+	2816	53,300
N160	55.5	2597	41,100	61.0+	2895	56,500
IMR-4831	54.0	2485	38,400	59.0+	2786	52,000
H-4831	56.0	2431	34,900	62.0+	2754	49,000
WXR	55.0	2401	34,200	61.0+	2720	47,800

Note: Loads shown in shaded panels are maximum.
Loads shown in bold designate potentially most accurate load.
\+ Designates a compressed powder charge.
In pressure column, C=Copper Units of Pressure (CUP); P=Pounds per Square Inch (PSI)

30-06 Springfield (7.62 x 63mm)

168 gr. Jacketed HPBT
3.300" OAL

BC: .462
SD: .253

Powder	Sugg Starting Grains	Velocity fps	Pressure	Max Load Grains	Velocity fps	Pressure
IMR-3031	42.0	2463	34,000	48.0	2808	49,600 C
IMR-4895	**43.0**	**2444**	**35,500**	48.5	2762	47,600 C
IMR-4064	45.0	2551	34,000	50.0	2816	46,700 C
AA2520	42.5	2383	36,200	48.0	2653	48,800 C
IMR-4320	45.0	2475	36,400	51.0	2816	50,000 C
RX15	47.5	2592	38,800	51.8	2832	50,000 C
AA2700	49.5	2475	41,800	55.0	2745	54,900 P
760	50.0	2541	37,300	57.0	2839	49,700 C
IMR-4350	52.0	2590	38,000	57.0+	2873	50,400 C
RX19	55.0	2489	37,800	60.9+	2810	50,000 C
N160	55.0	2559	39,800	61.0+	2891	58,800 P
IMR-4831	54.5	2544	35,700	60.6+	2847	47,800 C
H-4831	55.0	2564	37,300	61.5+	2873	47,200 C
WXR	55.0	2489	38,100	61.0+	2799	53,300 P

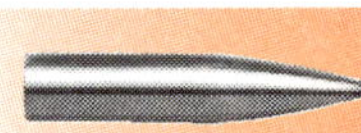

180 gr. Barnes "X"
3.275" OAL

BC: .511
SD: .271

Powder	Sugg Starting Grains	Velocity fps	Pressure	Max Load Grains	Velocity fps	Pressure
IMR-4895	41.5	2307	37,500	47.0	2578	48,100 C
IMR-4064	43.0	2447	34,000	48.5	2614	48,300 C
AA2520	40.0	2209	38,700	45.5	2467	48,500 C
IMR-4320	41.5	2178	35,700	47.5	2497	45,700 C
760	47.0	2325	37,100	53.3	2557	48,000 C
IMR-4350	51.0	2399	37,500	55.7+	2688	49,200 C
RX19	52.0	2336	37,200	**57.0+**	**2627**	**49,700 C**
N160	53.0	2492	45,600	58.5+	2733	59,600 P
IMR-4831	52.0	2374	34,100	57.6+	2717	48,700 C
XMR-3100	54.5	2419	40,500	59.8+	2625	48,400 C
H-4831	52.0	2390	38,200	57.4+	2649	48,100 C
WXR	55.0	2513	47,300	61.0+	2756	59,400 P
RX22	54.5	2441	38,200	61.5+	2715	48,200 C

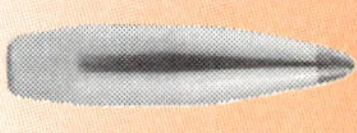

180 gr. Jacketed SBT
3.280" OAL

BC: .475
SD: .271

Powder	Sugg Starting Grains	Velocity fps	Pressure	Max Load Grains	Velocity fps	Pressure
IMR-4895	42.0	2386	36,000	47.5	2659	48,000 C
IMR-4064	43.0	2403	35,000	49.0	2710	47,200 C
AA2520	41.0	2275	35,400	47.3	2572	49,700 C
IMR-4320	41.0	2281	35,800	46.3	2584	49,900 C
760	45.0	2360	35,500	52.0	2666	49,000 C
IMR-4350	50.0	2469	38,300	**56.0+**	**2801**	**50,000 C**
RX19	52.0	2497	41,200	58.3+	2698	48,800 C
N160	53.0	2418	37,700	58.5+	2743	55,100 P
IMR-4831	52.5	2525	39,000	58.4+	2788	48,700 C
XMR-3100	53.5	2425	38,300	59.0+	2648	48,300 C
H-4831	55.0	2604	40,300	60.0+	2840	50,000 C
WXR	54.0	2311	33,600	60.0+	2674	50,300 P
RX22	55.5	2536	39,300	60.8+	2780	48,400 C

190 gr. Jacketed HPBT
3.280" OAL

BC: .533
SD: .286

Powder	Sugg Starting Grains	Velocity fps	Pressure	Max Load Grains	Velocity fps	Pressure
IMR-4895	41.0	2288	35,000	46.0	2569	50,000 C
IMR-4064	42.0	2309	32,500	47.5	2624	48,400 C
AA2520	41.0	2233	36,800	46.0	2469	48,500 C
IMR-4320	44.0	2341	34,500	**49.0**	**2631**	**48,800 C**
760	46.0	2350	36,200	53.0	2662	50,000 C
H414	47.0	2368	36,200	54.0	2694	49,800 C
IMR-4350	45.0	2205	31,500	51.0	2586	50,000 C
RX19	51.5	2341	37,300	56.0+	2606	48,300 C
N160	52.0	2383	39,800	58.0	2695	57,400 P
IMR-4831	51.0	2395	37,300	56.8+	2686	49,300 C
XMR-3100	53.5	2368	38,800	59.0+	2614	48,600 C
H-4831	53.0	2450	36,900	59.0+	2724	47,200 C
WXR	53.0	2341	38,200	58.0+	2647	54,100 P
RX22	55.0	2461	37,800	60.5+	2722	48,700 C

200 gr. Jacketed HPBT
3.280" OAL

BC: .565
SD: .301

Powder	Sugg Starting Grains	Velocity fps	Pressure	Max Load Grains	Velocity fps	Pressure
IMR-4895	41.0	2262	37,300	46.5	2512	48,400 C
IMR-4064	42.0	2309	38,100	47.0	2552	50,000 C
IMR-4320	43.0	2294	38,600	47.5	2502	47,600 C
760	45.0	2224	36,300	52.0	2565	50,000 C
H414	46.0	2258	35,000	53.0	2606	49,800 C
IMR-4350	49.0	2380	39,900	54.0+	2638	48,400 C
RX19	51.5	2302	36,500	55.6+	2526	45,800 C
N160	51.5	2340	38,800	57.0+	2655	58,000 P
IMR-4831	50.0	2323	35,800	55.5+	2587	47,400 C
XMR-3100	53.0	2336	39,000	58.8+	2573	48,600 C
H-4831	52.0	2343	36,400	58.0+	2620	46,300 C
WXR	52.0	2360	40,800	**58.0+**	**2659**	**57,600 P**
RX22	53.0	2340	40,900	59.0+	2650	56,600 P

220 gr. Jacketed RN
3.220" OAL

BC: .300
SD: .331

Powder	Sugg Starting Grains	Velocity fps	Pressure	Max Load Grains	Velocity fps	Pressure
IMR-4064	41.0	2183	37,300	46.5	2439	49,200 C
IMR-4320	41.0	2118	36,400	47.5	2421	49,200 C
760	44.5	2106	37,200	50.5	2381	49,400 C
H414	45.0	2153	40,200	51.0	2379	49,200 C
IMR-4350	49.0	2309	40,700	53.0+	2500	49,200 C
RX19	50.0	2127	34,600	**55.8+**	**2451**	**48,800 C**
N160	52.0	2303	46,400	58.0+	2463	57,900 P
IMR-4831	49.5	2192	35,700	54.6+	2458	47,100 C
XMR-3100	51.5	2172	37,800	57.5	2427	49,000 C
H-4831	51.0	2283	38,600	57.5+	2583	50,000 C
WXR	50.0	2208	38,500	56.0+	2490	55,100 P
RX22	53.0	2201	36,100	58.0+	2477	51,600 P
IMR-7828	51.0	2058	33,800	56.0+	2357	48,900 P

Note: Loads shown in shaded panels are maximum.
Loads shown in bold designate potentially most accurate load.
\+ Designates a compressed powder charge.
In pressure column, C=Copper Units of Pressure (CUP); P=Pounds per Square Inch (PSI)

30-06 Springfield (7.62 x 63mm)

#311359
115 gr. (#2 Alloy) 2.909" OAL

BC: .181
SD: .173

Powder	Sugg Starting Grains	Velocity fps	Pressure	Max Load Grains	Velocity fps	Pressure
Red Dot	11.0	1683	23,400	15.0	2000	36,900 C
700X	10.0	1610	22,000	14.0	1953	38,100 C
Green Dot	13.0	1792	28,800	16.0	2020	38,600 C
Unique	14.0	1960	27,000	19.0	2331	42,000C
SR-7625	12.5	1677	21,600	16.5	2008	42,400 C
Herco	14.5	1876	28,200	18.0	2132	39,000 C
XMP-5744	**24.0**	**1904**	**15,600**	32.0	2399	26,600 P

#311291
170 gr. (#2 Alloy) 3.013" OAL

BC: .202
SD: .256

Powder	Sugg Starting Grains	Velocity fps	Pressure	Max Load Grains	Velocity fps	Pressure
Unique	12.0	1508	24,000	16.5	1842	36,400 C
SR-7625	11.5	1363	23,400	14.5	1585	36,000 C
Herco	12.5	1488	26,400	16.0	1697	35,500 C
SR-4756	12.5	1418	22,200	16.5	1692	35,500 C
2400	18.5	1635	18,200	29.5	2296	43,100 C
SR-4759	20.0	1666	16,700	31.0	2325	46,900 C
XMP-5744	**26.0**	**1928**	**25,000**	35.0	2388	45,600 P
IMR-4198	25.0	1729	15,100	38.5	2501	45,100 C
RX7	**25.0**	**1733**	**15,100**	38.3	2480	47,600 C
IMR-3031	29.0	1710	13,900	42.0	2469	38,200 C

#311644
190 gr. (#2 Alloy) 3.075" OAL

BC: .272
SD: .286

Powder	Sugg Starting Grains	Velocity fps	Pressure	Max Load Grains	Velocity fps	Pressure
Unique	13.0	1404	21,800	17.2	1677	36,800 C
SR-7625	12.2	1295	23,400	15.2	1514	38,000 C
Herco	14.5	1437	20,500	18.0	1664	31,100 C
SR-4756	13.5	1378	23,600	17.0	1587	34,100 C
SR-4759	23.0	1716	21,000	28.2	1975	37,700 C
XMP-5744	27.0	1891	25,900	37.0	2385	49,900 P
RX7	**22.0**	**1650**	**21,000**	38.0	2250	40,500 P

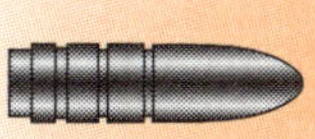

#311284
210 gr. (#2 Alloy) 3.063" OAL

BC: .332
SD: .316

Powder	Sugg Starting Grains	Velocity fps	Pressure	Max Load Grains	Velocity fps	Pressure
Unique	11.5	1326	26,400	15.5	1577	36,400 C
SR-7625	11.5	1242	30,000	13.0	1320	36,400 C
Herco	12.5	1335	30,500	15.5	1510	39,000 C
SR-4756	12.5	1264	26,400	15.5	1445	35,000 C
2400	19.0	1592	27,000	24.0	1827	32,000 C
SR-4759	22.0	1679	27,800	27.8	2004	45,300 C
XMP-5744	**28.0**	**1885**	**30,300**	38.5	2327	57,900 P
RX7	25.0	1685	22,200	38.2	2254	47,000 C

#311672
160 gr. (#2 Alloy) 3.050" OAL

BC: .245
SD: .241

Powder	Sugg Starting Grains	Velocity fps	Pressure P.S.I.	Max Load Grains	Velocity fps	Pressure P.S.I.
Unique	**18.0**	**1852**	**34,300**	24.0	2182	56,700
SR-4759	25.0	1950	24,700	33.0	2381	50,400
XMP-5744	26.0	1929	22,000	35.0	2396	37,800
IMR-4198	27.0	1940	19,600	37.0	2421	36,000
RX7	28.0	1904	17,600	38.0	2411	36,700
IMR-3031	32.0	1752	13,800	43.0	2384	29,800

#311041
173 gr. (#2 Alloy) 2.968" OAL

BC: .220
SD: .260

Powder	Sugg Starting Grains	Velocity fps	Pressure	Max Load Grains	Velocity fps	Pressure
Unique	12.0	1504	24,000	16.5	1811	35,500 C
SR-7625	11.5	1346	24,000	14.5	1570	36,900 C
Herco	13.0	1497	27,000	16.0	1689	36,000 C
SR-4756	12.5	1406	22,800	16.5	1653	36,400 C
2400	20.0	1751	22,200	27.0	2141	34,500 C
SR-4759	20.0	1642	13,900	31.5	2329	47,400 C
XMP-5744	**26.0**	**1929**	**24,300**	34.0	2393	43,000 P
IMR-4198	25.0	1717	13,100	38.5	2463	43,100 C
RX7	25.0	1661	12,800	40.5	2515	46,900 C
IMR-3031	30.0	1636	11,100	44.0	2522	40,800 C

#311299
200 gr. (#2 Alloy) 3.250" OAL

BC: .377
SD: .301

Powder	Sugg Starting Grains	Velocity fps	Pressure P.S.I.	Max Load Grains	Velocity fps	Pressure P.S.I.
Unique	12.7	1398	25,500	17.2	1673	38,300
2400	15.0	1376	15,400	21.0	1736	27,300
SR-4759	19.5	1626	23,100	27.0	2002	42,300
IMR-4227	21.0	1600	19,100	**29.0**	**1992**	**35,400**
XMP-5744	21.0	1609	19,500	29.5	1998	33,200
N130	23.0	1575	17,000	31.0	2020	34,200
IMR-4198	21.0	1601	19,100	30.0	1997	30,800
RX7	22.0	1588	17,700	30.5	1989	30,300
N133	25.0	1569	15,400	32.7	1981	27,700
Varget	30.0	1588	15,300	38.0	2017	27,800

Note: Loads shown in shaded panels are maximum.
Loads shown in bold designate potentially most accurate load.
In pressure column, C=Copper Units of Pressure (CUP); P=Pounds per Square Inch (PSI)

300 H & H Magnum (300 Holland & Holland) (7.63 x 72mm)

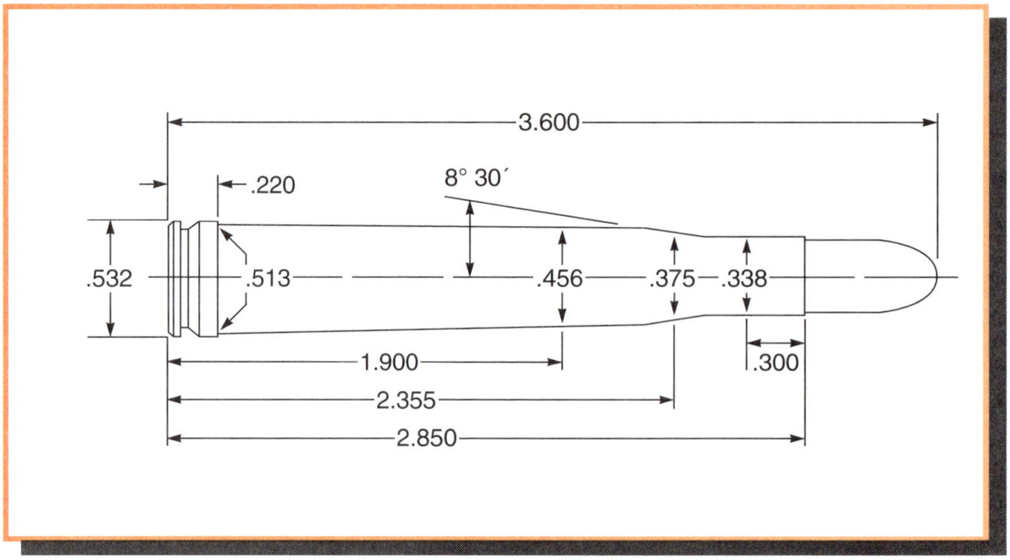

Comments:

The 300 Holland and Holland is the grandfather of the modern belted 30-caliber Magnums. In contrast to the later 300 Winchester and 300 Weatherby, this cartridge requires the belt for proper headspacing due to its long, sloping shape. Holland and Holland introduced their "Super 30" in the early 1920s. Its availability via only expensive imported or custom rifles limited its access to the average American hunter. The 300 H&H came to prominence in America when Ben Comfort won the 1000-yard stage of the National Matches in 1935. Winchester offered the chambering two years later in a Magnum-length Model 70. American shooters now had a reasonably priced 30-caliber Magnum rifle. The 300 H&H is suitable for most North American hunting and has seen extensive use on the African plains. It is still a potent cartridge but has been overtaken by later developments. The current generation of premium bullets are well suited to the performance levels of the H&H. IMR-4350 and IMR-4831 produced the best results in our tests.

Test Components:

Cases . . . . .Winchester
Trim-to Length . . . . .2.840"
Primers . . . . .WLR
Primer Size . . . . .Large Rifle
Lyman Shell Holder . . . . .No. 13
Jacketed Bullets Used . . . . . .Sierra HP #2110, 110 gr.
Hornady SP #3020, 130 gr.
Hornady SP #3031, 150 gr.
Nosler Ballistic Tip #30165, 165 gr.
Nosler Ballistic Tip #30180, 180 gr.
Nosler SP #35626, 200 gr.
Hornady RN #3090, 220 gr.
Cast Bullets Used . . . .(sized to .309" dia)
*gas check bullet *#311291, 170 gr.

Test Specifications: (Velocity Only)

Firearm Used . . . . .Remington Model 721
Barrel Length . . . . .26"
Twist . . . . .1-10"
Groove Dia. . . . . .308"

110 gr. Jacketed HP
3.600" OAL
BC: .177
SD: .166

Powder	Sugg Starting Grains	Velocity fps	Pressure	Max Load Grains	Velocity fps	Pressure
IMR-4064	61.0	3322	—	68.0	3708	—
760	64.0	2891	—	71.0	3485	—
IMR-4350	68.0	3047	—	76.0	3445	—
RX19	71.0	3045	—	79.0	3473	—
IMR-4831	70.5	3033	—	78.5	3488	—
XMR-3100	71.0	2954	—	79.0	3491	—
H-4831	73.0	3134	—	80.0+	3401	—

130 gr. Jacketed SP
3.600" OAL
BC: .295
SD: .196

Powder	Sugg Starting Grains	Velocity fps	Pressure	Max Load Grains	Velocity fps	Pressure
IMR-4064	60.0	3164	—	67.0	3497	—
760	62.0	2814	—	69.0	3346	—
IMR-4350	65.0	2954	—	72.0	3372	—
RX19	67.5	2936	—	**75.0**	**3370**	—
IMR-4831	66.5	2923	—	74.0	3392	—
XMR-3100	70.0	2953	—	78.0	3396	—
H-4831	73.0	3058	—	80.0+	3333	—

Note: Loads shown in shaded panels are maximum.
Loads shown in bold designate potentially most accurate load.
\+ Designates a compressed powder charge.

300 H & H Magnum (300 Holland & Holland) (7.63 x 72mm)

150 gr. Jacketed SP
3.600" OAL
BC: .338 **SD:** .226

Powder	Sugg Starting Grains	Velocity fps	Pressure	Max Load Grains	Velocity fps	Pressure
IMR-4064	59.0	3012	—	65.0	3268	—
760	61.0	2633	—	68.0	3146	—
IMR-4350	63.0	2799	—	**70.0**	**3169**	—
RX19	65.0	2779	—	72.5	3189	—
IMR-4831	65.0	2809	—	72.0	3218	—
XMR-3100	68.5	2792	—	76.0	3217	—
H-4831	72.0	2923	—	80.0+	3236	—

165 gr. Jacketed Ballistic Tip
3.600" OAL
BC: .475 **SD:** .248

Powder	Sugg Starting Grains	Velocity fps	Pressure	Max Load Grains	Velocity fps	Pressure
760	60.0	2720	—	66.5	3073	—
IMR-4350	62.0	2783	—	69.0	3135	—
RX19	64.0	2765	—	71.0	3123	—
IMR-4831	63.5	2752	—	**70.5**	**3131**	—
XMR-3100	67.5	2818	—	75.0	3157	—
H-4831	70.0	2793	—	77.0+	3095	—

180 gr. Jacketed Ballistic Tip
3.600" OAL
BC: .507 **SD:** .271

Powder	Sugg Starting Grains	Velocity fps	Pressure	Max Load Grains	Velocity fps	Pressure
760	58.5	2611	—	65.0	2972	—
IMR-4350	60.0	2680	—	67.0	3025	—
IMR-4831	62.0	2688	—	**69.0**	**3051**	—
XMR-3100	65.0	2697	—	72.5	3011	—
H-4831	66.0	2624	—	73.0+	2949	—
RX22	64.0	2613	—	71.0	2955	—

200 gr. Jacketed Spitzer
3.600" OAL
BC: .481 **SD:** .381

Powder	Sugg Starting Grains	Velocity fps	Pressure	Max Load Grains	Velocity fps	Pressure
IMR-4350	59.0	2538	—	65.0	2808	—
IMR-4831	60.0	2557	—	67.0	2862	—
XMR-3100	61.0	2443	—	68.0	2767	—
H-4831	64.0	2506	—	71.0	2801	—
RX22	62.0	2522	—	69.0	2793	—
RX25	66.0	2694	—	**73.5**	**2984**	—

220 gr. Jacketed RN
3.600" OAL
BC: .300 **SD:** .331

Powder	Sugg Starting Grains	Velocity fps	Pressure	Max Load Grains	Velocity fps	Pressure
IMR-4350	57.0	2450	—	64.0	2732	—
IMR-4831	58.5	2404	—	65.0	2663	—
XMR-3100	58.0	2233	—	64.0	2473	—
H-4831	62.0	2398	—	69.0	2710	—
RX22	60.0	2366	—	67.0	2611	—
RX25	66.0	2570	—	**73.0**	**2876**	—

#311291
170 gr. (#2 Alloy) 3.335" OAL
BC: .202 **SD:** .256

Powder	Sugg Starting Grains	Velocity fps	Pressure	Max Load Grains	Velocity fps	Pressure
SR-4759	27.0	1949	—	32.5	2185	—
XMP-5744	29.0	1905	—	34.0	2157	—
RX7	28.0	1801	—	36.2	2130	—

Note: Loads shown in shaded panels are maximum.
Loads shown in bold designate potentially most accurate load.
\+ Designates a compressed powder charge.

300 Remington Short Action Ultra Mag (SAUM)

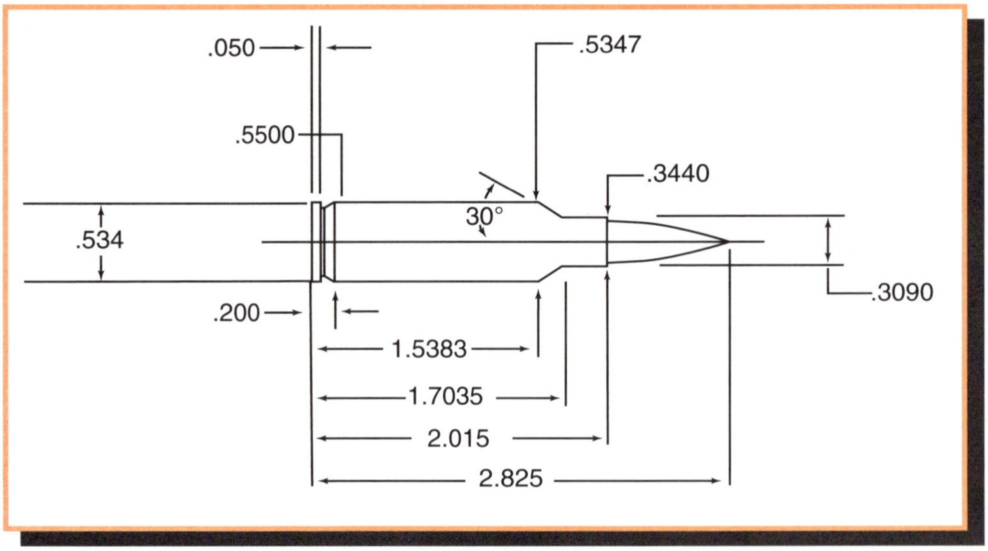

Comments:

Remington introduced their own concept of a rebated rim Short Magnum in 2002 as a short action adaptation of the popular Ultra Mag series. This was not however their first experience with short magnums. Remington first offered their belted 350 and 6.5mm Remington Magnums in the short action Model 600 rifle back in the mid-sixties. Neither cartridge enjoyed significant commercial success and both are long discontinued. Case length of the 300 SAUM is .085" shorter than the 300 WSM with a gentler 30-degree shoulder set approximately .125" farther back. The SAUM family of cartridges shares the appearance of the PPC with its short, fat disposition of powder. Factory specifications average around 150 to 200 feet per second over the 30-06 while using seven to eight percent more powder. Tests in Lyman's ballistic lab showed the 300 SAUM to average only around 30 to 50 feet below Winchester's short 300 while using three to four percent less powder. All propellants listed displayed fairly uniform ballistics throughout with IMR-4350, Reloder 19, Reloder 22, and the various VihtaVuori powders giving particularly consistent results. Cast bullets were selected for their capability to seat properly in the case while remaining within the maximum overall length. IMR-4198 and H4895 provided the most consistent results during cast bullet testing.

Test Components:

Cases . . . . .Remington
Trim-to Length . . . . .2.005"
Primers . . . . .Remington 9½ M & 9½
Primer Size . . . . .Large Rifle, Magnum & Standard
Lyman Shell Holder .No. 34
Jacketed Bullets Used . . . . .Swift Scirocco, 150 gr.
Swift Scirocco, 165 gr.
Nosler Ballistic Tip, #30180, 180 gr.
Nosler Partition, #35626, 200 gr.
Cast Bullets Used . . . .(sized to .309" dia)
*gas check bullet *#311672, 160 gr.
*#311644, 190 gr.

Test Specifications:
(Velocity & Pressure)

Firearm Used . . . . .Universal Receiver
Barrel Length . . . . .24"
Twist . . . . .1-10"
Groove Dia. . . . . .308"

150 gr. Jacketed Scirocco 2.800" OAL — **BC:** .430 **SD:** .226

Powder	Sugg Starting Grains	Velocity fps	Pressure P.S.I.	Max Load Grains	Velocity fps	Pressure P.S.I.
H414	61.5	3039	56,600	65.0+	3183	63,700
IMR-4350	61.5	2965	54,500	65.0+	3152	63,300
XMR-4350	62.5	2956	51,400	66.0+	3137	59,500
N160	64.5	3070	56,300	68.0+	3217	63,700
RX19	65.0	2930	53,400	**68.5+**	**3098**	**60,900**
MAGPRO	71.0	2944	52,900	75.0+	3146	61,800

165 gr. Jacketed Scirocco 2.800" OAL — **BC:** .470 **SD:** .248

Powder	Sugg Starting Grains	Velocity fps	Pressure P.S.I.	Max Load Grains	Velocity fps	Pressure P.S.I.
H414	58.5	2862	54,300	61.5	3025	62,700
IMR-4350	60.0	2870	57,200	63.0+	3002	62,800
XMR-4350	60.0	2852	53,600	**63.3+**	**3004**	**60,700**
N160	62.0	2922	56,500	65.5+	3042	62,600
IMR-4831	61.0	2857	56,000	64.5+	3006	63,100
RX19	63.0	2841	54,500	66.5+	3015	62,600
WXR	64.0	2816	54,000	67.5+	2996	62,900
RX22	65.5	2976	56,000	69.0+	3136	63,900
MAGPRO	70.0	2868	53,700	74.0+	3037	62,500

Note: Loads shown in shaded panels are maximum.
Loads shown in bold designate potentially most accurate load.
\+ Designates a compressed powder charge.

300 Remington Short Action Ultra Mag (SAUM)

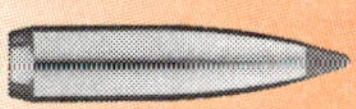

180 gr. Jacketed Ballistic Tip
2.800" OAL
BC: .507
SD: .271

Powder	Sugg Starting Grains	Velocity fps	Pressure P.S.I.	Max Load Grains	Velocity fps	Pressure P.S.I.
H414	58.0	2743	52,900	61.0	2932	64,300
IMR-4350	59.0	2778	57,300	62.0	2902	63,200
XMR-4350	60.5	2808	55,800	64.0+	2942	62,700
N160	61.0	2807	56,400	64.5+	2938	63,300
N560	63.5	2776	56,100	67.0+	2927	63,700
RX19	63.0	2747	54,500	66.5+	2919	63,200
IMR-4831	61.0	2761	55,500	64.5+	2931	64,000
RX22	63.5	2767	51,400	**67.0+**	**2938**	**59,400**
WXR	64.0	2731	54,400	67.5+	2904	63,200
N165	64.5	2649	45,300	68.0+	2829	53,300
MAGPRO	69.0	2802	53,600	73.0+	2945	60,100

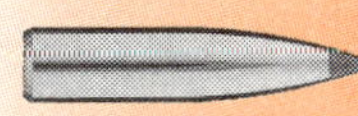

200 gr. Jacketed SP
2.825" OAL
BC: .481
SD: .301

Powder	Sugg Starting Grains	Velocity fps	Pressure P.S.I.	Max Load Grains	Velocity fps	Pressure P.S.I.
H414	56.5	2648	57,400	59.5	2785	63,700
N560	60.0	2595	53,500	63.5	2762	62,400
RX19	60.5	2651	56,000	64.0+	2794	64,000
IMR-4831	59.0	2636	55,100	62.0+	2787	62,900
RX22	61.0	2710	56,600	**64.5+**	**2846**	**63,700**
WXR	61.5	2654	56,500	65.0+	2798	63,900
H-4831	62.5	2666	59,200	66.0+	2774	63,700
IMR-7828	61.5	2611	54,300	65.0+	2770	62,600
MAGPRO	68.5	2789	57,900	72.0+	2886	62,400

***#311672**
160 gr. (#2 Alloy) 2.650" OAL
BC: .245
SD: .239

Powder	Sugg Starting Grains	Velocity fps	Pressure P.S.I.	Max Load Grains	Velocity fps	Pressure P.S.I.
SR-4759	21.5	1729	25,500	27.5	2099	36,900
XMP-5744	23.0	1708	24,800	30.0	2101	33,000
IMR-4227	23.5	1736	24,900	30.0	2105	33,700
IMR-4198	24.0	1701	22,900	31.0	2117	31,100
RX7	25.0	1728	23,500	32.5	2115	32,500
H-4895	32.5	1690	21,800	**40.0**	**2089**	**29,300**

***#311644**
190 gr. (#2 Alloy) 2.775" OAL
BC: .272
SD: .286

Powder	Sugg Starting Grains	Velocity fps	Pressure P.S.I.	Max Load Grains	Velocity fps	Pressure P.S.I.
SR-4759	20.5	1692	30,200	29.0	2113	48,700
XMP-5744	23.5	1708	27,600	31.0	2099	38,700
IMR-4227	23.0	1710	27,800	31.5	2120	42,200
IMR-4198	24.0	1705	26,300	**32.0**	**2095**	**37,000**
RX7	24.0	1668	26,100	32.5	2100	37,900
H-4895	30.0	1704	25,500	38.5	2101	33,400

Note: Loads shown in shaded panels are maximum.
Loads shown in bold designate potentially most accurate load.
\+ Designates a compressed powder charge.
* Designates the use of Remington 9½ primers.

300 WSM (Winchester Short Magnum)

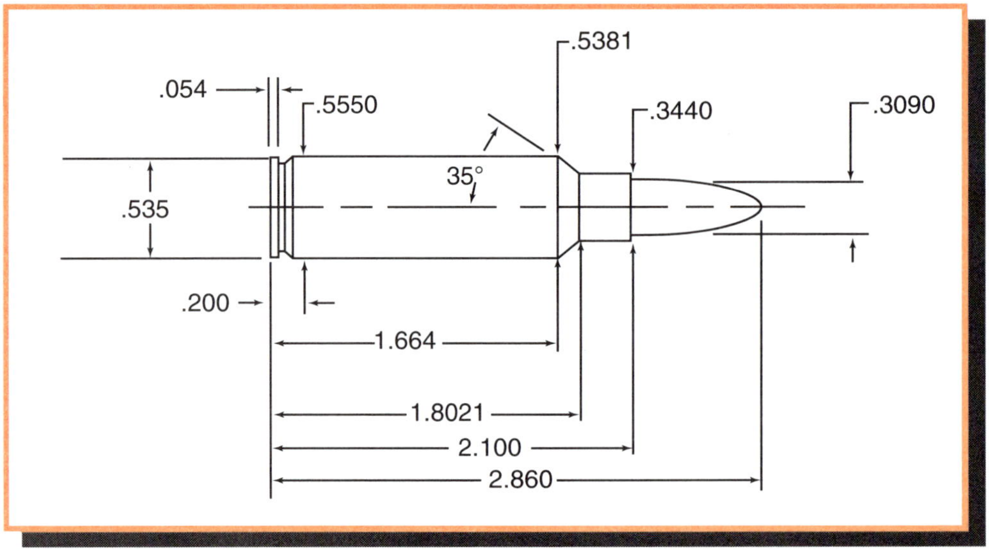

Comments:

Winchester launched their Short Magnum line of cartridges with the 300 WSM in 2001. The 300 WSM is Winchester's first completely new cartridge to come along in quite a few years as well as their first non-belted magnum. The appearance of the WSM is very similar to the PPC series of cartridges and features a slightly rebated rim. Benchrest shooters discovered long ago the inherent accuracy advantages of short, stubby cartridges. Many shooters report outstanding accuracy with this cartridge as well as lower perceived recoil. The absence of the belt found on traditional magnum cartridges offer smooth feeding and more positive head spacing. Its 2.100" case length, 35-degree shoulder, and pressure limit of 65,000 psi are designed to offer magnum level performance in a short action rifle.

Ballistics of the 300 WSM averaged 30 to 50 feet per second below the full-sized 300 Winchester Magnum while consuming around eight to ten percent less powder in our lab tests. Powders in the medium to slow burning range such as H-414, IMR-4350, and Reloder 19 all produced very good results. Cast bullets listed in the data section can be seated to within the maximum overall length while keeping the lube grooves and gas check contained within the neck. In addition to the listed accuracy loads, SR-4759 and Reloder 7 provided good results with cast bullet #311672. XMP-5744 and H-4895 worked well with cast bullet #311332. Initial public response has been quite positive and this cartridge will surely keep the wildcatters busy for years to come.

Test Components:

Cases . . . . .Winchester
Trim-to Length . . . . .2.090"
Primers . . . . .Winchester WLRM and WLR
Primer Size . . . . . . .Large Rifle Magnum and Standard
Lyman Shell Holder . . . . .No. 34
Jacketed Bullets Used . . . . .Swift Scirocco, 150 gr.
Nosler Ballistic Tip #30165, 165 gr.
Swift Scirocco, 180 gr.
Sierra HPBT #2210, 190 gr.
Nosler Partition #35626, 200 gr.
Sierra HPBT #2230, 200 gr.
Cast Bullets Used . . . .(sized to .309" dia)
*gas check bullet *#311672, 160 gr.
*#311332, 180 gr.

Test Specifications:
(Velocity & Pressure)

Firearm Used . . . . .Universal Receiver
Barrel Length . . . . .24"
Twist . . . . .1-10"
Groove Dia. . . . . .308"

150 gr. Jacketed Scirocco
2.850" OAL
BC: .430
SD: .266

Powder	Sugg Starting Grains	Velocity fps	Pressure P.S.I.	Max Load Grains	Velocity fps	Pressure P.S.I.
H-380	59.5	2895	48,600	66.0	3152	60,600
760	59.5	2903	48,400	66.0	3211	63,000
H-414	61.0	2873	44,600	67.5	3223	62,600
N160	63.0	2916	46,400	**70.0**	**3247**	**62,800**
IMR-4350	61.0	2807	44,000	67.5	3193	62,400
XMR-4350	61.0	2831	43,300	68.0	3198	60,800
RX19	63.0	2838	45,600	70.0	3169	60,200
WXR	64.0	2819	45,300	71.0+	3176	61,100
Magpro	72.0	2888	43,800	80.0+	3210	56,700

Note: Loads shown in shaded panels are maximum.
Loads shown in bold designate potentially most accurate load.
\+ Designates a compressed powder charge.

300 WSM (Winchester Short Magnum)

165 gr. Jacketed Ballistic Tip
2.850" OAL — **BC:** .475 **SD:** .248

Powder	Sugg Starting Grains	Velocity fps	Pressure P.S.I.	Max Load Grains	Velocity fps	Pressure P.S.I.
760	59.0	2812	49,800	65.5	3067	61,400
H-414	58.0	2822	51,700	64.7	3045	60,800
IMR-4350	59.0	2684	44,200	65.5	3049	62,500
XMR-4350	60.0	2730	44,500	**66.5**	**3053**	**60,900**
N160	61.5	2785	47,800	68.2	3073	61,600
RX19	62.0	2688	43,800	68.5	3046	60,900
IMR-4831	62.0	2688	45,000	68.7	3016	61,200
WXR	63.0	2740	45,800	70.0+	3055	61,600
H-4831	65.0	2775	49,700	72.0+	3024	62,400
Magpro	71.0	2821	44,100	79.0+	3103	55,800

190 gr. Jacketed HPBT
2.860" OAL — **BC:** .533 **SD:** .286

Powder	Sugg Starting Grains	Velocity fps	Pressure P.S.I.	Max Load Grains	Velocity fps	Pressure P.S.I.
760	56.5	2630	50,500	62.7	2864	62,600
H-414	55.0	2610	50,200	61.0	2844	62,800
IMR-4350	56.0	2574	46,300	62.5	2859	61,600
XMR-4350	57.0	2582	46,300	63.5	2858	61,300
N160	57.0	2611	49,400	63.5	2851	62,500
RX19	61.0	2609	47,300	67.5	2904	62,800
IMR-4831	59.0	2536	45,300	65.5	2865	63,200
WXR	60.5	2576	45,100	67.0	2879	61,500
H-4831	60.0	2541	46,300	66.5	2822	62,900
N165	62.0	2592	44,600	69.0+	2861	58,700
RX22	60.5	2692	50,000	**67.0**	**2947**	**63,800**
Magpro	67.5	2665	44,800	75.0+	2968	60,400

200 gr. Jacketed HPBT
2.860" OAL — **BC:** .565 **SD:** .301

Powder	Sugg Starting Grains	Velocity fps	Pressure P.S.I.	Max Load Grains	Velocity fps	Pressure P.S.I.
760	55.5	2561	51,100	61.5	2782	62,000
IMR-4350	55.5	2520	47,500	61.5	2783	62,300
XMR-4350	57.0	2549	48,600	63.0	2813	63,000
WXR	60.0	2576	49,100	66.5	2839	63,500
RX22	58.5	2631	52,100	**65.0**	**2842**	**62,800**
N165	62.0	2603	51,100	68.5+	2821	61,400

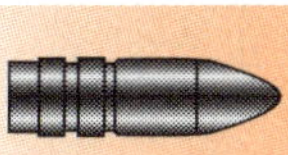

*#311332
180 gr. (#2 Alloy) 2.850" OAL — **BC:** .320 **SD:** .271

Powder	Sugg Starting Grains	Velocity fps	Pressure P.S.I.	Max Load Grains	Velocity fps	Pressure P.S.I.
SR-4759	**23.5**	**1804**	**25,600**	33.0	2241	45,500
XMP-5744	26.5	1792	22,500	36.0	2245	36,100
IMR-4198	27.0	1823	22,000	37.0	2248	36,600
N133	31.5	1794	19,200	41.0	2251	34,700
RX7	28.0	1822	21,300	37.0	2247	36,600
H-4895	34.0	1818	19,900	43.5	2234	30,200

180 gr. Jacketed Scirocco
2.860" OAL — **BC:** .520 **SD:** .271

Powder	Sugg Starting Grains	Velocity fps	Pressure P.S.I.	Max Load Grains	Velocity fps	Pressure P.S.I.
760	57.0	2670	49,900	63.2	2937	63,100
H-414	56.0	2665	49,800	62.0	2931	62,200
IMR-4350	58.0	2632	47,600	64.2	2921	62,900
XMR-4350	57.5	2606	45,000	64.0	2916	63,200
N160	57.5	2611	46,600	64.0	2880	62,500
RX19	61.5	2664	50,100	68.2	2952	63,800
IMR-4831	60.0	2599	47,800	66.7	2902	63,000
WXR	61.0	2671	49,500	68.0	2945	63,600
H-4831	62.0	2579	46,100	68.0+	2865	62,300
N165	63.0	2650	46,000	**70.0+**	**2936**	**60,500**
RX22	61.0	2697	48,900	67.0	2978	62,700
Magpro	68.0	2630	41,600	75.5+	2907	52,600

200 gr. Jacketed SP
2.860" OAL — **BC:** .481 **SD:** .301

Powder	Sugg Starting Grains	Velocity fps	Pressure P.S.I.	Max Load Grains	Velocity fps	Pressure P.S.I.
760	55.0	2523	50,500	61.0	2753	62,400
H-414	55.5	2534	52,500	61.5	2757	62,900
IMR-4350	55.0	2488	47,300	61.0	2734	61,200
XMR-4350	57.0	2529	48,600	63.0	2775	61,900
N160	57.0	2537	51,600	**63.0**	**2763**	**63,500**
RX19	60.0	2565	48,300	67.0	2826	63,300
IMR-4831	58.0	2445	45,300	64.0	2740	61,600
WXR	60.0	2542	48,700	66.0	2783	61,100
H-4831	58.5	2493	48,800	65.0	2719	62,300
N165	62.0	2539	48,400	68.0	2805	62,400
RX22	58.5	2590	50,100	65.0	2822	62,500
Magpro	67.5	2587	45,600	74.0+	2855	58,900

*#311672
160 gr. (#2 Alloy) 2.725" OAL — **BC:** .245 **SD:** .240

Powder	Sugg Starting Grains	Velocity fps	Pressure P.S.I.	Max Load Grains	Velocity fps	Pressure P.S.I.
SR-4759	21.5	1800	22,900	30.2	2250	40,300
XMR-5744	**25.0**	**1782**	**20,100**	34.3	2242	31,000
IMR-4198	25.5	1824	19,100	35.0	2263	31,200
N133	30.0	1791	16,800	39.0	2251	28,600
RX7	26.0	1805	19,100	36.0	2252	31,600
H4895	33.0	1806	17,000	42.5	2258	27,200

Note: Loads shown in shaded panels are maximum.
Loads shown in bold designate potentially most accurate load.
\+ Designates a compressed powder charge.
* Designates the use of Win. WLR primers.

300 Winchester Magnum

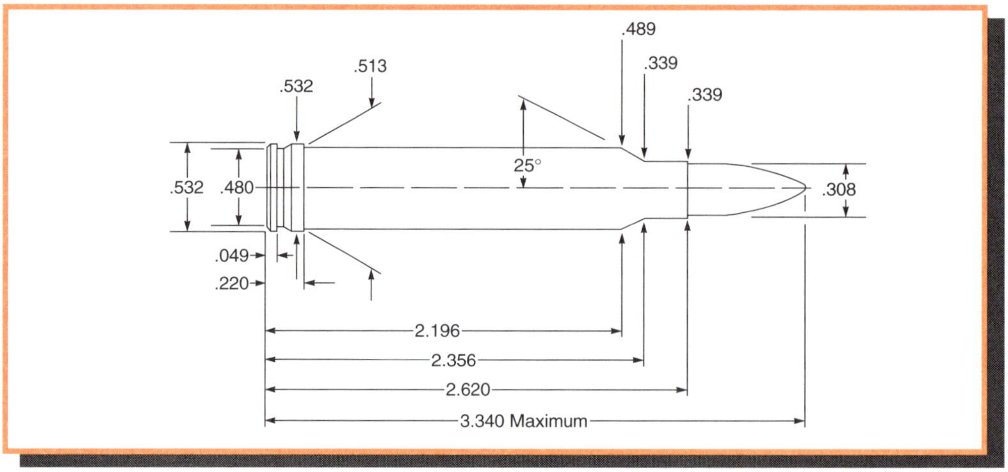

Comments:

This cartridge will duplicate the ballistics of the 300 H&H Magnum. Its main advantage is that it will work in a standard length action whereas the H&H round requires a magnum length action.

This cartridge tends to give the best results with Winchester primers. Our data was developed using the standard Winchester primer and the use of a magnum primer will cause a pressure increase. When using magnum type primers work up loads cautiously and stay at least one grain below the maximum listed loads.

Favored powders should be Hodgdon H4831 and IMR 7828. Strongly constructed bullets of 180 to 220 grains usually give the best balance of internal ballistic uniformity and accuracy

Test Components:

Cases . . . . .Winchester
Trim-to Length . . . . .2.610"
Primers . . . . .Winchester WLR
Primer Size . . . . .Large Rifle
Lyman Shell Holder . . . . .No. 13
Jacketed Bullets Used . . . .Hornady SP #3020, 130 gr.
Hornady SP #3031, 150 gr.
Nosler Partition #16330, 165 gr.
Sierra HPBT #2220, 180 gr.
Barnes X #30835, 180 gr.
Sierra HPBT #2210, 190 gr.
Sierra HPBT #2230, 200 gr.
Hornady RN #3090, 220 gr.
Cast Bullets Used . . . .(sized to .309" dia)
*gas check bullet *#311672, 160 gr.
*#311291, 170 gr.
*#311644, 190 gr.

Test Specifications: (Velocity & Pressure)

Firearm Used . . . . .Universal Receiver and Winchester Model 70
Barrel Length . . . . .24"
Twist . . . . .1-10"
Groove Dia. . . . . .308"

130 gr. Jacketed SP 3.215" OAL — **BC:** .295 **SD:** .196

Powder	Sugg Starting Grains	Velocity fps	Pressure	Max Load Grains	Velocity fps	Pressure
Varget	63.5	3176	49,300	70.0	3456	62.400 P
RX15	64.0	3169	48,200	70.0	3466	61,900 P
AA2700	67.5	3036	44,200	75.0	3401	61,800 P
760	69.0	3209	50,200	75.0	3404	61,500 P
H414	66.0	3036	43,500	75.0	3381	53,000 C
N160	**73.0**	**3121**	**45,200**	81.2+	3501	63,700 P
IMR-4350	71.0	3086	—	79.0+	3497	—
XMR-3100	72.0	2871	38,100	80.0+	3290	56,300 P
IMR-4831	73.0	3039	41,300	81.0+	3485	60,000 P
H4831	75.0	2932	—	83.0+	3290	—

150 gr. Jacketed SP 3.340" OAL — **BC:** .338 **SD:** .226

Powder	Sugg Starting Grains	Velocity fps	Pressure	Max Load Grains	Velocity fps	Pressure
AA2700	63.5	2876	47,500	70.5	3159	61,300 P
760	66.0	2958	51,300	71.0	3221	62,800 P
H414	60.0	2800	43,000	68.5	3111	52,200 C
N160	69.0	2950	47,000	77.0	3275	63,500 P
IMR-4350	68.0	2890	—	76.0	3290	—
RX19	70.0	2912	44,500	78.0	3285	62,700 P
IMR-4831	69.0	2886	43,300	**77.0**	**3299**	**63,300 P**
XMR-3100	70.0	2782	43,100	77.5+	3141	62,200 P
H4831	73.0	2793	—	81.0+	3194	—
RX22	74.0	2889	43,600	82.0+	3218	61,300 P
IMR7828	75.5	2944	38,600	81.0+	3268	52,500 C

Note: Loads shown in shaded panels are maximum.
Loads shown in bold designate potentially most accurate load.
\+ Designates a compressed powder charge.
In pressure column, C=Copper Units of Pressure(CUP); P=Pounds per Square Inch (PSI)

300 Winchester Magnum

165 gr. Jacketed SP
3.340" OAL — **BC:** .410 **SD:** .248

Powder	Sugg Starting Grains	Velocity fps	Pressure	Max Load Grains	Velocity fps	Pressure
H414	60.0	2781	48,400	67.0	3059	63,300 P
IMR-4350	67.0	2865	—	**74.0**	**3194**	**—**
N560	71.0	2761	46,200	78.5+	3098	61,500 P
RX19	68.5	2795	45,300	76.0+	3111	61,800 P
IMR-4831	67.5	2807	44,200	75.0	3164	63,200 P
XMR-3100	68.0	2758	45,100	75.5+	3071	63,200 P
H4831	71.0	2724	—	78.0+	3066	—
WXR	71.0	2838	45,900	78.5+	3059	63,100 P
RX22	72.5	2786	37,700	79.5+	3119	50,300 C
IMR-7828	74.0	2892	41,600	79.5+	3149	51,200 C

180 gr. Jacketed HPBT
3.340" OAL — **BC:** .475 **SD:** .271

Powder	Sugg Starting Grains	Velocity fps	Pressure	Max Load Grains	Velocity fps	Pressure
IMR-4350	65.0	2724	—	72.5	3048	—
RX19	68.0	2647	38,200	74.0	2962	49,700 C
IMR-4831	66.0	2674	43,700	73.0+	3035	63,000 P
XMR-3100	68.0	2562	37,200	75.5	2936	51,400 C
N560	67.5	2668	43,600	**75.0**	**3036**	**63,000 P**
H4831	70.0	2666	—	76.0+	2938	—
WXR	69.0	2703	45,000	77.0+	3048	63,300 P
RX22	70.5	2696	37,800	77.0+	3017	50,300 C
IMR-7828	69.0	2596	42,000	76.0+	2894	58,000 P
RX25	73.0	2741	46,200	81.0+	3084	63,200 P

180 gr. Barnes X
3.340" OAL — **BC:** .511 **SD:** .271

Powder	Sugg Starting Grains	Velocity fps	Pressure	Max Load Grains	Velocity fps	Pressure
IMR-4350	61.0	2509	36,700	68.0	2788	48,200 C
RX19	64.5	2587	37,000	72.5	2894	50,200 C
XMR-3100	64.0	2428	36,000	73.0	2794	50,600 C
N560	64.0	2616	48,500	71.0	2830	56,800 P
H4831	64.0	2567	39,700	74.0	2832	50,100 C
RX22	64.0	2575	38,000	73.5	2928	51,300 C
IMR-7828	69.0	2685	53,500	76.5+	2907	63,000 P
RX25	70.0	2787	54,400	78.0+	3009	63,100 P

190 gr. Jacketed HPBT
3.340" OAL — **BC:** .533 **SD:** .286

Powder	Sugg Starting Grains	Velocity fps	Pressure	Max Load Grains	Velocity fps	Pressure
IMR-4350	59.5	2583	42,800	68.0	2880	53,700 C
IMR-4831	65.0	2603	44,000	72.5	2936	62,500 P
N560	68.5	2599	45,100	75.0	2926	62,400 P
XMR-3100	67.5	2536	42,000	75.0+	2890	61,000 P
H4831	64.0	2522	37,600	73.0	2897	52,800 C
WXR	68.0	2633	46,800	75.5+	2933	62,500 P
RX22	69.0	2633	44,400	77.0+	2985	62,600 P
IMR-7828	69.0	2614	45,600	77.0+	2932	62,400 P
RX25	72.0	2633	44,000	**80.0+**	**2992**	**63,300 P**
H1000	76.0	2623	45,600	83.5+	2930	62,900 P
AA8700	78.0	2353	37,000	86.0+	2607	48,300 P

200 gr. Jacketed HPBT
3.340" OAL — **BC:** .565 **SD:** .301

Powder	Sugg Starting Grains	Velocity fps	Pressure	Max Load Grains	Velocity fps	Pressure
IMR-4350	62.0	2597	—	69.0	2849	—
IMR-4831	63.5	2569	44,800	70.5	2881	62,500 P
RX19	64.5	2480	35,600	73.0	2877	51,800 C
XMR-3100	64.0	2392	36,000	72.5	2751	51,000 C
H4831	67.0	2531	—	74.0+	2816	—
WXR	67.0	2593	46,500	74.0	2893	62,900 P
RX22	65.0	2489	35,800	74.5+	2878	51,500 C
IMR-7828	67.0	2476	37,700	74.5+	2828	51,800 C
RX25	71.0	2608	45,500	**78.0+**	**2938**	**62,900 P**
H1000	70.0	2442	36,000	80.0+	2798	49,900 C
AA8700	74.0	2253	35,000	82.0+	2483	44,500 P

220 gr. Jacketed RN
3.340" OAL — **BC:** .300 **SD:** .331

Powder	Sugg Starting Grains	Velocity fps	Pressure	Max Load Grains	Velocity fps	Pressure
IMR-4350	60.0	2463	—	67.5	2747	—
IMR-4831	63.0	2440	46,100	**70.0**	**2705**	**62,600 P**
RX19	64.0	2415	37,000	72.0	2730	50,900 C
XMR-3100	62.5	2389	45,000	69.5	2674	62,600 P
H4831	65.0	2415	—	72.0+	2688	—
WXR	65.0	2494	49,200	72.5+	2741	63,700 P
RX22	63.5	2374	36,700	72.5+	2717	51,400 C
IMR-7828	67.5	2445	45,000	71.0+	2747	63,300 P
RX25	69.0	2446	43,100	76.0+	2772	61,900 P
H1000	69.0	2337	36,800	78.0+	2653	49,400 C
AA8700	75.0	2278	38,300	83.0+	2552	51,300 P
H870	80.0	2460	44,000	84.0+	2641	51,400 P

Note: Loads shown in shaded panels are maximum.
Loads shown in bold designate potentially most accurate load.
\+ Designates a compressed powder charge.
In pressure column, C=Copper Units of Pressure(CUP); P=Pounds per Square Inch (PSI)

300 Winchester Magnum

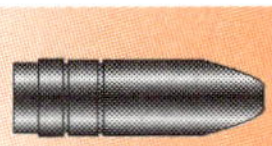

#311672
160 gr. (#2 Alloy) 3.200" OAL
BC: .245
SD: .240

Powder	Sugg Starting Grains	Velocity fps	Pressure P.S.I.	Max Load Grains	Velocity fps	Pressure P.S.I.
Unique	16.0	1621	21,700	28.0	2198	45,800
SR-4759	**22.0**	**1661**	**15,600**	38.5	2440	44,900
XMP-5744	25.5	1673	14,700	42.5	2466	36,100
RX7	28.0	1683	12,700	47.0	2427	33,800
H-4895	31.0	1694	13,200	50.0	2438	31,400
IMR-4831	40.0	1684	13,100	63.0	2386	27,000
XMR-3100	40.0	1659	14,100	65.0	2381	28,200

#311291
170 gr. (#2 Alloy) 3.135" OAL
BC: .202
SD: .256

Powder	Sugg Starting Grains	Velocity fps	Pressure	Max Load Grains	Velocity fps	Pressure
Unique	18.0	1705	29,400	22.0	1910	38,600 C
SR-7625	15.0	1480	27,600	18.0	1650	36,000 C
SR-4759	30.5	2075	28,400	40.0	2472	50,900 C
XMP-5744	**26.0**	**1648**	**16,300**	43.0	2415	40,000 P
RX7	37.0	2184	27,200	48.0	2575	46,100 C
H-4895	40.5	2157	25,900	56.0	2748	51,200 C
XMR-3100	39.5	1623	14,700	65.5	2398	34,000 P

#311644
190 gr. (#2 Alloy) 3.405" OAL
BC: .272
SD: .286

Powder	Sugg Starting Grains	Velocity fps	Pressure P.S.I.	Max Load Grains	Velocity fps	Pressure P.S.I.
Unique	18.0	1609	27,200	30.0	2144	56,900
SR-4759	23.0	1637	19,400	41.0	2367	56,100
XMP-5744	**26.0**	**1641**	**17,400**	44.0	2399	42,900
RX7	28.0	1672	16,300	48.0	2373	41,900
H-4895	32.0	1667	15,800	52.0	2391	38,300
IMR-4064	32.0	1640	15,300	53.0	2434	39,000
IMR-4831	40.0	1599	14,200	64.0	2444	35,200
XMR-3100	44.0	1695	16,200	68.0	2467	39,600

Note: Loads shown in shaded panels are maximum.
Loads shown in bold designate potentially most accurate load.
\+ Designates a compressed powder charge.
In pressure column, C=Copper Units of Pressure(CUP); P=Pounds per Square Inch (PSI)

300 Remington Ultra Mag

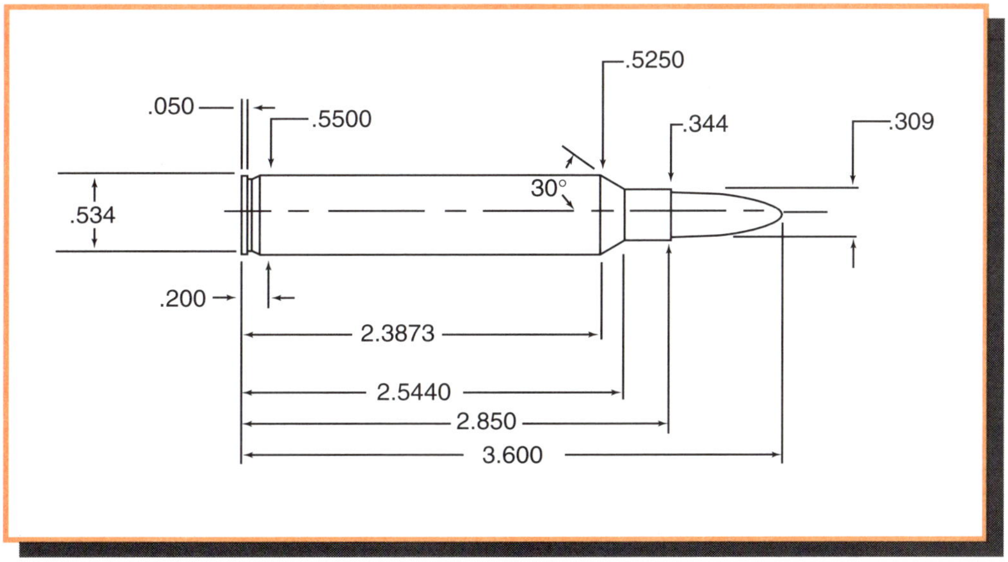

Comments:

Remington launched the Ultra Mag family with this, their first 30 caliber magnum in 1999. Remington broke with the long held magnum orthodoxy by basing the cartridge on the beltless 404 Jeffery case featuring a slightly rebated rim. The absence of the belt allows for increased capacity in comparison to the traditional belted Magnums while using the same diameter bolt face. Additional benefits include more positive headspacing and smoother feeding. Predictably, the new cartridge thrives on healthy charges of slow-burning powder and premium bullets. The 300 RUM is capable of some 200 feet per second over the 300 Winchester Magnum

Test Components:

Cases . . . . .Remington
Trim-to Length . . . . .2.840"
Primers . . . . .Remington 9½ M and 9½
Primer Size . . . . . .Large Rifle, Magnum and Standard
Lyman Shell Holder . . . . .No. 13
Jacketed Bullets Used . . .Barnes XLC #30854, 150 gr.
Nosler Ballistic Tip #30165, 165 gr.
Swift Scirocco, 180 gr.
Swift A-Frame, 200 gr.
Cast Bullets Used . . . .(sized to .309" dia)
*gas check bullet *#311644, 190 gr.

Test Specifications:
(Velocity & Pressure)

Firearm Used . . . . .Universal Receiver
Barrel Length . . . . .24"
Twist . . . . .1-10"
Groove Dia. . . . .308"

150 gr. Barnes XLC 3.545" OAL — **BC:** .428 **SD:** .226

Powder	Sugg Starting Grains	Velocity fps	Pressure P.S.I.	Max Load Grains	Velocity fps	Pressure P.S.I.
IMR-4831	86.0	3229	51,500	91.0	3476	63,900
WXR	89.0	3267	55,200	94.0	3422	63,100
XMR-3100	89.0	3143	51,000	94.0	3393	63,300
H-4831	89.0	3133	51,500	94.0	3348	62,600
RX22	**91.0**	**3201**	**50,500**	96.0	3467	63,300
IMR-7828	91.0	3220	53,500	96.0	3427	64,100
N165	95.0	3267	53,700	100.0	3480	64,800
H1000	97.0	3116	49,700	102.5+	3364	63,300
AA8700	106.0	3089	51,900	112.0+	3176	55,500

165 gr. Jacketed Ballistic Tip 3.600" OAL — **BC:** .475 **SD:** .248

Powder	Sugg Starting Grains	Velocity fps	Pressure P.S.I.	Max Load Grains	Velocity fps	Pressure P.S.I.
IMR-4831	84.5	3133	56,700	89.0	3270	62,800
WXR	87.0	3184	62,100	92.0	3250	63,400
XMR-3100	86.5	3033	52,000	90.5	3234	62,400
H-4831	87.0	3073	57,700	**92.0**	**3209**	**64,100**
RX22	89.0	3178	57,300	94.0	3316	63,800
IMR-7828	89.0	3111	56,000	94.0	3264	64,000
MAGPRO	92.0	3175	55,300	96.0	3302	60,200
RX25	92.0	3086	52,300	97.0	3296	62,900
H1000	94.5	3015	51,500	99.7	3211	62,100
N170	95.0	3164	60,300	100.0	3252	63,500
Retumbo	98.0	3145	50,900	103.0+	3310	58,800
AA8700	107.0	3120	57,300	113.0+	3232	62,200

Note: Loads shown in shaded panels are maximum.
Loads shown in bold designate potentially most accurate load.
\+ Designates a compressed powder charge.

300 Remington Ultra-Mag

180 gr. Jacketed Scirocco
3.600" OAL

BC: .520
SD: .271

Powder	Sugg Starting Grains	Velocity fps	Pressure P.S.I.	Max Load Grains	Velocity fps	Pressure P.S.I.
IMR-4831	80.0	2997	55,600	84.0	3152	63,700
WXR	85.0	2989	57,100	89.5	3112	62,300
XMR-3100	84.0	2950	53,600	89.0	3131	64,100
H-4831	84.0	3009	58,900	89.0	3123	64,600
RX22	86.0	3064	55,500	90.5	3222	63,500
IMR-7828	86.0	2987	54,100	**91.0**	**3171**	**64,200**
MAGPRO	89.0	3074	57,400	94.0	3192	62,100
RX25	89.0	2980	50,800	93.5	3196	62,400
H1000	92.0	2953	54,900	97.0	3120	63,700
N170	91.0	2999	56,000	96.0	3157	64,000
Retumbo	96.0	3093	55,000	101.0	3231	61,900
H50BMG	100.0	2971	55,600	106.0+	3121	63,400
H870	104.0	2983	50,900	110.0+	3091	55,400
AA8700	106.0	3024	55,200	112.0+	3179	62,800

200 gr. Jacketed A-Frame
3.535" OAL

BC: .444
SD: .301

Powder	Sugg Starting Grains	Velocity fps	Pressure P.S.I.	Max Load Grains	Velocity fps	Pressure P.S.I.
WXR	77.0	2705	50,500	81.0	2887	62,700
XMR-3100	81.0	2836	55,000	**85.5**	**2993**	**63,900**
RX22	83.0	2904	57,800	87.0	3018	63,700
IMR-7828	83.0	2872	58,700	87.0	2976	63,500
MAGPRO	80.0	2770	53,400	84.0	2901	60,000
N170	85.0	2778	55,000	89.0	2907	62,200
RX25	86.0	2884	55,700	91.0	3036	62,900
H1000	88.0	2816	54,200	93.0	2946	60,700
Retumbo	91.0	2941	56,100	96.0	3067	62,400
H50BMG	95.0	2877	60,500	100.5	2965	64,200
H870	104.0	3008	58,000	110.0+	3105	62,900
AA-8700	104.0	2990	60,600	110.0+	3066	63,100

***#311644**
190 gr. (#2 Alloy) 3.600" OAL

BC: .272
SD: .284

Powder	Sugg Starting Grains	Velocity fps	Pressure P.S.I.	Max Load Grains	Velocity fps	Pressure P.S.I.
SR-4759	23.0	1502	13,600	40.0	2198	39,000
XMP-5744	25.0	1484	12,500	**44.0**	**2208**	**30,800**
IMR-4198	25.0	1505	12,600	45.0	2208	32,500
RX7	27.5	1500	13,000	46.5	2200	30,800
Varget	33.0	1551	12,300	54.5	2207	27,600

Note: Loads shown in shaded panels are maximum.
Loads shown in bold designate potentially most accurate load.
* Designates use of 9½ primers.
\+ Designates a compressed powder charge.

300 Weatherby Magnum

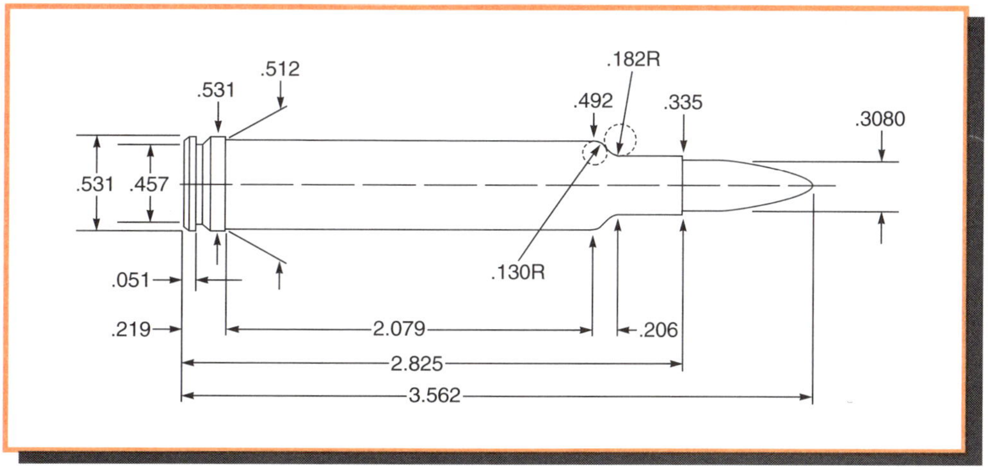

Comments:

This is probably the most popular–and one of the oldest–of all the calibers offered by Weatherby. Roy Weatherby based this 300 Magnum on a blown out 300 Holland and Holland case featuring the famous double radius shoulder in 1943. It has been the king of factory 30-caliber Magnums for many years being only recently challenged by a new generation of cartridges based on the 404 Jeffrey case as well as Weatherby's own 30-378. The 300 Weatherby can best the 300 Winchester Magnum by around 200 feet per second when properly loaded. The popularity of the 300 Weatherby has prompted most major manufacturers to offer rifles chambered for it over the years. H-4831 has been one of the more popular powders for loading this particular 300. Weatherby recommends the exclusive use of the Federal 215 Magnum primer for all jacketed loads. Cast bullet shooters should use the standard primers as indicated. All cast bullet propellants produced good results with XMP-5744 giving top ballistic uniformity. This data is intended for commercially manufactured and chambered rifles. It is not for use in custom chambered guns that may lack the free bore found in standard Weatherby chambers.

Test Components:

Cases . . . . Federal
Trim-to Length . . . . 2.825"
Primers . . . . Federal 215 and 210
Primer Size . . . . . . Large Rifle, Magnum and Standard
Lyman Shell Holder . . . . No. 13
Jacketed Bullets Used . . . Barnes XLC #30854, 150 gr.
Nosler Ballistic Tip #30165, 165 gr.
Barnes X #30835, 180 gr.
Swift A-Frame, 200 gr.
Hornady RN #3090, 220 gr.
Cast Bullets Used . . . . (sized to .309" dia)
*gas check bullet
*#311291, 170 gr.
*#311644, 190 gr.
*#311284, 210 gr.

Test Specifications: (Velocity Only)

Firearm Used . . . . Weatherby Mk V
Barrel Length . . . . 26"
Twist . . . . 1-10"
Groove Dia. . . . . .308"

150 gr. Barnes XLC
3.560" OAL
BC: .428
SD: .226

Powder	Sugg Starting Grains	Velocity fps	Pressure	Max Load Grains	Velocity fps	Pressure
IMR-4350	76.0	3217	—	81.0	3501	—
XMR-3100	80.5	3217	—	**85.0**	**3418**	—
RX22	85.0	3400	—	90.0+	3589	—
IMR-7828	85.0	3334	—	90.0+	3532	—
RX25	86.0	3301	—	91.0+	3511	—
H1000	85.0	3264	—	90.0+	3428	—

165 gr. Jacketed Ballistic Tip
3.560" OAL
BC: .475
SD: .248

Powder	Sugg Starting Grains	Velocity fps	Pressure	Max Load Grains	Velocity fps	Pressure
N160	78.0	3234	—	82.5	3411	—
IMR-4350	71.0	2988	—	75.0	3189	—
XMR-3100	80.0	3214	—	85.0	3385	—
H-4831SC	76.0	2914	—	80.0	3066	—
RX22	80.0	3292	—	85.0	3350	—
IMR-7828	81.0	3221	—	86.0	3376	—
H1000	**84.5**	**3063**	—	89.0	3220	—
RX25	85.5	3114	—	90.0	3445	—

Note: Loads shown in shaded panels are maximum.
Loads shown in bold designate potentially most accurate load.
\+ Designates a compressed powder charge.

300 Weatherby Magnum

180 gr. Barnes X
3.560" OAL

BC: .511
SD: .271

Powder	Sugg Starting Grains	Velocity fps	Pressure	Max Load Grains	Velocity fps	Pressure
N160	76.0	3028	—	80.0	3199	—
IMR-4350	68.5	2721	—	73.0	2994	—
XMR-3100	77.0	2949	—	81.0	3109	—
H-4831SC	77.0	2877	—	81.0	3054	—
RX22	78.0	2959	—	**82.0**	**3116**	—
IMR-7828	80.0	3033	—	85.0	3210	—
H1000	81.0	2818	—	85.0	2985	—
RX25	83.0	3108	—	87.5	3278	—

200 gr. Jacketed A-Frame
3.560" OAL

BC: .444
SD: .301

Powder	Sugg Starting Grains	Velocity fps	Pressure	Max Load Grains	Velocity fps	Pressure
XMR-3100	74.0	2771	—	78.0	2952	—
H-4831SC	73.5	2724	—	77.5	2867	—
RX22	74.0	2775	—	**78.0**	**2941**	—
IMR-7828	78.0	2919	—	82.0	3105	—
H1000	80.0	2740	—	85.0+	2967	—
RX25	80.0	2910	—	**84.0+**	**3092**	—
AA8700	88.0	2762	—	93.0+	2891	—
H870	88.0	2726	—	92.0	2833	—

220 gr. Jacketed RN
3.545" OAL

BC: .300
SD: .331

Powder	Sugg Starting Grains	Velocity fps	Pressure	Max Load Grains	Velocity fps	Pressure
XMR-3100	71.0	2624	—	75.0	2774	—
H-4831SC	73.0	2654	—	76.0	2800	—
RX22	72.0	2639	—	76.0	2795	—
IMR-7828	77.0	2811	—	81.0	2992	—
H1000	**76.0**	**2535**	—	79.5	2735	—
RX25	78.0	2795	—	82.0	3008	—
AA8700	88.0	2732	—	93.0	2849	—

***#311291**
170 gr. (#2 Alloy) 3.390" OAL

BC: .202
SD: .252

Powder	Sugg Starting Grains	Velocity fps	Pressure	Max Load Grains	Velocity fps	Pressure
SR-4759	**22.0**	**1645**	—	39.0	2393	—
XMP-5744	24.5	1616	—	40.0	2300	—
RX7	25.5	1642	—	39.0	2215	—

***#311644**
190 gr. (#2 Alloy) 3.530" OAL

BC: .272
SD: .284

Powder	Sugg Starting Grains	Velocity fps	Pressure	Max Load Grains	Velocity fps	Pressure
SR-4759	21.5	1594	—	35.5	2213	—
XMP-5744	24.5	1591	—	39.0	2219	—
RX7	**25.5**	**1613**	—	39.5	2155	—

***#311284**
210 gr. (#2 Alloy) 3.560" OAL

BC: .332
SD: .314

Powder	Sugg Starting Grains	Velocity fps	Pressure	Max Load Grains	Velocity fps	Pressure
SR-4759	21.0	1512	—	35.0	2125	—
XMP-5744	23.0	1488	—	**37.5**	**2102**	—
RX7	24.0	1521	—	39.0	2094	—

Note: Loads shown in shaded panels are maximum.
Loads shown in bold designate potentially most accurate load.
* Designates use of 210 primers.
\+ Designates a compressed powder charge.

30-378 Weatherby Magnum

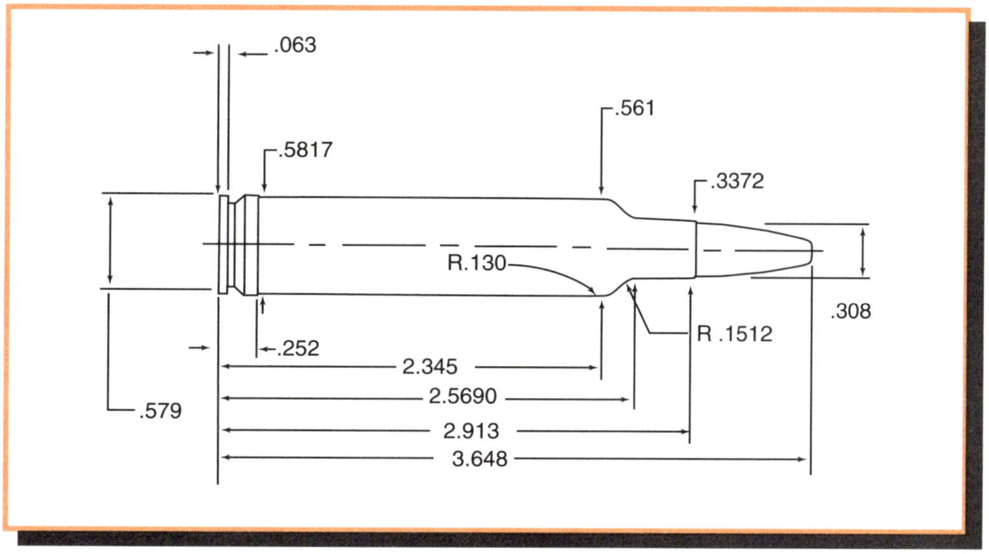

Comments:

Roy Weatherby first necked down his massive 378 case to 30-caliber in joint experiments with the United States Army during the 1950s. This cartridge survived in various wildcat incarnations since then, particularly with long-range benchrest shooters, before Weatherby formally introduced it as a standard chambering in their lineup in 1996. This is not a varmint cartridge. Shooters looking for a fast 30-caliber magnum rifle would be hard pressed to surpass this chambering. The latest generation of slow burning powders enable overbored cartridges such as this to perform to their full potential. While probably a bit more than many shooters would care to shoot extensively, this Weatherby is hard to beat — 3,400 feet per second with 180 grain bullets — for driving heavy bullets long distances.

Slow burning powder and premium bullets are in order for the 30-378. Powders well suited to this behemoth include H1000, H50BMG, and Reloder 25. The Swift Scirocco is ideal for the high velocities generated by this cartridge although shooters may have to experiment with seating depths for best accuracy. Limited cast bullet data is included for those shooters looking for off-season recreational shooting without the substantial recoil and blast of full house 30-378 loads. This data is intended for commercially produced and chambered rifles. It is not for use in custom chambered gun that may lack the free bore found in standard Weatherby chambers.

Test Components:

Cases . . . . .Weatherby
Trim-to Length . . . . .2.903"
Primers . . . . .Federal 215 and 210
Primer Size . . . . . . .Large Rifle, Magnum and Standard
Lyman Shell Holder . . . . .No. 17
Jacketed Bullets Used . . .Nosler Ballistic Tip, #30165, 165 gr.
Swift Scirocco, 180 gr.
Swift A-Frame, 200 gr.
Cast Bullets Used . . . . (sized to .309" dia)
*gas check bullet
*#311291, 170 gr.
*#311644, 190 gr.
*#311284, 210 gr.

Test Specifications: (Velocity Only)

Firearm Used . . . . .Weatherby Mk V
Barrel Length . . . . .26"
Twist . . . . .1-10"
Groove Dia. . . . . .308"

165 gr. Jacketed Ballistic Tip 3.630" OAL — **BC:** .475 **SD:** .248

Powder	Sugg Starting Grains	Velocity fps	Pressure	Max Load Grains	Velocity fps	Pressure
RX22	99.0	3329	—	104.0	3492	—
IMR-7828	100.0	3322	—	105.0	3520	—
RX25	103.0	3329	—	108.0	3537	—
H1000	106.0	3221	—	112.0	3481	—
AA8700	118.0	3364	—	124.0+	3540	—
H50BMG	116.0	3280	—	**123.0+**	**3456**	—
H870	116.0	3224	—	122.0+	3434	—

180 gr. Jacketed Scirocco 3.630" OAL — **BC:** .520 **SD:** .271

Powder	Sugg Starting Grains	Velocity fps	Pressure	Max Load Grains	Velocity fps	Pressure
RX22	96.0	3206	—	101.0	3372	—
IMR-7828	95.0	3142	—	101.0	3371	—
RX25	100.0	3194	—	106.0	3451	—
H1000	105.0	3209	—	110.0	3429	—
AA8700	115.0	3221	—	121.0	3406	—
H50BMG	114.0	3213	—	**120.0**	**3342**	—
H870	112.0	3147	—	118.0	3351	—

Note: Loads shown in shaded panels are maximum.
Loads shown in bold designate potentially most accurate load.
\+ Designates a compressed powder charge.

30-378 Weatherby Magnum

200 gr. Jacketed A-Frame 3.615" OAL — **BC:** .444 **SD:** .301

Powder	Sugg Starting Grains	Velocity fps	Pressure	Max Load Grains	Velocity fps	Pressure
RX25	97.0	3008	—	102.0	3195	—
H1000	100.0	2981	—	106.0	3174	—
AA8700	111.0	3002	—	117.0	3211	—
H50BMG	111.0	3116	—	**117.0**	**3270**	—
H870	110.0	3011	—	116.0	3205	—

***#311291** 170 gr. (#2 Alloy) 3.500" OAL — **BC:** .202 **SD:** .256

Powder	Sugg Starting Grains	Velocity fps	Pressure	Max Load Grains	Velocity fps	Pressure
SR-4759	25.0	1655	—	41.5	2306	—
XMP-5744	**28.0**	**1645**	—	45.5	2313	—
RX7	30.0	1711	—	49.0	2336	—

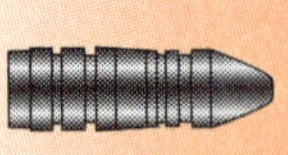

***#311644** 190 gr. (#2 Alloy) 3.630" OAL — **BC:** .272 **SD:** .284

Powder	Sugg Starting Grains	Velocity fps	Pressure	Max Load Grains	Velocity fps	Pressure
SR-4759	27.0	1710	—	43.0	2294	—
XMP-5744	**30.0**	**1695**	—	46.0	2272	—
RX7	28.0	1628	—	48.0	2248	—

***#311284** 210 gr. (#2 Alloy) 3.700" OAL — **BC:** .332 **SD:** .314

Powder	Sugg Starting Grains	Velocity fps	Pressure	Max Load Grains	Velocity fps	Pressure
SR-4759	26.0	1620	—	44.0	2258	—
XMP-5744	31.0	1701	—	**47.0**	**2261**	—
RX7	30.0	1620	—	48.0	2202	—

Note: Loads shown in shaded panels are maximum.
Loads shown in bold designate potentially most accurate load.
* Designates use of 210 primers.

7.62 x 39 Russian

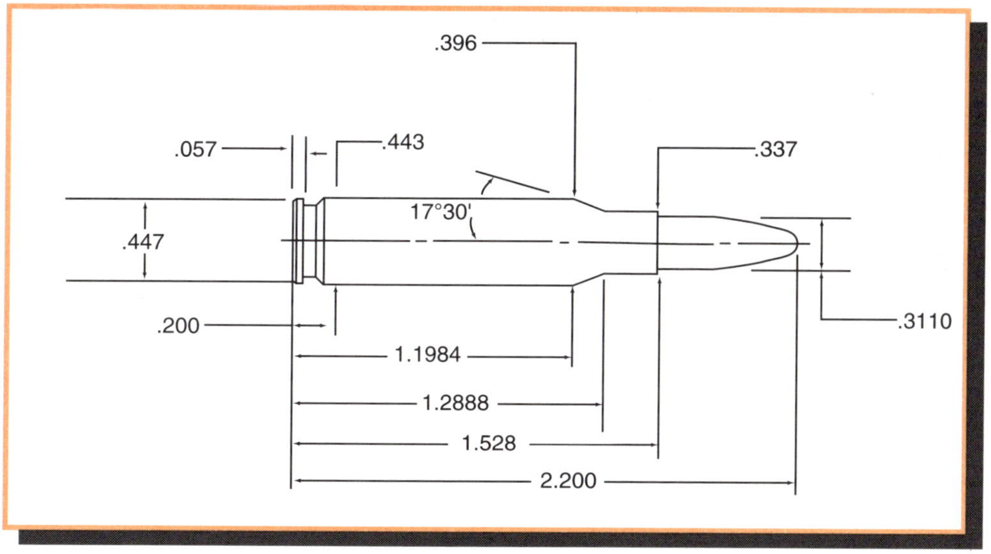

Comments:

The 7.62x39mm cartridge originated in the SKS rifle in the Soviet Union in the midst of World War Two. Some sources state that it was based on the German 7.92x33mm *Kurz* cartridge but it may have preceded the German development by a year or so. A fair number of Soviet pattern rifles including the SKS have been imported in recent years. Ruger offers the Mini-Thirty and bolt action rifles for this little cartridge. Ruger rifles utilize barrels with a .308" groove while those produced overseas use the standard .311". Lyman's die sets for the 7.62x39mm contain both size expander balls. Shooters are still advised to slug their bore to verify actual groove diameter before reloading.

Tons of surplus 7.62x39mm ammunition have been imported. Much of this is steel cased and/or berdan primed. Reloaders should properly discard these fired cases and use boxer primed brass cases for reloading purposes. Shooters should be advised that Winchester cases use Large Rifle primers whereas Remington cases use Small Rifle primers. Ballistically, this cartridge is usually lumped in with the 30-30. However, the 30-30 can be loaded with a wider range of bullet weights. Bullet selection is limited to the 110 to 130-grain range due to this cartridge's ballistic levels. Most semiautomatics chambered for this cartridge are doing pretty well if the shooter achieves three-inch groups at 100 yards. H-335 produced the best results in our lab tests. Listed starting loads for cast bullets may not function the action of semiautomatics.

Test Components:

Cases . . . . .Winchester
Trim-to Length . . . . .1.518"
Primers . . . . .Winchester WLR
Primer Size . . . . .Large Rifle
Lyman Shell Holder . . . . .No. 3
Jacketed Bullets Used . . . . . .Sierra HP #2110, 110 gr.
Hornady SP #3140, 123 gr. (.310")
Hornady SP #3020, 130 gr.

Cast Bullets Used . . . .(sized to .310" dia)
*gas check bullet — *#311359, 115 gr.
#311410, 130 gr.

Test Specifications:
(Velocity & Pressure)

Firearm Used . . . . .Universal Receiver
Barrel Length . . . . .20"
Twist . . . . .1-9½"
Groove Dia. . . . . .310"

110 gr. Jacketed HP 2.200" OAL — **BC:** .188 **SD:** .166

Powder	Sugg Starting Grains	Velocity fps	Pressure C.U.P.	Max Load Grains	Velocity fps	Pressure C.U.P.
IMR-4227	**21.0**	**2115**	**31,100**	23.0	2353	42,300
N130	26.5	2234	29,700	29.6	2468	39,900
AA1680	24.0	2245	29,400	26.7	2533	42,800
IMR-4198	22.6	2115	28,700	25.2	2416	40,200
RX7	26.5	2313	32,300	29.5	2558	43,400

123gr. Jacketed SP 2.200" OAL — **BC:** .252 **SD:** .183

Powder	Sugg Starting Grains	Velocity fps	Pressure C.U.P.	Max Load Grains	Velocity fps	Pressure C.U.P.
IMR-4227	20.0	1995	32,700	22.0	2204	43,000
N130	25.5	2167	35,400	28.6	2391	44,300
AA1680	23.0	2108	32,800	25.7	2412	44,600
IMR-4198	22.0	2064	33,800	24.6	2307	44,000
RX7	24.0	2116	32,100	26.8	2376	44,700
H-335	**30.5**	**2085**	**27,400**	34.5+	2363	43,300

Note: Loads shown in shaded panels are maximum.
Loads shown in bold designate potentially most accurate load.
\+ Designates a compressed powder charge.

7.62 x 39 Russian

130 gr. Jacketed SP
2.200" OAL

BC: .295
SD: .196

Powder	Sugg Starting Grains	Velocity fps	Pressure C.U.P.	Max Load Grains	Velocity fps	Pressure C.U.P.
IMR-4227	19.0	1936	30,800	21.5	2172	44,700
N130	25.0	2072	30,700	27.5+	2282	39,100
AA1680	22.0	1962	25,500	23.8	2259	40,100
IMR-4198	20.5	1932	28,200	23.4	2275	43,400
RX7	24.0	2086	29,800	26.0	2340	42,600
H-335	**29.5**	**2030**	**29,400**	33.2+	2239	36,800

#311359
115 gr. (#2 Alloy) 2.060" OAL

BC: .181
SD: .171

Powder	Sugg Starting Grains	Velocity fps	Pressure C.U.P.	Max Load Grains	Velocity fps	Pressure C.U.P.
Unique	8.0	1594	25,500	10.0	1816	36,500
SR-4759	16.7	1978	29,000	18.7	2186	40,200
IMR-4227	17.0	1906	25,800	19.2	2142	35,700
XMP-5744	**17.0**	**1744**	**22,100**	21.0	2107	42,600
IMR-4198	19.8	1968	25,800	22.6	2273	39,000
RX7	25.0	2189	28,100	30.0+	2561	41,100
AA2230	26.0	2092	32,000	32.0+	2445	44,200

#311410
130 gr. (#2 Alloy) 2.200" OAL

BC: .239
SD: .193

Powder	Sugg Starting Grains	Velocity fps	Pressure C.U.P.	Max Load Grains	Velocity fps	Pressure C.U.P.
Unique	8.0	1552	23,500	10.3	1786	36,900
SR-4759	16.5	1935	29,900	19.0	2139	37,600
IMR-4227	17.5	1883	25,800	19.5	2097	35,900
XMP-5744	**18.5**	**1804**	**22,700**	23.0	2175	43,400
IMR-4198	20.0	1897	25,500	22.0	2134	34,200
RX7	25.0	2164	30,600	29.0+	2442	38,800
AA-2230	25.0	2015	30,700	30.0+	2280	39,600

Note: Loads shown in shaded panels are maximum.
Loads shown in bold designate potentially most accurate load.
+ Designates a compressed powder charge.

7.62 x 54R Russian

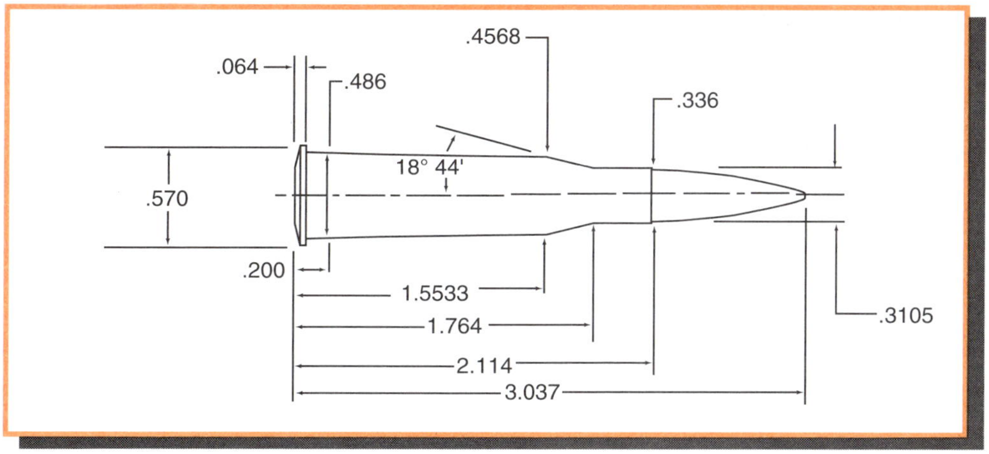

Comments:

This is currently the longest-lived military cartridge around. It was initially adopted by Imperial Russia in 1891 and continues to serve in the post-Soviet era as a light machine gun cartridge. It was the Russian mainstay of both World Wars and saw wide distribution among communist-bloc countries. Original ballistics yielded a 210-grain bullet traveling approximately 2,000 feet per second. Following the trend of the world's militaries, it was upgraded to a 150-grain bullet at 2,600 feet per second in 1909. American shooters first became familiar with the cartridge after World War One. Several U.S. companies produced Mosin-Nagant rifles under contract for the Czar but withheld delivery following the Bolshevik Revolution of 1917. Many of these brand new rifles chambered for the Russian cartridge subsequently turned up on the U.S. market. Many Mosin-Nagant rifles have recently been imported from various countries in condition ranging from mint to truly awful. Shooters should have their rifle examined by a qualified gunsmith before shooting. Winchester also produced the Model 95 in 7.62x54R for the Russian government, and these occasionally turn up.

Surplus ammunition is usually abundant but more often than not it is corrosive and berdan primed. Bullet diameter is nominally .310", however, large variations in groove diameters are the norm here. Our test data utilized .311" and .312" diameter bullets, as the dimensions in many rifles chambered for this caliber tend to run large. Some rifles will shoot .308" bullets just fine. Most of the bullets listed in our data are available from their manufacturers in .308" diameter should the shooter have a rifle with a smaller size bore. Use of .308" diameter bullets will require use of the smaller 308 caliber expanding button in the full-length resizing die for proper bullet tension. This can be an extremely accurate cartridge in a good rifle. It's been used throughout Scandinavia as a big-game cartridge and has been necked up and down to just about every conceivable bullet size in Finland. Norma and Lapua both offer boxer primed cases. Hodgdon's Varget and IMR-4350 provided the best ballistic uniformity with jacketed bullets. Data is provided for cast bullet #314299 but shooters can substitute cast bullet #311299 if the bore measures small.

Test Components:

Cases . . . . .Norma
Trim-to Length . . . . .2.105"
Primers . . . . .Federal 210
Primer Size . . . . .Large Rifle
Lyman Shell Holder .No. 17
Jacketed Bullets Used . . . .
Sierra SPT #2305, 125 gr. (.311")
Hornady SP # 3120, 150 gr. (.312")
Hornady RN #3130, 174 gr. (.312")
Sierra SPT #2310, 180 gr. (.311")
Cast Bullets Used . . . .(sized to .313" dia)
*gas check bullet *#314299, 200 gr.

Test Specifications:
(Velocity Only)

Firearm Used . . . . .SAKO Mosin-Nagant
Barrel Length . . . . .26"
Twist . . . . .1-9.5"
Groove Dia. . . . . .313"

125 gr. Jacketed SPT 2.800" OAL — **BC:** .274 **SD:** .185

Powder	Sugg Starting Grains	Velocity fps	Pressure	Max Load Grains	Velocity fps	Pressure
XMR-2015	41.5	2647	—	46.0	2894	—
IMR-4895	43.5	2517	—	48.0	2816	—
Varget	45.5	2677	—	**50.5**	**2994**	—
N140	45.0	2493	—	50.0	2824	—
RX15	46.0	2651	—	51.0	2973	—

150 gr. Jacketed SP 2.870" OAL — **BC:** .361 **SD:** .220

Powder	Sugg Starting Grains	Velocity fps	Pressure	Max Load Grains	Velocity fps	Pressure
XMR-2015	40.0	2456	—	44.5	2683	—
IMR-4895	42.5	2438	—	47.5	2748	—
Varget	43.0	2476	—	48.0	2772	—
N140	43.5	2395	—	48.5	2715	—
RX15	**43.5**	**2477**	—	48.5	2744	—

Note: Loads shown in shaded panels are maximum.
Loads shown in bold designate potentially most accurate load.

7.62 x 54R Russian

174 gr. Jacketed RN
2.815" OAL

BC: .262
SD: .255

Powder	Sugg Starting Grains	Velocity fps	Pressure	Max Load Grains	Velocity fps	Pressure
IMR-4895	41.0	2233	—	45.5	2496	—
XMR-2495	41.5	2300	—	46.0	2469	—
Varget	**42.0**	**2317**	—	46.5	2586	—
N140	42.0	2243	—	46.5	2484	—
RX15	42.0	2312	—	47.0	2540	—
N150	43.0	2287	—	48.0	2562	—
IMR-4350	47.5	2196	—	**52.5**	**2515**	—

180 gr. Jacketed SPT
2.975" OAL

BC: .411
SD: .266

Powder	Sugg Starting Grains	Velocity fps	Pressure	Max Load Grains	Velocity fps	Pressure
IMR-4895	41.0	2115	—	45.5	2488	—
XMR-2495	41.5	2314	—	46.0	2428	—
Varget	42.0	2261	—	**46.5**	**2542**	—
N140	42.0	2221	—	46.5	2471	—
RX15	42.0	2206	—	47.0	2499	—
N150	43.0	2238	—	48.0	2512	—
IMR-4350	**47.5**	**2160**	—	52.0	2504	—

#314299
200 gr. (#2 Alloy) 2.915" OAL

BC: .377
SD: .292

Powder	Sugg Starting Grains	Velocity fps	Pressure	Max Load Grains	Velocity fps	Pressure
2400	16.0	1456	—	22.0	1779	—
SR-4759	17.5	1524	—	29.0	2139	—
IMR-4227	20.0	1537	—	30.0	2053	—
XMP-5744	20.0	1513	—	**32.5**	**2153**	—
IMR-4198	20.0	1497	—	32.5	2108	—
N133	24.5	1533	—	34.0	2012	—
RX7	22.5	1525	—	35.0	2143	—

Note: Loads shown in shaded panels are maximum.
Loads shown in bold designate potentially most accurate load.

7.65 Argentine Mauser (7.65 x 53mm)

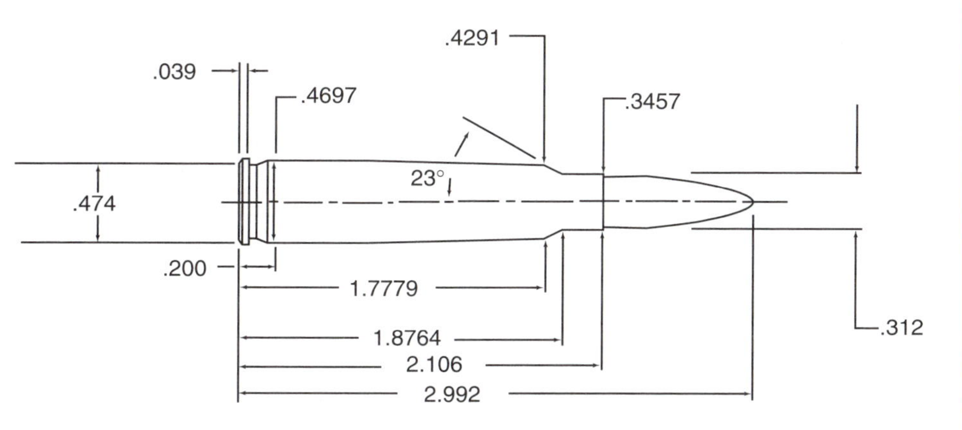

Comments:

The 7.65x53mm cartridge led a long, if not particularly illustrious, military career. This was Paul Mauser's first smokeless military cartridge. Its original incarnation in 1889 propelled a 215-grain round nose bullet at approximately 2,000 feet per second. However, this cartridge was not immune to the advances in ballistic technology at the turn of the twentieth century. Military loadings soon included a 182-grain bullet at 2,590 feet per second, and a 154-grain bullet at 2,750. Many rifles chambered for this caliber were sold throughout Latin America. Consequently, large numbers of foreign military surplus made their way to American shores over the years. The 7.65x53 features a shorter neck than typically found on other Mauser cartridges. This caliber never achieved the popularity with American shooters as did its 7x57 and 7.9x57 cousins. The 6.5x55 Swede has recently eclipsed its appeal on the surplus market. Remington and Winchester both offered rifles chambered in 7.65x53 and factory ammunition for several years between the World Wars. Norma still offers a soft-point factory load. Surplus ammunition is fairly common but it will be berdan primed and is often corrosive.

Ballistics of this Mauser cartridge are similar to the 308 Winchester. Powders suitable for the 308 will work here. The handloader also has a decent, if not overly broad selection of .311" and .312" diameter bullets to work with. There is no SAAMI pressure limit for this cartridge. Some shooters use reformed 30-06 cases so common sense is imperative when working up loads. Many of the recently imported rifles are in excellent condition and make good shooters. Those with an original Model 1889 should not use the listed maximum loads for jacketed bullets. Although well made, they do not have the inherent strength of later rifles built upon the stronger Model 98 action. Some Mauser rifles from South America have mismatched bolts. Like any surplus rifle, examination by a qualified gunsmith is in order. Shooters loading cast bullets should slug their bore and size bullets accordingly. Cast bullet #311291 may be too small if the rifle's groove diameter exceeds .312" If the bore exceeds .312" diameter, cast bullet #314299 can be substituted for #311299. XMP-5744 and SR-4759 provided the best results in cast bullet testing.

Test Components:

Cases . . . . .Norma
Trim-to Length . . . . .2.100"
Primers . . . . .Remington 9½
Primer Size . . . . .Large Rifle
Lyman Shell Holder . . . . .No. 2
Jacketed Bullets Used . . . .
Sierra SPT #2305, 125 gr. (.311")
Speer SP #2217, 150 gr. (.311")
Hornady RN #3130, 174 gr. (.312")
Sierra SPT #2310, 180 gr. (.311")
Cast Bullets Used . . . .(sized to .311" dia)
*gas check bullet *#311291, 170 gr.
*#311299, 200 gr.

Test Specifications: (Velocity Only)

Firearm Used . . . . .Argentine Mauser M1909
Barrel Length . . . . .29"
Twist . . . . .1-9.8"
Groove Dia. . . . . .311"

Note: Loads shown in shaded panels are maximum.
Loads shown in bold designate potentially most accurate load.

7.65 Argentine Mauser (7.65 x 53mm)

125 gr. Jacketed SPT
2.740" OAL
BC: .274 **SD:** .185

Powder	Sugg Starting Grains	Velocity fps	Pressure	Max Load Grains	Velocity fps	Pressure
IMR-3031	39.5	2527	—	44.0	2893	—
H-4895	41.5	2625	—	46.0	3001	—
748	42.0	2542	—	46.5	2934	—
N135	40.0	2632	—	44.5	2961	—
IMR-4064	43.0	2602	—	47.8	2983	—
Varget	42.5	2600	—	47.3	3005	—
IMR-4320	42.5	2618	—	**47.5**	**2943**	—
N140	43.5	2694	—	48.0	3017	—

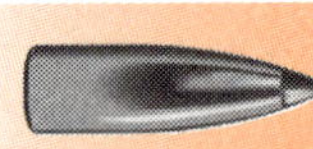

150 gr. Jacketed SP
2.815" OAL
BC: .411 **SD:** .222

Powder	Sugg Starting Grains	Velocity fps	Pressure	Max Load Grains	Velocity fps	Pressure
IMR-3031	38.0	2388	—	**42.5**	**2680**	—
H-4895	39.0	2375	—	43.5	2683	—
IMR-4064	40.5	2402	—	45.0	2715	—
Varget	40.0	2406	—	44.5	2707	—
IMR-4320	41.0	2429	—	45.5	2750	—
N140	40.5	2456	—	45.2	2721	—
RX15	41.0	2497	—	45.5	2740	—
760	44.5	2425	—	49.0	2635	—

174 gr. Jacketed RN
2.820" OAL
BC: .262 **SD:** .255

Powder	Sugg Starting Grains	Velocity fps	Pressure	Max Load Grains	Velocity fps	Pressure
IMR-4895	38.5	2255	—	42.7	2514	—
IMR-4064	39.5	2250	—	43.8	2510	—
Varget	39.5	2329	—	43.7	2576	—
IMR-4320	40.0	2314	—	44.5	2573	—
N140	39.0	2285	—	43.4	2519	—
RX15	**39.5**	**2333**	—	44.0	2572	—
760	43.0	2243	—	47.5	2495	—

180 gr. Jacketed SPT
2.875" OAL
BC: .411 **SD:** .266

Powder	Sugg Starting Grains	Velocity fps	Pressure	Max Load Grains	Velocity fps	Pressure
IMR-4895	38.0	2250	—	42.5	2545	—
IMR-4064	39.0	2269	—	43.5	2525	—
Varget	**39.0**	**2293**	—	43.5	2554	—
IMR-4320	40.0	2348	—	44.5	2592	—
N140	38.5	2270	—	43.0	2504	—
RX15	39.5	2330	—	44.0	2587	—
760	42.5	2245	—	47.0	2540	—

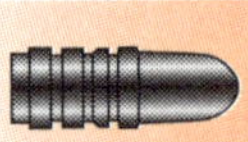

#311291
170 gr. (#2 Alloy) 2.820" OAL
BC: .202 **SD:** .249

Powder	Sugg Starting Grains	Velocity fps	Pressure	Max Load Grains	Velocity fps	Pressure
2400	15.0	1507	—	19.0	1787	—
SR-4759	15.5	1514	—	**22.0**	**1972**	—
IMR-4227	18.0	1547	—	23.5	1937	—
XMP-5744	18.0	1520	—	25.5	1983	—
IMR-4198	18.0	1528	—	25.5	1957	—
N133	24.0	1658	—	28.0	1909	—
RX7	19.0	1562	—	27.0	2004	—

#311299
200 gr. (#2 Alloy) 3.000" OAL
BC: .377 **SD:** .299

Powder	Sugg Starting Grains	Velocity fps	Pressure	Max Load Grains	Velocity fps	Pressure
2400	15.5	1474	—	19.5	1713	—
SR-4759	15.5	1448	—	22.0	1888	—
IMR-4227	19.0	1571	—	25.0	1923	—
XMP-5744	17.5	1444	—	**25.0**	**1920**	—
IMR-4198	18.5	1482	—	26.0	1916	—
N133	25.0	1515	—	29.5	1878	—
RX7	23.0	1606	—	27.0	1908	—

Note: Loads shown in shaded panels are maximum.
Loads shown in bold designate potentially most accurate load.

303 British

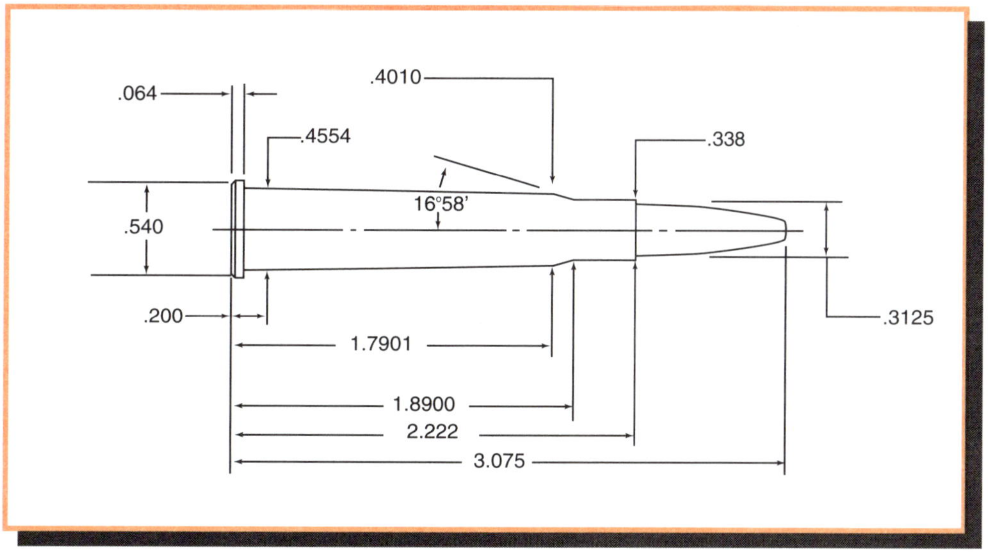

Comments:

The 303 British is another cartridge that owes its popularity to inexpensive, surplus rifles. This round should not be confused with the long obsolete 303 Savage. The 303 has seen extensive use as both a hunting and military cartridge around the world, especially in the vast reaches of the former British Empire. Great Britain adopted the 303 in 1888. Unlike most other small bore cartridges developed at this time, the 303 was originally loaded with compressed black powder and a 215-grain bullet around 2,000 feet per second. This loading changed to the smokeless Cordite propellant just four years later. The cartridge was altered again in 1910 with a 174-grain flat-base spitzer traveling approximately 2,550 feet per second. The newly designated MK VII cartridge served until its replacement by the 7.62 NATO in 1957.

The majority of 303 rifles encountered are of the Short Magazine Lee Enfield design (SMLE). Their fast-working action and 10-round magazine made them a formidable battle rifle throughout both World Wars. Unfortunately, "Enfields" as they are known are not an overly strong design. The rear locking lugs on these rifles do not provide the strong lockup as the front locking lugs found on firearms such as the Mauser 98 or Model 70 Winchester. The SAAMI Maximum Average Pressure (MAP) is set at 45,000 CUP in deference to this fact. Most major manufacturers currently offer factory ammunition with the 180-grain bullet being the most common. Surplus ammunition is quite plentiful but is usually berdan primed and corrosive. Ballistics of the 303 British are generally regarded as a step above the 30-40 Krag. Careful handloading increases the flexibility of the 303 as long as the shooter follows certain cautions. Persistent use of maximum loads in SMLE rifles will reduce case life expectancy considerably. Reloaders should pay close attention to the length of resized cases and trim when necessary. The generous chamber dimensions encountered in many SMLE rifles only compound this situation. Shooters should pay close attention for incipient case separation when sizing and trimming.

Enfields have been a popular surplus item over the years. Their production span covered over fifty years and four continents. Condition of these rifles vary from unissued to unusable and everywhere in between. Rifles should be checked over by a qualified gunsmith before firing. Rifles chambered in 303 sometimes contain extreme groove variation. While most rifles usually run around .312" or .313", some have measured as low as .308"and others as high as .317" Those reloading for an undersized bore can use cast bullet #311299 in place of #314299. Slug the bore and size accordingly. Varget and Reloder 15 produced very good accuracy with the heavier jacketed bullets in our test rifle. Lighter bullets responded to IMR-3031 but individual rifles may have different preferences. SR-4759 and XMP-5744 gave best results during cast bullet testing.

Test Components:

Cases . . . . .Remington
Trim-to Length . . . . .2.212"
Primers . . . . .Federal 210
Primer Size . . . . .Large Rifle
Lyman Shell Holder . . . . .No. 7
Jacketed Bullets Used . . . .
Sierra SPT #2305, 125 gr. (.311")
Hornady SP #3120, 150 gr. (.312")
Hornady RN #3130, 174 gr. (.312")
Sierra SPT #2310, 180 gr. (.311")
Cast Bullets Used . . . .(sized to .313" dia)
*gas check bullet *#314299, 200 gr.

Test Specifications:
(Velocity Only)

Firearm Used . . . . .Enfield #1 MK III*
Barrel Length . . . . .25¼"
Twist . . . . .1-10"
Groove Dia. . . . . .313"

303 British

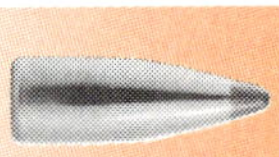

125 gr. Jacketed SPT
2.900" OAL

BC: .274
SD: .185

Powder	Sugg Starting Grains	Velocity fps	Pressure	Max Load Grains	Velocity fps	Pressure
IMR-3031	38.5	2305	—	**42.8**	**2720**	—
H4895	39.0	2322	—	43.5	2713	—
748	41.0	2253	—	45.3	2709	—
N135	39.0	2350	—	43.5	2714	—
IMR-4064	41.5	2343	—	46.0+	2709	—
N140	41.0	2410	—	45.5	2712	—
RX15	41.5	2469	—	46.3	2784	—

150 gr. Jacketed SP
2.975" OAL

BC: .361
SD: .220

Powder	Sugg Starting Grains	Velocity fps	Pressure	Max Load Grains	Velocity fps	Pressure
IMR-3031	37.0	2236	—	**41.5**	**2605**	—
H4895	38.0	2254	—	42.0	2572	—
748	39.0	2149	—	43.5	2523	—
IMR-4064	39.0	2212	—	44.0	2570	—
IMR-4320	39.5	2298	—	44.0	2604	—
N140	39.0	2173	—	43.3	2503	—
RX15	40.0	2335	—	44.5	2580	—
AA2700	43.0	2100	—	48.0	2432	—
H414	43.0	2211	—	48.0	2510	—

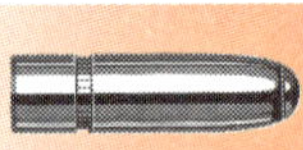

174 gr. Jacketed RN
2.945" OAL

BC: .262
SD: .255

Powder	Sugg Starting Grains	Velocity fps	Pressure	Max Load Grains	Velocity fps	Pressure
IMR-4895	37.0	2032	—	41.0	2341	—
IMR-4064	37.0	1981	—	41.5	2308	—
Varget	38.0	1993	—	42.0	2333	—
IMR-4320	38.0	2043	—	42.0	2361	—
N140	37.5	2021	—	41.8	2313	—
RX15	38.0	2107	—	**42.0**	**2367**	—
AA2700	41.0	1945	—	45.5	2233	—
H414	41.5	1978	—	46.3	2297	—

180 gr. Jacketed SPT
3.075" OAL

BC: .411
SD: .266

Powder	Sugg Starting Grains	Velocity fps	Pressure	Max Load Grains	Velocity fps	Pressure
IMR-4895	36.5	2020	—	40.5	2286	—
IMR-4064	37.5	2040	—	41.5	2307	—
Varget	37.5	1987	—	**41.7**	**2306**	—
IMR-4320	37.5	2046	—	41.5	2309	—
N140	37.5	2067	—	41.7	2317	—
RX15	38.0	2125	—	**42.0**	**2376**	—
AA2700	41.5	1994	—	45.5	2239	—
H414	41.5	2008	—	46.0	2263	—

#314299
200 gr. (#2 Alloy) 2.930" OAL

BC: .377
SD: .292

Powder	Sugg Starting Grains	Velocity fps	Pressure	Max Load Grains	Velocity fps	Pressure
Unique	11.0	1310	—	14.5	1549	—
2400	15.5	1404	—	20.0	1713	—
SR-4759	**18.0**	**1561**	—	24.5	1953	—
IMR-4227	20.0	1555	—	26.5	1953	—
XMP-5744	21.0	1566	—	27.5	1942	—
IMR-4198	21.0	1565	—	28.0	1946	—

Note: Loads shown in shaded panels are maximum.
Loads shown in bold designate potentially most accurate load.
\+ Designates a compressed powder charge.

32-20 Winchester (32 WCF)

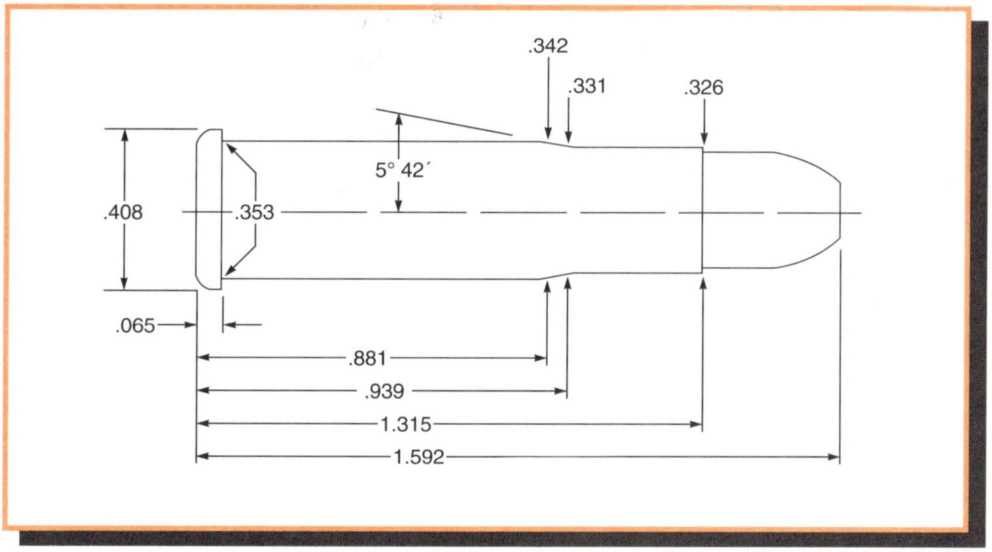

Comments:

The 32-20 dates back to 1882 when Winchester first offered it in the Model 1873 rifle. The cartridge has drifted in and out of popularity depending on who chambered it at any given time. Most manufacturers produced rifles in 32-20 at some point. It has been periodically produced in revolvers and more lately in the T/C Contender. Cowboy Action Shooting and Marlin's recent production of their Model 1894 in 32-20 have kept it going. The 32-20 also served as the parent cartridge for the 25-20 and the 218 Bee.

Factory ammunition with cases under the trim-to length has been encountered. This will cause no problem as long as they are segregated and the seating die adjusted to their particular length. Cartridges intended for tubular magazines should only be loaded with flat or blunt nosed bullets and crimped in place. Both bullets listed in data section are suitable for tubular magazines and are designed to perform at 32-20 velocity levels. Shooters need to be careful when seating bullets in the 32-20. Case neck walls are very thin, only .006" to .007" on average. An improperly adjusted seating die can lead to crushed cases. Seating and crimping in two separate steps will eliminate much of this aggravation. SAAMI established the Maximum Average Pressure (MAP) for the 32-20 at 16,000 CUP.

This data is not for use in older firearms originally designed for black powder or in handguns. Use only in modern firearms in good condition.

Test Components:

Cases . . . . .Winchester
Trim-to Length . . . . .1.305"
Primers . . . . .Winchester WSR
Primer Size . . . . .Small Rifle
Lyman Shell Holder . . . . .No. 10
Jacketed Bullets Used . . . . .Speer JHP #3981, 100 gr.
Cast Bullets Used . . . .(sized to .312" dia) #311008, 115 gr.

Test Specifications: (Velocity & Pressure)

Firearm Used . . . . .Universal Receiver
Barrel Length . . . . .14"
Twist . . . . .1-20"
Groove Dia. . . . . .311"

100 gr. Jacketed HP
1.560" OAL
BC: .167 **SD:** .147

Powder	Sugg Starting Grains	Velocity fps	Pressure C.U.P.	Max Load Grains	Velocity fps	Pressure C.U.P.
2400	6.8	1029	9,700	8.5	1280	15,800
IMR-4227	7.8	961	10,200	9.8	1181	15,200
AA1680	11.5	1154	8,000	12.5	1295	14,200
IMR-4198	**11.0**	**1122**	**10,100**	13.6+	1388	16,000
RX7	13.0	1086	6,700	16.3	1508	15,500

#311008
115 gr. (#2 Alloy) 1.590" OAL
BC: .154 **SD:** .169

Powder	Sugg Starting Grains	Velocity fps	Pressure C.U.P.	Max Load Grains	Velocity fps	Pressure C.U.P.
Unique	3.6	943	10,400	4.5	1117	16,000
2400	**6.6**	**1043**	**10,900**	8.3	1239	15,000
IMR-4227	7.8	972	11,400	9.7	1171	15,800
AA1680	11.0	1083	9,300	13.0	1332	15,600
IMR-4198	10.0	1019	8,900	12.4	1272	15,300
RX7	12.4	1038	6,500	15.5	1405	16,000

Note: Loads shown in shaded panels are maximum.
Loads shown in bold designate potentially most accurate load.
\+ Designates a compressed powder charge.

32 Winchester Special

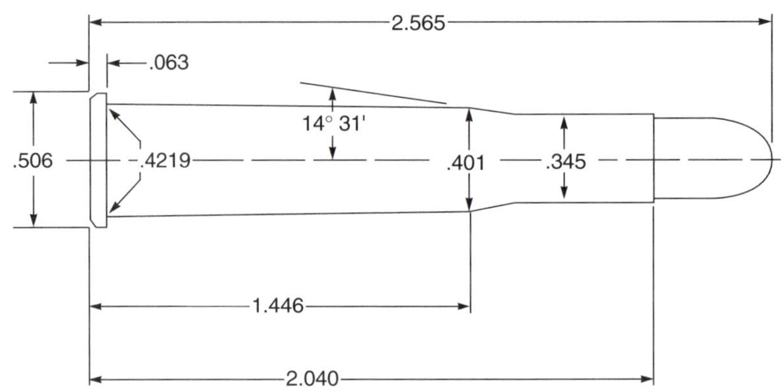

Comments:

The 32 Winchester Special is little more than the 30-30 necked up to .321" diameter bullets. It should not be confused with the 32 WCF (a.k.a. 32-20). The history of the 32 Special has been the subject of debate for some time. Smokeless powder was not yet widely available to reloaders in the late nineteenth century. Some sources argue the 32 Special was equipped with its 1-16" twist to enable factory cartridges to be reloaded with black powder. Others speculate the wide availability of 32-caliber rifle bullets in contrast to the then new 30-caliber led to these two very similar cartridges. Nevertheless, the 32 Special never achieved the universal acceptance enjoyed by the 30-30 and 32-caliber rifles gradually lost ground to 30-caliber cartridges. Winchester discontinued the 32 Special chambering many years ago but the cartridge still has its followers. Winchester and Federal both currently list a single 170-grain factory load. Shooters should carefully check all case lengths for uniformity before reloading. Any procedure applicable to loading the 30-30 will apply here.

Test Components:

Cases . . . . .Remington
Trim-to Length . . . . .2.030"
Primers . . . . .Remington 9½
Primer Size . . . . .Large Rifle
Lyman Shell Holder . . . . .No. 6
Jacketed Bullets Used . . . .Hornady FP #3210, 170 gr.

Test Specifications: (Velocity Only)

Firearm Used . . . . .Winchester Model 94
Barrel Length . . . . .20"
Twist . . . . .1-16"
Groove Dia. . . . . .321"

170 gr. Jacketed FP
2.605" OAL
BC: .249
SD: .236

Powder	Sugg Starting Grains	Velocity fps	Pressure	Max Load Grains	Velocity fps	Pressure
IMR-4198	24.0	1956	—	27.0	2150	—
RX7	**28.0**	**2123**	—	31.0	2283	—
IMR-3031	28.0	1912	—	31.0	2159	—
IMR-4895	28.0	1792	—	33.0	2136	—
IMR-4064	28.0	1685	—	33.0	2070	—
IMR-4320	31.0	1908	—	35.0	2169	—
H-380	36.0	1945	—	39.0	2109	—

Note: Loads shown in shaded panels are maximum.
Loads shown in bold designate potentially most accurate load.

8mm Mauser (8 x 57mm JS) (7.9 x 57mm)

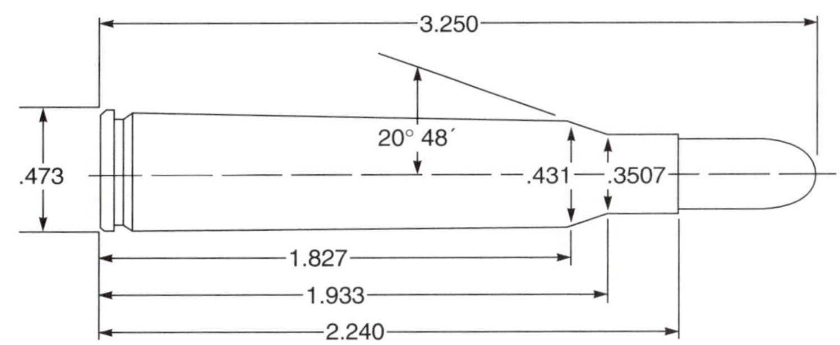

Comments:

This data is for use in rifles chambered only for the 8x57mm JS (a.k.a. 7.9x57 or 7.92x57) German military cartridge with the .323" diameter bullet. It is not for use in any rifle with the older .318" J-bore. The 8mm Mauser has caused no end of confusion to American shooters. This cartridge had its birth not from *Waffenfabrik Mauser* but from the Model 88 Commission rifle adopted by the German Infantry Board in 1888. This cartridge was originally loaded with a .318" diameter, 226-grain round nose bullet at 2100 feet per second and replaced the 11mm black powder cartridge in use since the 1870s. The German military revised it in 1905 with a .323" diameter bullet weighing 154-grains at nearly 2900 feet per second. This revolutionized small arms ballistics and prodded the United States Army to revise the just adopted 30-03 cartridge into what we now know as the 30-06. Military ordnance departments around the world scrambled to follow suit. All German military rifles manufactured since 1905 have the .323" diameter JS-bore. However, German gunmakers continued to produce .318" J-bore sporting rifles for the civil market in the post-World War One era due to the Versailles Treaty's restrictions of military arms production by German companies. Moreover, many Model 88 rifles reportedly were rechambered to accept the new .323" JS cartridge but retained the original .318" barrel — confused yet?

American shooters first became acquainted with the 8x57 immediately following World War Two when many GIs returned home bearing newly liberated K98 rifles. The lack of suitable boxer primed cases led to the development of the 8mm-06 wildcat, which remained popular for many years. Large numbers of surplus rifles produced by various manufacturers have also been imported even up to the present day. Surplus ammunition is still common but is usually berdan primed and corrosive. Eventually, the major suppliers produced 8x57 ammunition and reloading components, which are readily available. The 8x57 is often compared with the 30-06. However, SAAMI established a Maximum Average Pressure of 35,000 PSI for the 8x57 JS-bore in light of the real possibility of .323" cartridges being inadvertently chambered in .318" J-bore rifles. Ammunition offered by European manufacturers not confused by such designations has been loaded to the full ballistic potential that this cartridge is capable of.

Variations of the basic Model 98 Mauser were produced in very large quantities in different locales over a nearly fifty-year period. Quality of materials and workmanship will vary. Some are pristine while others show the rough finish of wartime production. Still others were produced by forced labor or under occupation. Shooters can encounter some specimens of dubious origin, which have manufacturers codes, or other identification marks removed. However, those with an 8x57 Mauser given a clean bill of health by a qualified gunsmith can get the most out of their rifles by careful handloading. Bullet selection is not nearly as extensive as for 30-caliber or even the 7mm but handloaders should be able to find a bullet to suit their needs. The 220-grain Hornady bullet was designed for the 8mm Remington Magnum and may not expand well at the 8x57's performance level. IMR-4895, IMR-4350 and IMR-4064 have all been used to good effect in the 8x57. Reloder 7, SR-4759, and XMP-5744 are good powder choices for cast bullets.

Test Components:

Cases . . . . Winchester and Federal
Trim-to Length . . . . 2.230"
Primers . . . . Remington 9½
and Federal 210
Primer Size . . . . Large Rifle
Lyman Shell Holder . . . . No. 2
Jacketed Bullets Used . . . . Hornady SP #3230, 125 gr.
Speer SP #2277, 150 gr.
Hornady RN #3235, 170 gr.
Hornady SP #3238, 220 gr.
Cast Bullets Used . . . . (sized to .323" dia)
*gas check bullet *#323470, 165 gr.

Test Specifications: (Velocity & Pressure)

Firearm Used . . . . Mauser 98
and Universal Receiver
Barrel Length . . . . Mauser 98 - 23"
Universal Receiver - 26"
Twist . . . . Mauser 98 - 1-9¼"
Universal Receiver - 1-9½"
Groove Dia. . . . . .323"

8mm Mauser (8 x 57mm JS) (7.9 x 57mm)

125 gr. Jacketed SP
2.875" OAL
BC: .246
SD: .171

Powder	Sugg Starting Grains	Velocity fps	Pressure	Max Load Grains	Velocity fps	Pressure
IMR-3031	46.0	2865	—	51.0+	3184	—
IMR-4895	50.0	2898	—	53.0+	3194	—
748	51.0	2813	32,800	58.0+	3063	35,100 C
N135	46.0	2793	35,200	**51.0**	**3059**	**46,600 P**
IMR-4064	49.0	2840	—	54.0+	3125	—
IMR-4320	50.0	2824	—	55.0+	3105	—
RX15	49.5	2811	37,300	55.0+	3098	48,100 P
H-380	50.5	2557	25,900	56.0+	2803	32,900 P
AA2700	52.0	2610	30,000	58.0+	2827	36,500 P
760	51.0	2493	27,500	58.0+	2739	32,400 C
H-414	50.0	2519	26,700	57.0+	2791	32,900 C
IMR-4350	50.0	2421	—	56.0+	2717	—

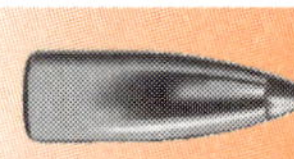

150 gr. Jacketed SP
2.945" OAL
BC: .369
SD: .205

Powder	Sugg Starting Grains	Velocity fps	Pressure	Max Load Grains	Velocity fps	Pressure
IMR-4895	44.0	2476	34,400	49.2	2795	47,200 P
748	47.0	2625	37,800	52.0	2858	47,400 P
IMR-4064	46.0	2522	35,600	51.0	2844	48,700 P
Varget	45.0	2562	37,000	50.0	2820	47,800 P
IMR-4320	48.0	2577	—	53.0+	2865	—
N140	45.5	2529	37,000	50.5	2771	46,600 P
RX15	45.0	2541	36,500	50.2	2816	48,100 P
H-380	53.0	2652	38,000	59.0+	2827	43,500 P
AA2700	52.0	2624	37,800	**58.0+**	**2818**	**44,400 P**
760	47.5	2316	28,200	54.0+	2605	36,800 C
H-414	47.5	2394	30,300	54.0+	2692	36,900 C
IMR-4350	50.0	2347	—	56.0+	2652	—

170 gr. Jacketed RN
2.850" OAL
BC: .217
SD: .233

Powder	Sugg Starting Grains	Velocity fps	Pressure	Max Load Grains	Velocity fps	Pressure
IMR-4895	42.5	2368	36,400	47.5	2633	48,500 P
748	40.5	2187	28,100	46.0	2437	34,400 C
IMR-4064	43.5	2313	33,600	48.5	2624	47,200 P
Varget	43.0	2415	38,000	48.0	2646	48,200 P
IMR-4320	46.0	2421	—	51.0+	2710	—
N140	43.0	2350	36,600	47.7	2586	47,500 P
RX15	43.5	2377	36,300	**48.5**	**2627**	**47,200 P**
N150	44.5	2430	39,100	49.5	2644	48,500 P
760	43.0	2100	28,100	49.0	2302	35,000 C
H-380	48.0	2371	34,500	53.0+	2594	43,300 P
IMR-4350	49.0	2237	—	54.0+	2518	—
XMR-3100	49.0	2053	29,000	54.5+	2323	39,000 P

220 gr. Jacketed SP
3.000" OAL
BC: .464
SD: .301

Powder	Sugg Starting Grains	Velocity fps	Pressure	Max Load Grains	Velocity fps	Pressure
IMR-4895	34.5	1841	31,800	39.0	2081	36,000 C
IMR-4064	35.0	1817	29,300	40.0	2087	35,800 C
IMR-4320	36.0	1867	35,200	41.0	2086	37,000 C
H-380	37.0	1810	27,000	42.0	2107	35,900 C
760	39.5	1798	27,200	45.0	2100	37,000 C
IMR-4350	39.5	1787	27,700	**45.0+**	**2056**	**35,800 C**
XMR-3100	44.0	1810	30,300	49.0+	2008	37,400 P
RX19	45.0	1970	35,300	50.0+	2178	42,800 P

#323470
165 gr. (#2 Alloy) 2.730" OAL
BC: .187
SD: .226

Powder	Sugg Starting Grains	Velocity fps	Pressure	Max Load Grains	Velocity fps	Pressure
SR-4759	20.5	1831	18,100	25.2	2139	30,200 C
XMP-5744	**19.0**	**1504**	**14,000**	34.0	2374	40,300 P
IMR-4198	27.0	1925	16,700	35.0	2482	34,900 C
RX7	28.0	1917	16,900	40.0	2527	33,700 C
H-4895	31.0	1957	18,600	37.5	2330	28,200 C

Note: Loads shown in shaded panels are maximum.
Loads shown in bold designate potentially most accurate load.
\+ Designates a compressed powder charge.
In pressure column, C=Copper Units of Pressure (CUP); P=Pounds per Square Inch (PSI)

8mm Remington Magnum

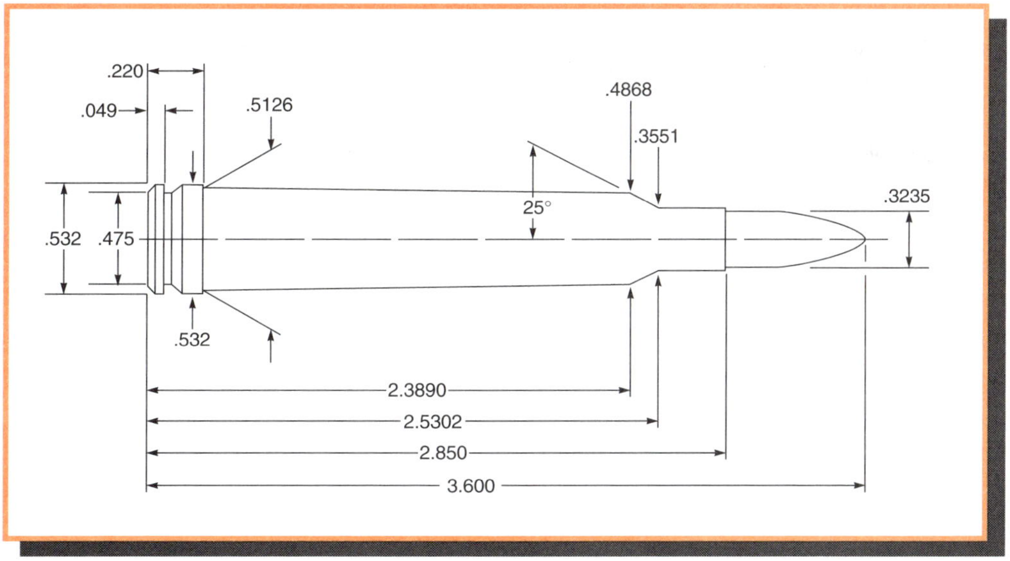

Comments:

This was Remington's answer to the 338 Winchester Magnum. With its larger case (a necked down 375 H&H) it offered greater potential; but nevertheless it never became popular. This may be due to the fact that its hunting use is limited to the heaviest North American game. Or perhaps as some found the original factory bullets were really too soft for use on heavy rugged game.

In any event when properly loaded with a good premium grade bullet the 8mm Remington is a cartridge that can handle almost any situation to be encountered in North America. And it will do nicely for much of the African plains game.

With standard bullets it will perform well in non-demanding situations.

This round is at its best with a good 220 grain bullet and loaded with Hercules Reloder 22.

Test Components:

Cases . . . . .Remington
Trim-to Length . . . . .2.840"
Primers . . . . .CCI 250
Primer Size . . . . .Large Rifle, Magnum
Lyman Shell Holder . . . . .No. 13
Jacketed Bullets Used . . . . . .Speer SP #2277, 150 gr.
Hornady RN #3235, 170 gr.
Sierra SPT #2410, 175 gr.
Nosler Ballistic Tip #32180, 180 gr.
Hornady SP #3238, 220 gr.
Cast Bullets Used . . . .(sized to .323" dia)
*gas check bullet *#323470, 165 gr.

Test Specifications: (Velocity & Pressure)

Firearm Used . . . . .Universal Receiver
Barrel Length . . . . .24"
Twist . . . . .1-10"
Groove Dia. . . . . .323"

150 gr. Jacketed SP
3.580" OAL
BC: .369
SD: .205

Powder	Sugg Starting Grains	Velocity fps	Pressure C.U.P.	Max Load Grains	Velocity fps	Pressure C.U.P.
IMR-4895	57.0	2821	35,200	64.0	3143	51,300
IMR-4320	57.5	2767	34,700	66.0	3126	52,300
H-380	65.0	2805	34,300	75.0	3198	53,600
760	71.0	3045	43,800	80.0	3294	52,900
H-414	68.0	3042	42,700	77.0	3272	51,700
IMR-4350	74.0	3102	41,800	**80.0**	**3348**	**50,700**
RX19	80.0	3070	41,600	89.0	3500	53,700
XMR-3100	81.0	3014	41,400	90.0+	3411	53,800
H-4831	80.0	3144	41,200	86.5	3382	50,000

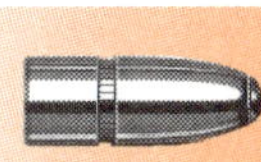

170 gr. Jacketed RN
3.440" OAL
BC: .217
SD: .233

Powder	Sugg Starting Grains	Velocity fps	Pressure C.U.P.	Max Load Grains	Velocity fps	Pressure C.U.P.
H-414	71.5	2955	44,300	78.0	3181	53,000
IMR-4350	70.0	2919	42,900	76.5	3177	52,000
RX19	77.0	2921	42,300	**85.5**	**3323**	**53,900**
XMR-3100	76.5	2806	37,700	85.7+	3206	53,200
H-4831	78.0	2979	43,000	84.5	3257	52,300
WXR	78.0	2854	42,400	86.5+	3241	53,900
RX22	81.5	3004	40,600	87.3+	3296	51,300
IMR-7828	83.0	2977	43,100	90.0+	3261	52,100
RX25	83.0	2822	39,000	92.0+	3227	50,000
AA8700	92.5	2908	41,400	102.4+	3155	50,200

Note: Loads shown in shaded panels are maximum.
Loads shown in bold designate potentially most accurate load.
\+ Designates a compressed powder charge.

8mm Remington Magnum

175 gr. Jacketed SPT
3.585" OAL

BC: .381
SD: .240

Powder	Sugg Starting Grains	Velocity fps	Pressure C.U.P.	Max Load Grains	Velocity fps	Pressure C.U.P.
H-414	69.0	2889	45,000	75.6	3099	52,400
IMR-4350	71.6	2871	41,100	77.5	3178	52,700
RX19	75.0	2852	42,100	**83.3+**	**3210**	**52,800**
XMR-3100	77.0	2851	41,000	83.5+	3159	52,800
H-4831	75.3	2843	42,100	82.0	3131	53,000
WXR	78.0	2805	40,300	86.0	3175	53,200
RX22	81.5	3034	42,500	87.2+	3305	52,200
IMR-7828	78.5	2812	39,100	86.3+	3216	52,500
RX25	84.0	2848	43,300	91.5+	3210	52,800
AA8700	92.5	2920	42,500	102.4+	3196	51,800

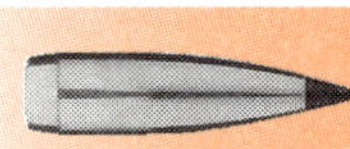

180 gr. Jacketed Ballistic Tip
3.595" OAL

BC: .394
SD: .247

Powder	Sugg Starting Grains	Velocity fps	Pressure C.U.P.	Max Load Grains	Velocity fps	Pressure C.U.P.
H-414	67.0	2706	42,300	74.0	3082	54,000
IMR-4350	70.0	2777	43,400	77.0	3164	54,000
RX19	74.0	2802	42,200	82.0	3157	53,600
XMR-3100	74.5	2721	41,100	82.5	3086	53,900
H-4831	73.0	2796	41,100	81.0	3196	53,600
WXR	77.0	2855	44,000	85.0	3197	54,000
RX22	78.0	2823	42,600	**86.0**	**3191**	**53,400**
IMR-7828	77.0	2749	42,100	85.0	3138	53,800
RX25	81.0	2869	44,100	90.0+	3246	54,000
H1000	83.0	2835	42,700	92.0+	3152	54,000
AA8700	88.0	2644	40,900	97.5+	2928	53,200

220 gr. Jacketed SP
3.597" OAL

BC: .464
SD: .301

Powder	Sugg Starting Grains	Velocity fps	Pressure C.U.P.	Max Load Grains	Velocity fps	Pressure C.U.P.
IMR-4350	65.4	2579	43,800	73.0	2819	52,800
RX19	70.0	2610	42,600	77.6	2917	53,800
XMR-3100	72.5	2600	42,800	79.0+	2860	53,600
H-4831	66.0	2473	40,700	74.0	2734	51,600
WXR	71.0	2606	43,700	79.0	2862	53,700
RX22	73.5	2673	40,600	**79.5+**	**2939**	**51,800**
IMR-7828	73.0	2596	40,300	80.0+	2888	50,800
RX25	75.0	2556	40,700	83.5+	2934	53,600
AA8700	87.0	2662	43,100	94.5+	2885	50,400
H870	84.5	2621	40,200	88.0+	2742	44,200

#323470
165 gr. (#2 Alloy) 3.398" OAL

BC: .187
SD: .226

Powder	Sugg Starting Grains	Velocity fps	Pressure C.U.P.	Max Load Grains	Velocity fps	Pressure C.U.P.
2400	24.0	1714	—	35.0	2313	—
SR-4759	**23.0**	**1712**	—	39.0	2469	—
IMR-4227	24.0	1729	—	40.0	2432	—
XMP-5744	24.5	1688	—	43.0	2532	—
IMR-4198	26.0	1707	—	44.0	2471	—
RX7	29.0	1655	—	49.0	2465	—
H-4895	32.0	1711	—	50.0	2488	—

Note: Loads shown in shaded panels are maximum.
Loads shown in bold designate potentially most accurate load.
\+ Designates a compressed powder charge.

338 Winchester Magnum

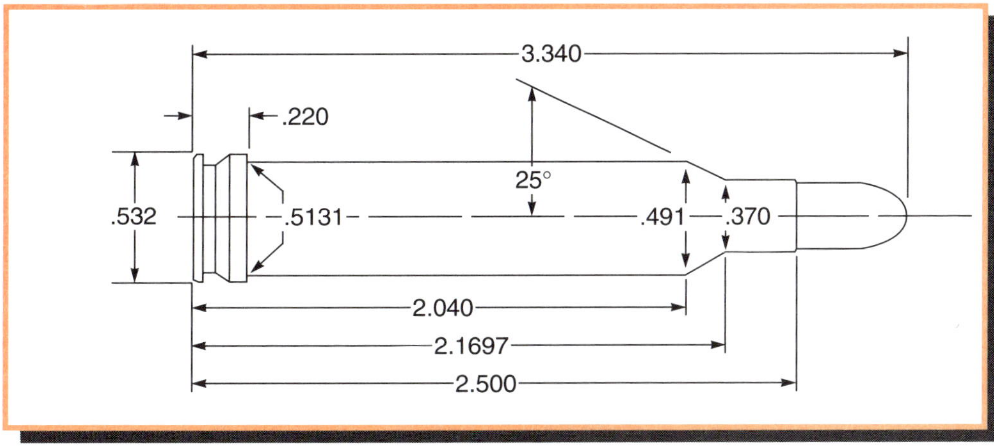

Comments:

This cartridge has attained classic status over the last 40 years. It started slow but gradually gained a momentum equal to its ability to handle most of the world's game.

This round is best served with a premium style bullet but for nondemanding situations many use popularly priced bullets.

Propellants that have proven very accurate include IMR 4831, Hercules Reloder 22, and IMR 7828 with all bullet weights. The shooter who really wants one cartridge for a wide variety of game, and who can see past the 30-06 mystique, would do well to learn to shoot a 338 Winchester Magnum. It has been used successfully on 30 pound duiker, 140 pound deer, giant Alaskan bear, and a lot of in between sized game around the world. It rivals the 375 H&H Magnum as being the most versatile of all big game cartridges. The 338 has the edge on lighter game whereas the 375 H&H can be pushed to the limit and used (carefully) for the likes of African cape buffalo.

Test Components:

Cases . . . . .Winchester
Trim-to Length . . . . .2.490"
Primers . . . . .Remington 9½ M
Primer Size . . . . .Large Rifle, Magnum
Lyman Shell Holder . . . . .No. 13
Jacketed Bullets Used . . . . . .Speer SP #2405, 200 gr.
Hornady SP #3320, 225 gr.
Hornady RN #3330, 250 gr.

Test Specifications:
(Velocity Only)

Firearm Used . . . . .Winchester Model 70
Barrel Length . . . . .24"
Twist . . . . .1-10"
Groove Dia. . . . . .336"

200 gr. Jacketed SP
3.340" OAL
BC: .448
SD: .250

Powder	Sugg Starting Grains	Velocity fps	Pressure	Max Load Grains	Velocity fps	Pressure
IMR-4350	69.0	2777	—	75.0+	3058	—
RX19	69.0	2634	—	77.0+	2968	—
IMR-4831	69.0	2664	—	**77.5+**	**3023**	—
XMR-3100	68.0	2513	—	76.0+	2805	—
H-4831	72.0	2652	—	80.0+	2923	—
RX22	71.0	2644	—	79.0+	2932	—

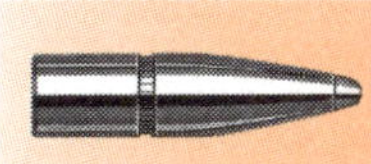

225 gr. Jacketed SP
3.310" OAL
BC: .397
SD: .281

Powder	Sugg Starting Grains	Velocity fps	Pressure	Max Load Grains	Velocity fps	Pressure
IMR-4350	67.0	2597	—	74.0+	2915	—
RX19	68.0	2533	—	75.0+	2795	—
IMR-4831	68.0	2524	—	75.5+	2801	—
XMR-3100	67.0	2386	—	74.0+	2616	—
H-4831	70.0	2493	—	78.0+	2797	—
IMR-7828	68.0	2306	—	76.0+	2602	—
RX22	69.0	2462	—	**77.0+**	**2747**	—

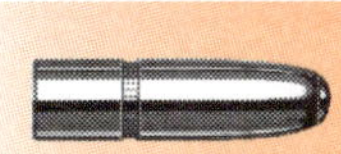

250 gr. Jacketed RN
3.300" OAL
BC: .291
SD: .313

Powder	Sugg Starting Grains	Velocity fps	Pressure	Max Load Grains	Velocity fps	Pressure
IMR-4350	65.0	2450	—	72.0+	2762	—
RX19	65.0	2297	—	72.0+	2591	—
IMR-4831	66.0	2367	—	73.5+	2686	—
XMR-3100	66.0	2267	—	72.0+	2509	—
H-4831	69.0	2403	—	76.0+	2666	—
IMR-7828	67.0	2225	—	74.0+	2523	—
RX22	68.0	2331	—	**76.0+**	**2652**	—

Note: Loads shown in shaded panels are maximum.
Loads shown in bold designate potentially most accurate load.
\+ Designates a compressed powder charge.

340 Weatherby Magnum

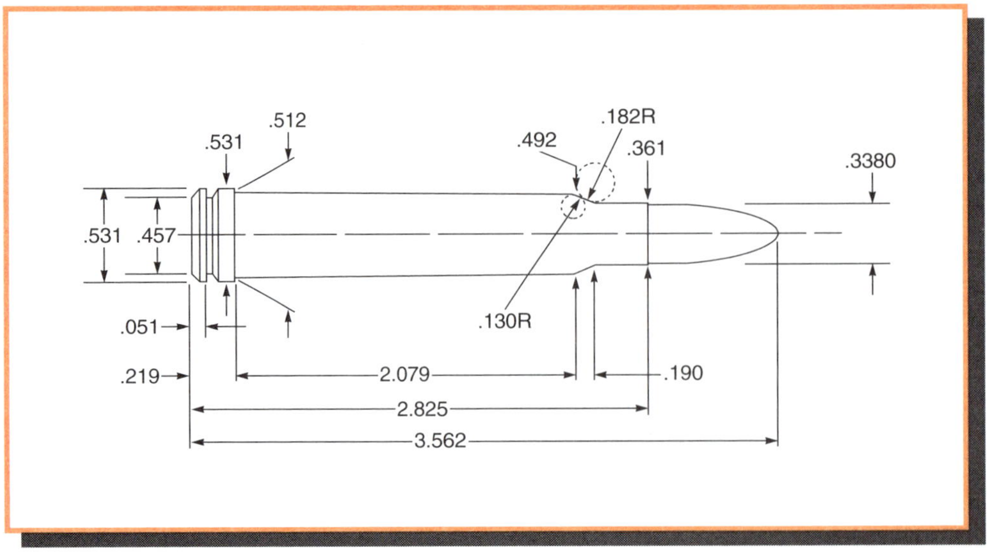

Comments:

This and the 338 Winchester were the only Magnum rifle chamberings available in the .338" diameter bore for many years. The 338 Winchester became quite popular while the Weatherby offering languished in proprietary status with a small, but loyal following. Only time will tell how the 340 fares against the new 338 Ultra Mag and its Weatherby stable mate, the 338-378. IMR-4350 and H-4831 often give best results. Weatherby recommends the exclusive use of Federal 215 primers. This data is intended for commercially produced and chambered rifles. It is not for use in custom guns that may lack the free bore found in standard Weatherby chambers.

Test Components:

Cases . . . . .Federal
Trim-to Length . . . . .2.815"
Primers . . . . .Federal 215
Primer Size . . . . .Large Rifle, Magnum
Lyman Shell Holder .No. 13
Jacketed Bullets Used . . . . . .Speer SP #2405, 200 gr.
Hornady SP #3320, 225 gr.
Hornady RN #3330, 250 gr.

Test Specifications:
(Velocity Only)

Firearm Used . . . . .Weatherby Mk V
Barrel Length . . . . .26"
Twist . . . . .1-10"
Groove Dia. . . . . .338"

200 gr. Jacketed SP
3.560" OAL
BC: .448
SD: .250

Powder	Sugg Starting Grains	Velocity fps	Pressure	Max Load Grains	Velocity fps	Pressure
IMR-4350	76.0	2724	—	84.0	3144	—
RX19	81.0	2977	—	**85.0**	**3112**	—
IMR-4831	79.0	2851	—	83.0	3025	—
XMR-3100	84.0	2964	—	88.0	3094	—
H-4831	81.0	2604	—	90.0+	3021	—
RX22	85.0	3004	—	89.5	3164	—
IMR-7828	86.0	3019	—	90.5	3147	—

225 gr. Jacketed SP
3.655" OAL
BC: .397
SD: .281

Powder	Sugg Starting Grains	Velocity fps	Pressure	Max Load Grains	Velocity fps	Pressure
IMR-4350	72.0	2538	—	80.0	2923	—
RX19	79.0	2894	—	83.0	3041	—
IMR-4831	77.0	2773	—	80.5	2920	—
XMR-3100	79.0	2795	—	83.0	2937	—
H-4831	78.0	2512	—	87.0+	2915	—
RX22	83.0	2954	—	**87.0+**	**3064**	—
IMR-7828	83.0	2939	—	87.0+	3035	—

250 gr. Jacketed RN
3.655" OAL
BC: .291
SD: .313

Powder	Sugg Starting Grains	Velocity fps	Pressure	Max Load Grains	Velocity fps	Pressure
IMR-4350	70.0	2418	—	77.0	2739	—
RX19	76.0	2713	—	80.0	2841	—
IMR-4831	75.0	2548	—	**79.0**	**2760**	—
XMR-3100	78.0	2635	—	82.0	2809	—
H-4831	79.0	2652	—	83.0	2813	—
RX22	80.0	2713	—	84.0	2876	—
IMR-7828	82.0	2791	—	**86.0**	**2934**	—

Note: Loads shown in shaded panels are maximum.
Loads shown in bold designate potentially most accurate load.
\+ Designates a compressed powder charge.

338 Remington Ultra Mag

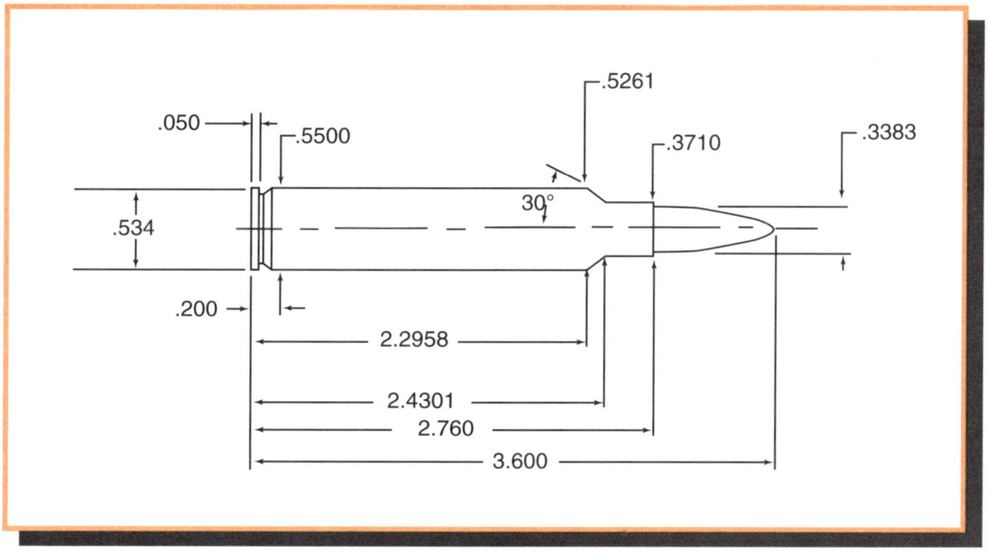

Comments:

Remington introduced the 338 member of the Ultra Mag family in 1999. This is Remington's first offering in the 338 caliber. The 338 RUM differs from the rest of its siblings by featuring a case body shortened by some .090". Barnes Bullets advises that data for their XLC coated bullets not be used for any uncoated bullets. Reloder 22 produced excellent results with a variety of different bullet weights.

Test Components:

Cases . . . . .Remington
Trim-to Length . . . . .2.750"
Primers . . . . .Remington 9½ M
Primer Size . . . . .Large Rifle, Magnum
Lyman Shell Holder . . . . .No. 13
Jacketed Bullets Used . . .Barnes XLC #33854, 185 gr.
Combined Tech. Ballistic Silvertip #51200, 200 gr.
Barnes XBT #33883, 210 gr.
Barnes X #33885, 225 gr.
Combined Tech. Fail Safe #53230, 230 gr.
Swift A-Frame, 250 gr.
Swift A-Frame, 275 gr.

Test Specifications:
(Velocity & Pressure)

Firearm Used . . . . .Universal Receiver
Barrel Length . . . . .24"
Twist . . . . .1-10"
Groove Dia. . . . . .338"

185 gr. Jacketed XLC
3.600" OAL
BC: .437
SD: .231

Powder	Sugg Starting Grains	Velocity fps	Pressure P.S.I.	Max Load Grains	Velocity fps	Pressure P.S.I.
N160	90.0	3102	54,400	94.5	3277	63,600
IMR-4350	87.5	3012	48,900	**92.0**	**3268**	**61,200**
RX19	91.5	3092	51,700	96.5	3318	63,500
IMR-4831	90.0	3119	54,200	94.5	3305	63,600
XMR-3100	94.5	3103	55,700	99.5+	3275	64,200
H-4831	93.0	2998	51,700	98.0	3200	62,800
WXR	94.0	3116	54,300	99.0	3294	63,300
N165	99.0	3196	56,200	104.0+	3360	64,500
RX22	95.5	3137	54,100	100.7+	3332	63,800
IMR-7828	95.0	3090	54,600	100.0+	3278	63,600

200 gr. Jacketed Silvertip
3.600" OAL
BC: .414
SD: .250

Powder	Sugg Starting Grains	Velocity fps	Pressure P.S.I.	Max Load Grains	Velocity fps	Pressure P.S.I.
IMR-4350	85.5	2794	52,500	90.0	3202	64,300
RX19	90.0	3015	50,800	94.5	3251	64,000
IMR-4831	88.0	3039	53,100	93.0	3239	64,400
XMR-3100	91.5	2984	53,200	96.5	3178	64,000
H-4831	91.0	2940	51,800	96.0	3136	63,700
WXR	92.0	3010	51,300	97.0	3200	62,200
N165	97.0	3026	51,200	102.0+	3210	61,400
RX22	93.0	3019	51,300	**98.0+**	**3244**	**63,200**
IMR-7828	94.0	3062	56,000	99.0+	3216	64,000
H1000	100.0	2999	53,500	105.0+	3174	62,600

Note: Loads shown in shaded panels are maximum.
Loads shown in bold designate potentially most accurate load.
\+ Designates a compressed powder charge.

338 Remington Ultra Mag

210 gr. Barnes XBT

3.575" OAL

BC: .471 **SD:** .263

Powder	Sugg Starting Grains	Velocity fps	Pressure P.S.I.	Max Load Grains	Velocity fps	Pressure P.S.I.
IMR-4350	82.0	2867	54,600	86.5	3039	63,800
RX19	87.0	2942	55,000	91.5	3114	64,500
IMR-4831	85.5	2907	52,500	90.0	3084	64,100
XMR-3100	88.0	2849	51,700	93.0	3047	62,900
H-4831	90.0	2866	53,200	94.5	3036	64,100
WXR	86.5	2869	53,800	91.0	3016	61,000
N165	88.5	2903	57,100	93.2	3035	63,700
RX22	90.0	2922	53,000	**95.2**	**3121**	**63,200**
IMR-7828	90.0	2905	54,500	94.7	3064	63,300
H1000	97.0	2881	52,800	102.0+	3049	62,200

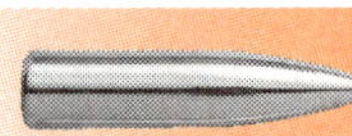

225 gr. Barnes X

3.600" OAL

BC: .482 **SD:** .281

Powder	Sugg Starting Grains	Velocity fps	Pressure P.S.I.	Max Load Grains	Velocity fps	Pressure P.S.I.
RX19	84.0	2831	57,500	88.7	2970	64,400
IMR-4831	82.0	2856	61,100	86.0	2946	63,600
XMR-3100	84.5	2737	53,900	89.0	2910	63,600
H-4831	84.0	2775	60,200	88.5	2863	63,700
WXR	83.5	2843	60,200	88.0	2939	64,200
N165	80.5	2699	55,900	85.0	2829	62,500
RX22	86.0	2852	56,600	**91.0**	**2990**	**63,000**
IMR-7828	84.0	2749	54,600	88.5	2894	61,700
H1000	92.0	2761	55,500	97.0+	2915	64,500
RX25	86.5	2748	51,500	91.0	2938	61,100

230 gr. Jacketed Fail-Safe

3.600" OAL

BC: .436 **SD:** .288

Powder	Sugg Starting Grains	Velocity fps	Pressure P.S.I.	Max Load Grains	Velocity fps	Pressure P.S.I.
IMR-4350	81.0	2699	51,400	85.7	2878	62,200
RX19	86.0	2801	54,000	91.0	2982	64,200
IMR-4831	84.0	2773	55,000	88.5	2921	63,700
XMR-3100	87.0	2739	53,500	91.5	2912	63,300
H-4831	87.0	2718	53,500	91.5	2876	62,900
WXR	86.0	2747	53,900	91.0	2899	63,100
RX22	89.5	2780	53,100	94.2	2973	63,800
IMR-7828	88.5	2771	54,900	**93.2**	**2936**	**64,200**
H1000	94.0	2796	56,200	99.0	2938	63,800
RX25	91.0	2769	53,000	95.7	2952	63,200
MagPro	93.0	2869	53,600	98.0	3032	62,600
Retumbo	97.5	2807	48,100	103.0+	2969	57,300

250 gr. Jacketed A-Frame

3.520" OAL

BC: .427 **SD:** .313

Powder	Sugg Starting Grains	Velocity fps	Pressure P.S.I.	Max Load Grains	Velocity fps	Pressure P.S.I.
RX19	83.0	2720	55,000	87.5	2869	64,200
IMR-4831	81.0	2684	54,600	85.0	2837	63,900
XMR-3100	83.0	2628	51,300	87.5	2809	63,600
H-4831	84.0	2651	55,200	88.5	2796	64,200
WXR	82.5	2699	57,300	87.0	2797	61,100
RX22	86.0	2752	55,300	**90.3+**	**2904**	**64,600**
IMR-7828	85.0	2703	55,500	89.5+	2844	64,000
H-1000	90.0	2660	54,000	95.0+	2811	63,300
RX25	87.0	2707	53,000	91.8+	2869	62,300
MagPro	89.0	2748	53,000	94.0	2873	59,000
Retumbo	96.0	2801	53,900	101.0+	2933	61,100

275 gr. Jacketed A-Frame

3.490" OAL

BC: .469 **SD:** .344

Powder	Sugg Starting Grains	Velocity fps	Pressure P.S.I.	Max Load Grains	Velocity fps	Pressure P.S.I.
RX19	79.0	2566	56,000	83.5	2696	63,800
XMR-3100	79.5	2449	50,400	83.7	2618	61,500
H-4831	80.0	2540	60,100	84.5	2629	64,000
RX22	81.0	2569	54,900	85.5	2718	64,400
IMR-7828	79.0	2519	55,800	84.0	2646	63,200
H1000	85.5	2482	51,700	**90.0**	**2654**	**63,800**
MagPro	85.5	2603	54,700	90.0	2718	61,300
Retumbo	89.0	2637	54,600	94.0	2726	59,400
AA-8700	96.0	2452	49,400	102.0+	2565	54,800

Note: Loads shown in shaded panels are maximum.
Loads shown in bold designate potentially most accurate load.
\+ Designates a compressed powder charge.

338-378 Weatherby Magnum

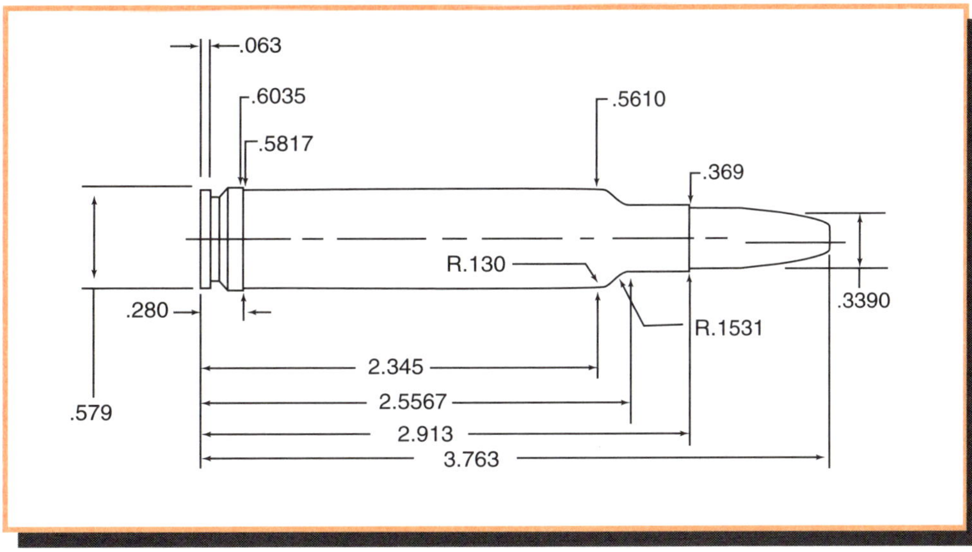

Comments:

This cartridge is similar to the old 338-378 Keith/Thompson wildcat, also based on the 378 Weatherby case. Powder capacity is approximately 15 percent more than the 338 Remington Ultra Mag and over 20 percent more than Weatherby's long time entry in the field, the 340 Weatherby Magnum. This cartridge can justifiably be called king of the factory 338s. Best performance requires large volumes of slow burning powder ignited by Federal 215 primers and premium bullets. IMR-7828 and H-1000 provided good ballistic uniformity with several different weight bullets. This number should amply handle anything in North America as well as much of Africa. This data is intended for commercially produced and chambered rifles. It is not intended for use in custom guns which may lack the free bore found in standard Weatherby chambers.

Test Components:

Cases . . . . .Weatherby
Trim-to Length . . . . .2.903"
Primers . . . . .Federal 215
Primer Size . . . . .Large Rifle, Magnum
Lyman Shell Holder . . . . .No. 17
Jacketed Bullets Used . . .Barnes XLC #33854, 185 gr.
Comb. Tech. Bal. Silvertip #51200, 200 gr.
Barnes XBT #33883, 210 gr.
Barnes X #33885, 225 gr.
Swift A-Frame, 250 gr.
Swift A-Frame, 275 gr.

Test Specifications: (Velocity Only)

Firearm Used . . . . .Weatherby Mk V
Barrel Length . . . . .26"
Twist . . . . .1-10"
Groove Dia. . . . . .338"

185 gr. Barnes XLC
3.700" OAL
BC: .437
SD: .231

Powder	Sugg Starting Grains	Velocity fps	Pressure	Max Load Grains	Velocity fps	Pressure
XMR-3100	103.5	3278	—	109.0	3537	—
H-4831 SC	104.0	3216	—	110.0	3421	—
RX22	105.5	3293	—	**111.0**	**3513**	—
IMR-7828	104.5	3277	—	110.0	3548	—
H1000	114.0	3366	—	120.0+	3556	—

200 gr. Jacketed Ballistic Tip
3.740" OAL
BC: .414
SD: .250

Powder	Sugg Starting Grains	Velocity fps	Pressure	Max Load Grains	Velocity fps	Pressure
XMR-3100	98.0	3052	—	103.5	3325	—
H-4831SC	99.0	3165	—	103.0	3288	—
RX22	99.0	3125	—	104.5	3317	—
IMR-7828	99.0	3061	—	**104.0**	**3323**	—
H1000	108.0	3129	—	113.5+	3355	—

Note: Loads shown in shaded panels are maximum.
Loads shown in bold designate potentially most accurate load.
\+ Designates a compressed powder charge.

338-378 Weatherby Magnum

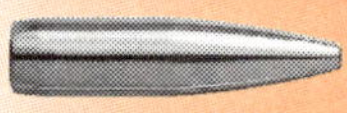

210 gr. Barnes XBT
3.740" OAL
BC: .471
SD: .263

Powder	Sugg Starting Grains	Velocity fps	Pressure	Max Load Grains	Velocity fps	Pressure
XMR-3100	99.0	3167	—	104.0	3377	—
H-4831SC	98.0	3119	—	**103.0**	**3219**	—
RX22	100.0	3208	—	105.5	3394	—
IMR-7828	101.0	3183	—	107.0	3445	—
H1000	108.0	3110	—	113.0	3309	—

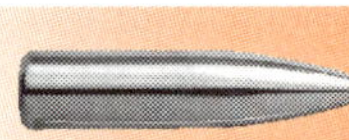

225 gr. Barnes X
3.720" OAL
BC: .482
SD: .281

Powder	Sugg Starting Grains	Velocity fps	Pressure	Max Load Grains	Velocity fps	Pressure
XMR-3100	98.0	2871	—	103.0	3082	—
H-4831SC	99.0	2904	—	104.0	3045	—
IMR-7828	100.0	2895	—	**105.0**	**3096**	—
H1000	105.5	2842	—	111.0	3026	—
RX25	107.0	3006	—	113.0	3242	—
H870	120.5	2992	—	127.0+	3116	—

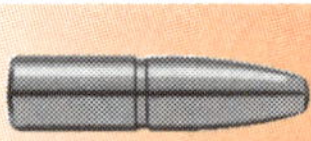

250 gr. A-Frame
3.675" OAL
BC: .427
SD: .313

Powder	Sugg Starting Grains	Velocity fps	Pressure	Max Load Grains	Velocity fps	Pressure
XMR-3100	96.0	2815	—	101.0	2988	—
H-4831SC	96.0	2832	—	101.0	2943	—
IMR-7828	97.0	2823	—	**102.0**	**3003**	—
H1000	103.5	2778	—	109.0	2980	—
RX25	102.0	2908	—	107.5	3105	—
H870	116.0	2868	—	122.0+	3059	—

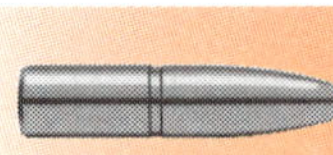

275 gr. A-Frame
3.640" OAL
BC: .469
SD: .344

Powder	Sugg Starting Grains	Velocity fps	Pressure	Max Load Grains	Velocity fps	Pressure
RX25	**98.0**	**2778**	—	103.0	2949	—
H1000	100.0	2746	—	105.0	2879	—
AA8700	110.0	2733	—	116.0+	2932	—
H50BMG	112.0	2789	—	117.0+	2894	—
H870	112.0	2779	—	118.0+	2909	—

Note: Loads shown in shaded panels are maximum.
Loads shown in bold designate potentially most accurate load.
\+ Designates a compressed powder charge.

348 Winchester

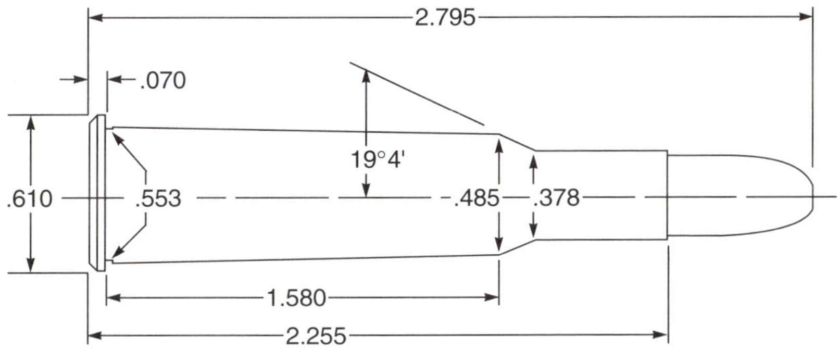

Comments:

The Model 71 Winchester is the only rifle made to ever chamber this cartridge. Conversely, this cartridge was offered in no other rifle but the Model 71. Winchester commenced Model 71 production in 1936 and discontinued it twenty years later. Browning resurrected the 348 when it ran a limited run of reproduction Model 71s during the 1980s. Factory ammunition included 150, 200 and 250-grain loadings at one time, however, Winchester currently lists only the 200-grain loading. The 348 is about the most firepower that has ever been packed into a lever-action rifle. The 348 retains a small but dedicated band of disciples, especially in the Alaskan bush. IMR-4895 is a good place to start for accuracy. Roll crimping is essential as on any tubular magazine.

Test Components:

Cases . . . . .Winchester
Trim-to Length . . . . .2.245"
Primers . . . . .Winchester WLR
Primer Size . . . . .Large Rifle
Lyman Shell Holder . . . . .No. 18
Jacketed Bullets Used . . . .Hornady FN #3410, 200 gr

Test Specifications: (Velocity& Pressure)

Firearm Used . . . . .Universal Receiver
Barrel Length . . . . .24"
Twist . . . . .1-12"
Groove Dia. . . . . .348"

200 gr. Jacketed FN
2.830" OAL
BC: .246
SD: .236

Powder	Sugg Starting Grains	Velocity fps	Pressure C.U.P.	Max Load Grains	Velocity fps	Pressure C.U.P.
IMR-4895	49.0	2340	35,000	**54.0**	**2546**	**39,400**
IMR-4064	48.0	2348	36,700	53.5	2523	39,700
Varget	50.0	2406	33,300	55.0	2557	39,400
IMR-4320	44.0	2202	36,800	49.0	2415	39,700
N140	43.0	2243	34,500	48.0	2432	39,200
RX15	44.0	2286	35,200	49.0	2472	39,500
N540	47.0	2310	34,700	52.0	2532	39,700
N150	44.0	2279	34,500	49.0	2487	39,500
N550	48.0	2197	34,900	53.0	2412	39,600
AA2700	50.0	2193	34,700	56.0	2449	38,300
IMR-4350	54.0	2194	34,600	60.0	2444	39,400
XMR-4350	56.0	2148	33,600	62.0	2432	40,000
H-4831	60.0	2166	33,100	67.0+	2519	39,100

Note: Loads shown in shaded panels are maximum.
Loads shown in bold designate potentially most accurate load.
\+ Designates a compressed powder charge.

9mm Luger (9mm Parabellum) (Rifle Data)

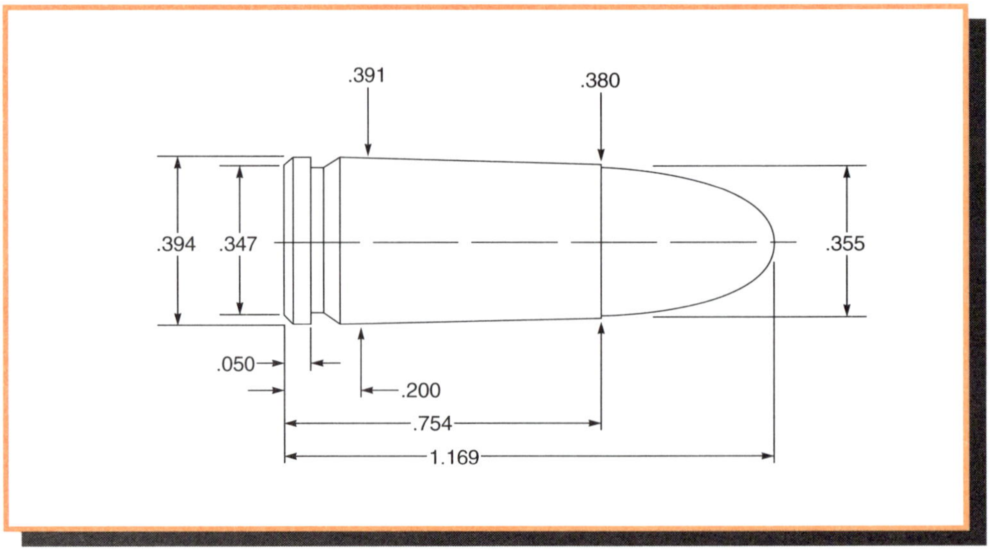

Comments:

The 9mm Luger has been one of the most popular pistol chamberings in the world. It has lost ground in recent years to the 40 Smith and Wesson. The data listed below was worked up in a pressure barrel and fired through a Ruger PC-9 Carbine. **Do not go below listed starting charges.** Doing so may cause a bullet to lodge in the rifle-length barrel. Firing a round with a bullet stuck in the barrel is extremely dangerous. Alliant's Power Pistol provided the best results in our testing.

Test Components:

Cases . . . . .Federal
Trim-to Length . . . . .751"
Primers . . . . .CCI 500
Primer Size . . . . .Small Pistol
Lyman Shell Holder . . . . .No. 12
Jacketed Bullets Used . .Hornady HP #35540, 115 gr.
Sierra HP #8125, 125 gr.
Sierra FMJ #8345, 130 gr.
Speer TMJ #4006, 147 gr.
Cast Bullets Used . . . .(sized to .355" dia)
#356242, 120 gr.
#356637, 147 gr.

Test Specifications:
(Velocity Only)

Firearm Used . . . . .Ruger Carbine, PC9
Barrel Length . . . . .16¼"
Twist . . . . .1-10"
Groove Dia. . . . . .354"

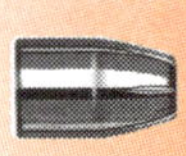

115 gr. Jacketed HP 1.090" OAL — **BC:** .129 **SD:** .130

Powder	Sugg Starting Grains	Velocity fps	Pressure	Max Load Grains	Velocity fps	Pressure
231	4.4	1117	—	4.9	1268	—
N340	4.5	1047	—	5.0	1238	—
Power Pistol	**5.9**	**1257**	—	6.5	1409	—
Blue Dot	6.8	1207	—	7.6+	1375	—

125 gr. Jacketed HP 1.075" OAL — **BC:** .137 **SD:** .142

Powder	Sugg Starting Grains	Velocity fps	Pressure	Max Load Grains	Velocity fps	Pressure
231	3.9	933	—	4.4	1106	—
N340	4.0	901	—	4.6	1048	—
Power Pistol	**5.1**	**1096**	—	5.7	1246	—
AA#5	5.4	1007	—	6.1	1188	—
Blue Dot	6.3	1001	—	7.1+	1270	—

Note: Loads shown in shaded panels are maximum.
Loads shown in bold designate potentially most accurate load.
\+ Designates a compressed powder charge.

9mm Luger (9mm Parabellum) (Rifle Data)

130 gr. FMJ
1.160" OAL

BC: .160
SD: .147

Powder	Sugg Starting Grains	Velocity fps	Pressure	Max Load Grains	Velocity fps	Pressure
Titegroup	3.8	1045	—	4.2	1109	—
231	3.9	1001	—	4.4	1107	—
N340	4.1	1013	—	4.6	1127	—
Power Pistol	**5.1**	**1130**	—	5.7	1262	—
AA#7	7.3	1126	—	8.2	1287	—

147 gr. TMJ
1.115" OAL

BC: .208
SD: .167

Powder	Sugg Starting Grains	Velocity fps	Pressure	Max Load Grains	Velocity fps	Pressure
Titegroup	3.2	919	—	3.6	1028	—
Unique	4.0	914	—	4.5	1050	—
N340	3.9	947	—	4.4	1060	—
Power Pistol	4.5	1048	—	**5.0**	**1121**	—
AA#7	6.4	1000	—	7.2	1119	—
Blue Dot	5.9	1071	—	6.6	1132	—

#356242
120 gr. (#2 Alloy) 1.065" OAL

BC: .154
SD: .136

Powder	Sugg Starting Grains	Velocity fps	Pressure	Max Load Grains	Velocity fps	Pressure
Titegroup	3.2	1105	—	3.8	1174	—
N340	4.0	1116	—	4.5	1242	—
AA#5	5.4	1143	—	6.0	1283	—
Power Pistol	**5.0**	**1265**	—	5.6	1369	—

#356637
147 gr. (#2 Alloy) 1.058" OAL

BC: .073
SD: .167

Powder	Sugg Starting Grains	Velocity fps	Pressure	Max Load Grains	Velocity fps	Pressure
Titegroup	2.5	863	—	2.8	927	—
Universal	3.0	885	—	3.4	989	—
N340	3.4	912	—	**3.8**	**987**	—
Power Pistol	4.1	997	—	4.6	1053	—
AA#7	6.3	976	—	7.1	1120	—

Note: Loads shown in shaded panels are maximum.
Loads shown in bold designate potentially most accurate load.

357 Magnum (Rifle Data)

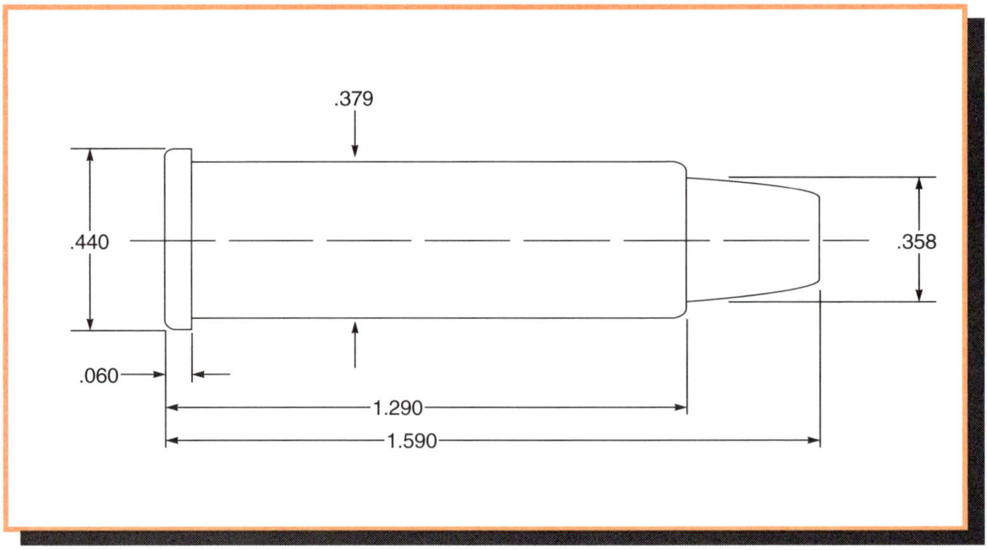

Comments:

Lever-action carbines chambered in revolver calibers have become quite popular recently. Part of this can be attributed to their lightweight and handling while the sport of Cowboy Action Shooting has also helped. All bullets used in rifles equipped with tubular magazines should be crimped in place. Powders that work well in 357 handgun loads, i.e. H110, 2400, N110, also worked well in the test rifle.

Test Components:

Cases . . . . Federal
Trim-to Length . . . . 1.280"
Primers . . . . CCI 550
Primer Size . . . . Small Pistol, Magnum
Lyman Shell Holder . . . . No. 1
Jacketed Bullets Used . . . . . Speer JHP #4007, 110 gr.
Hornady JHP #35710, 125 gr.
Speer JHP #4203, 140 gr.
Hornady JHP #35750, 158 gr.
Sierra JHC #8365, 170 gr.
Cast Bullets Used . . . . (sized to .357" dia)
*gas check bullet *#358156, 155 gr.
#358429, 170 gr.

Test Specifications:
(Velocity Only)

Firearm Used . . . . Winchester Model 94AE
Barrel Length . . . . 20"
Twist . . . . 1-16"
Groove Dia. . . . . .356"

110 gr. Jacketed HP
1.590" OAL
BC: .122
SD: .123

Powder	Sugg Starting Grains	Velocity fps	Pressure	Max Load Grains	Velocity fps	Pressure
Unique	7.4	1489	—	10.0	1873	—
Blue Dot	11.0	1609	—	14.6	2178	—
2400	17.0	2005	—	**20.3**	**2291**	—
N110	17.8	2069	—	19.8	2254	—
IMR-4227	16.0	1606	—	19.5	1921	—

125 gr. Jacketed HP
1.590" OAL
BC: .151
SD: .140

Powder	Sugg Starting Grains	Velocity fps	Pressure	Max Load Grains	Velocity fps	Pressure
Power Pistol	9.1	1731	—	10.2	1862	—
AA#5	10.6	1725	—	11.8	1858	—
Blue Dot	10.4	1534	—	13.3	2030	—
2400	15.0	1827	—	17.7	2048	—
H110	**21.0**	**2207**	—	22.0	2317	—
IMR-4227	16.0	1582	—	19.5	1920	—

Note: Loads shown in shaded panels are maximum.
Loads shown in bold designate potentially most accurate load.

357 Magnum (Rifle Data)

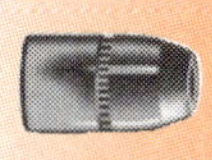

140 gr. Jacketed HP
1.590" OAL

BC: .152
SD: .157

Powder	Sugg Starting Grains	Velocity fps	Pressure	Max Load Grains	Velocity fps	Pressure
AA#5	9.5	1513	—	10.6	1639	—
Blue Dot	9.2	1329	—	11.9	1780	—
2400	14.0	1631	—	16.5	1889	—
N110	15.3	1805	—	**17.1**	**1955**	—
H110	17.3	1904	—	18.0	1931	—
IMR-4227	15.0	1510	—	17.8	1749	—

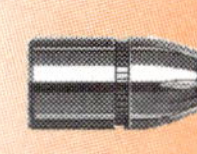

158 gr. Jacketed HP
1.590" OAL

BC: .206
SD: .177

Powder	Sugg Starting Grains	Velocity fps	Pressure	Max Load Grains	Velocity fps	Pressure
Blue Dot	9.6	1442	—	10.7	1610	—
AA#9	13.4	1639	—	**14.9**	**1774**	—
2400	12.0	1472	—	14.9	1735	—
N110	11.5	1327	—	14.0	1655	—
H110	16.3	1785	—	17.0	1829	—
IMR-4227	14.0	1391	—	16.1	1572	—

170 gr. Jacketed HC
1.590" OAL

BC: .176
SD: .191

Powder	Sugg Starting Grains	Velocity fps	Pressure	Max Load Grains	Velocity fps	Pressure
AA#7	9.8	1347	—	11.0	1440	—
AA#9	11.0	1347	—	12.6	1546	—
2400	10.5	1317	—	12.0	1436	—
N110	11.0	1318	—	13.5	1587	—
H110	14.0	1549	—	**15.5**	**1678**	—
IMR-4227	12.0	1202	—	13.7	1344	—

#358156
155 gr. (#2 Alloy) 1.590" OAL

BC: .213
SD: .174

Powder	Sugg Starting Grains	Velocity fps	Pressure	Max Load Grains	Velocity fps	Pressure
HS-6	8.8	1370	—	9.7	1491	—
AA#7	10.8	1493	—	12.0	1636	—
AA#9	13.0	1628	—	14.5	1785	—
2400	10.6	1358	—	14.0	1688	—
N110	12.0	1488	—	**15.0**	**1767**	—
IMR-4227	12.4	1257	—	15.2	1549	—

#358429
170 gr. (#2 Alloy) 1.553" OAL

BC: .286
SD: .191

Powder	Sugg Starting Grains	Velocity fps	Pressure	Max Load Grains	Velocity fps	Pressure
Blue Dot	8.3	1338	—	10.0	1592	—
AA#9	11.7	1514	—	**13.0**	**1666**	—
2400	11.5	1474	—	13.5	1677	—
N110	12.3	1552	—	13.7	1684	—
H110	14.4	1638	—	15.0	1704	—
IMR-4227	12.0	1268	—	14.5	1509	—

Note: Loads shown in shaded panels are maximum.
Loads shown in bold designate potentially most accurate load.

35 Remington

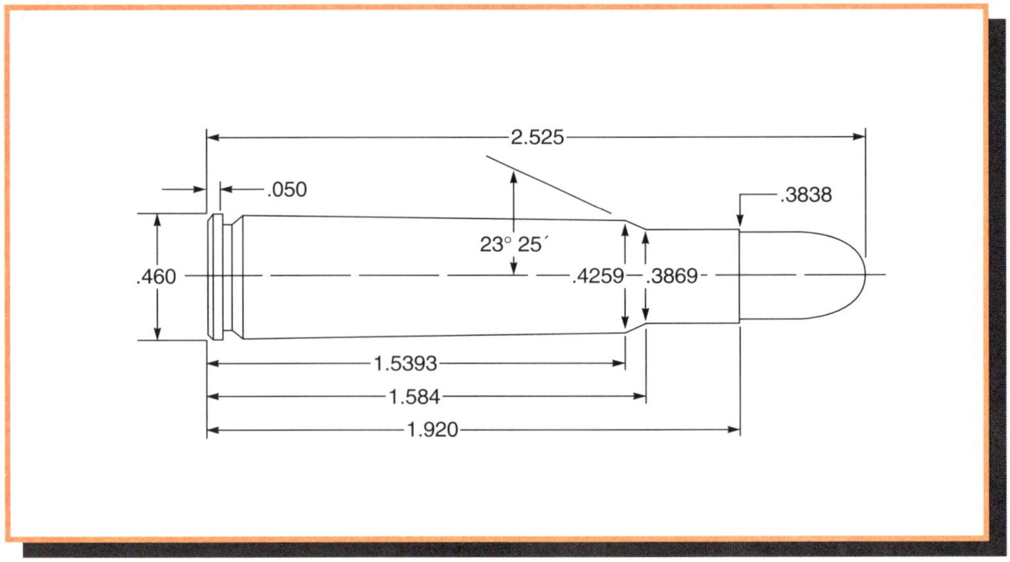

Comments:

The 35 Remington made its debut in 1906 as one of several chamberings (the others being the 25, 30, and 32 Remington) offered in Remington's new model 8 semiautomatic rifle. While the other calibers are long gone, the 35 Remington lives on today in the Marlin 336. Several different rifles have been produced in 35 Remington over the years. Rifles with tubular magazines are limited to flat or blunt nosed bullets. It's long been a popular brush cartridge out to 150 yards and is considered by many in the field to be superior to the 30-30.

The diminutive little Remington is the mildest of the 35 caliber rifle cartridges. Some current 358″ diameter bullets feature heavy construction designed for the higher velocities of other cartridges. Such bullets may not perform at 35 Remington levels. The Lyman staff selected the bullets listed in our data section based on their suitability for the 35 Remington's ballistic levels. Cast bullet # 358315 was originally designed for the 35 Remington. Shooters reloading for Marlin rifles with Micro-Groove® rifling should keep velocities below 1,600 feet per second best accuracy.

Test Components:

Cases . . . . .Remington
Trim-to Length . . . . .1.910"
Primers . . . . .Remington 9½
Primer Size . . . . .Large Rifle
Lyman Shell Holder . . . . .No. 2
Jacketed Bullets Used . . . . .Speer FSP #2435, 180 gr.
Sierra RN #2800, 200 gr.
Speer FSP #2439, 220 gr.
Cast Bullets Used . . . .(sized to .358" dia)
*gas check bullet #358430, 195 gr.
*#358315, 204 gr.

Test Specifications: (Velocity & Pressure)

Firearm Used . . . . .Universal Receiver
Barrel Length . . . . .22"
Twist . . . . .1-16"
Groove Dia. . . . . .357"

180 gr. Jacketed FSP
2.470" OAL
BC: .245
SD: .201

Powder	Sugg Starting Grains	Velocity fps	Pressure C.U.P.	Max Load Grains	Velocity fps	Pressure C.U.P.
N133	32.0	1941	28,300	35.5	2154	34,300
XMR-2015	**32.0**	**1904**	**27,800**	36.0	2135	34,900
RX7	29.0	1930	28,700	32.5	2121	34,600
IMR-3031	34.0	1829	28,700	38.5+	2158	35,000
H-4895	35.0	1812	29,000	39.0+	2073	34,900
748	38.0	1954	30,700	42.0	2131	34,500
IMR-4064	36.0	1838	29,500	40.5+	2101	34,800
Varget	37.0	1853	28,300	41.0+	2122	34,300

200 gr. Jacketed RN
2.475" OAL
BC: .148
SD: .223

Powder	Sugg Starting Grains	Velocity fps	Pressure C.U.P.	Max Load Grains	Velocity fps	Pressure C.U.P.
N133	31.0	1803	27,400	35.0	2060	34,600
XMR-2015	31.0	1780	28,500	**35.0**	**2007**	**34,600**
RX7	28.0	1860	31,000	31.0	1990	34,800
IMR-3031	33.0	1730	27,900	37.0	2029	35,000
H-4895	34.0	1740	27,600	38.5	2035	34,500
748	37.0	1875	31,000	41.0	2041	34,600
IMR-4064	34.0	1652	26,300	38.5	1947	34,700
Varget	36.0	1810	29,400	40.0	2031	34,700

Note: Loads shown in shaded panels are maximum.
Loads shown in bold designate potentially most accurate load.
\+ Designates a compressed powder charge.

220 gr. Jacketed FSP
2.470" OAL

BC: .316
SD: .245

Powder	Sugg Starting Grains	Velocity fps	Pressure C.U.P.	Max Load Grains	Velocity fps	Pressure C.U.P.
XMR-2015	30.0	1739	30,700	33.0	1884	34,700
RX7	27.0	1763	32,300	30.3	1936	35,000
IMR-3031	31.0	1641	28,900	35.0	1920	34,900
H-4895	33.0	1710	28,100	**36.5+**	**1947**	**34,700**
IMR-4064	34.0	1706	30,600	38.0+	1926	34,900
Varget	33.0	1588	27,300	37.2	1838	34,500

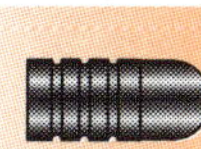

#358430
195 gr. (#2 Alloy) 2.215" OAL

BC: .145
SD: .217

Powder	Sugg Starting Grains	Velocity fps	Pressure C.U.P.	Max Load Grains	Velocity fps	Pressure C.U.P.
2400	15.0	1343	16,500	19.5	1731	26,600
SR-4759	16.0	1389	16,200	19.5	1745	26,400
IMR-4227	**18.5**	**1394**	**15,800**	22.5	1748	26,700
XMP-5744	18.0	1389	16,100	24.5	1788	29,300
IMR-4198	19.5	1365	15,500	25.5	1778	26,600
RX7	20.5	1329	15,000	26.5	1758	24,700

#358315
204 gr. (#2 Alloy) 2.470" OAL

BC: .165
SD: .227

Powder	Sugg Starting Grains	Velocity fps	Pressure C.U.P.	Max Load Grains	Velocity fps	Pressure C.U.P.
2400	15.0	1353	18,500	20.0	1707	29,000
SR-4759	15.5	1337	17,700	**21.0**	**1803**	**31,600**
IMR-4227	19.0	1375	17,000	24.2	1776	29,700
XMP-5744	19.0	1387	16,500	25.0	1778	29,000
IMR-4198	**20.0**	**1349**	**15,000**	26.5	1760	25,900
RX7	21.5	1364	14,900	27.7	1761	22,400
N133	25.0	1349	14,600	29.5	1749	23,500

Note: Loads shown in shaded panels are maximum.
Loads shown in bold designate potentially most accurate load.
\+ Designates a compressed powder charge.

358 Winchester (8.8 x 51mm)

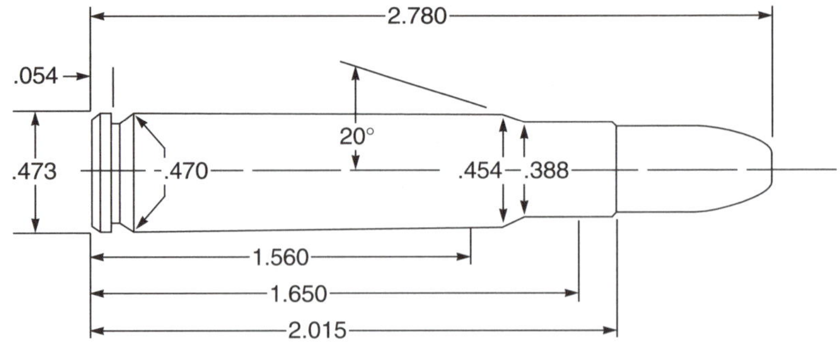

Comments:

The 358 Winchester is another promising cartridge that never achieved great commercial success. Winchester brought out this necked up 308 in 1955. It gained a reputation as a handy brush cartridge with a bit extra range to spare for open areas. 225 or 250-grain bullets and IMR-4064 usually give good results in the 358. Shooters loading ammunition from necked up GI brass should approach maximum loads cautiously due to the smaller capacity of military cases.

Test Components:

Cases . . . . .Winchester
Trim-to Length . . . . .2.005"
Primers . . . . .Remington 9½ and Winchester WLR
Primer Size . . . . .Large Rifle
Lyman Shell Holder . . . . .No. 2
Jacketed Bullets Used . . . . .Speer FSP #2435, 180 gr.
Sierra RN #2800, 200 gr.
Speer FSP #2439, 220 gr.
Hornady SP #3520, 250 gr.
Cast Bullets Used . . . .(sized to .359" dia)
*gas check bullet #358430, 195 gr.
*#358315, 204 gr.

Test Specifications: (Velocity & Pressure)

Firearm Used . . . . .Universal Receiver
Barrel Length . . . . .24"
Twist . . . . .1-12"
Groove Dia. . . . . .359"

180 gr. Jacketed FSP 2.680" OAL — **BC:** .245 **SD:** .201

Powder	Sugg Starting Grains	Velocity fps	Pressure C.U.P.	Max Load Grains	Velocity fps	Pressure C.U.P.
IMR-4198	37.0	2378	34,300	41.5	2667	49,300
RX7	40.0	2511	36,500	44.8	2752	49,400
N130	41.0	2501	35,200	**46.0**	**2788**	**49,800**
XMR-2015	44.0	2519	39,800	48.8+	2728	48,800
IMR-3031	46.0	2469	40,700	51.0	2728	49,100
H-335	48.0	2508	38,300	53.0+	2770	49,700
748	48.0	2340	30,800	53.0+	2521	36,900

200 gr. Jacketed RN 2.550" OAL — **BC:** .148 **SD:** .223

Powder	Sugg Starting Grains	Velocity fps	Pressure C.U.P.	Max Load Grains	Velocity fps	Pressure C.U.P.
RX7	34.0	2071	28,500	38.0	2309	44,500
IMR-4895	44.0	2220	35,900	49.0	2458	41,800
748	48.5	2227	36,700	53.5+	2507	47,900
N135	**46.0**	**2401**	**35,100**	51.5	2659	49,000
IMR-4064	46.0	2268	36,000	51.0+	2491	44,300
Varget	49.0	2340	34,800	54.0+	2612	45,500

Note: Loads shown in shaded panels are maximum.
Loads shown in bold designate potentially most accurate load.
+ Designates a compressed powder charge.

358 Winchester (8.8 x 51mm)

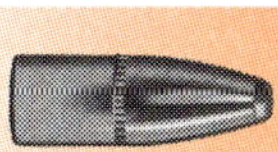

220 gr. Jacketed FSP 2.680" OAL — **BC:** .316 **SD:** .245

Powder	Sugg Starting Grains	Velocity fps	Pressure C.U.P.	Max Load Grains	Velocity fps	Pressure C.U.P.
IMR-3031	41.0	2199	39,600	46.0	2421	49,600
H-335	42.0	2184	35,500	47.0	2419	49,400
IMR-4895	44.0	2179	38,700	49.0	2438	49,600
748	47.5	2289	37,200	53.0	2499	49,100
N135	43.0	2274	36,400	47.5+	2483	48,800
IMR-4064	44.0	2174	37,700	**49.0+**	**2410**	**48,900**
Varget	47.0	2278	36,200	52.0	2514	48,700

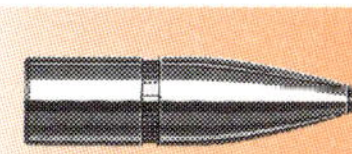

250 gr. Jacketed SP 2.810" OAL — **BC:** .375 **SD:** .279

Powder	Sugg Starting Grains	Velocity fps	Pressure C.U.P.	Max Load Grains	Velocity fps	Pressure C.U.P.
IMR-3031	36.0	1909	33,600	40.5	2198	49,400
H-335	38.0	1978	34,500	42.5	2182	49,600
IMR-4895	39.0	1979	40,200	43.0	2182	49,300
748	41.0	2003	33,600	46.0	2257	49,100
N135	39.0	2094	36,900	43.0	2283	49,200
Varget	41.0	2047	36,100	46.0	2284	49,300
RX15	39.0	1978	37,500	**43.3**	**2189**	**49,500**

***#358430** 195 gr. (#2 Alloy) 2.400" OAL — **BC:** .145 **SD:** .216

Powder	Sugg Starting Grains	Velocity fps	Pressure C.U.P.	Max Load Grains	Velocity fps	Pressure C.U.P.
SR-4759	17.5	1536	22,000	23.5	1965	31,300
IMR-4227	**20.0**	**1549**	**19,800**	27.0	1953	29,900
XMP-5744	21.0	1540	22,600	28.0	1912	30,000

***#358315** 204 gr. (#2 Alloy) 2.560" OAL — **BC:** .165 **SD:** .226

Powder	Sugg Starting Grains	Velocity fps	Pressure C.U.P.	Max Load Grains	Velocity fps	Pressure C.U.P.
2400	18.5	1628	22,000	25.0	1979	34,000
SR-4759	19.0	1601	21,700	24.5	1964	33,400
IMR-4227	22.5	1615	21,500	28.0	1967	29,600
XMP-5744	23.0	1627	23,200	**29.0**	**1975**	**33,400**
IMR-4198	25.0	1643	16,500	30.5	1981	27,600
N130	27.0	1627	17,900	32.5	1940	24,100
RX7	27.0	1618	16,500	32.0	2018	27,300

Note: Loads shown in shaded panels are maximum.
Loads shown in bold designate potentially most accurate load.
+ Designates a compressed powder charge.
* Designates the use of Win. WLR primers.

35 Whelen

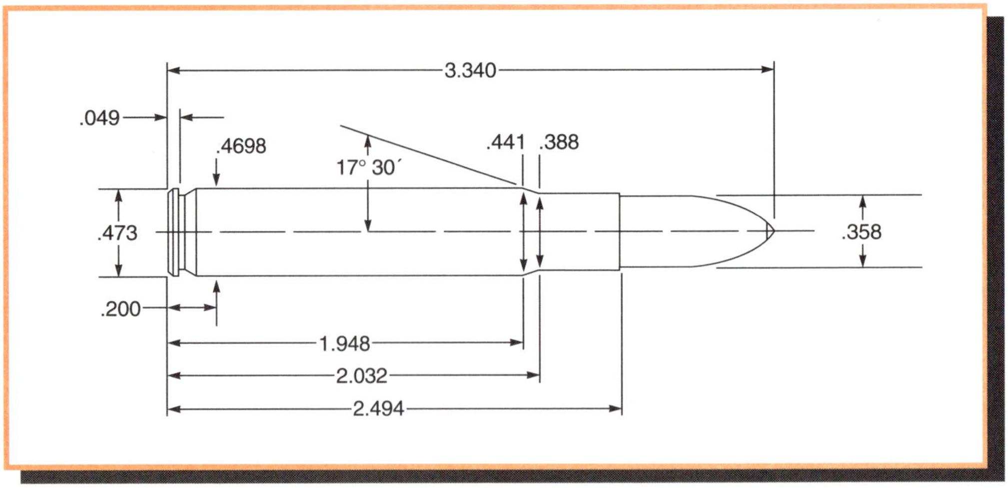

Comments:

This is yet another popular wildcat legitimized by Remington. The 35 Whelen is essentially a 30-06 necked up to accept .358″ diameter bullets. The cartridge had been around since the early 1920s before it became factory sanctioned in 1988. The Whelen ably fills a niche between the 30-06 and 416 Rigby for those shooters with an aversion to belted magnums. Its ability to be chambered in standard length actions and ready source of parent cases made it an economical alternative to the 375 H&H for many shooters. Credit for its creation has been the subject of some debate. Although the namesake of Colonel Townsend Whelen, the first rifles originated from the firm of Griffin and Howe. Factory ammunition and bulk cases are available from Remington but many shooters still choose to neck up 30-06 cases. Shooters reloading GI cases should approach maximum loads with caution due to the smaller capacity of military cases. IMR-4064 has long been a favorite for loading the Whelen. This data is intended for standard Remington chambers. Older, custom rifles may have minor dimensional variations in the chamber.

Test Components:

Cases . . . . .Remington
Trim-to Length . . . . .2.484"
Primers . . . . .Remington 9½
Primer Size . . . . .Large Rifle
Lyman Shell Holder . . . . .No. 2
Jacketed Bullets Used . . . . .Speer FSP #2435, 180 gr.
Sierra RN #2800, 200 gr.
Barnes X #35825, 225 gr.
Hornady SP #3520, 250 gr.
Hornady HP (Pistol) #35750, 158 gr.
Cast Bullets Used . . . .(sized to .358" dia)
*gas check bullet #358430, 195 gr.
*#358315, 204 gr.

Test Specifications: (Velocity & Pressure)

Firearm Used . . . . .Universal Receiver
Barrel Length . . . . .24"
Twist . . . . .1-16"
Groove Dia. . . . . .3575"

180 gr. Jacketed FSP
3.045" OAL
BC: .245
SD: .201

Powder	Sugg Starting Grains	Velocity fps	Pressure C.U.P.	Max Load Grains	Velocity fps	Pressure C.U.P.
IMR-3031	52.0	2438	39,100	57.5+	2740	49,900
AA2230	50.0	2345	34,900	55.7	2772	50,300
H-335	53.0	2301	32,700	59.0	2750	49,200
H-4895	54.0	2508	38,900	60.0+	2763	47,900
N135	54.0	2550	41,500	60.0+	2778	47,900
IMR-4064	55.5	2465	41,200	61.7+	2714	48,500
Varget	55.5	2577	41,700	61.7+	2814	48,500
AA2520	55.5	2628	41,800	61.5	2859	48,800
IMR-4320	53.0	2405	41,800	**58.8**	**2680**	**49,200**
RX15	57.0	2567	43,400	63.0+	2789	49,500

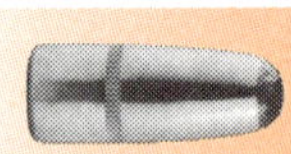

200 gr. Jacketed RN
3.045" OAL
BC: .148
SD: .223

Powder	Sugg Starting Grains	Velocity fps	Pressure C.U.P.	Max Load Grains	Velocity fps	Pressure C.U.P.
IMR-3031	51.0	2356	39,600	56.8+	2630	49,900
AA2230	49.5	2268	35,200	55.0	2641	48,400
H-335	53.0	2355	40,900	58.6	2657	49,700
H-4895	53.0	2411	37,500	59.0	2675	49,600
748	56.5	2461	35,600	63.0+	2540	47,200
N135	53.0	2492	42,400	**59.0+**	**2676**	**48,100**
IMR-4064	53.0	2313	38,900	59.0+	2594	49,300
Varget	55.0	2502	41,800	61.0+	2715	50,100
AA2520	52.0	2287	32,000	57.4	2659	47,700
IMR-4320	52.0	2257	37,800	58.0	2536	49,900
RX15	54.0	2428	41,700	60.0	2635	49,000

Note: Loads shown in shaded panels are maximum.
Loads shown in bold designate potentially most accurate load.
\+ Designates a compressed powder charge.

35 Whelen

225 gr. Barnes X
3.340" OAL

BC: .405
SD: .250

Powder	Sugg Starting Grains	Velocity fps	Pressure C.U.P.	Max Load Grains	Velocity fps	Pressure C.U.P.
IMR-3031	49.0	2354	37,500	54.4+	2617	48,400
H-335	50.0	2099	36,600	55.3	2419	49,500
H-4895	49.5	2218	38,300	55.0	2467	49,900
748	56.0	2356	37,200	62.0+	2456	43,800
IMR-4064	50.5	2145	39,900	56.0	2395	49,600
Varget	50.5	2204	40,000	56.0	2475	49,800
AA2520	50.0	2158	35,200	55.3	2506	50,100
IMR-4320	49.0	2077	40,600	54.0	2344	50,400
N140	52.0	2260	41,500	**58.0**	**2484**	**50,400**
RX15	50.5	2219	42,200	56.2	2424	48,400

250 gr. Jacketed SP
3.255" OAL

BC: .375
SD: .279

Powder	Sugg Starting Grains	Velocity fps	Pressure C.U.P.	Max Load Grains	Velocity fps	Pressure C.U.P.
IMR-3031	46.0	2237	35,400	51.3+	2485	48,200
H-4895	48.0	2104	34,300	**53.0**	**2377**	**50,300**
748	53.0	2252	40,200	59.0+	2382	45,200
IMR-4064	48.5	2088	38,000	54.0	2344	50,700
Varget	49.5	2178	40,300	55.0	2413	50,400
AA2520	48.0	2149	37,300	53.0	2410	49,300
IMR-4320	47.0	2024	38,900	52.5	2291	49,400
N140	51.0	2213	40,400	56.5+	2426	51,000
RX15	49.0	2095	37,400	54.0	2364	50,200

158 gr. Jacketed HP (Pistol Bullet) 2.787" OAL

BC: .206
SD: .177

Powder	Sugg Starting Grains	Velocity fps	Pressure C.U.P.	Max Load Grains	Velocity fps	Pressure C.U.P.
2400	26.7	1405	23,300	31.0	1546	40,800
SR-4759	22.0	1435	23,000	23.6	1564	39,000
XMP-5744	22.5	1411	22,000	**25.5**	**1570**	**38,500**

#358430
195 gr. (#2 Alloy) 2.805" OAL

BC: .145
SD: .217

Powder	Sugg Starting Grains	Velocity fps	Pressure C.U.P.	Max Load Grains	Velocity fps	Pressure C.U.P.
Unique	13.0	1546	20,600	19.2	1894	38,100
2400	22.5	1815	17,800	29.0	2255	43,600
SR-4759	**27.0**	**1935**	**20,200**	34.7	2337	42,400
IMR-4198	33.0	2060	21,800	44.0	2553	40,200
IMR-3031	39.0	2094	20,500	50.0+	2594	36,300

#358315
204 gr. (#2 Alloy) 3.045" OAL

BC: .165
SD: .227

Powder	Sugg Starting Grains	Velocity fps	Pressure C.U.P.	Max Load Grains	Velocity fps	Pressure C.U.P.
SR-4759	24.0	1707	28,800	31.0	2092	38,800
IMR-4227	30.0	1615	24,700	36.0	2096	31,900
XMP-5744	**28.0**	**1704**	**27,600**	37.0	2124	38,400
IMR-4198	30.0	1666	22,400	37.5	2097	31,200
RX7	36.0	1767	22,000	40.5	2131	30,000
IMR-3031	50.0	2558	34,600	58.5	2860	49,000

Note: Loads shown in shaded panels are maximum.
Loads shown in bold designate potentially most accurate load.
\+ Designates a compressed powder charge.

38-55 Winchester (38-55 Ballard)

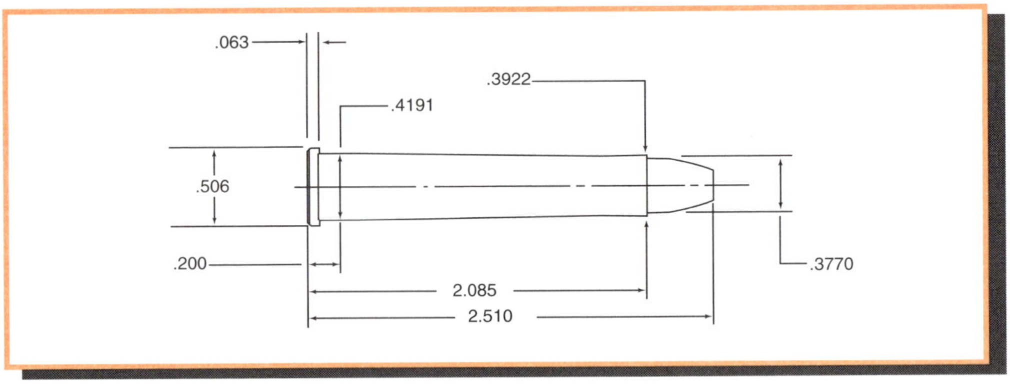

Comments:

Marlin first introduced the 38-55 in 1884 as a black powder cartridge for Ballard target rifles. Winchester produced rifles chambered in 38-55 shortly thereafter. This caliber survived into the smokeless era in the Model 94 along with several other makes and models before being finally discontinued in 1940. The cartridge developed a reputation for accuracy and effectiveness in the field that still serves it well today. The 38-55 has experienced a revival of late among the Black Powder Cartridge Silhouette and Cowboy Action Shooting establishments. U.S. Repeating Arms, Marlin, and several smaller custom manufacturers currently offer 38-55 chambered rifles. The 38-55 also served as the basis of several other cartridges brought out by Winchester over the years including the 30-30.

The 38-55 responds well to handloading, especially with the use of cast bullets. Many 38-55 caliber rifles show wide variations in groove diameters. Some older rifles measure as large as .382" Shooters should slug their bore and size bullets accordingly. Cast bullet #375248 was first introduced around 1905 and has been a favorite for the 38-55 ever since. It is also compatible for lever action rifles. Bullet #375674 exceeds the maximum overall length when seated and is for single shot rifles only. Shooters may occasionally encounter older brass that exceeds 2.085" in length. Maximum cartridge case length was originally 2.129" but has been established at 2.085" by SAAMI. All listed data has been worked up with the newer, shorter cases. SAAMI has also established a Maximum Average Pressure (MAP) of 30,000 CUP for the 38-55. Shooters should exercise caution if loading for older guns as many have developed excessive headspace. The following data is intended for use in modern firearms such as the Winchester 94. Shooters should never attempt to fire any 375 Winchester ammunition in any rifle chambered for the 38-55.

The 375 Winchester is dimensionally similar but loaded to substantially higher pressure. If in doubt, all older rifles should be examined by a qualified gunsmith before shooting.

Test Components:

Cases . . . . .Winchester
Trim-to Length . . . . .2.075"
Primers . . . . .Remington 9½
Primer Size . . . . .Large Rifle
Lyman Shell Holder . . . . .No. 6
Jacketed Bullets Used . . .Barnes FNSP #5510, 255 gr.
Cast Bullets Used . . . .(sized to .379" dia)
#375248, 249 gr.
#378674, 335 gr.

Test Specifications: (Velocity & Pressure)

Firearm Used . . . . .Universal Receiver
Barrel Length . . . . .30"
Twist . . . . .1-20"
Groove Dia. . . . . .379"

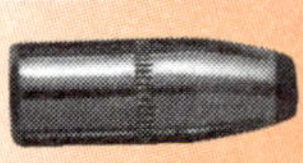

255 gr. Jacketed FNSP
2.480" OAL
BC: .290
SD: .259

Powder	Sugg Starting Grains	Velocity fps	Pressure C.U.P.	Max Load Grains	Velocity fps	Pressure C.U.P.
XMP-5744	**20.5**	**1229**	**14,700**	23.0	1487	24,600
IMR-4198	24.0	1379	16,200	27.0	1616	23,400
IMR-3031	31.5	1475	17,800	35.0+	1762	26,800
IMR-4895	31.5	1382	16,400	35.0+	1576	20,000

#375248
249 gr. (#2 Alloy) 2.550" OAL
BC: .290
SD: .248

Powder	Sugg Starting Grains	Velocity fps	Pressure C.U.P.	Max Load Grains	Velocity fps	Pressure C.U.P.
XMP-5744	**23.0**	**1490**	**17,400**	25.5	1679	24,400
IMR-4198	25.0	1590	21,200	28.0	1755	23,000
RX7	30.0	1664	16,200	35.0+	1998	23,600
IMR-3031	31.0	1506	17,000	35.0+	1812	25,600
IMR-4895	32.5	1574	18,400	36.0+	1729	23,200

#378674
335 gr. (#2 Alloy) *2.900" OAL
BC: .360
SD: .333

Powder	Sugg Starting Grains	Velocity fps	Pressure C.U.P.	Max Load Grains	Velocity fps	Pressure C.U.P.
XMP-5744	**21.0**	**1313**	**17,700**	24.0	1484	25,600
IMR-4198	24.0	1417	17,000	27.0	1598	25,400
RX7	29.0	1588	19,900	31.0	1633	21,900
IMR-3031	28.5	1347	16,700	32.0	1605	23,500
IMR-4895	30.0	1374	17,000	33.5+	1582	24,200

Note: Loads shown in shaded panels are maximum. Loads shown in bold designate potentially most accurate load.
\+ Designates a compressed powder charge.
* Exceeds SAAMI Maximum length. For use in single shot rifles only.

375 Winchester

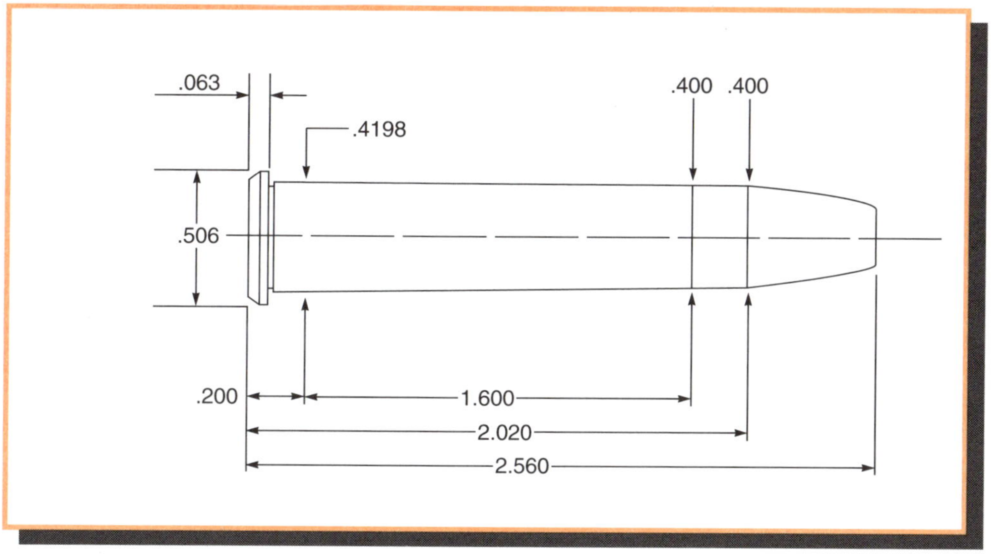

Comments:

This cartridge was the first attempt to beef up the ballistics that could be obtained with a 94 Winchester carbine. Basically it is the rebirth of the 38-55 loaded to higher pressure levels. The rifle was called the Big Bore.

Never put a 375 Winchester round into a 38-55 Winchester. Nor should you attempt to use 38-55 ammo in the 375 Winchester. Incompatible neck diameters can cause serious hazard.

Test Components:

Cases . . . . .Winchester
Trim-to Length . . . . .2.010"
Primers . . . . .Winchester WLR
Primer Size . . . . .Large Rifle
Lyman Shell Holder . . . . .No. 6
Jacketed Bullets Used . . . .Hornady FP #3705, 220 gr.
Barnes FP #375W20, 255 gr.
Cast Bullets Used . . . .(sized to .377" dia)
*gas check bullet #375248, 249 gr.
*#375449, 264 gr.

Test Specifications:
(Velocity & Pressure)

Firearm Used . . . . .Winchester Model 94 Big Bore
Universal Receiver
Barrel Length . . . . .Winchester; 20"
Universal Receiver; 24"
Twist . . . . .1-12"
Groove Dia. . . . . .376"

220 gr. Jacketed FP
2.500" OAL
BC: .217
SD: .223

Powder	Sugg Starting Grains	Velocity fps	Pressure C.U.P.	Max Load Grains	Velocity fps	Pressure C.U.P.
IMR-4198	29.0	1949	44,100	33.0	2145	50,900
RX7	34.0	2031	31,000	**38.0+**	**2259**	**42,100**
IMR-3031	30.3	1732	31,700	35.0+	1983	40,700
748	38.0	1705	25,200	41.5+	1879	30,800
H-4895	34.3	1868	30,000	38.0+	2069	36,600
AA2230	37.5	1951	34,900	42.0+	2178	44,500
AA2460	38.0	1991	33,800	42.0+	2162	41,700

255 gr. Jacketed FP
2.560" OAL
BC: .290
SD: .259

Powder	Sugg Starting Grains	Velocity fps	Pressure C.U.P.	Max Load Grains	Velocity fps	Pressure C.U.P.
IMR-4198	23.0	1542	34,500	25.7+	1743	48,700
RX7	25.2	1599	30,000	29.0+	1781	33,700
IMR-3031	26.8	1547	35,800	32.0+	1771	40,900
H-322	29.4	1736	36,500	**34.0+**	**1940**	**45,800**
H-4895	30.0	1632	27,300	34.0+	1838	35,500
AA2230	35.5	1844	37,700	40.5+	2045	48,400
AA2460	34.5	1796	37,800	39.0+	1977	47,100

Note: Loads shown in shaded panels are maximum.
Loads shown in bold designate potentially most accurate load.
+ Designates a compressed powder charge.

375 Winchester

***#375248**
249 gr. (#2 Alloy) 2.450" OAL
BC: .290
SD: .250

Powder	Sugg Starting Grains	Velocity fps	Pressure C.U.P.	Max Load Grains	Velocity fps	Pressure C.U.P.
2400	20.0	1595	—	24.0	1897	—
SR-4759	**21.0**	**1606**	—	26.0+	1866	—
IMR-4227	**22.0**	**1617**	—	26.0	1925	—
IMR-4198	24.0	1581	—	28.5+	1907	—
RX7	27.5	1614	—	32.0+	1811	—
IMR-3031	27.0	1563	—	31.4+	1771	—

***#375449**
264 gr. (#2 Alloy) 2.560" OAL
BC: .315
SD: .265

Powder	Sugg Starting Grains	Velocity fps	Pressure C.U.P.	Max Load Grains	Velocity fps	Pressure C.U.P.
2400	19.5	1503	—	24.3	1806	—
SR-4759	**20.0**	**1502**	—	24.0	1738	—
IMR-4227	21.0	1520	—	24.8	1807	—
IMR-4198	23.0	1496	—	25.5	1769	—
RX7	25.0	1507	—	32.0+	1762	—
IMR-3031	26.2	1484	—	32.0+	1770	—

Note: Loads shown in shaded panels are maximum.
Loads shown in bold designate potentially most accurate load.
\+ Designates a compressed powder charge.
* Fired in a Winchester Model 94.

375 H & H Magnum (375 Holland & Holland)

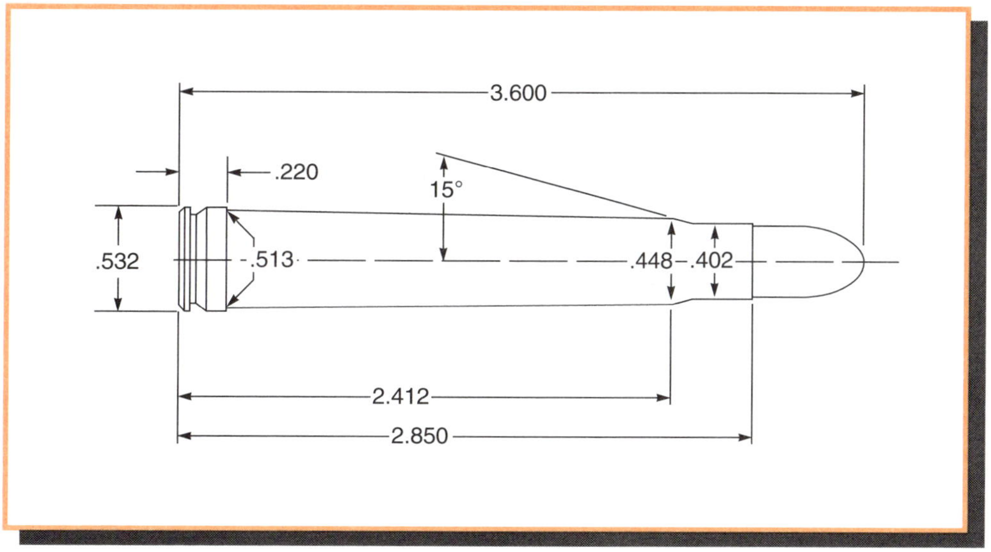

Comments:

This is the godfather of modern belted magnums. Introduced by Holland and Holland in 1912, it sired the 300 H&H in 1925. These two cartridges thus served as the basis for almost every popular belted magnum since. The 375 H&H made its reputation in Africa over many years. It may be considered a bit overpowered for much of the lower 48 but has seen quite a bit of use in Alaska. The 375 is one of the few magnums that actually uses the belt for proper headspacing due to the long, sloping configuration of the case. The length of this cartridge mandates use of a long, magnum length action. The big H&H properly loaded with the right bullets is also an exceptionally accurate, flat shooting cartridge. Do not attempt to crimp bullets without a cannelure. IMR-4350 and Reloder 15 are good powder selections for loading the 375.

Test Components:

Cases . . . . .Winchester
Trim-to Length . . . . .2.840"
Primers . . . . .Winchester WLRM
Primer Size . . . . .Large Rifle Magnum
Lyman Shell Holder . . . . .No. 13
Jacketed Bullets Used . . .Barnes XLC #37553, 235 gr.
Nosler Partition #44850, 260 gr.
Combined Tech. Fail-Safe, #53350, 270 gr.
Speer Grand Slam #2473, 285 gr.
Swift A-Frame, 300 gr.
Cast Bullets Used . . . .(sized to .378" dia)
*gas check bullet *#375449
#378674

Test Specifications: (Velocity Only)

Firearm Used . . . . .Winchester Model 70
Barrel Length . . . . .24"
Twist . . . . .1-12"
Groove Dia. . . . . .377"

235 gr. Barnes XLC 3.570" OAL — **BC:** .400 **SD:** .239

Powder	Sugg Starting Grains	Velocity fps	Pressure C.U.P.	Max Load Grains	Velocity fps	Pressure C.U.P.
IMR-3031	64.0	2591	—	71.0	2857	—
IMR-4895	**66.0**	**2593**	—	73.0	2864	—
IMR-4064	67.5	2583	—	75.0+	2845	—
N140	68.0	2626	—	76.0	2832	—
RX15	70.0	2690	—	78.0+	2974	—
H-380	73.0	2556	—	81.0+	2748	—

260 gr. Jacketed SP 3.570" OAL — **BC:** .314 **SD:** .264

Powder	Sugg Starting Grains	Velocity fps	Pressure C.U.P.	Max Load Grains	Velocity fps	Pressure C.U.P.
IMR-4895	63.0	2414	—	70.5	2719	—
IMR-4064	66.0	2433	—	73.0	2697	—
RX15	67.5	2572	—	75.0	2825	—
H-380	71.0	2481	—	78.5	2704	—
AA2700	69.0	2449	—	77.0	2694	—
N160	76.0	2496	—	**84.5+**	**2752**	—
XMR-4350	75.0	2401	—	84.0+	2680	—

Note: Loads shown in shaded panels are maximum.
Loads shown in bold designate potentially most accurate load.
\+ Designates a compressed powder charge.

375 H & H Magnum (375 Holland & Holland)

270 gr. Jacketed Fail-Safe 3.600" OAL — **BC:** .393 **SD:** .274

Powder	Sugg Starting Grains	Velocity fps	Pressure	Max Load Grains	Velocity fps	Pressure
IMR-4064	64.0	2403	—	71.5	2628	—
RX15	66.0	2488	—	73.5	2742	—
H-380	68.0	2347	—	76.0+	2541	—
AA2700	71.0	2448	—	79.0+	2537	—
N160	75.0	2344	—	**83.0+**	**2622**	—
IMR-4350	75.0	2398	—	83.0+	2707	—
XMR-4350	75.0	2351	—	83.5+	2629	—

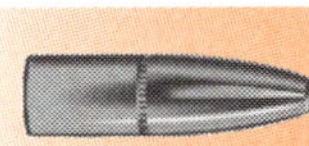

285 gr. Jacketed SP 3.600" OAL — **BC:** .354 **SD:** .290

Powder	Sugg Starting Grains	Velocity fps	Pressure	Max Load Grains	Velocity fps	Pressure
IMR-4895	57.0	2130	—	63.5	2360	—
IMR-4064	62.0	2254	—	69.0	2521	—
RX15	63.0	2324	—	70.0	2545	—
AA2700	66.0	2168	—	73.5	2446	—
N160	72.5	2303	—	80.5+	2595	—
XMR-4350	73.0	2255	—	81.0+	2529	—
RX19	72.0	2242	—	**80.0+**	**2506**	—
H-4831SC	76.5	2262	—	85.0+	2545	—

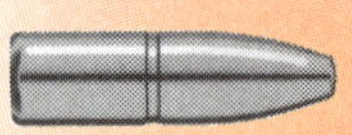

300 gr. Jacketed A-Frame 3.550" OAL — **BC:** .325 **SD:** .305

Powder	Sugg Starting Grains	Velocity fps	Pressure	Max Load Grains	Velocity fps	Pressure
IMR-4064	61.0	2219	—	68.0	2469	—
RX15	60.0	2244	—	67.0	2468	—
AA2700	67.5	2375	—	75.0	2556	—
N160	73.0	2298	—	81.5+	2570	—
XMR-4350	**71.0**	**2209**	—	79.0+	2508	—
IMR-4350	73.0	2369	—	81.0+	2629	—
RX19	70.0	2170	—	78.0+	2448	—
H-4831SC	76.0	2232	—	84.0+	2490	—

#375449 264 gr. (#2 Alloy) 3.440" OAL — **BC:** .315 **SD:** .264

Powder	Sugg Starting Grains	Velocity fps	Pressure	Max Load Grains	Velocity fps	Pressure
SR-4759	21.0	1372	—	32.0	1858	—
XMP-5744	27.0	1461	—	34.0	1736	—
IMR-4198	24.0	1387	—	**35.0**	**1813**	—
RX7	26.0	1422	—	34.0	1698	—

#378674 335 gr. (#2 Alloy) 3.570" OAL — **BC:** .360 **SD:** .335

Powder	Sugg Starting Grains	Velocity fps	Pressure	Max Load Grains	Velocity fps	Pressure
SR-4759	23.0	1353	—	32.5	1759	—
XMP-5744	26.0	1329	—	**36.0**	**1725**	—
IMR-4198	25.0	1314	—	36.0	1717	—
RX7	27.0	1367	—	37.5	1726	—

Note: Loads shown in shaded panels are maximum.
Loads shown in bold designate potentially most accurate load.
\+ Designates a compressed powder charge.

375 Remington Ultra Mag

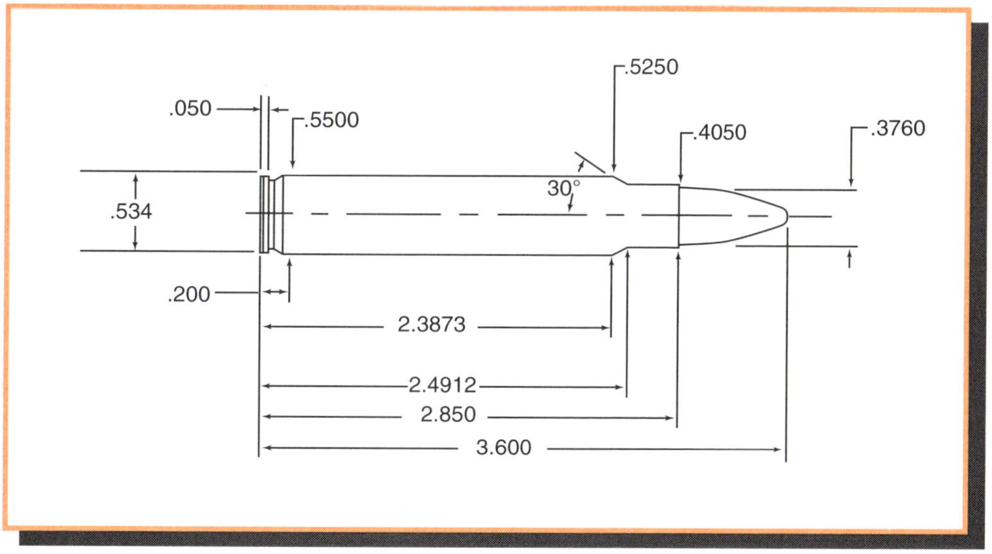

Comments:

Remington introduced their 375, the largest of the Ultra Mag family, in 2001. The 375 Ultra Mag offers approximately 25 percent more capacity than the old standby, the 375 Holland and Holland. The Ultra Mag produces on average some 300 feet per second increase in velocity over the H&H. A cartridge of the magnitude of the 375 Ultra Mag mandates the use of premium bullets such as the Barnes X, Speer Grand Slam, or Swift A-Frame. Do not attempt to roll crimp bullets without a cannelure. IMR-4831 produced the most uniform results with several different bullet weights. The 375 Ultra would seem to be somewhat over powered for the lower 48 but should make a name for itself in Africa.

Test Components:

Cases . . . . .Remington
Trim-to Length . . . . .2.840"
Primers . . . . .Federal 215
Primer Size . . . . .Large Rifle, Magnum
Lyman Shell Holder . . . . .No. 13
Jacketed Bullets Used . . . . . .Speer SP #2471, 235 gr.
Nosler Partition #44850, 260 gr.
Barnes X #37585, 270 gr.
Speer Grand Slam #2473, 285 gr.
Swift A-Frame, 300 gr.
Barnes RN Solid #37525, 300 gr.

Test Specifications:
(Velocity & Pressure)

Firearm Used . . . . .Universal Receiver
Barrel Length . . . . .26"
Twist . . . . .1-12"
Groove Dia. . . . . .375"

235 gr. Jacketed SP
3.550" OAL
BC: .317
SD: .239

Powder	Sugg Starting Grains	Velocity fps	Pressure P.S.I.	Max Load Grains	Velocity fps	Pressure P.S.I.
AA-2700	91.0	3038	54,700	99.0	3212	61,000
760	92.0	3087	57,100	100.0	3245	62,500
H-414	92.0	3111	59,300	100.5+	3262	64,400
N160	96.0	2989	47,400	**104.0+**	**3239**	**61,200**
IMR-4350	93.0	2952	47,200	101.0	3253	62,900
XMR-4350	95.0	2946	44,400	103.0+	3224	58,800
RX19	99.0	2977	47,700	108.0+	3245	61,000
IMR-4831	98.0	2965	48,300	106.0+	3235	62,300
H-4831	99.0	2906	46,400	108.0+	3148	59,400
WXR	99.0	2975	47,700	108.0+	3230	61,300
Mag Pro	103.0	2962	41,500	115.0+	3218	51,800

260 gr. Jacketed SP
3.565" OAL
BC: .314
SD: .264

Powder	Sugg Starting Grains	Velocity fps	Pressure P.S.I.	Max Load Grains	Velocity fps	Pressure P.S.I.
H-414	82.0	2677	48,500	89.0	2926	59,900
N160	86.0	2766	48,500	94.0	2980	60,200
IMR-4350	86.0	2734	45,800	93.5	3005	60,700
XMR-4350	86.0	2740	45,500	94.0	3008	60,600
RX19	90.0	2730	45,400	97.5	2995	59,300
IMR-4831	88.0	2715	45,600	**96.0**	**3003**	**61,100**
H-4831	93.0	2711	45,500	100.5	2973	61,600
WXR	89.0	2766	48,600	96.7	2978	59,100
Mag Pro	94.0	2662	37,600	104.0+	2985	52,100

Note: Loads shown in shaded panels are maximum.
Loads shown in bold designate potentially most accurate load.
\+ Designates a compressed powder charge.

375 Remington Ultra Mag

270 gr. Barnes X
3.565" OAL

BC: .503
SD: .275

Powder	Sugg Starting Grains	Velocity fps	Pressure P.S.I.	Max Load Grains	Velocity fps	Pressure P.S.I.
H-414	83.0	2669	47,700	90.0	2910	60,400
N160	88.0	2729	48,600	95.0	2954	61,100
IMR-4350	85.0	2687	46,900	92.5	2938	60,000
XMR-4350	88.0	2740	48,400	95.0	2983	61,700
IMR-4831	90.0	2730	49,300	**98.0**	**2968**	**60,100**
RX19	91.0	2739	48,500	**99.0**	**2989**	**61,500**
H-4831	93.0	2725	50,400	100.5+	2920	61,200
WXR	91.0	2745	49,200	99.0+	2984	61,300
IMR-7828	92.0	2682	46,200	100.0+	2923	57,000
Mag Pro	99.0	2763	43,700	110.0+	3032	55,860

285 gr. Jacketed SP
3.570" OAL

BC: .354
SD: .290

Powder	Sugg Starting Grains	Velocity fps	Pressure P.S.I.	Max Load Grains	Velocity fps	Pressure P.S.I.
H-414	81.0	2631	49,000	88.0	2850	60,000
N160	86.0	2650	46,600	93.5	2891	60,800
IMR-4350	85.0	2657	46,400	92.0	2921	61,700
XMR-4350	86.0	2623	44,000	93.0	2901	60,900
IMR-4831	88.0	2646	46,200	**96.0**	**2920**	**61,900**
RX19	90.0	2650	44,500	97.5	2925	60,100
XMR-3100	90.0	2623	45,900	97.2+	2875	60,000
H-4831	91.0	2630	46,400	98.5	2872	61,000
RX22	90.0	2762	50,700	97.0	2960	61,000
WXR	90.0	2676	46,400	98.0	2929	61,100
IMR-7828	92.0	2681	47,200	100.0+	2944	61,200

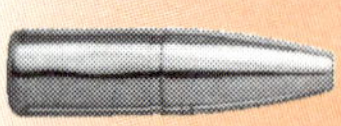

300 gr. Jacketed A-Frame
3.550" OAL

BC: .325
SD: .305

Powder	Sugg Starting Grains	Velocity fps	Pressure P.S.I.	Max Load Grains	Velocity fps	Pressure P.S.I.
N160	83.0	2542	46,400	90.5	2773	59,900
IMR-4350	83.0	2560	46,300	89.5	2802	60,300
XMR-4350	84.0	2552	45,200	90.5	2803	60,900
RX19	88.0	2578	45,700	95.0	2835	60,900
XMR-3100	88.0	2524	45,200	95.5	2784	60,700
H-4831	89.0	2575	48,200	**97.0**	**2794**	**61,600**
RX22	85.0	2638	50,000	92.0	2820	59,500
N165	89.0	2581	45,800	97.0	2805	58,900
WXR	86.0	2568	46,300	93.5	2809	60,500
IMR-7828	89.0	2559	45,800	97.0	2823	60,500

300 gr. RN Solid
3.590" OAL

BC: .307
SD: .305

Powder	Sugg Starting Grains	Velocity fps	Pressure P.S.I.	Max Load Grains	Velocity fps	Pressure P.S.I.
N160	89.0	2685	50,500	96.0	2861	60,100
IMR-4350	86.0	2681	50,300	93.5	2877	60,700
XMR-4350	89.0	2698	49,800	97.0+	2895	60,700
RX19	92.0	2661	47,000	100.0+	2893	59,300
XMR-3100	91.0	2585	45,800	99.0+	2830	59,300
H-4831	91.0	2619	48,400	99.0+	2806	59,000
RX22	91.0	2662	46,400	**99.0+**	**2893**	**58,100**
IMR-7828	91.0	2566	43,600	99.0+	2791	54,600
Mag Pro	94.5	2610	40,900	105.0+	2870	53,000

Note: Loads shown in shaded panels are maximum.
Loads shown in bold designate potentially most accurate load.
\+ Designates a compressed powder charge.

378 Weatherby Magnum

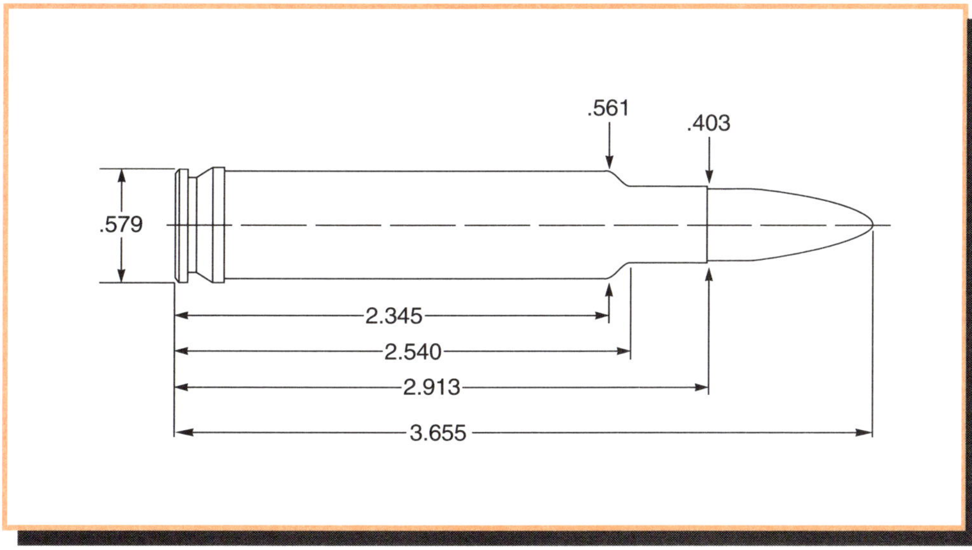

Comments:

This is one of the world's most powerful cartridges. It uses the 416 Rigby case necked down to accept a 0.375" bullet. Its recoil is in proportion to its enormous power. Few shooters can handle this cartridge unless the rifle is equipped with a muzzle brake. Due to extremely heavy recoil, the crimping of all bullets is urged.

With a muzzle brake and good reloads it is not difficult to get groups approaching one inch at 100 yards. Very strong premium grade bullets are required to withstand impact forces. Lesser bullets will simply disintegrate before penetration.

Popular priced bullets are satisfactory for range practice. Shooters will need practice to avoid flinching when using this round. It is the smallest diameter round that many serious African professional hunters will recommend for cape buffalo or elephant.

Magnum primers are essential to igniting the heavy powder charges. The Federal 215 primer is exclusively recommended by Weatherby.

The accompanying data is for use only in Weatherby Mark V rifles.

Test Components:

Cases . . . . .Weatherby
Trim-to Length . . . . .2.903"
Primers . . . . .Federal 215 & 210
Primer Size . . . . .Large Rifle, Magnum & Standard
Lyman Shell Holder . . . . .No. 17
Jacketed Bullets Used . . . . .Barnes X #37585, 270 gr.
Barnes RN Solid, #37525, 300 gr.
Cast Bullets Used . . . .(sized to .375" dia)
*gas check bullet #375248, 249 gr.
*#375449, 264 gr.

Test Specifications:
(Velocity Only)

Firearm Used . . . . .Weatherby Mk V
Barrel Length . . . . .26"
Twist . . . . .1-12"
Groove Dia. . . . . .375"

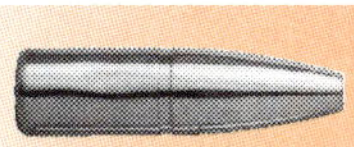

270 gr. Barnes X
3.625" OAL
BC: .503
SD: .275

Powder	Sugg Starting Grains	Velocity fps	Pressure	Max Load Grains	Velocity fps	Pressure
XMR-4350	98.0	2738	—	103.0	2946	—
RX19	**102.0**	**2842**	—	107.0	3016	—
H-4831	105.0	2788	—	110.0	2939	—
RX22	105.0	2817	—	110.0	2962	—
IMR-7828	105.0	2791	—	110.0	2956	—

300 gr. RN Solid
3.625" OAL
BC: .307
SD: .305

Powder	Sugg Starting Grains	Velocity fps	Pressure	Max Load Grains	Velocity fps	Pressure
RX19	100.0	2756	—	**105.0**	**2921**	—
XMR-3100	100.0	2655	—	105.0	2791	—
H-4831	104.0	2739	—	109.0	2862	—
RX22	104.0	2725	—	109.0	2842	—
IMR-7828	103.0	2719	—	108.0	2845	—

Note: Loads shown in shaded panels are maximum.
Loads shown in bold designate potentially most accurate load.

378 Weatherby Magnum

***#375248** 249 gr. (#2 Alloy) 3.325" OAL — **BC:** .290 **SD:** .253

Powder	Sugg Starting Grains	Velocity fps	Pressure	Max Load Grains	Velocity fps	Pressure
SR-4759	**26.0**	**1535**	—	35.0	1883	—
IMR-4227	30.0	1509	—	40.0	1899	—
XMP-5744	29.0	1525	—	42.0	1907	—
IMR-4198	29.0	1538	—	42.5	1936	—
RX7	32.0	1506	—	45.0	1920	—

***#375449** 264 gr. (#2 Alloy) 3.500" OAL — **BC:** .315 **SD:** .268

Powder	Sugg Starting Grains	Velocity fps	Pressure	Max Load Grains	Velocity fps	Pressure
SR-4759	28.0	1558	—	38.5	1941	—
IMR-4227	32.0	1562	—	46.0	1992	—
XMP-5744	**32.0**	**1574**	—	46.0	2014	—
IMR-4198	31.0	1560	—	45.0	1982	—
RX7	**33.0**	**1580**	—	50.0	1990	—

Note: Loads shown in shaded panels are maximum.
Loads shown in bold designate potentially most accurate load.
* Indicates use of Federal 210 primers.

40 Smith & Wesson (Rifle Data)

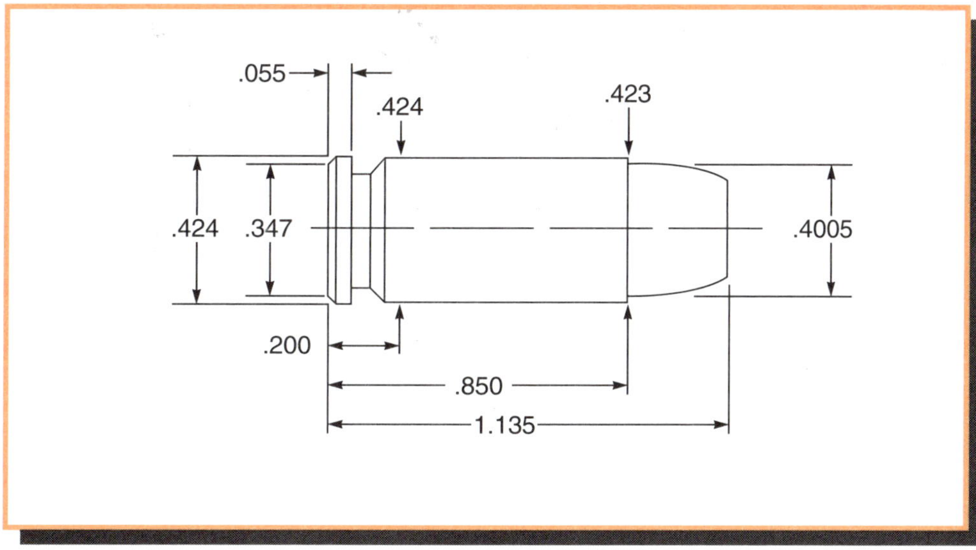

Comments:

The 40 Smith and Wesson has achieved widespread, popular acceptance since its introduction in 1992. It was only a matter of time before it was offered in a lightweight carbine. Loading for the 40 S &W is no different than any other auto loading pistol cartridge. The ballistics listed below were worked up in a pressure barrel and fired through a Ruger PC-4 Carbine. **Do not go below listed starting charges.** Doing so may cause a bullet to lodge in the rifle-length barrel. Firing a round with a bullet stuck in the barrel is extremely dangerous.

Test Components:

Cases . . . . .Winchester
Trim-to Length . . . . .847"
Primers . . . . .Winchester WSP
Primer Size . . . . .Small Pistol
Lyman Shell Holder . . . . .No. 15
Jacketed Bullets Used . . . .Winchester Silvertip, 155 gr.
Speer TMJ #4410, 165 gr.
Winchester Silvertip, 175 gr.
Sierra JHP #8460, 180 gr.
Cast Bullets Used . . . .(sized to .401" dia)
#401654, 150 gr.
#401638, 175 gr.

Test Specifications:
(Velocity Only)

Firearm Used . . . . .Ruger Carbine, PC-4
Barrel Length . . . . .16¼"
Twist . . . . .1-16"
Groove Dia. . . . . .401"

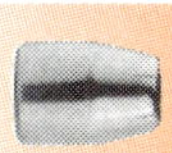

155 gr. Jacketed Silvertip
1.125" OAL
BC: .166
SD: .138

Powder	Sugg Starting Grains	Velocity fps	Pressure	Max Load Grains	Velocity fps	Pressure
Bullseye	5.1	1140	—	5.7	1257	—
231	5.3	1085	—	5.9	1204	—
AA#5	7.1	1184	—	8.0	1295	—
Unique	5.8	1113	—	6.5	1219	—
Power Pistol	7.0	1320	—	**7.8**	**1437**	—
N340	5.5	1008	—	6.2	1210	—
HS-6	7.6	1143	—	8.5	1288	—
800X	7.2	1228	—	8.0	1342	—
Blue Dot	8.2	1093	—	9.2	1235	—

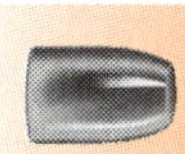

165 gr. Jacketed TMJ
1.120" OAL
BC: .135
SD: .147

Powder	Sugg Starting Grains	Velocity fps	Pressure	Max Load Grains	Velocity fps	Pressure
231	5.2	1105	—	5.8	1171	—
Unique	5.4	1092	—	**6.0**	**1182**	—
Power Pistol	6.3	1196	—	7.0	1319	—
N340	5.4	1057	—	6.1	1189	—
HS-6	7.6	1142	—	8.5	1276	—
800X	7.2	1251	—	8.0	1322	—
Blue Dot	8.2	1146	—	9.2	1300	—

Note: Loads shown in shaded panels are maximum.
Loads shown in bold designate potentially most accurate load.

40 Smith & Wesson (Rifle Data)

175 gr. Jacketed Silvertip
1.125" OAL

BC: .142
SD: .155

Powder	Sugg Starting Grains	Velocity fps	Pressure	Max Load Grains	Velocity fps	Pressure
Bullseye	4.7	1021	—	5.3	1156	—
231	5.1	1016	—	5.7	1116	—
Unique	5.3	984	—	5.9	1134	—
Power Pistol	6.0	1090	—	6.7	1229	—
Universal	4.9	955	—	5.5	1091	—
N340	5.4	1042	—	**6.0**	**1161**	—
WSF	5.8	1075	—	6.5	1164	—
HS-6	6.9	1006	—	7.7	1127	—
800X	6.8	1136	—	7.6	1242	—
Blue Dot	7.9	1066	—	8.8	1231	—

180 gr. Jacketed HP
1.115" OAL

BC: .140
SD: .161

Powder	Sugg Starting Grains	Velocity fps	Pressure	Max Load Grains	Velocity fps	Pressure
Bullseye	5.0	1095	—	5.6	1196	—
231	5.3	1082	—	5.9	1161	—
Unique	5.2	1030	—	5.8	1128	—
Power Pistol	6.0	1106	—	6.7	1231	—
Universal	4.9	1019	—	5.5	1119	—
N340	5.4	1087	—	6.0	1172	—
WSF	5.5	1051	—	6.2	1157	—
HS-6	7.2	1101	—	8.0	1236	—
800X	6.6	1145	—	**7.4**	**1238**	—
Blue Dot	7.9	1124	—	8.8	1273	—

#401654
150 gr. (#2 Alloy) 1.090" OAL

BC: .074
SD: .133

Powder	Sugg Starting Grains	Velocity fps	Pressure	Max Load Grains	Velocity fps	Pressure
Bullseye	5.4	1252	—	6.0	1338	—
AA#2	5.7	1238	—	6.4	1333	—
231	5.8	1265	—	6.5	1365	—
AA#5	6.7	1104	—	7.5	1221	—
Unique	5.6	1123	—	6.3	1264	—
Power Pistol	6.3	1253	—	**7.0**	**1344**	—
N340	5.3	1121	—	5.9	1240	—
HS-6	7.9	1221	—	8.8	1377	—
800X	7.2	1267	—	8.0	1364	—
Blue Dot	8.6	1256	—	9.6	1390	—

#401638
175 gr. (#2 Alloy) 1.100" OAL

BC: .088
SD: .155

Powder	Sugg Starting Grains	Velocity fps	Pressure	Max Load Grains	Velocity fps	Pressure
Bullseye	4.5	1071	—	5.0	1159	—
231	5.2	1124	—	5.8	1200	—
AA#5	6.2	1047	—	6.9	1159	—
Unique	5.2	1091	—	5.8	1184	—
Power Pistol	5.8	1136	—	**6.5**	**1251**	—
N340	4.8	1001	—	5.4	1160	—
HS-6	7.5	1211	—	8.4	1344	—
800X	6.8	1177	—	7.6	1298	—
Blue Dot	7.7	1184	—	8.6	1292	—

Note: Loads shown in shaded panels are maximum.
Loads shown in bold designate potentially most accurate load.

40-65 Winchester

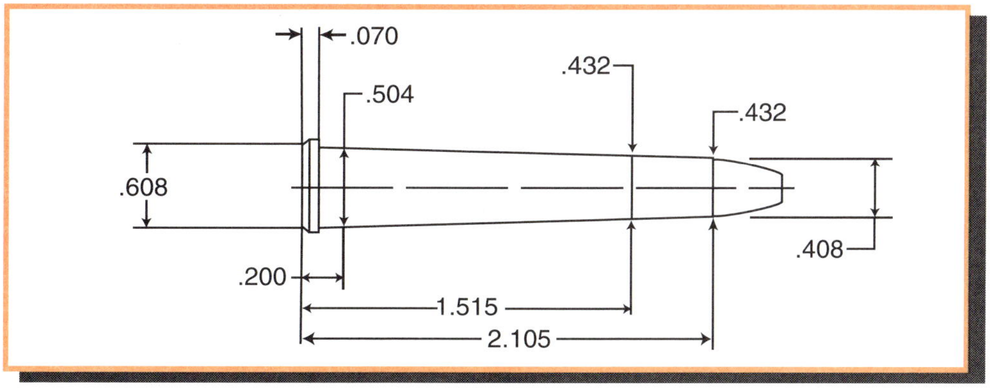

Comments:

Winchester introduced this cartridge in 1886 for use in both the 1885 single-shot and 1886 lever action. This was originally a black powder cartridge that survived in smokeless form before finally being dropped during the 1930s. This number has enjoyed resurgence in popularity in recent years in both Cowboy Action Shooting and Black Powder Cartridge Silhouette. The 40-65 has several factors in its favor; forming brass requires little more than running ever plentiful 45-70 Springfield cases through a full-length sizing die. Lyman offers several moulds producing bullets with good ballistic coefficients suitable long range shooting. Recoil with even the heaviest 400-grain bullets is mild thus enabling shooters to fire sixty-round matches without tiring out. Many older guns will show varying groove diameters, often around .406". Newly manufactured reproduction rifles utilize groove diameters of .408" to .409". All current Lyman die sets for the 40-65 are manufactured to accommodate .408"/ .409" bullet diameters. There is no SAAMI pressure limit for the 40-65. Lyman determined the maximum pressure limit used here by firing 40-65 cartridges loaded with compressed black powder in a special pressure barrel. Reloder 7 and XMP-5744 are excellent powder choices for this cartridge.

This data is intended for use in modern, newly manufactured guns rated for smokeless powder. This data is not for use in antique guns originally built for black powder.

Test Components:

Cases . . . . .Remington
Trim-to Length . . . . .2.100"
Primers . . . . .Remington 9½
Primer Size . . . . .Large Rifle
Lyman Shell Holder . . . . .No. 17
Cast Bullets Used . . . .(sized to .410" dia)
#410660, 385 gr.
#410655, 400 gr.
#410663, 400 gr.

Test Specifications:
(Velocity & Pressure)

Firearm Used . . . . .Universal Receiver
Barrel Length . . . . .36"
Twist . . . . .1-16"
Groove Dia. . . . . .409"

#410660 385 gr. (#2 Alloy) 2.940" OAL — **BC:** .352 **SD:** .327

Powder	Sugg Starting Grains	Velocity fps	Pressure C.U.P.	Max Load Grains	Velocity fps	Pressure C.U.P.
SR-4759	19.0	1270	11,200	21.5	1398	17,800
IMR-4198	21.5	1271	11,800	25.0	1431	16,600
XMP-5744	18.0	1135	11,900	**21.0**	**1274**	**16,000**
RX7	22.0	1298	10,200	28.0	1516	17,800
IMR-3031	28.0	1271	12,100	33.0	1525	18,200
IMR-4895	29.5	1265	12,100	34.0	1472	17,700
Varget	31.5	1294	11,800	36.5	1543	18,100

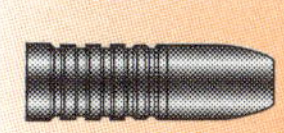

#410655 400 gr. (#2 Alloy) 2.780" OAL — **BC:** .350 **SD:** .340

Powder	Sugg Starting Grains	Velocity fps	Pressure C.U.P.	Max Load Grains	Velocity fps	Pressure C.U.P.
SR-4759	16.0	1113	11,200	19.0	1245	17,600
IMR-4198	19.0	1179	14,400	22.0	1293	17,300
XMP-5744	16.5	1073	13,800	20.0	1237	17,900
RX7	17.0	1064	13,000	20.5	1200	17,300
IMR-3031	25.0	1183	13,300	28.0	1334	16,900
IMR-4895	25.0	1215	15,400	**29.0**	**1298**	**17,500**
Varget	28.0	1173	10,400	32.5	1389	17,000

#410663 400 gr. (#2 Alloy) 2.860" OAL — **BC:** .435 **SD:** .340

Powder	Sugg Starting Grains	Velocity fps	Pressure C.U.P.	Max Load Grains	Velocity fps	Pressure C.U.P.
SR-4759	17.0	1140	8,900	21.0	1373	17,900
IMR-4198	21.0	1236	11,200	25.0	1442	18,700
XMP-5744	17.5	1104	11,600	22.0	1307	18,500
RX7	18.5	1208	15,300	23.5	1359	18,100
IMR-3031	28.0	1258	11,100	33.0	1488	18,200
IMR-4895	28.0	1249	15,000	**32.0**	**1483**	**17,800**
Varget	28.5	1235	8,200	33.5	1418	17,000

Note: Loads shown in shaded panels are maximum.
Loads shown in bold designate potentially most accurate load.

40-70 Sharps Straight

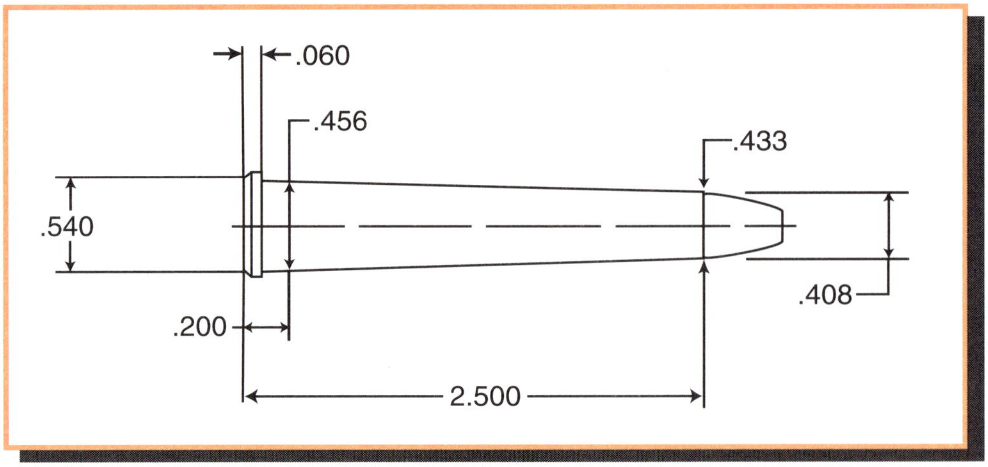

Comments:

This is yet another cartridge brought back from extinction by the Black Powder Cartridge Silhouette crowd. This cartridge dates to about 1876 and should not be confused with the bottlenecked 40-70 also produced by Sharps. Ballistics of the 40-70 Sharps Straight average around 70 to 100 feet per second over the 40-65 Winchester. Original Sharps specifications called for a (sometimes) paper-patched bullet of .403" diameter. Currently manufactured reproduction rifles feature groove diameters of .408"/. 409". All Lyman die sets for the 40-70 are designed for use with bullets sized to these diameters.

There is no SAAMI maximum average pressure for this cartridge. Lyman determined the maximum pressure limit of 20,000 CUP by firing 40-70 cartridges loaded with various granulations of compressed black powder in a special pressure test barrel. The cases used in load development were based on unformed, full-length (2½ inches long) 30-40 Krag cases with a Remington headstamp and were supplied by Buffalo Arms of Sandpoint, ID. Shooters loading cases of different manufacture may have different case volumes and should carefully work up loads.

This data is intended for use in modern, newly manufactured guns rated for smokeless powder. This data is not for use in antique guns originally built for black powder.

Test Components:

Cases . . . . Buffalo Arms
Trim-to Length . . . . 2.490"
Primers . . . . Remington 9½
Primer Size . . . . Large Rifle
Lyman Shell Holder . . . . No. 33
Cast Bullets Used . . . . (sized to .409" dia)
#410660, 385 gr.
#410655, 400 gr.
#410663, 400 gr.

Test Specifications:
(Velocity & Pressure)

Firearm Used . . . . Universal Receiver
Barrel Length . . . . 32"
Twist . . . . 1-16"
Groove Dia. . . . . .4075"

#410660 — 385 gr. (#2 Alloy) 3.250" OAL — **BC:** .352 **SD:** .327

Powder	Sugg Starting Grains	Velocity fps	Pressure C.U.P.	Max Load Grains	Velocity fps	Pressure C.U.P.
SR-4759	21.0	1162	8,300	24.0	1377	15,800
XMP-5744	21.0	1115	10,700	26.5	1368	16,900
IMR-4198	24.0	1269	12,000	30.0	1501	17,000
RX7	23.0	1129	9,100	29.5	1452	16,700
IMR-3031	31.0	1225	8,900	**37.5**	**1464**	**14,100**
Varget	36.0	1337	10,100	42.5	1613	17,000

#410655 — 400 gr. (#2 Alloy) 3.115" OAL — **BC:** .350 **SD:** .340

Powder	Sugg Starting Grains	Velocity fps	Pressure C.U.P.	Max Load Grains	Velocity fps	Pressure C.U.P.
SR-4759	19.0	1141	10,600	22.0	1332	17,300
XMP-5744	22.5	1224	14,100	**25.0**	**1320**	**17,300**
IMR-4198	23.0	1250	12,600	27.2	1420	16,900
RX7	22.0	1113	10,600	27.0	1375	17,100
Varget	31.0	1188	9,200	39.0	1535	17,200

#410663 — 400 gr. (#2 Alloy) 3.170" OAL — **BC:** .435 **SD:** .340

Powder	Sugg Starting Grains	Velocity fps	Pressure C.U.P.	Max Load Grains	Velocity fps	Pressure C.U.P.
SR-4759	19.0	1172	10,700	21.5	1326	16,600
XMP-5744	20.0	1086	11,700	25.0	1344	17,500
IMR-4198	21.0	1182	12,800	**26.5**	**1401**	**17,300**
RX7	21.0	1084	10,000	26.5	1377	16,900
IMR-3031	29.0	1211	11,400	36.0	1494	17,200
Varget	30.0	1218	9,600	38.5	1515	16,500

Note: Loads shown in shaded panels are maximum.
Loads shown in bold designate potentially most accurate load.

416 Rigby

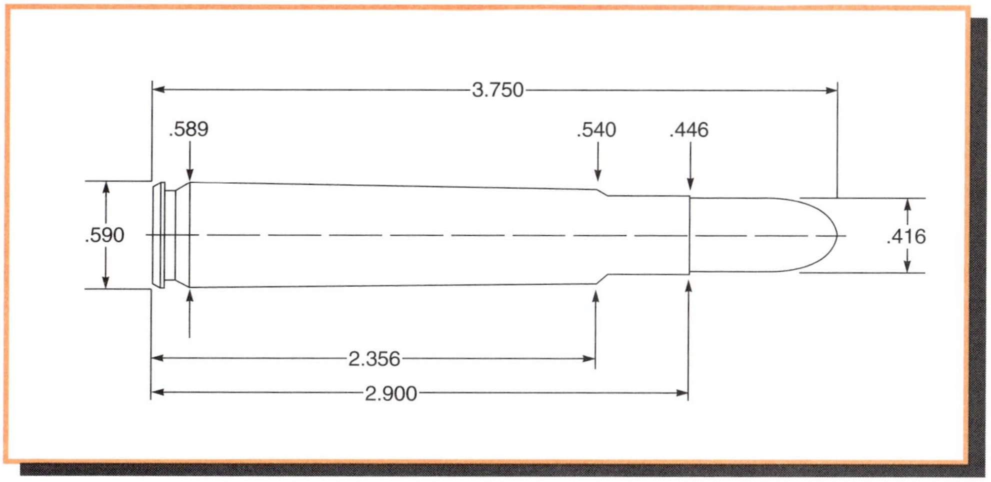

Comments:

The 416 Rigby has been a standard African safari cartridge for most of the twentieth century. The Rigby is a non-belted case that operates at fairly low (40,000 CUP) chamber pressure. Its positive feeding and easy case extraction offers piece of mind when the hunter is pursuing large, disagreeable game. This particular 416 also has a reputation for longer case life than belted cases. Many shooters had known of the 416 Rigby only through its formidable reputation for many years. The Rigby's availability had been restricted to expensive custom or double-rifles since its introduction in 1911. The cartridge has seen increased popularity in recent years due to its availability in Ruger's M77 and #1 rifles. IMR-4831 and 7828 are favored powders to fill this huge case. Lighter bullets such as the Barnes "X" can obtain on average two hundred feet per second over the standard 400-grain factory loading. The Rigby case has also served as the parent case for a number of new cartridges over the last several years.

Test Components:

Cases . . . . .Federal
Trim-to Length . . . . .2.890"
Primers . . . . .Federal 215
Primer Size . . . . .Large Rifle, Magnum
Lyman Shell Holder . . . . .No. 17
Jacketed Bullets Used . . . . .Barnes X #41682, 325 gr.
Barnes Solid RN #41628, 350 gr.
Barnes Solid RN #41660, 400 gr.

Test Specifications: (Velocity Only)

Firearm Used . . . . .Ruger Model 77
Barrel Length . . . . .23"
Twist . . . . .1-16½"
Groove Dia. . . . . .415"

325 gr. Barnes X
3.650" OAL

BC: .467
SD: .268

Powder	Sugg Starting Grains	Velocity fps	Pressure	Max Load Grains	Velocity fps	Pressure
IMR-4350	87.0	2300	—	97.0	2586	—
IMR-4831	90.0	2179	—	**100.0**	**2551**	—
XMR-3100	93.0	2234	—	103.0	2556	—
RX22	91.0	2305	—	101.5	2568	—
IMR-7828	96.0	2313	—	102.0	2549	—

350 gr. Solid RN
3.625" OAL

BC: .364
SD: .289

Powder	Sugg Starting Grains	Velocity fps	Pressure	Max Load Grains	Velocity fps	Pressure
IMR-4350	89.0	2268	—	95.0	2472	—
RX19	94.0	2334	—	100.0	2503	—
IMR-4831	94.0	2287	—	100.0	2496	—
XMR-3100	96.0	2284	—	102.0	2479	—
RX22	95.0	2308	—	**101.0**	**2483**	—
IMR-7828	96.0	2261	—	102.5	2490	—

400 gr. Solid RN
3.625" OAL

BC: .388
SD: .330

Powder	Sugg Starting Grains	Velocity fps	Pressure	Max Load Grains	Velocity fps	Pressure
IMR-4350	88.0	2201	—	94.0	2396	—
RX19	91.0	2241	—	96.5	2386	—
IMR-4831	93.0	2263	—	**98.5**	**2370**	—
XMR-3100	95.0	2233	—	101.5	2390	—
RX22	93.0	2275	—	99.0	2394	—
IMR-7828	98.0	2295	—	104.0	2410	—

Note: Loads shown in shaded panels are maximum.
Loads shown in bold designate potentially most accurate load.

416 Weatherby Magnum

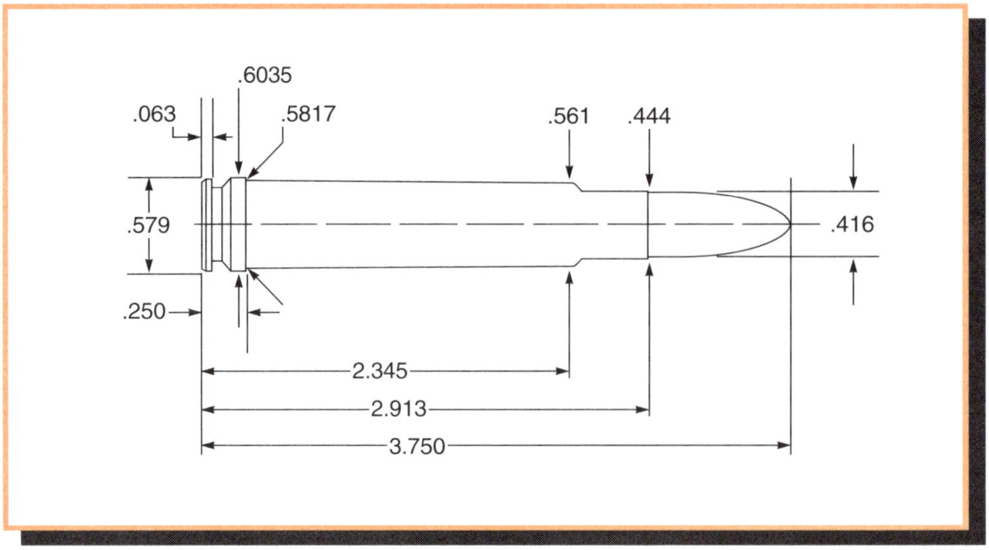

Comments:

This case is the 416 Rigby with a belt added and loaded to substantially higher pressure. The much higher chamber pressures result in almost 300 feet per second additional velocity with the 400 grain bullet compared to the 2400 feet per second velocity of factory 416 Rigby ammo. Recoil may exceed most shooters' ability. Of course the 416 Weatherby can easily be loaded down to Rigby ballistics which makes it a manageable round.

The slowest burning powders are ideal. For Rigby level ballistics IMR 4831 seems best. For full power loads, IMR 7828 is the wisest selection, though all the listed propellants are quite satisfactory. This is enough cartridge to make almost anyone believe in muzzle brakes. Even then full power loads are nothing for the timid shooter. Such loads would undoubtedly give an advantage when hunting elephant as compared to the Remington or Rigby 416's, but for almost any other purpose they might be overly potent.

Test Components:

Cases . . . . .Weatherby
Trim-to Length . . . . .2.903"
Primers . . . . .Federal 215
Primer Size . . . . .Large Rifle, Magnum
Lyman Shell Holder . . . . .No. 17
Jacketed Bullets Used . . . . .Barnes X #41682, 325 gr.
Barnes Solid RN #41628, 350 gr.
Barnes Solid RN #41660, 400 gr.

Test Specifications: (Velocity Only)

Firearm Used . . . . .Weatherby Mk V
Barrel Length . . . . .24"
Twist . . . . .1-14"
Groove Dia. . . . . .416"

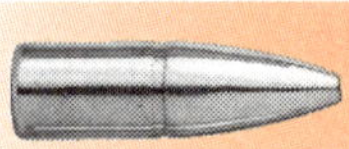

325 gr. Barnes X
3.700" OAL
BC: .467
SD: .268

Powder	Sugg Starting Grains	Velocity fps	Pressure C.U.P.	Max Load Grains	Velocity fps	Pressure C.U.P.
IMR-4350	102.0	2627	—	113.0+	2914	—
RX19	108.0	2613	—	120.0+	2902	—
IMR-4831	105.0	2603	—	117.0+	2877	—
AA 3100	107.0	2551	—	119.0+	2767	—
RX22	111.0	2586	—	124.0+	2887	—
IMR-7828	110.0	2536	—	**122.0+**	**2819**	—

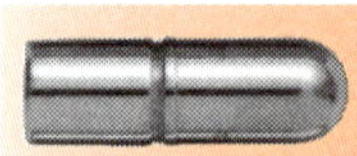

350 gr. Solid RN
3.640" OAL
BC: .364
SD: .289

Powder	Sugg Starting Grains	Velocity fps	Pressure C.U.P.	Max Load Grains	Velocity fps	Pressure C.U.P.
IMR-4350	98.0	2464	—	109.0+	2783	—
RX19	106.0	2533	—	118.0+	2822	—
IMR-4831	102.0	2460	—	114.0+	2766	—
AA3100	106.0	2469	—	118.0+	2687	—
RX22	109.0	2477	—	121.0+	2756	—
IMR-7828	107.0	2445	—	**119.0+**	**2727**	—

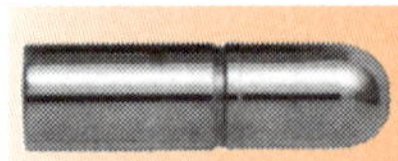

400 gr. Solid RN
3.750" OAL
BC: .388
SD: .330

Powder	Sugg Starting Grains	Velocity fps	Pressure C.U.P.	Max Load Grains	Velocity fps	Pressure C.U.P.
IMR-4350	94.0	2261	—	104.0+	2609	—
RX19	101.0	2307	—	108.0+	2650	—
IMR-4831	100.0	2309	—	**110.0+**	**2648**	—
RX22	104.0	2286	—	112.0+	2643	—
IMR-7828	104.0	2290	—	116.0+	2635	—

Note: Loads shown in shaded panels are maximum.
Loads shown in bold designate potentially most accurate load.
\+ Designates a compressed powder charge.

416 Remington Magnum

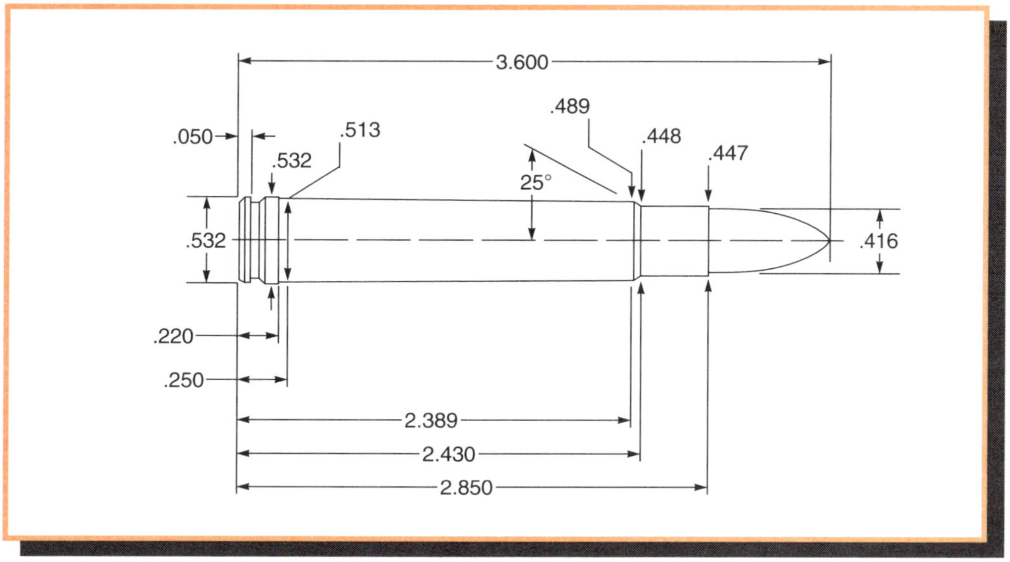

Comments:

This cartridge is a necked up 8mm Remington Magnum. It is ideal for big game applications from brown bear to elephant.

The 416 Remington Magnum is not much different from the wildcat 416 Hoffman. It offers all the power of the 416 Rigby (albeit at higher chamber pressure) and the advantage of more shells in the magazine because of its narrower case diameter.

The 400 grain bullet offers the most punch, but lighter bullets better fit the needs of most North American hunting. Hercules Reloder 15 is the single best propellant and 1" groups are possible with it and the 400 grain bullets.

Test Components:

Cases . . . . .Remington
Trim-to Length . . . . .2.840"
Primers . . . . .Remington 9½ M
Primer Size . . . . .Large Rifle, Magnum
Lyman Shell Holder . . . . .No. 13
Jacketed Bullets Used . . . . .Barnes X #41682, 325 gr.
Barnes Solid RN #41628, 350 gr.
Barnes Solid RN #41660, 400 gr.

Test Specifications:
(Velocity & Pressure)

Firearm Used . . . . .Universal Receiver
Barrel Length . . . . .24"
Twist . . . . .1-14"
Groove Dia. . . . . .4165"

325 gr. Barnes X 3.600" OAL — **BC:** .467 **SD:** .268

Powder	Sugg Starting Grains	Velocity fps	Pressure C.U.P.	Max Load Grains	Velocity fps	Pressure C.U.P.
XMR-2015	68.0	2345	42,900	76.0	2547	52,700
IMR-3031	74.5	2466	40,300	83.0+	2708	53,100
IMR-4895	75.5	2450	39,700	84.4+	2699	53,100
IMR-4064	77.0	2426	38,200	85.7+	2702	51,500
Varget	77.0	2445	37,400	**86.0+**	**2703**	**51,100**
IMR-4320	76.0	2428	39,800	84.2+	2651	51,400
RX15	80.5	2552	42,200	88.5+	2753	52,300
N140	78.0	2448	40,400	87.0+	2696	50,600
H-380	81.0	2389	38,300	91.0+	2644	48,500

350 gr. Solid RN 3.590" OAL — **BC:** .364 **SD:** .289

Powder	Sugg Starting Grains	Velocity fps	Pressure C.U.P.	Max Load Grains	Velocity fps	Pressure C.U.P.
XMR-2015	65.0	2171	37,700	72.0	2403	49,800
IMR-3031	74.0	2389	41,200	82.5+	2608	52,300
IMR-4895	75.0	2370	41,300	84.2+	2610	52,800
IMR-4064	76.5	2364	39,700	85.5+	2613	52,400
Varget	76.0	2371	38,200	84.0+	2587	48,100
IMR-4320	75.0	2341	40,300	84.0+	2565	49,900
RX15	79.0	2442	41,800	88.0+	2668	52,500
N140	76.0	2339	38,100	**84.0+**	**2577**	**49,500**
H-380	81.0	2329	39,800	91.0+	2567	48,700

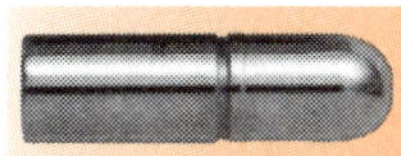

400 gr. Solid RN 3.590" OAL — **BC:** .388 **SD:** .330

Powder	Sugg Starting Grains	Velocity fps	Pressure C.U.P.	Max Load Grains	Velocity fps	Pressure C.U.P.
IMR-3031	67.5	2149	40,700	75.2+	2353	52,500
IMR-4895	69.0	2151	40,000	77.3+	2376	52,900
IMR-4064	71.0	2156	39,800	78.8+	2385	52,000
Varget	70.0	2131	37,700	77.0+	2359	49,400
IMR-4320	70.0	2106	39,000	77.2+	2320	50,800
RX15	71.5	2155	38,000	80.0+	2415	51,600
N140	70.0	2112	37,700	**78.0+**	**2327**	**48,400**
AA2700	78.0	2103	32,400	86.0+	2325	42,500
H-380	78.0	2200	42,200	87.0+	2403	50,400

Note: Loads shown in shaded panels are maximum.
Loads shown in bold designate potentially most accurate load.
\+ Designates a compressed powder charge.

44-40 Winchester (44 WCF)

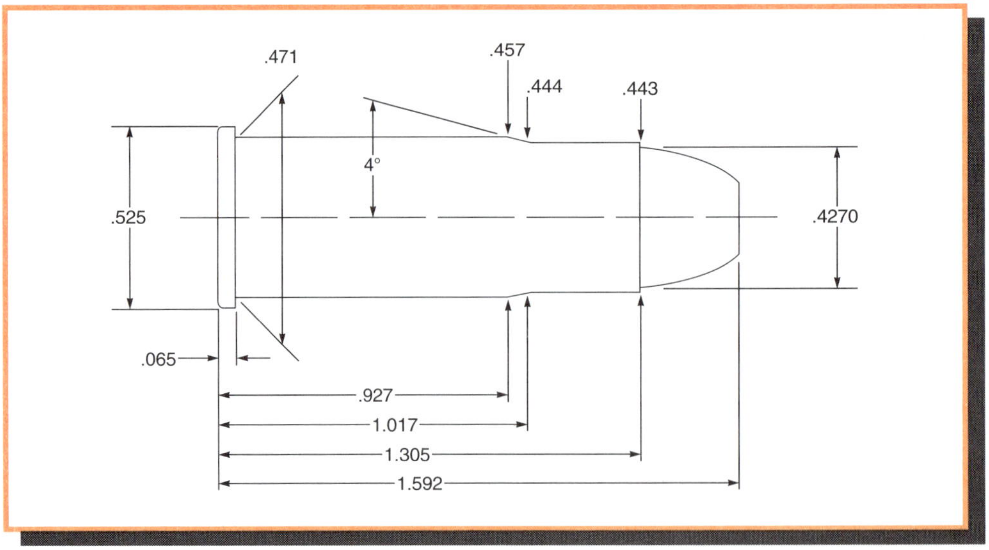

Comments:

These loads should not be used in handguns or in rifles that were designed for black powder.

Individual tolerances vary greatly in rifles chambered for this cartridge. Therefore, extreme care should be used in working up maximum loads. CAUTION: THE RELATIVE STRENGTH OF 44/ 40 FIREARMS VARIES GREATLY WITH ACTION DESIGN AND DATE OF MANUFACTURE. As a reference we have categorized actions into two groups:

GROUP I (Weaker Actions)

Winchester Model 1873 Lever Action
Whitney-Kennedy Lever Action
Colt-Burgess Lever Action
Marlin Model 1888 Lever Action
Colt "Lightning" Slide Action
Euroarms Replica 1873 Lever Action
Navy Arms Replica 1873 Lever Action
Remington No. 2 Rolling-Block S.S.
Ballard No. 2 Single Shot
Stevens Model 44 Single Shot

GROUP 2 (Strong Actions)

Winchester Model 1892 Lever Action
Marlin Model 1889 Lever Action
Marlin Model 1894 Lever Action
Remington-Keene Bolt Action
Remington Model 14 1/2 Slide Action
Winchester Single Shot Rifles
Remington No. 1 Rolling Block
S.S.Remington "Baby Carbine" S.S.
Stevens Model 44 1/2 Single Shot

For Group 1 we recommend not to exceed 13,700 CUP maximum pressure and for Group 2 not to exceed 22,000 CUP maximum pressure. Our data is divided up into loads for each category. In all cases use caution in working up to maximum loads.

Cases should be examined to determine if they are of the old balloon head construction. These cases should be considered unsafe for smokeless powder loads. Due to variations in groove diameters, it is recommended that you slug your barrel before reloading.

Test Components:

Cases . . . . .Winchester
Trim-to Length . . . . .1.295"
Primers . . . . .Winchester WLP
Primer Size . . . . .Large Pistol
Lyman Shell Holder .No. 14B
Jacketed Bullets Used . . . . .Speer JHP # 4425, 200 gr.
Cast Bullets Used . . . .(sized to .429" dia)
#427666, 200 gr.
#427098, 205 gr.

Test Specifications: (Velocity & Pressure)

Firearm Used . . . . .Universal Receiver
Navy Arms 1873 Carbine
Barrel Length . . . . .Universal Receiver; 24"
1873 Carbine; 19"
Twist . . . . .Universal Receiver; 1-36"
1873 Carbine; 1-20"
Groove Dia. . . . . .Universal Receiver; .429"
1873 Carbine; .427"

200 gr. Jacketed HP
1.600" OAL

BC: .122
SD: .155

Group 1

Powder	Sugg Starting Grains	Velocity fps	Pressure C.U.P.	Max Load Grains	Velocity fps	Pressure C.U.P.
Unique	7.6	865	10,600	8.5	981	13,300
2400	14.4	980	6,600	16.0	1183	11,900
IMR-4227	**16.5**	**895**	**4,800**	18.5	1212	11,600

Group 2

Powder	Sugg Starting Grains	Velocity fps	Pressure C.U.P.	Max Load Grains	Velocity fps	Pressure C.U.P.
Unique	8.5	981	13,300	10.2	1282	19,600
2400	18.0	1380	14,600	20.0	1638	19,000
IMR-4227	18.5	1212	11,600	20.5	1455	19,600

Note: Loads shown in shaded panels are maximum.
Loads shown in bold designate potentially most accurate load.

44-40 Winchester (44 WCF)

#427666
200 gr. (#2 Alloy) 1.580" OAL

BC: .149
SD: .155

Group 1 Powder	Sugg Starting Grains	Velocity fps	Pressure C.U.P.	Max Load Grains	Velocity fps	Pressure C.U.P.
Bullseye	5.6	989	7,700	7.0	1165	12,600
700X	5.1	993	7,400	6.4	1119	12,600
AA#2	6.0	994	7,400	6.5	1154	12,800
Unique	6.9	1056	6,700	8.6	1226	12,600
SR-4756	**7.5**	**904**	**5,900**	9.4	1086	12,300
2400	13.2	1035	7,900	16.5	1232	12,800
IMR-4227	15.0	933	7,700	17.0	1310	12,600

#427098
205 gr. (#2 Alloy) 1.592" OAL

BC: .103
SD: .159

Group 1 Powder	Sugg Starting Grains	Velocity fps	Pressure C.U.P.	Max Load Grains	Velocity fps	Pressure C.U.P.
Bullseye	5.2	952	—	6.5	1099	—
Red Dot	5.5	980	7,500	6.5	1125	13,600
700X	4.8	921	—	6.0	1086	—
Green Dot	7.0	1059	8,700	8.0	1194	12,900
AA#2	5.4	871	—	6.3	1060	—
Unique	6.3	895	—	7.9	1127	—
SR-4756	**7.4**	**848**	—	9.3	1050	—
2400	12.0	885	—	15.0	1088	—
IMR-4227	12.6	685	—	17.0	1083	—

Group 2 Powder	Sugg Starting Grains	Velocity fps	Pressure C.U.P.	Max Load Grains	Velocity fps	Pressure C.U.P.
Red Dot	7.0	1180	17,200	8.0	1290	21,900
Green Dot	8.0	1194	12,900	9.0	1340	19,300
Unique	9.5	1319	15,200	10.5	1410	19,700

Note: Loads shown in shaded panels are maximum.
Loads shown in bold designate potentially most accurate load.

44 Remington Magnum (Rifle Data)

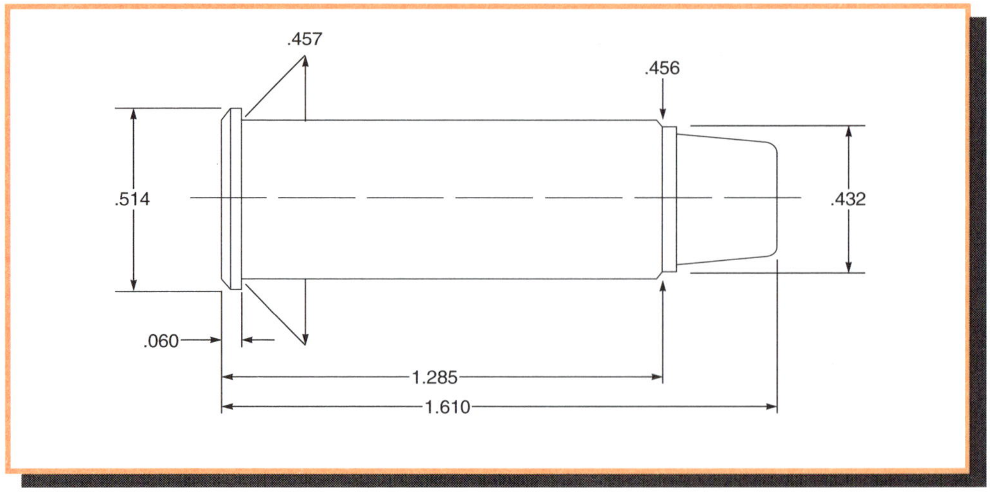

Comments:

Pressure limits for this cartridge are the same in rifles or handguns. High primers are an extraordinary danger in semi-automatic rifles. Be extra careful. Cast bullets tend to foul the gas piston in these same firearms and hence are not recommended for use in the semi-autos.

In rifles with shallow multi-groove rifling cast bullet velocity must be held to 1,600 fps. or less. The use of hard alloys and maximum sizing diameters will also help accuracy in these rifles.

Use the exact listed overall length to get correct seating depth with cast bullets. Bullet #429640 may not feed from the magazine in some Marlin rifles.

Use only flat nosed bullets in tubular magazines.

Test Components:

Cases . . . . .Remington
Trim-to Length . . . . .1.275"
Primers . . . . .CCI 300 & 350
Primer Size . . . . .Large Pistol, Std. & Magnum
Lyman Shell Holder . . . . .No. 7
Jacketed Bullets Used . . . .
Hornady HP/XTP #44100, 200 gr.
Speer JHP #4435, 225 gr.
Speer JHP #4453, 240 gr.
Hornady SP #4300, 265 gr.
Cast Bullets Used . . . .(sized to .429" dia)
*gas check bullet *#429215, 210 gr.
#429667, 240 gr.

Test Specifications: (Velocity Only)

Firearm Used . . . . .Winchester Model 94AE
Barrel Length . . . . .20"
Twist . . . . .1-26"
Groove Dia. . . . . .429"

200 gr. Jacketed HP/XTP 1.610" OAL — **BC:** .170 **SD:** .155

Powder	Sugg Starting Grains	Velocity fps	Pressure	Max Load Grains	Velocity fps	Pressure
AA#7	18.2	1695	—	20.2	1836	—
2400	**19.5**	**1603**	—	23.6	1880	—
*H110	27.7	2049	—	28.8+	2133	—
*296	27.0	2066	—	28.3+	2129	—
*H-4227	23.8	1715	—	26.5+	1870	—

225 gr. Jacketed HP 1.610" OAL — **BC:** .146 **SD:** .175

Powder	Sugg Starting Grains	Velocity fps	Pressure	Max Load Grains	Velocity fps	Pressure
Unique	10.8	1349	—	12.0	1475	—
Blue Dot	14.1	1420	—	15.7	1570	—
2400	19.8	1581	—	22.0	1753	—
N110	19.0	1578	—	**21.1**	**1717**	—
*H110	25.0	1834	—	26.0	1871	—
*296	24.0	1724	—	25.0	1822	—
*H-4227	22.5	1574	—	25.0	1750	—

Note: Loads shown in shaded panels are maximum.
Loads shown in bold designate potentially most accurate load.
* Designates the use of magnum primers.
\+ Designates a compressed powder charge.

44 Remington Magnum (Rifle Data)

240 gr. Jacketed HP
1.610" OAL

BC: .165
SD: .186

Powder	Sugg Starting Grains	Velocity fps	Pressure	Max Load Grains	Velocity fps	Pressure
Unique	10.8	1310	—	12.0	1445	—
Blue Dot	14.4	1457	—	16.0	1558	—
2400	18.4	1479	—	20.5	1602	—
N110	18.7	1555	—	20.8	1671	—
***H110**	23.5	1690	—	**24.5**	**1745**	—
*296	23.0	1654	—	24.0	1704	—
*H-4227	21.8	1504	—	24.2	1671	—

265 gr. Jacketed SP
1.610" OAL

BC: .189
SD: .205

Powder	Sugg Starting Grains	Velocity fps	Pressure	Max Load Grains	Velocity fps	Pressure
Blue Dot	12.6	1253	—	14.0	1425	—
2400	16.5	1318	—	18.3	1475	—
N110	16.6	1394	—	18.5	1523	—
*H110	20.6	1576	—	21.5	1623	—
***296**	21.0	1583	—	**22.0**	**1652**	—
*H-4227	19.3	1397	—	21.5	1538	—

#429215
210 gr. (#2 Alloy) 1.645" OAL

BC: .188
SD: .163

Powder	Sugg Starting Grains	Velocity fps	Pressure	Max Load Grains	Velocity fps	Pressure
Titegroup	9.0	1408	—	10.0	1490	—
HS-6	15.0	1621	—	16.7	1733	—
AA#7	17.2	1627	—	19.2	1805	—
Blue Dot	14.9	1588	—	16.6	1722	—
2400	19.8	1550	—	22.0	1705	—
N110	19.3	1683	—	21.5	1821	—
*H110	26.4	1929	—	27.5	1985	—
***296**	**25.9**	**1887**	—	27.0	1933	—

#429667
240 gr. (#2 Alloy) 1.645" OAL

BC: .149
SD: .186

Powder	Sugg Starting Grains	Velocity fps	Pressure	Max Load Grains	Velocity fps	Pressure
Titegroup	9.0	1356	—	10.0	1443	—
Unique	10.0	1332	—	11.7	1482	—
Herco	10.5	1362	—	11.7	1450	—
800X	12.1	1477	—	13.5	1597	—
Blue Dot	13.9	1447	—	15.5	1589	—
AA#9	18.2	1560	—	20.2	1710	—
2400	18.5	1574	—	20.6	1686	—
N110	18.0	1578	—	20.0	1695	—
*H110	22.5	1709	—	23.5	1765	—
IMR-4227	**21.6**	**1504**	—	24.0	1654	—

Note: Loads shown in shaded panels are maximum.
Loads shown in bold designate potentially most accurate load.
* Designates the use of magnum primers.

444 Marlin

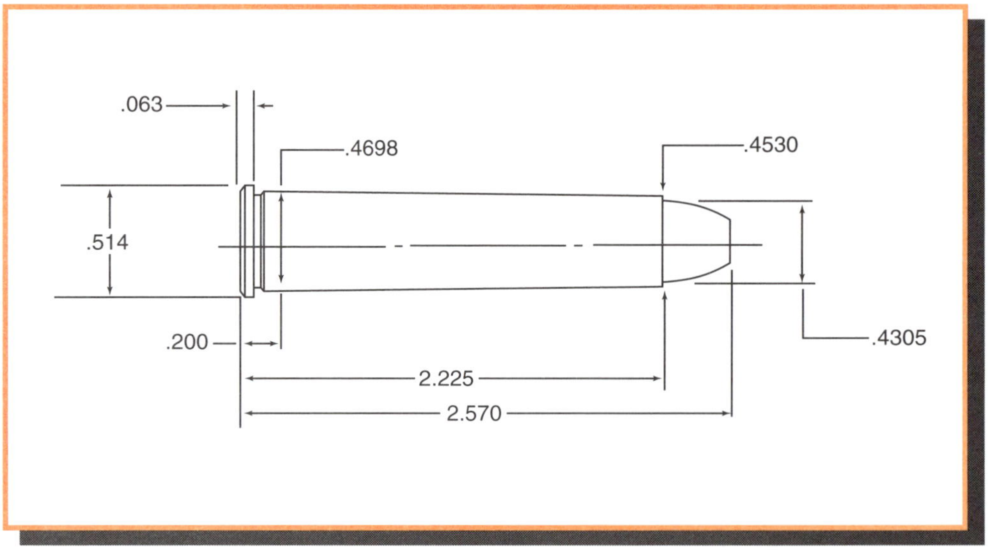

Comments:

The 444 Marlin has earned a reputation as a hard-hitting, short-range brush cartridge since its introduction in 1964. Some think of the "Triple-four" as simply a 44 Magnum stretched .940" to 2.225" case length. The performance — and appearance of the 444 is similar to the various 40-caliber black powder cartridges so prominent in the latter nineteenth and early twentieth century. The 225-grain bullet is best for plinking or off-season target practice. Hornady's 265-grain Flat-Point is constructed especially to feed and perform in the big Marlin. Many of the jacketed bullets loads are heavily compressed. Best results in loading may require seating and crimping in two separate operations. Those shooters loading cast bullets for Marlin rifles with Micro-Groove® rifling should keep velocities below 1,600 feet per second for best accuracy. A hard bullet alloy of at least 15 bhn will also help.

The 18-inch ported barrel of the Winchester Timber Carbine used in our tests yielded a dramatic difference in velocity compared to other published data. This difference is often upwards of two to three hundred feet per second. Most other sources of data for the 444 Marlin utilize unported 24-inch test barrels. Reloder 7 and IMR-4198 with jacketed bullets have worked well in the Marlin over the years.

Test Components:

Cases . . . . .Remington
Trim-to Length . . . . .2.215"
Primers . . . . .Remington 9½
Primer Size . . . . .Large Rifle
Lyman Shell Holder . . . . .No. 14B
Jacketed Bullets Used . . . . .Speer JHP #4435, 225 gr.
Speer JHP #4453, 240 gr.
Hornady SP #4300, 265 gr.
Hornady HP/XTC #44280, 300 gr.
Cast Bullets Used . . . .(sized to .431" dia)
*gas check bullet *#429215, 210 gr.
#429667, 240 gr.

Test Specifications:
(Velocity Only)

Firearm Used . . .Winchester Model 94 Timber Carbine
Barrel Length . . . . .18"
Twist . . . . .1-38"
Groove Dia. . . . . .431"

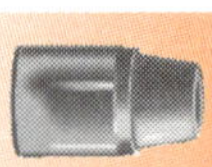

225 gr. Jacketed HP 2.550" OAL — **BC:** .146 **SD:** .175

Powder	Sugg Starting Grains	Velocity fps	Pressure	Max Load Grains	Velocity fps	Pressure
XMP-5744	36.0	1729	—	38.5	1848	—
N120	46.0	2184	—	51.0	2303	—
IMR-4198	41.0	1871	—	45.5	2124	—
H-4198	42.0	1702	—	46.5	2001	—
RX7	47.0	1883	—	52.0	2100	—
IMR-3031	49.5	1751	—	55.0+	2027	—
IMR-4895	51.5	1767	—	**57.0+**	**1995**	—

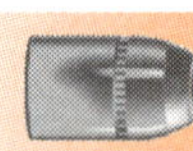

240 gr. Jacketed HP 2.505" OAL — **BC:** .165 **SD:** .186

Powder	Sugg Starting Grains	Velocity fps	Pressure	Max Load Grains	Velocity fps	Pressure
XMP-5744	33.0	1624	—	37.0	1812	—
N130	48.0	1954	—	53.0+	2278	—
IMR-4198	42.0	1944	—	**47.0**	**2212**	—
H-4198	44.0	1841	—	49.0	2184	—
RX7	46.0	1901	—	51.0+	2162	—
IMR-3031	49.5	1796	—	55.0+	2028	—
AA2230	50.0	1881	—	55.5	1961	—
IMR-4895	51.0	1799	—	57.0+	2016	—

Note: Loads shown in shaded panels are maximum.
Loads shown in bold designate potentially most accurate load.
\+ Designates a compressed powder charge.

444 Marlin

265 gr. Jacketed SP
2.560" OAL

BC: .189
SD: .205

Powder	Sugg Starting Grains	Velocity fps	Pressure	Max Load Grains	Velocity fps	Pressure
XMP-5744	31.5	1516	—	35.0	1683	—
N130	46.0	1879	—	51.0+	2169	—
IMR-4198	40.0	1799	—	**44.5**	**2077**	—
H-4198	42.0	1803	—	47.0	2118	—
RX7	42.0	1691	—	47.0	1934	—
AA2230	47.0	1769	—	52.5	1854	—
IMR-4895	48.0	1633	—	53.0	1873	—

300 gr. Jacketed HP/XTP
2.570" OAL

BC: .245
SD: .232

Powder	Sugg Starting Grains	Velocity fps	Pressure	Max Load Grains	Velocity fps	Pressure
XMP-5744	33.0	1551	—	36.5	1737	—
IMR-4198	35.0	1602	—	39.0	1798	—
H-4198	37.0	1579	—	41.5	1777	—
RX7	39.5	1614	—	44.0	1834	—
IMR-4895	47.0	1657	—	**52.0**	**1838**	—

#429215
210 gr. (#2 Alloy) 2.570" OAL

BC: .188
SD: .161

Powder	Sugg Starting Grains	Velocity fps	Pressure	Max Load Grains	Velocity fps	Pressure
SR-4759	26.0	1546	—	29.0	1752	—
XMP-5744	30.5	1569	—	34.0	1698	—
IMR-4198	31.5	1345	—	35.0	1587	—
RX7	39.0	1185	—	43.0	1513	—
IMR-3031	**40.0**	**1273**	—	44.0	1473	—

#429667
240 gr. (#2 Alloy) 2.570" OAL

BC: .149
SD: .185

Powder	Sugg Starting Grains	Velocity fps	Pressure	Max Load Grains	Velocity fps	Pressure
SR-4759	25.0	1596	—	28.0	1769	—
XMP-5744	29.0	1516	—	32.0	1634	—
IMR-4198	**32.5**	**1524**	—	36.0	1684	—
RX7	37.0	1494	—	41.0	1611	—
IMR-3031	40.5	1419	—	45.0	1596	—

Note: Loads shown in shaded panels are maximum.
Loads shown in bold designate potentially most accurate load.
\+ Designates a compressed powder charge.

45 Colt (Rifle Data)

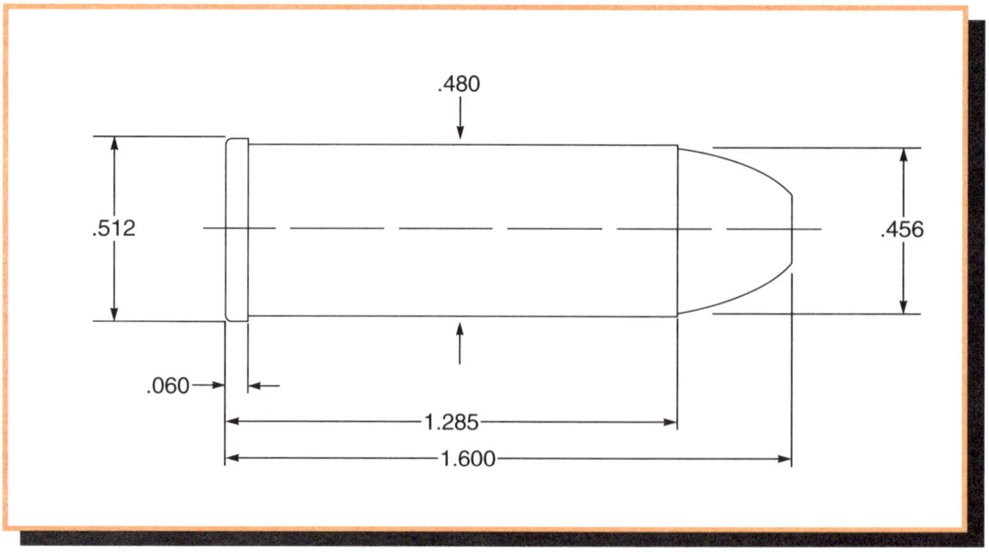

Comments:

While this cartridge has been chambered in revolvers since its inception, it's only recently been offered in lever action rifles. Shooters should be cautious when resizing this particular cartridge. The unusually small rim is prone to tearing off thus leaving the case stuck in the die. Powders favored for 45 Colt pistol loads will also work with the rifle although the longer 16-inch barrel may work better with slower burning powders such as 2400. Some Lyman customers ask if the 45 Colt can be loaded to the higher pressures of the 44 Remington Magnum. The 44 Magnum case is built to withstand the 40,000 CUP pressure level,...substantially more than the original loading for the Colt back in 1873. Those requiring 44 Magnum level performance should trade their rifle in for one chambered in 44 Magnum. A Model 94 in 45 Colt will otherwise be adequate as a short-range brush gun with the 250-grain XTP on deer sized game with proper shot placement. Cast bullet #452664 is our Cowboy Action Shooting bullet and produced excellent accuracy in the test rifle.

Test Components:

Cases . . . . .Winchester
Trim-to Length . . . . .1.275"
Primers . . . . .Winchester WLP
Primer Size . . . . .Large Pistol
Lyman Shell Holder . . . . .No. 11
Jacketed Bullets Used . . . . .Sierra JHP #8800, 185 gr.
Speer JHP #4479, 225 gr.
Sierra JHC #8820, 240 gr.
Hornady HP/XTC #45200, 250 gr.
Cast Bullets Used . . . .(sized to .452" dia)
#454190, 250 gr.
#452664, 250 gr.
#452424, 255 gr.

Test Specifications:
(Velocity Only)

Firearm Used . . . . .Winchester Model 94AE
Barrel Length . . . . .16"
Twist . . . . .1-38"
Groove Dia. . . . . .451"

185 gr. Jacketed HP
1.520" OAL
BC: .100
SD: .130

Powder	Sugg Starting Grains	Velocity fps	Pressure	Max Load Grains	Velocity fps	Pressure
700X	6.0	946	—	7.0	1108	—
N320	8.3	1102	—	9.3	1289	—
231	7.6	1021	—	8.5	1111	—
AA#5	13.0	1416	—	13.7	1493	—
Unique	7.5	944	—	10.0	1263	—
HS-7	13.0	1121	—	14.5	1238	—
Blue Dot	12.0	979	—	13.5	1116	—
IMR-4227	16.5	843	—	20.0	1145	—
XMP-5744	18.5	1099	—	**20.5**	**1194**	—

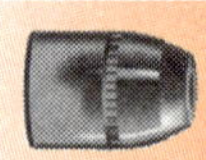

225 gr. Jacketed HP
1.557" OAL
BC: .169
SD: .158

Powder	Sugg Starting Grains	Velocity fps	Pressure	Max Load Grains	Velocity fps	Pressure
N320	7.2	900	—	8.0	1031	—
231	7.4	889	—	8.3	1047	—
AA#5	11.2	1170	—	11.8	1258	—
Unique	8.3	1049	—	10.8	1291	—
Power Pistol	8.5	1035	—	**9.5**	**1158**	—
SR-7625	7.0	741	—	9.3	1151	—
N340	9.1	1041	—	10.2	1194	—
HS-6	11.6	1096	—	13.6	1265	—
Blue Dot	12.4	1076	—	15.4	1310	—
IMR-4227	16.5	953	—	21.0	1265	—

Note: Loads shown in shaded panels are maximum.
Loads shown in bold designate potentially most accurate load.

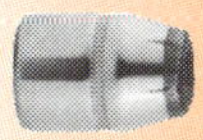

240 gr. Jacketed HC
1.575" OAL

BC: .150
SD: .168

Powder	Sugg Starting Grains	Velocity fps	Pressure	Max Load Grains	Velocity fps	Pressure
AA#5	10.9	1148	—	11.5	1210	—
Unique	7.5	952	—	9.5	1176	—
Power Pistol	8.2	1008	—	9.2	1127	—
N340	8.8	1012	—	9.8	1123	—
HS-6	10.0	906	—	11.0	1010	—
HS-7	12.0	1041	—	14.0	1185	—
2400	15.0	973	—	18.5	1321	—
IMR-4227	16.0	842	—	20.5	1215	—
XMP-5744	**16.5**	**958**	—	18.5	1054	—

250 gr. Jacketed HP/XTP
1.590" OAL

BC: .146
SD: .175

Powder	Sugg Starting Grains	Velocity fps	Pressure	Max Load Grains	Velocity fps	Pressure
Titegroup	5.6	828	—	6.3	935	—
231	6.5	824	—	7.3	902	—
AA#5	10.9	1124	—	11.5	1217	—
Unique	7.5	878	—	9.3	1102	—
Power Pistol	8.0	987	—	**9.0**	**1091**	—
N340	8.6	979	—	9.6	1097	—
2400	15.2	1054	—	17.5	1236	—
IMR-4227	17.0	949	—	20.3	1198	—
XMP-5744	16.0	927	—	18.0	1030	—

#454190
250 gr. (#2 Alloy) 1.600" OAL

BC: .269
SD: .175

Powder	Sugg Starting Grains	Velocity fps	Pressure	Max Load Grains	Velocity fps	Pressure
Red Dot	4.5	763	—	6.5	1002	—
700X	4.5	816	—	**6.0**	**992**	—
Titegroup	5.6	933	—	6.2	1004	—
N320	5.9	769	—	6.8	932	—
231	6.5	967	—	7.4	1069	—
Unique	6.0	809	—	9.0	1157	—
AA#5	10.5	1160	—	11.0	1237	—
SR-7625	6.0	767	—	8.5	1106	—
HS-6	9.5	894	—	10.5	994	—

#452664
250 gr. (#2 Alloy) 1.570" OAL

BC: .150
SD: .175

Powder	Sugg Starting Grains	Velocity fps	Pressure	Max Load Grains	Velocity fps	Pressure
Red Dot	5.8	928	—	6.5	1005	—
700X	5.4	936	—	**6.0**	**991**	—
Titegroup	5.6	966	—	6.2	1020	—
N320	6.0	879	—	6.8	976	—
Green Dot	6.3	970	—	7.0	1027	—
231	6.5	931	—	7.4	1031	—
PB	6.7	922	—	7.5	1010	—
AA#5	9.5	1062	—	10.5	1133	—
Unique	6.3	913	—	8.2	1101	—
SR-7625	7.0	882	—	8.5	1030	—
HS-6	8.5	882	—	10.5	1066	—

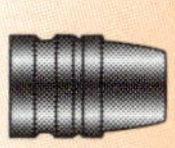

#452424
255 gr. (#2 Alloy) 1.575" OAL

BC: .210
SD: .178

Powder	Sugg Starting Grains	Velocity fps	Pressure	Max Load Grains	Velocity fps	Pressure
Red Dot	4.5	746	—	6.0	947	—
Titegroup	5.6	933	—	**6.2**	**1009**	—
700X	4.5	809	—	6.0	993	—
N320	5.8	790	—	6.7	941	—
231	6.5	930	—	7.2	996	—
AA#5	10.0	1119	—	10.7	1204	—
Unique	6.0	836	—	8.5	1110	—
SR-7625	6.0	775	—	8.0	1054	—
HS-6	8.8	832	—	10.3	990	—

Note: Loads shown in shaded panels are maximum.
Loads shown in bold designate potentially most accurate load.

45-70 Government (For 1873 Springfield)

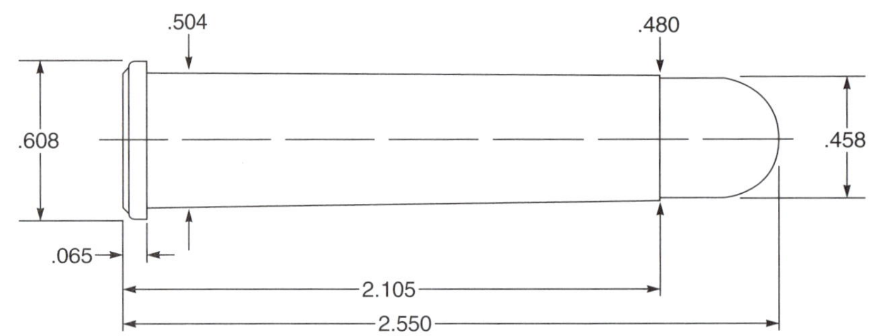

Comments:

Few rifle cartridges rival the longevity and versatility of the 45-70 Government. The United States Army adopted this cartridge in 1873 as a replacement for the short lived 50-70. The 45-70 has become synonymous with the Model 1873 Springfield rifle, nicknamed the "Trapdoor" due to the unique operation of its hinged breech. The cartridge was originally designated the 45-70-500 to indicate its 45-caliber bullet, 70-grains black powder, and 500-grain bullet composition. The Army subsequently developed a lighter load designated the 45-55-405 for use in cavalry carbines. The 30-40 Krag rifle and cartridge officially replaced the Trapdoor Springfield in 1892 but it soldiered on in various military units well into the twentieth century. It survived into the age of smokeless powder and is currently one of the more popular cartridges in the realm of handloading.

The three different loading levels of the 45-70 listed in *Lyman's 48th Edition* reflect this cartridge's versatility. The Trapdoor Springfield is not regarded as a strong action. It is perfectly adequate for its intended black powder pressure levels but shooters must exercise extreme caution when loading the modern smokeless powders. Overzealous reloaders have destroyed more than one Trapdoor through high-pressure loads. The following data is held to a pressure level of 18,000 CUP for use in the Model 1873 Trapdoor. It is also suitable for Remington Rolling Blocks, Sharps Rifles and replicas of any of these rifles as well as the Harrington & Richardson Trapdoor. The exclusive use of cast bullets is recommended for the older, original rifles. As one might expect, many of these rifles show variations in groove diameters. Shooters should slug the bore and size cast bullets accordingly for best accuracy. Never fire any older 45-70 unless it has been thoroughly checked over by a qualified gunsmith.

Older cases of unknown age or origin should not be used as they may be of the weak balloon head construction or have been fired with mercuric primers. The wide availability of current production cases renders use of such older cases unwise and unnecessary. Cast bullet #457125 duplicates the original 500-grain military bullet and has been in our product line for over 100 years. This has been a very accurate bullet in Black Powder Cartridge Silhouette competition. Reloder 7 has long been a favorite for loading both jacketed and cast bullets in the 45-70. XMP-5744 also works well with cast bullets.

Test Components:

Cases . . . . Remington
Trim-to Length . . . . 2.095"
Primers . . . . Remington 9½
Primer Size . . . . Large Rifle
Lyman Shell Holder . . . . No. 17
Jacketed Bullets Used . . . . Nosler PP #45325, 300 gr.
Remington SP, #B22899, 405 gr.
Cast Bullets Used . . . . (sized to .458" dia)
#457191, 292 gr.
#457122, 330 gr.
#457124, 385 gr.
#457193, 405 gr.
#457125, 500 gr.
#457658, 500 gr.
#457132, 535 gr.

Test Specifications: (Velocity & Pressure)

Firearm Used . . . . Universal Receiver
Barrel Length . . . . 24" & 30"
Twist . . . . 1-18"
Groove Dia. . . . . .457"

Note: Loads shown in shaded panels are maximum.
Loads shown in bold designate potentially most accurate load.

45-70 Government (For 1873 Springfield)

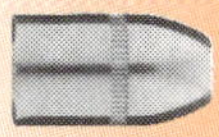

300 gr. Jacketed PP
2.500" OAL

BC: .199
SD: .204

Powder	Sugg Starting Grains	Velocity fps	Pressure C.U.P.	Max Load Grains	Velocity fps	Pressure C.U.P.
SR-4759	24.0	1290	14,700	27.0	1421	16,500
XMP-5744	28.5	1263	13,300	32.0	1423	17,200
N130	37.0	1432	14,200	41.0	1620	17,700
IMR-4198	31.0	1613	14,500	36.0	1692	16,300
RX7	36.0	1589	15,600	38.0	1652	17,000
IMR-3031	45.5	1580	14,500	**48.0**	**1661**	**16,300**
Varget	47.0	1447	13,300	52.5	1617	16,700

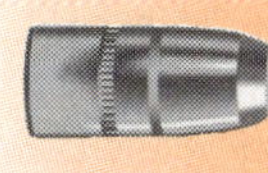

405 gr. Jacketed SP
2.550" OAL

BC: .251
SD: .277

Powder	Sugg Starting Grains	Velocity fps	Pressure C.U.P.	Max Load Grains	Velocity fps	Pressure C.U.P.
IMR-4198	33.0	1321	13,800	35.5	1433	16,700
XMR-2015	40.0	1349	13,100	44.0	1495	17,800
RX7	35.0	1337	13,900	**37.0**	**1440**	**16,000**
H322	41.0	1294	12,300	46.0	1485	17,300
IMR-3031	42.0	1280	13,700	44.5	1416	16,700
N135	42.0	1278	12,400	47.0	1450	16,300
Varget	44.0	1275	12,000	49.0	1467	16,600

#457191
292 gr. (#2 Alloy) 2.550" OAL

BC: .201
SD: .199

Powder	Sugg Starting Grains	Velocity fps	Pressure C.U.P.	Max Load Grains	Velocity fps	Pressure C.U.P.
SR-4759	26.0	1311	9,400	30.0	1602	14,600
XMP-5744	29.0	1376	10,500	**33.0**	**1502**	**14,200**
N130	38.0	1443	9,800	42.5	1659	15,400
XMR-2015	47.0	1608	11,800	53.0	1804	16,800
RX7	50.0	1779	11,900	55.0	2011	16,500
IMR-3031	48.0	1467	11,400	53.0+	1706	16,800
Varget	53.0	1591	13,000	59.0+	1799	17,100

#457122
330 gr. (#2 Alloy) 2.550" OAL

BC: .274
SD: .225

Powder	Sugg Starting Grains	Velocity fps	Pressure C.U.P.	Max Load Grains	Velocity fps	Pressure C.U.P.
SR-4759	26.5	1454	13,000	28.0	1533	15,200
XMP-5744	29.0	1357	13,700	32.5	1485	17,300
N130	40.0	1540	14,400	45.2	1680	16,500
IMR-4198	34.0	1440	12,100	38.0	1595	15,100
XMR-2015	41.0	1485	13,700	46.5	1654	17,700
RX7	40.0	1633	14,600	42.0	1711	16,600
IMR-3031	**43.0**	**1338**	**10,600**	47.5	1532	15,300
Varget	48.0	1379	9,500	53.0	1600	15,500

#457124
385 gr. (#2 Alloy) 2.540" OAL

BC: .299
SD: .262

Powder	Sugg Starting Grains	Velocity fps	Pressure C.U.P.	Max Load Grains	Velocity fps	Pressure C.U.P.
SR-4759	24.0	1252	10,400	**26.5**	**1426**	**16,600**
XMP-5744	26.5	1217	12,500	29.5	1324	16,000
IMR-4198	30.5	1298	11,300	34.0	1441	16,000
N130	36.0	1367	12,600	40.0	1509	16,000
XMR-2015	39.0	1403	14,400	43.5	1542	17,200
RX7	38.5	1543	14,400	41.0	1649	17,400
IMR-3031	40.0	1254	9,100	44.5	1449	15,800
Varget	44.0	1292	9,700	49.0	1504	15,700

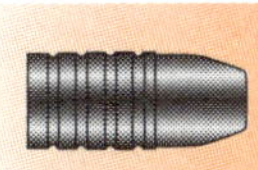

#457193
405 gr. (#2 Alloy) 2.550" OAL

BC: .307
SD: .276

Powder	Sugg Starting Grains	Velocity fps	Pressure C.U.P.	Max Load Grains	Velocity fps	Pressure C.U.P.
SR-4759	24.0	1258	12,600	26.5	1415	17,700
XMP-5744	26.0	1189	12,200	29.0	1313	16,100
IMR-4198	31.5	1312	12,400	**35.0**	**1463**	**17,700**
XMR-2015	36.0	1247	12,300	40.0	1399	16,600
RX7	37.0	1534	16,400	39.0	1578	17,900
IMR-3031	34.0	1161	11,100	38.5	1352	16,000
N135	42.0	1272	11,100	46.5	1510	17,400
Varget	42.0	1258	11,100	46.5	1436	15,800

Note: Loads shown in shaded panels are maximum.
Loads shown in bold designate potentially most accurate load.
* Fired in a 30" barrel.

45-70 Government (For 1873 Springfield)

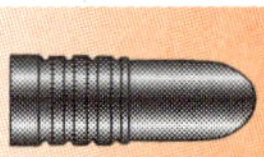

#457125 500 gr. (#2 Alloy) 2.835" OAL — **BC:** .391 **SD:** .341

Powder	Sugg Starting Grains	Velocity fps	Pressure C.U.P.	Max Load Grains	Velocity fps	Pressure C.U.P.
SR-4759	21.0	1108	13,400	24.0	1216	16,700
XMP-5744	**25.0**	**1088**	**14,200**	28.0	1183	17,400
IMR-4198	28.5	1107	12,300	31.5	1287	16,700
XMR-2015	32.0	1079	10,500	35.5	1243	17,500
RX7	30.0	1119	11,100	34.0	1284	15,100
IMR-3031	38.0	1075	12,400	42.0	1332	17,900
N135	39.0	1123	13,100	43.5	1369	17,700
Varget	40.5	1150	13,000	45.0	1356	16,700

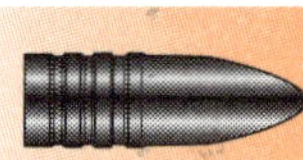

***#457658** 500 gr. (20 to 1) 2.990" OAL — **BC:** .372 **SD:** .341

Powder	Sugg Starting Grains	Velocity fps	Pressure C.U.P.	Max Load Grains	Velocity fps	Pressure C.U.P.
SR-4759	23.5	1127	12,100	27.5	1369	16,300
XMP-5744	**26.5**	**1253**	**15,600**	28.0	1317	17,900
IMR-4198	28.0	1354	10,400	32.0	1418	17,200
RX7	35.0	1220	15,200	40.0	1603	17,100
IMR-3031	37.0	1308	12,000	40.0	1412	16,400
IMR-4895	42.0	1376	16,300	45.0	1459	17,400
Varget	40.0	1248	12,900	43.5	1424	16,000

***#457132** 535 gr. (20 to 1) 2.930" OAL — **BC:** .402 **SD:** .364

Powder	Sugg Starting Grains	Velocity fps	Pressure C.U.P.	Max Load Grains	Velocity fps	Pressure C.U.P.
SR-4759	23.0	1064	7,600	26.5	1245	15,700
XMP-5744	**23.0**	**1101**	**13,000**	28.0	1310	17,900
IMR-4198	28.0	1221	13,300	31.5	1326	17,000
RX7	36.0	1230	10,600	41.0	1498	18,000
IMR-3031	36.0	1204	11,600	41.0	1409	18,000
IMR-4895	37.0	1274	11,200	42.0	1370	18,000
Varget	41.5	1308	10,100	46.5	1520	18,000

Note: Loads shown in shaded panels are maximum.
Loads shown in bold designate potentially most accurate load.
* Fired in a 30" barrel.

45-70 Government (For 1886 Winchester and 1895 Marlin Only)

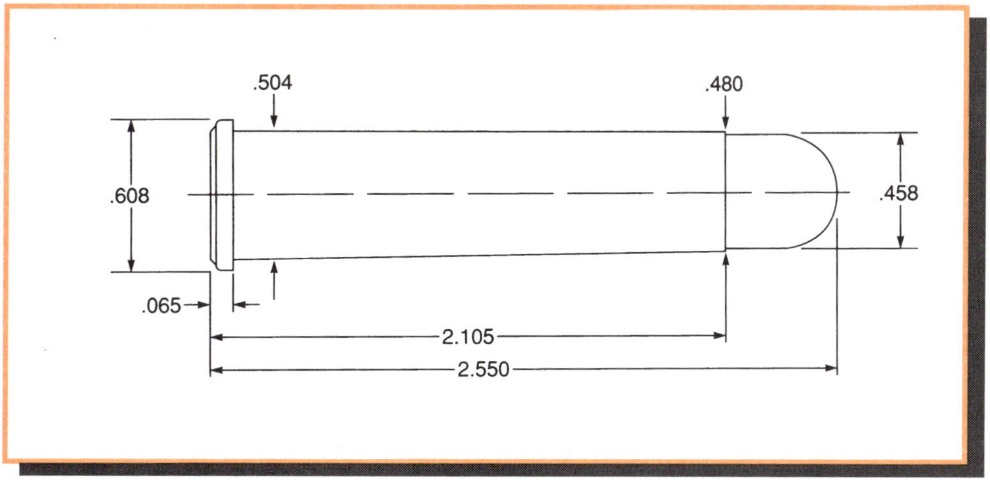

Comments:

The 45-70 Government has proven itself on most North American game over the last 130 years. Those looking for an effective, large caliber brush cartridge out to 150 yards would be hard pressed to beat it without going belted. The stronger lock up of the Winchester Model 1886 and Marlin 1895 lever action rifles allow the reloader to safely exceed the pressure levels associated with the elderly Trapdoor rifles. The data listed in this section is worked up to the SAAMI Maximum Average Pressure (MAP) of 28,000 CUP. Cartridges intended for lever action rifles should not exceed 2.550" overall length and must be crimped. Cast bullet # 457643 is specifically designed for lever-action rifles and gives excellent expansion when cast in a 20 to 1 alloy. Some Marlin rifles utilize shallow Micro-Groove® rifling. Bullet alloy should be at least 15 bhn and velocities should remain below 1,600 feet per second if cast bullet accuracy is to be acceptable in these rifles.

This data is not safe for Trapdoor Springfields or any of the rifles listed in the previous data section for 45-70. It is intended for use in Model 1886 Winchester/Browning rifles and the 1895 Marlin rifle. Use of this data in Trapdoor Rifles is potentially hazardous.

Test Components:

Cases . . . . .Remington
Trim-to Length . . . . .2.095"
Primers . . . . .Remington 9½
Primer Size . . . . .Large Rifle
Lyman Shell Holder . . . . .No. 17
Jacketed Bullets Used . . . .Nosler PP #45325, 300 gr.
Remington SP #B22899, 405 gr.
Cast Bullets Used . . . .(sized to .458" dia)
#457191, 292 gr.
#457122, 330 gr.
#457643, 400 gr.
#457193, 405 gr.

Test Specifications: (Velocity & Pressure)

Firearm Used . . . . .Universal Receiver
Barrel Length . . . . .24"
Twist . . . . .1-18"
Groove Dia. . . . . .457"

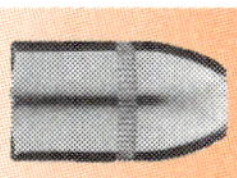

300 gr. Jacketed PP 2.500" OAL — **BC:** .199 **SD:** .204

Powder	Sugg Starting Grains	Velocity fps	Pressure C.U.P.	Max Load Grains	Velocity fps	Pressure C.U.P.
XMP-5744	37.0	1624	19,500	41.0	1804	26,800
IMR-4198	36.0	1688	17,000	41.0	1780	24,800
RX7	38.0	1628	17,000	45.0	1811	26,100
N130	49.0	1878	20,700	53.0	2101	27,400
XMR-2015	50.0	1847	21,200	55.0	2045	26,900
IMR-3031	48.0	1657	17,000	51.0	1881	25,300
H-322	51.0	1732	17,400	56.7	2038	27,400
H-4895	56.0	1713	16,700	**62.0+**	**2021**	**27,300**
IMR-4064	55.0	1611	15,900	61.0+	1852	24,600
Varget	57.0	1821	18,600	64.0+	2051	26,100

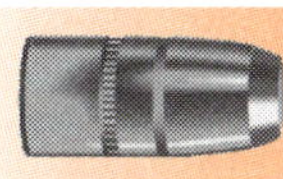

405 gr. Jacketed SP 2.550" OAL — **BC:** .251 **SD:** .277

Powder	Sugg Starting Grains	Velocity fps	Pressure C.U.P.	Max Load Grains	Velocity fps	Pressure C.U.P.
IMR-4198	37.0	1513	19,100	41.0	1661	25,700
RX7	41.0	1602	20,500	46.0	1795	27,100
N130	43.5	1556	19,300	48.5	1744	26,900
XMR-2015	44.5	1479	19,200	49.5	1679	27,700
IMR-3031	46.0	1469	17,700	**51.5**	**1702**	**26,200**
H-322	45.0	1430	15,100	50.0	1683	24,300
H-4895	49.0	1439	15,900	55.0+	1730	26,500
IMR-4064	49.0	1466	19,500	54.5+	1640	26,600
Varget	50.0	1481	16,500	56.0+	1752	26,400

Note: Loads shown in shaded panels are maximum.
Loads shown in bold designate potentially most accurate load.
\+ Designates a compressed powder charge.

45-70 Government (For 1886 Winchester and 1895 Marlin Only)

#457191 292 gr. (#2 Alloy) 2.550" OAL — **BC:** .201 **SD:** .199

Powder	Sugg Starting Grains	Velocity fps	Pressure C.U.P.	Max Load Grains	Velocity fps	Pressure C.U.P.
SR-4759	30.0	1602	14,600	34.0	1802	21,000
XMP-5744	38.0	1687	17,900	**42.0**	**1908**	**26,800**
IMR-4198	34.5	1613	16,600	49.7	2065	24,600
XMR-2015	54.5	1857	19,100	58.5	2065	25,700
N130	51.5	1857	15,800	57.5	2139	24,800
H322	55.0	1775	16,200	61.0+	2097	26,700

#457122 330 gr. (#2 Alloy) 2.550" OAL — **BC:** .274 **SD:** .225

Powder	Sugg Starting Grains	Velocity fps	Pressure C.U.P.	Max Load Grains	Velocity fps	Pressure C.U.P.
SR-4759	30.0	1637	17,200	33.5	1823	27,200
XMP-5744	34.0	1571	17,200	38.0	1726	24,700
IMR-4198	40.0	1784	20,900	44.0	1912	25,300
XMR-2015	50.0	1768	18,000	**55.0**	**1994**	**26,400**
N130	48.0	1798	18,200	54.0	2027	24,900
H322	49.0	1631	21,200	54.5	1911	24,800
RX7	49.0	1853	16,700	55.0	2096	25,100
IMR-3031	50.0	1665	16,900	55.0+	1900	26,000
Varget	56.0	1798	19,100	62.0+	2019	27,000

#457643 400 gr. (#2 Alloy) 2.530" OAL — **BC:** .280 **SD:** .272

Powder	Sugg Starting Grains	Velocity fps	Pressure C.U.P.	Max Load Grains	Velocity fps	Pressure C.U.P.
SR-4759	26.0	1405	17,600	29.0	1550	27,300
IMR-4227	31.0	1481	21,400	34.5	1597	26,500
XMP-5744	**31.5**	**1421**	**20,100**	35.0	1560	26,900
IMR-4198	35.5	1535	19,400	39.5	1699	27,500
XMR-2015	41.0	1457	17,700	45.5	1639	26,400
N130	41.0	1562	21,400	46.5	1786	26,200
H322	43.0	1462	16,200	47.5+	1717	26,800
Varget	48.5	1522	17,700	54.0+	1760	27,200

#457193 405 gr. (#2 Alloy) 2.550" OAL — **BC:** .307 **SD:** .276

Powder	Sugg Starting Grains	Velocity fps	Pressure C.U.P.	Max Load Grains	Velocity fps	Pressure C.U.P.
SR-4759	26.5	1429	19,000	29.5	1577	25,600
XMP-5744	32.0	1434	19,900	36.0	1600	27,000
IMR-4198	36.5	1533	18,300	40.5	1717	27,500
XMR-2015	42.5	1504	17,600	47.5	1712	26,100
N130	43.0	1612	19,000	48.0	1828	27,300
H322	43.5	1469	16,100	48.5	1740	27,100
Varget	**49.5**	**1591**	**18,400**	55.0+	1772	26,300

Note: Loads shown in shaded panels are maximum.
Loads shown in bold designate potentially most accurate load.
\+ Designates a compressed powder charge.

45-70 Government (For Ruger No. 1 and No. 3 Only)

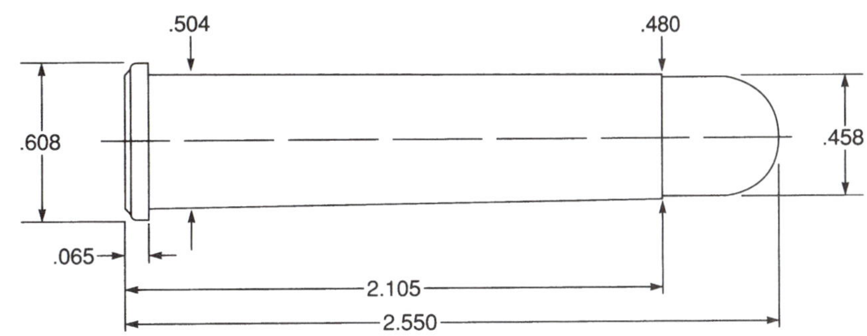

Comments:

The performance of the 45-70 Government is impeded only by the strength of the rifles for which it is chambered. 45-70 factory ammunition has always been loaded on the light side due to the large numbers of older, weaker designed rifles in circulation. Handloading is particularly advantageous when the shooter possesses a Ruger #1 or one of numerous Siamese Mausers converted to 45-70.

The following data is worked up to a Maximum Average Pressure (MAP) of 40,000 CUP. Some of the listed loads exceed the SAAMI maximum overall length of 2.550 inches but will not be a problem with single shot rifles. Heavily compressed loads should be crimped regardless of rifle type to prevent cartridges from "growing" in length after seating. Shooters using any of the following data should take careful precautions that no cartridges accidentally find their way into Trapdoor or lever action rifles. Those loading 45-70 ammunition to more than one pressure level for different rifles should mark the primer or bottom of the case head with a marker to identify one batch from the other. Reloaders should also consider using different brand cases to further reduce the chance of a high-pressure load being mistakenly fired in a weaker gun.

The following data should not be used in any other rifles than the Ruger No. 1, Ruger No. 3, or bolt-action rifle built upon a Model 98 action rifle that has been deemed safe by a qualified gunsmith. They should be loaded only in newly manufactured cases that have not been repeatedly fired. Firing any of the following loads in any weaker gun listed in a previous section will be dangerous.

Test Components:

Cases	Remington
Trim-to Length	2.095"
Primers	Remington 9½
Primer Size	Large Rifle
Lyman Shell Holder	No. 17
Jacketed Bullets Used	Nosler PP #45325, 300 gr.
	Speer SP #2478, 350 gr.
	Speer FN #2479, 400 gr.
	Hornady RN #4504, 500 gr.
Cast Bullets Used	(sized to .458" dia)
*gas check bullet	#457122, 292 gr.
	#457124, 385 gr.
	#457193, 405 gr.
	*#457671, 475 gr.
	#457125, 500 gr.
	#457132, 535 gr.

Test Specifications:
(Velocity & Pressure)

Firearm Used	Universal Receiver
Barrel Length	24" + 28"
Twist	1-18", 1-20"
Groove Dia.	.457", .456"

45-70 Government (For Ruger No. 1 and No. 3 Only)

300 gr. Jacketed PP — 2.550" OAL — **BC:** .199 **SD:** .204

Powder	Sugg Starting Grains	Velocity fps	Pressure C.U.P.	Max Load Grains	Velocity fps	Pressure C.U.P.
SR-4759	35.0	1786	23,000	42.0	2045	37,000
IMR-4227	32.0	1702	22,800	38.0	1974	36,300
*XMP-5744	41.0	1803	26,900	46.0	1984	39,800
IMR-4198	49.0	2047	24,000	55.0+	2347	35,200
*RX7	51.0	2092	32,900	57.0+	2267	38,700
***N130**	51.0	2001	29,800	**57.0+**	**2196**	**37,300**
*XMR-2015	52.0	1907	31,900	58.5+	2135	39,200
IMR-3031	55.0	1958	26,400	60.0+	2076	25,000
H-322	55.0	1984	23,900	62.0+	2076	22,400
*H-4895	58.0	1797	23,400	63.0+	2081	32,800

***350 gr. Jacketed SP** — 2.710" OAL — **BC:** .232 **SD:** .238

Powder	Sugg Starting Grains	Velocity fps	Pressure C.U.P.	Max Load Grains	Velocity fps	Pressure C.U.P.
IMR-4198	45.0	1881	31,500	50.0	2098	38,700
RX7	47.0	1942	31,900	53.0	2119	39,100
XMR-2015	48.0	1689	27,400	53.5	1881	36,900
IMR-3031	51.0	1709	27,900	57.0+	1975	35,700
Benchmark	51.0	1743	27,100	57.0	2025	36,600
H-322	51.0	1751	24,400	57.0	2051	36,700
N133	52.0	1914	29,200	**58.0+**	**2152**	**37,100**
H-335	59.0	1928	27,800	66.0+	2152	38,000

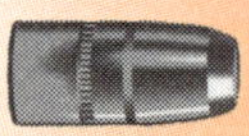

***400 gr. Jacketed FN** — 2.550" OAL — **BC:** .214 **SD:** .272

Powder	Sugg Starting Grains	Velocity fps	Pressure C.U.P.	Max Load Grains	Velocity fps	Pressure C.U.P.
XMR-2015	46.0	1625	26,200	51.0	1838	35,600
H-322	48.0	1630	25,800	53.0	1915	39,900
H-335	55.0	1730	29,500	**60.5**	**1875**	**36,700**
N135	50.0	1679	25,100	56.0	1924	37,500

500 gr. Jacketed RN — 2.930" OAL — **BC:** .287 **SD:** .341

Powder	Sugg Starting Grains	Velocity fps	Pressure C.U.P.	Max Load Grains	Velocity fps	Pressure C.U.P.
IMR-4198	35.0	1440	27,500	43.0	1657	38,400
RX7	37.0	1487	31,200	45.5	1738	39,100
*XMR-2015	44.0	1529	29,800	49.0	1695	37,000
IMR-3031	44.0	1444	22,300	53.1+	1742	32,700
*H-322	44.0	1460	27,200	49.0	1673	36,500
H-4895	50.0	1646	28,100	57.0	1879	39,000
***N135**	47.0	1543	28,000	**52.2**	**1735**	**38,800**
*Varget	49.0	1533	29,700	55.0	1759	39,600
*AA2520	50.5	1608	31,100	56.0	1768	39,600

***#457122 HP** — 330 gr. (#2 Alloy) 2.550" OAL — **BC:** .274 **SD:** .225

Powder	Sugg Starting Grains	Velocity fps	Pressure C.U.P.	Max Load Grains	Velocity fps	Pressure C.U.P.
XMP-5744	**39.5**	**1750**	**26,300**	44.0	1941	37,400
IMR-4227	39.8	1842	22,900	44.0	2035	38,800
IMR-4198	44.0	1862	24,500	49.0	2090	36,300
N130	51.5	1945	24,200	57.0+	2209	37,000
RX7	53.0	2016	23,200	59.0+	2255	35,500
XMR-2015	54.0	1843	24,000	60.0+	2094	36,900

***#457124** — 385 gr. (#2 Alloy) 2.540" OAL — **BC:** .299 **SD:** .262

Powder	Sugg Starting Grains	Velocity fps	Pressure C.U.P.	Max Load Grains	Velocity fps	Pressure C.U.P.
XMP-5744	36.0	1584	24,200	40.0	1744	37,800
IMR-4198	40.5	1712	25,100	45.0	1924	37,100
N130	47.5	1828	24,900	**53.0+**	**2057**	**37,700**
XMR-2015	49.5	1724	28,600	55.0+	1947	36,800
RX7	49.5	1896	25,000	55.0+	2115	38,100
H322	49.5	1750	25,600	55.0+	1993	38,300
Varget	53.0	1681	21,700	59.0+	1903	30,500

Note: Loads shown in shaded panels are maximum.
Loads shown in bold designate potentially most accurate load.
* Fired in a 24" barrel.
\+ Designates a compressed powder charge.

45-70 Government (For Ruger No. 1 and No. 3 Only)

***#457193**
405gr. (#2 Alloy) 2.550" OAL
BC: .307
SD: .276

Powder	Sugg Starting Grains	Velocity fps	Pressure C.U.P.	Max Load Grains	Velocity fps	Pressure C.U.P.
SR-4759	30.0	1595	25,500	34.0	1743	36,900
XMP-5744	35.0	1532	25,800	39.0	1715	38,900
IMR-4198	**40.0**	**1684**	**26,800**	44.5	1881	38,300
RX7	46.0	1785	23,600	51.0	1996	36,800
N130	46.0	1799	26,100	51.0	2009	38,700
XMR-2015	51.0	1833	29,100	57.0+	2063	38,700
H-322	47.0	1669	24,800	53.0	1909	37,000

***#457671**
475 gr. (#2 Alloy) 2.800" OAL
BC: .477
SD: .323

Powder	Sugg Starting Grains	Velocity fps	Pressure C.U.P.	Max Load Grains	Velocity fps	Pressure C.U.P.
SR-4759	28.0	1415	25,000	31.0	1556	37,000
XMP-5744	33.0	1413	25,200	37.0	1576	36,000
IMR-4198	38.0	1570	26,300	42.5	1735	37,300
RX7	43.0	1650	24,700	48.5	1819	34,900
XMR-2015	48.0	1693	27,000	54.0+	1909	39,600
IMR-3031	47.0	1657	32,400	**53.0+**	**1802**	**38,500**
H322	46.0	1539	21,800	51.0+	1811	38,800
N135	47.0	1495	21,100	52.0+	1741	29,500
Varget	52.0	1699	29,400	58.0+	1876	37,700

***#457125**
500 gr. (#2 Alloy) 2.835" OAL
BC: .391
SD: .341

Powder	Sugg Starting Grains	Velocity fps	Pressure C.U.P.	Max Load Grains	Velocity fps	Pressure C.U.P.
SR-4759	29.0	1415	26,900	**32.0**	**1537**	**39,700**
XMP-5744	33.0	1397	26,100	37.0	1540	35,200
RX7	44.0	1636	27,500	49.0	1797	38,200
XMR-2015	46.0	1600	24,600	52.0+	1795	37,400
H-322	46.0	1554	24,700	51.0	1753	36,700
N135	48.5	1571	22,900	54.0+	1801	38,200
Varget	50.0	1579	24,600	56.0+	1791	35,900

***#457132**
535 gr. (20 to 1) 2.930" OAL
BC: .402
SD: .364

Powder	Sugg Starting Grains	Velocity fps	Pressure C.U.P.	Max Load Grains	Velocity fps	Pressure C.U.P.
SR-4759	27.0	1328	26,600	30.0	1445	37,100
XMP-5744	33.0	1385	28,200	37.0	1519	37,600
IMR-4198	38.0	1499	27,500	42.0	1648	38,100
RX7	42.0	1501	24,500	45.5	1696	38,400
XMR-2015	**46.0**	**1534**	**24,000**	51.0+	1740	37,700
IMR-3031	45.0	1469	24,100	50.0	1674	37,000
H-322	44.0	1459	22,300	49.0	1685	37,100
N135	48.0	1551	24,300	53.5	1767	37,600
Varget	49.0	1552	25,900	54.0	1760	37,800

Note: Loads shown in shaded panels are maximum.
Loads shown in bold designate potentially most accurate load.
* Fired in a 24" barrel.
\+ Designates a compressed powder charge.

450 Marlin

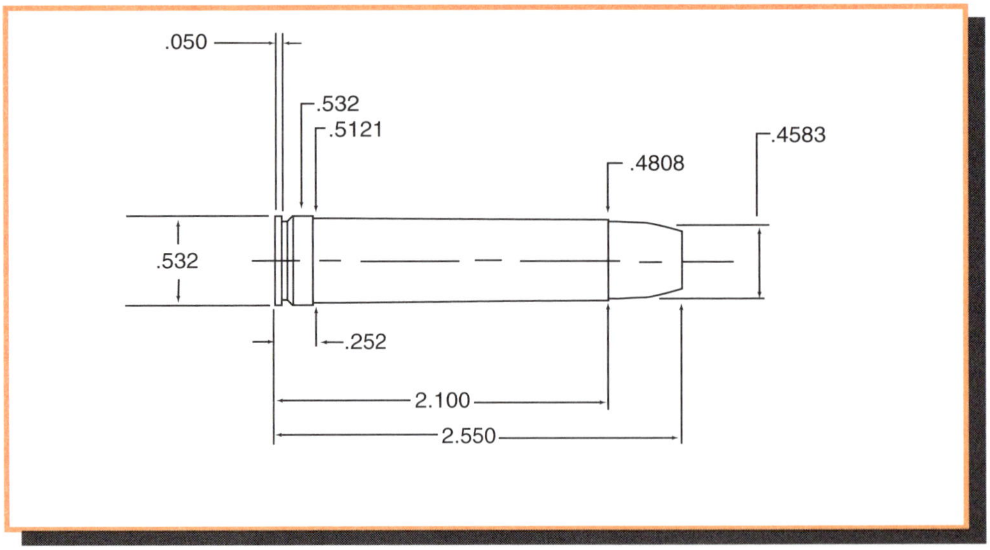

Comments:

The 450 Marlin evolved from a joint effort between Hornady and Marlin to produce a magnum version of the venerable old 45-70 Government. Readers of Lyman's reloading manual will find three different sections devoted to the 45-70 rifle with data appropriate to particular rifles. Shooters for many years now have hand-loaded 45-70 cartridges to higher than SAAMI approved levels for use in high-strength firearms such as the Ruger Number 1 and Mauser-based actions. The large number of older guns of weak or questionable design — notably the 1873 Trapdoor Springfield — prevented the 45-70 cartridge from ever being factory loaded to its full potential. The Marlin cartridge duplicates in standard factory form what was previously available only by handloading the 45-70. While below 458 Winchester Magnum level in performance, the 450 Marlin offers distinctly more punch than the 28,000 CUP level of current 45-70 factory ammo. The loading procedures for the Marlin are the same as the 45-70. Powders and components used in the 45-70 are also suitable for the Marlin.

Test Components:

Cases . . . . .Hornady
Trim-to Length . . . . .2.090"
Primers . . . . .Winchester WLR
Primer Size . . . . .Large Rifle
Lyman Shell Holder . . . . No. 13
Jacketed Bullets Used . . .Barnes XFN #45831, 250 gr.
Hornady HP #4500, 300 gr.
Hornady FP #4503, 350 gr.
Speer FSP #2479, 400 gr.
Cast Bullets Used . . . .(sized to .457" dia)
#457122, 330 gr.
#457643, 400 gr.

Test Specifications:
(Velocity & Pressure)

Firearm Used . . . . .Universal Receiver
Barrel Length . . . . .24"
Twist . . . . .1-20"
Groove Dia. . . . . .456"

250 gr. Barnes XFN
2.530" OAL
BC: .172
SD: .170

Powder	Sugg Starting Grains	Velocity fps	Pressure P.S.I.	Max Load Grains	Velocity fps	Pressure P.S.I.
AA1680	43.5	1946	28,900	48.5	2214	39,300
IMR-4198	44.0	1965	31,900	49.0	2246	41,700
RX7	45.5	1966	31,000	**50.7**	**2183**	**35,900**
N130	46.0	1977	30,200	51.0	2247	39,800
N133	51.0	2083	32,900	57.0+	2258	38,600
Benchmark	52.0	2024	31,100	58.0+	2184	35,700
H-322	54.0	2083	34,100	60.0+	2246	39,300

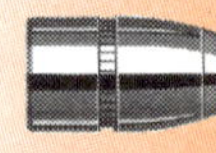

300 gr. Jacketed HP
2.530" OAL
BC: .197
SD: .204

Powder	Sugg Starting Grains	Velocity fps	Pressure P.S.I.	Max Load Grains	Velocity fps	Pressure P.S.I.
AA-1680	45.0	2029	31,600	50.5	2257	40,900
IMR-4198	44.5	1997	30,000	49.5	2260	42,000
RX7	46.0	2047	31,300	**51.5**	**2242**	**39,600**
XMR-2015	51.0	1919	31,000	56.5	2166	42,100
N133	52.0	2070	29,900	57.5	2318	41,500
Benchmark	55.0	2106	34,500	61.5+	2273	41,300
H-322	55.0	2128	35,500	61.0+	2288	42,600
IMR-3031	55.0	2075	35,200	61.0+	2238	42,800
748	58.0	1969	29,900	64.0+	2136	34,000

Note: Loads shown in shaded panels are maximum.
Loads shown in bold designate potentially most accurate load.
\+ Designates a compressed powder charge.

350 gr. Jacketed FP
2.530" OAL

BC: .193
SD: .238

Powder	Sugg Starting Grains	Velocity fps	Pressure P.S.I.	Max Load Grains	Velocity fps	Pressure P.S.I.
IMR-4198	41.0	1824	31,700	46.0	2028	41,100
RX7	43.0	1837	33,500	48.0	2040	42,400
XMR-2015	48.5	1807	34,400	54.0	2000	43,400
N133	48.0	1857	30,300	53.0	2089	40,700
Benchmark	49.5	1798	30,800	**55.0**	**2033**	**42,700**
H-322	50.0	1888	32,800	56.0+	2099	44,000
IMR-3031	49.5	1791	31,200	55.0	2035	41,900
AA2230	55.0	1991	36,400	61.0+	2188	43,600
748	56.0	1869	32,200	62.0+	2040	38,500

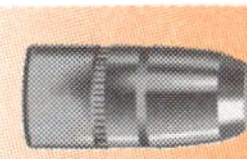

400 gr. Jacketed FSP
2.540" OAL

BC: .214
SD: .272

Powder	Sugg Starting Grains	Velocity fps	Pressure P.S.I.	Max Load Grains	Velocity fps	Pressure P.S.I.
RX7	40.5	1726	33,800	45.0	1893	42,100
XMR-2015	43.0	1599	31,000	48.0	1808	41,700
N133	44.5	1735	30,500	49.5	1935	40,400
H-322	46.0	1665	28,600	**51.0**	**1918**	**42,200**
IMR-3031	47.0	1692	32,800	52.0+	1915	42,900
H-4895	50.0	1815	35,000	56.0+	1975	41,500
748	52.0	1720	30,900	58.0+	1891	40,000

#457122
330 gr. (#2 Alloy) 2.535" OAL

BC: .274
SD: .226

Powder	Sugg Starting Grains	Velocity fps	Pressure P.S.I.	Max Load Grains	Velocity fps	Pressure P.S.I.
SR-4759	28.0	1652	22,800	35.0	1970	40,300
IMR-4227	33.0	1663	23,200	**41.0**	**2016**	**38,700**
XMP-5744	35.0	1662	24,300	44.0	2001	40,100
IMR-4198	37.0	1666	22,300	45.0	2022	34,600
RX7	39.0	1656	20,100	47.0	1992	31,500
XMR-2015	44.0	1645	23,400	52.0	2010	38,200

#457643
400 gr. (#2 Alloy) 2.505" OAL

BC: .280
SD: .274

Powder	Sugg Starting Grains	Velocity fps	Pressure P.S.I.	Max Load Grains	Velocity fps	Pressure P.S.I.
SR-4759	24.0	1408	23,700	**30.0**	**1675**	**41,000**
IMR-4227	27.0	1383	22,500	35.0	1713	38,000
XMP-5744	29.0	1408	24,400	37.0	1718	39,700
IMR-4198	30.5	1417	21,500	38.0	1714	33,000
RX7	31.0	1399	20,000	39.0	1721	33,000
XMR-2015	37.0	1398	22,600	44.0	1708	37,800

Note: Loads shown in shaded panels are maximum.
Loads shown in bold designate potentially most accurate load.
\+ Designates a compressed powder charge.

45-90 Winchester

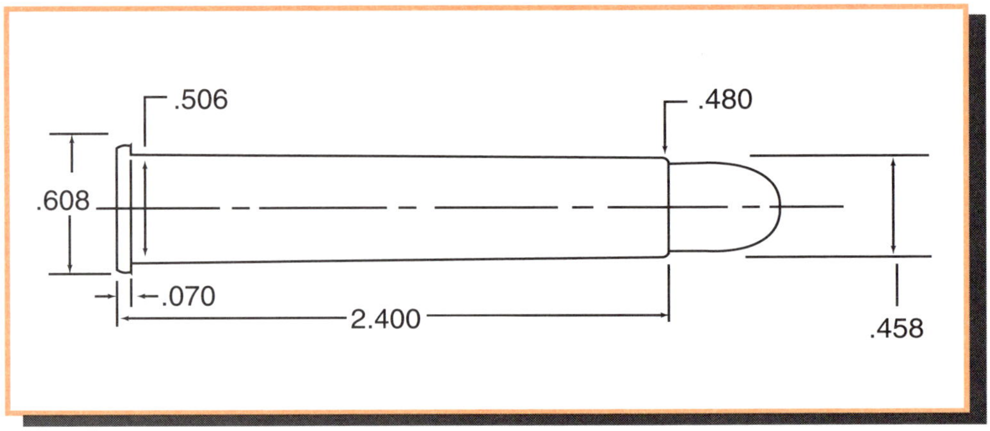

Comments:

The 45-90 had its origins with Winchester's 1885 High Wall and 1886 lever action. This was for many years about as big a cartridge as one could get in a lever gun. Newer and more modern developments rendered this cartridge obsolete by the 1930s when it was discontinued. This cartridge has made a comeback in recent years on the Black Powder Cartridge Silhouette circuit. Being originally a black powder cartridge, Lyman recommends the exclusive use of cast bullets. The 45-90 case is essentially a 45-70 case lengthened from 2.100" to 2.400". Tests in Lyman's ballistic lab showed an increase of around 200 feet per second over the 45-70 with Lyman cast bullet #457643.

This data is intended only for new or recently manufactured guns recommended for smokeless powder.

This data is not for use in antique rifles originally made for black powder.

Test Components:

Cases . . . . .Buffalo Arms
Trim-to Length . . . . .2.390"
Primers . . . . .Remington 9½
Primer Size . . . . .Large Rifle
Lyman Shell Holder . . . . .No. 17
Cast Bullets Used . . . .(sized to .458" dia)
#457124, 385 gr.
#457643, 400 gr.
#457125, 500 gr.
#457132, 535 gr.

Test Specifications: (Velocity & Pressure)

Firearm Used . . . . .Universal Receiver
Barrel Length . . . . .29½"
Twist . . . . .1-20"
Groove Dia. . . . . .4565"

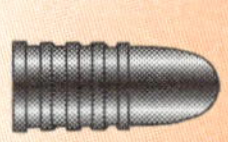

#457124
385 gr. (#2 Alloy) 2.850" OAL
BC: .299
SD: .262

Powder	Sugg Starting Grains	Velocity fps	Pressure C.U.P.	Max Load Grains	Velocity fps	Pressure C.U.P.
SR-4759	33.0	1675	15,400	36.5	1826	24,000
XMP-5744	**36.0**	**1568**	**13,700**	44.5	1866	24,400
IMR-4198	47.0	1861	19,400	51.7	2070	26,800
RX7	51.0	1944	18,800	57.0	2117	24,900
H-322	49.0	1685	15,300	61.0	2134	25,900
Varget	55.0	1636	12,800	65.0+	2071	25,000

#457643
400 gr. (#2 Alloy) 2.800" OAL
BC: .280
SD: .272

Powder	Sugg Starting Grains	Velocity fps	Pressure C.U.P.	Max Load Grains	Velocity fps	Pressure C.U.P.
SR-4759	31.0	1611	21,700	**35.0**	**1746**	**26,300**
XMP-5744	38.0	1637	22,900	42.5	1782	26,600
IMR-4198	38.0	1611	16,800	49.5	1948	26,800
RX7	40.0	1626	16,000	51.0	1957	25,500

#457125
500 gr. (#2 Alloy) 3.210" OAL
BC: .391
SD: .340

Powder	Sugg Starting Grains	Velocity fps	Pressure C.U.P.	Max Load Grains	Velocity fps	Pressure C.U.P.
SR-4759	33.5	1480	17,400	37.0	1590	25,100
XMP-5744	**39.0**	**1526**	**18,600**	43.0	1667	26,600
IMR-4198	42.0	1589	17,900	48.0	1770	25,600
RX7	46.0	1579	15,200	50.5	1772	24,800

#457132
535 gr. (20 to 1) 3.300" OAL
BC: .402
SD: .364

Powder	Sugg Starting Grains	Velocity fps	Pressure C.U.P.	Max Load Grains	Velocity fps	Pressure C.U.P.
SR-4759	35.0	1519	21,800	36.5	1546	25,600
XMP-5744	**40.0**	**1542**	**21,800**	43.0	1633	25,400
IMR-4198	40.0	1496	16,200	46.5	1699	24,900
RX7	44.0	1569	19,200	48.5	1707	25,500

Note: Loads shown in shaded panels are maximum.
Loads shown in bold designate potentially most accurate load.
\+ Designates a compressed powder charge.

45-100-2$\frac{6}{10}$ (45-100 Sharps)

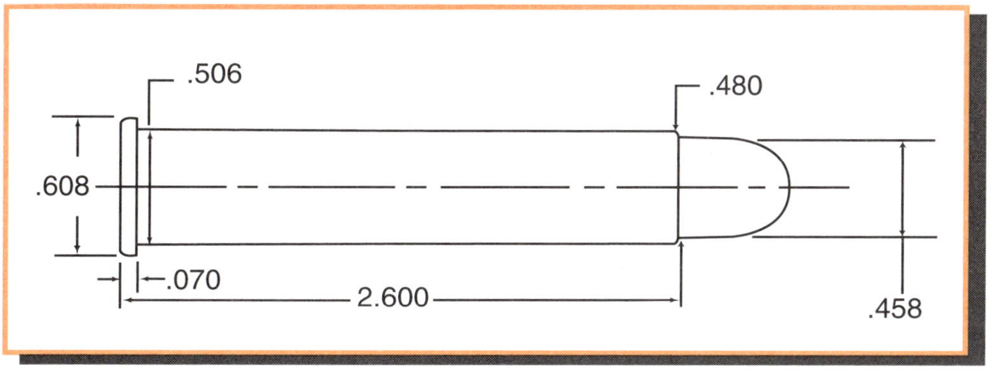

Comments:

This cartridge has been known under different designations with the primary difference being bullet weights and powder charges. The 45-100 saw limited use during the late 1870s before the demise of the Sharps Rifle Company. Renewed interest in Black Powder Cartridge Silhouette shooting have revived this number to an extent. Test results of the 45-100 in Lyman's ballistic lab showed little or no gain in velocity compared to the 45-90 due to the differing capacities of the parent cases. The 45-100 cases utilized here are based on unformed full-length 348 Winchester brass while cases used in our 45-90 data were a 45-70 case lengthened by .300". Loading the 45-100 is no different than for the 45-70. However, extreme care should be taken not to allow excessive amounts of case lubricant build upon the mid-section of the case during full-length resizing. The thin walls of this case are particularly prone to lube dents in this area. For best results, apply lube by hand — sparingly — to the neck area and lower portion of the case. If lube dents do occur, the inside of the full-length sizing body should be thoroughly cleaned with solvent before sizing additional cases.

Data listed here is intended for modern, new production guns intended for use with smokeless powder.

This data is not for use in antique rifles originally made for black powder.

Test Components:

Cases . . . . .Buffalo Arms
Trim-to Length . . . . .2.600"
Primers . . . . .Remington 9½ Magnum
Primer Size . . . . .Large Rifle, Magnum
Lyman Shell Holder . . . . .No. 17
Cast Bullets Used . . . .(sized to .458" dia)
#457193, 405 gr.
#457125, 500 gr.
#457658, 500 gr.
#457132, 535 gr.

Test Specifications:
(Velocity & Pressure)

Firearm Used . . . . .Universal Receiver
Barrel Length 29½
Twist . . . . .1-20"
Groove Dia. . . . . .4565"

#457193
405 gr. (#2 Alloy) 3.040" OAL
BC: .307
SD: .276

Powder	Sugg Starting Grains	Velocity fps	Pressure C.U.P.	Max Load Grains	Velocity fps	Pressure C.U.P.
SR-4759	29.0	1566	18,100	33.0	1724	26,600
XMP-5744	**36.0**	**1622**	**19,000**	40.0	1782	26,700
IMR-4198	38.0	1636	19,000	43.0	1846	26,700
RX7	38.0	1620	17,900	43.0	1830	26,000

#457125
500 gr. (#2 Alloy) 3.420" OAL
BC: .391
SD: .340

Powder	Sugg Starting Grains	Velocity fps	Pressure C.U.P.	Max Load Grains	Velocity fps	Pressure C.U.P.
SR-4759	30.0	1454	21,600	**34.0**	**1556**	**27,500**
XMP-5744	36.0	1479	20,100	40.0	1646	27,000
IMR-4198	39.0	1610	20,700	43.0	1721	27,200
RX7	39.0	1499	18,800	43.5	1640	27,200

#457658
500 gr. (20 to 1) 3.525" OAL
BC: .372
SD: .340

Powder	Sugg Starting Grains	Velocity fps	Pressure C.U.P.	Max Load Grains	Velocity fps	Pressure C.U.P.
SR-4759	32.5	1514	21,600	36.5	1621	26,300
XMP-5744	38.0	1541	18,900	43.0	1677	25,600
IMR-4198	43.0	1614	17,500	48.0	1787	26,000
RX7	**47.0**	**1672**	**20,500**	52.0	1829	26,600
XMR-2015	50.0	1614	19,000	56.0	1775	25,000

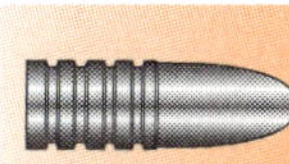

#457132
535 gr. (20 to 1) 3.550" OAL
BC: .402
SD: .364

Powder	Sugg Starting Grains	Velocity fps	Pressure C.U.P.	Max Load Grains	Velocity fps	Pressure C.U.P.
XMP-5744	**36.0**	**1479**	**20,300**	40.0	1632	27,900
IMR-4198	36.5	1450	20,500	40.5	1633	27,500
RX7	39.0	1481	16,200	47.0	1726	25,700

Note: Loads shown in shaded panels are maximum.
Loads shown in bold designate potentially most accurate load.

45-110-2⅞ (45-110 Sharps)

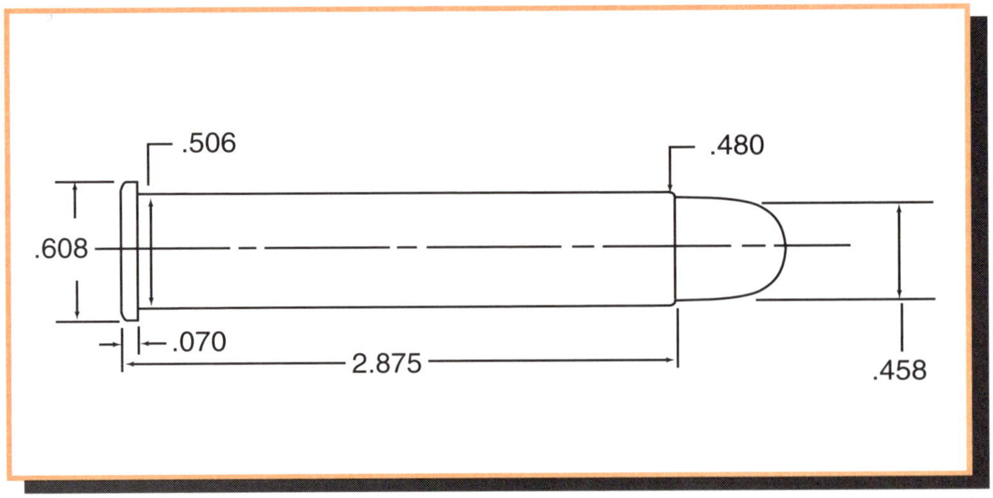

Comments:

This is another circa 1870s black powder cartridge that has gained some nostalgia appeal in recent years. The same procedures for resizing and loading the 45-100 are applicable here. The cases used in our lab tests were based on unformed, full-length 348 Winchester cases provided by Buffalo Arms of Sand Point, ID. Refer to the previous comments section for the 45-100 for information regarding resizing these cases. Lab results show a velocity gain of 125 to 175 feet per second over the 45-100.

This data is intended for use modern, newly manufactured guns rated for smokeless powder.

This data is not for use in antique guns originally built for black powder.

Test Components:

Cases . . . . .Buffalo Arms
Trim-to Length . . . . .2.865"
Primers . . . . .Remington 9½ M
Primer Size . . . . .Large Rifle, Magnum
Lyman Shell Holder . . . . .No. 17
Cast Bullets Used . . . .(sized to .458" dia)
#457193, 405 gr.
#457125, 500 gr.
#457132, 535 gr.

Test Specifications:
(Velocity & Pressure)

Firearm Used . . . . .Universal Receiver
Barrel Length . . . . .30"
Twist . . . . .1-20"
Groove Dia. . . . . .4565"

#457193
405 gr. (#2 Alloy) 3.320" OAL
BC: .307
SD: .276

Powder	Sugg Starting Grains	Velocity fps	Pressure C.U.P.	Max Load Grains	Velocity fps	Pressure C.U.P.
SR-4759	34.0	1578	14,200	42.5	1875	27,000
XMP-5744	38.0	1562	15,900	50.0	1911	26,900
IMR-4198	**40.0**	**1610**	**14,000**	53.0	1969	24,000
RX7	46.0	1585	10,400	60.0	2070	23,000

#457125
500 gr. (#2 Alloy) 3.750" OAL
BC: .391
SD: .340

Powder	Sugg Starting Grains	Velocity fps	Pressure C.U.P.	Max Load Grains	Velocity fps	Pressure C.U.P.
SR-4759	38.0	1524	18,400	42.0	1642	25,600
XMP-5744	**43.0**	**1569**	**18,200**	49.5	1746	25,600
IMR-4198	45.0	1569	14,700	53.0	1817	26,000
RX7	45.0	1526	15,500	59.0	1880	26,900
XMR-2015	49.0	1532	14,400	60.0	1829	24,500

#457132
535 gr. (20 to 1) 3.730" OAL
BC: .402
SD: .364

Powder	Sugg Starting Grains	Velocity fps	Pressure C.U.P.	Max Load Grains	Velocity fps	Pressure C.U.P.
SR-4759	**36.0**	**1470**	**17,200**	40.5	1600	26,900
XMP-5744	40.0	1473	16,500	46.0	1712	26,300
IMR-4198	42.5	1526	15,000	53.0	1805	26,700
XMR-2015	50.0	1539	15,900	57.5	1761	24,200

Note: Loads shown in shaded panels are maximum.
Loads shown in bold designate potentially most accurate load.

45-120-3¼ (45-120 Sharps)

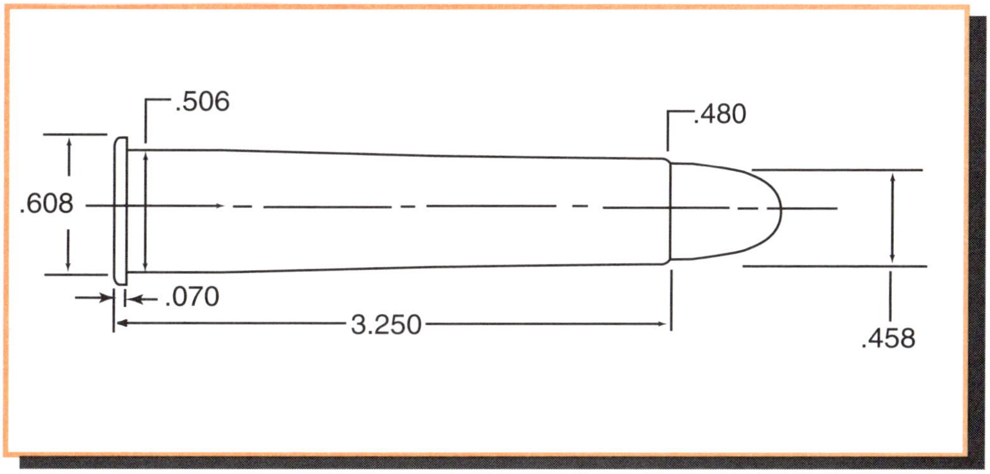

Comments:

This is yet another old-timer revived by the current popularity of Black Powder Cartridge Silhouette Shooting. This cartridge came on to the scene around 1879 shortly before the demise of the Sharps Rifle Company. Although widely considered one of the great buffalo cartridges, its exact history has been the subject of some debate. Many rifles from both domestic and foreign sources have come on to the market in recent years. Some of our consumer correspondence indicates that some rifles made overseas have groove diameters as small as .456". The 45-120 was originally designed to be loaded with compressed black powder and cast lead bullets. It is a very large case that operates at relatively low pressures. These circumstances limit the number of smokeless powders suitable for loading this cartridge. Those looking for a good first powder choice should try Accurate Arms XMP-5744. Accurate Arms formulated this propellant specifically for reduced loads and large cases such as the old Sharps cartridges. Shooters should slug their bores and size bullets accordingly.

This data is intended for newly manufactured guns rated for smokeless powder.

This data is not intended for use in antique guns originally built for black powder.

Test Components:

Cases . . . . .Bertram
Trim-to Length . . . . .3.240"
Primers . . . . .Federal 215M
Primer Size . . . . .Large Rifle Magnum
Lyman Shell Holder . . . . .No. 17
Cast Bullets Used . . . .(sized to .459" dia)
#457125, 500 gr.
#457658, 500 gr.
#457132, 535 gr.

Test Specifications:
(Velocity & Pressure)

Firearm Used . . . . .Universal Receiver
Barrel Length . . . . .26"
Twist . . . . .1-20"
Groove Dia. . . . . .458"

#457125
500 gr. (#2 Alloy) 3.960" OAL
BC: .391
SD: .339

Powder	Sugg Starting Grains	Velocity fps	Pressure C.U.P.	Max Load Grains	Velocity fps	Pressure C.U.P.
SR-4759	32.0	1383	18,000	36.5	1534	23,000
XMP-5744	**39.0**	**1532**	**18,300**	44.0	1648	22,200
IMR-4350	60.0	1562	19,500	66.5	1752	27,800

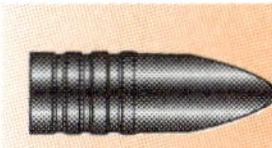

#457658
500 gr. (20 to 1) 4.125" OAL
BC: .372
SD: .339

Powder	Sugg Starting Grains	Velocity fps	Pressure C.U.P.	Max Load Grains	Velocity fps	Pressure C.U.P.
SR-4759	**36.0**	**1476**	**21,200**	40.0	1588	24,000
XMP-5744	41.0	1576	22,700	46.0	1712	25,800

#457132
535 gr. (20 to 1) 4.030" OAL
BC: .402
SD: .363

Powder	Sugg Starting Grains	Velocity fps	Pressure C.U.P.	Max Load Grains	Velocity fps	Pressure C.U.P.
XMP-5744	39.0	1495	21,300	43.5	1639	23,700

Note: Loads shown in shaded panels are maximum.
Loads shown in bold designate potentially most accurate load.

458 Winchester Magnum

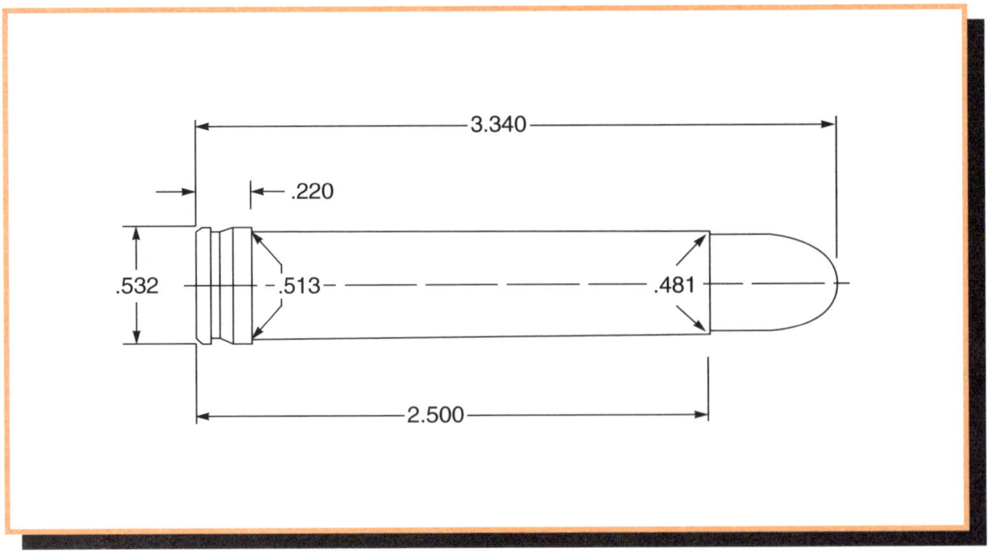

Comments:

Winchester threw its hat into the belted Magnum ring in 1956 with the 458. The cartridge has made a name for itself since then on the most dangerous game in Africa and around the world. The 458 Magnum has also developed a reputation for punishing recoil. Fortunately, it is a flexible cartridge which responds well to handloading. Shooters can load 500-grain bullets to full power or download it to 45-70 levels. Those loading the 458 should heed several cautions. Many maximum loads are heavily compressed. Use of Reloder 7 can minimize difficulties with powder compression, particularly with the long 500-grain bullets. Seating and crimping are best done in two separate operations to avoid crushed cases. Proper alignment of case mouth and bullet cannelure is crucial. Use of lighter weight bullets turn the 458 into a potent brush cartridge for much of North America.

Test Components:

Cases . . . . .Winchester
Trim-to Length . . . . .2.490"
Primers . . . . .Winchester WLR and WLRM
Primer Size . . . . . .Large Rifle, Standard and Magnum
Lyman Shell Holder .No. 13
Jacketed Bullets Used . . . . .Sierra FN #8900, 300 gr.
Speer SP #2478, 350 gr.
Remington SP #B22899, 405 gr.
Barnes X #45818, 450 gr.
Hornady RN #4504, 500 gr.
Cast Bullets Used . . . .(sized to .459" dia)
*gas check bullet — #457191, 292 gr.
#457124, 385 gr.
*#457671, 475 gr.

Test Specifications:
(Velocity & Pressure)

Firearm Used . . . . .Winchester Model 70
Ruger No.1
Universal Receiver
Barrel Length . . . . .Model 70; 22"
Ruger No.1; 24"
Universal Receiver; 24"
Twist . . . . .1-14"
Groove Dia. . . . . .Model 70: .457"
Ruger No.1: .459"
Universal Receiver: .457"

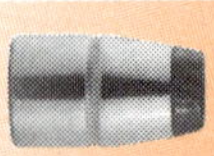

*** 300 gr. Jacketed FN**
2.940" OAL
BC: .120
SD: .204

Powder	Sugg Starting Grains	Velocity fps	Pressure C.U.P.	Max Load Grains	Velocity fps	Pressure C.U.P.
N130	72.0	2590	39,600	78.0	2789	52,900
IMR-4198	58.0	2265	28,800	**63.0**	**2514**	**45,600**
XMR-2015	70.0	2386	35,600	74.5+	2464	40,400
RX7	69.0	2563	33,700	75.0+	2757	52,400
IMR-3031	66.0	2133	20,100	72.0+	2391	31,500
H-322	74.0	2536	35,700	81.0+	2723	52,700

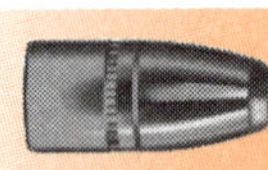

*** 350 gr. Jacketed SP**
3.105" OAL
BC: .232
SD: .238

Powder	Sugg Starting Grains	Velocity fps	Pressure C.U.P.	Max Load Grains	Velocity fps	Pressure C.U.P.
N130	69.0	2394	42,900	74.7	2550	52,700
IMR-4198	61.0	2287	43,100	66.3	2441	52,700
XMR-2015	72.0	2306	37,900	79.0+	2528	52,500
RX7	68.0	2350	41,600	74.0+	2529	52,700
IMR-3031	69.0	2158	30,300	74.0+	2353	42,200
H-322	**73.0**	**2354**	**37,700**	80.0+	2550	52,800

Note: Loads shown in shaded panels are maximum.
Loads shown in bold designate potentially most accurate load.
\+ Designates a compressed powder charge.
* Designates use of magnum primers.

458 Winchester Magnum

†† 405 gr. Jacketed SP 3.090" OAL — **BC:** .281 **SD:** .276

Powder	Sugg Starting Grains	Velocity fps	Pressure	Max Load Grains	Velocity fps	Pressure
IMR-3031	**64.0**	2036	—	**71.0+**	**2237**	—
IMR-4895	67.0	1988	—	75.0+	2217	—
H-335	74.0	1815	—	80.0+	1984	—
IMR-4064	67.0	1949	—	74.0+	2127	—
BL-C (2)	71.0	1733	—	77.0+	1919	—
IMR-4320	69.0	1988	—	77.0+	2247	—
H-380	72.0	1897	—	80.0+	2016	—

*** 450 gr. Barnes X** 3.280" OAL — **BC:** .488 **SD:** .306

Powder	Sugg Starting Grains	Velocity fps	Pressure	Max Load Grains	Velocity fps	Pressure
IMR-4198	49.0	1803	41,500	53.5	1948	52,800
XMR-2015	58.0	1844	36,200	63.3+	2034	52,400
RX7	55.0	1903	36,600	59.5	2052	52,600
IMR-3031	60.0	1875	41,600	65.0+	2052	52,800
IMR-4895	61.0	1852	40,600	67.0+	2026	53,000
H-335	68.0	1935	36,200	74.3+	2159	53,000
N135	**63.0**	**1979**	**43,700**	68.5+	2124	52,700

500 gr. Jacketed RN 3.340" OAL — **BC:** .287 **SD:** .341

Powder	Sugg Starting Grains	Velocity fps	Pressure C.U.P.	Max Load Grains	Velocity fps	Pressure C.U.P.
†† RX7	**60.0**	**1709**	—	65.0+	1937	—
†† IMR-3031	62.0	1890	—	69.0+	2053	—
†† IMR-4895	66.0	1886	—	73.0+	2066	—
†† H-335	73.0	1821	—	76.0+	2000	—
†† IMR-4064	64.0	1785	—	71.0+	1968	—
†† BL-C(2)	70.0	1736	—	75.0+	1912	—
†† IMR-4320	67.0	1858	—	75.0+	2066	—
**SR-4759	30.0	1216	18,000	34.5	1384	25,200
**XMP-5744	34.0	1212	16,000	39.5	1380	22,180

††† #457191 292 gr. (#2 Alloy) 2.995" OAL — **BC:** .201 **SD:** .198

Powder	Sugg Starting Grains	Velocity fps	Pressure	Max Load Grains	Velocity fps	Pressure
Unique	19.0	1460	—	23.0	1625	—
2400	28.0	1585	—	33.0	1765	—
SR-4759	32.0	1499	—	36.5	1762	—
IMR-4227	44.5	2100	—	50.5	2305	—
XMP-5744	37.0	1530	—	42.5	1744	—
IMR-4198	54.0	2135	—	58.0	2205	—
IMR-3031	**60.0**	**1885**	—	69.0	2220	—
IMR-4064	64.0	1865	—	72.0	2110	—

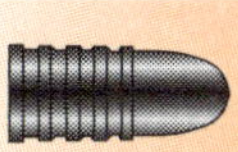

††† #457124 385 gr. (#2 Alloy) 3.012" OAL — **BC:** .299 **SD:** .261

Powder	Sugg Starting Grains	Velocity fps	Pressure	Max Load Grains	Velocity fps	Pressure
Unique	18.0	1315	—	22.0	1510	—
2400	25.0	1310	—	31.0	1565	—
SR-4759	**29.0**	**1431**	—	35.5	1723	—
IMR-4227	44.0	1935	—	50.0	2090	—
XMP-5744	33.5	1431	—	42.0	1706	—
IMR-4198	53.0	2010	—	57.0	2125	—
IMR-3031	59.0	1900	—	68.0+	2210	—
IMR-4064	62.0	1860	—	70.0+	2115	—

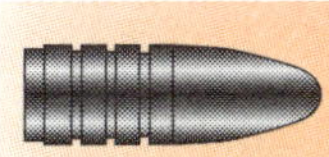

††† #457671 475 gr. (#2 Alloy) 3.145" OAL — **BC:** .477 **SD:** .322

Powder	Sugg Starting Grains	Velocity fps	Pressure	Max Load Grains	Velocity fps	Pressure
SR-4759	29.0	1428	—	35.0	1646	—
IMR-4227	35.5	1436	—	39.0	1648	—
XMP-5744	**34.5**	**1425**	—	40.5	1623	—
IMR-4198	37.5	1433	—	41.5	1623	—

Note: Loads shown in shaded panels are maximum.
Loads shown in bold designate potentially most accurate load.
\+ Designates a compressed powder charge.
* Designates use of magnum primers.
†† Fired in a Model 70.
††† Fired in a Ruger No.1.
** Designates a reduced load.

460 Weatherby Magnum

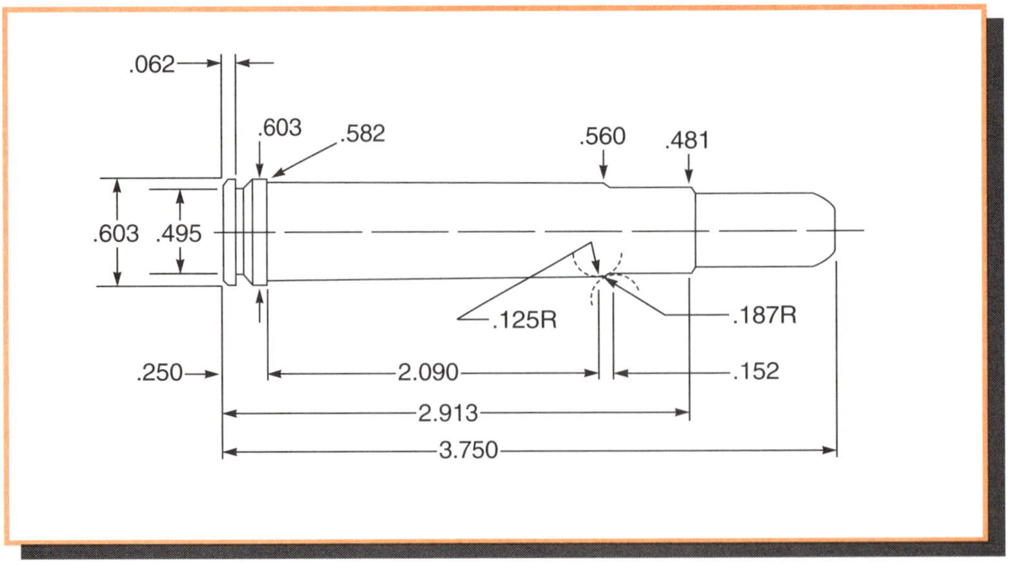

Comments:

The 460 Weatherby justifiably made its reputation as the last word in stopping power on dangerous game. It evolved as a necked up version of the 378 Weatherby in 1958 and was one of the earliest developments of that particular case. The 460 is designed to stop the most dangerous game on earth. By all accounts it does this very well. It's not especially useful for North America save maybe certain Alaskan situations but is a fun gun to shoot the first time around. Weatherby recommends the use of Federal 215 Magnum primers for all jacketed bullets. This data is intended for commercially produced and chambered rifles. It is not for use in custom guns that may lack the free bore found in standard Weatherby chambers.

Test Components:

Cases . . . . Weatherby
Trim-to Length . . . . 2.903"
Primers . . . . Federal 215 & 210
Primer Size . . . . Large Rifle, Magnum & Standard
Lyman Shell Holder . . . . No. 17
Jacketed Bullets Used. . .Remington SP #B22899, 405 gr.
Hornady RN #4504, 500 gr.
Cast Bullets Used . . . . (sized to .458" dia)
*gas check bullet *#457671, 475 gr.
#457125, 500 gr.

Test Specifications:
(Velocity Only)

Firearm Used . . . . Weatherby Mk V
Barrel Length . . . . 26"
Twist . . . . 1-16"
Groove Dia. . . . . .458"

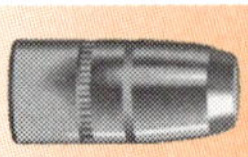

405 gr. Jacketed SP
3.485" OAL
BC: .281
SD: .276

Powder	Sugg Starting Grains	Velocity fps	Pressure	Max Load Grains	Velocity fps	Pressure
SR-4759	49.0	1868	—	**58.0**	**2101**	—
XMP-5744	56.0	1856	—	65.0	2091	—
IMR-4198	57.0	1865	—	69.0	2159	—

500 gr. RN
3.710" OAL
BC: .287
SD: .341

Powder	Sugg Starting Grains	Velocity fps	Pressure	Max Load Grains	Velocity fps	Pressure
H-414	110.0	2372	—	116.0	2494	—
IMR-4350	117.0	2545	—	123.0+	2682	—
XMR-4350	114.0	2401	—	120.0+	2506	—
RX19	**118.0**	**2521**	—	124.0+	2647	—
H-4831SC	119.0	2380	—	126.0+	2500	—
AA8700	128.0+	2004	—	135.0+	2119	—
**SR-4759	50.0	1720	—	59.0	1922	—
**XMP-5744	56.0	1706	—	65.0	1936	—

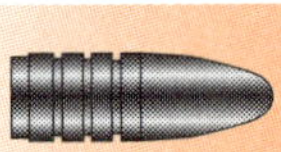

***#457671**
475 gr. (#2 Alloy) 3.700" OAL
BC: .477
SD: .323

Powder	Sugg Starting Grains	Velocity fps	Pressure	Max Load Grains	Velocity fps	Pressure
SR-4759	46.0	1717	—	56.0	2006	—
XMP-5744	**50.0**	**1700**	—	63.5	2039	—
RX7	57.5	1686	—	69.5	2015	—

***#457125**
500 gr. (#2 Alloy) 3.650" OAL
BC: .391
SD: .340

Powder	Sugg Starting Grains	Velocity fps	Pressure	Max Load Grains	Velocity fps	Pressure
SR-4759	46.0	1701	—	57.5	1959	—
XMP-5744	**52.0**	**1702**	—	62.5	1958	—
RX7	61.0	1745	—	68.0	1956	—

Note: Loads shown in shaded panels are maximum. Loads shown in bold designate potentially most accurate load.

\+ Designates a compressed powder charge.
* Designates the use of 210 primers.
**Designates a reduced load.

50-70 Government

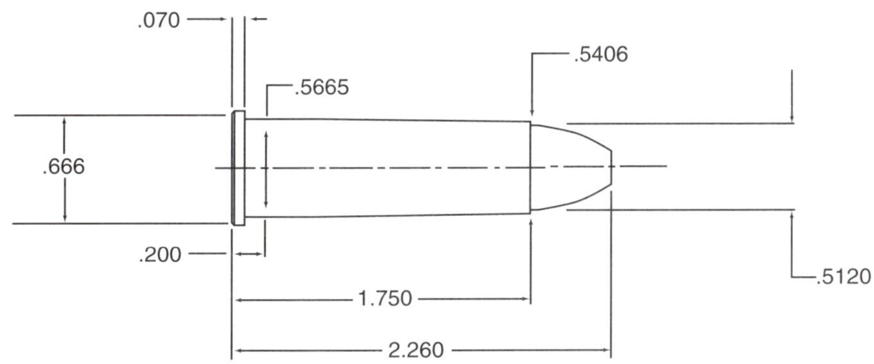

Comments:

The 50-70 Government led a short military career as the first general issue center-fire rifle cartridge in U.S. Army service. It served from 1866 until its replacement by the 45-70 in 1873. The 50-70 enjoyed fairly wide spread post-military usage throughout the remainder of the nineteenth century. The cartridge is most commonly encountered in rolling blocks and reportedly saw extensive use during the buffalo hunts of the late 1870s. The 50-70 has made a bit of a comeback recently among black powder cartridge aficionados.

This data applies only to modern, brass cases of recent manufacture. Do not use any old balloon head cases. This data is for use in newly manufactured replica rifles approved for smokeless powder. This data is not for use in antique firearms originally designed for black powder.

Test Components:

Cases . . . . .Bell
Trim-to Length . . . . .1.740"
Primers . . . . .Federal 215
Primer Size . . . . .Large Rifle, Magnum
Lyman Shell Holder . . . . .No. 22
Cast Bullets Used . . . .(sized to .511" dia)
#515141, 425 gr.

Test Specifications:
(Velocity Only)

Firearm Used . . . . .Shiloh Sharps
Barrel Length . . . . .22"
Twist . . . . .1-48"
Groove Dia. . . . . .511"

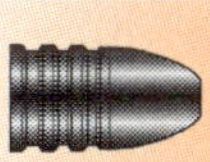

#515141
425 gr. (#2 Alloy) 2.312" OAL

BC: .250
SD: .232

Powder	Sugg Starting Grains	Velocity fps	Pressure	Max Load Grains	Velocity fps	Pressure
Unique	11.0	1003	—	14.8	1211	—
SR-4759	22.0	1114	—	26.5	1313	—
IMR-4227	22.0	1102	—	27.3	1333	—
IMR-4198	25.5	1129	—	30.0	1305	—

Note: Loads shown in shaded panels are maximum.
Loads shown in bold designate potentially most accurate load.

25 Automatic (25 ACP)

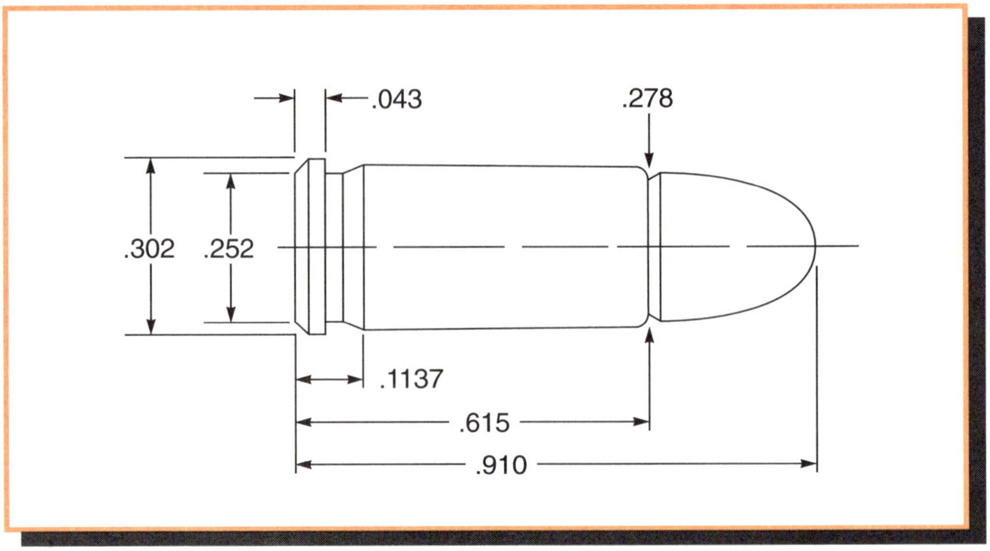

Comments:

The 25 ACP (Automatic Colt Pistol) came to prominence to American shooters in the Colt Vest Pocket pistol designed by none other than John M. Browning. This cartridge owes much of its popularity to the small size of the pistols chambered for it rather than its ballistics. Its marginal performance as a defensive round has drawn the ridicule of numerous gun writers over the years. The 25 ACP is most commonly encountered today through pocket pistols imported from overseas from sources such as Beretta. It has never been widely regarded as a handloader's cartridge and bullet selection is not extensive. Charge weights for the 25 ACP are quite small. Shooters should take care not to accidentally double-charge cases with powder. The 25 ACP case is semi-rimmed and can be lightly roll crimped. Bullseye and 231 are preferred powders.

Test Components:

Cases . . . . .Remington
Trim-to Length . . . . .610"
Primers . . . . .Winchester WSP
Primer Size . . . . .Small Pistol
Lyman Shell Holder . . . . .No. 32
Jacketed Bullets Used . . .Hornady HP #35450, 35 gr.
Hornady FMJ #3545, 50 gr.

Test Specifications:
(Velocity & Pressure)

Firearm Used . . . . .Universal Receiver
Barrel Length . . . . .2"
Twist . . . . .1-16"
Groove Dia. . . . . .251"

35 gr. Jacketed HP
.860" OAL
BC: .072
SD: .079

Powder	Sugg Starting Grains	Velocity fps	Pressure C.U.P.	Max Load Grains	Velocity fps	Pressure C.U.P.
Bullseye	**1.3**	**751**	—	1.5	863	—
Red Dot	1.2	774	—	1.4	955	—
700X	1.3	766	—	1.5	938	—
AA#2	1.3	737	—	1.5	796	—
HP-38	1.3	641	—	1.5	923	—
231	1.3	768	—	1.5	807	—

50 gr. Jacketed FMC
.900" OAL
BC: .116
SD: .113

Powder	Sugg Starting Grains	Velocity fps	Pressure C.U.P.	Max Load Grains	Velocity fps	Pressure C.U.P.
Bullseye	1.0	558	11,000	**1.4**	**754**	**17,300**
231	1.0	563	11,800	1.4	752	17,800
HP-38	0.8	565	12,200	1.1	710	17,500
Red Dot	1.0	562	11,800	1.3	749	18,100
700X	0.8	556	11,600	1.1	743	18,000
Unique	1.2	544	11,000	1.6	740	16,700

Note: Loads shown in shaded panels are maximum.
Loads shown in bold designate potentially most accurate load.

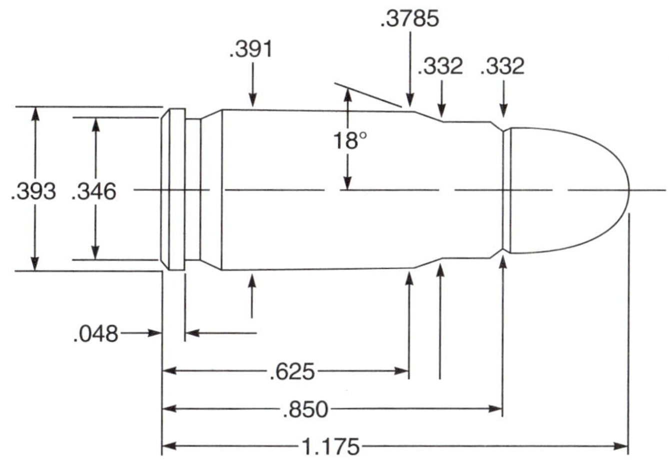

Comments:

Groove diameter variations are the norm in firearms chambered for this cartridge. In some handguns the starting load may not function the action. In such an instance work up the load cautiously.

For best results with cast bullets, size bullets to, or 0.001" above groove diameter. Cast bullet #313249 should be crimped on the leading edge of the first driving band.

Test Components:

Cases . . . . .Remington
Trim-to Length . . . . .845"
Primers . . . . .Remington 1½
Primer Size . . . . .Small Pistol
Lyman Shell Holder . . . . .No. 12
Jacketed Bullets Used . . . .Hornady FMJ #1419, 93 gr.
Cast Bullets Used . . . .(sized to .309" dia)
#313249, 85 gr.

Test Specifications:
(Velocity Only)

Firearm Used . . . . .Walther P-38
Barrel Length . . . . .4 15/16"
Twist . . . . .1-9.85"
Groove Dia. . . . . .3085"

93 gr. Jacketed FMJ
1.175" OAL
BC: .102
SD: .140

Powder	Sugg Starting Grains	Velocity fps	Pressure	Max Load Grains	Velocity fps	Pressure
700X	3.6	1065	—	**4.1**	**1171**	—
SR-4756	5.0	1038	—	5.5	1161	—
Bullseye	3.5	968	—	4.0	1067	—
Unique	4.5	1003	—	5.0	1115	—
231	3.7	935	—	4.2	1046	—
HP-38	3.5	1028	—	4.0	1115	—

#313249
85 gr. (#2 Alloy) 1.125" OAL
BC: .212
SD: .127

Powder	Sugg Starting Grains	Velocity fps	Pressure	Max Load Grains	Velocity fps	Pressure
700X	3.7	1143	—	4.2	1204	—
PB	4.0	1105	—	4.5	1177	—
SR-4756	4.7	966	—	5.6	1182	—
Bullseye	3.8	1106	—	4.5	1244	—
Red Dot	3.8	1116	—	4.3	1201	—
Green Dot	4.2	1111	—	4.7	1210	—
Unique	4.6	1110	—	5.2	1226	—
231	3.9	1054	—	**4.4**	**1144**	—

Note: Loads shown in shaded panels are maximum.
Loads shown in bold designate potentially most accurate load.

7.62 Tokarev/30 Mauser

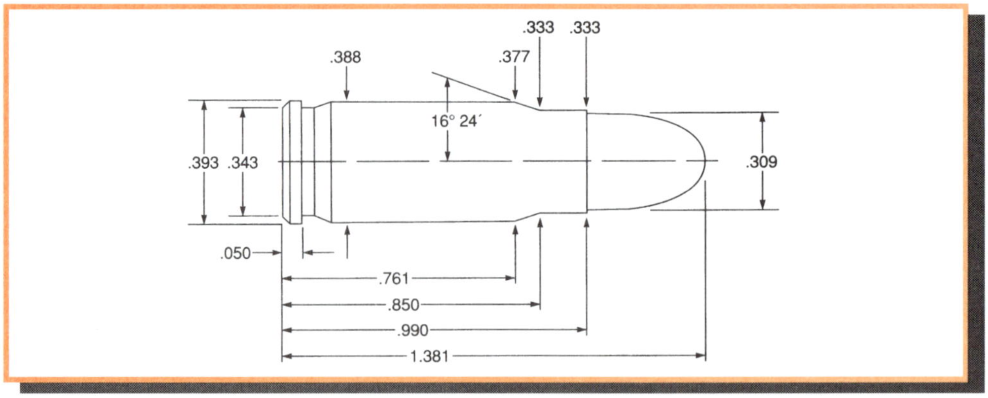

Comments:

The 7.62x25mm Tokarev and 30 Mauser are two different cartridges. However, their dimensions are so close that it has been common practice to use them interchangeably. The 30 Mauser had its origins in the Borchardt semiautomatic pistol developed in 1893. It became best known as the cartridge most often encountered in the Mauser C96 "Broomhandle" pistol. The Soviet Union adopted the 7.62x25 cartridge chambered in the Tokarev TT-30 during the early 1930s. The Soviets also produced large numbers of submachine guns chambered in this cartridge. The 7.62x25 became widely distributed throughout the Soviet sphere of influence during the Cold War and both pistols and ammunition have been commonplace on the U.S. market since the demise of the Eastern bloc. Much of this ammunition is useless to reloaders as it is steel cased.

Like any mass-produced military arm, groove diameters can vary. Cast bullets should be sized as close to groove diameter as possible. Make up a dummy round with a bullet of the chosen diameter to ensure the cartridge both chambers and falls freely. Bullets should be cast of a hard pistol alloy, preferably linotype. The listed starting loads may not function the action of some pistols. Do not use this data for the older 7.65 Borchardt pistol. Borchardt pistols operate at a lower chamber pressure. The Hornady bullets used in our data are available exclusively through The Old Western Scrounger, 12924 Highway A-12, Montague, CA, 96064.

Test Components:

Cases . . . . Starline
Trim-to Length . . . . .980"
Primers . . . . Remington 1½
Primer Size . . . . Small Pistol
Lyman Shell Holder . . . . No. 12
Jacketed Bullets Used . . . . Hornady FMJ #1502, 86 gr.
Hornady FMJ #1419, 93 gr.
Cast Bullets Used . . . . (sized to .311" dia)
#313249, 85 gr.

Test Specifications: (Velocity Only)

Firearm Used Cz52
Barrel Length . . . . 4.7"
Twist . . . . 1-12½"
Groove Dia. . . . . .310"

86 gr. FMJ
1.290" OAL
BC: .093
SD: .129

Powder	Sugg Starting Grains	Velocity fps	Pressure	Max Load Grains	Velocity fps	Pressure
Bullseye	4.8	1187	—	5.4	1308	—
700X	4.6	1190	—	5.7	1363	—
Red Dot	5.0	1240	—	**5.8**	**1395**	—
AA#2 Imp.	4.5	1125	—	5.9	1368	—
N320	4.8	1178	—	5.9	1375	—
HP-38	5.0	1157	—	6.0	1318	—
231	5.0	1177	—	6.1	1374	—
Unique	5.0	1165	—	6.2	1335	—
AA#5	6.0	1098	—	7.1	1278	—

93 gr. FMJ
1.310" OAL
BC: .102
SD: .140

Powder	Sugg Starting Grains	Velocity fps	Pressure	Max Load Grains	Velocity fps	Pressure
Bullseye	4.6	1118	—	5.3	1236	—
700X	4.5	1131	—	5.4	1274	—
Red Dot	4.9	1194	—	**5.7**	**1343**	—
AA#2 Imp.	4.5	1077	—	5.8	1276	—
N320	4.5	1074	—	5.8	1275	—
HP-38	4.9	1112	—	6.0	1289	—
231	5.0	1132	—	6.0	1283	—
Unique	5.0	1062	—	6.3	1303	—
AA#5	6.0	1042	—	7.1	1229	—

#313249
85 gr. (#2 Alloy) 1.290" OAL
BC: .212
SD: .125

Powder	Sugg Starting Grains	Velocity fps	Pressure	Max Load Grains	Velocity fps	Pressure
Titegroup	4.0	1222	—	5.0	1418	—
Bullseye	4.0	1184	—	5.2	1409	—
Red Dot	**4.1**	**1209**	—	5.0	1381	—
AA#2 Imp.	4.2	1221	—	5.2	1377	—
HP-38	4.7	1237	—	5.5	1374	—
231	4.5	1203	—	5.5	1379	—
Unique	4.6	1123	—	6.0	1376	—

Note: Loads shown in shaded panels are maximum.
Loads shown in bold designate potentially most accurate load.

32 Automatic (32 ACP)

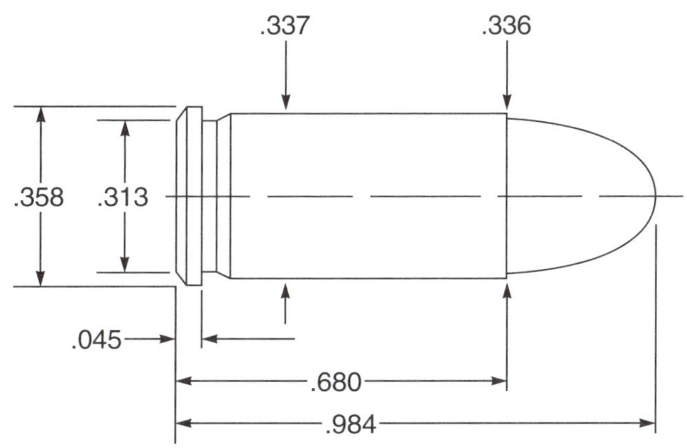

Comments:

The loads listed may be used with 0.311" jacketed bullets in pistols which have a groove diameter as small as 0.309 inches. Groove diameter variations are extensive in handguns chambered for this cartridge. Chamber limits make it unsafe to attempt to use larger diameter bullets, regardless of groove diameter measurement.

The very fast powders, such as Alliants Bullseye, Hodgdon HP-38, and Winchester 231 are best for this tiny cartridge.

Test Components:

Cases . . . . .Remington, Winchester
Trim-to Length . . . . .672"
Primers . . . . .Remington 1½, Winchester WSP
Primer Size . . . . .Small Pistol
Lyman Shell Holder . . . . .No. 23
Jacketed Bullets Used . . . . . .Sierra FMJ #8010, 71 gr.
Cast Bullets Used . . . .(sized to .309" dia) #311252, 75 gr.

Test Specifications: (Velocity & Pressure)

Firearm Used . . . . .Mauser HSc Universal Receiver
Barrel Length . . . . .Mauser; 3" Universal Receiver; 3¼"
Twist . . . . .1-16"
Groove Dia. . . . . .309"

71 gr. FMJ .984" OAL — **BC:** .120 **SD:** .105

Powder	Sugg Starting Grains	Velocity fps	Pressure C.U.P.	Max Load Grains	Velocity fps	Pressure C.U.P.
**700X	1.7	706	9,500	2.2	902	14,700
Bullseye	1.5	603	—	**2.2**	**825**	—
Red Dot	1.7	705	—	2.5	937	—
Unique	2.0	571	—	3.1	945	—
****HP-38**	1.7	733	11,100	**2.2**	**883**	**14,900**

#311252 75 gr. (#2 Alloy) .975" OAL — **BC:** .121 **SD:** .112

Powder	Sugg Starting Grains	Velocity fps	Pressure C.U.P.	Max Load Grains	Velocity fps	Pressure C.U.P.
700X	1.5	750	—	1.9	895	—
PB	1.6	665	—	2.2	870	—
SR-7625	1.7	675	—	2.3	860	—
Bullseye	1.5	685	—	2.0	830	—
Red Dot	1.5	695	—	2.3	940	—
Green Dot	1.6	705	—	2.4	895	—
Unique	1.8	625	—	2.6	860	—
****231**	1.5	620	7,200	**2.5**	**888**	**15,000**
**HP-38	1.5	688	9,200	2.0	824	13,800

Note: Loads shown in shaded panels are maximum.
Loads shown in bold designate potentially most accurate load.
** Designates use of Winchester cases and primers.

32 Smith & Wesson Long

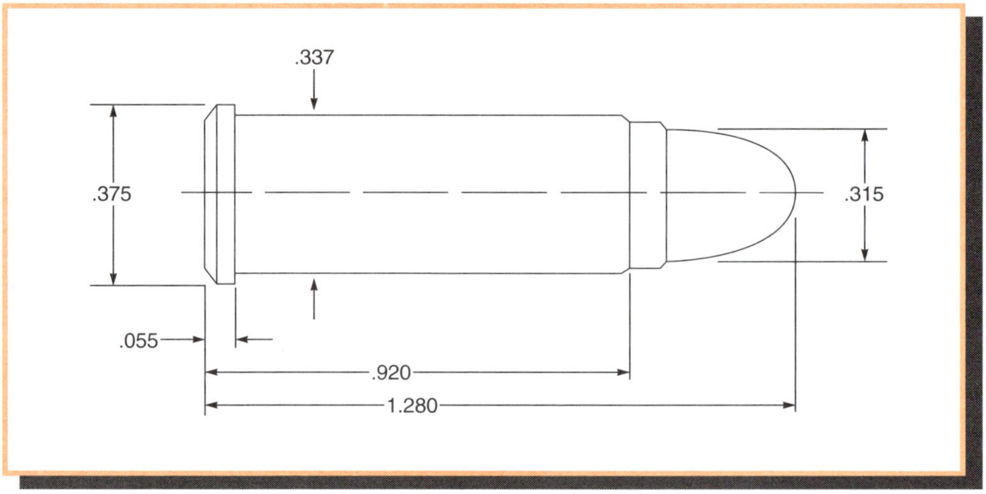

Comments:

This data is intended for use only in solid frame revolvers. We do not recommend use of these loads in older revolvers of the top break design due to their age and relative weakness. The 32 Smith & Wesson Long is dimensionally identical to the 32 Colt New Police. The main difference between the cartridges is the bullet weight of factory ammunition. Neither of these cartridges should be confused with the 32 Long Colt. Cartridge body diameter of the 32 Long Colt is nearly .020" smaller than these cartridges.

Cartridges bearing the 32 Long Colt headstamp should not be fired in any revolver marked 32 S&W Long or 32 Colt New Police. Ruptured cases will result.

The 32 S&W Long has developed a following among European target shooters due to its high accuracy potential. Internal barrel dimensions can vary widely from gun to gun. Shooters should slug their bore and size bullets to groove diameter or .001" larger. Make up a dummy round to ensure that your revolver will properly function with bullets of the chosen diameter. Fast propellants work best in these small cases. Alliant Bullseye or Winchester 231 are good choices.

Test Components:

Cases . . . . .Remington
Trim-to Length . . . . .910"
Primers . . . . .Remington 1½
Primer Size . . . . .Small Pistol
Lyman Shell Holder . . . . .No. 9
Jacketed Bullets Used . . . .
Hornady HP/XTP #32050, 85 gr.
Hornady HP/XTP #32070, 100 gr.
Cast Bullets Used . . . .(sized to .312" dia)
#311252, 75 gr.

Test Specifications:
(Velocity Only)

Firearm Used . . . . .Smith & Wesson Model 31
Barrel Length . . . . .4"
Twist . . . . .1-18¾"
Groove Dia. . . . . .312"

85 gr. Jacketed HP
1.160" OAL
BC: .145
SD: .125

Powder	Sugg Starting Grains	Velocity fps	Pressure	Max Load Grains	Velocity fps	Pressure
Bullseye	2.1	654	—	2.4	815	—
Titegroup	2.1	704	—	2.4	820	—
AA#2	2.1	742	—	2.4	909	—
HP-38	2.5	842	—	2.8	871	—
231	**2.6**	**752**	—	2.9	835	—
AA#5	3.2	655	—	3.8	799	—
Unique	2.7	709	—	3.0	837	—

100 gr. Jacketed HP
1.160" OAL
BC: .170
SD: .147

Powder	Sugg Starting Grains	Velocity fps	Pressure	Max Load Grains	Velocity fps	Pressure
Bullseye	2.0	600	—	2.3	718	—
AA#5	3.0	630	—	3.4	700	—
Unique	2.5	573	—	2.8	733	—
SR-4756	3.7	677	—	4.1	761	—
AA#9	5.3	626	—	5.9	767	—

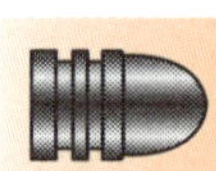

#311252
75 gr. (#2 Alloy) 1.115" OAL
BC: .121
SD: .110

Powder	Sugg Starting Grains	Velocity fps	Pressure	Max Load Grains	Velocity fps	Pressure
Bullseye	1.5	513	—	3.0	958	—
Red Dot	1.8	588	—	3.3	1028	—
700X	1.8	598	—	3.3	1098	—
Green Dot	1.9	563	—	3.4	993	—
231	2.0	677	—	3.5	1043	—
Unique	2.5	608	—	4.0	943	—
SR-7625	2.4	583	—	3.6	988	—

Note: Loads shown in shaded panels are maximum.
Loads shown in bold designate potentially most accurate load.

32 H&R Magnum

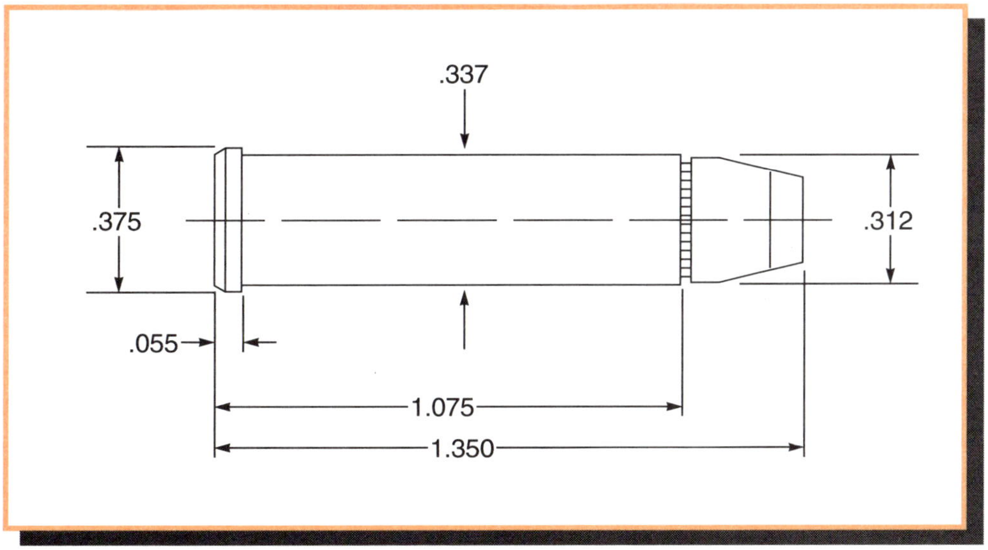

Comments:

Revolvers chambered for this cartridge will also fire 32 S&W, or 32 S&W Long ammo. However the reverse is not true.

Do not load 32 H&R data into any case except those so headstamped.

Alliants Unique, Hodgdon HP-38, and Winchester 231 will provide the best accuracy.

Test Components:

Cases . . . . .Federal
Trim-to Length . . . . .1.065"
Primers . . . . .Federal 100
Primer Size . . . . .Small Pistol
Lyman Shell Holder . . . . .No. 9
Jacketed Bullets Used . . . . . .Sierra FMJ #8010, 71 gr.
Hornady HP/XTP #32050, 85 gr.
Sierra JHC #8030, 90 gr.
Hornady HP/XTP #32070, 100 gr.
Cast Bullets Used . . . .(sized to .313" dia)
#313249, 85 gr.

Test Specifications: (Velocity & Pressure)

Firearm Used . . . . .Universal Receiver
Barrel Length . . . . .5"
Twist . . . . .1-16"
Groove Dia. . . . . .312"

71 gr. FMJ 1.350" OAL — **BC:** .120 **SD:** .105

Powder	Sugg Starting Grains	Velocity fps	Pressure C.U.P.	Max Load Grains	Velocity fps	Pressure C.U.P.
Bullseye	3.8	1098	15,500	4.3	1269	20,100
HP-38	3.0	902	14,900	3.9	1104	20,100
231	3.5	817	13,400	4.5	1132	20,000
Unique	**4.5**	**1004**	**14,400**	5.0	1169	18,200
SR-4756	4.7	900	12,400	5.5	1141	19,200
HS-6	5.4	1098	15,600	6.0	1310	20,100
800X	4.8	936	12,200	6.0	1174	19,400
Blue Dot	6.5	936	14,300	7.8	1188	20,600

85 gr. Jacketed HP 1.310" OAL — **BC:** .145 **SD:** .125

Powder	Sugg Starting Grains	Velocity fps	Pressure C.U.P.	Max Load Grains	Velocity fps	Pressure C.U.P.
Bullseye	3.3	914	16,800	3.7	1043	19,800
HP-38	2.3	755	13,700	3.5	1022	21,000
231	2.8	773	13,300	4.0	1054	20,800
Unique	**4.0**	**859**	**14,900**	4.5	1049	18,600
AA#5	4.0	788	13,000	5.4	1089	20,600
SR-4756	3.5	641	11,500	4.8	1073	20,700
HS-6	5.0	897	15,100	5.6	1124	20,000
800X	3.7	758	13,600	5.0	1005	20,500
Blue Dot	5.5	798	12,700	6.7	1044	19,200

Note: Loads shown in shaded panels are maximum.
Loads shown in bold designate potentially most accurate load.

32 H&R Magnum

90 gr. Jacketed HC
1.315" OAL

BC: .146
SD: .132

Powder	Sugg Starting Grains	Velocity fps	Pressure C.U.P.	Max Load Grains	Velocity fps	Pressure C.U.P.
Bullseye	3.1	844	16,700	3.5	1008	20,300
HP-38	2.0	612	12,100	3.3	941	20,600
231	2.5	679	12,500	**3.8**	**980**	**20,600**
Unique	3.8	869	13,600	4.3	1003	18,400
AA#5	4.0	723	11,900	5.4	1064	20,700
Herco	3.7	880	13,900	4.2	1005	19,400
SR-4756	3.2	549	9,800	4.6	1015	20,500
HS-6	5.4	863	14,300	6.0	1016	18,900
800X	3.8	805	13,400	5.0	1057	20,600
Blue Dot	5.0	719	11,800	6.5	1014	19,700
296	—	—	—	9.9	1177	19,400

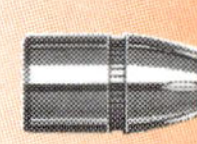

100 gr. Jacketed HP
1.315" OAL

BC: .170
SD: .147

Powder	Sugg Starting Grains	Velocity fps	Pressure C.U.P.	Max Load Grains	Velocity fps	Pressure C.U.P.
Bullseye	2.9	717	15,400	3.3	838	19,500
231	3.2	706	15,000	3.6	841	19,500
Unique	**3.6**	**781**	**15,900**	4.0	928	20,800
AA#5	4.5	766	15,300	5.1	931	20,100
Herco	3.7	740	15,200	4.2	895	19,700
HS-6	4.9	754	15,100	5.4	911	19,500
Blue Dot	5.1	711	14,300	5.7	935	20,100

#313249
85 gr. (#2 Alloy) 1.350" OAL

BC: .212
SD: .124

Powder	Sugg Starting Grains	Velocity fps	Pressure C.U.P.	Max Load Grains	Velocity fps	Pressure C.U.P.
Titegroup	3.1	992	15,900	3.5	1089	17,200
Bullseye	2.5	803	12,400	4.0	1115	20,900
Red Dot	2.5	801	12,800	3.8	1110	20,400
HP-38	2.0	738	10,900	3.5	1082	20,700
231	2.5	777	12,500	**4.1**	**1101**	**20,400**
Unique	3.0	747	10,800	4.5	1092	20,100
AA#5	4.0	820	11,500	5.4	1124	19,200
SR-7625	2.7	709	9,700	4.3	1083	19,800
Herco	3.5	845	12,500	4.9	1080	19,300
SR-4756	3.0	659	8,000	4.6	1091	18,900
800X	3.8	889	12,400	5.2	1136	19,300
Blue Dot	5.0	824	12,700	6.5	1105	19,000

Note: Loads shown in shaded panels are maximum.
Loads shown in bold designate potentially most accurate load.

32-20 Winchester (32-20 WCF)

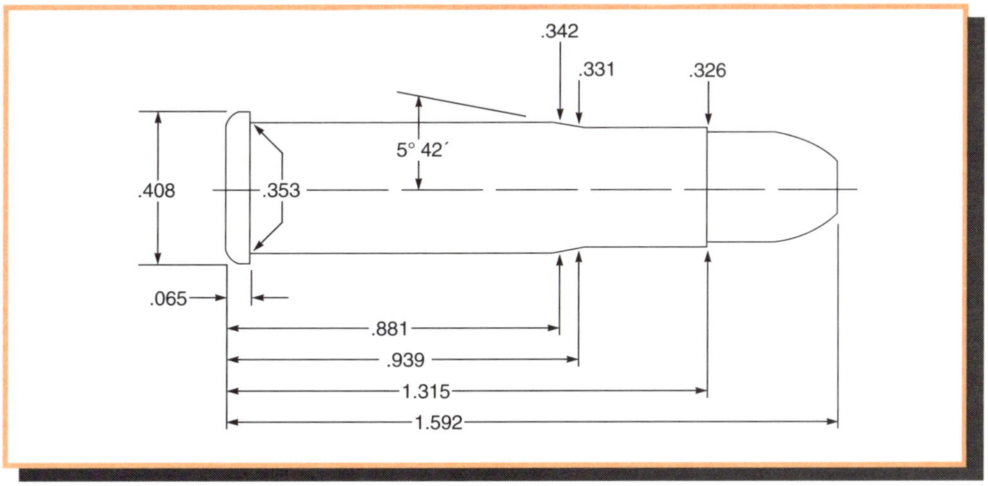

Comments:

This data is for revolvers made by Colt, Smith & Wesson and Ruger designed for smokeless powder 32-20 Winchester ammunition. Do not use this data for old guns designed for black powder or any of the old cheap Spanish copies of the S&W revolver.

The 32-20 Winchester has been chambered in revolvers for many years as a companion to the numerous lever action rifles which also chambered the cartridge.

This is a good game cartridge when fired in well-made revolvers.

Test Components:

Cases . . . . .Winchester
Trim-to Length . . . . 1.305"
Primers . . . . .Winchester WSP
Primer Size . . . . .Small Pistol
Lyman Shell Holder . . . . No. 10
Jacketed Bullets Used . . . .
Hornady HP/XTP #32050, 85 gr.
Sierra JHC #8030, 90 gr.
Cast Bullets Used . . . .(sized to .313" dia)
#311008, 115 gr.

Test Specifications:
(Velocity Only)

Firearm Used . . . . .Ruger Blackhawk
Barrel Length . . . . .6½"
Twist . . . . .1-16"
Groove Dia. . . . . .312"

85 gr. Jacketed HP
1.535" OAL
BC: .145
SD: .125

Powder	Sugg Starting Grains	Velocity fps	Pressure C.U.P.	Max Load Grains	Velocity fps	Pressure C.U.P.
231	3.2	745	—	4.7	1022	—
SR-7625	3.3	703	—	4.7	987	—
SR-4756	4.5	748	—	5.8	1101	—
Unique	4.1	742	—	**4.8**	**967**	—
AA#5	4.2	554	—	5.7	948	—

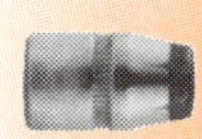

90 gr. Jacketed HP
1.535" OAL
BC: .112
SD: .132

Powder	Sugg Starting Grains	Velocity fps	Pressure C.U.P.	Max Load Grains	Velocity fps	Pressure C.U.P.
231	3.4	679	—	4.5	937	—
SR-7625	3.7	716	—	4.5	845	—
SR-4756	5.2	791	—	5.5	928	—
Unique	3.6	639	—	**4.8**	**864**	—
AA#5	4.7	697	—	5.4	875	—

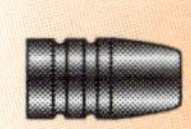

#311008
115 gr. (#2 Alloy) 1.530" OAL
BC: .154
SD: .168

Powder	Sugg Starting Grains	Velocity fps	Pressure C.U.P.	Max Load Grains	Velocity fps	Pressure C.U.P.
231	3.1	782	—	**3.8**	**905**	—
SR-7625	3.2	642	—	4.0	837	—
SR-4756	4.0	735	—	4.7	896	—
Unique	3.2	684	—	4.0	835	—
AA#5	3.8	649	—	4.5	825	—

Note: Loads shown in shaded panels are maximum.
Loads shown in bold designate potentially most accurate load.

380 Automatic (380 ACP) (9mm KURZ)

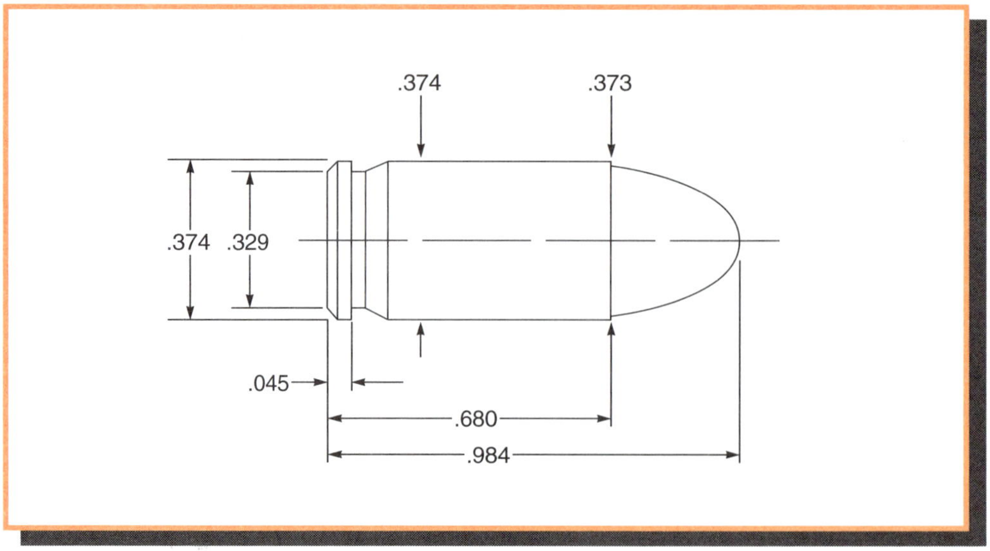

Comments:

Groove diameters have been measured as small as 0.355" to as large as 0.362 inch. Such variations can play havoc with accuracy. Because of chamber restrictions it is generally not safe to load bullets of a diameter larger than 0.355 inch.

Heavy (long) bullets of 0.355" diameter may cause case bulging due to the internal case taper. This may be ignored so long as the outside diameter of the case at the bulge does not exceed 0.374 inch.

Winchester 231 is an outstanding performer with all weight bullets in this cartridge.

Test Components:

Cases . . . . .Federal
Trim-to Length . . . . .677"
Primers . . . . .Federal 100 and Winchester WSP
Primer Size . . . . .Small Pistol
Lyman Shell Holder . . . . .No. 26
Jacketed Bullets Used . . . . . .Sierra JHC #8100, 90 gr.
Sierra FMJ #8105, 95 gr.
Hornady FMJ #35527, 100 gr.
Hornady HP/XTP #35540, 115 gr.
Cast Bullets Used . . . .(sized to .355" dia)
#356242, 90 gr.
#352242, 120 gr.

Test Specifications: (Velocity & Pressure)

Firearm Used . . . . .Universal Receiver
Barrel Length . . . . .3¾"
Twist . . . . .1-16"
Groove Dia. . . . . .355"

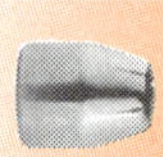

90 gr. Jacketed HC
.925" OAL
BC: .115
SD: .102

Powder	Sugg Starting Grains	Velocity fps	Pressure C.U.P.	Max Load Grains	Velocity fps	Pressure C.U.P.
Bullseye	2.0	713	9,100	**2.9**	**970**	**15,500**
Titegroup	2.4	828	13,300	2.7	914	15,600
Red Dot	1.9	691	8,400	2.8	983	15,700
700X	1.8	704	9,100	2.5	959	15,900
N320	2.6	769	11,300	2.9	907	15,300
AA#2	2.4	772	12,500	2.7	889	16,200
231	2.1	681	8,100	3.2	992	16,000
Unique	2.5	709	8,700	3.4	967	15,500
SR-7625	2.2	583	8,100	3.1	941	15,200
Power Pistol	3.6	881	14,300	4.0	953	16,300

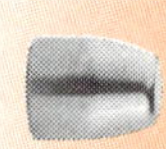

95 gr. FMJ
.900" OAL
BC: .116
SD: .108

Powder	Sugg Starting Grains	Velocity fps	Pressure C.U.P.	Max Load Grains	Velocity fps	Pressure C.U.P.
Bullseye	2.0	724	9,400	2.8	953	15,700
Titegroup	2.4	861	14,500	2.7	938	16,400
Red Dot	2.0	712	9,600	2.7	938	15,600
700X	1.7	693	9,500	2.5	935	15,700
N320	2.3	825	11,300	2.6	865	15,200
AA#2	1.9	694	9,900	2.9	922	15,600
231	2.1	704	9,700	**2.9**	**939**	**15,800**
AA#5	3.0	671	9,100	3.9	940	16,000
Unique	2.3	706	9,600	3.1	952	16,000
SR-7625	2.9	769	11,300	3.3	899	15,400
Power Pistol	3.5	856	13,500	3.9	933	15,600

Note: Loads shown in shaded panels are maximum.
Loads shown in bold designate potentially most accurate load.

380 Automatic (380 ACP) (9mm KURZ)

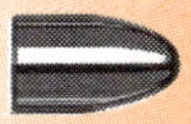

100 gr. FMJ
.955" OAL

BC: .115
SD: .113

Powder	Sugg Starting Grains	Velocity fps	Pressure C.U.P.	Max Load Grains	Velocity fps	Pressure C.U.P.
Bullseye	2.0	695	7,700	3.1	988	15,400
Titegroup	2.2	762	12,100	2.5	859	15,300
Red Dot	1.8	651	6,500	3.1	1026	15,900
700X	1.9	704	9,100	**2.5**	**889**	**14,700**
231	2.2	693	6,600	3.5	1024	15,600
Unique	2.6	725	8,600	3.6	1013	15,200
SR-7625	2.8	799	12,000	3.2	834	14,500
Power Pistol	3.5	843	13,700	3.9	930	16,500

115 gr. Jacketed HP
.956" OAL

BC: .129
SD: .130

Powder	Sugg Starting Grains	Velocity fps	Pressure C.U.P.	Max Load Grains	Velocity fps	Pressure C.U.P.
Bullseye	2.0	711	9,900	2.9	937	15,100
Red Dot	1.8	670	8,700	2.8	945	15,500
700X	1.6	658	9,100	**2.5**	**921**	**15,900**
231	2.0	664	8,300	3.1	934	15,300
SR-7625	2.2	667	8,700	3.1	928	15,800
Power Pistol	2.9	748	13,500	3.3	862	16,700
HS-6	3.1	677	8,700	4.5	945	15,200

#356242
90 gr. (#2 Alloy) .980" OAL

BC: .105
SD: .102

Powder	Sugg Starting Grains	Velocity fps	Pressure C.U.P.	Max Load Grains	Velocity fps	Pressure C.U.P.
*Bullseye	2.4	750	11,800	3.0	915	16,000
Titegroup	2.4	755	11,600	2.7	871	15,400
*Red Dot	2.0	690	11,000	2.6	880	15,200
*700X	1.9	695	11,100	2.5	900	15,800
*Green Dot	2.1	655	10,000	3.0	910	16,000
Amer. Select	2.7	843	13,100	3.1	919	16,400
231	2.3	701	7,900	**3.5**	**1000**	**15,600**
*PB	2.2	645	10,300	2.8	860	15,200
*Unique	3.1	755	11,800	3.7	920	16,000
SR-4756	2.8	633	6,900	4.1	970	15,600
Power Pistol	3.8	902	15,100	4.3	968	16,500
Blue Dot	4.0	667	7,000	5.5	961	14,600

#356242
120 gr. (#2 Alloy) .980" OAL

BC: .154
SD: .136

Powder	Sugg Starting Grains	Velocity fps	Pressure C.U.P.	Max Load Grains	Velocity fps	Pressure C.U.P.
*Bullseye	1.6	600	10,500	2.1	750	15,200
*Red Dot	1.6	615	11,000	2.1	770	15,800
*700X	1.4	610	11,300	1.9	750	15,800
*Green Dot	1.7	620	10,800	2.3	760	15,400
231	2.1	718	9,200	**3.2**	**946**	**15,800**
*PB	1.7	605	10,800	2.2	750	15,800
*Unique	2.2	630	11,300	2.8	780	15,600
SR-4756	2.3	662	7,800	3.5	927	15,800
Blue Dot	3.7	732	9,000	5.3	948	15,900

Note: Loads shown in shaded panels are maximum.
Loads shown in bold designate potentially most accurate load.
* Designates use of Super Vel cases.

9mm Luger (9mm Parabellum) (9 x 19mm)

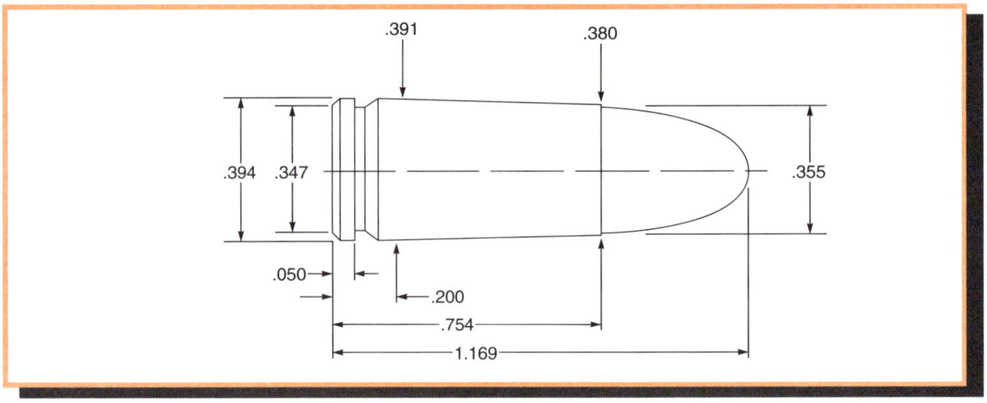

Comments:

The most common diameter for this cartridge is 0.354 inch. However, the industry standard has recently been established at 0.355 inch. All barrels of 0.354" to 0.356" are correctly matched with a bullet diameter of 0.355 inch. Barrels with larger groove diameters will often provide poor accuracy with a 0.355" bullet but chamber restrictions prevent the safe use of larger diameter bullets.

This cartridge headspaces from the mouth and therefore cases trimming must be uniform and accurate. Do not reduce cases below the trim-to length. Additionally do not roll crimp bullets as this will prevent the case from properly headspacing on its mouth. A modest taper crimp may be employed if found necessary.

The most popular cast bullet for this round has been #356402. However, there are many good choices based upon intended use. Bullet #356637 (147 gr.) closely duplicates the U.S. Military subsonic bullet design and has proven to be very accurate.

Alliants Bullseye, Winchester 231, and Alliants Unique are very good popular propellants for this cartridge.

With any handgun cartridge, it is important not to seat bullets to a shorter length than specified in the data. This is especially important with 9mm loads. Pressures can be raised dramatically with deep seating.

Test Components:

Cases . . . . .Federal
Trim-to Length . . . . .751"
Primers . . . . .CCI 500
Primer Size . . . . .Small Pistol
Lyman Shell Holder . . . . .No. 12
Jacketed Bullets Used . . . . . .Sierra JHP #8100, 90 gr.
Sierra FMJ #8105, 95 gr.
Hornady FMJ #35527, 100 gr.
Hornady HP/XTP #35540, 115 gr.
Sierra JHP #8125, 125 gr.
Sierra FMJ #8345, 130 gr.
Speer TMJ #4006, 147 gr.

Cast Bullets Used . . . .(sized to .356" dia)
*gas check bullet
#356242, 90 gr.
#356242, 120 gr.
#356402, 120 gr.
*#358093, 125 gr.
#356637, 147 gr.

Test Specifications:
(Velocity & Pressure)
Firearm Used . . . . .Universal Receiver
Barrel Length . . . . .4"
Groove Dia. . . . . .355"

90 gr. Jacketed HP — 1.010" OAL — **BC:** .115 **SD:** .102

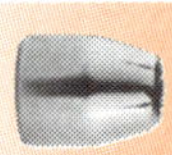

Powder	Sugg Starting Grains	Velocity fps	Pressure C.U.P.	Max Load Grains	Velocity fps	Pressure C.U.P.
Clays	3.0	982	22,000	3.7	1213	30,800
Titegroup	4.4	1170	24,100	4.9	1305	31,700
Bullseye	4.1	971	21,600	**5.2**	**1376**	**30,900**
700X	3.6	1027	25,000	4.5	1292	30,500
231	4.0	1156	20,600	5.2	1411	30,400
AA#5	5.2	835	16,600	6.5	1347	30,800
Unique	5.0	1089	26,900	6.3	1343	32,600
Power Pistol	6.3	1216	23,800	7.0	1331	30,400
N340	5.2	1161	24,400	5.8	1300	31,000
SR-4756	5.0	1028	24,000	6.2	1352	31,900
HS-6	5.9	858	20,200	7.4	1333	31,800
Blue Dot	7.0	1072	24,800	8.9	1341	31,500

95 gr. FMJ — 1.005" OAL — **BC:** .107 **SD:** .108

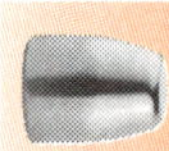

Powder	Sugg Starting Grains	Velocity fps	Pressure C.U.P.	Max Load Grains	Velocity fps	Pressure C.U.P.
Titegroup	4.3	1150	25,500	4.8	1264	30,800
Bullseye	3.5	893	19,400	5.0	1307	30,600
700X	3.2	893	19,600	4.5	1301	31,400
231	3.7	1068	19,200	5.1	1374	32,400
AA#5	5.2	915	18,100	6.5	1394	31,000
Unique	4.5	1007	22,000	5.8	1399	32,000
Power Pistol	6.2	1214	26,000	6.9	1337	31,800
Universal	4.7	939	15,600	5.3	1286	28,800
N340	**5.1**	**1162**	**25,200**	5.7	1291	32,400
SR-4756	4.5	983	20,700	5.8	1306	32,700
HS-6	5.8	952	20,100	7.0	1315	29,600
Blue Dot	6.8	1101	22,900	8.9+	1457	32,300

Note: Loads shown in shaded panels are maximum.
Loads shown in bold designate potentially most accurate load.

9mm Luger (9mm Parabellum) (9 x 19mm)

100 gr. FMJ
1.105" OAL

BC: .115
SD: .113

Powder	Sugg Starting Grains	Velocity fps	Pressure C.U.P.	Max Load Grains	Velocity fps	Pressure C.U.P.
Titegroup	**4.2**	**1131**	**25,800**	4.7	1222	30,200
Bullseye	4.0	976	22,000	5.0	1263	31,700
700X	3.5	870	20,200	4.4	1171	31,500
231	4.5	1061	22,900	5.0	1162	27,200
Unique	4.5	931	20,000	5.6	1256	31,400
Power Pistol	6.1	1143	24,100	6.8	1310	32,200
N340	5.1	1133	24,000	5.7	1255	31,000
SR-4756	4.5	681	13,000	5.7	1225	30,400
HS-6	6.0	801	15,900	7.4	1206	30,100
AA#7	8.0	884	18,600	9.6	1271	31,000
Blue Dot	6.5	973	19,800	7.8	1252	30,200

115 gr. Jacketed HP
1.090" OAL

BC: .129
SD: .130

Powder	Sugg Starting Grains	Velocity fps	Pressure C.U.P.	Max Load Grains	Velocity fps	Pressure C.U.P.
Titegroup	4.0	1046	27,100	4.5	1119	31,400
Bullseye	3.5	945	18,700	4.8	1184	31,700
700X	3.1	918	20,200	4.5	1171	31,700
231	3.5	959	20,700	4.9	1253	32,100
Unique	4.4	996	20,600	5.8	1233	30,700
Power Pistol	5.9	1102	27,100	**6.5**	**1212**	**31,500**
N340	4.5	999	24,300	5.0	1124	30,500
SR-4756	4.5	946	18,600	6.2+	1147	28,100
HS-6	5.7	956	17,400	7.2	1193	30,400
AA#7	6.8	871	20,700	8.5	1158	30,400
Blue Dot	6.8	1110	27,300	7.6+	1189	32,300

125 gr. Jacketed HP
1.075" OAL

BC: .137
SD: .142

Powder	Sugg Starting Grains	Velocity fps	Pressure C.U.P.	Max Load Grains	Velocity fps	Pressure C.U.P.
Clays	3.1	872	26,700	3.6	960	31,700
Titegroup	3.8	962	26,100	4.2	1045	31,200
Bullseye	3.9	979	27,900	4.5	1073	30,100
700X	3.8	930	22,900	4.2	1022	31,400
231	3.9	960	27,000	4.4	1043	32,000
AA#5	5.4	955	24,200	6.1	1078	31,400
Unique	4.5	997	27,300	5.0	1078	30,700
Power Pistol	5.1	979	23,100	5.7	1107	31,300
Universal	4.0	880	17,400	4.5	1042	29,300
N340	4.0	937	21,900	4.6	1060	31,200
HS-6	5.8	949	23,800	6.6	1080	30,400
AA#7	7.0	1008	26,700	**7.8**	**1119**	**31,600**
Blue Dot	6.3	1049	27,400	7.1+	1163	32,400

130 gr. FMJ
1.160" OAL

BC: .160
SD: .147

Powder	Sugg Starting Grains	Velocity fps	Pressure C.U.P.	Max Load Grains	Velocity fps	Pressure C.U.P.
Clays	2.5	690	19,200	3.6	933	31,500
Titegroup	3.8	936	24,400	4.2	1035	30,300
Bullseye	3.6	780	21,500	4.5	1002	31,000
700X	3.0	803	21,700	4.0	991	32,500
231	3.9	949	25,300	4.4	1039	30,600
Unique	3.5	710	18,200	4.8	1014	31,200
Power Pistol	5.1	1010	25,300	5.7	1115	30,700
Universal	4.0	904	18,300	4.5	1045	28,700
N340	4.1	941	23,300	4.6	1067	32,000
HS-6	5.0	752	18,000	6.5	1112	32,500
AA#7	6.5	833	20,500	8.2	1119	32,600
Blue Dot	5.4	852	21,700	6.8	1114	31,700

147 gr. TMJ
1.115" OAL

BC: .208
SD: .167

Powder	Sugg Starting Grains	Velocity fps	Pressure C.U.P.	Max Load Grains	Velocity fps	Pressure C.U.P.
Titegroup	3.2	914	27,700	3.6	983	31,900
Bullseye	3.4	871	23,500	3.9	971	29,900
700X	2.8	766	23,300	3.6	949	30,700
231	3.5	855	23,000	4.1	957	27,500
Unique	3.7	873	20,500	4.5	1039	29,100
Power Pistol	4.5	960	26,900	5.0	1043	32,400
Universal	3.1	778	19,000	3.5	924	29,200
N340	**3.9**	**935**	**28,500**	4.4	1012	32,200
WSF	3.5	790	21,500	4.4	983	30,800
HS-6	4.8	866	22,000	5.8	1046	30,600
AA#7	6.0	852	20,200	7.2	1014	29,000
Blue Dot	5.3	864	19,000	6.6	1052	28,600

#356242
90 gr. (#2 Alloys) 1.045" OAL

BC: .105
SD: .101

Powder	Sugg Starting Grains	Velocity fps	Pressure C.U.P.	Max Load Grains	Velocity fps	Pressure C.U.P.
Titegroup	4.3	1215	26,100	4.8	1299	30,600
Bullseye	**4.0**	**1060**	**19,800**	5.2	1315	32,000
700X	4.1	1125	22,800	5.1	1285	31,000
Red Dot	4.1	1120	22,800	5.5	1340	32,500
231	3.8	1093	20,500	5.1	1455	32,100
SR-7625	5.0	1180	23,400	5.9	1310	30,500
PB	4.8	1115	19,800	6.0	1360	33,000
Green Dot	4.3	1095	20,400	6.0	1325	31,000
N340	5.2	1201	23,800	5.8	1321	31,400
Unique	5.0	1065	19,800	6.8	1325	31,500
Power Pistol	6.3	1280	27,700	7.0	1389	31,900
HS-6	5.8	1011	19,500	7.2	1356	29,800
AA#5	6.6	1163	23,100	7.4	1284	29,300
SR-4756	5.3	1168	22,400	7.2	1380	30,500

Note: Loads shown in shaded panels are maximum.
Loads shown in bold designate potentially most accurate load.
\+ Designates a compressed powder charge.

9mm Luger (9mm Parabellum) (9 x 19mm)

#356242 — 120 gr. (#2 Alloy) 1.065" OAL — **BC:** .154 **SD:** .135

Powder	Sugg Starting Grains	Velocity fps	Pressure C.U.P.	Max Load Grains	Velocity fps	Pressure C.U.P.
Titegroup	3.2	1021	25,200	3.8	1091	30,000
Bullseye	3.4	939	23,600	4.2	1175	32,500
700X	3.0	940	26,000	3.7	1080	31,600
Red Dot	3.0	919	25,100	3.9	1140	32,000
231	3.0	1010	20,000	4.1	1148	30,200
SR-7625	3.0	852	21,400	3.9	1095	31,300
PB	3.2	872	24,500	4.1	1106	31,400
Green Dot	3.5	966	26,200	4.4	1152	32,800
N340	**4.0**	**1047**	**27,300**	4.5	1135	31,600
Unique	4.0	1013	26,500	5.0	1194	32,800
Power Pistol	5.0	1107	27,700	5.6	1190	31,500
HS-6	5.0	904	23,000	6.3	1182	31,700
AA#5	5.4	1021	24,600	6.0	1136	30,300

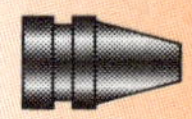

#356402 — 120 gr. (#2 Alloy) 1.110" OAL — **BC:** .146 **SD:** .135

Powder	Sugg Starting Grains	Velocity fps	Pressure C.U.P.	Max Load Grains	Velocity fps	Pressure C.U.P.
Titegroup	3.3	997	24,400	3.7	1109	32,200
Bullseye	3.8	1047	28,400	4.2	1124	31,200
700X	3.6	1031	26,600	4.0	1115	31,800
231	2.9	963	17,000	4.4	1264	29,600
Green Dot	3.8	1038	28,000	4.3	1136	31,400
N340	**4.0**	**986**	**23,300**	4.5	1098	30,200
Power Pistol	5.1	1124	28,200	5.7	1209	31,600
HS-6	5.0	910	20,000	6.2	1163	29,300
AA#5	5.4	1030	24,600	6.0	1150	31,100

#358093 — 125 gr. (#2 Alloy) 1.115" OAL — **BC:** .079 **SD:** .141

Powder	Sugg Starting Grains	Velocity fps	Pressure C.U.P.	Max Load Grains	Velocity fps	Pressure C.U.P.
Clays	2.2	863	25,100	3.1	1030	30,800
Titegroup	2.9	1002	26,800	3.3	1082	29,500
Bullseye	3.3	1011	28,700	3.9	1130	31,400
700X	2.8	948	25,000	3.5	1069	31,200
231	3.2	946	27,600	4.2	1116	30,000
Unique	3.4	1004	27,400	4.5	1195	32,600
N340	**3.6**	**1037**	**29,200**	4.0	1122	31,900
WSF	3.5	950	28,300	4.7	1158	30,900
SR-4756	3.4	918	25,100	4.5+	1110	31,600
AA#5	4.9	1063	26,500	5.5	1136	31,300
Power Pistol	4.8	1129	30,000	5.4	1202	32,300
HS-6	4.8	1028	27,100	5.8+	1175	32,600

#356637 — 147 gr. (#2 Alloy) 1.058" OAL — **BC:** .073 **SD:** .166

Powder	Sugg Starting Grains	Velocity fps	Pressure C.U.P.	Max Load Grains	Velocity fps	Pressure C.U.P.
Clays	1.9	669	21,000	2.8+	873	31,700
Titegroup	2.5	870	26,800	2.8	943	30,300
Bullseye	2.8	780	24,900	3.5	942	31,600
700X	2.7	803	26,500	3.3	921	32,200
Universal	3.0	875	26,300	3.4	969	31,000
Unique	3.2	810	25,600	3.8	956	31,800
N340	3.4	894	25,400	3.8	954	29,900
WSF	3.4	867	26,900	3.9	961	32,100
SR-4756	3.1	769	25,000	3.6	896	30,700
Power Pistol	4.1	939	26,100	**4.6**	**1033**	**30,200**
HS-6	4.4	846	26,700	5.0	968	31,900
AA#7	6.3	986	26,500	7.1	1061	29,300
Blue Dot	5.1	864	26,700	5.9	980	31,200

Note: Loads shown in shaded panels are maximum.
Loads shown in bold designate potentially most accurate load.
\+ Designates a compressed powder charge.

38 Super Automatic

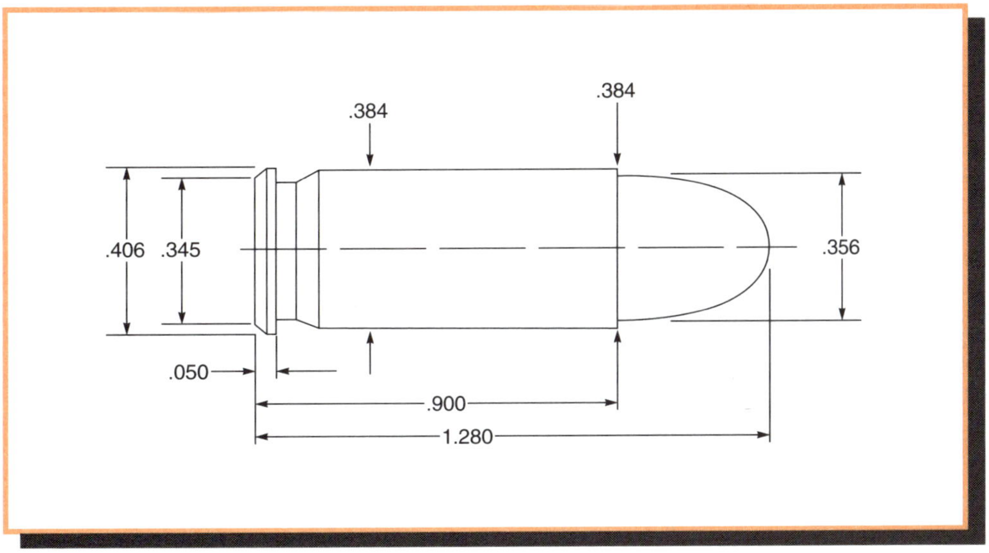

Comments:

Use this data only in pistols specifically marked 38 Super by the manufacturer. This data would be extremely dangerous if used in a pistol marked 38 Automatic. The 38 Automatic uses the same size case but loaded to much lower chamber pressures.

This has traditionally been a semi-rimless case that headspaces on the small rim. Some modern pistols and replacement barrels headspace on the case mouth. If your handgun headspaces on the case mouth do not roll crimp.

38 Super has become extremely popular with IPSC shooters due to its lower recoil which allows for faster recovery time.

Test Components:

Cases . . . . .Remington
Trim-to Length . . . . .895"
Primers . . . . .CCI 500
Winchester WSP (Cast Bullets Only)
Primer Size . . . . .Small Pistol
Lyman Shell Holder . . . . .No. 12
Jacketed Bullets Used . . . . . .Sierra JHP #8100, 90 gr.
Hornady FMJ #35527, 100 gr.
Hornady HP/XTP #35540, 115 gr.
Hornady FMJ/FP #35567, 124 gr.
Sierra FMJ #8345, 130 gr.
Speer TMJ #4006, 147 gr.

Cast Bullets Used . . . .(sized to .355" dia)
*gas check bullet #356242, 90 gr.
#356242, 120 gr.
#356402, 120 gr.
*#358093, 125 gr.
#356637, 147 gr.
#358311, 158 gr.

Test Specifications:
(Velocity & Pressure)

Firearm Used . . . . .Universal Receiver
Barrel Length . . . . .5"
Twist . . . . .1-16"
Groove Dia. . . . . .355"

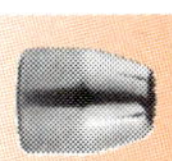

90 gr. Jacketed HP
1.125" OAL
BC: .095
SD: .102

Powder	Sugg Starting Grains	Velocity fps	Pressure C.U.P.	Max Load Grains	Velocity fps	Pressure C.U.P.
Titegroup	5.3	1382	26,500	5.9	1518	32,500
Bullseye	5.0	1222	17,600	6.8	1472	30,600
PB	5.5	1193	20,400	7.3	1482	29,900
231	5.5	1198	20,500	**6.9**	**1383**	**30,400**
Unique	6.0	1221	19,400	7.1	1409	30,000
Power Pistol	7.1	1393	24,600	7.9	1550	30,000
SR-7625	6.0	1203	21,200	7.8	1515	30,600
Herco	6.0	1193	17,500	8.2	1544	30,500
N340	6.5	1354	22,500	7.3	1499	32,000
HS-6	7.5	1210	20,000	9.4	1550	30,500

100 gr. FMJ
1.210" OAL
BC: .115
SD: .113

Powder	Sugg Starting Grains	Velocity fps	Pressure C.U.P.	Max Load Grains	Velocity fps	Pressure C.U.P.
Titegroup	5.0	1248	22,000	5.7	1368	29,200
Bullseye	5.7	1243	24,800	6.4	1364	29,800
700X	4.6	1136	21,800	6.5	1358	31,400
PB	5.3	1153	21,100	7.0	1333	30,400
231	5.5	1101	18,700	**6.5**	**1303**	**28,900**
Unique	5.8	1126	18,800	7.1	1343	31,000
Power Pistol	7.0	1300	22,600	7.8	1466	30,700
Herco	6.1	1141	19,600	7.9	1397	31,100
N340	6.3	1076	17,700	7.1	1413	32,200
HS-6	7.5	1172	17,800	9.3	1433	30,800

Note: Loads shown in shaded panels are maximum.
Loads shown in bold designate potentially most accurate load.

38 Super Automatic

115 gr. Jacketed HP
1.265" OAL

BC: .129
SD: .130

Powder	Sugg Starting Grains	Velocity fps	Pressure C.U.P.	Max Load Grains	Velocity fps	Pressure C.U.P.
Bullseye	5.0	1084	19,100	6.2	1182	31,300
700X	4.4	1032	20,200	6.0	1202	32,600
231	5.4	933	20,500	6.1	1209	31,200
Unique	**5.4**	**1039**	**19,900**	6.9	1186	30,700
Power Pistol	6.6	1067	21,500	7.4	1334	30,700
SR-7625	5.3	1042	19,200	6.6	1195	31,900
Herco	5.6	1039	19,600	6.9	1200	30,600
N340	5.8	1051	22,900	6.5	1295	31,600
WSF	6.3	1002	19,400	7.1	1296	31,300
HS-6	7.9	1020	22,700	8.8	1296	30,700

124 gr. FMJ/FP
1.165" OAL

BC: .174
SD: .141

Powder	Sugg Starting Grains	Velocity fps	Pressure C.U.P.	Max Load Grains	Velocity fps	Pressure C.U.P.
Bullseye	4.0	842	19,500	5.1	1125	29,600
231	4.4	879	22,400	**5.5**	**1132**	**31,300**
Unique	4.5	776	16,600	5.8	1123	31,200
AA#5	5.7	908	19,300	6.7	1175	29,400
Power Pistol	6.0	1000	24,700	6.9	1283	31,700
Herco	4.8	801	18,200	6.0	1123	30,900
WSF	5.5	961	23,100	6.2	1148	32,500
SR-4756	4.7	737	12,900	5.9	1156	31,000
HS-6	7.0	980	26,200	7.8	1184	29,800
HS-7	7.0	816	16,000	8.8	1212	29,900
AA#7	8.5	1051	22,700	9.5	1242	30,600
Blue Dot	6.5	904	20,400	8.3	1248	31,800
N105	8.2	983	20,100	9.2+	1303	31,000

130 gr. FMJ
1.280" OAL

BC: .160
SD: .147

Powder	Sugg Starting Grains	Velocity fps	Pressure C.U.P.	Max Load Grains	Velocity fps	Pressure C.U.P.
Bullseye	4.3	855	19,800	5.4	1095	30,300
231	4.5	847	21,300	5.7	1100	31,500
Unique	4.8	771	16,800	6.1	1089	31,300
Power Pistol	6.1	894	22,000	**6.8**	**1131**	**30,700**
Herco	5.0	766	17,900	6.2	1050	31,200
WSF	5.6	842	21,400	6.2	1119	31,800
SR-4756	5.0	747	13,500	6.1	1083	28,000
AA#7	7.8	814	14,600	9.4	1150	30,000
Blue Dot	6.6	790	15,700	8.3	1183	31,600
AA#9	10.8	1118	27,200	12.0+	1304	30,700

147 gr. TMJ
1.280" OAL

BC: .208
SD: .167

Powder	Sugg Starting Grains	Velocity fps	Pressure C.U.P.	Max Load Grains	Velocity fps	Pressure C.U.P.
Bullseye	4.3	972	22,000	5.0	1078	28,200
231	4.5	966	20,300	5.5	1098	30,200
Unique	4.7	962	21,300	5.5	1097	28,900
Power Pistol	5.6	831	24,300	6.3	1016	30,100
Universal	4.5	659	17,500	5.2	946	32,200
Herco	5.0	940	19,500	5.7	1080	27,800
WSF	**5.2**	**771**	**24,100**	5.8	939	30,500
SR-4756	5.0	895	16,300	5.8	1086	27,600
AA#7	8.3	1047	21,500	9.1	1160	28,600
Blue Dot	7.0	1051	24,900	7.8	1134	28,900
AA#9	9.3	883	22,000	10.5	1126	30,100
N105	7.8	896	21,600	8.7	1144	29,600

***#356242**
90 gr. (#2 Alloy) 1.175" OAL

BC: .105
SD: .102

Powder	Sugg Starting Grains	Velocity fps	Pressure C.U.P.	Max Load Grains	Velocity fps	Pressure C.U.P.
Bullseye	4.2	1022	19,500	5.6	1261	30,900
HP-38	4.1	1026	20,600	5.8	1246	30,500
231	4.7	1039	20,300	**6.3**	**1282**	**31,600**
PB	4.3	1009	18,900	5.7	1222	30,500
Unique	5.1	1061	20,400	6.3	1267	30,500
AA#5	8.3	1323	23,900	9.3	1443	30,200
Power Pistol	7.4	1363	25,100	8.3	1488	30,200
SR-4756	5.9	1089	19,700	7.1	1263	30,300

***#356242**
120 gr. (#2 Alloy) 1.160" OAL

BC: .154
SD: .136

Powder	Sugg Starting Grains	Velocity fps	Pressure C.U.P.	Max Load Grains	Velocity fps	Pressure C.U.P.
Bullseye	3.9	970	17,600	5.4	1214	29,200
700X	3.7	979	19,800	5.3	1206	29,200
231	4.3	988	17,700	5.8	1233	29,600
HP-38	3.9	975	17,800	5.6	1230	29,400
AA#5	6.9	1052	24,300	7.7	1302	32,200
Unique	4.9	1031	20,000	6.4	1281	28,800
Power Pistol	**6.1**	**1128**	**24,900**	6.8	1326	32,400
SR-7625	4.2	978	18,100	5.8	1230	29,300

Note: Loads shown in shaded panels are maximum.
Loads shown in bold designate potentially most accurate load.
* Designates use of Winchester primers.
\+ Designates a compressed powder charge.

38 Super Automatic

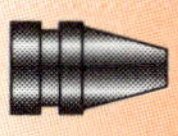

***#356402** 120 gr. (#2 Alloy) 1.230" OAL — **BC:** .146 **SD:** .136

Powder	Sugg Starting Grains	Velocity fps	Pressure C.U.P.	Max Load Grains	Velocity fps	Pressure C.U.P.
Bullseye	4.0	1005	18,600	5.3	1235	30,000
700X	4.0	1025	20,600	5.5	1252	29,900
231	5.0	1092	20,700	**6.5**	**1307**	**30,800**
HP-38	4.4	1063	20,900	5.9	1267	29,700
AA#5	6.9	1016	23,900	7.7	1227	29,800
Unique	5.1	1075	20,400	6.0	1255	31,000
Power Pistol	5.9	1041	24,100	6.6	1285	30,700
SR-7625	5.1	1099	20,600	6.6	1274	29,900

***#358093** 125 gr. (#2 Alloy) 1.250" OAL — **BC:** .079 **SD:** .142

Powder	Sugg Starting Grains	Velocity fps	Pressure C.U.P.	Max Load Grains	Velocity fps	Pressure C.U.P.
Bullseye	3.5	963	19,800	4.6	1155	30,000
700X	3.8	1042	22,800	5.0	1239	31,500
231	4.7	1134	23,600	5.8	1282	29,900
AA#5	5.0	1030	17,400	6.3	1267	31,100
Unique	4.0	1021	22,700	5.1	1236	31,500
Power Pistol	5.9	1140	28,700	6.6	1278	32,000
HS-6	6.0	1110	21,300	7.2	1287	29,600
Blue Dot	**6.8**	**1149**	**24,300**	8.0+	1287	30,600

***#356637** 147 gr. (#2 Alloy) 1.224" OAL — **BC:** .073 **SD:** .167

Powder	Sugg Starting Grains	Velocity fps	Pressure C.U.P.	Max Load Grains	Velocity fps	Pressure C.U.P.
Bullseye	4.3	793	24,300	4.8	940	31,300
Unique	4.5	791	22,900	5.2	954	31,300
Universal	4.2	702	21,700	4.7	876	31,500
Herco	4.9	830	25,200	5.7	928	31,200
HS-7	6.7	877	23,900	7.7	1048	31,400
AA#7	7.5	862	22,200	**8.9**	**1100**	**31,200**
Blue Dot	6.6	869	24,000	7.6	1061	31,200
AA#9	9.4	907	24,500	10.5+	1087	31,000

***#358311** 158 gr. (#2 Alloy) 1.245" OAL — **BC:** .228 **SD:** .179

Powder	Sugg Starting Grains	Velocity fps	Pressure C.U.P.	Max Load Grains	Velocity fps	Pressure C.U.P.
Unique	4.4	958	20,100	5.1	1058	28,100
Herco	4.5	920	21,000	5.4	1058	29,000
SR-4756	4.9	927	18,900	5.8	1085	29,800
AA#7	7.2	873	24,500	8.0	1035	32,000
HS-7	6.8	956	20,200	8.3	1100	28,200
Blue Dot	6.5	987	20,900	7.6	1133	28,400
AA#9	8.9	875	23,300	**9.9+**	**1037**	**28,200**

Note: Loads shown in shaded panels are maximum.
Loads shown in bold designate potentially most accurate load.
* Designates use of Winchester primers.
\+ Designates a compressed powder charge.

9 x 21 mm

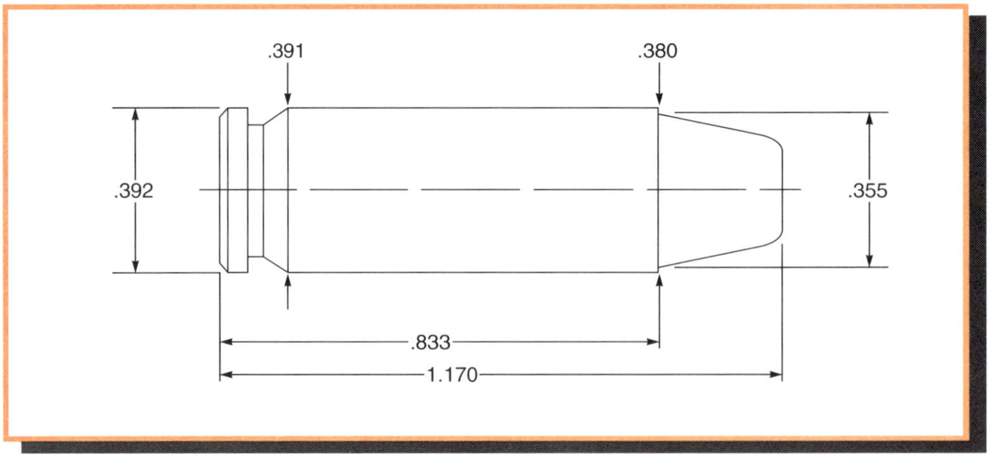

Comments:

The 9x21mm originated in Italy, due to a law which prohibits civilians from owning firearms chambering military calibers. Since this law banned the 9mm Luger (9x19mm), the answer for shooters desiring a 9mm was to lengthen the Luger case by 2mm creating the 9x21mm.

In this country, IPSC shooters have adopted this cartridge to make major power requirements which can not be safely reached with the 9mm Luger. Even though the 9x21mm case is longer than the 9mm Luger, the same overall cartridge length was maintained to allow for the easy conversion of 9mm Luger handguns. If this overall length is maintained, it is difficult to obtain major loads without exceeding the 33,000 c.u.p. S.A.A.M.I. pressure limit set for 9mm Luger firearms.

Since most IPSC shooters fire well-built custom handguns, they often exceed this limit, but we choose not to, as quality of the firearm to be used is not known to us.

Within this limit we have one 147 gr. jacketed bullet load which meets major requirements.

For lightweight bullets, we found 231 to work very well. For heavier bullets, try Blue Dot, HS-6 or AA 7.

Test Components:

Cases . . . . .Geco
Trim-to Length . . . . .830"
Primers . . . . .Winchester WSP
Primer Size . . . . .Small Pistol
Lyman Shell Holder . . . . .No. 12
Jacketed Bullets Used . . . . . .Sierra JHP #8100, 90 gr.
Sierra JHP #8110, 115 gr.
Hornady FMJ/FP #35567, 124 gr.
Sierra FMJ #8345, 130 gr.
Speer TMJ #4006, 147 gr.
Cast Bullets Used. . . . .(sized to .355" dia)
#356402, 120 gr.
#356637, 147 gr.

Test Specifications: (Velocity & Pressure)

Firearm Used . . . . .Universal Receiver
Barrel Length . . . . .5"
Twist . . . . .1-16"
Groove Dia. . . . . .355"

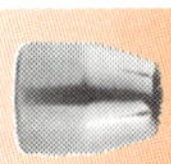

90 gr. Jacketed HP 1.063" OAL — **BC:** .115 **SD:** .102

Powder	Sugg Starting Grains	Velocity fps	Pressure C.U.P.	Max Load Grains	Velocity fps	Pressure C.U.P.
700X	4.5	1127	21,000	5.9	1419	31,800
SR-4756	6.9	1293	22,500	8.0+	1480	31,000
Bullseye	5.5	1290	26,000	7.2	1446	31,700
Unique	5.2	1219	23,200	6.6	1402	32,000
Blue Dot	8.7	1349	24,900	10.2+	1482	28,600
HS-6	7.7	1274	23,800	9.0	1411	31,700
AA#5	6.5	1195	20,300	7.8	1463	31,500
231	5.2	1234	24,000	**6.5**	**1456**	**30,800**

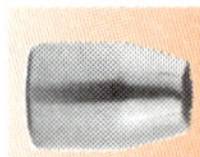

115 gr. Jacketed HP 1.063" OAL — **BC:** .155 **SD:** .130

Powder	Sugg Starting Grains	Velocity fps	Pressure C.U.P.	Max Load Grains	Velocity fps	Pressure C.U.P.
700X	4.0	882	22,000	5.5	1295	31,500
SR-4756	5.8	1119	24,600	7.0+	1323	32,800
Bullseye	4.5	935	23,000	5.8	1234	29,600
Unique	5.0	995	24,800	6.2	1317	31,000
Blue Dot	8.0	1180	25,100	**9.3+**	**1333**	**29,400**
HS-6	6.8	1048	24,400	8.1	1305	31,300
AA#7	8.0	979	20,500	10.0	1330	30,000
231	5.0	1028	25,200	6.3	1266	31,500

Note: Loads shown in shaded panels are maximum.
Loads shown in bold designate potentially most accurate load.
\+ Designates a compressed powder charge.

9 x 21 mm

124 gr. FMJ/FP 1.090" OAL — **BC:** .174 **SD:** .141

Powder	Sugg Starting Grains	Velocity fps	Pressure C.U.P.	Max Load Grains	Velocity fps	Pressure C.U.P.
700X	4.0	868	22,000	5.4	1235	31,500
SR-4756	5.0	887	20,800	6.5	1222	31,000
Bullseye	4.2	853	20,000	5.7	1242	29,400
Unique	4.5	909	20,500	5.8	1259	31,100
Blue Dot	7.3	1066	24,100	8.7+	1291	30,400
HS-6	6.5	937	21,500	7.8	1289	31,700
AA#7	7.3	928	19,800	9.8	1335	32,100
231	4.5	930	20,000	**6.1**	**1235**	**30,100**

130 gr. FMJ 1.169" OAL — **BC:** .160 **SD:** .147

Powder	Sugg Starting Grains	Velocity fps	Pressure C.U.P.	Max Load Grains	Velocity fps	Pressure C.U.P.
700X	3.7	761	18,500	5.1	1069	31,400
WSF	5.5	918	25,500	6.4	1128	31,500
Bullseye	4.0	811	19,800	5.6	1021	29,100
Unique	5.2	947	20,600	6.3	1179	29,200
Blue Dot	7.2	964	23,800	8.5	1247	31,100
HS-6	6.5	898	21,900	7.7	1173	29,400
AA#7	8.1	964	23,000	**9.7**	**1284**	**31,800**
231	4.6	873	23,500	5.6	1141	31,400

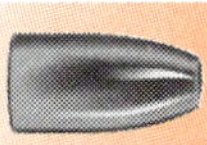

147 gr. TMJ 1.169" OAL — **BC:** .208 **SD:** .167

Powder	Sugg Starting Grains	Velocity fps	Pressure C.U.P.	Max Load Grains	Velocity fps	Pressure C.U.P.
WSF	5.1	856	24,700	6.2	1144	33,000
231	4.2	762	20,000	5.4	1047	31,400
Blue Dot	5.7	642	20,000	7.8	1100	31,400
HS-6	5.9	785	24,400	7.4	1159	33,000
AA#7	7.4	856	24,600	**9.2**	**1228**	**33,000**

#356402 120 gr. (#2 Alloy) 1.169" OAL — **BC:** .146 **SD:** .136

Powder	Sugg Starting Grains	Velocity fps	Pressure C.U.P.	Max Load Grains	Velocity fps	Pressure C.U.P.
700X	3.6	868	23,900	5.2	1292	34,000
SR-4756	4.8	826	18,100	6.0	1228	29,500
Bullseye	3.8	836	18,200	5.4	1242	29,700
Unique	4.3	888	21,000	5.6	1269	31,000
Blue Dot	7.0	1031	23,200	8.4+	1279	29,500
HS-6	6.1	934	20,500	7.4	1287	31,800
AA#5	5.4	951	20,800	6.5	1268	31,800
231	**4.1**	**870**	**20,000**	5.7	1252	30,300

#356637 147 gr. (#2 Alloy) 1.119" OAL — **BC:** .073 **SD:** .167

Powder	Sugg Starting Grains	Velocity fps	Pressure C.U.P.	Max Load Grains	Velocity fps	Pressure C.U.P.
SR-4756	4.0	767	23,300	4.7	930	31,800
Bullseye	3.4	700	22,600	4.5	1005	32,200
Unique	3.3	667	22,300	4.4	925	30,100
Blue Dot	5.1	775	18,900	7.0	1037	29,300
HS-6	**4.6**	**683**	**22,200**	6.0	1040	31,500
AA#7	6.0	705	18,800	8.0	1089	29,600
231	3.0	756	22,000	5.0	1065	31,300

Note: Loads shown in shaded panels are maximum.
Loads shown in bold designate potentially most accurate load.
\+ Designates a compressed powder charge.

357 SIG

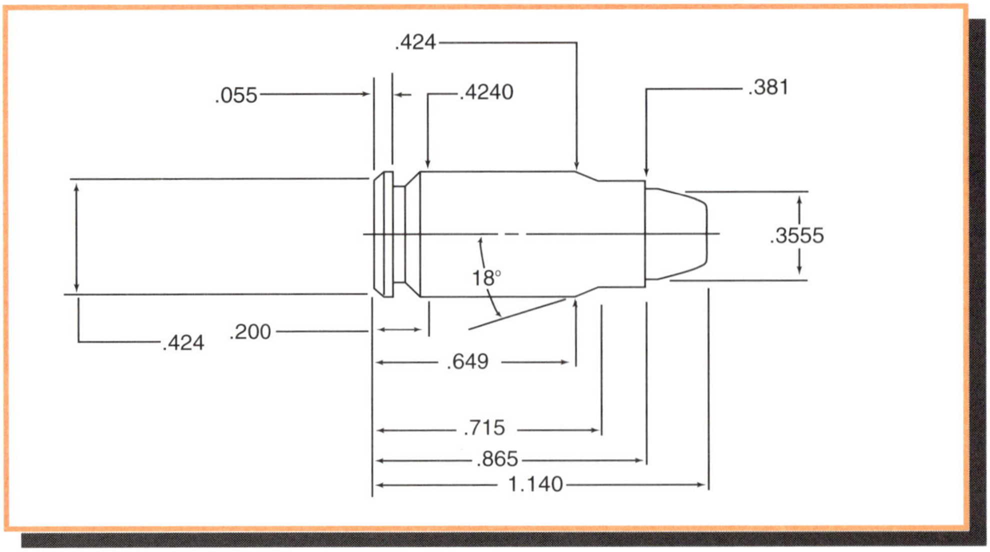

Comments:

The 357 SIG resulted from a joint effort between SIG Arms and Federal Cartridge to produce 357 Magnum level ballistics in a mid-size semiautomatic pistol. The simplest description of the 357 SIG is that of a 40 Smith & Wesson necked down to accept .355" diameter bullets. Reloaders should not however attempt to form brass by necking down 40 S&W cases, as they will be approximately .020" short. 10mm cases should also not be used due to their use of a large primer. Only .355" diameter bullets should be loaded for the SIG despite the 357 designation. Do not use .357" diameter bullets intended for use in the 38/357. The 357 SIG is a bottleneck cartridge but headspaces on the case mouth and cartridges should not be roll crimped. Uniform case length is imperative for proper functioning. Unfortunately, the short neck of the 357 SIG proved unsuitable for use with any of Lyman's cast bullets. The SAAMI Maximum Average Pressure (MAP) is 40,000 PSI.

Test Components:

Cases . . . . .Federal
Trim-to Length . . . . .860"
Primers . . . . .Federal 100
Primer Size . . . . .Small Pistol
Lyman Shell Holder . . . . .No. 15
Jacketed Bullets Used . . . . . .Sierra JHP #8100, 90 gr.
Sierra JHP #8110, 115 gr.
Speer Gold Dot #4360, 125 gr.
Speer TMJ #4006, 147 gr.

Test Specifications:
(Velocity & Pressure)

Firearm Used . . . . .Universal Receiver
Barrel Length . . . . .4"
Twist . . . . .1-16"
Groove Dia. . . . . .3553"

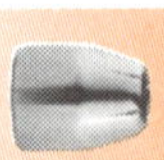

90 gr. Jacketed HP
1.140" OAL
BC: .095
SD: .102

Powder	Sugg Starting Grains	Velocity fps	Pressure P.S.I.	Max Load Grains	Velocity fps	Pressure P.S.I.
AA#2	6.3	1280	30,800	7.0	1423	38,800
HP-38	**7.3**	**1422**	**33,000**	8.0	1537	38,300
Amer. Select	7.9	1395	33,600	8.8	1487	37,400
AA#5	10.0	1422	30,200	11.0	1595	39,400
Unique	8.1	1420	31,700	9.0	1566	38,800
Universal	7.8	1343	26,600	8.7	1579	39,400
N340	7.6	1288	27,300	8.5	1518	38,100
Power Pistol	9.7	1488	31,600	10.8	1656	38,900
SR-4756	9.4	1451	34,300	10.5	1558	39,100
HS-6	10.0	1289	25,900	11.0	1529	35,700
AA#7	12.0	1419	29,100	13.5	1605	37,300
N105	12.1	1399	26,300	13.5+	1627	36,900

115 gr. Jacketed HP
1.140" OAL
BC: .127
SD: .130

Powder	Sugg Starting Grains	Velocity fps	Pressure P.S.I.	Max Load Grains	Velocity fps	Pressure P.S.I.
AA#5	8.8	1221	30,900	9.8	1364	38,600
Unique	6.9	1236	33,200	7.7	1331	38,800
Universal	6.4	1131	27,800	7.2	1305	38,400
N340	6.5	1186	31,800	7.3	1298	38,200
Power Pistol	8.1	1287	32,700	9.0	1416	39,300
SR-4756	8.2	1242	31,400	9.2	1372	39,300
800X	8.5	1315	33,800	9.5	1415	38,200
HS-6	9.0	1206	29,700	10.0	1369	38,600
AA#7	10.0	1268	31,900	11.5	1409	38,700
Blue Dot	9.2	1235	31,000	10.3	1384	38,500
N105	10.8	1331	31,100	12.0	1473	39,300
AA#9	12.4	1367	31,900	**14.0**	**1512**	**38,700**

Note: Loads shown in shaded panels are maximum.
Loads shown in bold designate potentially most accurate load.
\+ Designates a compressed powder charge.

125 gr. Jacketed HP
1.135" OAL

BC: .141
SD: .142

Powder	Sugg Starting Grains	Velocity fps	Pressure P.S.I.	Max Load Grains	Velocity fps	Pressure P.S.I.
AA#5	8.5	1197	31,800	9.5	1314	38,300
Unique	6.6	1157	33,800	7.3	1227	37,300
N340	6.4	1088	30,600	7.2	1229	38,100
Power Pistol	7.8	1196	32,000	8.7	1347	39,100
SR-4756	7.8	1202	34,100	8.7	1293	38,300
800X	8.5	1278	33,800	9.5	1394	39,300
HS-6	8.2	1119	28,700	9.3	1310	38,300
AA#7	10.0	1271	33,900	11.5	1389	39,500
Blue Dot	8.9	1160	30,900	9.9	1314	38,600
N105	10.1	1214	28,700	**11.3**	**1381**	**38,400**
AA#9	11.5	1293	31,900	13.0+	1419	38,200

147 gr. TMJ
1.135" OAL

BC: .208
SD: .167

Powder	Sugg Starting Grains	Velocity fps	Pressure P.S.I.	Max Load Grains	Velocity fps	Pressure P.S.I.
AA#5	7.3	968	32,600	8.2	1112	38,500
Power Pistol	6.7	1060	33,600	7.5	1198	38,400
SR-4756	6.2	1030	34,300	6.9	1099	39,000
800X	7.2	1141	35,200	8.0	1225	39,000
HS-6	7.1	880	31,500	7.9	1080	38,800
AA#7	9.0	1033	31,600	10.0	1227	39,500
Blue Dot	6.8	948	33,200	7.6	1086	37,200
N105	8.1	1092	31,300	9.0	1185	38,400
AA#9	9.7	1166	32,200	**11.0**	**1275**	**38,100**

Note: Loads shown in shaded panels are maximum.
Loads shown in bold designate potentially most accurate load.
\+ Designates a compressed powder charge.

38 Smith & Wesson

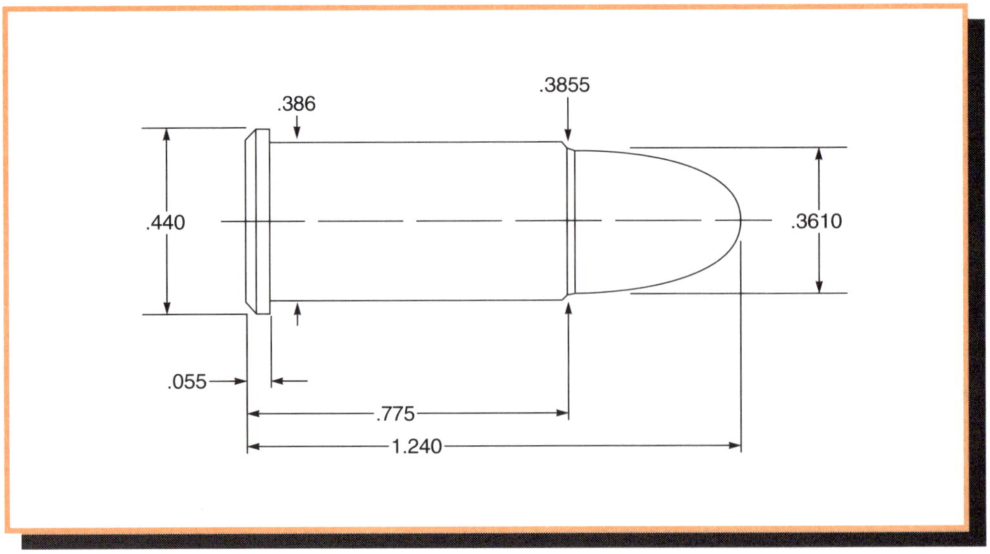

Comments:

The data listed is intended only for use in solid frame revolvers.

We do not recommend the use of reloads in revolvers of the top break design due to their age and relative weakness.

Because of wide internal barrel dimensions it is wise to slug your bore before beginning to reload. Size cast bullets to groove diameter or 0.001" larger. Then make up a dummy round to ensure that your revolver chambers will easily accept ammo with bullets of the chosen diameter.

The fastest burning propellants seem to work best. So, as with other small capacity cases, begin with either Bullseye or 231.

Test Components:

Cases . . . . .Remington, Winchester
Trim-to Length . . . . .765"
Primers . . . . .Winchester WSP
Primer Size . . . . .Small Pistol
Lyman Shell Holder . . . . .No. 21
Cast Bullets Used . . . .(sized to .360" dia)
#356242, 120 gr.
#358311, 160 gr.

Test Specifications:
(Velocity & Pressure)

Firearm Used . . . . .Universal Receiver
Smith & Wesson Model 33
Barrel Length . . . . .4"
Twist . . . . .1-18¾"
Groove Dia. . . . . .361"

#356242 — 120 gr. (#2 Alloy) 1.125" OAL — **BC:** .154 **SD:** .132

Powder	Sugg Starting Grains	Velocity fps	Pressure C.U.P.	Max Load Grains	Velocity fps	Pressure C.U.P.
Bullseye	2.0	560	—	3.0	805	—
Unique	3.5	700	—	4.5	945	—
700X	2.3	680	—	3.1	895	—
PB	2.6	575	—	3.4	825	—
SR-7625	2.7	580	—	3.5	830	—
231	2.0	610	6,900	**2.8**	**798**	**12,400**
Red Dot	2.4	650	—	3.2	855	—
Green Dot	2.5	615	—	3.3	815	—

#358311 — 160 gr. (#2 Alloy) 1.150" OAL — **BC:** .228 **SD:** .176

Powder	Sugg Starting Grains	Velocity fps	Pressure C.U.P.	Max Load Grains	Velocity fps	Pressure C.U.P.
Bullseye	1.8	545	—	2.6	730	—
Red Dot	2.1	605	—	2.8	770	—
Green Dot	2.2	595	—	2.9	745	—
Unique	3.2	715	—	3.7	805	—
700X	2.1	665	—	2.6	785	—
PB	2.3	620	—	2.8	730	—
SR-7625	2.4	660	—	2.9	770	—
HS-7	3.6	619	8,100	4.5	749	11,900
231	1.5	486	4,600	**2.5**	**727**	**12,400**

Note: Loads shown in shaded panels are maximum.
Loads shown in bold designate potentially most accurate load.

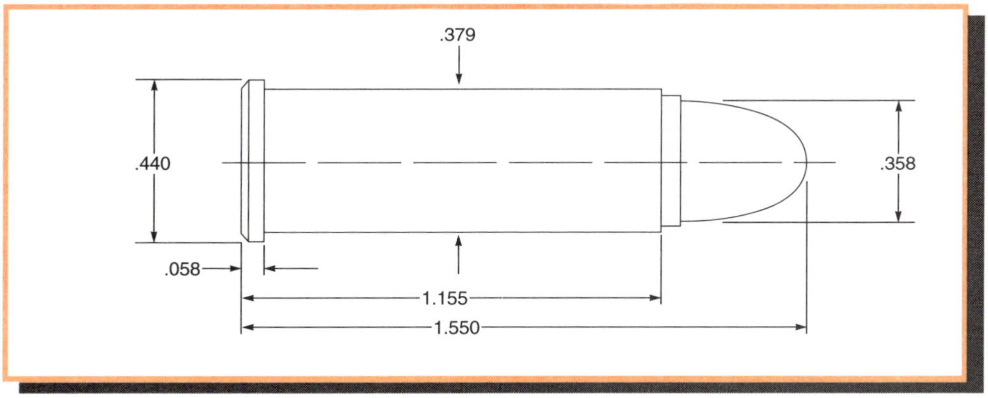

Comments:

Loads designated with a +P symbol are higher pressure loads and must not be used in standard 38 Special revolvers unless the firearm manufacturer has suggested the use of such ammunition in your specific revolver. This caution should not be ignored. Under no circumstance should +P (plus P) ammo be used in revolvers with aluminum frames and/or aluminum cylinders.

Do not use charge weights below the starting load to prevent bullet jackets from half jacket bullets from becoming lodged in the barrel (the lead cores may exit the muzzle and strike the target). To crimp half jacket bullets, form the crimp at the junction of the jacket and exposed lead nose.

For target (mid-range) loads we suggest bullet #358091 with the suggested starting grains load. Bullet #358311 duplicates the factory lead round nose 158 grain bullet very closely. Bullet #358429 closely duplicates the factory 158 grain semi-wadcutter. This is the Elmer Keith design bullet and makes an excellent choice for hunting.

Handguns can vary in groove diameter and it is wise to slug your barrel before sizing cast bullets.

Bullseye and 231 consistently provide the best accuracy with all bullets.

Test Components:

Cases . . . . .Federal
Trim-to Length . . . . .1.149"
Primers . . . . .CCI 500
Primer Size . . . . .Small Pistol
Lyman Shell Holder . . . . No. 1
Jacketed Bullets Used . . . . .Speer JHP #4007, 110 gr.
Hornady JHP #35710, 125 gr.
Speer JHP #4203, 140 gr.
Speer JHP #4205, 146 gr.
Hornady JHP #35750, 158 gr.

Cast Bullets Used . . . .(sized to .358" dia)
*gas check bullet
#356242, 90 gr.
#356242, 120 gr.
#358091, 150 gr.
*#358156, 155 gr.
#358665, 158 gr.
#358311, 160 gr.
#358429, 170 gr.

Test Specifications: (Velocity & Pressure)

Firearm Used . . . . .Universal Receiver
Barrel Length . . . . .4"
Twist . . . . .1-18¾"
Groove Dia. . . . . .357"

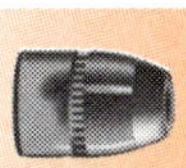

110 gr. Jacketed HP
1.430" OAL
BC: .122
SD: .123

Powder	Sugg Starting Grains	Velocity fps	Pressure C.U.P.	Max Load Grains	Velocity fps	Pressure C.U.P.
Bullseye	4.0	836	12,200	4.5	895	14,300
				(+P)5.2	**1054**	**18,600**
700X	4.1	875	13,700	4.6	983	16,800
231	5.1	844	12,700	5.7	949	15,100
				(+P)6.3	1062	18,400
Unique	5.8	864	13,400	6.5	1007	15,700
				(+P)7.1	1101	17,900
Power Pistol	5.7	795	11,300	6.4	980	15,600
				(+P)6.7	1098	18,700
SR-4756	5.8	675	9,100	7.3	980	16,400
				(+P)7.5	1027	17,500
HS-6	6.5	593	7,700	9.0	978	16,200
				(+P)9.5	1043	18,000
Blue Dot	7.9	822	12,400	8.8	938	14,900

125 gr. Jacketed HP
1.470" OAL
BC: .151
SD: .140

Powder	Sugg Starting Grains	Velocity fps	Pressure C.U.P.	Max Load Grains	Velocity fps	Pressure C.U.P.
Clays	3.2	719	11,800	3.8	870	15,900
Bullseye	3.2	568	10,200	**4.4**	**860**	**16,500**
				(+P5.0	920	18,300
700X	2.8	473	8,100	4.1	814	16,500
				(+P)4.5	877	17,300
AA#2	4.3	751	11,800	5.2	899	16,200
231	3.9	589	10,600	5.1	820	15,600
				(+P)5.4	882	18,300
Unique	4.0	498	8,000	6.0	895	16,700
Power Pistol	5.4	776	12,400	6.0	897	16,100
				(+P)6.4	1005	18,800
SR-4756	4.8	535	7,400	6.3	874	16,200
HS-6	5.6	600	9,400	7.4	862	16,100
Blue Dot	7.6	718	11,400	8.5	876	15,400

Note: Loads shown in shaded panels are maximum.
Loads shown in bold designate potentially most accurate load.
(+P) Designates higher pressure loads. See Comments.

38 Special

140 gr. Jacketed HP
1.470" OAL
BC: .152
SD: .157

Powder	Sugg Starting Grains	Velocity fps	Pressure C.U.P.	Max Load Grains	Velocity fps	Pressure C.U.P.
Bullseye	3.5	533	11,900	4.3	714	16,500
231	3.9	496	12,100	**4.7**	**684**	**15,800**
Unique	4.3	537	11,200	5.3	709	15,900
SR-7625	4.2	505	10,800	5.1	715	16,100
HS-6	5.6	526	11,200	7.2	777	16,700
Blue Dot	7.2	688	13,600	8.0	838	16,300

146 gr. Jacketed HP
1.408" OAL
BC: .159
SD: .163

Powder	Sugg Starting Grains	Velocity fps	Pressure C.U.P.	Max Load Grains	Velocity fps	Pressure C.U.P.
Unique	4.3	741	11,200	5.1	745	16,800
				(+P)5.4	**795**	**17,600**
Power Pistol	4.7	756	14,500	5.0	811	16,200
Herco	5.0	711	15,100	5.2	744	15,900
SR-4756	5.0	633	13,400	5.6	778	16,600
HS-6	6.2	691	14,800	6.7	782	16,000
Blue Dot	6.7	595	12,500	7.4	772	16,600
2400	8.9	690	12,900	9.4	836	16,100

158 gr. Jacketed HP
1.480" OAL
BC: .206
SD: .177

Powder	Sugg Starting Grains	Velocity fps	Pressure C.U.P.	Max Load Grains	Velocity fps	Pressure C.U.P.
Clays	2.8	648	14,500	3.2	712	16,600
Unique	4.7	645	14,400	5.0	710	16,600
				(+P)5.2	**761**	**18,200**
AA#5	5.0	674	12,100	5.8	821	16,300
Power Pistol	4.3	659	14,800	4.8	700	15,600
SR-4756	4.9	611	13,800	5.4	705	16,800
HS-6	5.5	572	12,400	6.5	726	16,600
Blue Dot	6.0	607	13,500	6.7	711	15,700
2400	8.4	645	13,500	9.4	745	16,400

#356242
90 gr. (Linotype) 1.456" OAL
BC: .105
SD: .100

Powder	Sugg Starting Grains	Velocity fps	Pressure C.U.P.	Max Load Grains	Velocity fps	Pressure C.U.P.
700X	3.0	675	5,700	**5.0**	**1107**	**15,700**
				(+P)5.3	1168	18,300
Bullseye	3.3	765	7,600	5.3	1110	15,700
				(+P)5.6	1168	17,600
Red Dot	3.8	830	8,800	5.2	1108	16,300
				(+P)5.5	1154	17,700
231	3.5	689	5,900	5.7	1104	16,300
				(+P)6.0	1163	17,800
Titegroup	5.7	1006	10,300	6.4	1216	15,700

#356242
120 gr. (Linotype) 1.450" OAL
BC: .154
SD: .134

Powder	Sugg Starting Grains	Velocity fps	Pressure C.U.P.	Max Load Grains	Velocity fps	Pressure C.U.P.
700X	2.5	647	7,600	3.9	940	16,200
				(+P)4.2	987	17,900
Bullseye	2.8	690	8,600	**4.6**	**1001**	**16,400**
				(+P)4.9	1045	18,400
Red Dot	3.3	749	10,100	4.6	974	16,200
				(+P)5.0	1047	18,400
SR-7625	4.0	788	10,600	5.0	988	16,200
231	3.4	693	8,800	5.1	986	16,600
				(+P)5.4	1036	17,900
Unique	4.0	725	9,000	5.7	1012	16,200
Titegroup	3.8	900	12,700	4.3	1014	15,400

#358091
150 gr. (Linotype) 1.317" OAL
BC: .038
SD: .167

Powder	Sugg Starting Grains	Velocity fps	Pressure C.U.P.	Max Load Grains	Velocity fps	Pressure C.U.P.
700X	2.9	854	13,300	3.3	917	15,500
Bullseye	3.1	837	13,100	3.5	925	16,400
Titegroup	**3.1**	**854**	**13,600**	3.5	942	16,000
AA#2	3.1	829	12,700	3.5	899	15,800
231	3.7	872	14,400	4.2	942	16,200
Unique	3.7	820	13,000	4.2	894	14,300

Note: Loads shown in shaded panels are maximum.
Loads shown in bold designate potentially most accurate load.
(+P) Designates higher pressure loads. See Comments.

#358156 155 gr. (Linotype) 1.460" OAL

BC: .213
SD: .173

Powder	Sugg Starting Grains	Velocity fps	Pressure C.U.P.	Max Load Grains	Velocity fps	Pressure C.U.P.
Bullseye	2.8	638	10,000	**4.0**	**858**	**16,000**
				(+P)4.4	915	18,100
PB	3.0	584	8,700	4.2	825	16,200
231	3.5	680	11,100	4.7	885	16,800
				(+P)5.0	934	18,000
AA#5	5.0	793	13,300	5.6	870	15,600
Unique	3.4	601	8,900	5.1	895	16,100
				(+P)5.4	954	18,000
Universal	4.3	758	12,600	4.8	869	15,600
				(+P)5.1	931	17,400
SR-4756	4.2	630	9,500	5.5	898	16,700
HS-6	4.4	564	8,000	6.7	894	16,600
Blue Dot	6.0	686	10,400	7.2	867	15,700

#358665 158 gr. (Linotype) 1.445" OAL

BC: .267
SD: .176

Powder	Sugg Starting Grains	Velocity fps	Pressure C.U.P.	Max Load Grains	Velocity fps	Pressure C.U.P.
Bullseye	3.0	786	14,900	3.4	855	16,600
231	3.6	769	13,100	**4.0**	**837**	**15,900**
Unique	4.0	791	13,100	4.5	871	16,000
Universal	4.2	760	12,500	4.7	867	15,800
AA#5	4.9	781	13,100	5.5	857	15,700
SR-4756	4.9	785	13,000	5.5	872	15,800
Blue Dot	5.9	765	12,200	6.6	855	15,700

#358311 160 gr. (Linotype) 1.550" OAL

BC: .228
SD: .178

Powder	Sugg Starting Grains	Velocity fps	Pressure C.U.P.	Max Load Grains	Velocity fps	Pressure C.U.P.
Bullseye	**3.2**	**796**	**13,200**	3.6	863	15,800
				(+P)4.1	936	18,300
Titegroup	3.2	775	11,100	3.6	877	16,200
231	3.5	675	9,600	4.9	906	16,800
				(+P)5.2	956	18,500
Unique	4.2	801	11,400	4.7	905	15,900
				(+P)5.3	982	18,500
SR-4756	4.2	620	8,700	5.5	874	16,000
AA#5	5.2	813	12,000	5.8	904	16,100
HS-6	4.5	610	8,200	6.7	901	16,200
Blue Dot	6.3	809	13,200	7.0	883	15,800

#358429 170 gr. (Linotype) 1.537" OAL

BC: .286
SD: .189

Powder	Sugg Starting Grains	Velocity fps	Pressure C.U.P.	Max Load Grains	Velocity fps	Pressure C.U.P.
Clays	2.3	722	11,500	2.9	824	15,900
Bullseye	2.8	781	13,600	3.2	833	15,500
Titegroup	3.0	808	14,100	**3.4**	**864**	**16,500**
231	3.0	599	9,300	4.6	855	16,900
Unique	3.7	761	12,200	4.2	841	15,500
SR-7625	3.4	621	9,700	4.6	821	16,300
AA#5	4.3	755	11,500	5.1	894	16,100
SR-4756	4.2	645	10,000	5.4	836	16,400
HS-6	5.0	677	11,000	6.4	856	16,800
Blue Dot	6.0	780	12,700	6.7	851	15,000

Note: Loads shown in shaded panels are maximum.
Loads shown in bold designate potentially most accurate load.
(+P) Designates higher pressure loads. See Comments.

357 Magnum

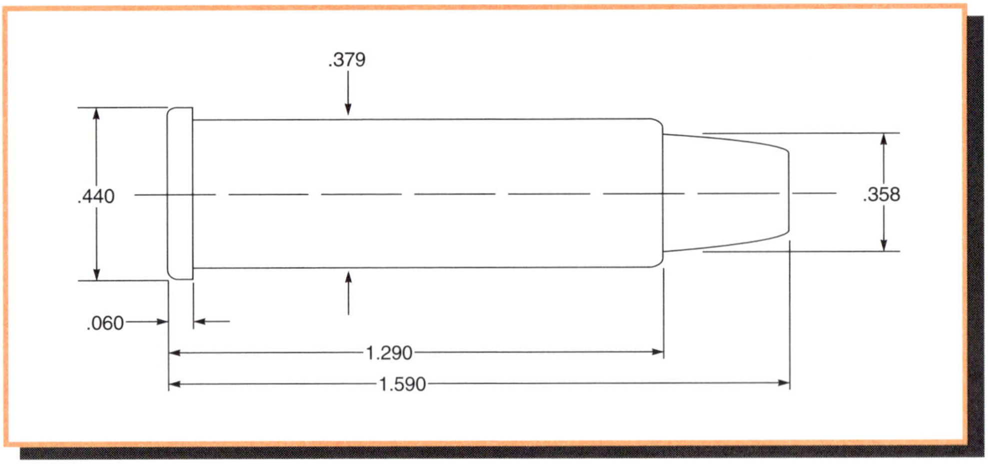

Comments:

Never use 357 Magnum Loads in 38 Special cases as very dangerous pressure will result.

Handguns can vary in groove diameter and it is wise to slug your barrel before sizing cast bullets.

In order to maintain a maximum overall cartridge length of 1.590", it is sometimes necessary to crimp cast bullets on the forward edge of the first driving band.

When using half jacketed bullets velocities must be kept above 750 fps. to prevent bullet jackets from becoming lodged in the barrel (the lead cores may exit the muzzle and strike the target). To crimp half jacket bullets, form the crimp at the junction of the jacket and exposed lead nose.

Bullet #358156 is extremely popular for heavy loads. Bullet #358429 closely duplicates the factory 158 grain semi-wadcutter. This is the Elmer Keith design bullet and makes an excellent choice for hunting.

For light loads Bullseye works best. Heavy loads work well with a wide range of powders but you might want to try 2400 for your first accuracy tests.

Test Components:

Cases . . . . .Federal
Trim-to Length . . . . .1.280"
Primers . . . . .CCI 550
Primer Size . . . . .Small Pistol, Magnum
Lyman Shell Holder . . . . .No. 1
Jacketed Bullets Used . . . . .Speer JHP #4007, 110 gr.
Hornady JHP #35710, 125 gr.
Speer JHP #4203, 140 gr.
Speer JHP #4205, 146 gr.
Hornady JHP #35750, 158 gr.
Hornady JTC SIL #3577, 180 gr.

Cast Bullets Used . . . .(sized to .357" dia)
*gas check bullet
#356242, 90 gr.
#356242, 120 gr.
#356402, 120 gr.
*#358093, 125 gr.
#358477, 150 gr.
*#358156, 155 gr.
#358311, 160 gr.
#358429, 170 gr.

Test Specifications:
(Velocity & Pressure)

Firearm Used . . . . .Universal Receiver
Barrel Length . . . . .4"
Twist . . . . .1-18¾"
Groove Dia. . . . . .356"

110 gr. Jacketed HP
1.590" OAL
BC: .122
SD: .123

Powder	Sugg Starting Grains	Velocity fps	Pressure C.U.P.	Max Load Grains	Velocity fps	Pressure C.U.P.
Bullseye	7.0	1030	20,500	9.7	1471	40,300
231	8.9	1142	29,300	9.5	1328	37,500
Unique	7.4	1052	17,300	10.0	1392	36,400
Blue Dot	10.0	930	12,400	14.6	1545	39,400
2400	13.5	1045	17,400	20.3+	1561	39,100
N110	**17.8**	**1568**	**33,900**	19.8	1693	41,300
IMR-4227	14.8	1100	20,700	19.5+	1383	30,100

125 gr. Jacketed HP
1.590" OAL
BC: .151
SD: .140

Powder	Sugg Starting Grains	Velocity fps	Pressure C.U.P.	Max Load Grains	Velocity fps	Pressure C.U.P.
Bullseye	6.5	864	21,000	8.6	1315	38,400
Unique	7.0	990	18,900	9.7	1359	41,600
AA#5	10.6	1223	31,100	11.8	1471	39,300
Power Pistol	9.0	1238	32,100	10.0	1479	38,600
N340	8.1	1147	31,800	9.0	1366	40,800
HS-7	10.9	1098	21,400	14.4	1403	39,200
Blue Dot	10.4	1113	20,000	13.3	1402	36,700
2400	13.0	1159	24,900	**17.7**	**1478**	**40,600**
N110	14.5	1157	26,000	17.5+	1371	38,500
H110	21.0	1357	33,500	22.0+	1506	42,600
IMR-4227	15.6	1232	31,200	19.5+	1427	37,400

Note: Loads shown in shaded panels are maximum.
Loads shown in bold designate potentially most accurate load.
+ Designates a compressed powder charge.

140 gr. Jacketed HP
1.590" OAL

BC: .152
SD: .157

Powder	Sugg Starting Grains	Velocity fps	Pressure C.U.P.	Max Load Grains	Velocity fps	Pressure C.U.P.
Bullseye	5.7	705	20,300	8.0	1116	39,200
Unique	7.0	949	20,800	8.9	1233	40,000
AA#5	9.4	1035	30,600	10.5	1218	36,000
HS-6	8.4	1076	30,200	9.5	1257	37,500
Blue Dot	9.2	917	17,900	11.9	1272	35,200
2400	12.5	1047	26,500	**16.5**	**1343**	**40,300**
N110	15.3	1258	32,200	17.1	1505	39,900
H110	17.3	1287	31,600	18.0	1324	35,700
IMR-4227	15.0	1106	28,800	17.8	1372	41,400

146 gr. Jacketed HP
1.590" OAL

BC: .159
SD: .163

Powder	Sugg Starting Grains	Velocity fps	Pressure C.U.P.	Max Load Grains	Velocity fps	Pressure C.U.P.
Bullseye	5.7	759	23,300	7.5	1077	39,900
Unique	6.3	928	25,900	8.1	1200	41,100
N340	7.6	1066	34,600	8.5	1212	45,000
AA#5	9.5	1034	32,100	10.6	1208	38,500
Blue Dot	9.0	943	21,300	11.5	1273	39,700
AA#7	10.8	1027	26,400	12.0	1337	41,700
2400	11.0	895	20,800	**15.1**	**1242**	**39,100**
N110	14.5	1075	27,700	16.2	1358	37,200
H110	16.3	1336	36,000	17.0	1380	41,900
IMR-4227	14.5	1045	29,500	16.0	1196	36,000

158 gr. Jacketed HP
1.590" OAL

BC: .206
SD: .177

Powder	Sugg Starting Grains	Velocity fps	Pressure C.U.P.	Max Load Grains	Velocity fps	Pressure C.U.P.
Unique	6.4	887	18,900	8.3	1185	38,300
Blue Dot	9.6	929	27,600	10.7	1158	39,800
AA#7	10.2	1005	31,200	11.5	1220	38,700
N110	11.5	851	21,800	14.0	1139	38,200
2400	11.3	938	20,700	**14.9**	**1279**	**41,800**
AA#9	13.4	1158	32,900	14.9	1357	42,900
H-110	16.3	1178	31,700	17.0	1309	38,400
IMR-4227	12.2	944	21,500	16.1	1249	41,400

180 gr. JTC SIL
1.585" OAL

BC: .232
SD: .202

Powder	Sugg Starting Grains	Velocity fps	Pressure C.U.P.	Max Load Grains	Velocity fps	Pressure C.U.P.
Blue Dot	8.4	782	27,300	9.4	970	39,100
N110	10.8	814	26,400	13.5+	999	40,800
2400	10.0	597	23,100	12.6	949	40,500
AA#9	11.2	857	29,200	12.5	1069	39,000
H-110	13.9	947	30,300	**14.5**	**1101**	**38,100**
296	—	—	—	14.9	1096	42,000
IMR-4227	11.7	602	22,900	14.6	933	40,000

#356242
90 gr. (Linotype) 1.585" OAL

BC: .105
SD: .101

Powder	Sugg Starting Grains	Velocity fps	Pressure C.U.P.	Max Load Grains	Velocity fps	Pressure C.U.P.
700X	6.0	1206	19,700	**8.3**	**1581**	**41,500**
Bullseye	7.5	1365	25,900	9.5	1630	39,900
SR-7625	7.5	1262	20,700	9.8	1613	40,900
231	8.1	1329	25,600	10.1	1629	41,200
AA#5	10.0	1106	14,800	12.5	1642	30,900
Power Pistol	10.5	1301	25,300	11.7	1495	41,100

#356242
120 gr. (Linotype) 1.585" OAL

BC: .154
SD: .135

Powder	Sugg Starting Grains	Velocity fps	Pressure C.U.P.	Max Load Grains	Velocity fps	Pressure C.U.P.
Bullseye	5.7	1045	16,500	8.0	1320	35,800
Unique	6.9	1088	21,100	**9.1**	**1409**	**42,000**
SR-7625	5.5	974	17,500	8.0	1335	41,100
AA#5	8.6	989	19,600	10.8	1450	37,300
Power Pistol	9.9	1308	33,100	11.0	1425	41,200
HS-6	11.2	1108	25,600	12.5	1376	40,100
HS-7	10.0	1053	18,400	13.2	1436	40,300
2400	13.0	1178	25,700	16.4	1494	41,700

Note: Loads shown in shaded panels are maximum.
Loads shown in bold designate potentially most accurate load.
\+ Designates a compressed powder charge.

357 Magnum

#356402
120 gr. (Linotype) 1.590" OAL
BC: .146
SD: .135

Powder	Sugg Starting Grains	Velocity fps	Pressure C.U.P.	Max Load Grains	Velocity fps	Pressure C.U.P.
Bullseye	**6.5**	**994**	**18,800**	8.3	1390	37,600
Unique	7.2	1134	15,800	9.4	1482	41,300
SR-7625	7.0	1205	23,000	8.8	1429	41,600
Power Pistol	9.9	1326	33,200	11.0	1449	41,000
HS-6	10.8	1325	30,400	12.0	1532	38,400
Blue Dot	10.8	1156	17,000	13.0	1564	41,400
2400	13.5	1140	17,300	19.5+	1565	37,200

#358093
125 gr. (Linotype) 1.585" OAL
BC: .079
SD: .140

Powder	Sugg Starting Grains	Velocity fps	Pressure C.U.P.	Max Load Grains	Velocity fps	Pressure C.U.P.
Bullseye	**5.5**	**890**	**26,800**	6.8	1158	38,200
Unique	6.0	769	16,600	7.8	1290	38,900
Herco	6.5	903	23,000	8.0	1275	41,000
Power Pistol	7.8	1065	32,100	8.7	1360	40,000
AA#5	7.5	932	22,300	9.4	1359	40,000
HS-6	9.4	987	30,500	10.5	1230	38,600
Blue Dot	9.0	854	19,800	10.5	1339	39,600
2400	11.5	1026	26,400	14.0	1429	38,900

#358477
150 gr. (Linotype) 1.510" OAL
BC: .286
SD: .168

Powder	Sugg Starting Grains	Velocity fps	Pressure C.U.P.	Max Load Grains	Velocity fps	Pressure C.U.P.
Bullseye	4.6	772	17,300	7.0	1114	36,900
Herco	6.2	972	17,700	7.8	1225	41,900
Blue Dot	8.2	950	13,000	10.8	1356	41,200
HS-6	9.0	1103	32,600	10.0	1301	40,100
AA#7	10.8	1068	25,300	12.0	1390	40,700
AA#9	12.1	1096	31,100	13.5	1331	40,600
2400	11.0	998	17,900	15.0	1362	41,400
N110	13.2	1152	32,400	14.7	1374	39,900
H110	16.3	1245	31,300	17.0	1351	39,900
IMR-4227	11.8	1044	21,700	**16.0+**	**1333**	**40,200**

#358156
155 gr. (Linotype) 1.590" OAL
BC: .213
SD: .174

Powder	Sugg Starting Grains	Velocity fps	Pressure C.U.P.	Max Load Grains	Velocity fps	Pressure C.U.P.
Bullseye	4.8	828	18,000	7.0	1122	39,400
Herco	5.0	785	16,300	7.5	1151	41,000
HS-6	8.8	1069	29,100	9.7	1271	40,200
Blue Dot	8.5	975	21,500	10.5	1277	40,800
AA#7	10.8	893	28,300	12.0	1146	41,300
AA#9	13.0	998	29,500	14.5	1221	39,600
2400	10.6	999	24,900	**14.0**	**1299**	**41,900**
N110	12.0	1074	27,400	15.0+	1299	41,800
H110	15.0	1310	35,800	15.7	1363	40,300
IMR-4227	11.4	973	24,100	15.2	1254	41,300

#358311
160 gr. (Linotype) 1.590" OAL
BC: .228
SD: .179

Powder	Sugg Starting Grains	Velocity fps	Pressure C.U.P.	Max Load Grains	Velocity fps	Pressure C.U.P.
Bullseye	4.9	821	17,500	6.9	1119	41,100
Herco	6.3	963	18,700	7.9	1203	41,000
HS-6	8.5	1046	32,600	9.7	1225	41,000
Blue Dot	8.2	888	11,600	10.9	1316	39,200
AA#7	10.2	1004	27,400	11.5	1213	41,000
AA#9	12.3	1123	28,700	13.7	1350	41,800
2400	11.4	1024	20,200	15.5	1344	39,700
N110	12.6	1054	26,400	14.0	1293	39,000
H110	15.8	1343	34,500	16.5	1463	40,300
IMR-4227	11.8	977	19,600	**17.0+**	**1345**	**40,600**

#358429
170 gr. (Linotype) 1.553" OAL
BC: .286
SD: .190

Powder	Sugg Starting Grains	Velocity fps	Pressure C.U.P.	Max Load Grains	Velocity fps	Pressure C.U.P.
Bullseye	4.1	813	16,500	6.1	976	39,300
Herco	5.6	885	17,900	7.1	1105	39,000
Blue Dot	8.3	970	18,400	10.0	1233	39,200
AA#9	11.7	1028	33,800	13.0	1231	40,500
2400	9.7	879	15,900	13.5	1242	41,100
N110	12.3	1053	33,700	13.7	1212	40,000
H110	14.4	1172	33,500	**15.0**	**1285**	**40,900**
IMR-4227	9.8	835	14,100	14.5+	1233	40,800

Note: Loads shown in shaded panels are maximum.
Loads shown in bold designate potentially most accurate load.
+ Designates a compressed powder charge.

9mm Makarov

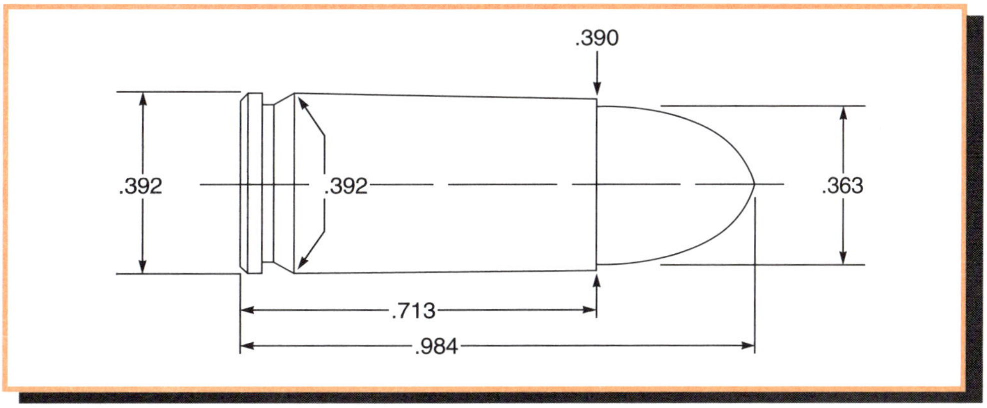

Comments:

The 9mm Makarov is the current standard caliber for Russian military handguns. Other former Soviet Bloc countries also use this cartridge. We have added this cartridge due to the demand for data to satisfy the recent influx of surplus arms.

While called a 9mm this cartridge actually uses a .363 inch (9.2mm) diameter bullet. As such, finding suitable bullets used to take some looking. Recently, however a number of suitable bullets have come onto the market. The jacketed bullets used were obtained from Schroeder Bullets, 1421 Thermal Ave., San Diego, CA., 92154. Other bullet makers, such as Sierra and Hornady, have introduced bullets as more handguns have become available

Cases were obtained from Starline, 1300 W. Henry St., Sedalia, MO., 65301, for our test work. Cases can also be formed from 9mm Luger brass trimmed to the proper length and fireformed. The volume of these cases is often less, however so approach maximum loads with caution.

231 and 700x gave us our most uniform results throughout our testing.

Test Components:

Cases . . . . .Starline
Trim-to Length . . . . .710"
Primers . . . . .Remington 1½
Primer Size . . . . .Small Pistol
Lyman Shell Holder . . . . .No. 12
Jacketed Bullets Used . . . . .Schroeder JHP, 90 gr.
Schroeder FMJ, 95 gr.
Schroeder JHC, 100 gr.

Test Specifications: (Velocity Only)

Firearm Used . . . . .East German PM
Barrel Length . . . . .3.6"
Twist . . . . .1-19"
Groove Dia. . . . . .366"

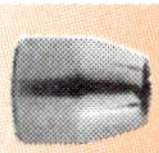

90 gr. Jacketed HP
.970" OAL
BC: .105
SD: .097

Powder	Sugg Starting Grains	Velocity fps	Pressure C.U.P.	Max Load Grains	Velocity fps	Pressure C.U.P.
Bullseye	2.3	567	—	3.5	948	—
AA#2	2.5	581	—	3.8	955	—
231	2.6	621	—	**3.8**	**962**	—
700X	2.3	644	—	3.4	882	—
AA#5	4.0	667	—	5.0	955	—
Unique	3.3	651	—	4.3	966	—

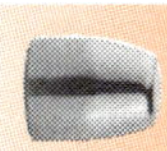

95 gr. FMJ
.970" OAL
BC: .184
SD: .103

Powder	Sugg Starting Grains	Velocity fps	Pressure C.U.P.	Max Load Grains	Velocity fps	Pressure C.U.P.
Bullseye	2.2	587	—	3.4	872	—
AA#2	2.4	670	—	3.7	945	—
231	2.5	613	—	3.7	900	—
700X	2.3	616	—	**3.4**	**900**	—
AA#5	3.8	574	—	5.0	909	—
Unique	3.2	672	—	4.2	901	—

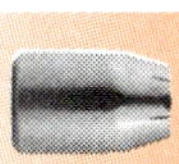

100 gr. Jacketed HP
.970" OAL
BC: .130
SD: .108

Powder	Sugg Starting Grains	Velocity fps	Pressure C.U.P.	Max Load Grains	Velocity fps	Pressure C.U.P.
Bullseye	2.0	573	—	**3.2**	**864**	—
AA#2	2.2	540	—	3.4	801	—
231	2.3	542	—	3.5	820	—
700X	2.1	501	—	3.3	868	—
AA#5	3.6	616	—	4.8	825	—
Unique	3.0	639	—	4.1	887	—

Note: Loads shown in shaded panels are maximum.
Loads shown in bold designate potentially most accurate load.

38-40 Winchester

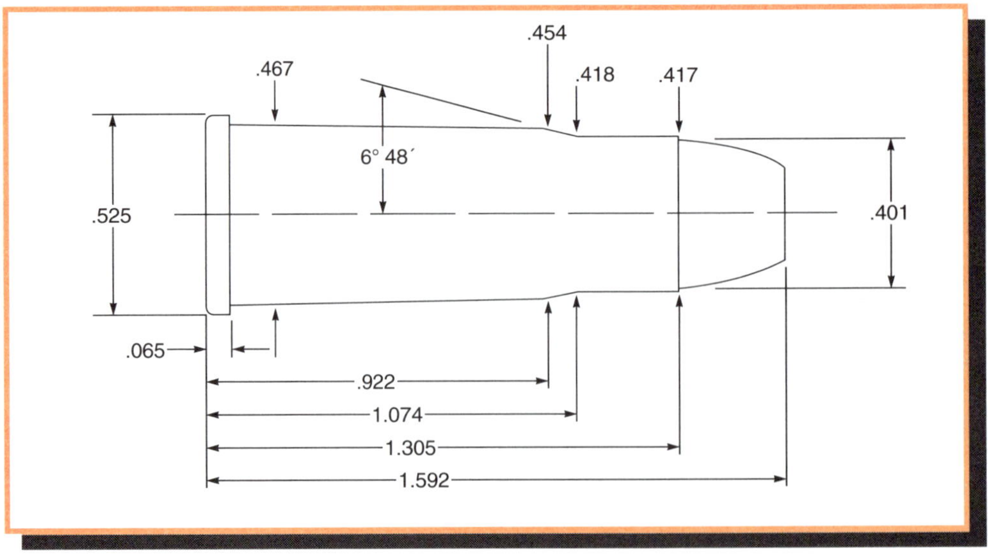

Comments:

This is a very old cartridge and care should be taken not to use the data listed in any handgun which has deteriorated. Additionally this data should not be used in conjunction with any firearm designed for black powder cartridges.

Individual tolerances vary greatly in firearms chambered for this cartridge even among the same make and model. Therefore exercise extreme caution when developing loads as you work up from the suggested starting grains listing.

Never load old brass as it may be of the weaker balloon head types or have been fired at one time with corrosive primers.

Test Components:

Cases . . . . .Winchester
Trim-to Length . . . . .1.300"
Primers . . . . .Winchester WLP
Primer Size . . . . .Large Pistol
Lyman Shell Holder .No. 14B
Cast Bullets Used . . . .(sized to .401" dia)
#401043, 175 gr.

Test Specifications:
(Velocity & Pressure)

Firearm Used . . . . .Universal Receiver
Barrel Length . . . . .6½"
Twist . . . . .1-36"
Groove Dia. . . . . .401"

#401043
175 gr. (#2 Alloy) 1.575" OAL
BC: .098
SD: .155

Powder	Sugg Starting Grains	Velocity fps	Pressure C.U.P.	Max Load Grains	Velocity fps	Pressure C.U.P.
Bullseye	4.0	740	7,100	5.9	965	13,200
Unique	**8.0**	**931**	**8,600**	10.0	1115	13,500
2400	11.0	740	6,600	14.0	1111	13,900

Note: Loads shown in shaded panels are maximum.
Loads shown in bold designate potentially most accurate load.

40 Smith & Wesson

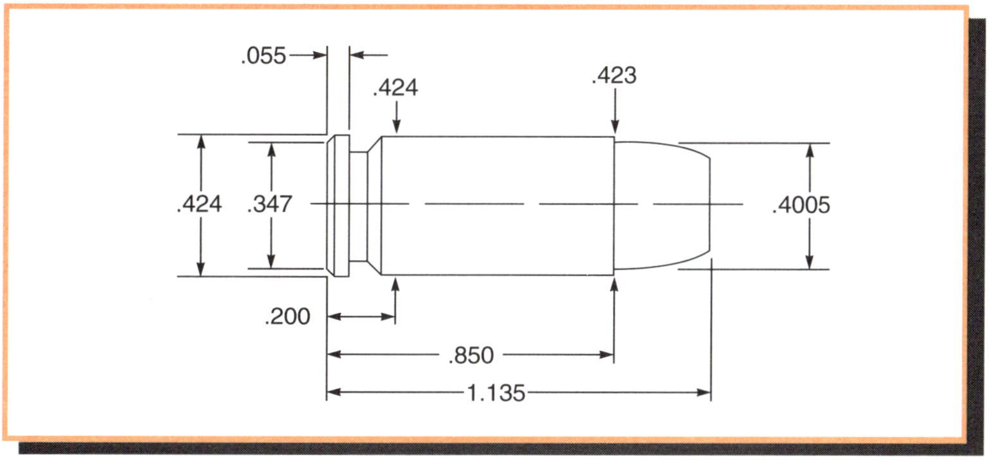

Comments:

The 40 S&W has quickly become a very popular cartridge as it is a good choice between the 9mm and 45 ACP. It has lighter recoil and greater magazine capacity than the 45 ACP and is more powerful than the 9mm. Accuracy in most handguns is superb, especially with Bullseye, Winchester 231 and Accurate Arms No. 5.

This cartridge headspaces from the mouth and therefore case trimming must be uniform and accurate. Do not reduce cases below the trim-to length. Additionally do not roll crimp bullets as this will prevent the case from properly headspacing on its mouth. A modest taper crimp may be employed if found necessary.

It is important not to seat bullets to a shorter length than listed in any handgun cartridge; but especially so in this one. Pressures can be raised dramatically with deep seating.

Test Components:

Cases . . . . .Winchester
Trim-to Length . . . . .845"
Primers . . . . .Winchester WSP
Primer Size . . . . .Small Pistol
Lyman Shell Holder .No. 15
Jacketed Bullets Used . . . .Nosler JHP #44838, 135 gr.
Sierra JHP #8430, 150 gr.
Winchester Silvertip #SHP40, 155 gr.
Speer TMJ #4410, 165 gr.
Winchester Silvertip #SHP10, 175 gr.
Sierra JHP #8460, 180 gr.
Sierra FPJ #8480, 190 gr.
Cast Bullets Used . . . .(sized to .401" dia)
#401654, 150 gr.
#401043, 175 gr.
#401638, 175 gr.

Test Specifications:
(Velocity & Pressure)

Firearm Used . . . . .Universal Receiver
Barrel Length . . . . .4"
Twist . . . . .1-16"
Groove Dia. . . . . .401"

135 gr. Jacketed HP
1.085" OAL
BC: .093
SD: .121

Powder	Sugg Starting Grains	Velocity fps	Pressure C.U.P.	Max Load Grains	Velocity fps	Pressure C.U.P.
Bullseye	**5.5**	**1070**	**19,600**	6.3	1203	23,200
AA#2	5.0	1014	19,300	6.0	1168	23,500
231	5.5	1005	17,800	6.7	1185	23,300
Unique	5.8	1025	18,500	6.8	1198	23,000
AA#5	7.2	1026	17,500	8.3	1221	23,500
Power Pistol	7.3	1027	19,200	8.2	1225	23,200
N340	6.3	1020	19,300	7.0	1193	23,000
HS-6	8.0	1022	17,600	9.0	1171	22,000
Blue Dot	9.0	1058	19,500	10.3	1162	23,100

150 gr. Jacketed HP
1.100" OAL
BC: .130
SD: .134

Powder	Sugg Starting Grains	Velocity fps	Pressure C.U.P.	Max Load Grains	Velocity fps	Pressure C.U.P.
Clays	3.5	858	17,100	4.4	1025	21,600
Bullseye	4.9	740	18,200	5.7	890	22,500
231	**5.0**	**799**	**17,900**	6.2	1054	22,800
Unique	5.5	982	17,800	6.6	1132	23,300
AA#5	6.9	998	18,400	7.7	1101	23,300
WST	5.5	968	19,100	6.4	1096	23,400
Power Pistol	6.7	954	19,000	7.5	1105	22,300
N340	5.7	922	18,800	6.3	1091	23,100
HS-6	7.6	990	17,100	8.6	1093	22,300
800X	6.9	999	18,500	8.4	1109	23,200
Blue Dot	8.6	999	18,800	10.0	1140	23,100

Note: Loads shown in shaded panels are maximum.
Loads shown in bold designate potentially most accurate load.

155 gr. Jacketed Silvertip
1.125" OAL

BC: .166 **SD:** .138

Powder	Sugg Starting Grains	Velocity fps	Pressure C.U.P.	Max Load Grains	Velocity fps	Pressure C.U.P.
Bullseye	5.1	771	19,500	5.7	885	22,500
231	5.3	877	18,000	5.9	1053	22,300
Unique	5.8	847	18,300	6.5	1011	22,800
AA#5	7.1	916	18,400	8.0	1054	22,500
Power Pistol	7.0	942	20,400	**7.8**	**1082**	**23,300**
N340	5.5	893	19,800	6.2	993	22,600
HS-6	7.6	924	18,900	8.5	1089	22,800
800X	7.2	988	19,800	8.0	1056	22,400
Blue Dot	8.2	881	18,800	9.2	1010	22,100

165 gr. TMJ
1.120" OAL

BC: .135 **SD:** .147

Powder	Sugg Starting Grains	Velocity fps	Pressure C.U.P.	Max Load Grains	Velocity fps	Pressure C.U.P.
Bullseye	4.8	733	19,200	5.4	825	22,600
231	5.2	937	19,300	5.8	1050	22,100
Unique	5.4	952	19,300	6.0	1034	21,700
AA#5	6.9	934	18,400	7.6	1025	21,400
Power Pistol	6.3	855	17,900	**7.0**	**953**	**22,000**
N340	5.5	873	18,700	6.1	997	23,000
HS-6	7.6	833	17,400	8.5	979	22,600
800X	7.2	891	19,700	8.0+	1004	22,600
Blue Dot	8.3	922	19,900	9.2	1018	22,400

175 gr. Jacketed Silvertip
1.125" OAL

BC: .142 **SD:** .156

Powder	Sugg Starting Grains	Velocity fps	Pressure C.U.P.	Max Load Grains	Velocity fps	Pressure C.U.P.
Bullseye	4.7	735	20,000	5.3	843	23,400
WST	5.1	880	19,800	**5.7**	**991**	**22,800**
231	5.1	864	19,900	5.7	983	22,500
Universal	4.9	889	19,400	5.5	973	21,800
Unique	5.3	932	19,500	5.9	1044	23,000
Power Pistol	6.0	844	18,900	6.7	981	22,200
N340	5.4	891	20,600	6.0	1004	23,000
WSF	5.8	930	20,700	6.5	1001	22,500
HS-6	6.9	843	18,000	7.7	970	22,500
800X	6.8	887	20,200	7.6	975	22,900
Blue Dot	7.9	952	19,500	8.8	1057	22,400

180 gr. Jacketed HP
1.115" OAL

BC: .170 **SD:** .161

Powder	Sugg Starting Grains	Velocity fps	Pressure C.U.P.	Max Load Grains	Velocity fps	Pressure C.U.P.
Clays	3.4	551	17,900	4.0	912	21,900
Bullseye	4.9	628	18,200	5.6	844	23,500
231	5.0	927	20,400	5.6	1015	22,800
Universal	4.8	815	18,500	5.5	941	22,800
Unique	4.9	839	18,800	5.6	916	23,100
Power Pistol	6.0	876	19,300	6.7	974	22,900
N340	5.4	904	20,800	6.0	1014	23,100
WSF	5.0	853	17,700	6.2	993	23,400
HS-6	6.9	860	19,600	8.0	1048	23,500
800X	5.9	906	18,700	**7.4**	**1013**	**23,400**
Blue Dot	7.3	754	18,000	8.8	990	23,300

190 gr. Jacketed FP
1.135" OAL

BC: .190 **SD:** .170

Powder	Sugg Starting Grains	Velocity fps	Pressure C.U.P.	Max Load Grains	Velocity fps	Pressure C.U.P.
Bullseye	3.9	520	16,200	5.0	722	23,100
231	4.2	540	16,800	5.3	723	23,200
Universal	4.7	730	14,600	5.3	846	23,000
Unique	4.5	535	15,900	5.3	720	23,100
Power Pistol	5.9	720	19,900	6.5	816	23,000
AA#5	5.5	545	16,000	6.5	745	23,200
N340	5.0	773	18,900	**5.6**	**870**	**23,200**
WSF	4.9	530	16,100	5.8	702	22,700
HS-6	6.4	594	16,200	7.2	734	22,300
800X	5.2	573	17,200	6.3	737	22,200
Blue Dot	6.7	567	16,800	8.1	724	22,600

#401654
150 gr. (#2 Alloy) 1.090" OAL

BC: .074 **SD:** .133

Powder	Sugg Starting Grains	Velocity fps	Pressure C.U.P.	Max Load Grains	Velocity fps	Pressure C.U.P.
Bullseye	5.0	1056	18,600	6.0	1202	23,100
231	5.5	1079	19,000	6.5	1195	22,500
AA#2	**5.0**	**1041**	**18,600**	6.4	1192	22,900
Unique	5.0	980	16,200	6.3	1196	23,000
Power Pistol	6.3	971	18,700	7.0	1112	23,300
AA#5	6.3	969	16,200	7.5	1195	23,500
N340	5.3	911	17,800	5.9	1057	22,500
HS-6	7.5	1038	18,000	8.8	1224	23,500
800X	6.8	1085	19,800	8.0	1180	23,300
Blue Dot	8.3	1033	18,400	9.6	1171	22,800

Note: Loads shown in shaded panels are maximum.
Loads shown in bold designate potentially most accurate load.
\+ Designates a compressed powder charge.

40 Smith & Wesson

#401043
175 gr. (#2 Alloy) 1.125" OAL

BC: .098
SD: .155

Powder	Sugg Starting Grains	Velocity fps	Pressure C.U.P.	Max Load Grains	Velocity fps	Pressure C.U.P.
Bullseye	4.2	665	17,100	5.1	812	22,000
231	4.3	863	15,400	5.8	977	22,500
Universal	4.8	832	15,300	5.5	1006	21,900
Unique	4.8	911	15,900	5.8	1023	23,300
Power Pistol	5.9	908	18,900	6.6	998	21,600
AA#5	6.1	903	19,300	6.9	1066	23,000
N340	5.2	918	18,700	**5.8**	**1038**	**23,000**
HS-6	7.0	940	17,000	8.2	1047	23,300
800X	5.6	853	16,500	7.4	974	22,400
Blue Dot	7.3	897	17,900	9.0	1123	23,500

#401638
175 gr. (#2 Alloy) 1.100" OAL

BC: .088
SD: .155

Powder	Sugg Starting Grains	Velocity fps	Pressure C.U.P.	Max Load Grains	Velocity fps	Pressure C.U.P.
Clays	3.3	641	17,800	4.0	992	20,900
Bullseye	4.0	654	16,900	5.0	842	23,100
231	4.3	796	16,000	5.8	970	22,700
Universal	4.7	871	17,400	5.4	1009	22,900
Unique	4.7	898	18,600	5.8	971	23,200
Power Pistol	5.8	873	18,900	6.5	996	22,900
AA#5	6.0	844	18,600	6.9	1047	22,600
N340	4.8	873	18,800	**5.4**	**1008**	**22,400**
HS-6	7.0	976	18,600	8.4	1129	22,800
800X	5.8	878	17,000	7.6	996	22,900
Blue Dot	6.9	979	18,100	8.6	1059	23,200

Note: Loads shown in shaded panels are maximum.
Loads shown in bold designate potentially most accurate load.

10mm Automatic

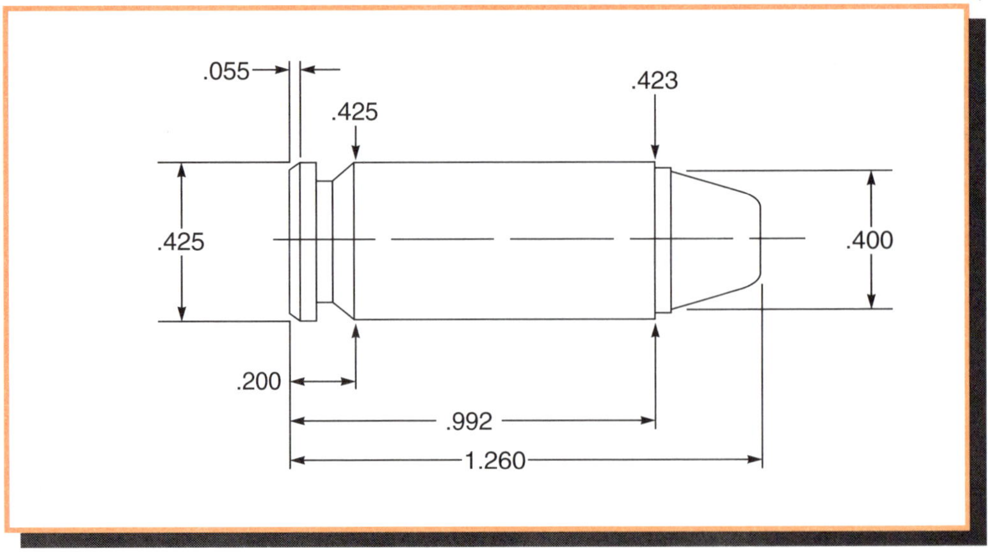

Comments:

This is a high performance semi-auto pistol cartridge with applications in law enforcement and hunting.

The 10mm Auto cartridge headspaces on the mouth and therefore case trimming must be uniform and accurate. Do not reduce cases below the recommended trim-to length. Additionally do not roll crimp bullets as this will prevent the case from properly headspacing on its mouth. A modest taper crimp is also suggested.

It is important not to seat bullets to a shorter length than listed in any handgun cartridge; but especially so in this one. Pressures can be raised dramatically with deep seating.

Winchester 231 frequently provided the best accuracy though not the highest velocities when used in this cartridge.

Test Components:

Cases . . . . .Winchester, Norma
Trim-to Length . . . . .982"
Primers . . . . .CCI 300
Primer Size . . . . .Large Pistol
Lyman Shell Holder . . . . .No. 15
Jacketed Bullets Used . . . .Nosler JHP #44838, 135 gr.
Sierra JHP #8430, 150 gr.
Winchester Silvertip #SHP40, 155 gr.
Speer TMJ #4410, 165 gr.
Winchester Silvertip #SHP10, 175 gr.
Sierra JHP #8460, 180 gr.
Sierra FPJ #8480, 190 gr.
Hornady HP/XTP #40060, 200 gr.
Cast Bullets Used . . . .(sized to .401" dia)
#401654, 150 gr.
#401043, 175 gr.
#401638, 175 gr.

Test Specifications: (Velocity & Pressure)

Firearm Used . . . . .Universal Receiver
Barrel Length . . . . .5"
Twist . . . . .1-16"
Groove Dia. . . . . .400"

***135 gr. Jacketed HP**
1.225" OAL
BC: .093
SD: .121

Powder	Sugg Starting Grains	Velocity fps	Pressure C.U.P.	Max Load Grains	Velocity fps	Pressure C.U.P.
Bullseye	6.0	1163	23,300	7.3	1335	29,300
Red Dot	5.5	1063	17,100	7.2	1375	28,800
Unique	6.8	1127	19,800	8.3	1382	29,700
700X	5.5	1094	19,800	7.1	1345	29,700
231	**6.0**	**1097**	**20,300**	7.7	1327	29,100
AA#5	8.0	1102	15,700	9.8	1397	28,200

***150 gr. Jacketed HP**
1.250" OAL
BC: .130
SD: .134

Powder	Sugg Starting Grains	Velocity fps	Pressure C.U.P.	Max Load Grains	Velocity fps	Pressure C.U.P.
Bullseye	5.5	1049	20,900	7.0	1263	28,500
Red Dot	4.7	864	19,300	6.5	1243	28,800
Unique	6.5	994	21,800	8.1	1265	29,500
700X	5.0	937	20,700	6.8	1282	30,000
800X	7.0	1029	20,200	9.0	1301	28,900
231	5.7	1053	20,700	**7.5**	**1292**	**29,800**
HS-6	8.3	959	17,900	9.9	1278	27,200
AA#5	7.0	986	18,100	9.0	1278	29,800

Note: Loads shown in shaded panels are maximum.
Loads shown in bold designate potentially most accurate load.
* Designates the use of Norma cases.

10mm Automatic

155 gr. Jacketed Silvertip
1.250" OAL

BC: .166
SD: .138

Powder	Sugg Starting Grains	Velocity fps	Pressure C.U.P.	Max Load Grains	Velocity fps	Pressure C.U.P.
231	6.5	1105	25,500	7.3	1229	28,900
Universal	6.5	973	14,200	7.3	1156	24,100
Unique	6.7	1085	22,000	7.5	1201	28,900
N340	6.8	1058	20,700	7.6	1212	29,600
WSF	7.2	1061	20,100	8.0	1197	28,000
Power Pistol	8.0	1140	23,500	8.9	1275	29,000
800X	8.8	1221	25,500	9.8	1315	28,000
HS-6	9.5	1139	23,700	10.6	1308	29,800
AA#7	11.2	1183	23,000	12.5	1324	29,800
Blue Dot	9.7	1158	23,500	10.8	1303	29,700
AA#9	14.0	1240	22,300	**15.5**	**1343**	**27,900**

165 gr. TMJ
1.250" OAL

BC: .135
SD: .147

Powder	Sugg Starting Grains	Velocity fps	Pressure C.U.P.	Max Load Grains	Velocity fps	Pressure C.U.P.
231	6.3	1045	22,500	7.1	1182	28,600
Universal	6.3	963	14,500	7.1	1118	23,700
Unique	6.5	1012	18,800	7.3	1172	27,300
N340	6.5	1004	17,900	7.3	1172	28,300
WSF	6.9	1057	21,700	7.7	1182	26,200
Power Pistol	7.8	1157	23,000	8.7	1265	29,000
800X	8.3	1133	21,600	9.3	1266	28,900
HS-6	9.3	1098	21,500	10.4	1235	28,300
AA#7	10.3	1035	18,300	11.5	1174	25,000
Blue Dot	9.1	1087	20,900	10.2	1230	27,400
AA#9	13.5	1183	20,900	**15.0**	**1318**	**29,100**

175 gr. Jacketed Silvertip
1.250" OAL

BC: .142
SD: .156

Powder	Sugg Starting Grains	Velocity fps	Pressure C.U.P.	Max Load Grains	Velocity fps	Pressure C.U.P.
231	5.9	1004	23,500	6.6	1109	28,900
Universal	5.8	930	18,800	6.5	1065	26,900
Unique	6.3	1011	23,600	7.0	1107	28,300
N340	6.1	997	23,300	6.8	1096	28,800
WSF	6.4	984	22,900	7.2	1129	28,800
Power Pistol	7.2	1076	24,100	8.1	1199	29,700
800X	8.1	1086	24,900	9.0	1209	29,000
HS-6	8.6	1050	22,200	9.6	1186	29,300
AA#7	10.2	1079	22,600	11.4	1201	28,600
Blue Dot	8.7	1048	21,400	9.7	1162	27,300
AA#9	12.5	1088	22,900	**13.9**	**1212**	**27,000**

*180 gr. Jacketed HP
1.250" OAL

BC: .170
SD: .161

Powder	Sugg Starting Grains	Velocity fps	Pressure C.U.P.	Max Load Grains	Velocity fps	Pressure C.U.P.
700X	4.5	892	17,300	5.8	1053	28,800
Bullseye	5.5	998	21,000	6.5	1088	29,300
231	5.9	1037	26,300	6.6	1125	28,700
Unique	5.8	1018	21,800	6.7	1064	28,100
Power Pistol	7.2	1107	24,800	**8.0**	**1209**	**28,900**
800X	6.8	831	20,300	8.4	1073	28,000
HS-6	8.1	1057	21,000	9.6	1225	28,700
AA#7	9.5	984	18,000	11.5	1087	28,300
Blue Dot	8.1	939	18,000	10.3	1133	30,000

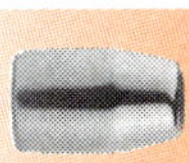

*190 gr. Jacketed FP
1.250" OAL

BC: .190
SD: .170

Powder	Sugg Starting Grains	Velocity fps	Pressure C.U.P.	Max Load Grains	Velocity fps	Pressure C.U.P.
Clays	4.3	878	22,900	5.6	1071	29,800
700X	4.8	951	20,300	5.8	1086	29,800
Bullseye	4.7	879	18,400	6.1	1088	28,200
231	5.8	980	22,900	6.5	1075	29,000
Unique	5.5	896	19,200	6.5	1001	27,600
Power Pistol	7.0	1060	25,000	**7.8**	**1168**	**28,800**
800X	6.4	987	22,200	8.5	1189	29,300
HS-6	7.7	744	20,700	9.2	1037	30,000
AA#7	9.0	932	20,600	10.5	1106	28,000
Blue Dot	8.3	968	19,200	9.9	1175	28,600

*200 gr. Jacketed HP
1.250" OAL

BC: .199
SD: .179

Powder	Sugg Starting Grains	Velocity fps	Pressure C.U.P.	Max Load Grains	Velocity fps	Pressure C.U.P.
700X	4.4	936	19,900	5.8	1080	29,500
Bullseye	4.8	910	20,400	5.7	1059	26,500
231	5.6	984	24,600	6.3	1067	28,400
Unique	4.8	836	17,700	6.1	1071	29,000
Power Pistol	6.3	1000	22,700	7.1	1092	28,200
800X	5.9	963	22,600	**7.6**	**1119**	**29,200**
HS-6	7.5	987	22,600	8.6	1126	29,500
AA#7	9.0	987	19,900	10.6	1145	28,200
Blue Dot	8.0	1015	22,900	9.6	1148	29,900

Note: Loads shown in shaded panels are maximum.
Loads shown in bold designate potentially most accurate load.
* Designates the use of Norma cases.

10mm Automatic

***#401654**
150 gr. (#2 Alloy) 1.240" OAL

BC: .074
SD: .133

Powder	Sugg Starting Grains	Velocity fps	Pressure C.U.P.	Max Load Grains	Velocity fps	Pressure C.U.P.
700X	5.5	1167	24,000	6.5	1310	30,000
Red Dot	5.5	1144	21,800	6.5	1300	29,700
Bullseye	6.2	1205	24,600	7.2	1336	29,000
231	6.5	1206	23,800	**7.4**	**1320**	**28,500**
Unique	6.3	1133	20,300	7.5	1323	29,000
AA#5	8.0	1205	21,000	9.0	1343	28,900
800X	8.4	1265	23,700	9.4	1381	28,600
Blue Dot	9.7	1218	23,600	11.0	1369	29,400

***#401043**
175 gr. (#2 Alloy) 1.260" OAL

BC: .098
SD: .155

Powder	Sugg Starting Grains	Velocity fps	Pressure C.U.P.	Max Load Grains	Velocity fps	Pressure C.U.P.
700X	5.2	1102	24,700	5.8	1170	28,200
Red Dot	5.5	1087	24,500	6.2	1194	29,300
Bullseye	5.5	1069	22,200	**6.5**	**1199**	**28,300**
231	5.5	1023	20,300	6.7	1189	28,800
Unique	6.1	1097	23,300	7.0	1224	28,800
N340	6.4	1127	24,200	7.2	1214	28,300
AA#5	7.0	1038	18,100	8.2	1236	28,400
Power Pistol	7.4	1155	23,300	8.3	1265	28,400
800X	7.0	1090	20,300	8.2	1242	27,600
HS-6	8.0	1071	20,300	9.4	1236	29,000
Blue Dot	9.0	1100	21,100	10.4	1245	28,100

***#401638**
175 gr. (#2 Alloy) 1.253" OAL

BC: .088
SD: .155

Powder	Sugg Starting Grains	Velocity fps	Pressure C.U.P.	Max Load Grains	Velocity fps	Pressure C.U.P.
Red Dot	5.4	1113	23,900	6.0	1152	28,700
Bullseye	5.9	1115	23,500	**6.4**	**1197**	**28,400**
231	5.9	1115	24,000	6.5	1181	29,400
Unique	6.1	1104	22,900	6.8	1180	28,900
WSF	6.2	1094	22,800	7.0	1176	28,600
N340	6.3	1052	21,700	7.1	1190	29,100
AA#5	7.2	1138	24,000	7.6	1196	28,700
Power Pistol	7.3	1139	23,300	8.2	1239	28,100
800X	7.7	1092	23,300	8.6	1269	28,300
Blue Dot	9.3	1175	23,800	10.4	1275	29,900

Note: Loads shown in shaded panels are maximum.
Loads shown in bold designate potentially most accurate load.
* Designates the use of Norma cases.

41 Remington Magnum

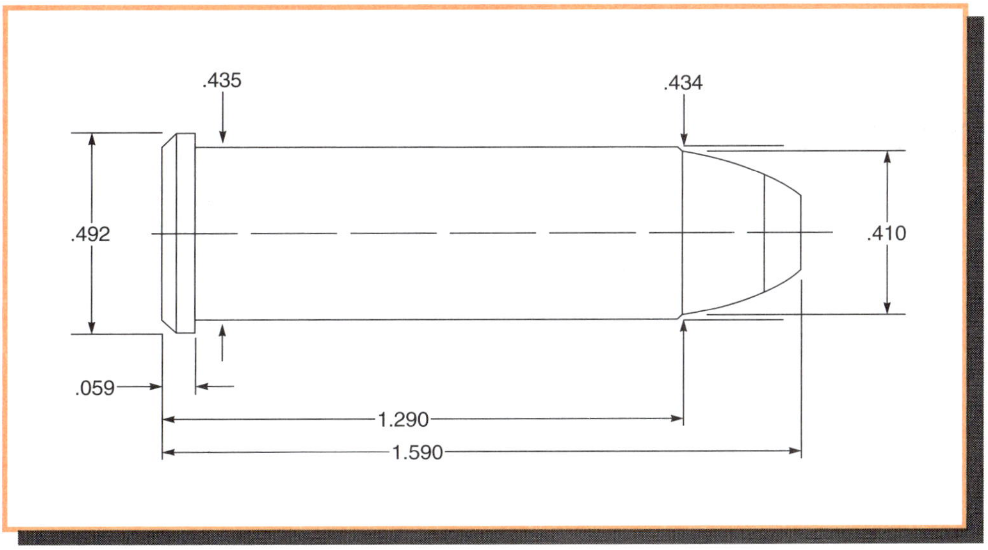

Comments:

Remington and Smith & Wesson introduced the 41 Magnum in the Model 57 revolver in 1964. The cartridge never gained a great deal of acceptance with the law enforcement market for which it was intended. Nor did it ever become terribly popular with the civil market outside of a dedicated handful of shooters, notably handgun hunters. The 41 Magnum offers shooters flatter trajectory with a bit less muzzle blast than the 44 Magnum. Those reloading for the 41 Magnum have a decent, if not extensive, selection of bullets to choose from. Pressure levels of the 41 Magnum require cast bullet #410610 be cast of a fairly hard alloy such as Linotype or equivalent. Alliant's 2400 and AA#9 produced the best results in our testing. This data can be used in the T/C Contender.

Test Components:

Cases . . . . .Winchester
Trim-to Length . . . . .1.280"
Primers . . . . .Winchester WLP
Primer Size . . . . .Large Pistol
Lyman Shell Holder . . . . .No. 30
Jacketed Bullets Used . . . . .Sierra JHC #8500, 170 gr.
Speer JHP #4405, 200 gr.
Hornady HP/XTP #41000, 210 gr.
Speer JSP #4417, 220 gr.
Cast Bullets Used . . . .(sized to .410" dia)
*gas check bullet *#410610, 215 gr.

Test Specifications:
(Velocity & Pressure)

Firearm Used . . . . .Universal Receiver
Barrel Length . . . . .4"
Twist . . . . .1-18¾"
Groove Dia. . . . . .409"

170 gr. Jacketed HC
1.580" OAL
BC: .123
SD: .144

Powder	Sugg Starting Grains	Velocity fps	Pressure C.U.P.	Max Load Grains	Velocity fps	Pressure C.U.P.
Unique	9.9	1185	27,800	11.0	1381	38,000
Blue Dot	13.7	1113	24,700	15.3	1534	37,500
AA#9	18.0	1240	25,200	20.0	1597	37,000
2400	17.5	1279	29,200	**19.5**	**1467**	**36,400**
H110	24.1	1582	29,900	25.2	1700	36,800
IMR-4227	20.0	1243	31,300	22.5	1392	36,600

200 gr. Jacketed HC
1.590" OAL
BC: .113
SD: .170

Powder	Sugg Starting Grains	Velocity fps	Pressure C.U.P.	Max Load Grains	Velocity fps	Pressure C.U.P.
Unique	8.5	979	29,200	9.5	1157	38,900
Blue Dot	11.7	1024	28,300	13.0	1296	38,700
AA#9	15.7	1149	29,300	17.5	1397	36,100
2400	16.0	1169	33,200	**17.8**	**1392**	**38,500**
H110	20.0	1314	32,400	21.0	1533	37,600
IMR-4227	17.5	994	29,000	19.5	1225	39,500

Note: Loads shown in shaded panels are maximum.
Loads shown in bold designate potentially most accurate load.

41 Remington Magnum

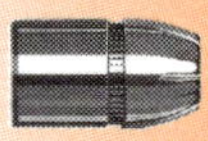

210 gr. Jacketed HP
1.585" OAL

BC: .182
SD: .178

Powder	Sugg Starting Grains	Velocity fps	Pressure C.U.P.	Max Load Grains	Velocity fps	Pressure C.U.P.
Unique	9.0	1021	33,400	10.0	1178	38,700
Blue Dot	11.9	952	25,900	13.3	1207	37,000
AA#9	15.9	1060	26,500	**17.7**	**1323**	**36,900**
2400	16.0	1143	32,200	17.8	1262	35,700
H110	20.5	1387	35,400	21.5	1473	37,000
IMR-4227	18.0	1017	30,600	20.0	1212	38,800

220 gr. Jacketed SP
1.590" OAL

BC: .137
SD: .178

Powder	Sugg Starting Grains	Velocity fps	Pressure C.U.P.	Max Load Grains	Velocity fps	Pressure C.U.P.
Unique	7.9	924	30,100	8.8	1060	38,600
Blue Dot	10.8	929	27,800	12.0	1111	36,500
AA#9	15.0	1083	30,700	**16.7**	**1285**	**37,900**
2400	14.5	1023	30,100	16.2	1198	37,000
H110	18.6	1268	33,500	19.4	1397	39,100
IMR-4227	16.2	920	30,900	18.0	1098	38,700

#410610
215 gr. (#2 Alloy) 1.575" OAL

BC: .158
SD: .183

Powder	Sugg Starting Grains	Velocity fps	Pressure C.U.P.	Max Load Grains	Velocity fps	Pressure C.U.P.
Unique	**7.8**	**960**	**25,800**	8.7	1111	36,900
SR-4756	9.0	938	27,000	10.0	1114	36,800
Blue Dot	10.2	941	24,600	11.4	1121	34,200
AA#7	13.0	1105	30,400	14.5	1276	35,700
AA#9	15.0	1110	29,200	16.7	1343	38,000
2400	14.5	1044	29,000	16.2	1312	38,100
IMR-4227	15.7	891	26,800	17.5	1105	35,700

Note: Loads shown in shaded panels are maximum.
Loads shown in bold designate potentially most accurate load.

44-40 Winchester (44WCF)

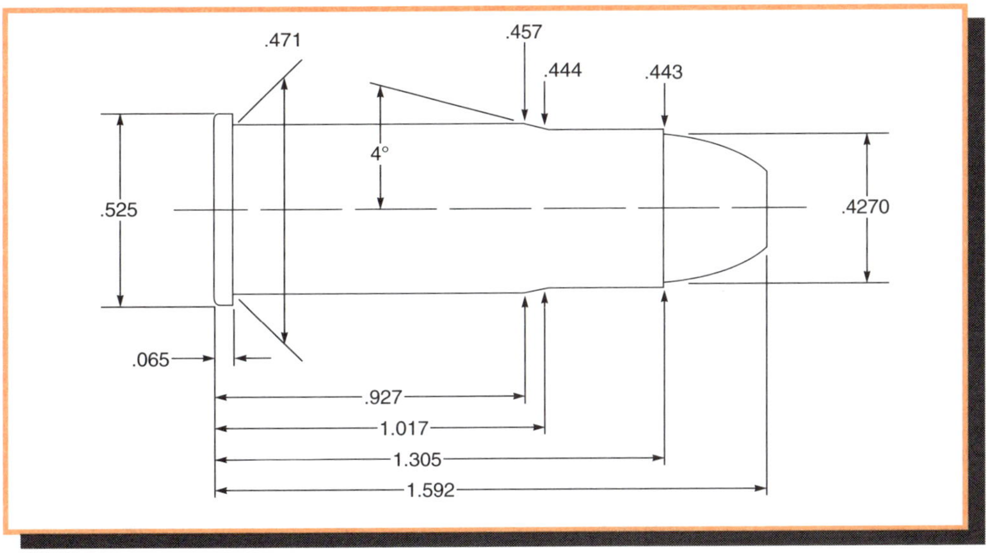

Comments:

This is a very old cartridge and care should be taken not to use the data listed in any handgun which has deteriorated. Additionally this data should not be used in conjunction with any firearm designed for black powder cartridges.

Individual tolerances vary greatly in firearms chambered for this cartridge, even among the same make and model.

Therefore exercise extreme caution when developing loads as you work up from the suggested starting grains listing.

Never load old brass as it may be of the weaker balloon head type or have been fired at one time with corrosive primers.

Test Components:

Cases . . . . Winchester
Trim-to Length . . . . 1.295"
Primers . . . . Winchester WLP
Primer Size . . . . Large Pistol
Lyman Shell Holder . . . . No. 14B
Cast Bullets Used . . . . (sized to .429" dia)
#427666, 200 gr.
#427098, 205 gr.

Test Specifications: (Velocity & Pressure)

Firearm Used . . . . Universal Receiver
Barrel Length . . . . 6"
Twist . . . . 1-20"
Groove Dia. . . . . .428"

#427666 200 gr. (#2 Alloy) 1.580" OAL — **BC:** .149 **SD:** .155

Powder	Sugg Starting Grains	Velocity fps	Pressure C.U.P.	Max Load Grains	Velocity fps	Pressure C.U.P.
700X	**5.0**	**808**	**7,400**	6.4	979	12,600
AA#2	6.0	824	7,400	6.5	996	12,800
Bullseye	5.6	852	7,700	7.0	1020	12,600
Unique	6.9	853	6,700	8.6	1044	12,600
SR-4756	7.5	727	5,900	9.4	988	12,300
2400	13.0	885	7,900	16.5	1077	12,800
IMR-4227	15.0	910	7,700	17.0	984	12,600

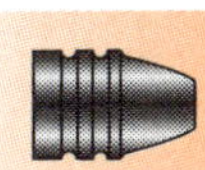

#427098 205 gr. (#2 Alloy) 1.592" OAL — **BC:** .103 **SD:** .159

Powder	Sugg Starting Grains	Velocity fps	Pressure C.U.P.	Max Load Grains	Velocity fps	Pressure C.U.P.
700X	**4.8**	**789**	**7,200**	6.0	936	12,100
AA#2	5.4	773	6,300	6.3	900	10,600
Bullseye	5.2	833	8,200	6.5	962	12,700
Unique	6.3	739	5,700	7.9	980	12,000
SR-4756	7.4	773	5,700	9.3	961	12,400
2400	12.0	751	5,100	15.0	974	12,400
IMR-4227	12.5	699	5,300	15.8	915	12,000

Note: Loads shown in shaded panels are maximum.
Loads shown in bold designate potentially most accurate load.

44 Smith & Wesson Special

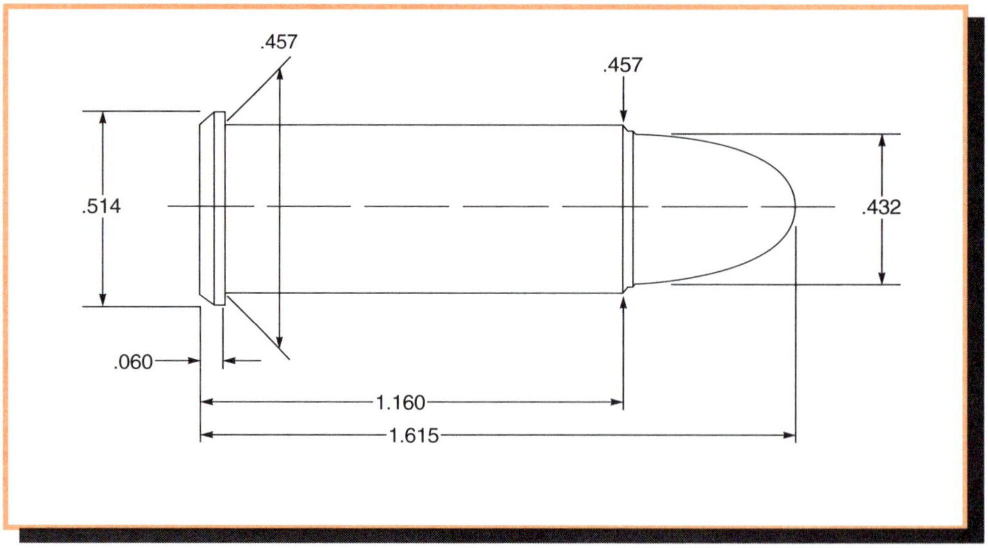

Comments:

The 44 Special came into being in 1907 as a lengthened version of the old 44 Russian cartridge in order to better utilize smokeless powder. The cartridge developed a sterling reputation for accuracy in the following decades. The large number of older revolvers of weaker construction kept commercially loaded ammunition on the mild side. Reloaders with a good heavy framed revolver could significantly improve over the factory ballistics by careful handloading. Elmer Keith used the 44 Special extensively in pursuit of high-velocity handgun loads. His efforts culminated in the introduction of the 44 Remington Magnum in 1955. The 44 Special subsequently fell by the wayside for quite a few years but interest has picked up recently. Two factors contributed to its recent popularity: Cowboy Action Shooting, and new generation compact revolvers chambered exclusively for the 44 Special. Cast bullets should be sized to either groove diameter or .001" over. Shooters should slug their bore before beginning to reload, especially if they have an older revolver. Shooters should make up a dummy round to ensure that the finished cartridge will chamber easily in all cylinders. Elmer Keith designed bullet #429421, which has been very popular in this cartridge for many years. 231, Unique, and 2400 produce good results in the 44 Special.

Test Components:

Cases . . . . .Winchester
Trim-to Length . . . . .1.150"
Primers . . . . .CCI 300
Primer Size . . . . .Large Pistol
Lyman Shell Holder . . . . .No. 7
Jacketed Bullets Used . . . . .Sierra JHC #8600, 180 gr.
Hornady HP/XTP #44100, 200 gr.
Speer JHP #4435, 225 gr.
Speer JHP #4453, 240 gr.
Cast Bullets Used . . . .(sized to .429" dia)
*gas check bullet
#427098, 205 gr.
*#429215, 210 gr.
#429383, 245 gr.
#429421, 245 gr.
*#429244, 255 gr.

Test Specifications:
(Velocity & Pressure)

Firearm Used . . . . .Universal Receiver
Barrel Length . . . . .4" & 7½"
Twist . . . . .1-20"
Groove Dia. . . . . .429"

180 gr. Jacketed HC
1.470" OAL
BC: .140
SD: .139

Powder	Sugg Starting Grains	Velocity fps	Pressure C.U.P.	Max Load Grains	Velocity fps	Pressure C.U.P.
Red Dot	6.4	793	11,600	7.1	850	13,700
231	6.5	689	9,300	7.5	832	13,600
2400	**14.0**	**674**	**8,100**	15.8	851	13,300
IMR-4227	15.1	685	9,400	16.8	862	13,100

200 gr. Jacketed HP
1.495" OAL
BC: .170
SD: .155

Powder	Sugg Starting Grains	Velocity fps	Pressure C.U.P.	Max Load Grains	Velocity fps	Pressure C.U.P.
Red Dot	5.5	660	9,000	6.3	776	13,800
231	6.4	707	11,100	7.1	797	13,800
2400	**12.0**	**599**	**8,000**	13.5	838	13,700
IMR-4227	14.3	657	10,200	15.8	882	13,900

Note: Loads shown in shaded panels are maximum.
Loads shown in bold designate potentially most accurate load.

44 Smith & Wesson Special

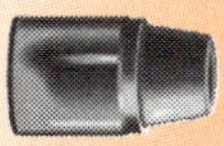

225 gr. Jacketed HP
1.615" OAL

BC: .146
SD: .175

Powder	Sugg Starting Grains	Velocity fps	Pressure C.U.P.	Max Load Grains	Velocity fps	Pressure C.U.P.
Red Dot	4.9	590	10,900	5.7	767	13,900
231	5.0	585	11,300	5.8	729	13,900
2400	11.6	653	10,600	13.3	790	13,800
IMR-4227	**12.3**	**640**	**10,600**	14.0	749	13,400

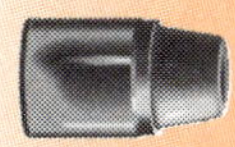

240 gr. Jacketed SP
1.175" OAL

BC: .165
SD: .186

Powder	Sugg Starting Grains	Velocity fps	Pressure C.U.P.	Max Load Grains	Velocity fps	Pressure C.U.P.
Red Dot	5.0	671	10,800	5.6	713	13,600
231	5.0	654	8,300	5.6	759	13,900
2400	12.1	604	10,300	13.5	723	13,000
IMR-4227	**13.3**	**616**	**11,100**	14.8	711	13,300

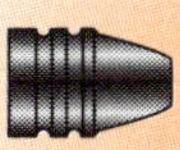

#427098
205 gr. (#2 Alloy) 1.537" OAL

BC: .103
SD: .159

Powder	Sugg Starting Grains	Velocity fps	Pressure C.U.P.	Max Load Grains	Velocity fps	Pressure C.U.P.
Unique	7.0	736	10,500	**8.2**	**869**	**14,000**
SR-7625	6.8	761	9,700	7.7	838	12,600
HS-7	10.0	732	10,100	11.6	874	13,600
Blue Dot	10.5	747	10,000	12.0	855	13,300
2400	13.3	734	10,800	15.2	870	13,500

#429215
210 gr. (#2 Alloy) 1.500" OAL

BC: .188
SD: .163

Powder	Sugg Starting Grains	Velocity fps	Pressure C.U.P.	Max Load Grains	Velocity fps	Pressure C.U.P.
Unique	6.3	729	9,500	**7.7**	**884**	**13,600**
SR-7625	6.2	732	10,900	6.8	801	13,500
HS-7	9.0	659	8,100	11.0	841	12,600
Blue Dot	9.5	689	8,800	11.0	885	13,400
2400	13.0	792	11,300	14.5	905	14,000

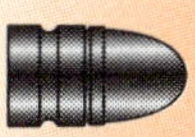

***#429383**
245 gr. (#2 Alloy) 1.570" OAL

BC: .191
SD: .190

Powder	Sugg Starting Grains	Velocity fps	Pressure C.U.P.	Max Load Grains	Velocity fps	Pressure C.U.P.
231	**4.6**	**703**	**9,200**	5.7	843	13,600
Universal	5.7	675	7,400	7.0	938	13,700
Unique	5.2	771	10,100	6.5	905	14,000
SR-4756	6.4	704	8,300	8.0	898	14,000
HS-7	7.6	729	8,700	9.5	910	13,700
2400	10.0	706	7,600	12.5	935	13,900
IMR-4227	12.6	784	7,500	14.5	879	13,700

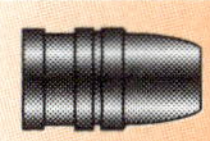

#429421
245 gr. (#2 Alloy) 1.570" OAL

BC: .209
SD: .190

Powder	Sugg Starting Grains	Velocity fps	Pressure C.U.P.	Max Load Grains	Velocity fps	Pressure C.U.P.
231	4.4	679	7,700	5.5	821	13,100
Unique	6.0	649	10,300	**6.9**	**767**	**13,300**
SR-4756	6.8	661	9,000	8.1	780	13,800
HS-7	9.1	728	11,300	10.4	828	14,000
2400	11.4	704	10,400	13.2	797	13,800
IMR-4227	11.7	674	7,100	14.7	872	13,900

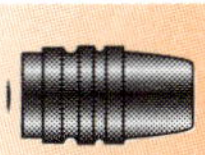

#429244
255 gr. (#2 Alloy) 1.550" OAL

BC: .201
SD: .198

Powder	Sugg Starting Grains	Velocity fps	Pressure C.U.P.	Max Load Grains	Velocity fps	Pressure C.U.P.
Unique	5.7	635	10,200	6.6	747	13,600
SR-4756	6.2	584	8,400	7.5	765	14,000
HS-7	8.5	606	8,700	**10.0**	**743**	**13,400**
2400	10.5	602	8,500	12.3	742	12,700

Note: Loads shown in shaded panels are maximum.
Loads shown in bold designate potentially most accurate load.
* Designates fired in a 7½" barrel.

44 Remington Magnum

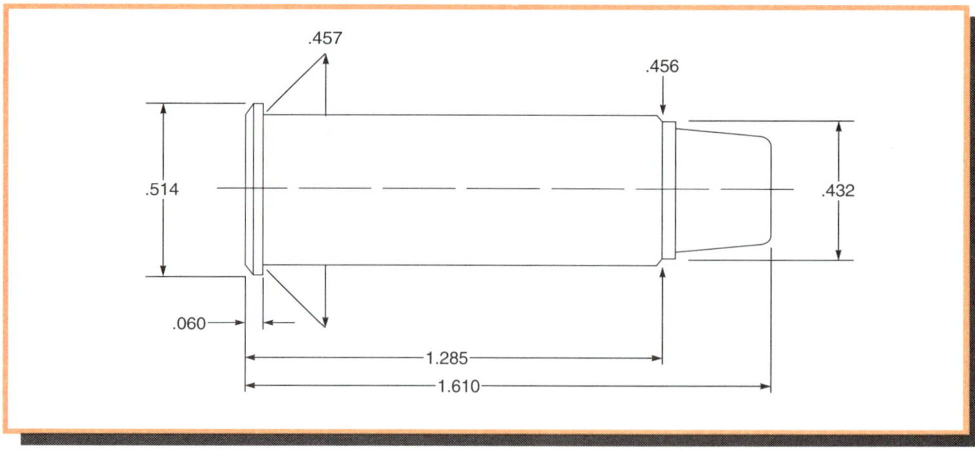

Comments:

The 44 Remington Magnum is the product of Elmer Keith's experimentation with high-pressure loadings in the 44 Special during the early 1950s. Smith & Wesson first produced revolvers for it in 1955 and Sturm, Ruger & Co. soon followed. The 44 Magnum has since been offered by all of the major handgun manufacturers and most of the smaller ones as well. It has also proven an effective short-range brush cartridge when chambered in various lever action and semiautomatic carbines. The 44 Magnum requires practice for the average shooter to shoot well due to its noticeable recoil and muzzle blast. Many new owners of 44 Magnums fire 44 Specials in their revolvers to acclimate themselves to the gun or for use as a general purpose target load.

This cartridge offers a lot to the hand loader. It has a high degree of accuracy, an extensive selection of bullet designs and weights, and is well suited to cast bullets. The loads listed for cast bullet #429215, #429303, 429244, #429421, #429650, and #429667 exceed the standard maximum cartridge length. However, the cylinders of most newer revolvers chambered for 44 Magnum will accept the listed overall length. All bullets must be well crimped to prevent bullet movement under recoil. Bullet # 429215 features a gas check and is a good general purpose bullet. Elmer Keith designed bullet # 429421 and many handloaders consider it to be *the* bullet for loading the 44 Magnum. Cast bullets should be made of a hard alloy, preferably Linotype or an equivalent of 22 bhn. Alliant 2400 is a favored powder among those loading the 44 Magnum. Winchester 296 and Hodgdon's H-110 are also popular for full power loads. This data can be used for the T/C Contender. Shooters loading for lever action rifles should refer to the rifle data section for more approximate velocity levels.

Test Components:

Cases . . . . .Remington
Trim-to Length . . . . .1.275"
Primers . . . . .CCI 300 & 350
Primer Size . . . . .Large Pistol, Std. & Magnum
Lyman Shell Holder . . . . .No. 7
Jacketed Bullets Used . . . . .Sierra JHC #8600, 180 gr.
Hornady HP/XTP #44100, 200 gr.
Speer JHP #4435, 225 gr.
Sierra JHC #8610, 240 gr.
Speer JHP #4453, 240 gr.
Hornady FP #4300, 265 gr.
Sierra JSP #8630, 300 gr.
Cast Bullets Used. . . . .(sized to .429" dia)
*gas check bullet *#429303, 200 gr.
*#429215, 210 gr.
#429667, 240 gr.
#429421, 245 gr.
*#429244, 255 gr.
*#429650, 300 gr.

Test Specifications:
(Velocity & Pressure)

Firearm Used . . . . .Universal Receiver
Barrel Length . . . . .4"
Twist . . . . .1-20"
Groove Dia. . . . . .429"

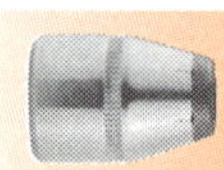

180 gr. Jacketed HC
1.610" OAL

BC: .130
SD: .139

Powder	Sugg Starting Grains	Velocity fps	Pressure C.U.P.	Max Load Grains	Velocity fps	Pressure C.U.P.
Unique	12.0	1061	25,100	13.6	1301	35,700
231	10.2	1077	22,100	13.0	1267	36,300
SR-4756	11.4	970	17,400	13.9	1262	34,400
AA#5	12.8	976	16,400	16.0	1344	36,000
HS-7	17.8	1111	24,200	20.6	1341	36,300
Blue Dot	17.8	1126	25,800	20.2	1404	36,400
2400	21.3	1103	24,800	**25.5**	**1370**	**36,400**
*H-110	29.0	1541	31,700	30.0	1587	34,400
*296	29.0	1498	32,400	30.0	1563	36,700
IMR-4227	24.2	1136	27,600	27.0	1317	34,800
*H-4227	25.0	1306	30,600	27.8	1440	37,600

Note: Loads shown in shaded panels are maximum.
Loads shown in bold designate potentially most accurate load.
* Designates the use of magnum primers.
\+ Designates a compressed powder charge.

44 Remington Magnum

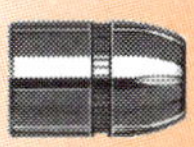

200 gr. Jacketed HP 1.610" OAL — **BC:** .170 **SD:** .155

Powder	Sugg Starting Grains	Velocity fps	Pressure C.U.P.	Max Load Grains	Velocity fps	Pressure C.U.P.
Unique	**10.5**	**905**	**21,100**	13.2	1148	35,100
SR-4756	11.0	894	18,500	13.6	1164	36,300
800X	12.0	861	20,200	14.9	1239	36,100
AA#7	18.0	1163	28,800	20.0	1365	38,600
Blue Dot	15.7	942	19,600	18.7	1245	35,100
N110	21.0	1142	25,500	24.0+	1381	38,700
2400	19.5	972	23,100	23.6	1227	35,000
*H-110	27.7	1444	34,600	28.8	1492	39,800
*296	27.0	1432	31,700	28.3	1493	37,000
IMR-4227	22.8	1000	22,800	26.4	1261	36,600
*H-4227	23.8	1221	31,000	26.5	1383	39,000

225 gr. Jacketed HP 1.610" OAL — **BC:** .146 **SD:** .175

Powder	Sugg Starting Grains	Velocity fps	Pressure C.U.P.	Max Load Grains	Velocity fps	Pressure C.U.P.
Unique	**10.8**	**1021**	**29,900**	12.0	1200	39,300
SR-4756	10.2	865	20,800	12.6	1123	36,500
800X	12.0	958	27,700	14.7	1178	36,300
AA#7	15.5	836	23,300	19.0	1227	37,300
Blue Dot	14.0	982	28,700	15.7	1264	37,500
N110	19.0	1168	30,800	21.1	1354	38,000
2400	17.7	935	22,300	22.0	1199	36,000
*H-110	25.0	1348	34,100	26.0	1415	39,300
*296	24.0	1112	27,600	25.0	1235	33,100
IMR-4227	20.4	1010	24,800	24.0	1205	35,800
*H-4227	22.5	1027	33,600	25.0	1182	38,800

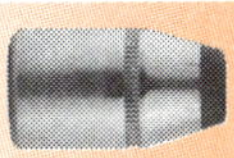

240 gr. Jacketed HC 1.610" OAL — **BC:** .172 **SD:** .186

Powder	Sugg Starting Grains	Velocity fps	Pressure C.U.P.	Max Load Grains	Velocity fps	Pressure C.U.P.
Unique	10.3	957	31,400	11.5	1074	39,000
SR-4756	9.9	854	23,700	12.3	1072	35,800
800X	11.0	776	23,700	13.7	1079	37,300
Blue Dot	13.4	939	29,300	14.9	1155	38,600
N110	17.3	999	25,500	19.5+	1189	37,400
2400	17.5	993	30,400	19.5	1179	38,400
AA#9	19.0	829	22,300	22.2	1135	33,500
*H-110	22.5	1169	34,200	23.5	1271	38,800
*296	22.5	1168	34,700	23.5	1264	39,200
IMR-4227	19.3	921	22,600	**23.1+**	**1177**	**36,800**
*H-4227	20.5	1032	30,100	23.0+	1167	37,800

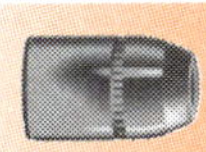

240 gr. Jacketed HP 1.610" OAL — **BC:** .165 **SD:** .186

Powder	Sugg Starting Grains	Velocity fps	Pressure C.U.P.	Max Load Grains	Velocity fps	Pressure C.U.P.
Unique	10.8	951	30,500	12.0	1084	38,700
800X	11.2	807	24,700	13.9	1068	36,500
Blue Dot	14.4	960	29,000	16.0	1160	38,400
N110	18.7	1052	31,000	20.8	1217	39,400
2400	18.4	1028	30,600	20.5	1154	37,700
AA#9	19.0	815	22,100	22.0	1195	37,100
*H-110	23.5	1196	35,900	24.5	1233	37,400
*296	23.0	1193	34,000	24.0	1292	38,500
IMR-4227	19.3	884	23,100	**23.1**	**1109**	**36,000**
*H-4227	21.8	1021	30,500	24.2	1181	38,600

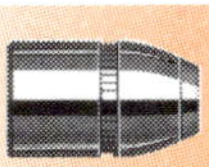

265 gr. Jacketed FP 1.610" OAL — **BC:** .189 **SD:** .205

Powder	Sugg Starting Grains	Velocity fps	Pressure C.U.P.	Max Load Grains	Velocity fps	Pressure C.U.P.
800X	10.7	745	25,400	13.4	999	36,700
Blue Dot	12.6	883	29,500	14.0	1051	38,600
N110	16.6	965	29,800	18.5	1133	38,900
2400	**16.5**	**937**	**31,800**	18.3	1105	38,700
AA#9	18.0	823	25,300	21.3	1066	35,600
*H-110	20.6	1036	33,200	21.5	1140	38,400
*296	21.1	1092	35,000	22.0	1150	38,700
IMR-4227	18.5	870	23,800	22.0	1088	34,800
*H-4227	19.3	950	29,600	21.5	1116	37,400

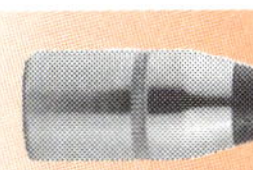

300 gr. Jacketed SP 1.720" OAL — **BC:** .250 **SD:** .232

Powder	Sugg Starting Grains	Velocity fps	Pressure C.U.P.	Max Load Grains	Velocity fps	Pressure C.U.P.
800X	10.0	785	25,200	12.5	968	35,200
Blue Dot	12.4	765	26,800	13.5	923	36,800
N110	16.0	870	27,000	17.5	965	33,600
2400	15.7	820	30,300	17.5	934	37,800
AA#9	17.5	895	27,200	19.3	1020	35,200
*H-110	19.6	909	34,500	20.5	982	38,300
*296	19.8	892	32,400	20.7	966	36,800
IMR-4227	18.5	790	22,000	**21.0**	**1046**	**36,100**
*H-4227	18.7	839	29,600	20.8	985	38,600

Note: Loads shown in shaded panels are maximum.
Loads shown in bold designate potentially most accurate load.
* Designates the use of magnum primers.
\+ Designates a compressed powder charge.

44 Remington Magnum

#429303 — 200 gr. (Linotype) 1.692" OAL — **BC:** .105 **SD:** .155

Powder	Sugg Starting Grains	Velocity fps	Pressure C.U.P.	Max Load Grains	Velocity fps	Pressure C.U.P.
Titegroup	9.2	1107	32,600	10.2	1223	38,500
231	9.1	913	19,600	**12.6**	**1231**	**36,900**
Unique	11.0	1004	22,100	13.2	1265	36,700
Herco	11.4	1035	24,300	13.8	1245	36,700
SR-4756	12.3	1043	22,600	15.5	1289	36,700
HS-6	14.8	1086	28,900	16.5	1327	38,400
AA#7	16.9	1065	26,200	18.8	1273	37,500
Blue Dot	15.8	1062	22,800	19.8	1363	36,800
N110	19.8	1232	30,200	22.0+	1438	39,300
2400	19.0	991	20,400	24.0+	1314	34,600
*H-110	26.6	1472	36,700	27.8+	1507	37,300
IMR-4227	22.0	1038	22,100	24.5+	1201	29,600

#429215 — 210 gr. (Linotype) 1.645" OAL — **BC:** .188 **SD:** .163

Powder	Sugg Starting Grains	Velocity fps	Pressure C.U.P.	Max Load Grains	Velocity fps	Pressure C.U.P.
Titegroup	9.0	1023	30,700	10.0	1143	37,400
231	**9.3**	**914**	**21,000**	12.8	1191	37,000
Unique	10.0	930	21,000	13.2	1221	36,000
Herco	11.4	984	22,800	13.5	1205	36,000
SR-4756	12.2	1036	24,900	15.2	1228	36,100
HS-6	15.0	1105	29,700	16.7	1290	37,400
AA#7	17.2	1094	27,500	19.2	1305	37,100
800X	12.0	932	23,900	15.1	1274	36,200
Blue Dot	14.9	1036	26,700	16.6	1324	38,500
N110	19.3	1180	21,200	21.5	1413	38,200
2400	19.8	1191	31,600	22.0	1348	37,100
*H-110	26.4	1298	27,500	27.5+	1447	33,800
*296	25.5	1340	29,500	27.0+	1508	36,400
IMR-4227	21.8	1017	22,100	25.5+	1227	32,800

#429667 — 240 gr. (Linotype) 1.645" OAL — **BC:** .149 **SD:** .186

Powder	Sugg Starting Grains	Velocity fps	Pressure C.U.P.	Max Load Grains	Velocity fps	Pressure C.U.P.
Titegroup	9.0	1003	33,900	10.0	1115	39,300
Unique	10.0	937	25,300	11.7	1133	37,900
Herco	10.5	958	28,400	11.7	1101	37,500
800X	12.0	1046	30,600	**13.5**	**1198**	**38,600**
Blue Dot	14.0	933	24,500	15.5	1159	36,000
AA#9	18.2	1009	25,700	20.2	1217	35,000
N110	18.0	1064	28,700	20.0	1283	37,900
2400	18.5	1101	34,700	20.6	1258	38,900
*H-110	22.5	1204	31,800	23.5	1266	35,700
IMR-4227	21.5	1036	29,000	24.0	1263	37,800
*H-4227	20.7	1032	27,500	23.0	1248	37,300

#429421 — 245 gr. (Linotype) 1.710" OAL — **BC:** .209 **SD:** .190

Powder	Sugg Starting Grains	Velocity fps	Pressure C.U.P.	Max Load Grains	Velocity fps	Pressure C.U.P.
Titegroup	8.5	915	29,400	9.5	1028	37,100
Unique	9.8	912	20,800	13.0	1147	36,800
Herco	10.9	918	22,900	13.1	1090	34,600
800X	10.6	812	20,700	13.3	1113	34,200
Blue Dot	14.5	1034	29,800	16.0	1194	36,600
AA#9	17.5	711	15,100	21.7	1229	37,000
N110	18.0	1034	25,900	20.0	1300	38,800
2400	18.5	1087	30,300	**20.6**	**1248**	**37,200**
*H-110	24.0	1218	31,900	25.0	1301	35,300
IMR-4227	20.2	938	21,300	24.0+	1196	34,600
*H-4227	21.5	1083	29,400	23.8	1281	38,400

#429244 — 255 gr. (Linotype) 1.680" OAL — **BC:** .201 **SD:** .198

Powder	Sugg Starting Grains	Velocity fps	Pressure C.U.P.	Max Load Grains	Velocity fps	Pressure C.U.P.
Titegroup	8.1	873	31,800	9.0	945	37,700
Unique	8.9	805	17,500	12.1	1100	36,400
SR-4756	10.5	892	21,500	12.2	1119	35,300
Herco	11.0	927	24,500	12.4	1085	35,400
800X	10.6	842	25,300	13.2	1075	36,600
HS-7	14.5	921	22,600	17.5	1160	37,000
Blue Dot	14.0	911	20,300	17.4	1160	34,800
AA#9	17.5	750	17,000	21.7	1223	34,600
2400	18.2	915	21,600	**22.2**	**1165**	**35,300**
*H-110	21.8	1097	31,500	22.8	1186	36,400
*296	22.5	1150	33,400	23.5	1223	37,800
IMR-4227	20.2	903	20,600	24.0+	1135	33,700
*H-4227	19.8	992	29,900	22.0	1137	36,400

#429650 — 300 gr. (Linotype) 1.700" OAL — **BC:** .165 **SD:** .233

Powder	Sugg Starting Grains	Velocity fps	Pressure C.U.P.	Max Load Grains	Velocity fps	Pressure C.U.P.
SR-4756	8.5	796	24,800	10.0	907	34,900
Herco	8.1	793	30,100	9.0	850	35,400
800X	8.5	794	23,800	10.5	948	36,000
Blue Dot	10.5	761	22,400	11.7	895	33,200
AA#9	15.3	875	25,700	18.0	1075	36,300
N110	14.2	873	24,300	16.0	1029	34,800
2400	14.0	857	27,600	**15.7**	**1001**	**37,200**
*H-110	17.7	976	34,300	18.5	1036	37,900
*296	18.2	982	34,500	19.0	1064	38,300
IMR-4227	16.5	837	24,000	19.0+	1008	33,600
*H-4227	17.0	916	31,100	19.0	1036	37,900

Note: Loads shown in shaded panels are maximum.
Loads shown in bold designate potentially most accurate load.
* Designates the use of magnum primers.
\+ Designates a compressed powder charge.

45 Automatic (45 ACP)

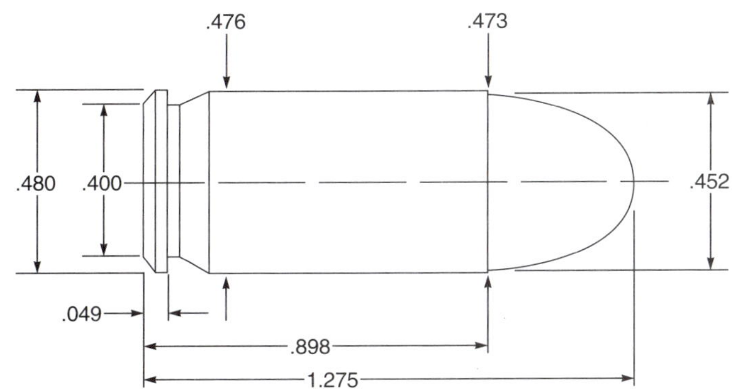

Comments:

The 45 ACP (Automatic Colt Pistol) originated in the Colt Model 1905 pistol developed by John Moses Browning. The United States Army sought a 45-caliber pistol to replace the 38 Long Colt that proved semi-disastrous during the Philippine Campaign. John Browning subsequently chambered his masterpiece, the Model 1911 pistol, in 45 ACP and the legend was born. The original military loading consisted of a 200-grain bullet traveling just over 900 feet per second. This soon changed to the now standard 230-grain bullet at 850 feet per second. The Model 1911 and 1911A1 pistol served as the standard U.S. military sidearm for over 70 years. Although officially replaced by the 9mm Beretta in 1985, the 45 ACP cartridge remains in use with Special Operations units. Law enforcement agencies have shown increased interest in the cartridge lately. Several European arms makers are now also producing 45 ACP chambered pistols.

The 45 ACP has long been one of the most accurate and popular pistol cartridges around and shows no sign of slowing down despite its age. It is not however a cartridge that one can shoot once or twice a year and remain proficient with. The wide availability of components and bullet designs make the 45 ACP a great candidate for handloaders. Bullseye, Unique, and 231 often give best results with most bullet weights in this cartridge. The 45 ACP headspaces on the case mouth so cartridges must not be roll crimped. The roll-crimping feature in Lyman seating dies for the 45 ACP is for use with the 45 Auto Rim cartridges used in revolvers. A slight taper crimp can be used on the 45 ACP if necessary. Reloaders should trim cases only if necessary and chamfer the case mouth only enough to break the burr. Shooters should adhere to the listed cartridge overall lengths. Variations in overall lengths may result in excessive pressures or difficulty feeding.

Many 1911 pattern pistols will not function well with anything but round nose bullets unless a qualified gunsmith alters the feed ramp. Cast bullet #452374 has been in the Lyman line since 1924 and is ideal for original, unaltered pistols. This bullet closely duplicates the shape and weight of the original military full metal-jacketed load. Cast bullet #452460 has been a popular semi wad cutter bullet since 1950. Cast bullet #452630 is a bevel-based design. This is a relatively new addition to our line and makes for a good mid-range practice load. Sizing this bullet through a lube-sizer such as our Lyman Model 4500 will often leave a small deposit of bullet lubricant around the bevel. Many shooters simply wipe the base of the bullet on a clean sheet of paper after sizing. Groove diameters can vary from .450" to .453" but the shooter should not size cast bullets larger than .451" due to chamber dimensions.

Test Components:

Cases . . . . .Remington, Federal
Trim-to Length . . . . .888"
Primers . . . . .Remington 2½, CCI 300
Primer Size . . . . .Large Pistol
Lyman Shell Holder . . . . .No. 2
Jacketed Bullets Used . .Remington SWC/FMJ, 185 gr.
Hornady HP/XTP #45100, 185 gr.
Speer Gold Dot HP #4478, 200 gr.
Speer JHP #4479, 225 gr.
Speer TMJ #4480, 230 gr.
Cast Bullets Used . . . .(sized to .451" dia)
#452460, 200 gr.
#452630, 200 gr.
#452374, 225 gr.

Test Specifications:
(Velocity & Pressure)

Firearm Used . . . . .Universal Receiver
Barrel Length . . . . .5"
Twist . . . . .1-16"
Groove Dia. . . . . .450"

45 Automatic (45 ACP)

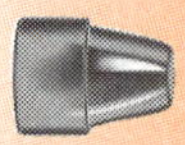

185 gr. Jacketed SWC
1.135" OAL — **BC:** .068 **SD:** .130

Powder	Sugg Starting Grains	Velocity fps	Pressure C.U.P.	Max Load Grains	Velocity fps	Pressure C.U.P.
700X	3.4	675	7,900	5.0	967	17,800
Bullseye	3.5	678	8,000	**5.6**	**975**	**16,600**
231	4.1	680	9,100	6.1	981	16,600
SR-7625	4.4	706	9,100	6.5	972	16,800
Unique	4.7	661	7,700	7.5	1055	18,000
SR-4756	5.7	666	7,900	7.7	1002	17,000
HS-6	6.6	689	8,000	9.1	1038	18,000
Blue Dot	7.3	666	8,000	10.2	1015	17,800

185 gr. Jacketed HP
1.175" OAL — **BC:** .139 **SD:** .130

Powder	Sugg Starting Grains	Velocity fps	Pressure C.U.P.	Max Load Grains	Velocity fps	Pressure C.U.P.
Clays	4.1	766	13,500	4.6	841	16,600
700X	3.5	612	7,500	5.5	933	17,800
Bullseye	3.5	610	7,300	6.0	976	18,000
231	4.4	640	7,500	6.1	937	15,700
HP-38	3.5	599	7,500	**5.8**	**960**	**17,900**
Red Dot	3.5	607	7,200	6.1	942	17,600
N320	5.7	812	10,500	6.4	988	17,300
PB	4.6	660	9,100	6.9	968	17,700
SR-7625	4.9	633	7,100	7.0	960	17,900
Unique	4.8	630	7,800	7.8	985	17,800
AA#5	7.4	720	10,700	9.2	1008	17,700
Power Pistol	7.4	917	13,400	8.3	1024	16,800
WSF	7.2	881	12,300	8.0	1047	17,500

200 gr. Jacketed HP
1.178" OAL — **BC:** .138 **SD:** .140

Powder	Sugg Starting Grains	Velocity fps	Pressure C.U.P.	Max Load Grains	Velocity fps	Pressure C.U.P.
Clays	3.9	713	13,100	4.4	790	16,500
700X	3.3	597	7,400	4.9	871	17,300
Bullseye	3.4	604	7,100	**5.7**	**942**	**17,800**
231	5.5	772	11,900	6.2	897	16,700
N320	5.1	650	12,500	5.7	857	16,200
SR-7625	4.5	601	7,300	6.7	935	17,400
Unique	4.0	604	7,300	6.5	927	17,700
SR-4756	5.4	588	7,100	7.4	961	18,000
AA#5	6.8	623	9,400	8.5	900	16,800
HS-6	5.7	566	6,500	9.0	951	17,600
Power Pistol	6.6	818	12,900	7.4	935	16,600
WSF	6.5	832	13,600	7.3	955	17,900
Blue Dot	8.7	747	10,400	9.7	899	15,800

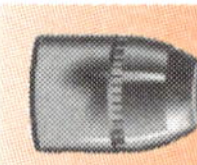

225 gr. Jacketed HP
1.243" OAL — **BC:** .169 **SD:** .158

Powder	Sugg Starting Grains	Velocity fps	Pressure C.U.P.	Max Load Grains	Velocity fps	Pressure C.U.P.
700X	3.1	549	7,500	4.8	796	17,200
Bullseye	3.1	554	7,300	5.3	840	17,400
SR-7625	4.4	587	7,400	6.3	844	17,400
Unique	4.1	548	6,800	**6.7**	**883**	**17,500**
SR-4756	4.8	552	6,700	7.3	882	17,300
N340	6.0	689	11,700	6.7	830	16,300
Power Pistol	6.5	776	14,500	7.3	876	17,100
WSF	6.0	717	14,100	6.7	807	16,500
HS-7	6.7	564	6,800	10.3	884	17,600

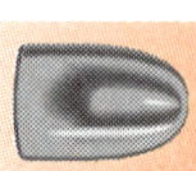

230 gr. TMJ
1.275" OAL — **BC:** .153 **SD:** .162

Powder	Sugg Starting Grains	Velocity fps	Pressure C.U.P.	Max Load Grains	Velocity fps	Pressure C.U.P.
700X	3.6	667	11,200	**4.8**	**867**	**17,900**
Bullseye	3.8	662	10,400	5.3	878	17,100
SR-7625	4.9	678	10,800	5.9	838	15,800
Unique	5.9	749	12,500	6.6	825	15,800
SR-4756	6.0	695	11,200	7.0	866	17,400
N340	5.9	742	12,700	6.6	838	16,500
Power Pistol	6.4	775	13,500	7.2	858	16,600
Blue Dot	8.3	684	11,000	9.2	806	15,100

#452460
200 gr. (#2 Alloy) 1.161" OAL — **BC:** .076 **SD:** .140

Powder	Sugg Starting Grains	Velocity fps	Pressure C.U.P.	Max Load Grains	Velocity fps	Pressure C.U.P.
*Titegroup	4.5	813	13,800	5.1	895	17,300
700X	4.0	745	10,300	5.2	940	17,300
Bullseye	3.5	645	6,900	5.6	869	15,700
Red Dot	4.0	695	8,400	5.3	895	14,700
*N320	4.6	774	13,400	5.2	889	17,100
*WST	4.2	750	13,300	4.7	804	15,300
Green Dot	4.5	715	9,000	5.8	895	14,400
****231**	4.0	694	9,200	**6.0**	**987**	**18,000**
SR-7625	5.0	735	9,000	6.2	945	17,000
Unique	5.0	670	7,700	7.5	980	16,600
**SR-4756	5.3	704	9,100	7.3	993	17,000
*Power Pistol	6.3	816	13,300	7.0	919	16,800
**AA#5	6.4	686	10,000	8.0	970	17,000
**HS-6	6.0	690	8,600	8.7	1016	17,700
**Blue Dot	7.1	701	8,900	10.6	1012	17,200

Note: Loads shown in shaded panels are maximum.
Loads shown in bold designate potentially most accurate load.
* Designates use of CCI primers.
** Designates use of Federal cases.

#452630 — 200 gr. (#2 Alloy) 1.235" OAL — **BC:** .063 **SD:** .140

Powder	Sugg Starting Grains	Velocity fps	Pressure C.U.P.	Max Load Grains	Velocity fps	Pressure C.U.P.
Clays	3.9	701	10,200	4.4	839	15,500
*Titegroup	4.8	801	12,200	5.4	920	17,100
700X	4.4	676	10,000	5.5	884	17,500
Bullseye	4.9	840	12,900	**6.0**	**909**	**17,000**
Red Dot	4.4	713	12,400	5.5	894	17,600
*N320	5.1	746	10,000	5.7	865	14,800
*WST	4.4	735	11,100	4.9	840	16,300
Green Dot	5.1	727	11,900	6.4	894	16,900
*231	5.4	769	12,700	6.1	885	16,300
SR-7625	5.4	653	8,600	6.7	900	16,800
*WSF	6.3	815	12,200	7.0	904	16,600
Unique	6.0	717	11,400	7.5	913	17,400
SR-4756	6.4	660	9,300	8.0	933	16,700
*Power Pistol	6.6	810	12,700	7.4	955	17,500
AA#5	7.0	697	10,500	8.5	927	16,800
*HS-6	8.5	795	12,800	9.5	912	16,800
Blue Dot	8.5	679	9,800	10.6	972	17,300

#452374 — 225 gr. (#2 Alloy) 1.272" OAL — **BC:** .158 **SD:** .158

Powder	Sugg Starting Grains	Velocity fps	Pressure C.U.P.	Max Load Grains	Velocity fps	Pressure C.U.P.
**Clays	3.6	683	10,600	4.3	818	17,100
*Titegroup	4.5	735	11,200	5.1	841	16,100
700X	4.0	695	10,500	5.0	855	17,300
Bullseye	4.0	680	10,100	5.0	815	14,400
Red Dot	4.3	705	11,100	5.3	835	15,300
*WST	4.2	707	11,500	4.7	794	17,100
Green Dot	4.8	725	11,400	5.8	845	15,100
****231**	4.0	661	9,200	**5.8**	**902**	**17,500**
SR-7625	5.0	675	9,000	6.0	850	15,200
*WSF	6.0	749	12,300	6.7	839	15,100
*N340	5.9	730	10,300	6.6	864	16,300
Unique	5.5	695	10,100	7.3	905	16,500
SR-4756	5.5	662	8,800	7.5	955	18,000
*Power Pistol	6.3	764	11,900	7.0	865	16,300
**AA#5	6.6	625	9,200	8.2	874	17,300
**HS-6	6.2	664	8,400	8.6	921	16,600
**Blue Dot	7.5	660	8,600	10.7	964	17,300

Note: Loads shown in shaded panels are maximum.
Loads shown in bold designate potentially most accurate load.
* Designates use of CCI primers.
** Designates use of Federal cases.

45 Auto Rim

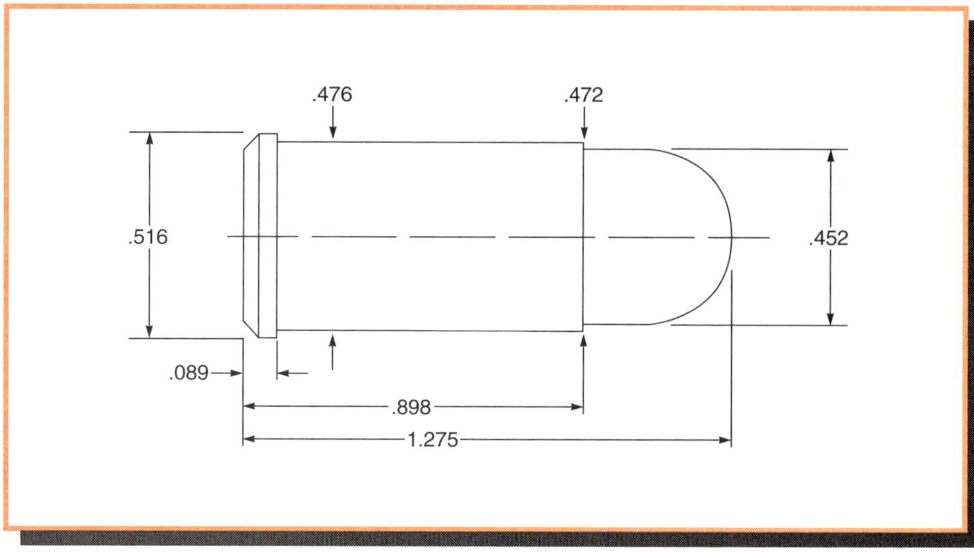

Comments:

The 45 Auto Rim was designed for use in revolvers that were chambered for the 45 Automatic. The 45 Auto Rim cartridge did away with the need to use the cumbersome half moon clips in such revolvers. The rim of this case both headspaced the round and afforded a purchase for the revolvers extractor. Bullseye, 231 or Unique should give the best results.

Test Components:

Cases . . . . .Remington
Trim-to Length . . . . .888"
Primers . . . . .Remington 2½
Primer Size . . . . .Large Pistol
Lyman Shell Holder .No. 14A
Jacketed Bullets Used...Hornady HP/XTP #45100,185 gr.
Speer GDHP #4478, 200 gr.
Speer JHP #4479, 225 gr.
Cast Bullets Used . . . .(sized to .451" dia)
#452460, 200 gr.
#452374, 225 gr.

Test Specifications:
(Velocity & Pressure)

Firearm Used . . . . .Universal Receiver
Smith & Wesson Model 25
Barrel Length . . . . .Universal Receiver 4½"
Model 25; 6½"
Twist . . . . .1-16"
Groove Dia. . . . . .451"

185 gr. Jacketed HP
1.236" OAL
BC: .139
SD: .130

Powder	Sugg Starting Grains	Velocity fps	Pressure C.U.P.	Max Load Grains	Velocity fps	Pressure C.U.P.
700X	4.5	713	8,200	5.2	838	11,900
SR-7625	6.0	709	7,500	7.2	849	13,700
Bullseye	4.6	722	8,600	5.8	915	14,800
Red Dot	4.8	711	8,600	**6.4**	**933**	**14,700**
Unique	5.0	585	5,100	7.6	905	13,200
231	4.4	618	6,400	6.1	889	13,700

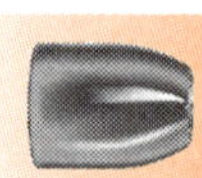

200 gr. Jacketed HP
1.180" OAL
BC: .138
SD: .140

Powder	Sugg Starting Grains	Velocity fps	Pressure C.U.P.	Max Load Grains	Velocity fps	Pressure C.U.P.
700X	3.4	491	5,000	4.9	799	13,400
SR-7625	4.8	608	7,200	6.0	809	12,400
Bullseye	4.0	614	7,800	5.4	838	13,700
Red Dot	4.0	615	7,700	5.4	852	14,600
Unique	5.8	719	10,100	7.1	882	14,500
231	4.0	570	7,800	**5.7**	**852**	**14,800**
HS-6	6.1	558	6,400	8.7	846	13,600

Note: Loads shown in shaded panels are maximum.
Loads shown in bold designate potentially most accurate load.

45 Auto Rim

225 gr. Jacketed HP
1.255" OAL

BC: .169
SD: .158

Powder	Sugg Starting Grains	Velocity fps	Pressure C.U.P.	Max Load Grains	Velocity fps	Pressure C.U.P.
SR-7625	4.4	518	6,100	6.2	812	14,900
Unique	5.0	609	8,200	7.0	834	14,500
Blue Dot	8.5	718	10,200	10.0	864	14,500
231	3.7	496	6,500	**5.2**	**773**	**13,800**
HS-6	6.7	634	8,800	8.3	812	13,600

#452460
200 gr. (#2 Alloy) 1.161" OAL

BC: .076
SD: .140

Powder	Sugg Starting Grains	Velocity fps	Pressure C.U.P.	Max Load Grains	Velocity fps	Pressure C.U.P.
700X	4.0	675	—	5.0	875	—
PB	4.5	650	—	5.5	840	—
SR-7625	5.0	660	—	6.1	850	—
Bullseye	3.0	510	—	**4.7**	**775**	—
Red Dot	4.0	645	—	5.2	845	—
Green Dot	4.5	660	—	5.6	840	—
Unique	5.0	610	—	6.8	845	—
231	4.0	694	7,200	5.5	936	14,100

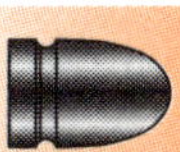

#452374
225 gr. (#2 Alloy) 1.265" OAL

BC: .158
SD: .158

Powder	Sugg Starting Grains	Velocity fps	Pressure C.U.P.	Max Load Grains	Velocity fps	Pressure C.U.P.
700X	4.0	625	—	4.8	770	—
PB	4.5	565	—	5.4	735	—
SR-7625	5.0	560	—	5.9	750	—
Bullseye	3.0	475	—	4.5	690	—
Red Dot	4.0	625	—	5.0	760	—
Green Dot	4.5	615	—	5.4	750	—
Unique	5.0	550	—	6.6	785	—
231	4.0	660	6,900	**5.5**	**873**	**13,800**
HS-6	7.3	799	9,400	8.6	922	13,800

Note: Loads shown in shaded panels are maximum.
Loads shown in bold designate potentially most accurate load.

45 Colt

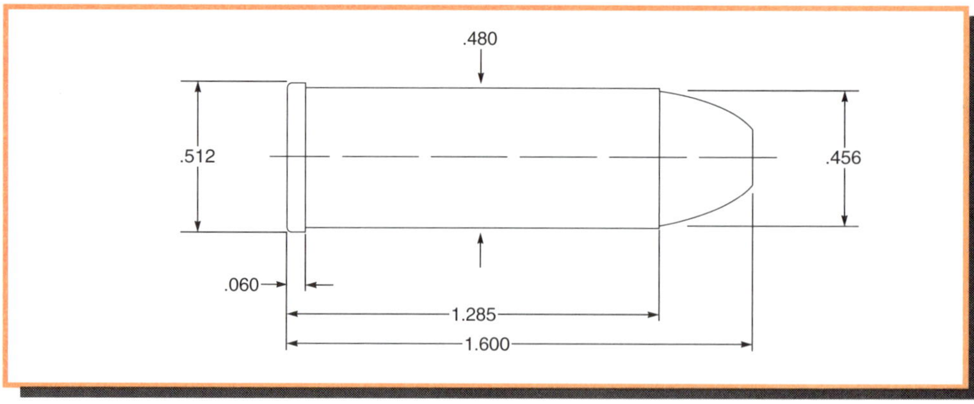

Comments:

The year 1873 was a milestone in development of metallic cartridges. 1873 marked the introduction of the 44-40 (44 WCF), the 45-70 Government, and the 45 Colt. All three cartridges are still going strong 130 years after their debut. Widespread use in Cowboy Action shooting is also contributing to their current popularity. Colt introduced their 45 cartridge in the famed Model 1873 Single Action Army. The original loading propelled a 255-grain lead bullet with 40-grains of black powder around 900 feet per second. The 45 Colt is also known—mistakenly–as the 45 Long Colt. In contrast to several other offerings from Colt, i.e. 32 Short/Long, 38 Short/Long, 41 Short/Long, there never was a 45 Short Colt.

Several caveats apply when reloading the 45 Colt. Shooters must pay particular attention in regard to the cartridge's small diameter rim. This rim provides less bearing surface to the shell holder in comparison to other cartridges and tears easily if undue force is applied during the reloading process. The grooves of Pre-World War II revolvers normally measure .454" in diameter. Later production revolvers and lever action rifles are built with .451" diameter grooves. Those loading for older revolvers with the original .454" groove should stick with cast bullet #454190. This is our original bullet for the 45 Colt and dates to the nineteenth century. If in doubt, slug the bore. Cast bullet #452424 is an Elmer Keith design and has long been an accurate performer. Bullet #452664 is designed for Cowboy Action shooters. It will function through lever action rifles and produced excellent accuracy with Hodgdon's Titegroup. Unique and 231 are both long time favorites for loading the 45 Colt.

Test Components:

Cases . . . . .Winchester
Trim-to Length . . . . .1.275"
Primers . . . . .Winchester WLP
Primer Size . . . . .Large Pistol
Lyman Shell Holder . . . . .No. 11
Jacketed Bullets Used . . . . .Sierra JHP #8800, 185 gr.
Speer GDHP #4478, 200 gr.
Speer JHP #4479, 225 gr.
Sierra JHC #8820, 240 gr.
Hornady HP/XTP #45200, 250 gr.
Cast Bullets Used . . . .(sized to .452" dia)
#454190, 250 gr.
#452664, 250 gr.
#452424, 255 gr.

Test Specifications: (Velocity Only)

Firearm Used . . . . .Ruger Blackhawk
Barrel Length . . . . .7½"
Twist . . . . .1-16"
Groove Dia. . . . . .451"

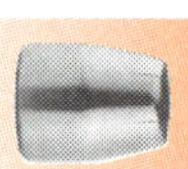

185 gr. Jacketed HP
1.520" OAL — **BC:** .110 **SD:** .130

Powder	Sugg Starting Grains	Velocity fps	Pressure C.U.P.	Max Load Grains	Velocity fps	Pressure C.U.P.
700X	6.0	658	—	7.0	884	—
N320	8.3	873	—	9.3	1123	—
231	7.6	866	—	8.5	1006	—
Unique	7.5	615	—	**10.0**	**934**	—
AA#5	10.8	876	—	12.0	946	—
Blue Dot	12.0	692	—	13.5	826	—
HS-7	13.0	700	—	14.5	908	—
IMR-4227	16.5	711	—	20.0	875	—
XMP-5744	18.4	842	—	20.5	978	—

200 gr. Jacketed HP
1.557" OAL — **BC:** .138 **SD:** .140

Powder	Sugg Starting Grains	Velocity fps	Pressure C.U.P.	Max Load Grains	Velocity fps	Pressure C.U.P.
N320	7.7	843	—	8.6	1031	—
Unique	**8.5**	**847**	—	9.5	934	—
SR-7625	7.0	581	—	9.5	894	—
AA#5	10.3	820	—	11.5	913	—
SR-4756	9.5	679	—	11.8	950	—
HS-6	10.3	740	—	13.3	993	—
Blue Dot	11.6	560	—	12.8	835	—

Note: Loads shown in shaded panels are maximum.
Loads shown in bold designate potentially most accurate load.

45 Colt

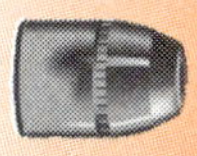

225 gr. Jacketed HP
1.575" OAL

BC: .169
SD: .158

Powder	Sugg Starting Grains	Velocity fps	Pressure C.U.P.	Max Load Grains	Velocity fps	Pressure C.U.P.
N320	7.2	764	—	8.0	862	—
231	7.4	780	—	8.3	954	—
Unique	8.0	772	—	9.0	969	—
SR-7625	8.4	777	—	9.3	899	—
Power Pistol	**8.5**	**837**	—	9.5	974	—
N340	9.0	901	—	10.0	1078	—
AA#5	10.0	787	—	11.2	908	—
HS-6	10.7	737	—	11.9	874	—
Blue Dot	11.7	756	—	13.0	888	—

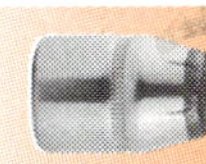

240 gr. Jacketed HC
1.575" OAL

BC: .170
SD: .168

Powder	Sugg Starting Grains	Velocity fps	Pressure C.U.P.	Max Load Grains	Velocity fps	Pressure C.U.P.
Unique	8.0	827	—	8.8	917	—
Power Pistol	**8.2**	**861**	—	9.2	957	—
N340	8.8	870	—	9.8	982	—
AA#5	9.3	771	—	10.4	895	—
HS-6	10.0	705	—	11.0	875	—
2400	15.0	765	—	17.0	997	—
XMP-5744	16.5	790	—	18.5	873	—

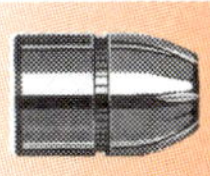

250 gr. Jacketed HP
1.590" OAL

BC: .146
SD: .175

Powder	Sugg Starting Grains	Velocity fps	Pressure C.U.P.	Max Load Grains	Velocity fps	Pressure C.U.P.
Titegroup	5.6	653	—	6.3	803	—
231	6.5	681	—	7.3	813	—
Unique	7.8	755	—	8.7	877	—
Power Pistol	8.0	754	—	9.0	893	—
N340	8.5	850	—	9.6	969	—
AA#5	**10.0**	**777**	—	11.0	925	—
2400	14.8	755	—	16.5	904	—
SR-4759	15.5	742	—	17.5	940	—
XMP-5744	16.0	725	—	18.0	822	—

#454190
250 gr. (#2 Alloy) 1.600" OAL

BC: .269
SD: .175

Powder	Sugg Starting Grains	Velocity fps	Pressure C.U.P.	Max Load Grains	Velocity fps	Pressure C.U.P.
700X	4.5	535	—	6.0	800	—
Titegroup	5.6	803	—	6.2	876	—
Red Dot	4.5	535	—	6.5	835	—
N320	6.0	685	—	6.8	822	—
Green Dot	4.5	505	—	7.0	835	—
231	6.5	789	—	7.4	894	—
PB	5.0	545	—	7.5	820	—
SR-7625	6.0	555	—	8.5	890	—
Unique	6.0	595	—	**9.0**	**875**	—
AA#5	9.5	703	—	10.5	885	—
HS-6	9.5	748	—	10.5	851	—

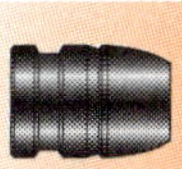

#452664
250 gr. (#2 Alloy) 1.570" OAL

BC: .150
SD: .175

Powder	Sugg Starting Grains	Velocity fps	Pressure C.U.P.	Max Load Grains	Velocity fps	Pressure C.U.P.
700X	5.4	744	—	6.0	809	—
Titegroup	**5.6**	**814**	—	6.2	881	—
Red Dot	5.8	752	—	6.5	836	—
N320	6.0	754	—	6.8	833	—
Green Dot	6.3	750	—	7.0	837	—
231	6.5	796	—	7.4	902	—
PB	6.7	730	—	7.5	822	—
SR-7625	7.0	721	—	8.5	897	—
Unique	6.3	754	—	8.2	931	—
AA#5	9.5	803	—	10.5	886	—
HS-6	8.5	640	—	10.5	859	—

#452424
255 gr. (#2 Alloy) 1.575" OAL

BC: .210
SD: .178

Powder	Sugg Starting Grains	Velocity fps	Pressure C.U.P.	Max Load Grains	Velocity fps	Pressure C.U.P.
700X	4.5	535	—	6.0	785	—
Titegroup	5.6	816	—	6.2	896	—
Red Dot	4.5	550	—	6.0	780	—
N320	5.8	692	—	6.7	841	—
Green Dot	4.5	500	—	6.5	765	—
231	6.5	797	—	7.2	879	—
PB	5.0	530	—	7.0	750	—
SR-7625	6.0	555	—	8.0	835	—
Unique	**6.0**	**590**	—	8.5	845	—
AA#5	9.3	737	—	10.4	845	—
HS-6	8.8	676	—	10.3	838	—

Note: Loads shown in shaded panels are maximum.
Loads shown in bold designate potentially most accurate load.

454 Casull

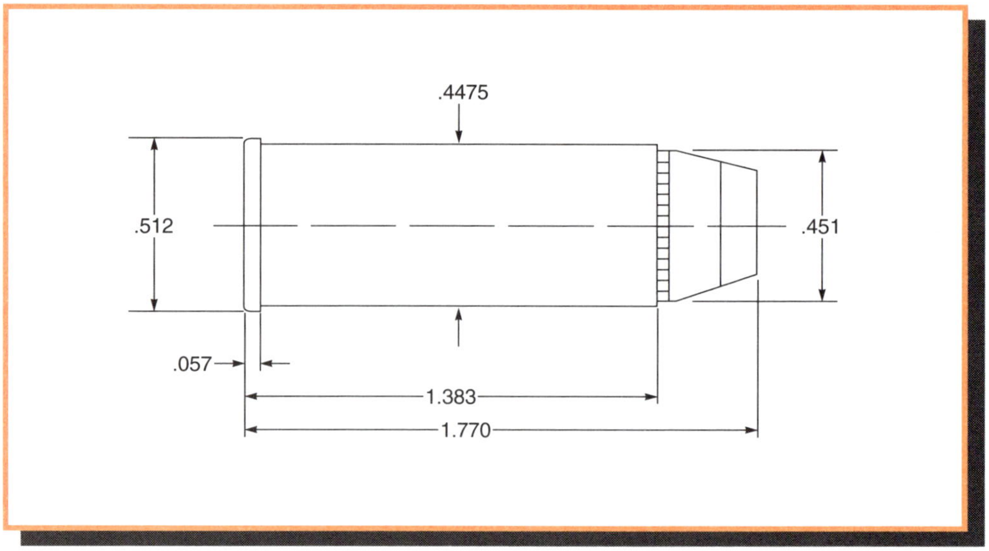

Comments:

The 454 Casull is a very powerful cartridge designed by Dick Casull for handgun hunting. in fact, it is one of, if not the most, powerful production built revolver/cartridge combinations available. The 454 Casull is based on a lengthened and thickened 45 Colt case. It uses small rifle primers, preferably magnum primers, to ignite the heavy charges of slow-burning powder.

The loads shown in this manual should only be used in Casull cases and fired in Casull pistols.

Cast bullets should be cast in linotype or a similar strength alloy and sized to .451". Larger bullets will not allow the cartridge to be used due to the tight tolerances found in the chambers.

IMR-4227 and AA1680 will give very good results with this cartridge.

Test Components:

Cases . . . . .Freedom Arms
Trim-to Length . . . . .1.378"
Primers . . . . .CCI 450
Primer Size . . . . .Small Rifle, Magnum
Lyman Shell Holder . . . . .No. 11
Jacketed Bullets Used . . . . . .Speer HP #4479, 225 gr.
Hornady HP/XTP #45200, 250 gr.
Hornady HP/XTP #45230, 300 gr.
Cast Bullets Used . . . .(sized to .451" dia)
*gas check bullet *#452490, 255 gr.
*#452651, 325 gr.

Test Specifications:
(Velocity & Pressure)

Firearm Used . . . . .Universal Receiver
Barrel Length . . . . .7½"
Twist . . . . .1-24"
Groove Dia. . . . . .451"

225 gr. Jacketed HP
1.690" OAL
BC: .169
SD: .158

Powder	Sugg Starting Grains	Velocity fps	Pressure C.U.P.	Max Load Grains	Velocity fps	Pressure C.U.P.
Unique	12.7	1348	20,100	15.2	1513	33,900
2400	25.7	1602	34,200	29.5	1877	31,900
AA#9	27.0	1634	25,500	31.0	1890	37,700
IMR-4227	29.7	1550	25,300	**33.5+**	**1823**	**42,000**

250 gr. Jacketed HP
1.690" OAL
BC: .146
SD: .175

Powder	Sugg Starting Grains	Velocity fps	Pressure C.U.P.	Max Load Grains	Velocity fps	Pressure C.U.P.
Unique	11.8	1260	22,500	13.5	1420	37,100
2400	21.8	1490	28,500	26.5	1709	40,400
AA#9	24.6	1523	26,800	29.0	1769	39,900
IMR-4227	26.2	1381	23,300	**30.5+**	**1666**	**39,800**

Note: Loads shown in shaded panels are maximum.
Loads shown in bold designate potentially most accurate load.
\+ Designates a compressed powder charge.

300 gr. Jacketed HP
1.760" OAL

BC: .180
SD: .210

Powder	Sugg Starting Grains	Velocity fps	Pressure C.U.P.	Max Load Grains	Velocity fps	Pressure C.U.P.
Unique	10.5	1049	19,300	13.0	1267	34,900
2400	20.0	1146	23,000	24.5	1384	39,800
AA#9	22.0	1325	23,300	25.8	1556	37,000
IMR-4227	24.2	1275	25,000	**28.2**	**1530**	**41,300**

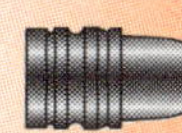

#452490
255 gr. (Linotype) 1.760" OAL

BC: .160
SD: .179

Powder	Sugg Starting Grains	Velocity fps	Pressure C.U.P.	Max Load Grains	Velocity fps	Pressure C.U.P.
Bullseye	9.7	1176	20,500	13.0	1395	38,000
Unique	11.6	1237	20,000	14.2	1424	36,400
2400	21.7	1441	25,300	27.0	1747	41,700
AA#9	24.4	1547	28,500	28.0	1723	38,100
IMR-4227	25.8	1393	23,700	**30.0**	**1663**	**40,200**

#452651
325 gr. (Linotype) 1.755" OAL

BC: .191
SD: .228

Powder	Sugg Starting Grains	Velocity fps	Pressure C.U.P.	Max Load Grains	Velocity fps	Pressure C.U.P.
Unique	9.5	1015	23,400	11.5	1161	35,800
2400	17.0	1161	22,500	20.0	1354	34,800
AA#9	18.5	1166	20,500	22.0	1378	36,800
IMR-4227	20.0	1084	22,200	24.0	1334	38,600
AA-1680	24.0	1092	19,800	**28.0**	**1324**	**37,400**

Note: Loads shown in shaded panels are maximum.
Loads shown in bold designate potentially most accurate load.

480 Ruger

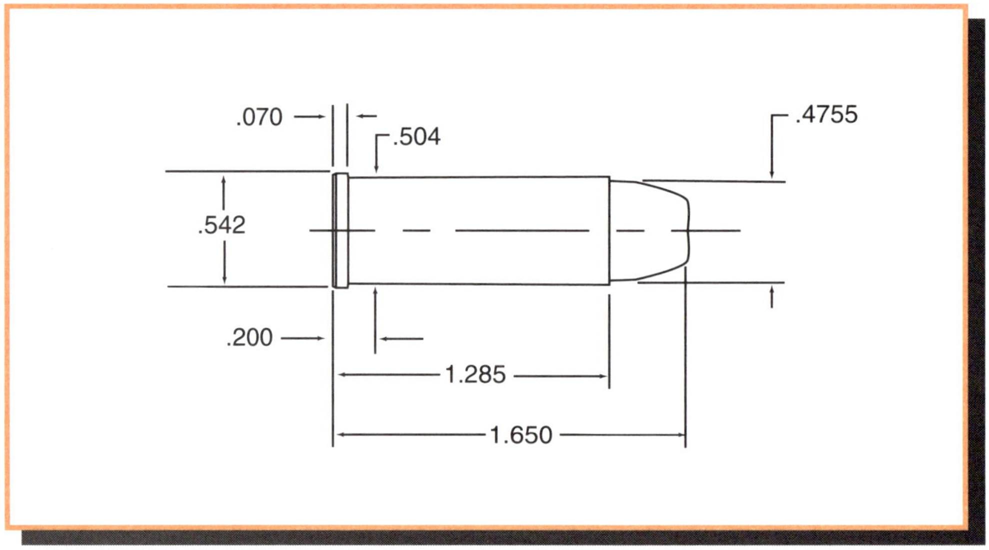

Comments:

The 480 Ruger is the result of a joint effort between Hornady and Sturm, Ruger & Co, to develop a cartridge filling the ballistic niche between the 44 Remington Magnum and the 475 Linebaugh. Like the Linebaugh, the 480 case is also based on the 45-70 featuring a smaller rim. Case length of the 480 measures .120" shorter than the Linebaugh and is best described as a "Special" version of the 475 much like the 44 Special's relationship to the 44 Magnum or the 38 Special to the 357. The 480s lower maximum average pressure of 48,000 PSI also allow revolvers to chamber six rounds, instead of the five normally found with the higher 55,000 PSI pressure level of the 475. Loading for the big Ruger is no different than other straight-walled pistol cartridges. Not surprisingly, the cartridge worked best with the slower pistol powders one would normally associate with magnum pistol cartridges. The Lyman technical staff test-fired a Ruger Super Redhawk and none found it particularly unpleasant to shoot. Its 9 1/2 inch barrel and 3 lb. 10 oz weight no doubt contributed to its shooting characteristics.

Test Components:

Cases . . . . .Hornady
Trim-to Length . . . . .1.275"
Primers . . . . .Winchester WLP
Primer Size . . . . .Large Pistol
Lyman Shell Holder . . . . .No. 17
Jacketed Bullets Used . . . .
Hornady HP/XTP #47500, 325 gr.
Hornady HP/XTP #47550, 400 gr.

Test Specifications:
(Velocity & Pressure)

Firearm Used . . . . .Universal Receiver
Barrel Length . . . . .7½"
Twist . . . . .1-18"
Groove Dia. . . . . .475"

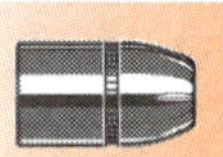

325 gr. Jacketed HP
1.620" OAL
BC: .149
SD: .206

Powder	Sugg Starting Grains	Velocity fps	Pressure P.S.I.	Max Load Grains	Velocity fps	Pressure P.S.I.
2400	20.5	1269	35,800	23.1	1432	46,200
N110	21.0	1321	35,600	23.7	1434	47,200
AA#9	22.0	1350	36,600	24.5	1500	45,900
H110	26.1	1448	40,500	27.0	1506	44,300
296	27.1	1503	43,900	28.0	1550	46,800
IMR-4227	24.0	1216	35,000	26.8	1378	46,200
H-4227	24.5	1312	37,100	**27.2**	**1453**	**46,100**
Lil'Gun	23.0	1388	38,800	25.3	1510	45,800

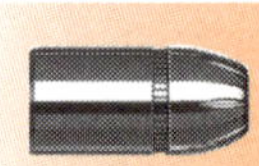

400 gr. Jacketed HP
1.620" OAL
BC: .182
SD: .253

Powder	Sugg Starting Grains	Velocity fps	Pressure P.S.I.	Max Load Grains	Velocity fps	Pressure P.S.I.
2400	15.5	1036	36,200	17.5	1175	46,400
N110	16.5	1072	33,900	18.3	1184	45,600
AA#9	17.3	1147	34,800	**19.3**	**1247**	**46,700**
H110	20.1	1204	40,500	20.8	1255	44,900
296	20.8	1211	40,500	21.5	1261	45,700
IMR-4227	19.0	1005	37,900	21.0	1142	46,800
H-4227	19.0	1070	36,500	21.0	1186	44,400
AA1680	21.5	937	34,600	24.0+	1045	43,000

Note: Loads shown in shaded panels are maximum.
Loads shown in bold designate potentially most accurate load.
\+ Designates a compressed powder charge.

50 Action Express (50 AE)

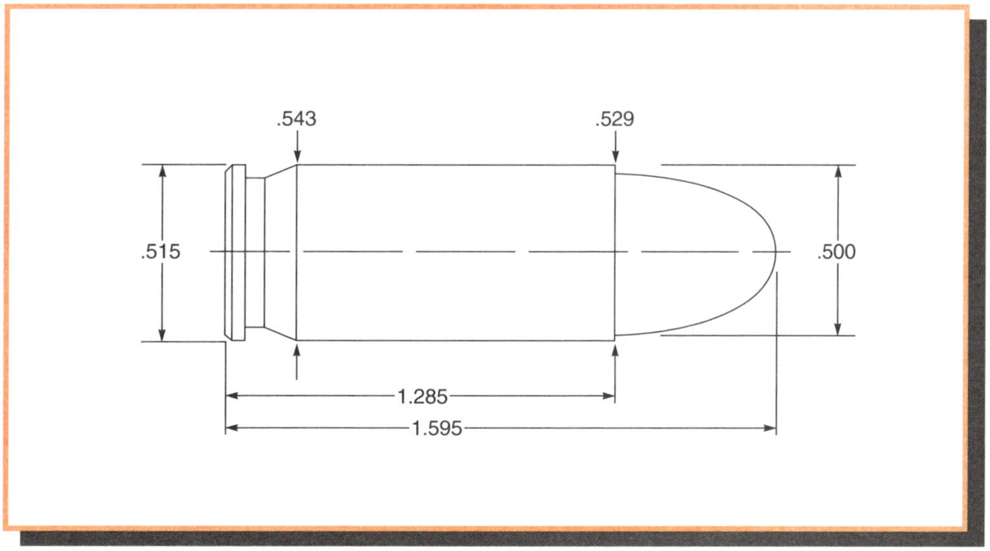

Comments:

The 50 Action Express is a very powerful cartridge designed to be fired in the semi-automatic Desert Eagle pistol. The 50 AE case uses a rebated rim of the same diameter as the 44 Magnum, and has a tapered body. The 50 AE fires 325 gr. .500" diameter bullets at nearly 1400 fps, making this an excellent hunting handgun.

Recoil is heavy, but due to the weight of the handgun and its gas-operating system, recoil is controllable.

Slow-burning powders are required to generate enough gas volume to properly cycle the gas-operated action. H110, IMR-4227 and AA1680 all gave us very consistent results during our testing.

Test Components:

Cases . . . . .Speer
Trim-to Length . . . . .1.280"
Primers . . . . .CCI 350
Primer Size . . . . .Large Pistol, Magnum
Lyman Shell Holder . . . . .No. 7
Jacketed Bullets Used . . . . . .Speer HP #4495, 325 gr.

Test Specifications:
(Velocity Only)

Firearm Used . . . . .Desert Eagle
Barrel Length . . . . .6"
Twist . . . . .1-19"
Groove Dia. . . . . .501"

325 gr. Jacketed HP
1.575" OAL

BC: .149
SD: .186

Powder	Sugg Starting Grains	Velocity fps	Pressure	Max Load Grains	Velocity fps	Pressure
2400	22.0	1037	—	23.4	1202	—
AA#9	22.5	967	—	23.8	1077	—
IMR-4227	29.5	1223	—	**31.0**	**1323**	—
H-110	31.2	1300	—	32.5	1372	—
AA-1680	34.0	1092	—	37.5+	1289	—

Note: Loads shown in shaded panels are maximum.
Loads shown in bold designate potentially most accurate load.
\+ Designates a compressed powder charge.

T/C Contender & Encore

Thompson/Center Contender and Encore Data

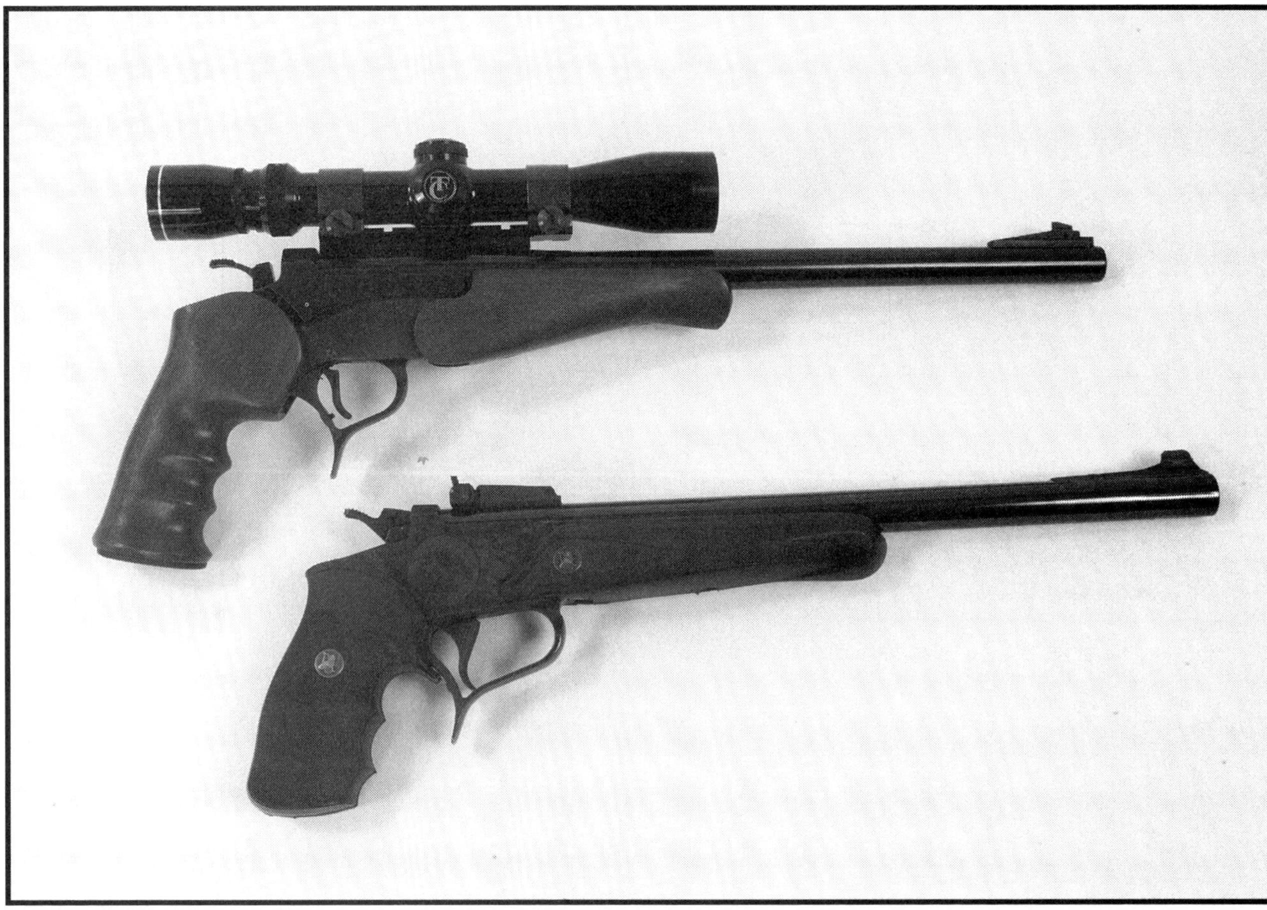

Introduction

This data was developed for use in the T/C Contender and Encore handguns. It can also be used in other single shot or bolt-action handguns such as the Remington XP-100 and Savage Striker. Do not use these loads in any other type of handgun.

Single shot and/or bolt-action handguns have become extremely popular with handgun hunters and silhouette shooters. They are available in a wide variety of calibers, many of which are best known as rifle cartridges. These range from the 22 Hornet up to the 45-70 Government. We have included reduced loads for several of the cartridges listed if a lower velocity and recoil load is desired.

Maximum overall lengths as suggested by SAAMI can be exceeded in the Contender or Encore. We recommend using the exact cartridge overall length shown for each specific bullet in this data section.

The data in this section cannot be used without a full understanding of what it represents or without complete reloading knowledge. Any attempts to use this data without the necessary cautions, procedures and background knowledge could be extremely dangerous, causing personal and property injury, even death.

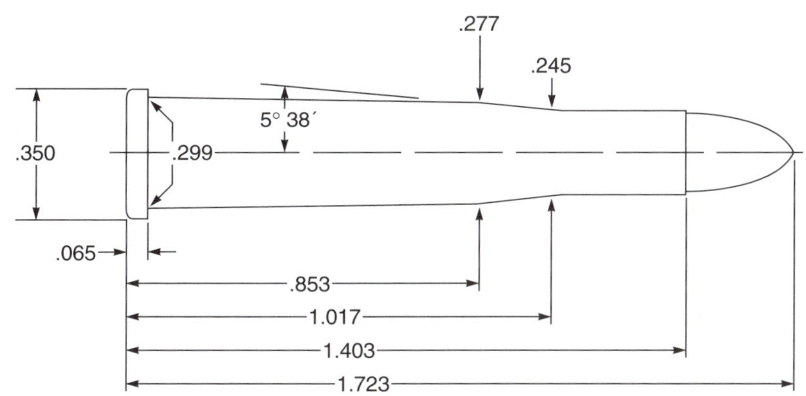

Comments:

The Hornet ia a fine short range varmint cartridge for the Contender and noise level is kept to a minimum. Trajectory is quite flat for 100 yard shooting.

IMR 4227 is the first choice for accuracy. Do not forget to use thin jacketed bullets expressly designed for the Hornet.

Test Components:

Cases . . . . .Winchester
Trim-to Length . . . . .1.393"
Primers . . . . .Winchester WSR
Primer Size . . . . .Small Rifle
Lyman Shell Holder . . . . .No. 4
Jacketed Bullets Used . . . . . .Sierra JSP #1200, 40 gr.
Hornady JSP #2230, 45 gr.
Hornady V-Max #22261, 50 gr.
Hornady A-Max #22492, 52 gr.
Cast Bullets Used . . . .(sized to .224" dia)
*gas check bullet *#225438, 44 gr.
*#225415, 55 gr.

Test Specifications:
(Velocity Only)

Firearm Used . . . . .Thompson/Center Contender
Barrel Length . . . . .10"
Twist . . . . .1-14"
Groove Dia. . . . . .224"

40 gr. Jacketed SP
1.722" OAL

BC: .122
SD: .114

Powder	Sugg Starting Grains	Velocity fps	Pressure	Max Load Grains	Velocity fps	Pressure
2400	7.2	1953	—	8.0	2150	—
N110	8.2	2077	—	9.2	2239	—
H-110	10.5	2230	—	11.0	2265	—
IMR-4227	10.0	2105	—	**11.7+**	**2348**	—
AA-1680	12.6	2118	—	14.0+	2285	—

45 gr. Jacketed SP
1.722" OAL

BC: .202
SD: .128

Powder	Sugg Starting Grains	Velocity fps	Pressure	Max Load Grains	Velocity fps	Pressure
2400	7.0	1920	—	8.3	2238	—
N110	7.9	2005	—	8.8	2107	—
H-110	9.8	2280	—	10.2	2348	—
IMR-4227	10.0	2062	—	**11.3**	**2280**	—
AA-1680	11.2	2002	—	12.5	2145	—

Note: Loads shown in shaded panels are maximum.
Loads shown in bold designate potentially most accurate load.
\+ Designates a compressed powder charge.

22 Hornet

50 gr. Jacketed V-Max
1.723" OAL

BC: .242
SD: .142

Powder	Sugg Starting Grains	Velocity fps	Pressure	Max Load Grains	Velocity fps	Pressure
2400	6.3	1695	—	7.0	1854	—
N110	**7.3**	**1840**	—	8.2	1963	—
H-110	9.0	2107	—	9.5	2187	—
IMR-4227	9.9	2128	—	11.0+	2236	—
AA-1680	9.9	1833	—	11.0+	2005	—

52 gr. Jacketed A-Max
1.800" OAL

BC: .247
SD: .148

Powder	Sugg Starting Grains	Velocity fps	Pressure	Max Load Grains	Velocity fps	Pressure
2400	6.1	1607	—	6.8	1771	—
N110	7.2	1790	—	8.0	1899	—
H-110	8.5	2055	—	9.0	2133	—
IMR-4227	9.5	1922	—	**10.5**	**2077**	—
AA-1680	9.7	1792	—	10.8	1944	—

#225438
44 gr. (#2 Alloy) 1.675" OAL

BC: .094
SD: .125

Powder	Sugg Starting Grains	Velocity fps	Pressure	Max Load Grains	Velocity fps	Pressure
Unique	4.2	1715	—	4.7	1836	—
2400	5.4	1594	—	6.0	1693	—
N110	6.3	1668	—	7.0	1847	—
IMR-4227	7.2	1732	—	**8.0**	**1890**	—
AA-1680	9.4	1909	—	10.5	1874	—

#225415
55 gr. (#2 Alloy) 1.694" OAL

BC: .162
SD: .157

Powder	Sugg Starting Grains	Velocity fps	Pressure	Max Load Grains	Velocity fps	Pressure
700X	2.0	1084	—	3.7	1558	—
Green Dot	**2.5**	**1195**	—	4.2	1635	—
Unique	3.0	1281	—	4.6	1673	—
SR-7625	3.0	1138	—	4.4	1601	—
SR-4759	5.6	1403	—	7.5	1848	—

Note: Loads shown in shaded panels are maximum.
Loads shown in bold designate potentially most accurate load.
\+ Designates a compressed powder charge.

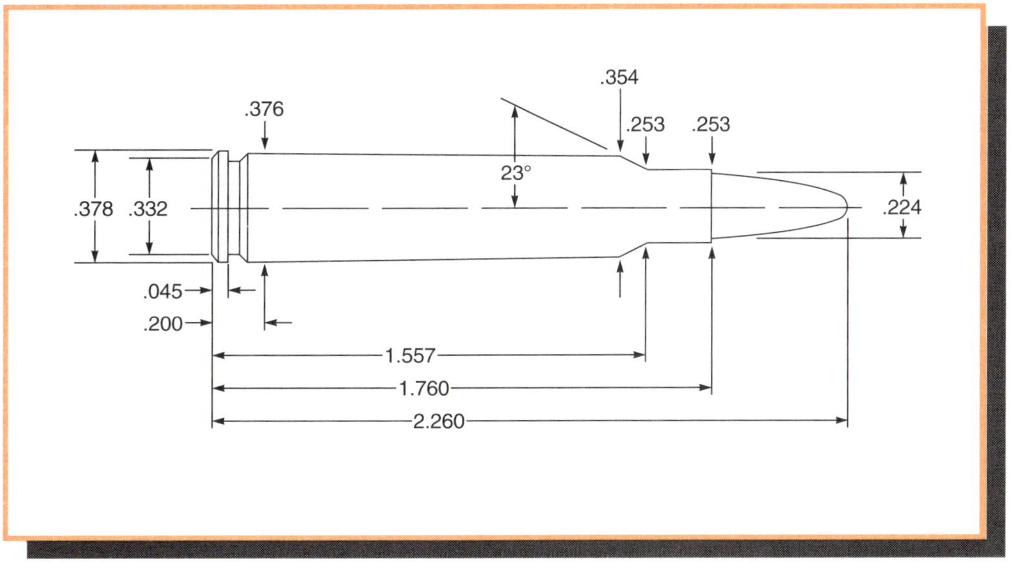

Comments:

The 223 Remington has seen extensive use with heavy bullets in recent years in conjunction with faster twist barrels. However, the 1-14" twist of the T/C Contender requires use of lighter weight varmint style bullets for peak performance. Both the 40 and 50-grain jacketed bullets used showed a four to 500 foot per second reduction in velocity compared to a 24-inch barrel. Both listed bullets will ably perform at these velocities due to their construction. IMR-3031 has usually worked well for jacketed bullets in the 223. XMP-5744 produced good accuracy with both cast bullets. Approach all maximum loads with caution and back off powder charges at any sign of high pressure.

Test Components:

Cases . . . . .Remington
Trim-to Length . . . . .1.750"
Primers . . . . .Remington 7½
Primer Size . . . . .Small Rifle
Lyman Shell Holder . . . . .No. 26
Jacketed Bullets Used . . .Hornady V-Max #22241, 40 gr.
Sierra Blitz #1340, 50 gr.
Cast Bullets Used . . . .(sized to .224" dia)
*gas check bullet *#225415, 55 gr.
*#225646, 55 gr.

Test Specifications: (Velocity Only)

Firearm Used . . . . .Thompson/Center Contender
Barrel Length . . . . .14"
Twist . . . . .1-14"
Groove Dia. . . . . .224"

40 gr. Jacketed V-Max 2.215" OAL — **BC:** .200 **SD:** .114

Powder	Sugg Starting Grains	Velocity fps	Pressure	Max Load Grains	Velocity fps	Pressure
IMR-4198	20.5	2892	—	22.9	3143	—
XMR-2015	24.5	2882	—	27.0+	3171	—
RX7	21.5	2848	—	24.0	3151	—
IMR-3031	23.5	2480	—	26.0+	2931	—
H-322	24.3	2780	—	**27.0**	**3141**	—
AA-2230	24.3	2808	—	27.0	3116	—

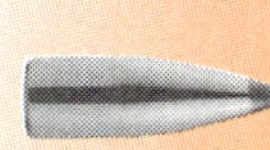

50 gr. Jacketed SP 2.235" OAL — **BC:** .222 **SD:** .142

Powder	Sugg Starting Grains	Velocity fps	Pressure	Max Load Grains	Velocity fps	Pressure
IMR-4198	**20.0**	**2694**	—	22.0	2868	—
XMR-2015	23.0	2542	—	25.0	2884	—
RX7	21.5	2721	—	24.0	3043	—
IMR-3031	22.0	2229	—	25.5+	2798	—
Benchmark	23.5	2529	—	26.5	2930	—
AA-2230	23.5	2654	—	26.5	3098	—

223 Remington

#225415
55 gr. (#2 Alloy) 2.060" OAL

BC: .162
SD: .157

Powder	Sugg Starting Grains	Velocity fps	Pressure	Max Load Grains	Velocity fps	Pressure
Unique	6.0	1703	—	**8.0**	**2029**	—
SR-4759	11.0	1834	—	13.0	2094	—
IMR-4227	10.5	1745	—	12.5	2086	—
XMP-5744	11.0	1582	—	14.5	2011	—
748	21.0	1839	—	23.0	2080	—

#225646
55 gr. (#2 Alloy) 2.260" OAL

BC: .155
SD: .157

Powder	Sugg Starting Grains	Velocity fps	Pressure	Max Load Grains	Velocity fps	Pressure
Unique	**6.0**	**1605**	—	9.0	2068	—
SR-7625	6.7	1639	—	9.5	2049	—
AA#9	7.8	1534	—	11.0	1997	—
2400	8.0	1524	—	11.5	2003	—
SR-4759	9.5	1527	—	11.5	1971	—
IMR-4227	10.0	1595	—	13.0	2037	—
XMP-5744	10.0	1552	—	14.0	1935	—
AA-1680	11.0	1629	—	14.3	2065	—

Note: Loads shown in shaded panels are maximum.
Loads shown in bold designate potentially most accurate load.
\+ Designates a compressed powder charge.

22-250 Remington

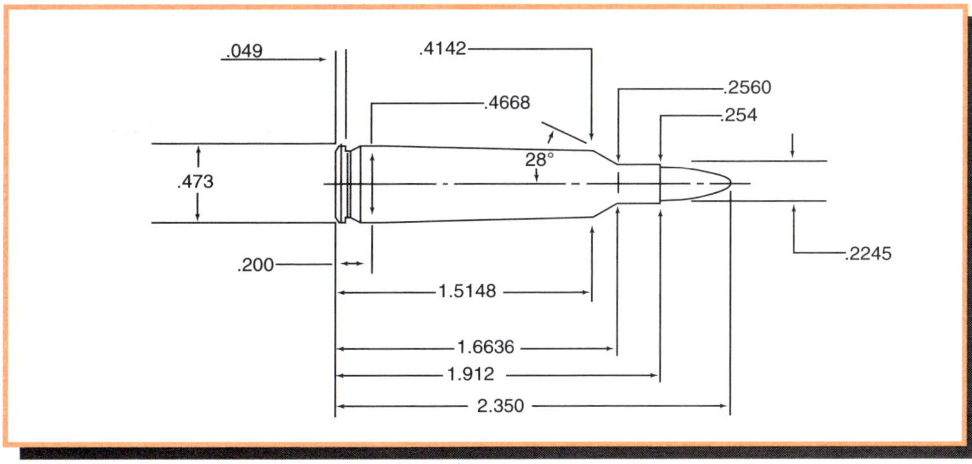

Comments:

The 22-250 Remington has been a very popular varmint round for some time. The full power jacketed loads fired in the 15-inch barrel of the Thompson/Center Encore produce around 400 feet per second less than with a full-length 24-inch rifle barrel. These velocity levels with the Encore are still about the same as they would be if fired in a 223 Remington with a standard 24-inch barrel with equivalent weight bullets. Lubing the nose grooves on cast bullet #225646 is optional. Approach all maximum loads with caution and back off powder charges at any sign of high pressure

Test Components:

Cases . . . . .Winchester
Trim-to Length . . . . .1.902"
Primers . . . . .Winchester WLR
Primer Size . . . . .Large Rifle
Lyman Shell Holder . . . . .No. 2
Jacketed Bullets Used . . . . . .Sierra SPT #1310, 45 gr.
Hornady V-Max #22271, 55 gr.
Cast Bullets Used . . . .(sized to .224" dia)
*gas check bullet *#225646, 55 gr.

Test Specifications: (Velocity Only)

Firearm Used . . . . .Thompson/Center Encore
Barrel Length . . . . .15"
Twist . . . . .1-12"
Groove Dia. . . . . .224"

45 gr. Jacketed SPT
2.310" OAL
BC: .210
SD: .128

Powder	Sugg Starting Grains	Velocity fps	Pressure	Max Load Grains	Velocity fps	Pressure
IMR-3031	31.0	2925	—	35.5	3350	—
AA-2230	33.3	3469	—	36.5	3674	—
IMR-4895	**32.5**	**2859**	—	36.5	3331	—
AA-2460	34.0	3238	—	37.2	3472	—
Varget	36.0	3071	—	39.5	3551	—
RX15	33.5	2949	—	37.0	3279	—
*SR-4759	15.0	2138	—	19.5	2775	—
*XMP-5744	18.0	2133	—	23.5	2646	—

55 gr. Jacketed V-Max
2.345" OAL
BC: .255
SD: .157

Powder	Sugg Starting Grains	Velocity fps	Pressure	Max Load Grains	Velocity fps	Pressure
IMR-4895	32.0	2869	—	35.5	3156	—
AA-2460	30.0	2741	—	34.8	3142	—
IMR-4064	**32.0**	**2682**	—	35.5	3172	—
Varget	34.0	2912	—	37.0	3263	—
RX15	31.7	2649	—	35.5	3040	—
N150	32.7	2775	—	36.5	3173	—
**SR-4759	16.0	2261	—	20.0	2702	—
**XMP-5744	18.5	2148	—	24.5	2721	—

#225646
55 gr. (#2 Alloy) 2.350" OAL
BC: .155
SD: .157

Powder	Sugg Starting Grains	Velocity fps	Pressure	Max Load Grains	Velocity fps	Pressure
Red Dot	7.7	1790	—	10.4	2105	—
700X	8.0	1851	—	11.0	2236	—
PB	9.0	1828	—	11.0	2104	—
SR-7625	**9.2**	**1825**	—	11.7	2121	—
SR-4756	10.2	1925	—	12.6	2158	—
SR-4759	16.5	2241	—	18.5	2525	—
IMR-4227	16.5	2073	—	21.0	2449	—
IMR-4198	18.5	2051	—	24.6	2653	—
XMP-5744	16.0	1921	—	18.0	2126	—

Note: Loads shown in shaded panels are maximum.
Loads shown in bold designate potentially most accurate load.
**Reduced load.

243 Winchester

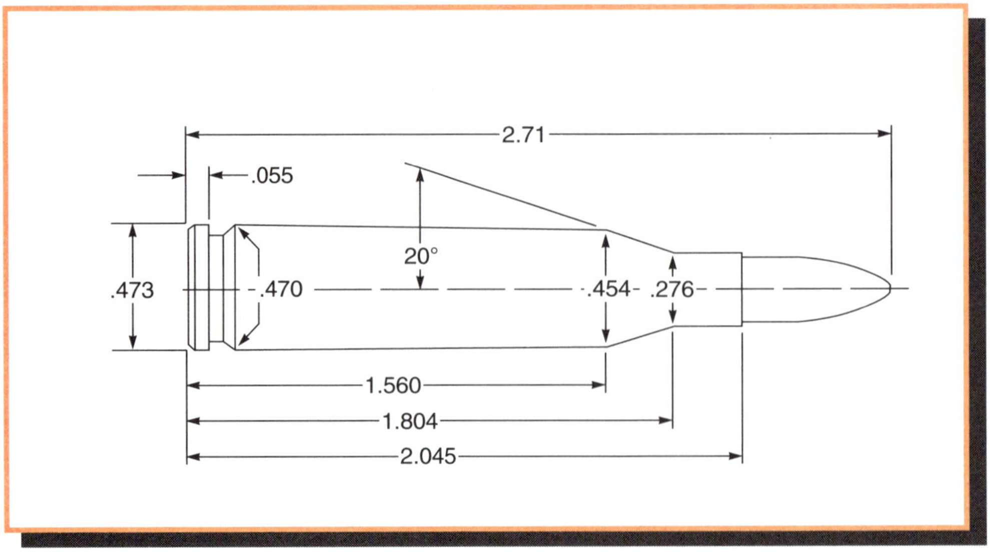

Comments:

The 80-grain jacketed bullet is intended for varmint-sized game. This bullet showed about 400 foot per second difference when compared to a 24-inch barrel. Loads listed for cast bullet #245496 showed particularly good ballistic uniformity.

Approach all maximum loads with caution and back off powder charges at any sign of high pressure.

Test Components:

Cases . . . . .Remington
Trim-to Length . . . . .2.035"
Primers . . . . .Remington 9½
Primer Size . . . . .Large Rifle
Lyman Shell Holder . . . . .No. 2
Jacketed Bullets Used . . . . . . .Speer SP #1211, 80 gr.
Cast Bullets Used . . . .(sized to .243" dia)
*gas check bullet *#245496, 84 gr.

Test Specifications:
(Velocity Only)

Firearm Used . . . . .Thompson/Center Encore
Barrel Length . . . . .15"
Twist . . . . .1-10"
Groove Dia. . . . . .243"

80 gr. Jacketed SP
2.655" OAL
BC: .365
SD: .194

Powder	Sugg Starting Grains	Velocity fps	Pressure	Max Load Grains	Velocity fps	Pressure
IMR-4198	27.5	2531	—	31.5	2769	—
XMR-2015	26.0	2035	—	32.0	2412	—
IMR-3031	33.0	2622	—	37.5	2858	—
IMR-4064	34.0	2433	—	39.5	2804	—
IMR-4320	36.0	2619	—	**40.5**	**2934**	—
**XMP-5744	19.5	1866	—	22.0	2048	—

#245496
84 gr. (#2 Alloy) 2.480" OAL
BC: .202
SD: .203

Powder	Sugg Starting Grains	Velocity fps	Pressure	Max Load Grains	Velocity fps	Pressure
700X	8.5	1613	—	10.5	1802	—
SR-7625	**9.5**	**1541**	—	11.5	1730	—
SR-4756	11.0	1666	—	13.0	1841	—
SR-4759	15.0	1783	—	20.0	2194	—
XMP-5744	13.0	1377	—	17.5	1739	—
IMR-4227	15.5	1670	—	20.0	2015	—
IMR-4198	16.0	1650	—	22.7	2121	—

Note: Loads shown in shaded panels are maximum.
Loads shown in bold designate potentially most accurate load.
**Reduced load.

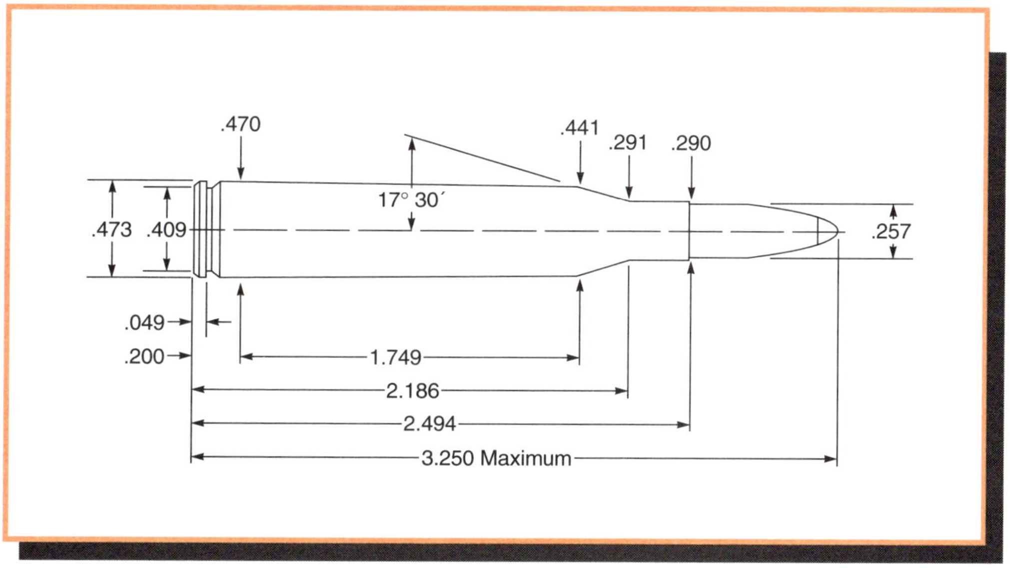

Comments:

The 87-grain jacketed bullet showed approximately a 400 foot per second difference when compared to a full-length rifle barrel. Cast bullet #257420 was originally designed for use in the 25-20 and is suited to the velocity levels shown with the 15-inch barrel of the 25-06 Encore. Approach all maximum loads with caution and back off powder charges at any sign of high pressure.

Test Components:

Cases . . . . .Remington
Trim-to Length . . . . .2.484"
Primers . . . . .Winchester WLR
Primer Size . . . . .Large Rifle
Lyman Shell Holder . . . . .No. 2
Jacketed Bullets Used . . . . .Hornady SP #2530, 87 gr.
Cast Bullets Used . . . .(sized to .257" dia)
*gas check bullet *#257420, 65 gr.

Test Specifications: (Velocity Only)

Firearm Used . . . . .Thompson/Center Encore
Barrel Length . . . . .15"
Twist . . . . .1-10"
Groove Dia. . . . . .257"

87 gr. Jacketed SP
3.096" OAL
BC: .322
SD: .188

Powder	Sugg Starting Grains	Velocity fps	Pressure	Max Load Grains	Velocity fps	Pressure
IMR-3031	42.0	2667	—	47.0	3007	—
IMR-4064	44.0	2617	—	48.5	2883	—
IMR-4320	44.0	2610	—	**49.0**	**3014**	—
RX15	43.0	2560	—	48.0	2875	—
RX19	50.5	2382	—	57.0	2845	—
**SR-4759	17.0	1833	—	19.0	1947	—
**XMP-5744	22.5	1866	—	25.0	2031	—

#257420
65 gr. (#2 Alloy) 2.815" OAL
BC: .129
SD: .140

Powder	Sugg Starting Grains	Velocity fps	Pressure	Max Load Grains	Velocity fps	Pressure
Unique	12.8	1966	—	17.0	2341	—
SR-7625	11.0	1685	—	15.0	2091	—
SR-4756	12.0	1692	—	16.0	2090	—
SR-4759	18.0	1745	—	24.0	2378	—
IMR-4227	19.0	1577	—	25.0	2060	—
XMP-5744	**20.0**	**1759**	—	27.0	2199	—
IMR-4198	19.0	1531	—	26.0	1972	—

Note: Loads shown in shaded panels are maximum.
Loads shown in bold designate potentially most accurate load.
**Designates a reduced load.

270 Winchester

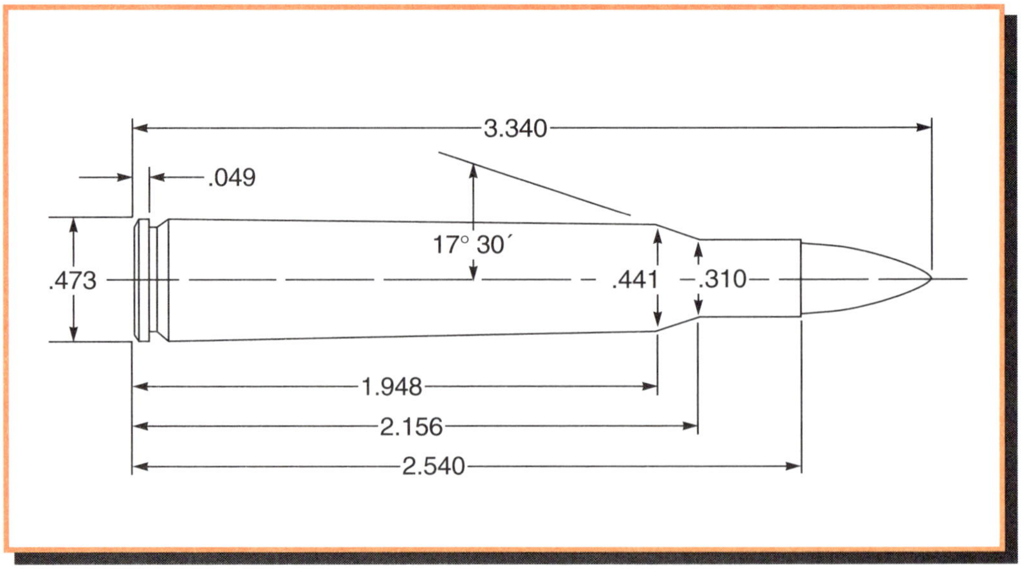

Comments:

Loads for both the 100 and 130-grain bullets showed a nearly 500 foot per second difference when compared to a full-length barrel. Cast bullet # 280642 displayed excellent ballistic uniformity, as did the reduced loads for the 130-grain jacketed bullet. Lube grooves on the nose of the cast bullet can be lubed by hand if the shooter desires. Approach all maximum loads with caution and back off powder charges at any sign of high pressure.

Test Components:

Cases . . . . .Winchester
Trim-to Length . . . . .2.530"
Primers . . . . .Winchester WLR
Primer Size . . . . .Large Rifle
Lyman Shell Holder . . . . .No. 2
Jacketed Bullets Used . . . .Hornady SP #2710, 100 gr.
Sierra SBT #1820, 130 gr.

Cast Bullets Used . . . .(sized to .277" dia)
*gas check bullet *#280642, 150 gr.

Test Specifications:
(Velocity Only)

Firearm Used . . . . .Thompson/Center Encore
Barrel Length . . . . .15"
Twist . . . . .1-10"
Groove Dia. . . . . .277"

100 gr. Jacketed SP
3.175" OAL
BC: .307 **SD:** .186

Powder	Sugg Starting Grains	Velocity fps	Pressure	Max Load Grains	Velocity fps	Pressure
IMR-4895	46.5	2491	—	50.0	2828	—
IMR-4064	46.0	2484	—	51.5	2811	—
Varget	47.0	2517	—	52.0	2851	—
RX15	**48.0**	**2601**	—	53.5	2909	—
**SR-4759	23.0	2037	—	27.5	2317	—
**XMP-5744	27.0	1958	—	31.3	2184	—

#280642
150 gr. (#2 Alloy) 3.225" OAL
BC: .260 **SD:** .279

Powder	Sugg Starting Grains	Velocity fps	Pressure	Max Load Grains	Velocity fps	Pressure
Unique	12.0	1405	—	19.0	1845	—
SR-7625	13.0	1404	—	18.0	1713	—
SR-4756	13.0	1370	—	19.0	1745	—
SR-4759	16.0	1395	—	**23.0**	**1842**	—
XMP-5744	18.0	1346	—	24.5	1721	—
AA-1680	20.0	1428	—	28.0	1841	—
RX7	19.0	1376	—	27.0	1812	—

130 gr. Jacketed SBT
3.250" OAL
BC: .418 **SD:** .242

Powder	Sugg Starting Grains	Velocity fps	Pressure	Max Load Grains	Velocity fps	Pressure
IMR-4895	42.0	2324	—	46.5	2576	—
IMR-4064	44.0	2358	—	**48.0**	**2569**	—
Varget	42.0	2267	—	46.0	2442	—
RX15	44.0	2359	—	47.0	2477	—
**SR-4759	24.0	1943	—	28.5	2219	—
**XMP-5744	27.0	1877	—	31.0	2106	—

Note: Loads shown in shaded panels are maximum.
Loads shown in bold designate potentially most accurate load.
**Designates a reduced load.

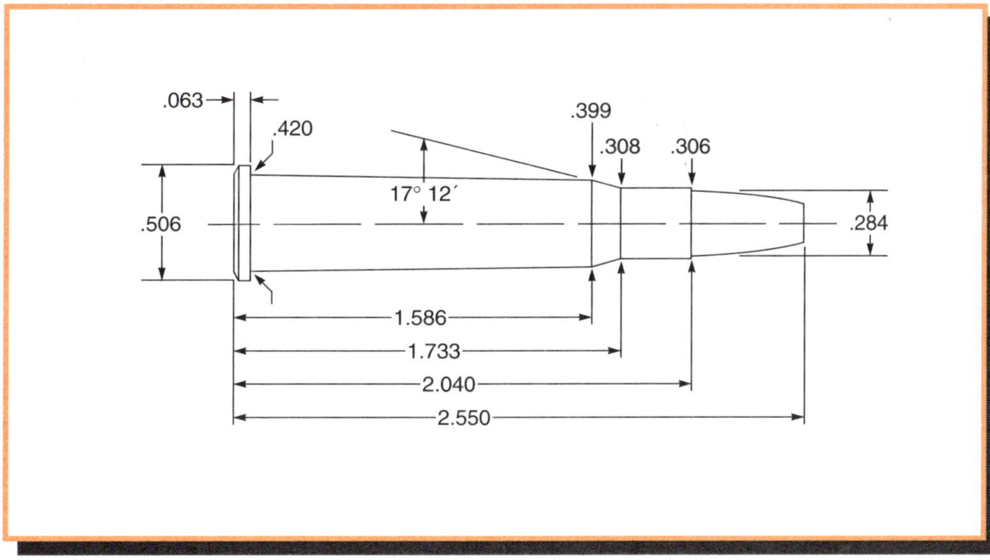

Comments:

This is a modified version of the 30-30 Win. that has gained some popularity, perhaps because of the recent trend toward 7mm cartridges and bullets.

The cartridge is suitable for light big game at modest ranges. The Thompson/Center Contender, being a single shot design, allows the use of Spitzer bullets. Good powder choices are Winchester 748 and IMR 3031.

Test Components:

Cases . . . . .Federal
Trim-to Length . . . . .2.030"
Primers . . . . .CCI 200
Primer Size . . . . .Large Rifle
Lyman Shell Holder . . . . .No. 6
Jacketed Bullets Used . . . . . .Sierra SP #1900, 120 gr.
Sierra SP #7250, 130 gr.
Sierra SBT #1905, 140 gr.
Sierra SBT #1913, 150 gr.
Sierra HPBT #1925, 160 gr.
Cast Bullets Used . . . .(sized to .285" dia)
*gas check bullet *#287346, 135 gr.

Test Specifications:
(Velocity Only)

Firearm Used . . . . .Thompson/Center Contender
Barrel Length . . . . .14"
Twist . . . . .1-9"
Groove Dia. . . . . .284"

120 gr. Jacketed SP 2.550" OAL — **BC:** .326 **SD:** .213

Powder	Sugg Starting Grains	Velocity fps	Pressure	Max Load Grains	Velocity fps	Pressure
IMR-3031	30.5	2063	—	34.5+	2350	—
H-335	31.0	2096	—	34.8	2398	—
AA-2460	29.5	1974	—	33.8	2267	—
BL-C (2)	32.3	1955	—	**36.8**	**2316**	—
IMR-4895	31.8	2068	—	35.8+	2312	—
748	34.0	2073	—	39.5+	2430	—

130 gr. Jacketed SP 2.550" OAL — **BC:** .321 **SD:** .230

Powder	Sugg Starting Grains	Velocity fps	Pressure	Max Load Grains	Velocity fps	Pressure
IMR-3031	29.5	2081	—	33.6+	2254	—
H-335	29.5	1912	—	**33.5**	**2249**	—
AA-2460	28.3	1918	—	32.3	2124	—
BL-C (2)	31.7	1897	—	35.9	2212	—
IMR-4895	30.0	1950	—	34.6+	2201	—
748	31.0	1778	—	35.0+	2171	—

Note: Loads shown in shaded panels are maximum.
Loads shown in bold designate potentially most accurate load.
\+ Designates a compressed powder charge.

7-30 Waters

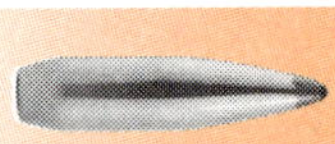

140 gr. Jacketed SBT
2.550" OAL

BC: .418
SD: .248

Powder	Sugg Starting Grains	Velocity fps	Pressure	Max Load Grains	Velocity fps	Pressure
IMR-3031	27.5	1856	—	32.6+	2175	—
H-335	28.7	1818	—	**33.0**	**2192**	—
AA-2460	29.0	1914	—	32.6	2152	—
BL-C(2)	31.4	1920	—	35.0	2141	—
IMR-4895	28.8	1839	—	33.6+	2173	—
748	31.5	1920	—	34.5	2134	—

150 gr. Jacketed SBT
2.550" OAL

BC: .430
SD: .266

Powder	Sugg Starting Grains	Velocity fps	Pressure	Max Load Grains	Velocity fps	Pressure
IMR-3031	26.2	1850	—	30.7+	2060	—
H-335	27.5	1865	—	32.0	2129	—
AA-2460	26.7	1806	—	31.6	2024	—
BL-C(2)	30.3	1858	—	33.7	2070	—
IMR-4895	28.2	1831	—	32.5+	2112	—
748	31.6	1927	—	**34.8+**	**2180**	—

160 gr. Jacketed HPBT
2.550" OAL

BC: .394
SD: .283

Powder	Sugg Starting Grains	Velocity fps	Pressure	Max Load Grains	Velocity fps	Pressure
IMR-3031	23.6	1688	—	27.8+	1963	—
H-335	27.0	1768	—	31.4	2029	—
AA-2460	26.7	1786	—	30.8	1960	—
BL-C(2)	29.7	1789	—	33.3	2021	—
IMR-4895	25.8	1787	—	29.7+	1968	—
748	31.5	1875	—	**34.8+**	**2111**	—

#287346
135 gr. (#2 Alloy) 2.475" OAL

BC: .235
SD: .237

Powder	Sugg Starting Grains	Velocity fps	Pressure	Max Load Grains	Velocity fps	Pressure
SR-4759	16.5	1592	—	20.0	1916	—
XMP-5744	20.0	1694	—	22.5	1872	—
RX7	23.0	1901	—	26.9	2069	—
IMR-3031	25.5	1793	—	31.0+	2139	—
H-335	26.2	1811	—	32.0	2161	—
AA-2460	26.0	1847	—	**32.0**	**2123**	—
748	29.5	1827	—	33.5+	2123	—

Note: Loads shown in shaded panels are maximum.
Loads shown in bold designate potentially most accurate load.
\+ Designates a compressed powder charge.

7mm-08 Remington

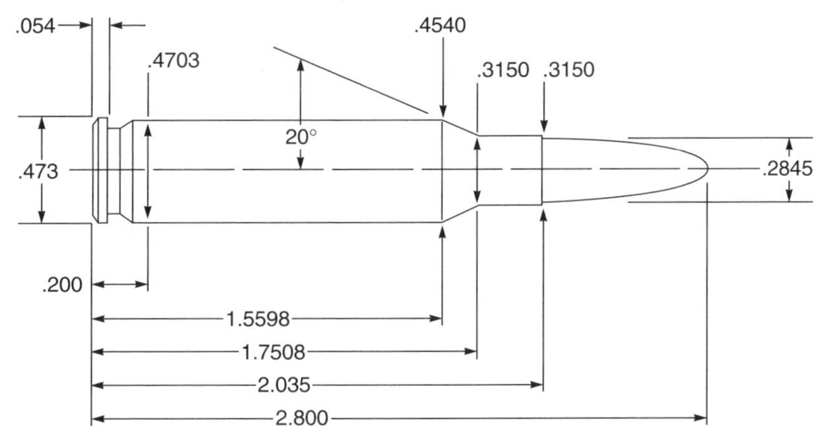

Comments:

The 7mm-08 has become popular for High-Power, hunting, and metallic silhouette applications. The loads listed below were worked up in a pressure barrel and fired through both a 12 and 15-inch barrel. The 15-inch barrel yielded velocities around 300 feet per second below those listed for the full-length 24-inch barrel. The 12-inch barrel showed a loss of one hundred to 175 feet per second from the 15-inch barrel. Unique, SR-4759, and XMP-5744 all produced excellent ballistic uniformity with both cast bullets. Approach all loads with caution and back off powder charges at any sign of high pressure.

Test Components:

Cases . . . . .Remington
Trim-to Length . . . . .2.025"
Primers . . . . .Remington 9½
Primer Size . . . . .Large Rifle
Lyman Shell Holder . . . . .No. 2
Jacketed Bullets Used . . . .Hornady SP #2820, 139 gr.
Hornady SP #2830, 154 gr.
Cast Bullets Used . . . .(sized to .284" dia)
*gas check bullet *#287346, 135 gr.
*#287641, 160 gr.

Test Specifications: (Velocity Only)

Firearm Used . . . . .Thompson/Center Encore
Barrel Length . . . . .12" & 15"
Twist . . . . .1-9"
Groove Dia. . . . . .284"

139 gr. Jacketed SP
2.750" OAL — BC: .392 SD: .246

Powder	Sugg Starting Grains	12" Bbl. Velocity fps	15" Bbl. Velocity fps	Max Load Grains	12" Bbl. Velocity fps	15" Bbl. Pressure fps
IMR-3031	37.5	2162	2305	40.0	2345	2524
AA2460	37.0	2140	2398	41.0	2348	2542
IMR-4895	38.0	2160	2257	**42.0**	**2332**	**2502**
IMR-4064	38.0	2062	2173	43.0	2407	2584
**SR-4759	20.0	1717	1813	26.0	2102	2213
**XMP-5744	23.0	1676	1777	28.5	1990	2123

154 gr. Jacketed SP
2.800" OAL — BC: .433 SD: .273

Powder	Sugg Starting Grains	12" Bbl. Velocity fps	15" Bbl. Velocity fps	Max Load Grains	12" Bbl. Velocity fps	15" Bbl. Pressure fps
IMR-3031	34.5	2029	2096	39.0	2296	2376
AA-2460	35.0	2038	2062	40.0	2307	2425
IMR-4895	37.0	2051	2148	**41.0**	**2293**	**2411**
IMR-4064	36.5	1989	2072	41.8	2330	2437
**SR-4759	22.0	1762	1868	25.5	1993	2089
**XMP-5744	24.0	1698	1777	27.5	1893	2012

Note: Loads shown in shaded panels are maximum.
Loads shown in bold designate potentially most accurate load.
**Reduced load.

7mm-08 Remington

#287346
135 gr. (#2 Alloy) 2.600" OAL

BC: .235
SD: .239

Powder	Sugg Starting Grains	12" Bbl. Velocity fps	15" Bbl. Velocity fps	Max Load Grains	12" Bbl. Velocity fps	15" Bbl. Pressure fps
Unique	**12.0**	**1496**	**1546**	15.5	1743	1807
SR-4759	15.5	1390	1505	23.5	1961	2097
IMR-4227	15.5	1208	1358	24.0	1774	1909
XMP-5744	**16.5**	**1265**	**1365**	25.0	1775	1922
IMR-4198	16.5	1202	1243	25.5	1755	1935
RX7	17.5	1196	1346	26.5	1724	1926

#287641
160 gr. (#2 Alloy) 2.705" OAL

BC: .382
SD: .283

Powder	Sugg Starting Grains	12" Bbl. Velocity fps	15" Bbl. Velocity fps	Max Load Grains	12" Bbl. Velocity fps	15" Bbl. Pressure fps
Unique	12.0	1378	1438	15.5	1608	1660
SR-4759	21.0	1708	1812	**26.0**	**1976**	**2074**
IMR-4227	21.0	1549	1627	27.0	1900	1993
XMP-5744	18.0	1316	1448	**26.0**	**1766**	**1905**
IMR-4198	23.0	1608	1719	29.5	1973	2110
RX7	22.0	1491	1625	28.0	1803	1988

Note: Loads shown in shaded panels are maximum.
Loads shown in bold designate potentially most accurate load.

30-30 Winchester

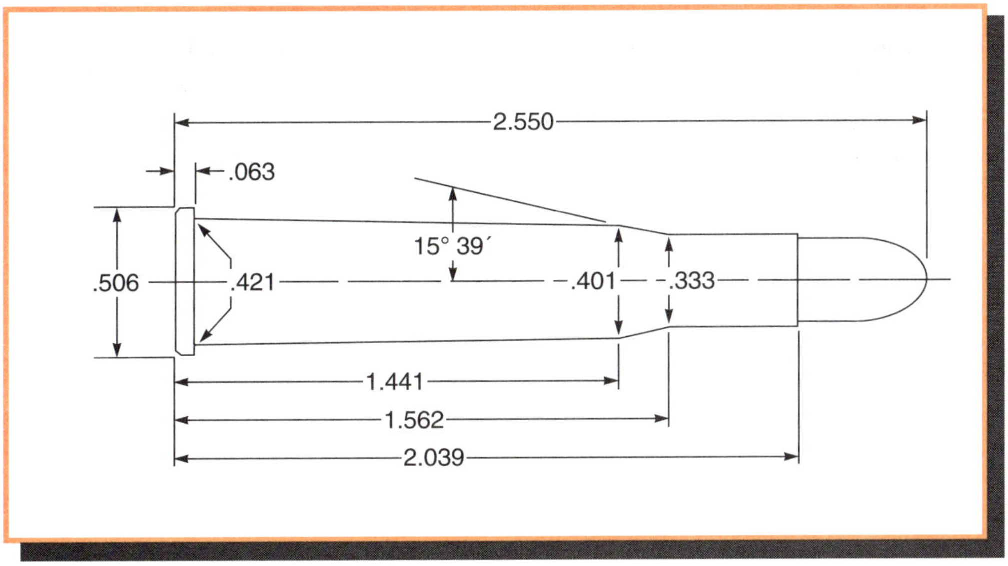

Comments:

When used in the Contender this is an extremely accurate cartridge.

The 110 grain to 130 grain bullets are for varmints and plinking. This cartridge works very well with a number of powders including IMR-3031, Winchester 748 and Reloder 7.

It is also very good with cast bullets, particularly #311291.

Test Components:

Cases . . . . .Winchester
Trim-to Length . . . . .2.029"
Primers . . . . .Winchester WLR
Primer Size . . . . .Large Rifle
Lyman Shell Holder . . . . .No. 6
Jacketed Bullets Used . . . .Hornady SP #3010, 110 gr.
Sierra SP #2120, 125 gr.
Hornady SP #3020, 130 gr.
Hornady SP #3031, 150 gr.
Cast Bullets Used . . . .(sized to .309" dia)
*gas check bullet *#311359, 115 gr.
*#311291, 170 gr.
*#311041, 173 gr.

Test Specifications:
(Velocity Only)

Firearm Used . . . . .Thompson/Center Contender
Barrel Length . . . . .14"
Twist . . . . .1-14"
Groove Dia. . . . . .3087"

110 gr. Jacketed SP
2.530" OAL
BC: .256
SD: .166

Powder	Sugg Starting Grains	Velocity fps	Pressure	Max Load Grains	Velocity fps	Pressure
IMR-4198	25.0	2038	—	29.0	2346	—
IMR-3031	28.5	1854	—	33.5	2166	—
748	35.5	1892	—	**41.0+**	**2222**	—

125 gr. Jacketed SP
2.600" OAL
BC: .277
SD: .188

Powder	Sugg Starting Grains	Velocity fps	Pressure	Max Load Grains	Velocity fps	Pressure
IMR-4198	24.0	1909	—	26.0	2159	—
IMR-3031	28.0	1893	—	32.0	2239	—
748	35.5	1907	—	**41.0+**	**2210**	—

Note: Loads shown in shaded panels are maximum.
Loads shown in bold designate potentially most accurate load.
\+ Designates a compressed powder charge.

30-30 Winchester

130 gr. Jacketed SP
2.620" OAL

BC: .295
SD: .196

Powder	Sugg Starting Grains	Velocity fps	Pressure	Max Load Grains	Velocity fps	Pressure
IMR-4198	23.0	1768	—	25.5	2095	—
IMR-3031	27.5	1779	—	31.5	2183	—
748	35.0	1851	—	**40.5+**	**2166**	—

150 gr. Jacketed SP
2.745" OAL

BC: .338
SD: .226

Powder	Sugg Starting Grains	Velocity fps	Pressure	Max Load Grains	Velocity fps	Pressure
IMR-4198	19.5	1457	—	23.5	1749	—
RX7	21.0	1595	—	28.0	1934	—
IMR-3031	25.5	1602	—	28.5	1847	—
748	33.0	1779	—	**38.5+**	**2077**	—
IMR-4064	29.0	1699	—	31.5	1950	—

#311359
115 gr. (#2 Alloy) 2.440" OAL

BC: .181
SD: .172

Powder	Sugg Starting Grains	Velocity fps	Pressure	Max Load Grains	Velocity fps	Pressure
Unique	7.5	1303	—	**11.4**	**1740**	—
SR-4759	18.0	1754	—	20.0	2027	—
RX7	20.0	1656	—	30.5	2039	—
IMR-3031	26.0	1805	—	32.0	2116	—
748	24.5	1531	—	36.5	1981	—

#311291
170 gr. (#2 Alloy) 2.540" OAL

BC: .202
SD: .254

Powder	Sugg Starting Grains	Velocity fps	Pressure	Max Load Grains	Velocity fps	Pressure
Unique	7.0	1065	—	10.6	1416	—
SR-4756	7.0	962	—	10.5	1283	—
2400	15.0	1459	—	18.5	1704	—
SR-4759	14.8	1395	—	**18.5**	**1650**	—
IMR-4227	16.5	1423	—	22.0	1812	—
XMP-5744	18.0	1456	—	24.0	1856	—
IMR-4198	18.0	1409	—	24.0	1870	—
RX7	19.0	1761	—	28.0	1936	—
IMR-3031	22.5	1439	—	28.5	1927	—
748	23.8	1429	—	37.3+	2025	—

#311041
173 gr. (#2 Alloy) 2.510" OAL

BC: .220
SD: .259

Powder	Sugg Starting Grains	Velocity fps	Pressure	Max Load Grains	Velocity fps	Pressure
Unique	7.0	1084	—	10.6	1419	—
SR-4756	7.0	980	—	10.5	1309	—
2400	14.5	1425	—	19.0	1786	—
SR-4759	15.5	1442	—	17.7	1661	—
IMR-4227	17.0	1449	—	22.0	1829	—
XMP-5744	**17.0**	**1384**	—	22.0	1721	—
IMR-4198	18.0	1393	—	23.5	1835	—
RX7	20.0	1486	—	28.6	2088	—
IMR-3031	21.5	1309	—	27.0	1775	—
748	25.0	1482	—	35.0	1978	—

Note: Loads shown in shaded panels are maximum.
Loads shown in bold designate potentially most accurate load.
\+ Designates a compressed powder charge.

308 Winchester

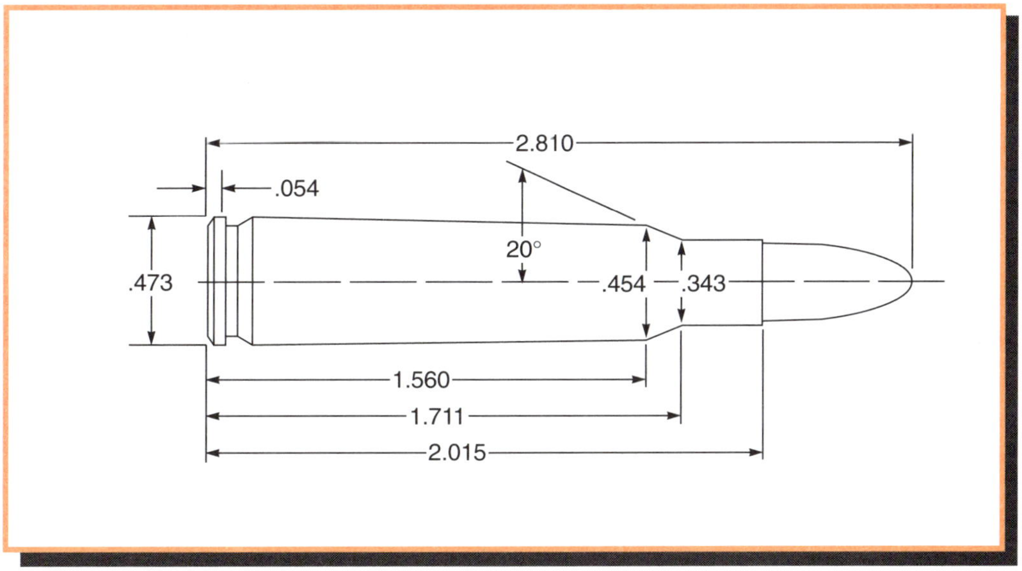

Comments:

Cast bullet #311359 produced excellent ballistic uniformity with all powders listed. Cast bullet #311674 and #311332 are ideal candidates for metallic silhouette. XMP-5744 also produced good results with the reduced load for the 165-grain Partition and bullet #311332. Approach all maximum loads with caution and back off powder charges at any sign of high pressure.

Test Components:

Cases . . . . .Remington
Trim-to Length . . . . .2.005"
Primers . . . . .Remington 9½
Primer Size . . . . .Large Rifle
Lyman Shell Holder . . . . .No. 2
Jacketed Bullets Used . . . .Hornady SP #3020, 130 gr.
Speer SP #2025, 150 gr.
Nosler Partition #16330, 165 gr.
Cast Bullets Used . . . .(sized to .308" dia)
*gas check bullet *#311359, 115 gr.
*#311672, 160 gr.
*#311332, 180 gr.

Test Specifications:
(Velocity Only)

Firearm Used . . . . .Thompson/Center Encore
Barrel Length . . . . .15"
Twist . . . . .1-10"
Groove Dia. . . . . .308"

130 gr. Jacketed SP
2.690" OAL
BC: .295
SD: .196

Powder	Sugg Starting Grains	Velocity fps	Pressure	Max Load Grains	Velocity fps	Pressure
IMR-4198	35.0	2442	—	38.0	2702	—
XMR-2015	41.0	2419	—	45.0	2724	—
Benchmark	**41.5**	**2529**	—	46.0	2774	—
IMR-3031	42.0	2489	—	45.0+	2806	—
IMR-4895	43.0	2395	—	47.0+	2694	—
Varget	46.0	2558	—	50.0	2829	—
**SR-4759	26.0	2256	—	29.0	2464	—
**XMP-5744	31.0	2220	—	33.5	2385	—

150 gr. Jacketed SP
2.735" OAL
BC: .301
SD: .226

Powder	Sugg Starting Grains	Velocity fps	Pressure	Max Load Grains	Velocity fps	Pressure
IMR-4198	33.0	2368	—	36.0	2580	—
XMR-2015	38.0	2281	—	42.0	2490	—
IMR-3031	40.0	2328	—	43.5	2596	—
AA-2230	40.5	2447	—	44.5	2618	—
IMR-4895	40.0	2201	—	45.0	2525	—
Varget	42.5	2240	—	**47.0**	**2505**	—
**SR-4759	24.5	2031	—	27.5	2207	—
**XMP-5744	28.0	1986	—	31.0	2143	—

Note: Loads shown in shaded panels are maximum.
Loads shown in bold designate potentially most accurate load.
\+ Designates a compressed powder charge.
** Reduced load.

308 Winchester

165 gr. Jacketed Partition 2.780" OAL — **BC:** .410 **SD:** .248

Powder	Sugg Starting Grains	Velocity fps	Pressure	Max Load Grains	Velocity fps	Pressure
IMR-3031	37.0	2085	—	41.5	2454	—
AA-2460	38.0	2126	—	42.5	2421	—
IMR-4895	37.5	1990	—	42.5	2338	—
Varget	41.0	2133	—	46.0	2420	—
N150	42.0	2228	—	**47.0+**	**2484**	—
**SR-4759	23.0	1856	—	26.0	2080	—
**XMP-5744	27.0	1877	—	30.0	2051	—

#311359 115 gr. (#2 Alloy) 2.430" OAL — **BC:** .181 **SD:** .173

Powder	Sugg Starting Grains	Velocity fps	Pressure	Max Load Grains	Velocity fps	Pressure
Red Dot	10.0	1547	—	14.0	1866	—
Green Dot	11.0	1605	—	15.0	1917	—
Unique	12.0	1668	—	**17.0**	**2053**	—
SR-7625	12.0	1599	—	15.0	1852	—
Herco	14.0	1805	—	17.0	2027	—

#311672 160 gr. (#2 Alloy) 2.635" OAL — **BC:** .245 **SD:** .241

Powder	Sugg Starting Grains	Velocity fps	Pressure	Max Load Grains	Velocity fps	Pressure
Unique	9.8	1317	—	13.0	1554	—
SR-7625	10.7	1326	—	14.0	1564	—
2400	15.5	1432	—	19.5	1707	—
SR-4759	18.0	1549	—	27.5	2158	—
IMR-4227	20.0	1506	—	**30.0**	**2178**	—
XMP-5744	20.0	1492	—	31.0	2148	—
IMR-4198	21.0	1493	—	31.7	2178	—
RX7	25.0	1675	—	34.5	2274	—

#311332 180 gr. (#2 Alloy) 2.705" OAL — **BC:** .320 **SD:** .271

Powder	Sugg Starting Grains	Velocity fps	Pressure	Max Load Grains	Velocity fps	Pressure
Unique	12.2	1429	—	15.5	1633	—
SR-7625	11.5	1327	—	14.0	1494	—
SR-4756	13.0	1404	—	15.5	1562	—
2400	17.0	1519	—	24.0	1968	—
SR-4759	23.5	1825	—	26.3	1991	—
IMR-4227	25.0	1764	—	29.0	2068	—
XMP-5744	21.0	1522	—	**32.5**	**2158**	—
IMR-4198	26.5	1797	—	32.0	2144	—

Note: Loads shown in shaded panels are maximum.
Loads shown in bold designate potentially most accurate load.
+ Designates a compressed powder charge.
** Designates a reduced load.

30-06 Springfield

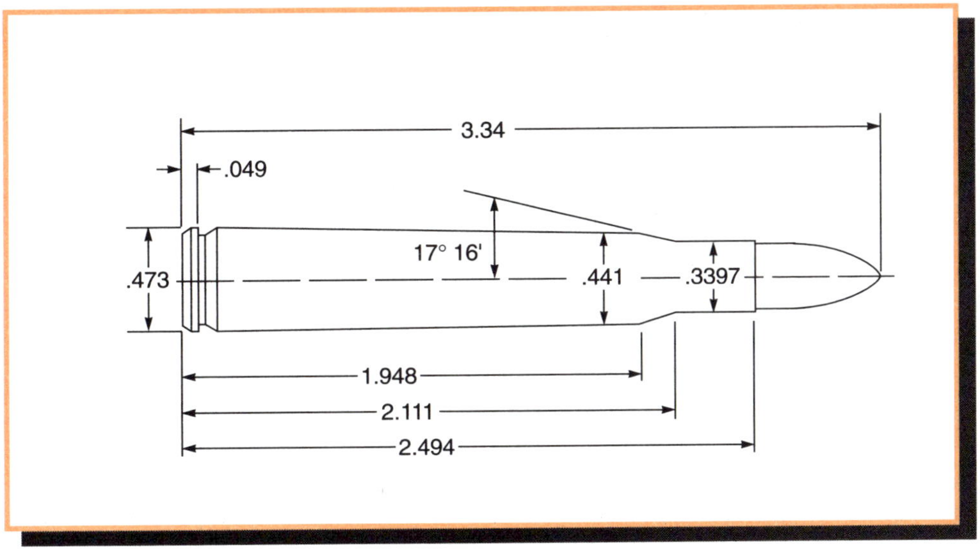

Comments:

A member of Lyman's technical staff described recoil of the Thompson/Center Encore with the 15-inch 30-06 barrel as "moderately heavy" with full power jacketed loads. Those seeking milder practice loads can utilize the reduced loads with SR-4759 or XMP-5744. Data listed for the 170-grain Hornady FN bullet will duplicate, and sometimes exceed, the ballistics of a 30-30 Winchester with a 24" barrel. Cast bullet #311359 is a good choice for casual plinking. Cast bullet #311672 and #311644 are good choices for metallic silhouette shooting. Approach all maximum loads with caution and back off powder charges at any sign of high pressure.

Test Components:

Cases . . . . .Winchester
Trim-to Length . . . . .2.484"
Primers . . . . .Winchester WLR
Primer Size . . . . .Large Rifle
Lyman Shell Holder . . . . .No. 2
Jacketed Bullets Used . . . . .Sierra RN #2135, 150 gr.
Hornady FN #3060, 170 gr.
Cast Bullets Used . . . .(sized to .308" dia)
*gas check bullet
*#311359, 115 gr.
*#311672, 160 gr.
*#311644, 190 gr.

Test Specifications:
(Velocity Only)

Firearm Used . . . . .Thompson/Center Encore
Barrel Length . . . . .15"
Twist . . . . .1-10"
Groove Dia. . . . . .3075"

150 gr. Jacketed RN
2.970" OAL
BC: .227
SD: .226

Powder	Sugg Starting Grains	Velocity fps	Pressure	Max Load Grains	Velocity fps	Pressure
XMR-2015	43.0	2399	—	48.0	2600	—
IMR-3031	44.0	2408	—	49.0	2614	—
IMR-4895	45.0	2342	—	49.0	2563	—
****SR-4759**	**30.0**	**2208**	—	32.5	2341	—
**XMP-5744	32.5	2030	—	35.5	2187	—

170 gr. Jacketed FN
3.000" OAL
BC: .189
SD: .256

Powder	Sugg Starting Grains	Velocity fps	Pressure	Max Load Grains	Velocity fps	Pressure
IMR-3031	41.0	2180	—	46.0	2392	—
AA2460	42.0	2221	—	46.5	2350	—
IMR-4895	42.0	2139	—	47.0	2414	—
748	44.0	2208	—	49.0	2308	—
**SR-4759	27.0	1937	—	31.5	2179	—
****XMP-5744**	**30.5**	**1875**	—	34.0	2058	—

Note: Loads shown in shaded panels are maximum.
Loads shown in bold designate potentially most accurate load.
**Designates a reduced load.

30-06 Springfield

#311359 — 115 gr. (#2 Alloy) 2.909" OAL — **BC:** .181 **SD:** .173

Powder	Sugg Starting Grains	Velocity fps	Pressure	Max Load Grains	Velocity fps	Pressure
Green Dot	13.0	1654	—	16.0	1861	—
Unique	14.0	1720	—	**19.0**	**2021**	—
Herco	14.5	1713	—	18.0	1931	—
SR-7625	12.5	1525	—	16.5	1796	—
XMP-5744	24.0	1678	—	32.0	2146	—

#311672 — 160 gr. (#2 Alloy) 3.050" OAL — **BC:** .245 **SD:** .241

Powder	Sugg Starting Grains	Velocity fps	Pressure	Max Load Grains	Velocity fps	Pressure
Unique	18.0	1773	—	24.0	2074	—
SR-4759	25.0	1892	—	**33.0**	**2272**	—
XMP-5744	26.0	1689	—	35.0	2130	—
IMR-4198	27.0	1696	—	37.0	2206	—
RX7	28.0	1660	—	38.0	2173	—

#311644 — 190 gr. (#2 Alloy) 3.075" OAL — **BC:** .272 **SD:** .286

Powder	Sugg Starting Grains	Velocity fps	Pressure	Max Load Grains	Velocity fps	Pressure
Unique	13.0	1373	—	17.2	1615	—
SR-4756	13.5	1335	—	17.0	1534	—
SR-4759	23.0	1691	—	**28.0**	**1939**	—
XMP-5744	27.0	1675	—	37.0	2179	—
RX7	22.0	1358	—	38.0	2229	—

Note: Loads shown in shaded panels are maximum.
Loads shown in bold designate potentially most accurate load.
* Designates a reduced load.

357 Magnum

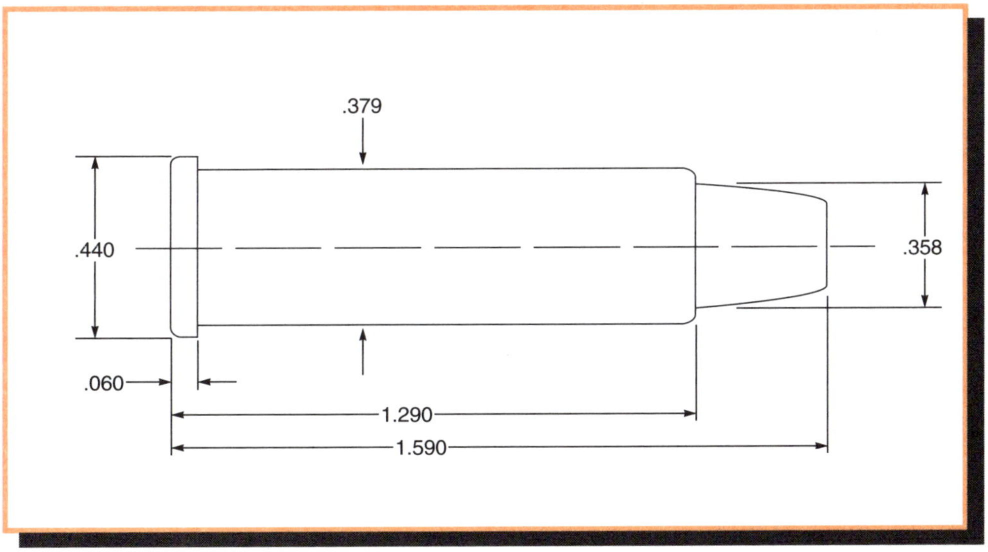

Comments:

Lacking the cylinder gap of a revolver, the Contender pistol will deliver higher velocity from identical barrel length.

Do not use 357 Magnum Loads in 38 Special cases as very dangerous pressure will result.

Bullet #358156 is extremely popular for heavy loads. Bullet #358156 closely duplicates the factory 158 grain semi-wadcutter. This is the Elmer Keith design bullet and makes an excellent choice for hunting.

Heavy loads work well with a wide range of powders but you might want to try 2400 for your first accuracy tests. Bullets may be seated in tune with the individual Contenders rifling since there is no other constraint with regard to maximum over-all cartridge length.

Test Components:

Cases . . . . .Federal
Trim-to Length . . . . .1.280"
Primers . . . . .CCI 550
Primer Size . . . . .Small Pistol, Magnum
Lyman Shell Holder . . . . .No. 1
Jacketed Bullets Used . . . .
Hornady HP/XTP #35700,110 gr.
Hornady HP/XTP #35710, 125 gr.
Speer JHP -SWC #4205, 146 gr.
Sierra JSP #8340, 158 gr.
Hornady JTC-SIL, #35771, 180 gr.
Cast Bullets Used . . . .(sized to .357" dia)
*gas check bullet
#356242, 120 gr.
*#358156, 155 gr.
#358429, 170 gr.
*#358315, 205 gr.

Test Specifications:
(Velocity Only)

Firearm Used . . . . .Thompson/Center Contender
Barrel Length . . . . .10"
Twist . . . . .1-18"
Groove Dia. . . . . .356"

110 gr. Jacketed HP
1.590" OAL
BC: .131
SD: .123

Powder	Sugg Starting Grains	Velocity fps	Pressure	Max Load Grains	Velocity fps	Pressure
Unique	7.4	1334	—	10.0	1737	—
AA#5	11.0	1832	—	13.0	2079	—
Blue Dot	10.0	1342	—	14.6	1968	—
2400	13.5	1427	—	20.3+	1952	—
IMR-4227	14.8	1446	—	**19.5+**	**1733**	—

125 gr. Jacketed HP
1.590" OAL
BC: .151
SD: .140

Powder	Sugg Starting Grains	Velocity fps	Pressure	Max Load Grains	Velocity fps	Pressure
Unique	7.0	1261	—	9.7	1643	—
AA#5	10.6+	1525	—	11.8	1704	—
Blue Dot	10.4	1355	—	13.3	1807	—
N110	14.5	1346	—	17.5	1801	—
2400	13.0	1302	—	17.7	1778	—
H-110	21.0	1907	—	22.0	2019	—
IMR-4227	15.6	1435	—	**19.5+**	**1749**	—

Note: Loads shown in shaded panels are maximum.
Loads shown in bold designate potentially most accurate load.
\+ Designates a compressed powder charge.

357 Magnum

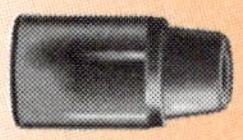

146 gr. Jacketed HP
1.590" OAL

BC: .159
SD: .164

Powder	Sugg Starting Grains	Velocity fps	Pressure	Max Load Grains	Velocity fps	Pressure
Unique	6.3	1108	—	8.1	1416	—
AA#5	9.5	1364	—	10.6	1507	—
Blue Dot	9.5	1335	—	11.5	1580	—
AA#7	10.8	1277	—	12.0	1497	—
N110	14.5	1509	—	16.0	1683	—
2400	11.0	1183	—	15.0	1546	—
IMR-4227	14.4	1167	—	**16.0**	**1355**	—

158 gr. Jacketed SP
1.590" OAL

BC: .161
SD: .177

Powder	Sugg Starting Grains	Velocity fps	Pressure	Max Load Grains	Velocity fps	Pressure
Unique	6.4	1102	—	8.3	1378	—
Blue Dot	9.6	1283	—	10.7	1441	—
AA#9	13.4	1484	—	14.9	1626	—
2400	11.3	1093	—	**14.9**	**1491**	—
H-110	16.3	1522	—	16.7	1540	—
IMR-4227	12.2	1130	—	16.1	1430	—

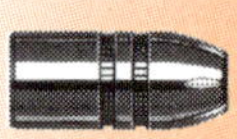

180 gr. Jacketed SIL
1.585" OAL

BC: .230
SD: .202

Powder	Sugg Starting Grains	Velocity fps	Pressure	Max Load Grains	Velocity fps	Pressure
Blue Dot	8.4	1140	—	9.4	1316	—
AA#9	11.2	1225	—	**12.5**	**1394**	—
2400	10.0	1129	—	12.6	1412	—
H-110	13.9	1305	—	14.5	1389	—
IMR-4227	11.7	1093	—	14.6	1383	—

#356242
120 gr. (#2 Alloy) 1.640" OAL

BC: .154
SD: .135

Powder	Sugg Starting Grains	Velocity fps	Pressure	Max Load Grains	Velocity fps	Pressure
Unique	6.9	1241	—	**9.1**	**1580**	—
Power Pistol	9.9	1685	—	11.0	1795	—
AA#5	8.6	1481	—	10.8	1771	—
HS-6	10.8	1376	—	12.0	1657	—
2400	13.0	1381	—	16.4	1669	—

#358156
155 gr. (#2 Alloy) 1.578" OAL

BC: .213
SD: .174

Powder	Sugg Starting Grains	Velocity fps	Pressure	Max Load Grains	Velocity fps	Pressure
HS-6	8.8	1280	—	9.7	1382	—
Blue Dot	8.5	1187	—	10.5	1443	—
AA#7	10.8	1357	—	12.0	1516	—
2400	10.6	1138	—	**14.0**	**1466**	—
AA#9	13.0	1505	—	14.5	1604	—
H110	15.0	1485	—	15.7	1508	—

#358429
170 gr. (#2 Alloy) 1.647" OAL

BC: .286
SD: .190

Powder	Sugg Starting Grains	Velocity fps	Pressure	Max Load Grains	Velocity fps	Pressure
Blue Dot	8.3	1141	—	10.0	1368	—
2400	9.7	998	—	**13.5**	**1422**	—
N110	12.3	1305	—	13.7	1503	—
AA#9	11.7	1323	—	13.0	1492	—
H-110	14.4	1368	—	15.0	1469	—

#358315
204 gr. (#2 Alloy) 1.880" OAL

BC: .165
SD: .229

Powder	Sugg Starting Grains	Velocity fps	Pressure	Max Load Grains	Velocity fps	Pressure
Blue Dot	9.0	1225	—	11.0	1410	—
2400	12.0	1318	—	14.5	1490	—
H-110	15.8	1442	—	**16.5**	**1515**	—

Note: Loads shown in shaded panels are maximum.
Loads shown in bold designate potentially most accurate load.

44 Remington Magnum

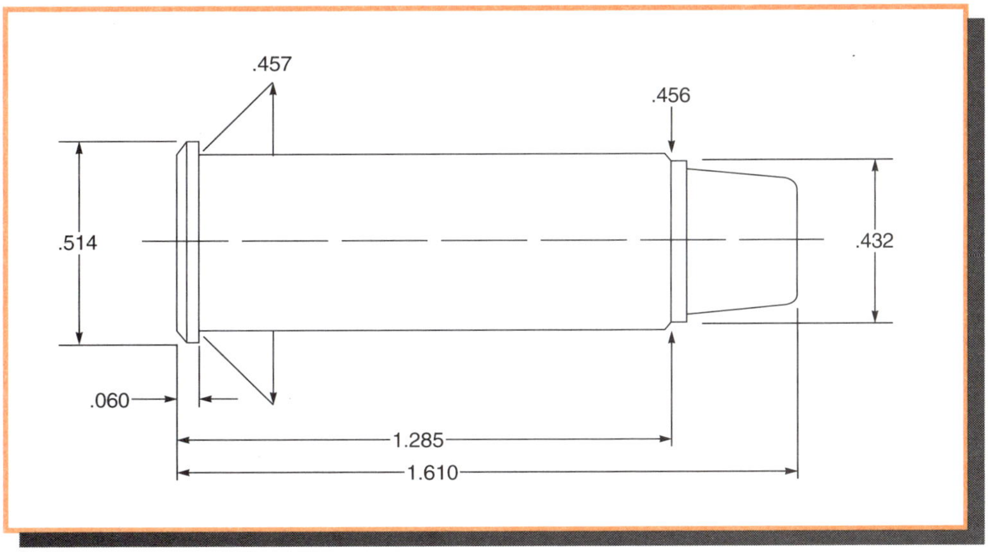

Comments:

This is one of the more popular Contender cartridges and it will fill a broad range of applications.

Bullet #429215 is extremely popular. However, all cast bullets designs generally work well.

IMR 4227 and Alliant 2400 are suggested for combining maximum accuracy with full power loads.

Test Components:

Cases . . . . .Remington
Trim-to Length . . . . .1.275"
Primers . . . . .CCI 300 & 350
Primer Size . . . . .Large Pistol, Std. & Magnum
Lyman Shell Holder . . . . .No. 7
Jacketed Bullets Used . . . . .Sierra HC #8600, 180 gr.
Hornady HP/XTP #44100, 200 gr.
Sierra JHC #8610, 240 gr.
Hornady FP #4300, 265 gr.
Sierra SP #8630, 300 gr.
Cast Bullets Used . . . .(sized to .430" dia)
*gas check bullet
*#429215, 210 gr.
*#429244, 255 gr.
*#429650, 300 gr.

Test Specifications:
(Velocity Only)

Firearm Used . . . . .Thompson/Center Contender
Barrel Length .10" and 14"
Twist . . . . .1-22"
Groove Dia. . . . . .430"

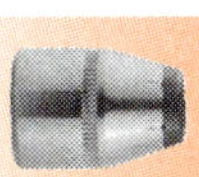

180 gr. Jacketed HC
1.610" OAL
BC: .130
SD: .139

Powder	Sugg Starting Grains	10" Bbl. Velocity fps	14" Bbl. Velocity fps	Max Load Grains	10" Bbl. Velocity fps	14" Bbl. Pressure fps
Unique	12.0	1474	1505	13.6	1634	1679
AA#5	12.8	1448	1497	16.0	1744	1835
Blue Dot	17.8	1570	1639	20.2	1770	1823
2400	21.3	1424	1481	25.5	1683	1773
***H-110**	29.0	1719	1859	**30.0+**	**1952**	**2098**
IMR-4227	24.2	1437	1503	27.0	1633	1741

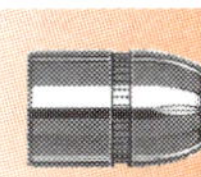

200 gr. Jacketed HP
1.610" OAL
BC: .170
SD: .155

Powder	Sugg Starting Grains	10" Bbl. Velocity fps	14" Bbl. Velocity fps	Max Load Grains	10" Bbl. Velocity fps	14" Bbl. Pressure fps
Unique	10.5	1280	1321	13.0	1498	1567
800X	12.0	1412	1488	14.9	1646	1702
Blue Dot	15.7	1388	1421	18.7	1608	1691
AA#7	18.0	1503	1524	20.0	1678	1706
2400	19.5	1324	1444	**23.6**	**1562**	**1656**
*H-110	27.7	1807	1894	28.8	1876	1991
IMR-4227	22.8	1353	1446	26.4	1630	1709

Note: Loads shown in shaded panels are maximum.
Loads shown in bold designate potentially most accurate load.
* Designates the use of a magnum primer.

44 Remington Magnum

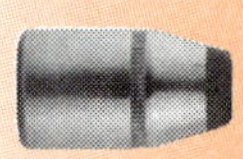

240 gr. Jacketed HC
1.610" OAL

BC: .185
SD: .186

Powder	Sugg Starting Grains	10" Bbl. Velocity fps	14" Bbl. Velocity fps	Max Load Grains	10" Bbl. Velocity fps	14" Bbl. Pressure fps
Unique	10.3	1202	1259	11.5	1247	1386
Blue Dot	13.4	1248	1287	14.9	1400	1466
2400	17.5	1304	1364	**19.5**	**1469**	**1512**
*H-110	22.5	1509	1561	23.5	1554	1644
IMR-4227	19.3	1160	1212	23.1	1413	1511

265 gr. Jacketed FP
1.650" OAL

BC: .189
SD: .205

Powder	Sugg Starting Grains	10" Bbl. Velocity fps	14" Bbl. Velocity fps	Max Load Grains	10" Bbl. Velocity fps	14" Bbl. Pressure fps
Unique	8.0	920	963	11.5	1255	1320
800X	10.7	1181	1248	13.4	1375	1436
Blue Dot	12.6	1151	1172	14.0	1290	1313
N110	16.6	1224	1252	18.5	1360	1398
2400	16.5	1228	1256	**18.3**	**1359**	**1396**
AA#9	18.0	1336	1377	21.3	1516	1583
***H-110**	**20.5**	**1368**	**1413**	21.5	1439	1478
IMR-4227	17.5	1015	1037	22.0	1361	1387

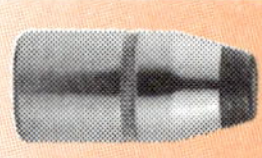

300 gr. Jacketed SP
1.720" OAL

BC: .230
SD: .232

Powder	Sugg Starting Grains	10" Bbl. Velocity fps	14" Bbl. Velocity fps	Max Load Grains	10" Bbl. Velocity fps	14" Bbl. Pressure fps
Blue Dot	12.4	1103	1067	13.5	1226	1231
2400	15.7	1142	1137	**17.5**	**1244**	**1270**
AA#9	17.5	1239	1243	19.3	1351	1370
*H-110	19.5	1293	1298	20.5	1361	1395
IMR-4227	18.5	1160	1169	21.0	1373	1448

#429215
210 gr. (#2 Alloy) 1.645" OAL

BC: .188
SD: .162

Powder	Sugg Starting Grains	10" Bbl. Velocity fps	14" Bbl. Velocity fps	Max Load Grains	10" Bbl. Velocity fps	14" Bbl. Pressure fps
Unique	10.0	1263	1323	13.2	1543	1624
800X	12.0	1426	1499	15.1	1639	1747
AA#7	17.2	1405	1506	19.2	1549	1679
Blue Dot	14.9	1372	1443	16.6	1553	1679
N110	19.3	1398	1542	21.5	1552	1681
2400	19.8	1479	1524	**22.0**	**1600**	**1720**
IMR-4227	21.8	1314	1403	25.5+	1569	1680

#429244
255 gr. (#2 Alloy) 1.680" OAL

BC: .201
SD: .197

Powder	Sugg Starting Grains	10" Bbl. Velocity fps	14" Bbl. Velocity fps	Max Load Grains	10" Bbl. Velocity fps	14" Bbl. Pressure fps
Unique	8.9	1111	1155	12.1	1360	1434
800X	10.6	1282	1325	13.2	1488	1546
Blue Dot	14.0	1255	1304	17.4	1494	1575
2400	18.0	1252	1327	**20.0**	**1529**	**1604**
AA#9	17.5	1337	1430	21.7	1544	1645
*H-110	21.8	1408	1545	22.8	1478	1594
IMR-4227	20.2	1227	1322	24.0	1497	1575

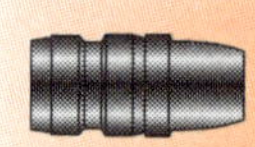

#429650
300 gr. (#2 Alloy) 1.700" OAL

BC: .165
SD: .232

Powder	Sugg Starting Grains	10" Bbl. Velocity fps	14" Bbl. Velocity fps	Max Load Grains	10" Bbl. Velocity fps	14" Bbl. Pressure fps
Herco	8.1	968	1012	9.0	1033	1095
Blue Dot	10.5	1032	1089	11.7	1128	1201
2400	**14.0**	**1110**	**1188**	15.7	1229	1299
AA#9	15.3	1213	1289	18.0	1366	1414
IMR-4227	16.5	1117	1185	19.0+	1299	1356

Note: Loads shown in shaded panels are maximum.
Loads shown in bold designate potentially most accurate load.
\+ Designates a compressed powder charge.
* Designates the use of a magnum primer.

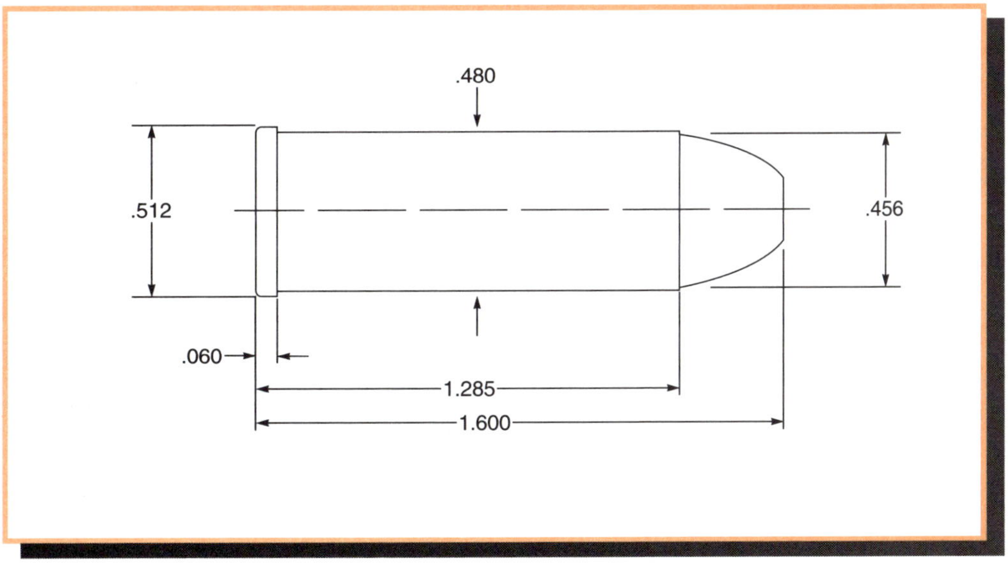

Comments:

Shooters must be careful in regard to this cartridges small diameter rim. The rim may tear off if excessive force is applied during the reloading process. Unique is a long time favorite for loading the 45 Colt, especially with cast bullets. Data fired through the 15-inch Encore barrel showed little or no velocity gain compared to the 10-inch Contender barrel due to the long chamber configuration to accommodate the 410 shotgun shell. **Shooters using the 15-inch 45 Colt/410 barrel must remove the internal choke before firing 45 Colt ammunition.** Bullet #452424 is the Elmer Keith design and has proven very popular over the years.

Test Components:

Cases . . . . Winchester
Trim-to Length . . . . 1.275"
Primers . . . . Winchester WLP
Primer Size . . . . Large Pistol
Lyman Shell Holder . . . . No. 11
Jacketed Bullets Used .Speer Gold Dot #4478, 200 gr.
Sierra JHC #8820, 240 gr.
Hornady HP/XTP #45200, 250 gr.
Cast Bullets Used . . . .(sized to .452" dia)
*gas check bullet #452424, 255 gr.
*#452651, 300 gr.

Test Specifications:
(Velocity Only)

Firearm Used . .Thompson/Center Contender and Encore
Barrel Length . . . . .10"-Contender and 15" Encore
Twist . . . . .1-24" Contender and 1-16" Encore
Groove Dia. . . . . .452"

200 gr. Jacketed HP
1.580" OAL
BC: .138
SD: .140

Powder	Sugg Starting Grains	10" Bbl. Velocity fps	15" Bbl. Velocity fps	Max Load Grains	10" Bbl. Velocity fps	15" Bbl. Pressure fps
Unique	11.2	1279	1169	12.2	1391	1231
Blue Dot	12.8	1030	855	15.8	1305	1087
2400	19.0	1161	1029	**21.5**	**1366**	**1502**
IMR-4227	18.8	1080	855	23.5	1341	1099

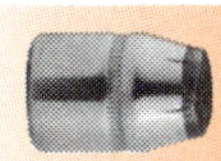

240 gr. Jacketed HC
1.580" OAL
BC: .150
SD: .168

Powder	Sugg Starting Grains	10" Bbl. Velocity fps	15" Bbl. Velocity fps	Max Load Grains	10" Bbl. Velocity fps	15" Bbl. Pressure fps
Unique	9.5	1096	1055	**12.5**	**1354**	**1378**
Blue Dot	11.8	973	754	14.8	1214	1127
2400	18.5	1155	1207	23.0	1472	1483
IMR-4227	20.5	1179	1061	25.0	1453	1325

Note: Loads shown in shaded panels are maximum.
Loads shown in bold designate potentially most accurate load.

45 Colt

250 gr. Jacketed HP
1.590" OAL

BC: .146
SD: .175

Powder	Sugg Starting Grains	10" Bbl. Velocity fps	15" Bbl. Velocity fps	Max Load Grains	10" Bbl. Velocity fps	15" Bbl. Pressure fps
Unique	9.3	1027	966	**12.2**	**1324**	**1285**
Blue Dot	12.0	946	796	14.5	1154	1016
2400	17.5	1060	1065	22.0	1410	1428
IMR-4227	20.3	1196	971	24.0	1362	1159

#452424
255 gr. (#2 Alloy) 1.630" OAL

BC: .210
SD: .178

Powder	Sugg Starting Grains	10" Bbl. Velocity fps	15" Bbl. Velocity fps	Max Load Grains	10" Bbl. Velocity fps	15" Bbl. Pressure fps
700X	6.3	955	890	**8.0**	**1127**	**1073**
Red Dot	6.5	917	890	8.3	1055	1071
Unique	8.5	1013	946	10.2	1162	1105
SR-7625	8.0	984	804	9.7	1153	938

#452651
300 gr. (#2 Alloy) 1.585" OAL

BC: .191
SD: .210

Powder	Sugg Starting Grains	10" Bbl. Velocity fps	15" Bbl. Velocity fps	Max Load Grains	10" Bbl. Velocity fps	15" Bbl. Pressure fps
700X	5.5	632	790	6.7	768	901
Red Dot	5.5	600	769	6.9	763	902
Unique	6.5	655	792	**8.5**	**900**	**978**
SR-7625	7.0	722	781	8.2	841	895

Note: Loads shown in shaded panels are maximum.
Loads shown in bold designate potentially most accurate load.

454 Casull

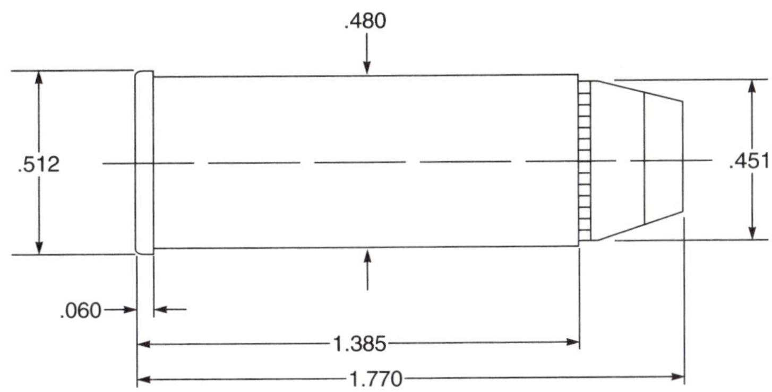

Comments:

Shooters who like the 44 Remington Magnum should love the 454 Casull. The following data was worked up in a pressure barrel and fired through a Thompson/Center Encore for velocity readings. This provided on average an increase of 50 to 150 feet per second. Cast bullets should be produced from linotype or its equivalent. Refer to the 454 Casull data in the previous section for additional information.

Test Components:

Cases . . . . .Freedom Arms
Trim-to Length . . . . .1.380"
Primers . . . . .CCI 450
Primer Size . . . . .Small Rifle, Magnum
Lyman Shell Holder . . . . .No. 11
Jacketed Bullets Used . . . . . .Speer HP #4479, 225 gr.
Hornady HP/XTP #45200, 250 gr.
Hornady HP/XTP #45230, 300 gr.
Cast Bullets Used . . . .(sized to .451" dia)
*gas check bullet *#452490, 255 gr.
*#452651, 300 gr.

Test Specifications:
(Velocity Only)

Firearm Used . . . . .Thompson/Center Encore
Barrel Length . . . . .12"
Twist . . . . .1-16"
Groove Dia. . . . . .451"

225 gr. Jacketed HP
1.690" OAL
BC: .169
SD: .158

Powder	Sugg Starting Grains	Velocity fps	Pressure	Max Load Grains	Velocity fps	Pressure
2400	25.5	1653	—	29.5	1947	—
AA#9	27.0	1849	—	31.0	2007	—
IMR-4227	29.5	1595	—	**33.5+**	**1885**	—

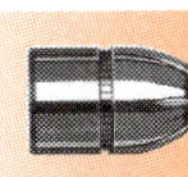

250 gr. Jacketed HP
1.690" OAL
BC: .146
SD: .175

Powder	Sugg Starting Grains	Velocity fps	Pressure	Max Load Grains	Velocity fps	Pressure
2400	22.0	1505	—	26.5	1763	—
AA#9	24.5	1685	—	29.0	1944	—
IMR-4227	26.0	1399	—	**30.5+**	**1731**	—

Note: Loads shown in shaded panels are maximum.
Loads shown in bold designate potentially most accurate load.
\+ Designates a compressed powder charge.

454 Casull

300 gr. Jacketed HP
1.760" OAL

BC: .180
SD: .210

Powder	Sugg Starting Grains	Velocity fps	Pressure	Max Load Grains	Velocity fps	Pressure
2400	20.0	1340	—	24.5	1589	—
AA#9	22.0	1459	—	25.8	1730	—
IMR-4227	24.0	1262	—	**28.2**	**1567**	—

#452490
255 gr. (Linotype) 1.760" OAL

BC: .160
SD: .179

Powder	Sugg Starting Grains	Velocity fps	Pressure	Max Load Grains	Velocity fps	Pressure
2400	21.7	1489	—	27.0	1846	—
AA#9	24.5	1739	—	**28.0**	**1905**	—
IMR-4227	25.8	1434	—	30.0	1699	—

#452651
300 gr. (Linotype) 1.755" OAL

BC: .191
SD: .211

Powder	Sugg Starting Grains	Velocity fps	Pressure	Max Load Grains	Velocity fps	Pressure
2400	17.0	1270	—	20.0	1412	—
AA#9	18.5	1319	—	22.0	1589	—
IMR-4227	20.0	1178	—	24.0	1407	—
AA-1680	24.0	1107	—	**28.0**	**1382**	—

Note: Loads shown in shaded panels are maximum.
Loads shown in bold designate potentially most accurate load.

45-70 Government

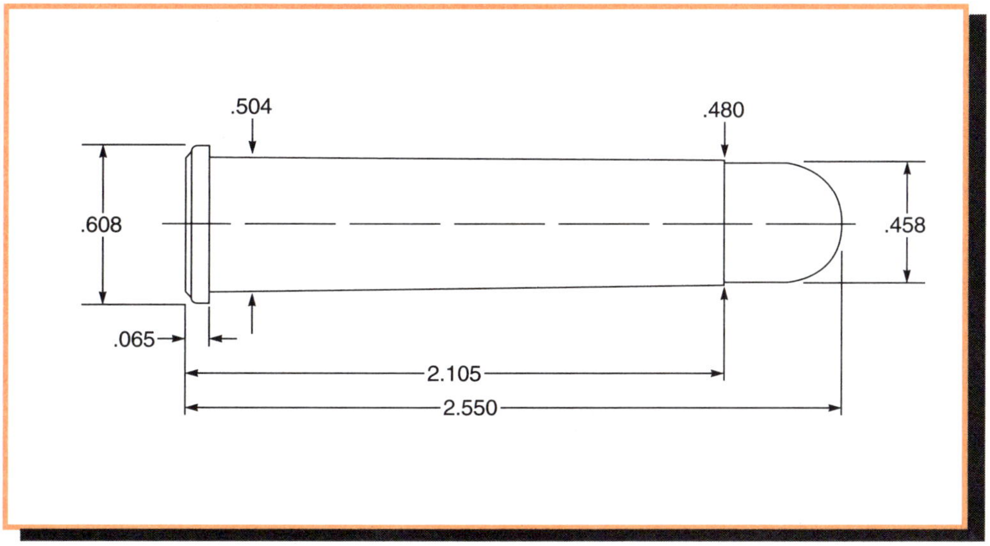

Comments:

This is not a cartridge for handgunners of faint heart. Recoil is very severe when shot in the T/C Contender. Alliant 2400 and Reloder 7 are good choices.

Test Components:

Cases .... Remington
Trim-to Length .... 2.095"
Primers .... Remington 9½
Primer Size .... Large Rifle
Lyman Shell Holder .... No. 17
Jacketed Bullets Used ... Hornady HP #4500, 300 gr.
Speer SP #2479, 400 gr.
Cast Bullets Used .... (sized to .459" dia)
#457191, 292 gr.
#457193, 405 gr.

Test Specifications:
(Velocity Only)

Firearm Used .... Thompson/Center Contender
Barrel Length .... 16"
Twist .... 1-14"
Groove Dia. .... .458"

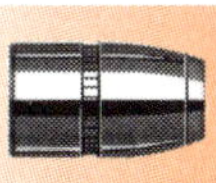

300 gr. Jacketed HP
2.550" OAL
BC: .197
SD: .204

Powder	Sugg Starting Grains	Velocity fps	Pressure	Max Load Grains	Velocity fps	Pressure
2400	24.0	1346	—	**30.0**	**1685**	—
IMR-4227	27.0	1327	—	32.0	1625	—
IMR-4198	31.0	1248	—	36.0	1449	—
RX7	33.0	1347	—	38.0	1600	—
IMR-3031	38.0	1206	—	43.0	1410	—

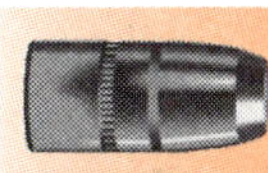

400 gr. Jacketed FSP
2.550" OAL
BC: .214
SD: .272

Powder	Sugg Starting Grains	Velocity fps	Pressure	Max Load Grains	Velocity fps	Pressure
Unique	11.0	746	—	15.0	1014	—
2400	22.0	1200	—	27.0	1426	—
IMR-4227	25.0	1211	—	30.0	1397	—
IMR-4198	28.0	1042	—	33.0	1293	—
RX7	29.0	1191	—	**34.0**	**1354**	—
IMR-3031	34.0	1018	—	39.0	1283	—

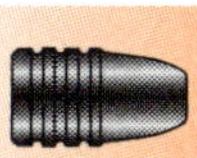

#457191
292 gr. (#2 Alloy) 2.550" OAL
BC: .201
SD: .198

Powder	Sugg Starting Grains	Velocity fps	Pressure	Max Load Grains	Velocity fps	Pressure
SR-4759	26.0	932	—	30.0	1471	—
IMR-4198	28.5	1122	—	**34.5**	**1481**	—
IMR-3031	34.0	961	—	44.0	1433	—

#457193
405 gr. (#2 Alloy) 2.550" OAL
BC: .307
SD: .275

Powder	Sugg Starting Grains	Velocity fps	Pressure	Max Load Grains	Velocity fps	Pressure
IMR-4198	21.5	733	—	28.5	1195	—
RX7	33.0	1386	—	**41.0**	**1578**	—
IMR-3031	34.0	1061	—	38.5	1304	—

Note: Loads shown in shaded panels are maximum.
Loads shown in bold designate potentially most accurate load.

Section 6

Appendix A

Bullet Type Abbreviations

BTSP: boat tail soft point

FMJ: full metal jacket

FMTBT: full metal jacket boat tail

FN: flat nose

FP: flat point

FSP: flat soft point

GDHP: Gold Dot hollow point (Speer)

GSSP: Grand Slam soft point (Speer)

HP: hollow point

HPBT: hollow point boat tail

HP/XTP: Hollow Point – Extreme Terminal Performance (Hornady)

JHC: jacketed hollow cavity

JSP: jacketed soft point

PP: Protected Point (Nosler)

SBT: spitzer boat tail

Sil: silhouette

SMP: semi-pointed

SP: soft point

Sp SP spitzer soft point

SPSX: Soft Point Super Explosive (Hornady)

SPT: spitzer

TMJ: Totally Metal Jacket (Speer)

WC: wadcutter

Appendix B

Useful formulas for reloaders

Energy in foot-pounds

$$\frac{\text{bullet weight in grains x velocity}^2}{450400} = \text{ft-lbs}$$

Example: 150gr. bullet at 2700 fps

$$\frac{150 \times 2700^2}{450400} = 2428 \text{ ft-lbs}$$

Sectional Density

A bullet's weight in pounds divided by the square of its diameter in inches.

$$\frac{\frac{W}{7000}}{D^2} = SD$$

Example: 150 gr. bullet, .308" in diameter

$$\frac{\frac{150}{7000}}{.308^2} = .226$$

Sight Correction

This formula will show how much change in inches will be needed to correctly zero your sights.

$$M = \left(\frac{D}{R}\right) \times S$$

Where M = movement in inches needed to zero sights.

D = impact deviation of bullet in inches.
(how far off from the aim point that the bullet is hitting.)
R = range in inches.
S = sight radius. (distance between sights)

Example: firearm is hitting 6" low at 50 yards.

D = 6" R = 50 x 36 = 1800" S = 20"

$$M = \frac{6''}{1800} \times 20''$$

$$M = .067''$$

.067" of sight adjustment is needed to zero firearm.

Free Recoil Energy

Using this formula, you can calculate the amount of recoil for any given gun and load.

$$\text{Free Recoil Energy (ft-lbs)} = \frac{(W1\ Vp + 4700\ W2)^2}{64.348\ Wg}$$

Where: W1 = weight of bullet in pounds. (grains divided by 7000 equals pounds)
W2 = weight of powder charge in pounds.
Wg = weight of gun in pounds.
Vp = muzzle velocity of bullet in feet per second.

Example: W1 for a 150 gr. bullet = 150 divided by 7000 = .021
W2 for a 50 gr. charge = 50 divided by 7000 = .007
Wg for a 9 pound rifle = 9
Vp for a velocity of 2700 fps = 2700

$$\frac{(.021 \times 2700 + 4700 \times .007)^2}{64.389 \times 9}$$

$$\frac{(56.7 + 32.9)^2}{579.132} = 13.86 \text{ ft-lbs}$$

Appendix C

Shellholder Charts

Rifle Cartridges	Lyman	Hornady	RCBS	Redding
17 Remington	26	16	10	10
22 Hornet	4	3	12	14
218 Bee	10	7	1	3
222 Remington	26	16	10	10
223 Remington	26	16	10	10
22 PPC	3	6	32	12
222 Remington Mag	26	16	10	10
225 Winchester	5	4	11	4
224 Weatherby Mag	3	1	na	12
22-250 Remington	2	1	3	1
220 Swift	5	4	11	4
6mm PPC	3	6	32	12
243 Winchester	2	1	3	1
6mm/244 Remington	2	1	3	1
240 Weatherby Mag.	2	1	3	1
25-20 Winchester	10	7	1	3
250 Savage	2	1	3	1
257 Roberts	2	1	3	1
25-06 Remington	2	1	3	1
257 Weatherby Mag.	13	5	4	6
6.5x55 Swed. Mau.	27	19	2	1
260 Remington	2	1	3	1
264 Winchester	13	5	4	6
270 Winchester	2	1	3	1
270 WSM	34	35	43	6
270 Weatherby Mag.	13	5	4	6
7-30 Waters	6	2	2	2
7mm-08 Remington	2	1	3	1
7mm Mauser	2	1	3	1
280 Remington	2	1	3	1
284 Winchester	2	1	3	1
7mm WSM	34	35	43	6
7mm Rem. SAUM	13	5	38	6
7mm Remington Mag	13	5	4	6
7mm Weatherby Mag	13	5	4	6
7mm STW	13	5	4	6
7mm Rem Ultra Mag	13	5	38	6
30 M1 Carbine	19	22	17	22
30-30 Winchester	6	2	2	2
300 Savage	2	1	3	1
30-40 Krag	7	11	7	8
308 Winchester	2	1	3	1

Rifle Cartridges	Lyman	Hornady	RCBS	Redding
30-06 Springfield	2	1	3	1
300 Rem SAUM	13	5	38	6
300 WSM	34	35	43	6
300 H&H Mag	13	5	4	6
300 Winchester Mag.	13	5	4	6
300 Rem Ultra Mag.	13	5	38	6
300 Weatherby Mag.	13	5	4	6
30-378 Wthby Mag.	17	14	14	18
7.62x39	3	6	32	12
7.62x54R	17	23	13	15
7.65 Arg. Mauser	2	24	3	1
303 British	7	11	7	8
32-20 Winchester	10	7	1	3
32 Winchester Spec.	6	2	2	2
8mm Mauser	2	1	3	1
8mm Remington Mag	13	5	4	6
338 Winchester Mag.	13	5	4	6
338 Rem Ultra Mag.	13	5	38	6
338/378 Wthby Mag.	17	14	14	18
340 Weatherby Mag.	13	5	4	6
348 Winchester	na	25	5	20
35 Remington	2	26	9	1
358 Winchester	2	1	3	1
35 Whelen	2	1	3	1
38-55 Winchester	6	2	2	2
375 Winchester	6	2	2	2
375 H&H Mag.	13	5	4	6
375 Rem. Ultra Mag.	13	5	38	6
378 Weatherby Mag.	17	14	14	18
416 Remington Mag.	13	5	4	6
416 Rigby	17	38	37	18
416 Weatherby Mag.	17	14	14	18
40-65 Winchester	17	14	14	18
40-70 Sharps Str.	33	na	na	8
44-40 Winchester	14B	9	35	9
444 Marlin	14B	27	28	19
45-70 Government	17	14	14	18
450 Marlin	13	5	4	6
45-90 Winchester	17	14	14	18
45-100	17	14	14	18
45-110	17	14	4	18
45-120	17	14	14	18
458 Winchester Mag.	13	5	4	6
460 Weatherby Mag.	17	14	14	18
50-70 Government	22	na	31	na

Pistol Cartridges	Lyman	Hornady	RCBS	Redding
25 ACP	32	37	29	27
7mm T/C-U	26	16	10	10
30 Luger	12	8	16	13
30 Mauser	12	na	16	13
32 ACP	23	22	17	22
32 S&W Long	9	36	23	10
32 H&R Mag	9	36	23	10
38 S&W	21	28	6	12
380 Auto	26	16	10	10
9mm Luger	12	8	16	13
357 Sig	15	10	27	5
38 Super	12	8	39	5
38 Special	1	6	6	12
357 Magnum	1	6	6	12
9mm Makarov	12	8	16	13
40 S&W	15	10	27	5
10mm Auto	15	10	27	5
38-40 Winchester	14B	9	35	9
41 Remington Mag	30	29	30	21
44 Special	7	30	18	19
44 Remington Mag	7	30	18	19
45 ACP	2	1	3	1
45 Auto Rim	na	31	8	17
45 Winchester Mag	2	1	3	7
45 Colt	11	32	20	23
454 Casull	11	32	20	23
480 Ruger	17	14	40	na
50 Action Express	7	40	33	na

Appendix D

Relative Burn Rate Chart

Listed in order from fastest to slowest

1) R-1, Norma
2) N310, VihtaVuori
3) Bullseye, Alliant
4) Titewad, Hodgdon
5) Red Dot, Alliant
6) Clays, Hodgdon
7) 700X, IMR
8) Titegroup, Hod.
9) Amer. Select, All.
10) Solo 1000, Acc.
11) No.2 Imp., Accu.
12) Green Dot, Alliant
13) International, Hod.
14) N320, VihtaVuori
15) PB, IMR
16) WST, Winchester
17) HP-38, Hodgdon
18) 231, Winchester
19) SR-7625, IMR
20) Unique, Alliant
21) Universal, Hod.
22) N330, VihtaVuori
23) Herco, Alliant
24) No. 5, Accurate
25) WSF, Winchester
26) HS-6, Hodgdon
27) Power Pistol, All.
28) N340, VihtaVuori
29) SR-4756, IMR
30) 800X, IMR
31) N105, VihtaVuori
32 No. 7, Accurate
33) HS-7, Hodgdon
34) Blue Dot, Alliant
35) 2400, Alliant
36) No. 9, Accurate
37) N110, VihtaVuori
38) H-110, Hodgdon
39) 296, Winchester
40) SR-4759, IMR
41) N120, VihtaVuori
42) IMR-4227, IMR
43) H-4227, Hodgdon
44) N130, VihtaVuori
45) 1680, Accurate
46) Lil'Gun, Hodgdon
47) IMR-4198, IMR
48) H-4198, Hodgdon
49) N133, VihtaVuori
50) XMR-2015, Acc.
51) Reloder 7, Alliant
52) H322, Hodgdon
53) Benchmark, Hod.
54) IMR-3031, IMR
55) 2230, Accurate
56) H335, Hodgdon
57) H-4895, Hodgdon
58) IMR-4895, IMR
59) 2460, Accurate
60) BL-C(2), Hodgdon
61) XMR-2495, Acc.
62) 748, Winchester
63) N135, VihtaVuori
64) IMR-4064, IMR
65) Varget, Hodgdon
66) 2520, Accurate
67) N540, VihtaVuori
68) N140, VihtaVuori
69) Reloder 15, Alliant
70) IMR-4320, IMR
71) H380, Hodgdon
72) 2700, Accurate
73) N150, VihtaVuori
74) N550, VihtaVuori
75) 760, Winchester
76) H-414, Hodgdon
77) IMR-4350, IMR
78) H-4350, Hodgdon
79) XMR-4350, Acc.
80) N160, VihtaVuori
81) N560, VihtaVuori
82) IMR-4831, IMR
83) 3100, Accurate
84) Reloder 19, Alliant
85) H-4831, Hodgdon
86) WXR, Winchester
87) Reloder 25, Alliant
88) N165, VihtaVuori
89) Reloder 22, Alliant
90) IMR-7828, IMR
91) N170, VihtaVuori
92) H-1000, Hodgdon
93) MagPro, Accurate
94) Retumbo, Hod.
95) H50BMG, Hod.
96) H-870, Hodgdon
97) 8700, Accurate

LYMAN RELOADING RECORDS LOG (ONE)

Data Loaded	Caliber	Brass Brand Lot No.	Primer Brand Type Lot No.	Powder Brand Type Lot No. Charge	TImes Fired	Times Trimmed	Bullet Brand Type Weight Lot No.

LYMAN RELOADING RECORDS LOG (ONE)

Data Loaded	Caliber	Brass Brand Lot No.	Primer Brand Type Lot No.	Powder Brand Type Lot No. Charge	Times Fired	Times Trimmed	Bullet Brand Type Weight Lot No.

LYMAN RELOADING RECORDS LOG (TWO)

Date	Firearm	Comments

LYMAN RELOADING RECORDS LOG (TWO)

Date	Firearm	Comments

NOTES

NOTES

NOTES

NOTES

NOTES

Shooters are Switching to Butch's Patches

Butch's "Triple Twill" Patches have "Drop Zones" to capture fouling!

Butch's "Triple Twill" Patches

Last year "Butch's Bore Shine" took the shooting market by storm. Customers told us that after trying Butch's, they will never use another bore cleaner.

We were so pleased with the success of Butch's we set out to find the perfect patch to go along with Butch's solvent and oil. Like Butch's bore shine, we wanted a better product, a patch that would hold more liquid and remove fouling faster than ordinary cotton flannel patches. By chance, we discovered Triple Twill, a fabric used by laboratories and cleanrooms. Triple Twill's special weave has hundreds of tiny "ridges and drop zones" which scrub and capture fouling residues far better than flannel patches. Butch's Triple Twill won't shed lint like other patches can. They also have been specially treated to gain the absolute maximum in liquid absorbency.

Butch's Triple Twill Patches are the highest quality patches you can buy anywhere!

Butch's Patches the Perfect Compliment to World Famous Butch's Bore Shine!

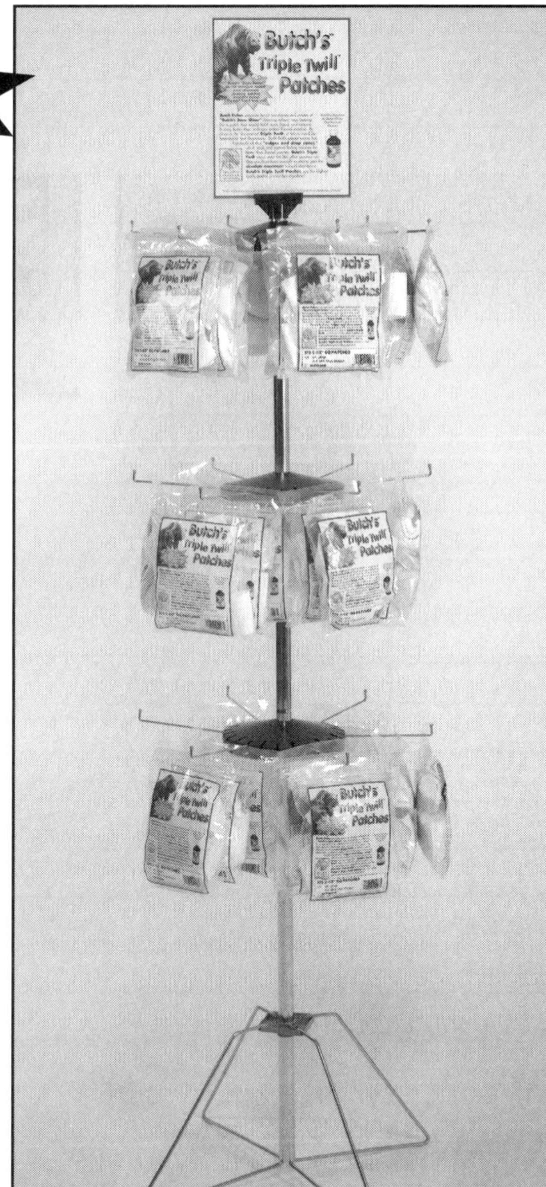

New Dealer Profit Center! Build a Tree with Butch's Patches!

Butch's "Triple Twill" Patches

Part No.	Description	
#02901	1⅛" SQ. 22-270 Cal	(Bag of 1000)
#02904	1¾" SQ. 7mm - 35 Cal Rifle	(Bag of 750)
#02906	2¼" SQ. 38 CAL - 45 Cal Rifle 10mm - 45 Cal Pistol	(Bag of 500)
#02908	2½" SQ. 45-58 Cal Rifle .410 - 20 Ga. Shotgun	(Bag of 375)
#02911	3" SQ. 10,12 & 16 Ga.Shotgun	(Bag of 300)

Butch's Gun Care Products
The New Choice of Competitive Shooters!

Butch Fisher - Bench Rest Competitor and originator of Butch's Gun Care Products

Butch's GUN CARE PRODUCTS

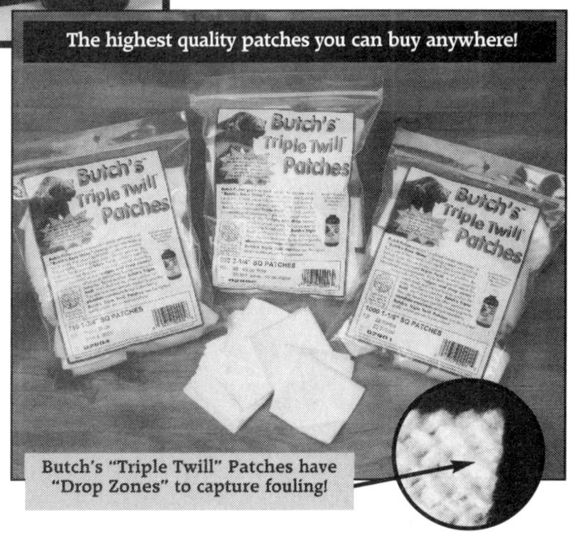

Butch's "Triple Twill" Patches have "Drop Zones" to capture fouling!

Butch's Gun Care Products have been developed by well known Bench Rest Competitor Butch Fisher. Always looking for a competitive edge, Butch created Bore Shine to clean his rifle faster and cleaner than any other solvent. Butch's Bore Shine was used by most competitors at this years National Bench Rest Championship. Now Butch has introduced Butch's Patches made from a special "Triple Twill" fabric that cleans and removes fouling better and faster. Triple Twill Fabric has pockets called Drop Zones that grab and remove fouling. The patches are specially treated for maximum absorption putting more solvent to work.

Butch's Bore Shine and Triple Twill Patches... the Deadly Duo for a clean barrel!

Lyman Products Corp
475 Smith Street
Middletown, CT 06457
860-632-2020 • FAX 860-632-1699
www.lymanproducts.com

To Order Call 800-22-LYMAN

Butch's "Triple Twill" Patches

We were so pleased with the success of Butch's we set out to find the perfect patch to go along with Butch's solvent and oil. Like Butch's bore shine, we wanted a better product, a patch that would hold more liquid and remove fouling faster than ordinary cotton flannel patches. By chance, we discovered Triple Twill, a fabric used by laboratories and cleanrooms. Triple Twill's special weave has hundreds of tiny "ridges and drop zones" which scrub and capture fouling residues far better than flannel patches. Butch's Triple Twill won't shed lint like other patches can. They also have been specially treated to gain the absolute maximum in liquid absorbency.

Butch's "Triple Twill" Patches

Part No.	Description	Price
#02901	1⅛" SQ. 22-270 Cal. (Bag of 1000)	**$11.50**
#02913	1½" SQ. 6mm Bench Rest (Bag of 1000)	**$11.50**
#02904	1¾" SQ. 7mm - 35 Cal Rifle (Bag of 750)	**$10.50**
#02906	2¼" SQ. 38 CAL - 45 Cal Rifle (Bag of 500) 10mm - 45 Cal Pistol	**$11.50**
#02908	2½" SQ. 45-58 Cal Rifle (Bag of 375) .410 - 20 Ga. Shotgun	**$10.50**
#02911	3" SQ. 10,12 & 16 Ga. (Bag of 300) Shotgun	**$8.75**

Butch's Gun Care Products are distributed exclusively by Lyman Products Corp. If you would like more information on Butch's products or want us to recommend a distributor that carries Butch's call 1-800-22-Lyman.

New Products for Shooters from Lyman and Pachmayr

Lyman Electronic Scale Funnel Pan

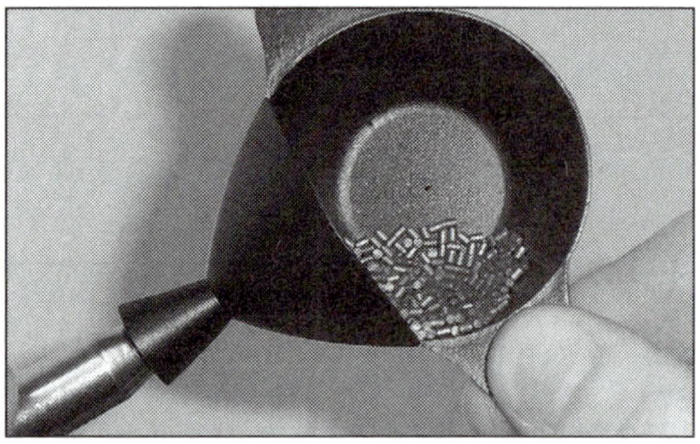

Pour directly from the Funnel Pan into the case! Saves time. Combines 2 steps into one. (See Order Form)

Lyman Electronic Digital Trigger Pull Gauge

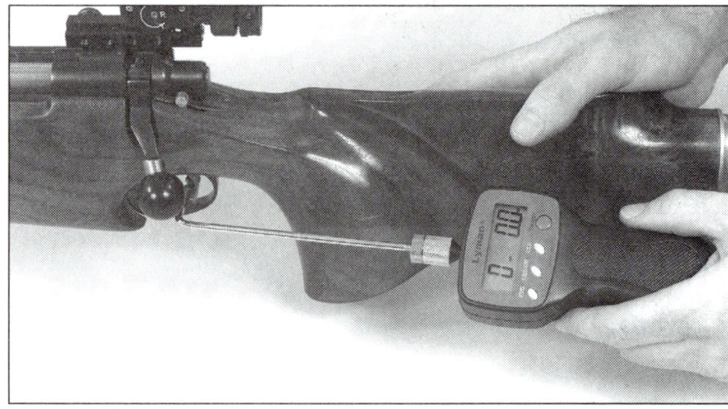

State of the art technology makes this the fastest and most accurate gauge on the market! Accuracy to 1/10 ounce over working range of 0 to 12 pounds.

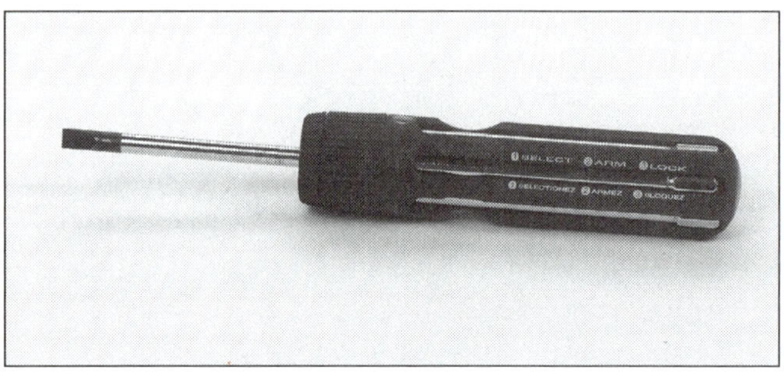

Pachmayr 1911 Tool Kit

A true All-in-One Tool Kit for 1911 or similar style guns. Change blades quickly with a turn of the dial! 1911 Tool Kit includes:

- 5/32" Pin Punch
- .150" Small Screwdriver
- .325" Large Screwdriver
- 5/32"Hex Drive
- .210" Medium Screwdriver
- 3/32" Hex Drive

Lyman Shooters Eyeglass Diopter

Instantly improves eye focus for shooters with prescription glasses. Eliminates target fuzzing and distortion. (See Order Form)

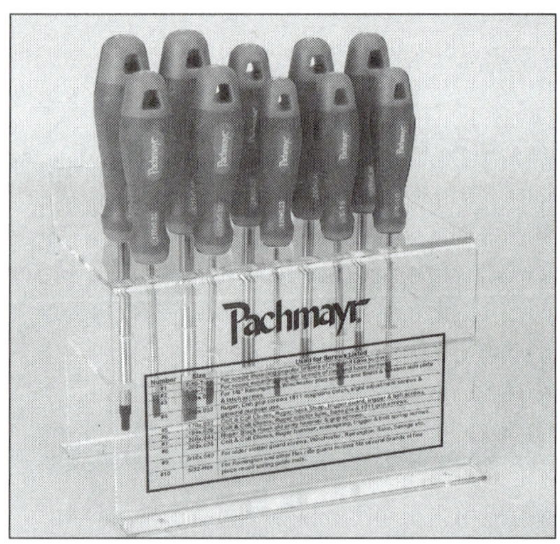

Pachmayr Master Gunsmith 10 Piece Screwdriver Set

10 classic screwdrivers designed specifically for gunsmiths and firearms enthusiasts. Magnetized with rubber coated handle with "caddy". Tips are magnetized and parallel ground. They are the exact size to fit all Colt, S&W, Ruger pistols and revolver screws. Torx bits are included for all scope rings and bases.
(See Order Form)

Pachmayr® A-ZOOM®

Precision Metal Snap-Caps

THE WORLD'S BEST PRACTICE AMMO!

SHOTGUN

ITEM #	DESCRIPTION	PACK	RETAIL
* 12210	10 GAUGE	2	9.98
* 12211	12 GAUGE	2	9.98
* 12212	16 GAUGE	2	9.98
* 12213	20 GAUGE	2	9.98
* 12214	28 GAUGE	2	9.98
* 12215	410 GAUGE	2	9.98

REVOLVER

ITEM #	DESCRIPTION	PACK	RETAIL
16130	32 H & R	6	19.98
16137	32 H & R Mag	6	19.98
16160	32 S & W Short	6	19.98
16135	32 S & W Long	6	19.98
16112	32-20 Win	6	19.98
* 16119	357 Magnum	6	18.98
16131	38 Blank	6	19.98
16125	38 S & W	6	19.98
* 16118	38 Special	6	18.98
16128	38-40 Win	6	19.98
16127	41 Magnum	6	19.98
16141	44 Colt	6	19.98
* 16120	44 Magnum	6	18.98
16140	44 Russian	6	19.98
* 16121	44 Special	6	18.98
* 16123	44-40 Win	6	18.98
* 16124	45 Colt	6	18.98
16126	454 Casull	6	19.98
16136	455 Webley	6	19.98

PISTOL

ITEM #	DESCRIPTION	PACK	RETAIL
* 15117	10mm Auto	5	15.98
* 15152	25 Auto	5	15.98
* 15153	32 Auto	5	15.98
* 15159	357 Sig	5	15.98
15158	38 Super	5	15.98
* 15113	380 Auto	5	15.98
* 15114	40 S&W	5	15.98
15138	40 Super	5	16.98
15134	400 Cor Bon	5	16.98
15150	41 AE	5	16.98
* 15115	45 Auto	5	15.98
15155	45 Magnum	5	16.98
15151	50 AE	5	16.98
15133	7.62 Tokarev	5	16.98
15129	7.63 Mauser	5	16.98

PISTOL CONTINUED

ITEM #	DESCRIPTION	PACK	RETAIL
15110	8mm Auto Blank	5	16.98
* 15116	9mm Luger	5	15.98
15132	9mm Makarov	5	16.98
15142	9mm Auto Steyr	5	16.98

RIFLE GROUP A

ITEM #	DESCRIPTION	PACK	RETAIL
* 12236	22 Hornet	2	9.98
12255	220 Swift	2	11.98
* 12238	222 Rem	2	9.98
* 12254	22-250 Rem	2	9.98
* 12222	223 Rem	2	9.98
* 12223	243 Win	2	9.98
12256	25-06 Rem	2	9.98
12287	260 Rem	2	11.98
* 12224	270 Win	2	9.98
* 12225	30 Carbine	2	9.98
* 12237	300 Win Mag	2	9.98
* 12227	30-06 Sprg	2	9.98
* 12226	303 British	2	9.98
* 12229	30-30 Win	2	9.98
* 12228	308 Win,	2	9.98
12296	300WSM	2	9.98
* 12230	338 Win Mag	2	9.98
12270	444 Marlin	2	11.98
* 12231	45-70 Govt	2	9.98
* 12251	6.5 x 55 Swedish	2	9.98
* 12252	7mm Rem Mag	2	9.98
12277	7mm STW	2	11.98
* 12234	7.62 x 39	2	9.98
12233	7.62 x 54 Russ	2	9.98
12247	7mm-08 Rem	2	9.98
* 12232	7 x 57 Mauser	2	9.98
* 12235	8 x 57 Mauser	2	9.98

RIFLE GROUP B

ITEM #	DESCRIPTION	PACK	RETAIL
12290	250 Savage	2	11.98
12286	25-20 Win	2	11.98
12257	25-35 Win	2	11.98
12258	257 Roberts	2	11.98
12246	280 Rem	2	11.98
12239	30 R Blaser	2	11.98
12284	300 Wea Mag	2	11.98
12272	300 Savage	2	11.98
12263	300 Rem Ultra Mag	2	11.98

RIFLE GROUP B CONTINUED

ITEM #	DESCRIPTION	PACK	RETAIL
12220	32-40 Win	2	11.98
12250	338 Lapua Mag	2	11.98
12273	340 Weatherby	2	11.98
12274	358 Win	2	11.98
12248	375 HH Mag	2	9.98
12249	375 Flgd Mag NE	2	11.98
12260	375 Win	2	11.98
12261	38-55 Win	2	11.98
12266	404 Jeffery	2	11.98
12221	40-65	2	11.98
12288	416 Rem Mag	2	11.98
12289	416 Wea Mag	2	11.98
12267	458 Win Mag	2	11.98
12295	5.6 x 50R Mag	2	11.98
12241	5.6 x 52 R	2	11.98
12276	6mm Rem	2	11.98
12291	6.5 Carcano	2	11.98
12262	6.5 x 54 MS	2	11.98
12294	6.5 x 57 R	2	11.98
12278	7mm Dakota	2	11.98
12293	7 x 64 Brenneke	2	9.98
12281	7.5 x 55 Swiss	2	11.98
12264	7.7 x 58 JAP	2	11.98
12242	7 x 57 R	2	9.98
12244	7 x 65 R	2	9.98
12253	8mm Rem Mag	2	11.98
12265	8 x 57 R	2	11.98
12282	9 x 57 Mauser	2	11.98
12240	9.3 x 62 Mauser	2	11.98
12280	9.3 x 72 R	2	11.98
12269	9.3 x 74 R	2	9.98

GROUP C LARGE RIFLE/NITRO EXPRESS

ITEM #	DESCRIPTION	PACK	RETAIL
11413	30-378 WB Mag	1	6.98
11401	416 Rigby	1	6.98
11403	450/400 x 3.25 NE	1	6.98
11405	470 NE	1	6.98
11406	475 #2 NE	1	6.98
11407	500 Jeffery	1	6.98
11408	500/465 NE	1	6.98
11410	505 Gibbs	1	6.98
11411	577 NE	1	6.98
11412	600 NE	1	6.98
11414	700 NE	1	6.98
* 11451	50 BMG	1	6.98

***Indicates Best Seller**

SEND TO COMPONENT MANUFACTURER

Gentlemen:

I have read about your products in Lyman's RELOADING HANDBOOK, 48th Ed. Please send me a current catalog showing your reloading components products.

Name: ______________________________

Address: ______________________________

____

Zip Code: ______________________________

48th Ed. 1

SPECIAL ORDER FORM

Congratulations! This Lyman purchase qualifies you to obtain our specially selected new products at special pricing below. Please complete the following and mail to Lyman Products.

Name: ______________________________

Address: ______________________________

City/State/Zip: ______________________________

Tel.#: (If we have a question) ______________

Method of Payment (Check one)

❑ VISA ❑ MasterCard ❑ Discover ❑ Check Enclosed

Account number ______________ / Exp. Date

____

Signature (required for credit card orders)

Product Description	Part No.	Qty.	Retail Price	Promo Price	Total
A-Zoom Precision Metal Snap Caps (Specify Caliber & Part No)			See Chart	Save $3.00 off Retail	
Butchs Gun Care Products					
Bore Shine 4oz	02937		6.65	4.99	
Bore Shine 16 oz	02941		19.95	16.95	
Gun Oil	02948		5.95	4.75	
Black Powder Bore Shine	02949		6.65	4.99	
Butch's Triple Twill Patches					
1 ⅛" Square – 22-270 1000 p/bag	02901		11.65	9.95	
1 ½" Square – 6mm 1000 p/bag	02913		11.65	9.95	
1 ¾" Square – 270-35 cal. 750 p/bag	02904		10.75	8.95	
2 ¼" Square - 35-45 cal 500 p/bag	02906		11.65	9.95	
2 ½" Square - 45-58 cal. 375 p/bag	02908		10.75	8.95	
3" Square - 10, 12, & 16 ga. 300 p/bag	02911		8.95	7.75	
Pachmayr Master Gunsmith 10 Piece Screwdriver Set	03050		44.95	39.95	
Pachmayr 1911 Tool Kit	03041		32.50	24.95	
Lyman Electronic Digital Trigger Pull Gauge	7832248		59.95	49.95	
"Powder Pal"Funnel Pan	7752433		5.95	4.75	
Shooters Eyeglass Diopter	3112020		19.95	14.95	

Shipping/Handling Charge	**2.00**
CT Residents Only: (Add 6% Sales Tax)	
Total Amount	

Questions? Call Toll Free 1-800-22-LYMAN
Offer valid only in U.S. and Canada. ***Good thru 2003.***

PLACE
STAMP
HERE

PRINT NAME AND ADDRESS OF MANUFACTURER BELOW

See **Special Order Form** on other side. Tear or cut off card, where shown. When folded and secured, card can be used as mailer by MC/Visa charge customers. For check or money order, use envelope. Any questions, please call toll-free **1-800-22-LYMAN** (except foreign customers, who may call (860) 632-2020). Hours: 9:00a.m. - 3:30 p.m. Eastern time, Monday-Friday.

Fold here and staple or tape order card enclosed. Do not send open!

PLACE
STAMP
HERE

LYMAN PRODUCTS CORP
DEPT 48HB - HB OFFERS
475 SMITH STREET
MIDDLETOWN, CT 06457